ATOMS, MOLECULES
AND **CLUSTERS** IN
ELECTRIC FIELDS

Theoretical Approaches to the
Calculation of Electric Polarizability

Series in Computational, Numerical and Mathematical Methods in Sciences and Engineering   ISSN: 1793-3439

Series Editor: Theodore E Simos *(Univ. of Peloponnese, Greece)*

ATOMS, MOLECULES AND **CLUSTERS** IN **ELECTRIC FIELDS**

Theoretical Approaches to the Calculation of Electric Polarizability

editor

George Maroulis
University of Patras, Greece

Imperial College Press

Published by

Imperial College Press
57 Shelton Street
Covent Garden
London WC2H 9HE

Distributed by

World Scientific Publishing Co. Pte. Ltd.
5 Toh Tuck Link, Singapore 596224
USA office: 27 Warren Street, Suite 401-402, Hackensack, NJ 07601
UK office: 57 Shelton Street, Covent Garden, London WC2H 9HE

British Library Cataloguing-in-Publication Data
A catalogue record for this book is available from the British Library.

ATOMS, MOLECULES AND CLUSTERS IN ELECTRIC FIELDS
Theoretical Approaches to the Calculation of Electric Polarizability
Series in Computational, Numerical and Mathematical Methods in Sciences and Engineering — Vol. 1

ISBN   1-86094-676-3

Printed in Singapore by B & JO Enterprise

PREFACE

ATOMS, MOLECULES AND CLUSTERS IN ELECTRIC FIELDS. THEORETICAL APPROACHES TO THE CALCULATION OF ELECTRIC POLARIZABILITY

Homage to D.M.Bishop's contribution.

This volume presents a collection of papers focusing on the theoretical determination of the electric polarizability of atoms, molecules and clusters. This is currently a vigorously expanding research field. The book aims at a wide readership, ranging from specialists and workers in field to graduate and undergraduate students in the molecular sciences. All chapters are written by scientists with extensive experience in the field. Thus, this volume offers a comprehensive account of recent progress in the field, and the trends and emerging perspectives.

Electric polarizability is routinely associated with fundamental characteristics of atomic and molecular systems: hardness, softness and hypersoftness, stiffness and compressibility. The basic theory of electric polarizability and hyperpolarizability is of central importance to the rational approach to the description and interpretation of a wide range of phenomena. These include nonlinear optics, scattering and phenomena induced by intermolecular interactions. It is also present as a key element in the rigorous analysis of spectroscopic observations. In recent years the theoretical determination of electric polarizability and hyperpolarizability is making decisive contributions to new fields with important potential for advanced technological applications. Such fields are the molecular simulation and modeling of fundamental processes and the search for new optical materials. One should also add here the powerful emergence of pharmacology, where electric polarizability and hyperpolarizability is used as descriptors in QSAR and drug design studies.

It is easily understood that the increasing demand for accurate polarizability and hyperpolarizability values has resulted in intense investigations on all computational aspects of their determination. At a relatively early stage the agreement between theory and experiment for these quantities was a major issue. The careful analysis of the observed discrepancies has brought about an effective rapprochement between theory and experiment. The development of theoretical methods of high predictive capability and reliable computational strategies has transformed computational quantum chemistry into a true vehicle of progress for modern molecular science.

In his review of the atomic static dipole polarizabilities Schwerdtfeger offers a critical evaluation of the available values for all elements from Z =1 to 119. His compilation of data is certainly a valuable reference and will be of great help to working scientists.

Quinet, Champagne and Kirtman present detailed analysis of the zero-point vibrational averaging (ZPVA) correction to the first hyperpolarizability of mono-substituted benzenes. Reliable estimates of the ZPVA correction are relatively rare for polyatomic systems. Their study offers valuable insights into the determination of this important quantity.

Pouchan, Zhang and Bégué propose a computationally advanced studied of the polarizability and hyperpolarizability of small silicon clusters. There are several studies of the polarizability of these important systems, but very little is known of their hyperpolarizability.

Fuentealba's contribution focuses on the calculation of the static polarizability of lithium, sodium, silicon and copper clusters. The polarizability of copper clusters has been a matter of some controversy in recent years.

The extensive review of Gu, Imamura and Aoki presents in detail a very promising contribution to the field. The elongation method for polymers paves the way to a new type of investigation with important potential applications to nonlinear optics.

Torii discusses the response of molecular vibrations to intermolecular electrostatic interactions. The author presents a comprehensive account of their effects. Illustrative examples conclude this important review.

The following chapter treats in an exemplary manner the highly non-trivial problem of the determination of the polarizability and hyperpolarizability of liquid water. Kongsted, Osted, Mikkelsen and Christiansen rely on Coupled Cluster/Molecular Mechanics response theory. The determination of electric properties of molecules in condensed phases constitutes a major challenge to Computational Quantum Chemistry.

Jensen and van Duijnen present the Discrete Solvent Reaction Field Model. This powerful QM/MM based method offers a promising tool for the determination of the electric properties of molecules in condensed phases.

Wu, Li, Li and Sun show that dipole bound anions as $(HF)_n^-$, n=2, 3, 4, are characterized by very large first hyperpolarizabilities.

The long review of Nakano reports an extensive investigation of the hyperpolarizability of open-shell molecules. Very little is known about the electric properties of such systems. This comprehensive review offers an invaluable account of this difficult subject, illustrated by many examples.

Coutinho and Canuto present a sequential Monte Carlo/Quantum Mechanics approach to the dipole polarizability of atomic liquids. Their treatment of liquid argon shows that the polarizability in the liquid is slightly larger than in the gas phase.

Bancewicz, Le Duff and Godet show that the analysis of interaction-induced light-scattering spectra can lead to the determination of multipolar polarizabilities of the interacting molecules. Their results for CH_4, CF_4 and SF_6 are in very good agreement with the available **ab initio** values.

Buldakov and Cherepanov have obtained polarizability functions for N_2 and O_2. Their method allows the determination of polarizability curves for a wide range of internuclear separations. In addition, they present dipole polarizabilities for the dimers $(N_2)_2$ and $(O_2)_2$.

A critical compilation of atomic polarizabilities and hyperpolarizabilities is offered by Thakkar and Lupinetti. With the notable exception of the work of Stiehler and Hinze (Numerical Hartree-

Fock hyperpolarizabilities for atoms from Z=1 to 36), critical presentations of atomic hyperpolarizability data are very rare in the literature.

Yan, Zhang and Li present a review of accurate methods for the calculation of the polarizability of few-body atomic and molecular systems. They also review polarizability data for He, Li, Be, H_2^+, H_2 and the positronic system PsH.

Wu reviews recent theoretical advances in the calculation of electric hyperpolarizabilities of transition metal clusters. His review presents novel results, shows clearly the computational difficulty of similar investigations and offers valuable insights into this promising subject.

Maroulis and Haskopoulos present new results for the interaction polarizability and hyperpolarizability in the complexes of N_2, CO_2, H_2O, $(H_2O)_2$ and O_3 with He.

Chandrakumar, Ghanty and Ghosh review recent theoretical results for the polarizability of lithium and sodium clusters. Due to the extensive experimental investigations reported in recent years, the electric properties of these systems have been extensively studied. This important review will be of great value to future investigations.

Last, Senet, Yang and van Alsenoy present a study of the charge distribution and polarizability in water clusters up to the icosamer. With the exception of the monomer, no experimental values are known for the polarizability of water clusters.

This Book is dedicated to David M.Bishop in celebration of his contribution to the field, on the occasion of his 70th birthday.

George Maroulis
Department of Chemistry
University of Patras
Patras, Greece

To my beloved wife Anastasia and our glorious children Penthesileia, Neoptolemos and Ianthe

Contents

CHAPTER 1

ATOMIC STATIC DIPOLE POLARIZABILITIES

Peter Schwerdtfeger

Theoretical and Computational Chemistry Centre,
Institute of Fundamental Sciences, Massey University (Albany Campus),
Private Bag 102904, North Shore MSC, Auckland, New Zealand.
Email: p.a.schwerdtfeger@massey.ac.nz

A review over the latest calculated and experimental static dipole polarizabilities of the neutral atoms is given. Periodic trends are analyzed and discussed. It is concluded that there is much room for improvement on the current available polarizability data, especially for open-shell systems and the heavier elements.

1. Introduction

The accurate determination of atomic and molecular response properties to an external electric or magnetic field is currently a very active and also challenging field in theoretical chemistry and physics. In an atom or molecule, the lowest order response of an electron cloud to an external electric field is described by its polarizability.[1,2][†] The static dipole polarizability, α, is a linear response property, and as defined as the second-derivative of the total electronic energy with respect to the external homogeneous electric field (quadratic Stark effect), it is very sensitive to basis set, electron correlation and relativistic effects, and to the vibrational structure in the case of a molecule. It is by no means an easy task to accurately calculate or measure static or dynamic dipole polarizabilities or higher moments of atoms or molecules (for excellent reviews on the theoretical

---

[†] The polarizability of the nucleus is very small and can safely be neglected.[3,4] For example the polarizability of a neutron is approximately 10^{-3} fm^3.[5]

background see Dalgarno[1] or Lazzeretti[6]), which play such an important role in the treatment of weakly interacting systems,[7,8] scattering processes, optical and dielectric properties, atom cooling and trapping, or in core-polarization and core-valence correlation effects.[9-15] Only in the last decade have accurate relativistic procedures, including many-body perturbation theory or coupled cluster techniques, become available for the calculation of various electric and magnetic response properties.[6]

The last extensive review on the measurement of atomic dipole polarizabilities for electronic ground states has been given by Miller and Bederson in 1988, and together with their 1977 review, can be considered as classics in this field.[16,17] More recently, Stiehler and Hinze reviewed calculations of dipole polarizabilities for the lighter elements up to Kr.[18] The purpose of the current review is to present an updated list of atomic polarizabilities determined by quantum theory including the most recent calculations for the heavier elements. This can be compared with the list of polarizabilities updated regularly by Miller in the Handbook of Chemistry and Physics, which also contains experimentally determined values.[16,19] There one finds a comprehensive list of formulae used for the experimental determination of dipole polarizabilities. More recent reviews on experimental techniques have been given by van Wijngaarden.[20,21] For molecular polarizabilities see refs.22-24 and other chapters in this book.

2. Theory

The atomic or molecular Hamiltonian in a non-uniform (inhomogeneous) static (time-independent) electric field is[25,26]

$$H = H_0 - \mu_\alpha F_\alpha - \tfrac{1}{3}\theta_{\alpha\beta}F_{\alpha\beta} - \tfrac{1}{15}\omega_{\alpha\beta\gamma}F_{\alpha\beta\gamma}\cdots \qquad (1)$$

where H_0 is the unperturbed atomic or molecular Hamiltonian of the system (nonrelativistic or relativistic), and we work in a stationary system-fixed coordinate system with $\alpha, \beta, \gamma = x, y$ or z. The Einstein sum convention is used. F_α, $F_{\alpha\beta}$, $F_{\alpha\beta\gamma}$, ... are the electric field, electric field gradient ($F_{\alpha\beta}=V_\alpha F_\beta$), and higher order (hyper) gradients (e.g. $F_{\alpha\beta\gamma}=V_\alpha V_\beta F_\gamma$), respectively. μ_α, $\theta_{\alpha\beta}$, $\omega_{\alpha\beta\gamma}$... are the multipolar tensor operators (dipole, quadrupole, octupole etc.). Using perturbation theory we obtain the response of the total energy of the system with respect to the electric field components,

$$
\begin{aligned}
E = E_0 &- \mu_\alpha^0 F_\alpha - \tfrac{1}{2}\alpha_{\alpha\beta}F_\alpha F_\beta - \tfrac{1}{6}\beta_{\alpha\beta\gamma}F_\alpha F_\beta F_\gamma \\
&- \tfrac{1}{24}\gamma_{\alpha\beta\gamma\delta}F_\alpha F_\beta F_\gamma F_\delta - \tfrac{1}{120}\delta_{\alpha\beta\gamma\delta\varepsilon}F_\alpha F_\beta F_\gamma F_\delta F_\varepsilon - \cdots \\
&- \tfrac{1}{3}\theta_{\alpha\beta}^0 F_{\alpha\beta} - \alpha_{\alpha,\beta\gamma}F_\alpha F_{\gamma\beta} - \alpha_{\alpha,\beta\gamma}F_\alpha F_{\delta\gamma\beta} - \cdots \\
&- \tfrac{1}{2}\alpha_{\alpha\beta,\gamma\delta}F_{\beta\alpha}F_{\delta\gamma} - \tfrac{1}{2}\alpha_{\alpha\beta,\gamma\delta\varepsilon}F_{\beta\alpha}F_{\varepsilon\delta\gamma} - \cdots
\end{aligned}
\tag{2}
$$

where E_0 is the energy in the absence of the electric field perturbation, i.e. $H_0\Psi_0 = E_0\Psi_0$ and μ_α^0 and $\theta_{\alpha\beta}^0$ are the permanent dipole and quadrupole moments respectively (for details see Buckingham[25]). For symmetric cases the number of tensor components in (2) can reduce significantly.[25,27] In this review we focus mainly on static dipole-polarizabilities of atoms which according to eq. (2) are,

$$
\alpha_{\alpha\beta} = -\left.\frac{\partial^2 E}{\partial F_\alpha \partial F_\beta}\right|_{\vec{F}=0} = -\left.\frac{\partial \mu_\alpha}{\partial F_\beta}\right|_{\vec{F}=0} = -\left.\frac{\partial \mu_\beta}{\partial F_\alpha}\right|_{\vec{F}=0}
\tag{3}
$$

From perturbation theory we get the static dipole polarizability for the ground state of an atom or molecule (in a.u.),

$$
\alpha_{\alpha\beta} = 2\sum_m \frac{\langle 0|r_\alpha|m\rangle\langle m|r_\beta|0\rangle}{E_m - E_0}
\tag{4}
$$

where for a multi-electron system r_α is understood as the dipole operator for the N-electron system, $r_\alpha = \sum_i r_{\alpha,i}$. Expression (4) is called the *oscillator strength* or *sum-over-state expression* (SOS) for the static dipole polarizability. A similar expression is obtained for the frequency dependent (dynamic) dipole polarizability obtained for time-dependent external field perturbations (see ref.10 for more details),

$$\alpha_{\alpha\beta}(\omega) = -\left\langle\left\langle r_\alpha; r_\beta\right\rangle\right\rangle_\omega = 2\sum_{m\neq 0}\frac{(E_m - E_0)\left\langle 0\left|r_\alpha\right|m\right\rangle\left\langle m\left|r_\beta\right|0\right\rangle}{(E_m - E_0)^2 - \omega^2} \tag{5}$$

and we introduced the notation usually used in response theory. Eq.(5) has poles at $\omega = E_m - E_0$ identical to atomic absorption frequencies (the substitution $\omega \to i\omega$ removes these poles). Equations (3) to (5) are all used for the calculation of dipole polarizabilities. Similar expressions can be obtained for higher order (hyper) polarizabilities.[6]

For a spherically symmetric system we have $\alpha_{xx} = \alpha_{yy} = \alpha_{zz} = \alpha$. For atoms in a $S=0$ or $J=0$ state we can simply rewrite (5) to

$$\alpha(\omega) = -\left\langle\left\langle z; z\right\rangle\right\rangle_\omega = \frac{2}{3}\sum_{m\neq 0}\frac{(E_m - E_0)}{(E_m - E_0)^2 - \omega^2}\left|\left\langle 0\left|\vec{r}\right|m\right\rangle\right|^2 \tag{6}$$

and only one diagonal component of the polarizability tensor is required, that is α reduces to a scalar property. For small frequencies the Cauchy formula is often used,

$$\alpha(\omega) = \sum_k S(-2k-2)\,\omega^{2k} \tag{7}$$

where $S(-2k-2)$ are the Cauchy moments.

Applying a static external electric field (without loss of generality) in z-direction (Stark perturbation), will however break the atomic symmetry to $C_{\infty v}$. Hence, L is not a good quantum number anymore and the degenerate M_L states split into states of $\Lambda = |M_L|$. It is therefore more convenient to change from a Cartesian to a spherical tensor basis [28,29] where the field is applied in z-direction (without loss of generality). As a result, we obtain $(L+1)$ components for the dipole polarizability. For example, for the ^3P state of carbon we obtain two components of the dipole polarizability. For closed shell atoms ($^1S_0$), or in general states of S-symmetry, this presents no problem and most calculations are therefore concentrated on elements in Groups 1, 2, 11, 12 and 18 of the periodic table. There are only a few calculations for atoms in higher L-states.[18,30] For these the M_L-averaged (scalar) dipole polarizabilities for a specific L-state are usually given,[31]

$$\bar{\alpha}_L = \frac{1}{2L+1} \sum_{|M_L|=0}^{L} (2 - \delta_{\Lambda 0}) \, \alpha_L(M_L) \qquad (8)$$

$\alpha_L(M_L) = \alpha_L(-M_L)$ can be determined by multi-reference techniques. The differences between different M_L-values for the polarizability can be quite substantial.[18]

For the relativistic (spin-orbit coupled) case we can always assume from eq.(3) that the Stark perturbation in z-direction is small compared to spin-orbit-coupling. For example, the ^3P state of carbon now splits into $^3P_0$, $^3P_1$ and $^3P_2$ due to spin-orbit splitting. The quadratic Stark effect results into one, two and three polarizability components for these states respectively and relativistic multiconfiguration techniques are needed to calculate these components. Khadjavi *et al.*[32] and Angel and Sandars[33] derived a useful relation for open-shell systems using perturbation theory and basic vector algebra,

$$\langle JM_J | \hat{\alpha}_{zz} | JM_J \rangle = \alpha_J(M_J) = \bar{\alpha} + \alpha_a \frac{3M_J^2 - J(J+1)}{J(2J-1)} \qquad (9)$$

where $\alpha_J(M_J) = \alpha_J(-M_J)$. The subscript a stands for the spherical anisotropy component and $J > 1/2$.[‡] This equation is obtained from lowest-order perturbation theory,[34] hence if the number of possible M_J states exceeds two for a specific J-state (for the two independent

---

[‡] The most common notation for $\bar{\alpha}$ and α_a is α_0 and α_2 respectively, where α_0 is called the scalar and α_2 the tensor polarizability.

polarizability components), eq.(9) is valid only if α_a is small compared to $\bar{\alpha}$. This is not always the case as we will see. For $J=1/2$ the anisotropy component vanishes as one expects.

Similarly we get for the nonrelativistic case (*LS*-coupling, valid only if spin-orbit coupling and α_a are small) for $L > 0$,

$$\alpha_L(M_L) = \bar{\alpha} + \alpha_a \frac{3M_L^2 - L(L+1)}{L(2L-1)} \tag{10}$$

Again, this relation is not strictly fulfilled if the number of possible M_L states exceeds two, but is helpful for estimating the individual M_L components of the spherical atomic polarizability. For a more detailed discussion see Bonin and Kresin.[10] Note that in this definition we have $\alpha_J(J) = \bar{\alpha} + \alpha_a$ (or $\alpha_L(L) = \bar{\alpha} + \alpha_a$), and $\alpha_0(0) = \bar{\alpha}$ as well as $\alpha_{\frac{1}{2}}(\frac{1}{2}) = \bar{\alpha}$. We point out that α_a does not have to be positive in sign. The anisotropy component vanishes for $L=0$ (*S*-states). These relations are in use for the measurements of Stark shifts in electronic transitions,[20,35] where the anisotropy component α_a can easily be obtained[36] (for more details see Bederson and Robinson[37]).

The determination of $\alpha(M_L)$ values for a specific atomic state is quite challenging from a computational point of view, and it is therefore not surprising that accurate polarizabilities for heavier open-shell atoms are currently not available. Note that for hyperfine levels (J, M_J) can simply be replaced by (F, M_F) in eq.(9). In the case of small spin-orbit coupling the average polarizabilities $\bar{\alpha}$ defined in eqs. (9) and (10) should not deviate substantially from each other, i.e. $\bar{\alpha}_L = \bar{\alpha}_J = \bar{\alpha}_F$. In this case relations between the anisotropy components can be derived using simple angular momentum algebra.[33] The Cartesian tensor components can be obtained from $\bar{\alpha}$ and α_a through basic vector algebra.[10] Relations between the different sets of quantum numbers (L, M_L), (J, M_J) and (F, M_F) are given by Brink and Satchler[38] or Angel and Sandars.[33] In the case of a *P*-state the situation simplifies because α_{xx} for $M_L = 0$ is identical to α_{zz} for $M_L = \pm 1$ and vice versa. In general, if the electric field coincides with the axis of symmetry (*z*-axis), the off-diagonal elements in α vanish and $\alpha_{xx} = \alpha_{yy}$. That is diagonalization of the polarizability tensor for a P-state leads to the two components $\alpha_1(0)$ and $\alpha_1(1)$, and $\bar{\alpha} = tr(\alpha)/3$ does not change anyway.

Polarizabilities from eq.(3) can be determined either numerically through finite field perturbations (FFP)[39-41] (either by using a homogeneous electric field perturbation, or by a dipole field using point charges at large distances to minimize inhomogeneity effects), or analytically by using coupled perturbed methods such as the coupled perturbed Hartree-Fock (CPHF) or coupled perturbed configuration interaction (CPHF) methods (for details see ref.42). Alternatively, linear response theory can be used at high levels of theory (for an excellent review on this subject see Christiansen *et al.*[43]). Implementation of response theory into a 4-component relativistic framework is currently in progress.[44]

3. Hydrogenic Systems

It is useful to review the current status on the polarizability for hydrogenic systems (see also ref.45 for a more detailed review). Waller,[46] Wentzel[47] and Epstein[48] already determined the exact nonrelativistic static dipole polarizability in 1926 for all (n,l,m_l) states, which is $\alpha(Z) = 4.5\ Z^{-4}$ a.u. for the electronic ground state. Precise values of all nonrelativistic static polarizabilities and hyperpolarizabilities up to sixth order can be found in a paper by Bishop and Pipin.[49] For the relativistic case (Dirac equation) we do not get exact analytical solutions. However, in 1969 Bartlett and Power obtained a relativistic expression for the dipole polarizability up to order $(Z^2 c^{-2})$ using the Dirac equation,[50]

$$\alpha(Z) = \tfrac{9}{2} Z^{-4} \left[1 - \tfrac{28}{27}(Z^2 c^{-2}) \right] \overset{Z=1}{=} 4.4997515...\quad \text{a.u.} \tag{11}$$

Here the velocity of light is $c = 137.03599$ a.u. Kaneko extended the work to higher relativistic multipole polarizabilities and shielding factors.[51] Goldman and co-workers obtained more precise values by accurate relativistic variational calculations and a fit to a power series in $(Zc^{-1})^m$ (refs. 52-54, see also ref.55),

$$\alpha(Z) = Z^{-4} \sum_i \lambda_i \left(Z^2 c^{-2} \right)^i \overset{Z=1}{=} 4.49975149514292 \text{ a.u.} \tag{12}$$

with $\lambda_0 = +9/2$, $\lambda_1 = -14/3$, $\lambda_2 = +2584/4889$, $\lambda_3 = +101/3000$, $\lambda_4 = +22/5000$. Note the strong Z-dependence ($\sim Z^{-4}$) of the dipole polarizability. Fig.1 shows Goldman's results dependent on the nuclear charge Z.[52-54] It demonstrates that relativistic effects soon become important even for the lighter elements (Z^2-dependence). Another important result is that relativistic effects decrease the dipole polarizability (negative sign of λ_1), a direct consequence of the relativistic 1s-contraction. At higher charges Z, QED corrections may become important which have not been investigated yet for the hydrogen atom.

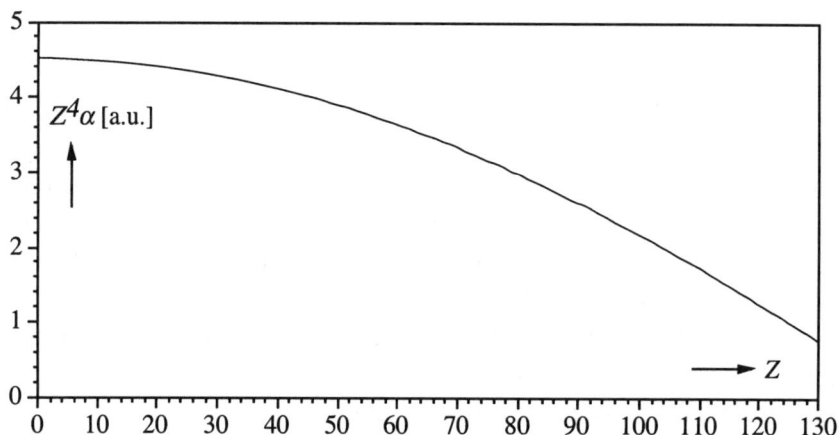

Figure 1. Dependence of the static dipole polarizability (scaled by Z^4) on the nuclear charge Z for hydrogenic atoms. The data are taken from refs. 52-54.

More recently, Szmytkowski studied static polarizabilities of the relativistic hydrogenlike atom using a Sturmian expansion of the Dirac-Coulomb Green function.[56-58] Dynamic polarizabilities of hydrogenic atoms have been investigated as early as 1963 by Karplus and Kolker,[59] and later by Tang and Chan,[60] and this is still an active field of research.[61-64] The dipole polarizability of the hydrogen atom confined in a sphere of a certain radius (at high pressures) has been studied recently by Sen *et al.*,[65] Dutt *et al.*[66] and Laughlin.[67] Interestingly, excited electronic state polarizabilities of hydrogenlike atoms have not been investigated in detail yet.[68] Here the hydrogen atom represents a special case, since the degeneracy in the states with principal quantum number $n > 1$ already results in a first-order (linear) Stark term.[69]

4. Multi-Electron Atoms. The Static Dipole Polarizabilities from Z=1 to 119

Calculated static dipole polarizabilities for the neutral atoms up to Z=119 are listed in Table 1 further below. Only the most accurate calculations are listed. For all other atomic values we refer to the estimated relativistic LDA results obtained from linear response theory by Doolen[70] using the method described in refs.71 and 72, as listed in Miller's compilation.[19] Experimental values are also included for comparison. For conversion to other units for the polarizability: 1 a.u. = 0.14818474 $Å^3$ = 0.16487776 $\times 10^{-40}$ C m^2/V.

It is well known that electron correlation effects are important for electric or magnetic properties. It is however clear from Fig.1 that relativistic effects for dipole polarizabilities cannot be neglected for the heavier elements, and therefore have to be included for precise calculations. It is now well established that relativistic effects are important not only for inner atomic shells, but through relativistic direct, indirect and spin-orbit effects for valence shells as well.[73]

The first relativistic calculations for neutral atoms up to Z=92 based on density functional theory (DFT) using the Slater exchange were carried out by Feiock and Johnson in 1969.[74] Nonrelativistic data based on the Schofield-Pople approximation are available from Fraga.[75] These two data bases are not directly comparable to extract relativistic effects, but in 1981 Desclaux *et al.*[76] and Sin Fai Lam[77] demonstrated for a number of heavy atoms that relativistic effects

Table 1. Static dipole polarizabilities (in a.u.) for the neutral atoms (see end of table for details).

Z	Atom	α_D	comments	Refs.
1	H	4.4997515	R, Dirac, variational, Slater basis	[52]
2	He	1.383191	R, Breit-Pauli+QED, mass pol., corr. basis	[78]
		1.383	*exp.*	[79]
3	Li	164.05	NR, r_{12}, + R DK correction	[80,81]
		164.0(3.4)	*exp.*	[82]
4	Be	37.755	NR, r_{12}	[80]
5	B	20.5	NR, PNO-CEPA, ^2P, M_L res.	[83]
		20.43	NR, CCSD(T), ^2P, M_L res.	[84]
		20.53/20.54	R, Dirac, MRCI, $^2P_{1/2}$/$^2P_{3/2}$, M_J res.	[85]
6	C	11.0	NR, CASPT2, ^3P, M_L res.	[86]
		11.67	NR, CCSD(T), ^3P, M_L res.	[84]
7	N	7.43	NR, PNO-CEPA, ^4S	[83]
		7.41	R, DK, CASPT2, ^4S	[87]
		7.26	NR, CCSD(T), ^4S	[84]
		7.6±0.4	*exp.*	[82,88]
8	O	6.04	NR, PNO-CEPA, ^3P, M_L res.	[83]
		6.1	NR, CASPT2, ^3P, M_L res.	[86]
		5.24	NR, CCSD(T), ^3P, M_L res.	[84]
9	F	3.76	NR, PNO-CEPA, ^2P, M_L res.	[83]
		3.76	NR, CASPT2, ^2P, M_L res.	[89]
		3.70	NR, CCSD(T), ^2P, M_L res.	[84]
10	Ne	2.68	NR, CCSD(T)	[90]
		2.665	NR, CC3	[91]
		2.666	R, CC3+FCI+DK3 correction	[91-93]
		2.670±0.005	*exp.*	[94]
11	Na	162.6	R, SD all orders + exp. data	[95]
		162.7(8)	*exp.*	[96]
12	Mg	71.7	NR, MBPT4	[97]
		71.8	NR, MBPT4	[98]
		70.9	R, DK, CASPT2	[99]
13	Al	56.3	NR, PNO-CEPA, ^2P	[100]
		62.0	NR, numerical MCSCF, ^2P, M_L res.	[18]
		57.74	NR, CCSD(T), ^2P, M_L res.	[101]
		55.4/55.9	R, Dirac, MRCI, $^2P_{1/2}$/$^2P_{3/2}$, M_J res.	[85]
		46±2	*exp.*	[102,103]
14	Si	36.7	NR, PNO-CEPA, ^3P, M_L res.	[100]
		36.5	NR, CASPT2, ^3P, M_L res.	[86]
		37.4	NR, CCSD(T), ^3P, M_L res.	[104]
		37.17	NR, CCSD(T), ^3P, M_L res.	[101]
15	P	24.7	NR, PNO-CEPA, ^4S	[100]
		24.6	NR, CASPT2, ^4S	[86]
		24.9	R, DK, CASPT2, ^4S	[87]
		24.93	NR, CCSD(T), ^4S	[101]

Z	Atom	α_D	comments	Refs.
16	S	19.6	NR, PNO-CEPA, ^3P, M_L res.	[100]
		19.6	NR, CASPT2, ^3P, M_L res.	[86]
		19.6	NR, CASPT2, ^3P, M_L res.	[89]
		19.37	NR, CCSD(T) , ^3P, M_L res.	[101]
17	Cl	14.7	NR, PNO-CEPA, ^2P, M_L res.	[100]
		14.6	NR, CASPT2, ^2P, M_L res.	[86]
		14.73	NR, CASPT2, ^2P, M_L res.	[89]
		14.57	NR, CCSD(T) , ^2P, M_L res.	[101]
18	Ar	11.10	NR, PNO-CEPA	[100]
		11.084	NR, CCSD(T)	[105]
		11.1	R, DK, CASPT2	[87]
		11.10	R, CCSD(T) + DK3 correction	[93,105]
		11.070(7)	*exp.*	[106,107]
19	K	289.1	R, SD all orders + exp. data	[95]
		291.1	R, DK, CCSD(T)	[108]
		293±6	*exp.*	[82]
20	Ca	160	R, CI, MBPT	[109]
		152.0	R, MVD, CCSD+T	[110]
		163	R, DK, CASPT2	[99]
		158.6	R, DK+SO, CCSD(T)	[111]
		169±17	*exp.*	[112]
21	Sc	120	R, LDA	[70]
		107	NR, small CI, VPA	[31]
		142.28	NR, MCPF	[113]
22	Ti	99	R, LDA	[70]
		92	NR, small CI, VPA	[31]
		114.34	NR, MCPF	[113]
23	V	84	R, LDA	[70]
		81	NR, small CI, VPA	[31]
		97.34	NR, MCPF	[113]
24	Cr	78	R, LDA	[70]
		94.72	NR, MCPF	[113]
		78.4	DK, CASPT2	[114]
25	Mn	63	R, LDA	[70]
		65	NR, small CI, VPA	[31]
		75.52	NR, MCPF	[113]
		66.8	DK, CASPT2	[114]
26	Fe	57	R, LDA	[70]
		58	NR, small CI, VPA	[31]
		63.93	NR, MCPF	[113]
		62.65	NR, GGA(PW86)	[115]
27	Co	51	R, LDA	[70]
		53	NR, small CI, VPA	[31]
		57.71	NR, MCPF	[113]
28	Ni	46	R, LDA	[70]
		48	NR, small CI, VPA	[31]
		51.10	NR, MCPF	[113]

Z	Atom	α_D	comments	Refs.
29	Cu	53.44	NR, MCPF	[113]
		45.0	R, PP, QCISD(T)	[116]
		46.5	R, DK, CCSD(T)	[117]
		40.7	DK, CASPT2	[114]
30	Zn	42.85	NR, MCPF	[113]
		39.2	NR, CCSD(T), MP2 basis correction	[118]
		38.0	R, PP, CCSD(T)	[119]
		37.6	R, MVD, CCSD(T)	[120]
		38.4	DK, CASPT2	[114]
		38.8±0.3	*exp.*	[118]
31	Ga	54.9	NR, PNO-CEPA, ^2P, M_L res.	[121]
		49.0	R, DK, MRCI, ^2P, M_L res.	[85]
		47.7/49.6	R, Dirac, MRCI, $^2P_{1/2}/^2P_{3/2}$, M_J res.	[85]
32	Ge	41.0	NR, PNO-CEPA, ^3P, M_L res.	[121]
33	As	29.1	NR, PNO-CEPA, ^4S	[121]
		29.8	R, DK, CASPT2, ^4S	[87]
34	Se	26.24	R, MVD, CASPT2, ^3P, M_L res.	[89]
35	Br	21.9	R, DK, SO-CI, $^2P_{1/2}$	[122]
		21.8	R, DK, SO-CI, $^2P_{3/2}$, M_J res.	[122]
		21.03	R, MVD, CASPT2, M_L res., ^2P	[89]
36	Kr	16.8	R, DK3, CCSD(T)	[106]
		16.6	R, DK, CASPT2	[87]
		17.075	*exp.*	[106]
37	Rb	318.6	R, SD all orders + exp. data	[95]
		316.2	R, DK, CCSD(T)	[108]
		316±6	*exp.*	[82]
38	Sr	199	R, CI, MBPT	[109]
		199.4	R, DK+SO, CCSD(T)	[111]
		186±15	*exp.*	[112]
39	Y	153	R, LDA	[70]
40	Zr	121	R, LDA	[70]
41	Nb	106	R, LDA	[70]
42	Mo	86	R, LDA	[70]
		72.5	DK, CASPT2	[114]
43	Tc	77	R, LDA	[70]
		80.4	DK, CASPT2	[114]
44	Ru	65	R, LDA	[70]
45	Rh	58	R, LDA	[70]
46	Pd	32	R, LDA	[70]
47	Ag	52.2	R, PP, QCISD(T)	[116]
		52.5	R, DK, CCSD(T)	[117]
		36.7	DK, CCSD(T)	[114]

Z	Atom	α_D	comments	Refs.
48	Cd	46.3	R, PP, CCSD(T)	[119]
		46.8	R, MVD, CCSD(T)	[120]
		46.9	DK,CASPT2	[114]
		49.65±1.46	*exp.*	[123]
49	In	65.2	R, DFT, $^2P_{1/2}$	[124]
		69.6	R, DK, MRCI, ^2P, M_L res.	[85]
		67.2/74.2	R, Dirac, MRCI, $^2P_{1/2}/^2P_{3/2}$, M_J res.	[85]
		68.7±8.1	*exp.*	[125]
50	Sn	52	R, LDA	[70]
		53.3	R. PP. MBPT	[126]
		61.3	R, PP, PW91	[126]
51	Sb	45	R, LDA	[70]
		42.2	R, DK, CASPT2	[87]
52	Te	37	R, LDA	[70]
53	I	35.1	R, DK, SO-CI, $^2P_{1/2}$	[122]
		34.6	R, DK, SO-CI, $^2P_{3/2}$, M_J res.	[122]
54	Xe	27.06	R, DK3, CCSD(T)	[93]
		27.36	R, SOPP, CCSD(T) + MP2 basis set correction	[127]
		26.7	R, DK, CASPT2	[87]
		27.815	*exp.*	[106]
55	Cs	399.9	R, SD all orders + exp. data	[95]
		396.0	R, DK, CCSD(T)	[108]
		401.0±0.6	*exp.*	[128]
56	Ba	273	R, CI, MBPT	[109]
		273.5	R, DK+SO, CCSD(T)	[111]
		268±22	*exp.*	[112]
57	La	210	R, LDA	[70]
58	Ce	200	R, LDA	[70]
59	Pr	190	R, LDA	[70]
60	Nd	212	R, LDA	[70]
61	Pm	203	R, LDA	[70]
62	Sm	194	R, LDA	[70]
63	Eu	187	R, LDA	[70]
64	Gd	159	R, LDA	[70]
65	Tb	172	R, LDA	[70]
66	Dy	165	R, LDA	[70]
67	Ho	159	R, LDA	[70]
68	Er	153	R, LDA	[70]
69	Tm	147	R, LDA	[70]
70	Yb	142	R, LDA	[70]

Z	Atom	α_D	comments	Refs.
71	Lu	148	R, LDA	[70]
72	Hf	109	R, LDA	[70]
73	Ta	88	R, LDA	[70]
74	W	75	R, LDA	[70]
75	Re	65	R, LDA	[70]
		61.1	DK, CASPT2	[114]
76	Os	57	R, LDA	[70]
77	Ir	51	R, LDA	[70]
78	Pt	44	R, LDA	[70]
79	Au	35.1	R, PP, QCISD(T)	[116]
		36.1	R, DK, CCSD(T)	[117]
		27.9	DK, CASPT2	[114]
80	Hg	34.4	R, PP, CCSD(T)	[119]
		31.2	R, MVD, CCSD(T)	[120]
		33.3	DK,CASPT2	[114]
		33.91±0.34	*exp.*	[129]
81	Tl	67.4	R, DK, MRCI, ^2P, M_L res.	[85]
		51.2/80.6	R, Dirac, MRCI, $^2P_{1/2}/^2P_{3/2}$, M_J res.	[85]
		51	*exp.*	[19]
82	Pb	46	R, LDA	[70]
		51.0	R, SOPP, CCSD(T)	[130]
83	Bi	50	R, LDA	[70]
		48.6	R, DK, CASPT2	[87]
84	Po	46	R, LDA	[70]
85	At	45.6	R, DK, SO-CI, $^2P_{1/2}$	[122]
		43.0	R, DK, SO-CI, $^2P_{3/2}$, M_J res.	[122]
86	Rn	33.18	R, DK3, CCSD(T)	[93]
		34.33	R, SOPP, CCSD(T) + MP2 basis set correction	[127]
		32.6	R, DK, CASPT2	[87]
87	Fr	317.8	R, SD all orders + experimental data	[95]
		315.2	R, DK, CCSD(T)	[108]
88	Ra	246.2	R, DK+SO, CCSD(T)	[111]
89	Ac	217	R, LDA	[70]
90	Th	217	R, LDA	[70]
91	Pa	171	R, LDA	[70]
92	U	152.7	R, LR	[70]
		168±10	*exp.*, $^5L_6$, M_J res.	[131]
93	Np	167	R, LDA	[70]
94	Pu	165	R, LDA	[70]
95	Am	157	R, LDA	[70]
		116	DK, CASPT2	[132]
96	Cm	155	R, LDA	[70]

Z	Atom	α_D	comments	Refs.
97	Bk	153	R, LDA	[70]
98	Cf	138	R, LDA	[70]
99	Es	133	R, LDA	[70]
100	Fm	161	R, LDA	[70]
101	Md	123	R, LDA	[70]
102	No	118	R, LDA	[70]
112		25.8	R, PP, CCSD(T)	[119]
		28.7	R, SOPP, CCSD(T)	[130]
114		34.4	R, SOPP, CCSD(T)	[130]
118		52.4	R, SOPP, CCSD(T)	[130]
119		163.8	R, DK, CCSD(T)	[108]

Table 1. Static dipole pPolarizabilities (in a.u.) for the neutral atoms. If not otherwise stated by the state symmetry, $M_L(M_J)$ -averaged polarizabilities are listed. Abbreviations: exp.: experimentally determined value; NR: nonrelativistic; R: Relativistic, DK: Douglas-Kroll; MVD: mass-velocity-Darwin; SO: Spin-orbit coupled; PP: pseudopotential; LDA: local density approximation; PW91: Perdew-Wang 91 functional; MBPT: many-body perturbation theory; CI: configuration interaction; CCSD(T): coupled cluster singles doubles (SD) with perturbative triples; CEPA: coupled electron pair approximation; MR: multi-reference; CAS: complete active space; VPA: variational perturbation approach [1]. For all other abbreviations see text or references. M_L res. denotes that the polarizability for each M_L state can be found in the reference given. For all other abbreviations see the literature reference given. Experimental values are given in italics.

are most important. For example, for mercury the nonrelativistic Hartree-Fock (NRHF) value is α= 80 a.u., while the relativistic Dirac-Hartree-Fock (DHF) value is almost halved to 43 a.u.[76] This is due to the relativistic *6s*-shell contraction. For closed p-shells relativistic effects are less pronounced, i.e. Sin Fai Lam gives for Rn α(NRHF)= 47.6 a.u. and α(DHF)= 46.4 a.u.[77] Relativistic effects can change the overall trend in the periodic table as will be discussed in the next section.

Helium has been the subject of very precise calculations.[45] Accurate nonrelativistic correlated values are given by Bishop and Pipin (α= 1.383192 a.u.),[49] Dalgarno and co-workers (α= 1.38319217440 a.u.)[133] and Pachucki and Sapirstein (α= 1.383192174455(1) a.u.,[78] see also ref.134). Relativistic and mass polarization corrections were considered by Weinhold in 1982 within a Breit-Pauli formalism, and found to be quite small, $\Delta\alpha=$ 1.03×10^{-5} a.u.[135] This agrees with the experimental value, which is 1.383223 ± 0.000067 a.u., obtained in 1980 by Gugan and Michel through dielectric constant gas thermometry.[136] One should mention that the individual relativistic and mass-polarization contributions are quite large (some of the order of 10^{-4} a.u.), but accidentally cancel out.[135] Pachucki and Sapirstein[78] and more recently Cencek et al.[137] improved Weinhold's result, and including also QED corrections[78] one obtains $\alpha=$ $1.383191(2)$ a.u. The latest accurate QED results for the dipole polarizability of ^{4}He including recoil corrections gives $1.38376079(23)$ a.u. (the nonrelativistic value is 1.383809986).[138] Both values include the mass polarization term.

Johnson and Cheng studied static dipole polarizabilities for two-electron systems.[139] It is interesting to compare relativistic effects for both the one- and two-electron series, Fig.2. To a very good approximation the hydrogenic values follow a Z^2-increase in relativistic effects as discussed before. After an initial strong increase in relativistic effects, an approximately Z^2-behaviour is also found for He. As pointed out by Johnson and Cheng, in the high Z-limit $\Delta_R\alpha \rightarrow$ $-0.486 \times 10^{-3}/Z^2$ a.u. This is approximately twice the value for a hydrogenic atom.

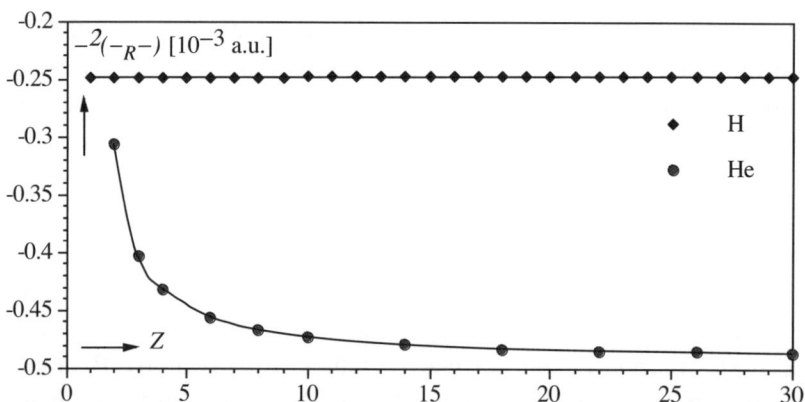

Figure 2. Relativistic change in the static dipole polarizability (scaled by Z^2) with respect to the nuclear charge Z for hydrogenic atoms. The data are taken from refs.52-54,139.

Dalgarno and co-workers give a precise nonrelativistic value for lithium (α= 164.111 a.u.),[133] which compares well with the experimental value of 164.0(3.4) a.u.[82] Komasa obtained α= 164.11171 a.u. using correlated Gaussian functions at the nonrelativistic level of theory.[80] Relativistic effects for Li have been estimated recently *($\Delta_R\alpha$= −0.06 a.u.)* which brings the polarizability to 164.05 a.u.[81]

For beryllium the most accurate value is again the nonrelativistic result of Komasa with α= 37.755 a.u.[80] or Tunega et al. with α= 37.73 a.u.[140] Relativistic effects will certainly lower this value. A recent DK/CASPT2 result of Roos and co-worker gave 37.2 a.u.[99]

For the following open-shell systems only very few accurate calculations are available, and we will not discuss these in detail here (see table 1).[18] For the first and second *p*-row elements a detailed discussion is given by Thakkar and co-workers,[84,101] and for Al by Fuentealba.[141] It is however interesting to compare the open-shell results obtained by Stiehler and Hinze[18] with the ones obtained from the approximation used in eq.(10). For example, for a *D*-state we have

$$
\begin{aligned}
M_L&= 0: & \alpha_2(0) &= \bar{\alpha} - \alpha_a \\
M_L&=\pm 1: & \alpha_2(1) &= \bar{\alpha} - \tfrac{1}{2}\alpha_a \\
M_L&=\pm 2: & \alpha_2(2) &= \bar{\alpha} + \alpha_a
\end{aligned}
\tag{13}
$$

For the Si ^1D state Stiehler and Hinze get $\alpha_2(0)$= 49.785 a.u. and $\alpha_2(2)$= 26.900 a.u. which gives $\bar{\alpha}$ = 42.775 a.u. and α_a= −7.011 a.u. This gives $\alpha_2(1)$= 45.780 a.u. in good agreement with the calculated value of 45.896 a.u.[18] For an *F*-state we have

$$
\begin{aligned}
M_L&= 0: & \alpha_3(0) &= \bar{\alpha} - \tfrac{4}{5}\alpha_a \\
M_L&=\pm 1: & \alpha_3(1) &= \bar{\alpha} - \tfrac{3}{5}\alpha_a \\
M_L&=\pm 2: & \alpha_3(2) &= \bar{\alpha} \\
M_L&=\pm 3: & \alpha_3(3) &= \bar{\alpha} + \alpha_a
\end{aligned}
\tag{14}
$$

For the V ^4F state Stiehler and Hinze get $\alpha_3(2)$= 113.189 a.u. and $\alpha_3(3)$= 111.351 a.u. which gives $\bar{\alpha}$ = 113.189 a.u. and α_a= −1.838 a.u. This gives $\alpha_3(0)$= 114.659 a.u. and $\alpha_3(1)$= 114.292 a.u. in reasonable agreement with the calculated values of 114.951 a.u. and 114.297 respectively.[18] Stiehler and Hinze also give M_L-resolved MCSCF polarizabilities from Sc to V and Fe to Ni.[18] We note that for Ga the Stark electric field splitting of the P-state becomes quite sizable with $\alpha(M_L=0)$= 80.67 a.u. and $\alpha(M_L=\pm 1)$= 42.04 a.u.[121]

Neon has been investigated quite recently and the most accurate values come from Jørgensen and co-workers using linear response theory at the coupled cluster level. Their most precise value is an iterative triples CC3 result together with a d-aug-cc-pV6Z basis set which gives 2.665 a.u.[91] Including corrections accounting for deviations to the full CI result estimated at the CC3 d-aug-pVDZ level of theory brings α down to 2.662 a.u.[92] This is already in excellent agreement with experiment (2.670±0.005 a.u.).[94] There may be a small correction due to relativistic effects. Indeed, Nicklass *et al.* pointed out that scalar relativistic effects increase the Ne polarizability by 0.007 a.u. bringing the value now very close to the experimental one.[142] This was confirmed recently by Douglas-Kroll calculations of Nakajima and Hirao.[93] Spin-orbit effects will most likely further increase this value (relativistic expansion of the $p_{3/2}$ spinor). Direct relativistic perturbation theory used by W. Klopper *et al.* of relativistic effects gives a correction of 0.00443 a.u.[143] They report a static dipole polarizability for Ne at the CCSD(T) level of theory 2.66312 a.u. Accurate values for the static second hyperpolarizability of neon are also available.[143,144]

Accurate polarizabilities for the heavier elements are only available for the Group 1, 2, 11, 12 and 18 elements, and more recently[86] for the Group 15 elements. They are listed in Table 1. The most precise values for the heavier elements are the ones for the Group 1 atoms from Na to Fr by Derevianko *et al.*[95] using a relativistic single-double all-order[145] within a sum-over-states approach, and using experimental values (if available) for energy levels. For these one-valence s-electron systems the sum-over-states approach converges fast with increasing principle quantum number of the excited p-states, and the continuum does not seem to be of any importance. For sodium the HF limit is estimated at α= 190.2 a.u., and coupled cluster calculations give 165.1 a.u.[146] A recent nonrelativistic CCSD(TQ) calculation gives 166.1 a.u.[81] Relativistic effects lower the polarizability by 1.0 a.u. This gives a value still about 2 a.u. above the Derevianko *et al.* result, which is in accordance with the latest atom-interferometry experiment by Ekstrom *et al.*[96] of α= 162.7(8) a.u. More recently Thakkar and Lupinetti recommend 162.88±0.6 a.u. for the dipole polarizability of Na from relativistic coupled cluster calculations.[147] The most recent high precision experiment comes from Amini and Gould for Cs in good agreement with theoretical predictions.[128] Roos and co-workers published non-relativistic polarizabilities for the first-row transition elements using a modified coupled pair functional (MCPF),[113] but these values are not M_L-resolved. They also published relativistic ANO basis sets for Group 1 and 2 elements and list polarizabilities at the Douglas-Kroll CASPT2 level from Be down to Ra and for a number of transition elements.[99,114] For Zn, Cd and Hg accurate refractive index measurements by Goebel and Hohm[118,123,129] are available. Here, the theoretical calculations (table 1) do not seem to reach the 1% accuracy level yet, mainly because of basis set limitations and approximations in the relativistic and electron correlation procedures used. For mercury, Goebel and Hohm give the first three Cauchy moments according to eq.(7).[129]

Very recently Fleig has derived scalar relativistic and spin-orbit coupled polarizabilities for the group 13 atoms in the periodic table (see table 1).[85] This work is particularly interesting since one can test the different coupling schemes, i.e. *LS* and *jj*. If spin-orbit coupling is small one approximately gets the useful relation $\bar{\alpha}_L = \bar{\alpha}_J \ (= \bar{\alpha}_F)$.[10] Fig.3 shows the scalar polarizabilities of the Group 13 elements in both coupling schemes. For boron we have $\bar{\alpha}_L = 20.59$ a.u., $\bar{\alpha}_J = 20.53$ a.u. for $J=1/2$ and $\bar{\alpha}_J = 20.54$ a.u. for $J=3/2$, in close agreement with each other. Figure 3 shows that the deviation between these values increase with nuclear charge Z and becomes very large for thallium, where spin-orbit splitting is as large as 0.96 eV between the $^2P_{1/2}$ and $^2P_{3/2}$ states. Eq.(3) now allows the J-averaging of polarizabilities in complete analogy to J-averaging of energies. Hence we get $\bar{\bar{\alpha}}_{L=1} = (\bar{\alpha}_{J=1/2} + 2\bar{\alpha}_{J=3/2})/3$ which gives $\bar{\bar{\alpha}}_{L=1} = 71.4$ a.u. for Tl in reasonable agreement with the scalar relativistic result of 70.0 a.u.

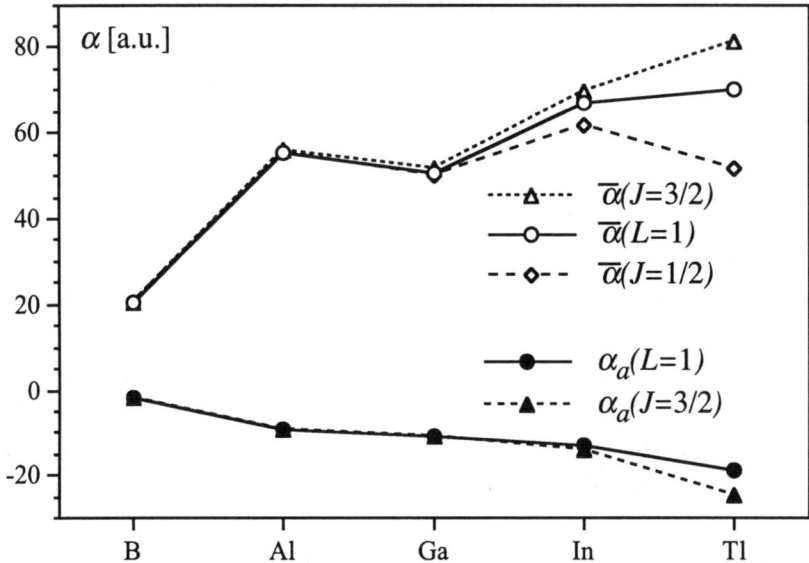

Figure 3. Average and anisotropy component of the static dipole polarizabilities of the Group 13 elements (in a.u.). The values are derived from the work of Fleig.[85]

Concerning the anisotropy component α_a one has the following useful relation between the two different coupling schemes

$$\alpha_a(J) = f(L,S,J)\alpha_a(L) \tag{15}$$

with

$$f(L,S,J) = (-1)^{J+L+S} \left(\frac{(2J+1)(2J-1)(2L+3)(2L+1)(L+1)}{(2J+3)(J+1)L(2L-1)} \right)^{1/2} \begin{Bmatrix} L & L & 2 \\ J & J & S \end{Bmatrix} \tag{16}$$

Here the expression in curly brackets $\{...\}$ denotes the well-known Wigner 6j-symbol. This implies that the sign of $\alpha_a(J)$ is determined by $\alpha_a(L)$ and $f(L,S,J)$ or vice versa. For the Group 13 elements we have $f(1,1/2,1/2)=0$ and $f(1,1/2,3/2)=1$. Hence for $J=3/2$ we have $\alpha_a(J)=\alpha_a(L)$. Again, for boron this is a good approximation, i.e.

$\alpha_a(J)=-2.00$ a.u. and $\alpha_a(L)=-1.97$ a.u. Figure 3 shows that the deviation between the two values is largest for thallium as expected.

Polarizabilities for spin-orbit-split states of Group 17 elements have been investigated recently by Fleig and Sadlej.[122] For example, Fleig and Sadlej obtain for the $^2P_{1/2}$ state of At $\alpha= 44.88$ a.u. at the two-component Douglas-Kroll CI level of theory. The Stark effect splits the $^2P_{3/2}$ state into two states resulting in two polarizability components, $M_J=3/2$ with $\alpha= 45.21$ a.u. and $M_J=1/2$ with $\alpha= 38.04$ a.u.[122] The $J= 3/2$, $M_J=3/2$ state shows the highest polarizability component and the anisotropy component α_a is now positive with +3.6 a.u. Hence the sign is opposite compared to the $^2P_{3/2}$ state of thallium, which is easily explained because for the Group 17 elements we have a hole state in the p-shell, and the $p_{3/2}$ orbital is more extended in space and more polarizable compared to the $p_{1/2}$ spinor.[122] The scalar relativistic result is 36.0 a.u.[122]

A recent measurement for the polarizability of the $[Rn]5f^36d^17s^2$ ($^5L_6$) state of uranium using a light-force technique should be mentioned,[131] Table 1. This technique should be applicable to a wide range of atoms. The uncertainty in the α_a value (eq.(9)) is still quite high and accurate theoretical calculations are required here.

5. Trends and Correlation with other Properties

Fig.4 shows the atomic polarizabilities together with the ionization potentials of the neutral elements. It is evident and intuitive that an increasing ionization potential IP implies a decreasing polarizability, $IP \sim 1/\alpha$, as has been pointed out by Fricke in 1985.[148] This also explains the zig-zag trend for the Group 13 elements in Fig.3, which mirrors inversely the trend in the ionization potentials. Pyykkö recently published a detailed analysis on periodic trends in alkali and alkaline earth metal ionization potentials.[149] Much of what is discussed in there can be transferred to polarizabilities through the $IP \sim 1/\alpha$ relation. The main $\alpha(Z)$ behaviour can be explained through shell-closure effects, that is the inert gases have small polarizabilities, and the Group 1 elements have the largest polarizabilities at the beginning of a new shell. The polarizability then decreases within one period of the periodic table.   Some other anomalies seen in Fig. 4, which cannot be explained by shell-closure effects, may indeed be due to large errors in the calculated or experimental polarizabilities.

For a dielectric sphere or shell the polarizability $\alpha \sim R^3$ (R is the radius of the sphere),[150] which establishes the relationship between the polarizability and the "atomic volume". It is therefore of no surprise that a number of papers are dealing with the correlation between polarizabilities and other properties such as the ionization potential,[151] atomic radius,[152,153] electronegativity,[154,155] hardness,[156,157] etc.[158] It was argued before that a useful relationship is $\alpha \sim 1/IP^2$, although there are   other arguments for a $\alpha \sim 1/IP^3$ relationship especially for highly ionized atoms with closed shells.[159] Fig.5 shows a plot of atomic dipole polarizabilities versus the ionization potential squared.   The trend is clearly visible.   Most of the elements deviating substantially from the linear curve are open-shell systems like the Group 13 elements.

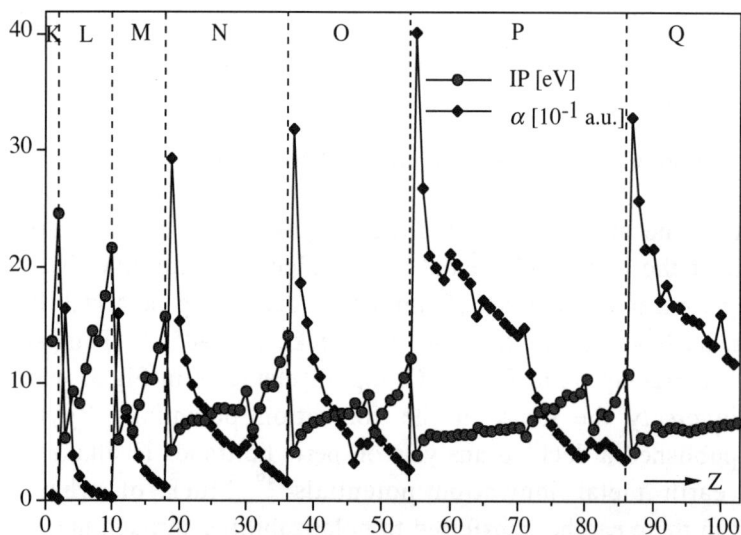

Figure 4. Static dipole polarizability (in 10^{-1} a.u.) and ionization potentials for the neutral atoms. The polarizabilities published by Miller are used.[19]

As mentioned before, relativistic effects can change the trend in polarizabilities within a Group of the periodic table, which are expected to increase monotonically with nuclear charge Z. Fig.6 shows nonrelativistic and relativistic dipole polarizabilities for the Group 1, 2, 11 and 12 elements of the periodic table. Relativistic dipole polarizabilities of Group 18 elements are also included. Relativistic effects increase approximately like Z^2 for these elements and are large enough to change the monotonically increasing trend at Fr and Ra for the s-block elements.[160] This change occurs one period earlier for the Group 11 and 12 atoms as pointed out in 1981 by Desclaux *et al.*[76] Here both Au and Hg have a lower polarizability than Cu and Zn, respectively, which underlines the importance of relativistic effects for these elements.[161]

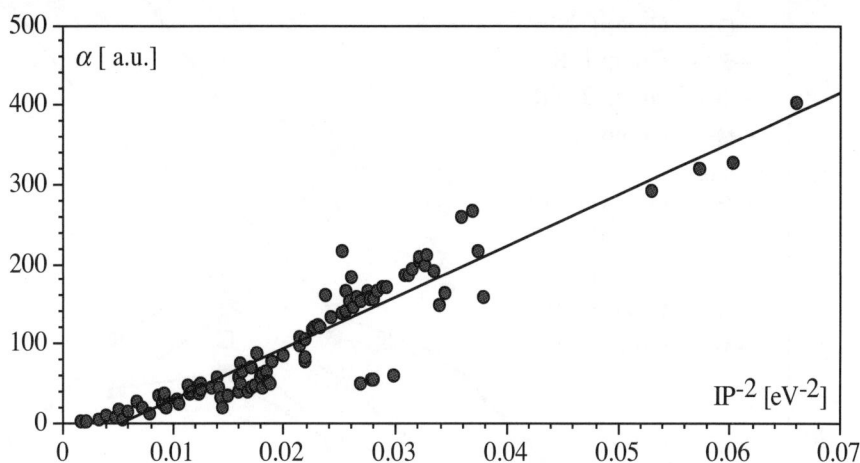

Figure 5. Static dipole polarizability versus the inverse of the ionization potential squared. The polarizabilities published by Miller are used.[19]

Relativistic effects in the rare gas systems have been studied at the scalar relativistic pseudopotential[142] and all-electron Douglas-Kroll third-order level (DK3).[93] Interestingly, for Ne and Ar relativistic effects lead to a slight increase in the polarizability, while for Ar, Xe and Rn the polarizability decreases. For Xe and Rn relativistic effects become sizable, i.e. $\Delta_R\alpha= -0.23$ a.u. for Xe, and $\Delta_R\alpha= -1.04$ a.u. for Rn.[93] Spin-orbit effects have not been included in these studies, which can substantially alter these values. This is indicated by the spin-orbit coupled pseudopotential results of Runeberg and Pyykkö.[127] Their results lie above the scalar relativistic results of Nakajima and Hirao.[93] As a final note, it would be interesting to study the relativistic change in the dipole polarizability of eka-Rn, the superheavy element with Z=118, where the $p_{3/2}$ spin-orbit expansion is quite substantial.

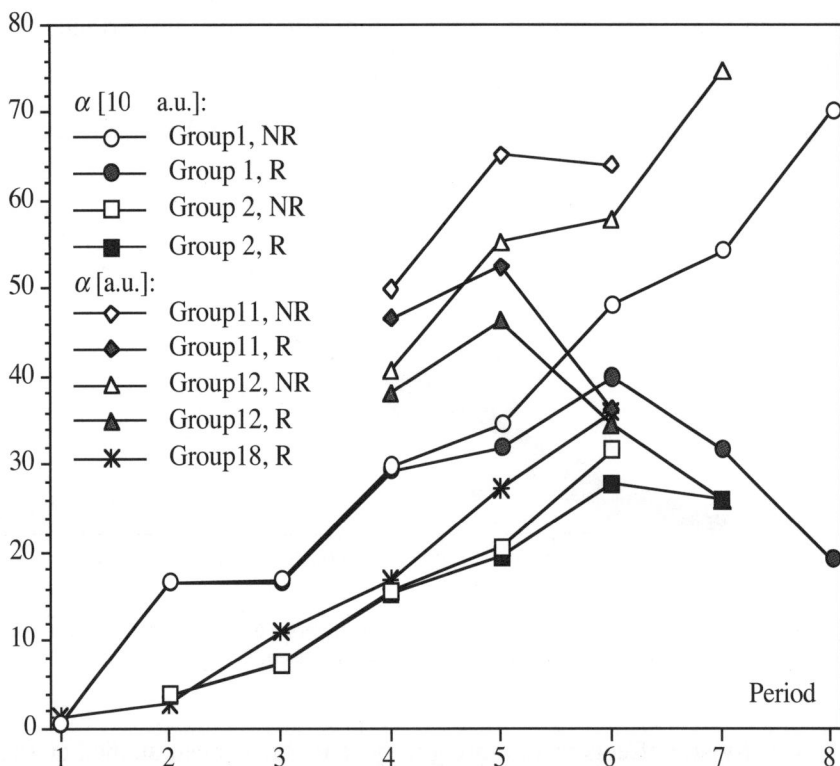

Figure 6. Calculated nonrelativistic and relativistic static dipole polarizability for Group 1, 2, 11, 12 and 18 elements of the periodic table. Note that the Group 1 and 2 polarizabilities are scaled by a factor of 0.1 in order to make the comparison to the other elements more visible.

6. Atomic Dipole Polarizabilities from Density Functional Theory

Perhaps with the exception of frequency dependent dipole polarizabilities[162] and higher order terms in eq.(2), for atomic static dipole polarizabilities density functional theory (DFT) does not seem to be the method of choice. More precise values are currently obtained from *ab-initio* theory. For example, Stott and Zaremba reported an LDA value of α= 1.89 a.u. for He,[163] which is by far too high compared to the experimental value of 1.3832 a.u.[136] Using a large uncontracted *(10s/6p/5d/4f/2g)* Gaussian basis set we calculate the following nonrelativistic values (in a.u.) for the He polarizability using eq.(3): *ab-initio* results 1.322 (Hartree-Fock), 1.362 (second-order many-body perturbation theory), 1.385 (configuration interaction); DFT results using various functionals: 1.686 ($X\alpha$), 1.644 (LDA), 1.558 (BPW91), and 1.505 (B3LYP). The LDA and B3LYP results are basically identical to the ones published previously published by Ionnou *et al.*[162] A recent more systematic DFT study on atomic dipole polarizabilities for a number of different elements has been performed by Cucinotta and Ballone.[164] For example, for the ^4S state of nitrogen DFT gives 8.17 a.u. (LDA), 8.11 a.u. (PBE), and 8.74 a.u. (meta-GGA),[164] which compares to the best available CCSD(T) value of 7.26 a.u. by Das and Thakkar.[84] For the heavier inert gases Ne to Xe similar large deviations are obtained.[165] The reason for the general overestimation of polarizabilities by DFT is well understood and can be found in the wrong asymptotic behaviour of the exchange-correlation potential. A recent study of Ren and Hsue of atomic polarizabilities uses the optimized effective potential (OEP) method for the rare gases and the earth alkaline atoms.[166] They obtain for Xe 29.6 a.u. and for Ba 301 a.u.,[166] which compares to the experimental values of 27.8 a.u. for Xe[106] and 268 a.u. for Ba.[112] See also the recent work by Hirata *et al.*[167] who got much improved values for the rare gases using a time-dependent OEP method. Another interesting comparison can be obtained from linear response four-component DFT calculations for the Hg atom[168] using large uncontracted basis sets (in a.u.): Hartree-Fock 44.81, LDA

33.72, BLYP 34.77 and B3LYP 35.75 (exp.[129] 33.92). For the future development of density functionals, the troublesome long-range region of the charge density can be tested from polarizability calculations of negatively charged atoms.[166] Nevertheless, atomic DFT polarizabilities give at least some estimate of the accuracy of the density functional applied in subsequent molecular calculations, where DFT can be more useful compared to the computer time consuming *ab-initio* methods. Also, for metallic clusters where the HOMO/LUMO gap is very small and single-reference methods fail, DFT seems to be the only choice.

7. Conclusion

The accurate determination of dipole polarizabilities for neutral atoms will remain an active field for a long time.[169] Note that only 20% of the listed dipole polarizabilities compiled by Miller are from experimental work.[19] There is a need from both experiment and theory to improve on the current values and produce precise data of better than 1% accuracy. This has not been achieved for most elements in the periodic table. Especially open shell systems need more careful investigations. The sum-over-states approach or linear response theory seems to be the preferred choice for few-electron systems or one-valence electron systems, where only a manageable amount of excited states are needed to produce accurate polarizabilities (see for example the recent work of Safronova *et al.*[170]). For more complicated systems the finite field approach seems to be the preferred method. Recent accurate experimental measurements come from Stark effects in excited electronic states where little theoretical data are available.[20,35] If accurate ground state polarizabilities are known, excited state polarizabilities can be obtained from quadratic Stark effect measurements in electronic transitions. There seems to be little theoretical work for excited electronic state polarizabilities, which can be very large.[171,172] Finally we note that for future work relativistic effects should be included within a four-component framework[168] including Breit interactions.

Acknowledgment

The author likes to thank Ivan Lim and Nicola Gaston (Auckland), Uwe Hohm (Braunschweig), Antonio Rizzo (Pisa), Jürgen Hinze (Bielefeld), Gary Doolen (Los Alamos National Laboratory), Dirk Andrae (Bielefeld), Vitaly Kresin (Los Angeles), Timo Fleig (Düsseldorf), Ajit Thakkar (Fredericton), Pekka Pyykkö (Helsinki) and Keith Bonin (Winston-Salem) for helpful discussions. Financial support from Marsden funding by the Royal Society of New Zealand is gratefully acknowledged.

References

1. A. Dalgarno, *Adv. Phys.* **11**, 281 (1962).
2. J. van Vleck, *The Theory of Electric and Magnetic Susceptibilities* (Oxford University Press, Oxford, 1932).
3. J. Schmiedmayer, P. Riehs, J. A. Harvey and N. W. Hill, *Phys. Rev. Lett.* **66**, 1015 (1999).
4. V. Bernard, N. Kaiser, and V.-G. Meissner, *Phys. Rev. Lett.* **67**, 1515 (1991).
5. J. Christensen, F. X. Lee, W. Wilcox, and L. Zhou, *Nucl. Phys. B* **119**, 269 (2003).
6. P. Lazzeretti, in *Handbook of Molecular Physics and Quantum Chemistry*, (S. Wilson, P. F. Bernath, R. McWeeney (ed.)), Vol.3, Wiley, Chichester (2003); pg. 53.
7. R. A. Buckingham, *Proc. Roy. Soc. (London)* **160**, 94 (1937).
8. R. A. Buckingham, *Proc. Roy. Soc. (London)* **160**, 113 (1937).
9. I. G. Kaplan, in *Handbook of Molecular Physics and Quantum Chemistry*, Eds. S. Wilson, P. F. Bernath and R. McWeeny (Wiley, Chichester, Vol.3, 2003); pg. 161.
10. K. D. Bonin and V. V. Kresin, *Electric-Dipole Polarizabilities of Atoms, Molecules and Clusters* (World Scientific, Singapore, 1997).
11. K. D. Bonin and M. A. Kadar-Kallen, *Int. J. Mod. Phys. Rev. B* **24**, 3313 (1994).
12. M. P. Bogaard and B. J. Orr, in *Physical Chemistry, Vol.2, Molecular Stucture and Properties*, Ed. A. D. Buckingham (Butterworths, London, 1975); pg.149
13. D. C. Hanna, M. A. Yuratich, and D. Cotter, *Nonlinear Optics of Free Atoms and Molecules* (Springer, Berlin, 1979).
14. W. Müller, J. Flesch, and W. Meyer, *J. Chem. Phys.* **80**, 3297 (1984).
15. H. Metcalf and P. van der Straten, *Phys. Rep.* **244**, 203 (1994).
16. T. M. Miller and B. Bederson, *Adv. At. Mol. Opt. Phys.* **25**, 37 (1988).
17. T. M. Miller and B. Bederson, *Adv. At. Mol. Phys.* **13**, 1 (1977).
18. J. Stiehler and J. Hinze, *J. Phys. B* **28**, 4055 (1995).

19. T. M. Miller, in *CRC Handbook of Chemistry and Physics*, Ed. D. R. Lide (CRC Press, New York, 2002).
20. W. A. van Wijngaarden, *Adv. At. Mol. Opt. Phys.* **36**, 141 (1996).
21. W. A. van Wijngaarden, *At. Phys.* **16**, 305 (1999).
22 G. Maroulis, in *Reviews of Modern Quantum Chemistry*, Ed. K. D. Sen (World Scientific, New Jersey, Vol. 1, 2002); pg.320.
23. P. Dahle, K. Ruud, T. Helgaker and P. R. Taylor, in *Pauling's Legacy, Modern Modelling of the Chemical Bond*, Eds. Z. B. Maksic, W. J. Orville-Thomas (Theoretical and Computational Chemistry Series, Elsevier, Amsterdam, Vol.6, 1999); pg. 147.
24. Special issue on *Computational aspects of electric polarizability calculations: Atoms, Molecules and clusters*, G. Maroulis (ed.), *J. Comput. Meth. Sci. Eng.* **4**, 235-764 (2005).
25. A. D. Buckingham, *Adv. Chem. Phys.* **12**, 107 (1967).
26. A. D. Buckingham, *Quart. Rev.* **13**, 183 (1959).
27. A. D. Buckingham and M. J. Stephen, *Trans. Faraday Soc.* **53**, 884 (1957).
28. W. Happer and B. S. Mathur, *Phys. Rev.* **163**, 12 (1967).
29. M. Weisbluth, *Atoms and Molecules* (Academic Press, New York, 1978).
30. K. Andersson and A. Sadlej, *Phys. Rev. A* **46**, 2356 (1992).
31. G. S. Chandler and R. Glass, *J. Phys. B* **20**, 1 (1987).
32. A. Khadjavi, A. Lurio, and W. Happer, *Phys. Rev.* **167**, 128 (1968).
33. J. R. P. Angel and P. G. H. Sandars, *Proc. Roy. Soc. A* **305**, 125 (1968).
34. L. D. Landau and E. M. Lifschitz, Lehrbuch der Theoretischen Physik. Quantenmechanik, Vol. III, Akademie Verlag, Berlin (1984); pg. 273.
35. J. Walls, J. Clarke, S. Cauchi, G. Karkas, H. Chen, and W. A. van Wijngaarden, *Eur. Phys. J. D* **14**, 9 (2001).
36. N. J. Martin, P. G. H. Sandars, and G. K. Woodgate, *Proc. Roy. Soc. A* **305**, 139 (1968).
37. B. Bederson and E. J. Robinson, *Adv. Chem. Phys.* **10**, 1 (1966).
38. D. M. Brink and G. R. Satchler, *Angular Momentum* (Oxford University Press, Oxford, 1994).
39. H. D. Cohen and C. C. J. Roothaan, *J. Chem. Phys.* **43**, 34 (1965).
40. J. A. Pople, J. W. Melver, and N. S. Ostlund, *J. Chem. Phys.* **49**, 2961 (1968).
41. G. H. F. Diercksen, B. O. Roos, and A. J. Sadlej, *Int. J. Quantum Chem. Symp.* **17**, 265 (1983).
42. Y. Yamaguchi, Y. Osamura, J. D. Goddard, and H. F. Schaefer III, *A New Dimension to Quantum Chemistry, Analytic Derivative Methods in Ab Initio Molecular Electronic Structure Theory* (Oxford University Press, Oxford, 1994).
43. O. Christiansen, P. Jørgensen, and C. Hättig, *Int. J. Quantum Chem.* **68**, 1 (1998).
44. T. Saue and H. J. Aa. Jensen, *J. Chem. Phys.* **118**, 522 (2002).
45. D. M. Bishop, in *Pauling's Legacy, Modern Modelling of the Chemical Bond*, Eds. Z. B. Maksic, W. J. Orville-Thomas (Theoretical and Computational Chemistry Series, Elsevier, Amsterdam, Vol.6, 1999); pg. 129.
46. I. Waller, *Z. Physik* **38**, 635 (1926).
47. G. Wentzel, *Z. Physik* **38**, 518 (1926).
48. P. S. Epstein, *Phys. Rev.* **28**, 695 (1926).
49. D. M. Bishop and J. Pipin, *Chem. Phys. Lett.* **236**, 15 (1995).
50. M. L. Bartlett and E. A. Power, *J. Phys. A* **2**, 419 (1969).

51. S. Kaneko, *J. Phys. B* **10**, 3347 (1977).
52. S. P. Goldman, *Phys. Rev. A* **39**, 976 (1989).
53. G. W. F. Drake and S. P. Goldman, *Phys. Rev. A* **23**, 2093 (1981).
54. G. W. F. Drake and S. P. Goldman, *Adv. At. Mol. Phys.* **25**, 393 (1988).
55. W. R. Johnson, S. A. Blundell, and J. Sapirstein, *Phys. Rev. A* **37**, 307 (1988).
56. R. Szmytkowski, *J. Phys. B; At. Mol. Opt. Phys.* **30**, 825 (1997).
57. R. Szmytkowski, *Phys. Rev. A* **65**, 012503 (2001).
58. R. Szmytkowski and K. Mielewczyk, *J. Phys. B; At. Mol. Opt. Phys.* **37**, 3961 (2004).
59. M. Karplus and H. J. Kolker, *J. Chem. Phys.* **39**, 1493 (1963).
60. A. Z. Tang and F. T. Chang, *Phys. Rev. A* **33**, 3671 (1986).
61. L. A. Thu, L. V. Hoang, L. I. Komarov, and T. S. Romanova, *J. Phys. B* **27**, 4083 (1994).
62. L. A. Thu, L. V. Hoang, L. I. Komarov, and T. S. Romanova, *J. Phys. B* **29**, 2897 (1996).
63. G. Figari and V. Magnasco, *Chem. Phys. Lett.* **342**, 99 (2001).
64. J. Pipin and D. M. Bishop, *J. Phys. B* **25**, 17 (1992).
65. K. D. Sen, B. Mayer, P. C. Schmidt, J. Garza, R. Vargas, and A. Vela, *Int. J. Quantum Chem.* **90**, 491 (2002).
66. R. Dutt, A. Mukherjee, and Y. P. Varshni, *Phys. Lett.* **280**, 318 (2001).
67. C. Laughlin, *J. Phys. B* **37**, 4085 (2004).
68. M. I. Bhatti, K. D. Coleman, and W. F. Perger, *Phys. Rev. A* **68**, 044503 (2003).
69. C. Cohen-Tannoudji, B. Diu and F. Laloé, *Méchanique Quantique* (Hermann Éditeurs des Sciences et des Arts, Paris, Vol.II, 1986).
70. G. Doolen, personal communication and values taken from ref.19.
71. A. Zangwill and P. Soven, *Phys. Rev. A* **21**, 1561 (1980).
72. G. Doolen and D. A. Liberman, *Phys. Scr.* **36**, 77 (1987).
73. P. Pyykkö, *Chem. Rev.* **88**, 563 (1988).
74. F. D. Feiock and W. R. Johnson, *Phys. Rev.* **187**, 39 (1969).
75. S. Fraga, J. Karwoski, and K. M. Saxena, *Handbook of Atomic Data* (Elsevier, Amsterdam, 1976).
76. J. P. Desclaux, L. Laaksonen, and P. Pyykkö, *J. Phys. B* **14**, 419 (1981).
77. L. T. Sin Fai Lam, *J. Phys. B* **14**, 3543 (1981).
78. K. Pachucki and J. Sapirstein, *Phys. Rev. A* **63**, 012504 (2001).
79. A. C. Newell and R. D. Baird, *J. Appl. Phys.* **36**, 3751 (1965).
80. J. Komasa, *Phys. Rev. A* **65**, 012506 (2002).
81. I. S. Lim, M. Pernpointner, M. Seth, J. K. Laerdahl, and P. Schwerdtfeger, *Phys. Rev. A* **60**, 2822 (1999).
82. R. W. Molof, H. L. Schwartz, T. M. Miller, and B. Bederson, *Phys. Rev. A* **10**, 1131 (1974).
83. H.-J. Werner and W. Meyer, *Phys. Rev. A* **13**, 13 (1976).
84. A. K. Das and A. J. Thakkar, *J. Phys. B, At. Mol. Opt. Phys.* **31**, 2215 (1998).
85. T. Fleig, to be published.
86. K. Anderson and A. J. Sadlej, *Phys. Rev. A* **46**, 2356 (1992).
87. B. O. Roos, R. Lindh, P.-A. Malmqvist, V. Veryazov, and P.-O. Widmark, *J. Phys. Chem.* **108**, 2851 (2004).
88. R. D. Alpher and D. R. White, *Phys. Fluids* **2**, 153 (1959).
89. M. Medved, P. W. Fowler, and J. M. Hutson, *Mol. Phys.* **7**, 453 (2000).

90. J. E. Rice, G. E. Scuseria, T. J. Lee, P. R. Taylor, and J. Almlöf, *Chem. Phys. Lett.* **191**, 23 (1992).
91. K. Hald, F. Pawlowski, P. Jørgensen, and C. Hättig, *J. Chem. Phys.* **118**, 1292 (2003).
92. H. Larsen, J. Olsen, C. Hättig, P. Jørgensen, O. Christiansen, and J. Gauss, *J. Chem. Phys.* **111**, 1917 (1999).
93. T. Nakajima and K. Hirao, *Chem. Lett.* 766 (2001).
94. R. H. Orcutt, and R. H. Cole, *J. Chem. Phys.* **46**, 697 (1967).
95. A. Derevianko, W. R. Johnson, M. S. Safronova, and J. F. Babb, *Phys. Rev. Lett.* **82**, 3589 (1999).
96. C. R. Ekstrom, J. Schmiedmayer, M. S. Chapman, T. D. Hammond, and D. E. Pritchard, *Phys. Rev. A* **51**, 3883 (1995).
97. E. F. Archibong and A. J. Thakkar, *Phys. Rev. A* **44**, 5478 (1991).
98. A. Sadlej and M. Urban, *J. Mol. Struct. (Theochem)* **234**, 147 (1991).
99. B. O. Roos, V. Veryazov, and P.-O. Widmark, *Theor. Chem. Acta* **111**, 345 (2004).
100. E. A. Reinsch and W. Meyer, *Phys. Rev. A* **14**, 915 (1976).
101. C. Lupinetti and A. J. Thakkar, *J. Chem. Phys.* **122**, 044301 (2005).
102. P. Milani, I. Moullet, and W. A. de Heer, *Phys. Rev. A* **42**, 5150 (1990).
103. All theoretical values yield significantly larger values compared to the experimental results of ref.102 which casts some doubt on the accuracy of this experiment.
104. G. Maroulis and C. Pouchan, *J. Phys. B, At. Mol. Opt. Phys.* **36**, 2011 (2003).
105. P. Soldán, E. P. F. Lee, and T. G. Wright, *Phys. Chem. Chem. Phys.* **3**, 4661 (2001).
106. U. Hohm and K. Kerl, *Mol. Phys.* **69**, 803 (1990).
107. T. K. Bose and R. H. Cole, *J. Chem. Phys.* **46**, 697 (1966).
108. I. Lim, P. Schwerdtfeger, B. Metz and H. Stoll, *J. Chem. Phys.* **122**, 104103 (2005).
109. S. Porsev and A. Derevianko, *Phys. Rev. A* **65**, 020701 (2002).
110. A. J. Sadlej, M. Urban, and O. Gropen, *Phys. Rev. A* **44**, 5547 (1991).
111. I. Lim and P. Schwerdtfeger, *Phys. Rev. A* **70**, 062501 (2004).
112. H. L. Schwartz, T. M. Miller, and B. Bederson, *Phys. Rev. A* **10**, 1924 (1974).
113. R. Pou-Amérigo, M. Merchán, I. Nebot-Gil, P-O.Widmark, and B. O. Roos, *Theor. Chim. Acta* **92**, 149 (1995).
114. B. O. Roos, R. Lindh, P-Å. Malmqvist, V. Veryazov, and P-O. Widmark, *J. Phys. Chem. A* **109**, 6575 (2005).
115. P. Calaminici, *Chem. Phys. Lett.* **387**, 253 (2004).
116. P. Schwerdtfeger and G. A. Bowmaker, *J. Chem. Phys.* **100**, 4487 (1994).
117. P.Neogrady, V. Kellö, M. Urban, and A. J. Sadlej, *Int. J. Quant. Chem.* **63**, 557 (1997).
118. D. Goebel, U. Hohm, and G. Maroulis, *Phys. Rev. A* **54**, 1973 (1996).
119. M. Seth, P. Schwerdtfeger, and M. Dolg, *J. Chem. Phys.* **106**, 3623 (1997).
120. V. Kellö and A. J. Sadlej, *Theor. Chim. Acta* **84**, 353 (1995).
121. E. A. Reinsch and W. Meyer, published in ref.18.
122. T. Fleig and A. J. Sadlej, *Phys. Rev. A* **65**, 032506 (2002).
123. D. Goebel and U. Hohm, *Phys. Rev. A* **52**, 3691 (1995).
124. D. A. Liberman and A. Zangwill, theoretical value published as a personal communication in ref.125.

125. T. P. Guella, T. M. Miller, B. Bederson, J. A. D. Stockdale, B. Jaduszliwer, *Phys. Rev. A* **29**, 2977 (1984).
126. D. Benker, P. Schwerdtfeger, to be published.
127. N. Runeberg and P. Pyykkö, *Int. J. Quantum Chem.* **66**, 131 (1998).
128. J. M. Amini and H. Gould, *Phys. Rev. Lett.* **91**, 153001 (2003).
129. D. Goebel and U. Hohm, *J. Phys. Chem.* **100**, 7710 (1996).
130. C. S. Nash, TAN03: *International Conference on the Chemistry and Physics of the Transactinide Elements* (Napa, California, 2003).
131. M. A. Kadar-Kallen and K. D. Bonin, *Phys. Rev. Lett.* **72**, 828 (1994).
132. B. O. Roos, R. Lindh, P-Å. Malmqvist, V. Veryazov, and P-O.Widmark, *Chem. Phys. Lett.* **409**, 295 (2005).
133. Z.-C. Yan, J. F. Babb, A. Dalgarno, and G. W. F. Drake, *Phys. Rev. A* **54**, 2824 (1996).
134. M. Masili and A. F. Starace, *Phys. Rev. A* **68**, 012508 (2003).
135. F. Weinhold, *J. Phys. Chem.* **86**, 1111 (1982).
136. D. Gugan and G. W. H. Michel, *Mol. Phys.* **39**, 783 (1980).
137. W. Cencek, K. Szalewicz, and B. Jeziorski, *Phys. Rev. Lett.* **86**, 5675 (2001).
138. G. Lach, B. Jeziorski, and K. Szalewicz, *Phys. Rev. Lett.* **92**, 233001 (2004).
139. W. R. Johnson and K. T. Cheng, *Phys. Rev. A* **53**, 1375 (1996).
140. D. Tunega, J. Noga, and W. Klopper, *Chem. Phys. Lett.* **269**, 435 (1997).
141. P. Fuentealba, *Chem. Phys. Lett.* **397**, 459 (2004).
142. A. Nicklass, M. Dolg, H. Stoll, and H. Preuss, *J. Chem. Phys.* **102**, 8942 (1995).
143. W. Klopper, S. Coriani, T. Helgaker, and P. Jørgensen, *J. Phys. B: At. Mol. Opt. Phys.* **37**, 3753 (2004).
144. F. Pawlowsky, P. Jørgensen, and C. Hättig, *Chem. Phys. Lett.* **391**, 27 (2004).
145. M. S. Safronova, A. Derevianko, and W. R. Johnson, *Phys. Rev. Lett.* **58**, 1016 (1998).
146. G. Maroulis, *Chem. Phys. Lett.* **334**, 207 (2001).
147. A. J. Thakkar and C. Lupinetti, *Chem. Phys. Lett.* **402**, 270 (2005).
148. B. Fricke, *J. Chem. Phys.* **84**, 862 (1986).
149. P. Pyykkö, *Int. J. Quantum Chem.* **85**, 18 (2001).
150. P. Lambin, A. A. Lucas, and J. P. Vigneron, *Phys. Rev. B* **46**, 1794 (1992).
151. D. Ik and P. Gi, *Phys. Scr.* **27**, 402-406 (1983).
152. P. Politzer, P. Jin, and J. S. Murray, *J. Chem. Phys.* **117**, 8197 (2002).
153. T. K. Ghanty and S. K. Ghosh, *J. Phys. Chem.* **100**, 17429 (1996).
154. J. K. Nagle, *J. Am. Chem. Soc.* **112**, 4741 (1990).
155. S. Hati and D. Datta, *J. Phys. Chem.* **99**, 10742 (1995).
156. S. Hati and D. Datta, *J. Phys. Chem.* **98**, 10451 (1994).
157. T. K. Ghanty and S. K. Ghosh, *J. Phys. Chem.* **97**, 4951 (1993).
158. U. Hohm, *Chem. Phys. Lett.* **183**, 304 (1991).
159. I. K. Dmitrieva and G. I. Plindov, *Phys. Scr. A* **27**, 402 (1983).
160. V. Kellö, A. J. Sadlej and K. Faegri Jr., *Phys. Rev. A* **47**, 1715 (1993).
161. P. Schwerdtfeger, in *Strength from Weakness: Structural Consequences of Weak Interactions in Molecules, Supermolecules and Crystals*, Eds. A. Domenico and I. Hargittai (Kluwer, Netherlands, 2002); p. 169.
162. A. G. Ioannou, S. M. Colwell, and R. D. Amos, *Chem. Phys. Lett.* **278**, 278 (1997).
163. M. J. Stott and E. Zaremba, *Phys. Rev. A* **21**, 12 (1980).
164. C. Cucinotta and P. Ballone, *Phys. Scr.* **T109**, 166 (2004).

165. A. Zangwill, *J. Chem. Phys.* **78**, 5926 (1983).
166. C.-Y. Ren and C.-S. Hsue, *Chin. J. Phys.* 42, 162 (2004).
167. S. Hirata, S. Ivanov, R. J. Bartlett, and I. Grabowski, *Phys. Rev. A* **71**, 032507 (2005).
168. P. Saalek, T. Helgaker, and T. Saue, *Chem. Phys.* **311**, 187 (2005).
169. U. Hohm, *Vacuum* **58**, 117 (2000).
170. M. S. Safronova, C. J. Williams, and C. W. Clark, *Phys. Rev. A* **69**, 022509 (2004).
171. C.-H. Li, S. M. Rochester, M. G. Kozlov and D. Budker, *Phys. Rev. A* **69**, 042507 (2004).
172. T. Kondo, D. Angom, I. Endo, A. Fukumi, T. Horiguchi, M. Iinuma, and T. Takahashi, *Eur. Phys. J. D* **25**, 103 (2003).

CHAPTER 2

First-order ZPVA correction to first hyperpolarizabilities of mono-substituted benzene molecules

Olivier Quinet [a], Benoît Champagne [a] and Bernard Kirtman [b]

[a] *Laboratoire de Chimie Théorique Appliquée, Facultés Universitaires Notre-Dame de la Paix, B-5000 Namur, Belgium*
[b] *Department of Chemistry and Biochemistry, University of California, Santa Barbara, California 93106 USA*

The first-order Zero-Point Vibrational Average (ZPVA) correction to the dynamic electronic first hyperpolarizability β^e has been computed at the time-dependent Hartree-Fock level for five mono-substituted benzene molecules. In most cases the difference between 6-31G and 6-31G* basis set results is sufficiently small so that qualitative structure-property relationships can be drawn. With one exception the ZPVA correction is less than about 10% of the corresponding β^e and may lead either to an increase or decrease in the total magnitude. Although the frequency dispersion of β^e is systematic that is not true of the ZPVA correction which, in addition, changes to a lesser degree. Substituent motions are typically the most important and electrical anharmonicity is often, though not always, the major contributor to the ZPVA.

1. Introduction

Materials exhibiting a large second- (and third-) order nonlinear optical (NLO) response are of crucial interest for communication technologies that employ frequency doublers or electro-optical switches. They are also used in probes relying on chiral-sensitive sum-frequency [1] or vibrational sum-frequency generation [2] to study molecular or material properties. Over the past three decades, a lot of research has been devoted to the synthesis and characterization of new efficient chromophores as well as to their processing in order to obtain suitable materials for use in NLO devices. This quest for efficient NLO compounds has been accompanied by theoretical modeling of the underlying property, the frequency-dependent hyperpolarizabilities, β (and γ) [3,4]. Not only is one interested in the magnitude of the

33

NLO properties for atoms, molecules, polymers, and solids but also in their structure-property relationships. Both aspects are addressed in the work presented here.

Calculating the (hyper)polarizabilities of polyatomic molecules remains a challenge due to the combined importance of electron correlation and vibrational effects. In the case of materials, or molecules in solution, taking into account the surroundings constitutes another difficult issue[5,6]. This article deals with a vibrational effect, namely the frequency-dependent zero-point vibrational averaging (ZPVA) correction, particularly for the first hyperpolarizability of mono-substituted benzene molecules .

According to the clamped nucleus treatment of electronic and nuclear motions, in the presence of applied electric fields [7] the polarization response may be decomposed into the sum of an electronic component evaluated at the equilibrium geometry, a pure vibrational term, and the ZPVA correction. As far as the pure vibrational hyperpolarizabilities are concerned, following the general time-dependent perturbation treatment of Bishop and Kirtman, [8,9] many works have shown that this contribution can be comparable in magnitude to the electronic hyperpolarizability (or larger) and, therefore, of major importance for the design of nonlinear optical (NLO) materials [4,10,11,12]. In particular, it has been shown that the pure vibrational contribution strongly depends on the NLO process, much more than on the UV/visible frequency. Finite field (FF) methods (or their analytical equivalent) have been developed which streamline evaluation of the pure vibrational component for systems of practical interest as NLO materials [13,14,15,16,17]. A further simplification, within the FF approach, has been afforded by the introduction of Field-Induced Coordinates (FICs) [18,19,20,16] that utilize only the essential vibrational information. These FF methods were initially applicable only to the case of static fields or in the infinite optical frequency limit[13], with the latter being especially useful because of the weak frequency-dependence in the UV/visible region (and below electronic resonances). However, promising attempts to take into account the frequency dispersion have now been made as well [21,22].

The ZPVA correction has received less attention than the pure vibrational terms (or, at least, the pure vibrational terms that are lowest-order in electrical and mechanical anharmonicity). It is defined as the difference between the average electronic property in the vibrational ground state and the property evaluated at the equilibrium geometry.

$$\Delta\beta^{\mathrm{ZPVA}} = \langle\Psi^{\mathrm{v}}|\beta(R)|\Psi^{\mathrm{v}}\rangle - \beta(R_0) \qquad (1)$$

Consequently, the ZPVA correction exhibits a frequency-dependence that is more similar to the corresponding electronic response than the pure vibrational response. In the FF methods noted above the static ZPVA correction is used to obtain pure vibrational contributions that are beyond those of lowest-order (the so-called curvature contributions). However, the frequency-dependent ZPVA is much more difficult to determine. Owing to the substantial computational task (*vide supra*), calculations of ZPVA corrections to hyperpolarizabilities have been mostly limited either to the static response or to small systems. In the latter category, dynamic ZPVA corrections to electric-field-induced second harmonic generation, $\Delta\gamma^{\text{ZPVA}}(-2\omega; \omega, \omega, 0)$, have been determined for methane using the random phase approximation (RPA) and the multi-configurational random phase approximation (MCRPA) [23], while the dc-Pockels effect $\Delta\beta^{\text{ZPVA}}(-\omega; \omega, 0)$ and second harmonic generation $\Delta\beta^{\text{ZPVA}}(-2\omega; \omega, \omega)$, have been calculated for methane, tetrafluoromethane and tetrachloromethane at the Hartree-FocK and MP2 levels of approximation [24]. In both instances, a numerical finite displacement scheme was adopted to obtain derivatives of the first hyperpolarizability with respect to the vibrational normal coordinates, Q (see Eqs 3-4 below). Similar approaches have been followed by Norman *et al.*[25], as well as by Ingamells *et al.*[26], for determining the dynamic $\Delta\gamma^{\text{ZPVA}}$ of methanol and ethylene, respectively.

In the above studies [23,24,25,26] the ZPVA correction was evaluated, as in most investigations on polyatomic molecules, to first-order in electrical and mechanical anharmonicity following the general variation-perturbation treatments of Kern, Raynes, Spackman and co-workers : [27,28,29], i.e.

$$\Delta P^{\text{ZPVA}} \equiv [P]^{0,1} + [P]^{1,0} \qquad (2)$$

Here the exponents n, m in $[P]^{n,m}$ refer to the order of perturbation theory in electrical and mechanical anharmonicity, respectively and $P = \alpha, \beta, \gamma$. This approach requires the computationally intensive evaluation of first and, especially, second derivatives of the hyperpolarizability with respect to each normal coordinate Q_a as well as the determination of the cubic force contants, F_{abb}. For the first hyperpolarizability, the terms in Eq. 2 are

$$[\beta]^{1,0} = \frac{1}{4} \sum_a \frac{\partial^2 \beta^{\text{e}}(-\omega_\sigma; \omega_1, \omega_2)/\partial Q_a^2}{\omega_a} \qquad (3)$$

$$[\beta]^{0,1} = -\frac{1}{4} \sum_a \left(\sum_b \frac{F_{abb}}{\omega_b} \right) \frac{\partial \beta^{\text{e}}(-\omega_\sigma; \omega_1, \omega_2)/\partial Q_a}{\omega_a^2} \qquad (4)$$

As indicated earlier FF/analytical strategies have been developed to reduce the computational cost of calculating the derivatives, but only for static properties. Using FICs, Luis *et al.*[20] simplified the evaluation of the first-order mechanical anharmonicity contribution, $[P]^{0,1}$. Although this part of their procedure could be extended to dynamic properties that is not the case for their treatment of the $[P]^{1,0}$ term. Another method for determining the static ZPVA contribution [30,31] is based on expanding the property about some reference geometry. By making an appropriate choice of that geometry one can eliminate the analogue of the $[P]^{1,0}$ term in Eq. 2. In order to do so, however, third derivatives of the property with respect to vibrational displacements must be evaluated. Finally, a non-perturbative procedure, which yields the exact static ZPVA to all orders, has very recently been proposed [17]. It utilizes vibrational SCF and post-SCF methodology. With regard to the magnitude of the static ZPVA correction, a study on a representative set of conjugated molecules found that in several cases it amounted to more than 20% of the static electronic response [32].

Much less has been done for the ZPVA correction to dynamic hyperpolarizabilities. The key new consideration that arises is the evaluation of first and second derivatives of the frequency-dependent property with respect to Cartesian coordinates in Eqs 3-4. Analytical approaches are preferred for this purpose and a first step in that direction has recently been achieved with a formulation for $\Delta\beta^{\text{ZPVA}}$ at the time-dependent Hartree-Fock (TDHF) level [33]. The latter treatment makes use of procedures akin to those employed in deriving the $2n+1$ rule[34] and the interchange relations[35] to reduce the computational task. Although electron-correlated schemes should be developed, results at the Hartree-Fock level can still be informative. So far, however, even the TDHF approach has been applied only for a few small molecules, namely H_2O, NH_3, and CH_4[33], and to the prototype *p*-nitroaniline compound [36].

The purpose of this paper is to augment our body of knowledge concerning dynamic ZPVA first hyperpolarizabilities by carrying out TDHF calculations on a set of five mono-substituted benzene molecules. Apart from the quantitative values, we are primarily interested in the structure-property relationships that they reveal and in comparing behavior with the corresponding pure electronic properties [37]. In the next section we summarize the important methodological steps which lead to the TDHF expressions for $\Delta\beta^{\text{ZPVA}}$ and we describe the computational approach. Then, the results are presented and discussed.

2. Methodology

The TDHF expression for the electronic contribution to the dynamic first hyperpolarizabilty ($\beta^e_{\xi\zeta\eta}(-\omega_\sigma;\omega_1,\omega_2)$) may be written in matrix form as

$$\beta^e_{\xi\zeta\eta}(-\omega_\sigma;\omega_1,\omega_2) = -\text{Tr}\left[H^\xi D^{\zeta\eta}(\omega_1,\omega_2)\right] \tag{5}$$

where H^ξ is the dipole moment matrix and $D^{\zeta\eta}(\omega_1,\omega_2)$ is the second-order perturbed density matrix associated with the dynamic electric fields $F_\zeta(\omega_1)$ and $F_\eta(\omega_2)$. For the sake of clarity, and to avoid confusion with geometrical indices, the exponent 'e' is eliminated in the following formalism. The first and second derivatives of $\beta_{\xi\zeta\eta}(-\omega_\sigma;\omega_1,\omega_2)$ with respect to the a and/or b atomic Cartesian coordinates, which are needed to evaluate Eqs 3-4, are given by[38,39],

$$\beta^a_{\xi\zeta\eta}(-\omega_\sigma;\omega_1,\omega_2) = \frac{\partial}{\partial a}\left(\beta_{\xi\zeta\eta}(-\omega_\sigma;\omega_1,\omega_2)\right)$$
$$= -\text{Tr}\left[\begin{array}{l} H^{\xi,a} D^{\zeta\eta}(\omega_1,\omega_2) \\ + H^\xi D^{\zeta\eta,a}(\omega_1,\omega_2) \end{array}\right] \tag{6}$$

$$\beta^{ab}_{\xi\zeta\eta}(-\omega_\sigma;\omega_1,\omega_2) = \frac{\partial^2}{\partial a\partial b}\left(\beta_{\xi\zeta\eta}(-\omega_\sigma;\omega_1,\omega_2)\right)$$
$$= -\text{Tr}\left[\begin{array}{l} H^\xi D^{\zeta\eta,ab}(\omega_1,\omega_2) + H^{\xi,a} D^{\zeta\eta,b}(\omega_1,\omega_2) \\ + H^{\xi,b} D^{\zeta\eta,a}(\omega_1,\omega_2) + H^{\xi,ab} D^{\zeta\eta}(\omega_1,\omega_2) \end{array}\right] \tag{7}$$

In addition to the second-order field-perturbed density matrix, Eqs 6-7 require the determination of third- and fourth-order perturbed density matrices or, equivalently, third- and fourth-order LCAO coefficient matrices.

Various schemes to determine the desired quantities are based on a multiple perturbation theory treatment of the TDHF equation,

$$FC - i\left(\frac{\partial SC}{\partial t}\right) = SC\epsilon; \tag{8}$$

subject to the normalization condition,

$$C^\dagger SC = 1; \tag{9}$$

and using the density matrix,

$$D = CnC^\dagger \tag{10}$$

The matrices involved are expanded as a power series in the simultaneous perturbations due to the dynamic electric fields $\lambda^\zeta = E_\zeta e^{+i\omega_\zeta t}$ and the atomic Cartesian coordinates $\lambda_a = x_a$. This leads to a set of TDHF perturbation equations obtained by equating terms of the same order and time-dependence.

Iterative schemes. One may solve the perturbation equations iteratively for the mixed (i.e. different fields and/or geometrical coordinates) derivatives of the density matrix. This is the procedure followed in the CPHF [40], TDHF [41,42,43] and extended TDHF [44,38] methodologies. The main idea in implementing the iterative schemes is to express the perturbed LCAO coefficients (i.e. the coefficient derivatives) in terms of the unperturbed ones. For example,

$$C_{ri}^{\zeta\eta,a}(\omega_1,\omega_2) = \sum_{j}^{\text{MO}} C_{rj}^{0} U_{ji}^{\zeta\eta,a}(\omega_1,\omega_2) \tag{11}$$

Introduction of the $U_{ji}^{\zeta\eta,a}(\omega_1,\omega_2)$ matrix permits us to rewrite the derivative of the density matrix in three parts, i.e. as

$$D_{rs}^{\zeta\eta,a}(\omega_1,\omega_2) = 2 \sum_{i}^{\text{occ}} \sum_{j}^{\text{virt}} C_{rj}^{0} U_{ji}^{\zeta\eta,a}(\omega_1,\omega_2) C_{is}^{0\dagger} + C_{ri}^{0} U_{ij}^{\zeta\eta,a\dagger}(-\omega_1,-\omega_2) C_{js}^{0\dagger}$$

$$+ 2 \sum_{i,j}^{\text{occ}} C_{rj}^{0} \left(U_{ji}^{\zeta\eta,a}(\omega_1,\omega_2) + U_{ji}^{\zeta\eta,a\dagger}(-\omega_1,-\omega_2) \right) C_{is}^{0\dagger}$$

$$+ \text{terms lower-order in U} \tag{12}$$

First of all, the sum of $U^{\zeta\eta,a}$ with $U^{\zeta\eta,a\dagger}$ in the second line of Eq. 12 can be related to terms lower-order in U by invoking the derivative of the normalization condition. Because the elements of the (occ,occ) block of $U^{\zeta\eta,a}(\omega_1,\omega_2)$ and $U^{\zeta\eta,a}(-\omega_1,-\omega_2)$ matrices are coupled in this way they are said to occur in the form of *nonindependent pairs* [40]. On the other hand, the elements of the (occ-virt) block, which appear in the first line of Eq. 12, must be determined directly from the corresponding third order TDHF equation. Within the non-canonical treatment [40] the off-diagonal (occ,virt) elements of $U^{\zeta\eta,a}$ read as

$$U_{ij}^{\zeta\eta,a}(\omega_1,\omega_2) = \frac{G_{ij}^{\zeta\eta,a}(\omega_1,\omega_2) + Q_{ij}^{\zeta\eta,a}(\omega_1,\omega_2)}{\epsilon_j^0 - \epsilon_i^0 - \omega_\sigma} \tag{13}$$

where $G^{\zeta\eta,a}$ is the derivative of the Fock matrix in the molecular orbital representation and $Q^{\zeta\eta,a}$ gathers contributions due to products of lower-order terms. The solution of this equation is obtained iteratively because $G^{\zeta\eta,a}$ depends on the density matrix derivative on the lhs of Eq. 12

Non-iterative schemes. The iterative determination of derivatives of the U matrix can be time-consuming especially second derivatives w.r.t. atomic

Cartesian coordinates . Fortunately, this step can be avoided by using the $2n + 1$ rule [34]. However, the resulting expression for the second derivatives of β w.r.t. atomic Cartesian coordinates then involves corresponding second derivatives of the coefficient matrix. In order to avoid the time-consuming computations that would be entailed in evaluating the latter we use interchange relations [45,35], which introduce third derivatives of the coefficient matrix w.r.t. electric fields. Although the $2n + 1$ rule is abandoned in this regard the computational effort is considerably reduced.

After several steps and mathematical rearrangements, the final noniterative expressions for the first- and second-order derivatives of the dynamic first hyperpolarizability are given by

$$
\begin{aligned}
\beta^a_{\xi\zeta\eta}&(-\omega_\sigma;\omega_1,\omega_2) \\
= -\sum_{r,s}^{\mathrm{AO}} \quad & D^{\zeta\eta}_{sr}(\omega_1,\omega_2)\mathcal{F}^{\xi a}_{rs}(-\omega_\sigma) + D^{\xi\zeta}_{sr}(-\omega_\sigma,\omega_1)\mathcal{F}^{\eta a}_{rs}(\omega_2) \\
& + D^{\xi\eta}_{sr}(-\omega_\sigma,\omega_2)\mathcal{F}^{\zeta a}_{rs}(\omega_1) \\
-2\sum_{i,j}^{\mathrm{MO}} \quad & \mathcal{U}^{\xi a}_{ij}(-\omega_\sigma)G^{\zeta\eta}_{ji}(\omega_1,\omega_2) + \mathcal{U}^{\eta a}_{ij}(\omega_2)G^{\xi\zeta}_{ji}(-\omega_\sigma,\omega_1) \\
& + \mathcal{U}^{\zeta a}_{ij}(\omega_1)G^{\xi\eta}_{ji}(-\omega_\sigma,\omega_2) + \mathcal{U}^{\zeta\eta a}_{ij}(\omega_1,\omega_2)G^{\xi}_{ji}(-\omega_\sigma) \\
& + \mathcal{U}^{\xi\eta a}_{ij}(-\omega_\sigma,\omega_2)G^{\zeta}_{ji}(\omega_1) + \mathcal{U}^{\xi\zeta a}_{ij}(-\omega_\sigma,\omega_1)G^{\eta}_{ji}(\omega_2) \\
& + \mathcal{U}^{\xi\zeta\eta}_{ij}(-\omega_\sigma,\omega_1,\omega_2)G^{a}_{ji} \\
-2\sum_{i}^{\mathrm{occ}} \quad & -\omega_\sigma S^{\xi\zeta\eta a}_{-\omega_\sigma,i}(-\omega_\sigma,\omega_1,\omega_2) + \omega_1 S^{\xi\zeta\eta a}_{\omega_1,i}(-\omega_\sigma,\omega_1,\omega_2) \\
& + \omega_2 S^{\xi\zeta\eta a}_{\omega_2,i}(-\omega_\sigma,\omega_1,\omega_2) + \mathcal{X}^{\xi\zeta\eta a}_{ii}(-\omega_\sigma,\omega_1,\omega_2)\epsilon^0_i \\
-2\sum_{i,j}^{\mathrm{occ}} \quad & X^{\xi a}_{ij}(-\omega_\sigma)\epsilon^{\zeta\eta}_{ji}(\omega_1,\omega_2) + X^{\eta a}_{ij}(\omega_2)\epsilon^{\xi\zeta}_{ji}(\omega_\sigma,\omega_1) \\
& + X^{\zeta a}_{ij}(\omega_1)\epsilon^{\xi\eta}_{ji}(-\omega_\sigma,\omega_2) + \mathcal{X}^{\zeta\eta a}_{ij}(\omega_1,\omega_2)\epsilon^{\xi}_{ji}(-\omega_\sigma) \\
& + \mathcal{X}^{\xi\eta a}_{ij}(\omega_\sigma,\omega_2)\epsilon^{\zeta}_{ji}(\omega_1) + \mathcal{X}^{\xi\zeta a}_{ij}(-\omega_\sigma,\omega_1)\epsilon^{\eta}_{ji}(\omega_2) \\
& + \mathcal{X}^{\xi\zeta\eta}_{ij}(-\omega_\sigma,\omega_1,\omega_2)\epsilon^{a}_{ji}
\end{aligned}
$$

$$(14)$$

and

$$\beta^{ab}_{\xi\zeta\eta}(-\omega_\sigma;\omega_1,\omega_2)$$

$$= -\sum_{r,s}^{\text{AO}} \quad D^{\xi\zeta\eta}_{sr}(-\omega_\sigma,\omega_1,\omega_2)\mathcal{F}^{ab}_{rs} + D^{\xi\eta}_{sr}(-\omega_\sigma,\omega_2)\mathcal{F}^{\zeta ab}_{rs}(\omega_1)$$
$$+ D^{\xi\zeta}_{sr}(-\omega_\sigma,\omega_1)\mathcal{F}^{\eta ab}_{rs}(\omega_2) + D^{\zeta\eta}_{sr}(\omega_1,\omega_2)\mathcal{F}^{\xi ab}_{rs}(-\omega_\sigma)$$

$$-2\sum_{i,j}^{\text{MO}} \quad \mathcal{U}^{ab}_{ij}G^{\xi\zeta\eta}_{ji}(-\omega_\sigma,\omega_1,\omega_2)$$
$$+ \mathcal{U}^{\xi\eta b}_{ij}(-\omega_\sigma,\omega_2)G^{\zeta a}_{ji}(\omega_1) + \mathcal{U}^{\xi\eta a}_{ij}(-\omega_\sigma,\omega_2)G^{\zeta b}_{ji}(\omega_1)$$
$$+ \mathcal{U}^{\xi\zeta b}_{ij}(-\omega_\sigma,\omega_1)G^{\eta a}_{ji}(\omega_2) + \mathcal{U}^{\xi\zeta a}_{ij}(-\omega_\sigma,\omega_1)G^{\eta b}_{ji}(\omega_2)$$
$$+ \mathcal{U}^{\zeta\eta b}_{ij}(\omega_1,\omega_2)G^{\xi a}_{ji}(-\omega_\sigma) + \mathcal{U}^{\zeta\eta a}_{ij}(\omega_1,\omega_2)G^{\xi b}_{ji}(-\omega_\sigma)$$
$$+ \mathcal{U}^{\xi ab}_{ij}(-\omega_\sigma)G^{\zeta\eta}_{ji}(\omega_1,\omega_2) + \mathcal{U}^{\eta ab}_{ij}(\omega_2)G^{\xi\zeta}_{ji}(-\omega_\sigma,\omega_1)$$
$$+ \mathcal{U}^{\zeta ab}_{ij}(\omega_1)G^{\xi\eta}_{ji}(-\omega_\sigma,\omega_2)$$
$$+ \mathcal{U}^{\xi\zeta\eta b}_{ij}(-\omega_\sigma,\omega_1,\omega_2)G^a_{ji} + \mathcal{U}^{\xi\zeta\eta a}_{ij}(-\omega_\sigma,\omega_1,\omega_2)G^b_{ji}$$
$$+ \mathcal{U}^{\xi\eta ab}_{ij}(-\omega_\sigma,\omega_2)G^\zeta_{ji}(\omega_1) + \mathcal{U}^{\xi\zeta ab}_{ij}(-\omega_\sigma,\omega_1)G^\eta_{ji}(\omega_2)$$
$$+ \mathcal{U}^{\zeta\eta ab}_{ij}(\omega_1,\omega_2)G^\xi_{ji}(\omega_\sigma)$$

$$-2\sum_i^{\text{occ}} \quad -\omega_\sigma \mathcal{S}^{\xi\zeta\eta ab}_{-\omega_\sigma,i}(-\omega_\sigma,\omega_1,\omega_2) + \omega_1 \mathcal{S}^{\xi\zeta\eta ab}_{\omega_1,i}(-\omega_\sigma,\omega_1,\omega_2)$$
$$+ \omega_2 \mathcal{S}^{\xi\zeta\eta ab}_{\omega_2,i}(-\omega_\sigma,\omega_1,\omega_2) + \mathcal{X}^{\xi\zeta\eta ab}_{ii}(-\omega_\sigma,\omega_1,\omega_2)\epsilon^0_i$$

$$-2\sum_{i,j}^{\text{occ}} \quad X^{ab}_{ij}\epsilon^{\xi\zeta\eta}_{ji}(-\omega_\sigma,\omega_1,\omega_2) + \mathcal{X}^{\xi ab}_{ij}(-\omega_\sigma)\epsilon^{\zeta\eta}_{ji}(\omega_1,\omega_2)$$
$$+ \mathcal{X}^{\eta ab}_{ij}(\omega_2)\epsilon^{\xi\zeta}_{ji}(\omega_\sigma,\omega_1) + \mathcal{X}^{\zeta ab}_{ij}(\omega_1)\epsilon^{\xi\eta}_{ji}(\omega_\sigma,\omega_2)$$
$$+ X^{\xi\eta b}_{ij}(-\omega_\sigma,\omega_2)\epsilon^{\zeta a}_{ji}(\omega_1) + X^{\xi\eta a}_{ij}(-\omega_\sigma,\omega_2)\epsilon^{\zeta b}_{ji}(\omega_1)$$
$$+ X^{\xi\zeta b}_{ij}(-\omega_\sigma,\omega_1)\epsilon^{\eta a}_{ji}(\omega_2) + X^{\xi\zeta a}_{ij}(-\omega_\sigma,\omega_1)\epsilon^{\eta b}_{ji}(\omega_2)$$
$$+ X^{\zeta\eta b}_{ij}(\omega_1,\omega_2)\epsilon^{\xi a}_{ji}(-\omega_\sigma) + X^{\zeta\eta a}_{ij}(\omega_1,\omega_2)\epsilon^{\xi b}_{ji}(-\omega_\sigma)$$
$$+ \mathcal{X}^{\xi\eta ab}_{ij}(-\omega_\sigma,\omega_2)\epsilon^\zeta_{ji}(\omega_1) + \mathcal{X}^{\xi\zeta ab}_{ij}(-\omega_\sigma,\omega_1)\epsilon^\eta_{ji}(\omega_2)$$
$$+ \mathcal{X}^{\zeta\eta ab}_{ij}(\omega_1,\omega_2)\epsilon^\xi_{ji}(-\omega_\sigma) + \mathcal{X}^{\xi\zeta\eta b}_{ij}(-\omega_\sigma,\omega_1,\omega_2)\epsilon^a_{ji}$$
$$+ \mathcal{X}^{\xi\zeta\eta a}_{ij}(-\omega_\sigma,\omega_1,\omega_2)\epsilon^b_{ji}$$

$$\text{(15)}$$

where the definition of the various matrices not previously defined can be found in Refs. [46,44,47]. The important thing to note here is that these matrices and, hence, Eq. 15, involves no second derivatives of the LCAO coefficient matrix w.r.t. atomic Cartesian coordinates (C^{ab}, $C^{\zeta ab}(\omega_1)$, ...). This expression is symmetric with respect to permutation of the electric field components (with a simultaneous switch of optical frequencies) and/or the atomic Cartesian coordinates. The procedures used to compute β^a and β^{ab} are partially non-iterative, but not completely so because they require iterative determination of lower-order quantities: i.e. first, second, and third-order derivatives with respect to the electric field ; first derivatives with respect to the geometrical displacements ; and mixed second derivatives with respect to both the electric field and the geometrical perturbations.

3. Applications

Our computational method to obtain the TDHF ZPVA correction for β has been implemented in the GAMESS quantum chemistry package[48]. In the present application to a set of mono-substituted benzenes (Ph-R with R = CH_3, CHO, OH, NH_2, and NO_2), this fully analytical, partially non-iterative, procedure was used to evaluate the electronic hyperpolarizabilities as well as their first and second derivatives with respect to geometrical coordinates. On the other hand, the anharmonic force constants (F_{abb}) required for Eq. 4 were obtained by numerical differentiation of the Hessian. For that purpose we computed the Hessian at geometries displaced from equilibrium along the different normal coordinates. The displacement amplitudes were ± 0.050 and ± 0.100 a.u. and a Romberg procedure[49] was employed for increased numerical accuracy. In order to study basis set effects both the 6-31G and 6-31G* basis sets were utilized. Although one would like to consider larger basis sets the storage requirements become prohibitively large under our current circumstances. Indeed, for benzaldehyde, almost 2Gb of memory is needed for the 6-31G* basis set, while for Ph-NO_2 slightly more than 2Gb is needed. Larger calculations would require a computer with larger memory and/or implementation of a direct algorithm. For consistency the geometry optimizations must be done at the Hartree-Fock level and, for accuracy, a very tight threshold was used on the residual atomic forces (2×10^{-6} a.u.). Table 1 shows the equilibrium conformation for the different molecules calculated with both basis sets. In the case of Ph-NH_2 the 6-31G* basis set yields a non-planar geometry. This is an artifact of the Hartree-Fock method, which disappears when electron correlation is included at the MP2 level. Although toluene is listed as non-planar due to the methyl group hydrogens it should be noted that the carbon atom of the methyl group is co-planar with the benzene ring. Finally, we checked that the longitudinal β^e values calculated here reproduce the results of Ref. [37].

Tables 2-11 contain the electronic and ZPVA results at $\lambda = 1064$nm ($\hbar\omega = 0.0428227$ a.u. $= 1.16527$ eV) for $\beta_{//}$ and β^K. The latter quantities are related to the experimentally determined electric field-induced second harmonic generation and to the electro-optic Kerr effect, respectively. $\beta_{//}$ is proportional to the projection of β in the direction of the permanent dipole moment which defines the z axis:

$$\beta_{//}(-2\omega;\omega,\omega) = \frac{1}{5}\sum_{\eta}\left(\begin{array}{c}\beta_{z\eta\eta}(-2\omega;\omega,\omega) + \beta_{\eta z\eta}(-2\omega;\omega,\omega) + \\ \beta_{\eta\eta z}(-2\omega;\omega,\omega)\end{array}\right) \quad (16)$$

whereas β^{K} is defined as

$$\beta^{K}(-\omega;\omega,0) = \frac{3}{10}\sum_{\eta}3\beta_{\eta z\eta}(-\omega;\omega,0) - \beta_{\eta\eta z}(-\omega;\omega,0) \quad (17)$$

We also present static field values which are, in fact, the same for $\beta_{//}$ and β^{K}. According to Eqs.3 and 4 the ZPVA response can be written as a sum over the normal modes Q_a. The two most important normal mode contributions are given in the tables for each symmetry species. In addition, the decomposition into electrical and mechanical anharmonicity is reported. For most of the molecules the difference between the 6-31G and 6-31G* basis sets is fairly small, which justifies the qualitative conclusions drawn below. In the case of aniline, there is an unrealistic loss of planarity when using 6-31G*.

From the bottom line of the tables we see that the ZPVA correction to the static and dynamic β is always less than about 10% of the electronic value except in the case of toluene where the electronic and ZPVA terms are of similar magnitude. Even when the magnitude of the ratio is small, however, there is considerable variation depending upon the particular substituent. The exceptional behavior of toluene seems to be more due to the small size of β^{e} than the large size of the ZPVA correction since the electronic hyperpolarizability of this molecule is calculated to be roughly a factor of seven, or more, smaller than that of all of the other molecules studied here. The extent to which these results can be generalized remains to be seen. On the other hand, although our experience is limited [16,32] we expect that the relative importance of the ZPVA correction will decrease when electron correlation is taken into account.

For the molecules examined in this paper it is about equally likely that the ZPVA correction will either increase or decrease the magnitude of β. In fact, for nitrobenzene at the 6-31G level (Table 10) it depends upon whether one is dealing with SHG or the Kerr effect and static limit. With respect to frequency dispersion β^{e} always increases in magnitude between the static limit and SHG at 1064 nm, in some cases by over 40%. The ZPVA correction behaves less systematically and the changes are usually smaller.

Symmetry, when present, plays a key role in the distribution of the ZPVA corrections between electrical and mechanical anharmonicity. In the absence of degenerate vibrations the anharmonic force constants F_{abb} vanish unless Q_a is totally symmetric. Thus, the ZPVA correction associated with non-totally symmetric modes is due entirely to electrical anharmonicity. For that reason the electrical anharmonicity often, though not always, is the major contributor to the total $[\beta]^I$. The situation is complicated by the fact that mechanical anharmonicity can be more important than electrical anharmonicity for the totally symmetric vibrations and, in general, the sign of each individual term may be either positive or negative. Partly because of sign variations, the contribution from the 2-3 most important vibrations of each symmetry often differs markedly from the total. Although there are a couple of exceptions, for each vibration the relative importance of mechanical vs. electrical anharmonicity is about the same for all three processes.

It is noteworthy that the particular vibration(s) making the largest contribution to the ZPVA correction is(are) almost always associated with the substituent group. From the results in tables 2, 4, 6, 8, and 10 we identify these vibrations as the methyl group wag (1583 cm^{-1}) and symmetric CH stretch (3185 cm^{-1}) in Ph-CH$_3$; the aldehyde CH stretch (1892 cm^{-1}) and a ring deformation similar to the Kekulé vibrational mode (1392 cm^{-1}) in Ph-CHO; the OH stretch (4048 cm^{-1}) and the CO stretch coupled with an aromatic/quinod ring deformation (1386 cm^{-1}) in Ph-OH; the NH$_2$ torsion (344 cm^{-1}) and the symmetric N-H out-of-plane bend (382 cm^{-1}) in Ph-NH$_2$; and the in-plane NO$_2$ twist (1482 cm^{-1}) in Ph-NO$_2$. The predominant role of the substituent group may be explained by the fact that its presence destroys the center of symmetry and allows for beta to be non-zero.

Acknowledgment

This paper is dedicated to Prof. D.M. Bishop who has been a pioneer in developing the theory of molecular nonlinear optical properties in general and the effect of vibration on these properties in particular. [50,51,52,53,3].

O.Q. and B.C. thank the Belgian National Fund for Scientific Research for their Postdoctoral Researcher and Senior Research Associate positions, respectively. The calculations were performed thanks to the Interuniversity Scientific Computing Facility (ISCF), installed at the Faculs Universitaires Notre-Dame de la Paix (Namur, Belgium), for which the authors gratefully acknowledge the financial support of the FNRS-FRFC and the "Loterie Nationale" for the convention n 2.4578.02, and of the FUNDP.

References

1. P. Fischer, D. S. Wiersma, R. Righini, B. Champagne, and A. D. Buckingham, Phys. Rev. Lett. **85**, 4253 (2000).
2. A. A. Mani, Z. D. Schultz, Y. Caudano, B. Champagne, C. Humbert, L. Dreesen, A. A. Gewirth, J. O. White, P. A. Thiry, and A. Peremans, J. Phys. Chem. B **108**, 16135 (2004).
3. D. M. Bishop and P. Norman, in *Handbook of Advanced Electronic and Photonic Materials and Devices*, edited by H. S. Nalwa (Academic Press, New York, 2001), Vol. 9, Chap. 1, p. 1.
4. B. Champagne and B. Kirtman, in *Handbook of Advanced Electronic and Photonic Materials and Devices*, edited by H. S. Nalwa (Academic Press, New York, 2001), Vol. 9, Chap. 2, p. 63.
5. D. M. Bishop, Int. Rev. Phys. Chem. **13**, 21 (1994).
6. B. Champagne and D. M. Bishop, Adv. Chem. Phys. **126**, 41 (2003).
7. D. M. Bishop, B. Kirtman, and B. Champagne, J. Chem. Phys. **107**, 5780 (1997).
8. D. M. Bishop and B. Kirtman, J. Chem. Phys. **95**, 2646 (1991).
9. D. M. Bishop and B. Kirtman, J. Chem. Phys. **97**, 5255 (1992).
10. B. Kirtman and B. Champagne, Int. Rev. Phys. Chem. **16**, 389 (1997).
11. B. Champagne, Th. Legrand, E. A. Perpète, O. Quinet, and J. M. André, Coll. Czech. Chem. Commun. **63**, 1295 (1998).
12. B. Champagne and B. Kirtman, Chem. Phys. **245**, 213 (1999).
13. D. M. Bishop, M. Hasan, and B. Kirtman, J. Chem. Phys. **103**, 4157 (1995).
14. J. M. Luis, J. Martí, M. Duran, J. L. Andrés, and B. Kirtman, J. Chem. Phys. **108**, 4123 (1998).
15. B. Kirtman, J. M. Luis, and D. M. Bishop, J. Chem. Phys. **108**, 10008 (1998).
16. M. Torrent-Sucarrat, M. Sola, M. Duran, J. M. Luis, and B. Kirtman, J. Chem. Phys. **118**, 711 (2003).
17. M. Torrent-Sucarrat, J.M. Luis, and B. Kirtman, J. Chem. Phys. , accepted.
18. J. M. Luis, M. Duran, J. L. Andrés, B. Champagne, and B. Kirtman, J. Chem. Phys. **111**, 875 (1999).
19. J. M. Luis, M. Duran, B. Champagne, and B. Kirtman, J. Chem. Phys. **113**, 5203 (2000).
20. J. M. Luis, B. Champagne, and B. Kirtman, Int. J. Quantum Chem. **80**, 471 (2000).
21. D. M. Bishop and B. Kirtman, J. Chem. Phys. **109**, 9674 (1998).
22. J. M. Luis, M. Duran, and B. Kirtman, J. Chem. Phys. **115**, 4473 (2001).
23. D. M. Bishop and S. P. A. Sauer, J. Chem. Phys. **107**, 8502 (1997).
24. D. M. Bishop and P. Norman, J. Chem. Phys. **111**, 3042 (1999).
25. P. Norman, Y. Luo, and H. Ågren, J. Chem. Phys. **109**, 3580 (1998).
26. V. E. Ingamells, M. G. Papadopoulos, and S. G. Raptis, Chem. Phys. Lett. **307**, 484 (1999).
27. W. C. Ermler and C. W. Kern, J. Chem. Phys. **55**, 4851 (1971).
28. W. T. Raynes, P. Lazzaretti, and R. Zanasi, Mol. Phys. **64**, 1061 (1988).
29. A. J. Russell and M. A. Spackman, Mol. Phys. **84**, 1239 (1995).

30. P. O. Åstrand, K. Ruud, and D. Sundholm, Theor. Chem. Acc. **103**, 365 (2000).
31. V. E. Ingamells, M. Papadopoulos, and A. J. Sadlej, J. Chem. Phys. **112**, 1645 (2000).
32. M. Torrent-Sucarrat, M. Sola, M. Duran, J. M. Luis, and B. Kirtman, J. Chem. Phys. **116**, 5363 (2002).
33. O. Quinet, B. Champagne, and B. Kirtman, J. Chem. Phys. **118**, 505 (2003).
34. T. S. Nee, R. G. Parr, and R. J. Bartlett, J. Chem. Phys. **64**, 2216 (1976).
35. B. Kirtman, J. Chem. Phys. **49**, 3895 (1968).
36. O. Quinet, B. Champagne, and B. Kirtman, J. Mol. Struct. **633**, 199 (2003).
37. B. Champagne, Int. J. Quantum Chem. **65**, 689 (1997).
38. O. Quinet and B. Champagne, J. Chem. Phys. **117**, 2481 (2002), publisher's note : **118**, 5692 (2003).
39. O. Quinet and B. Champagne, J. Chem. Phys. **118**, 5692 (2003), publisher's note.
40. Y. Yamaguchi, Y. Osamura, J. D. Goddard, and H. F. Schaefer III, *A New Dimension to Quantum Chemistry: Analytic Derivative Methods in Ab Initio Molecular Electronic Struture Theory* (Oxford University Press, Oxford, 1994).
41. H. Sekino and R. J. Bartlett, J. Chem. Phys. **85**, 976 (1986).
42. S. P. Karna, M. Dupuis, E. Perrin, and P. N. Prasad, J. Chem. Phys. **92**, 7418 (1990).
43. S. P. Karna and M. Dupuis, J. Comp. Chem. **12**, 487 (1991).
44. O. Quinet and B. Champagne, J. Chem. Phys. **115**, 6293 (2001).
45. A. Dalgarno and A. L. Stewart, Proc. Roy. Soc. (London) **A242**, 245 (1958).
46. O. Quinet, Ph.D. thesis, *Elaboration of quantum chemical procedures for determining mixed derivatives with respect to Cartesian coordinates and oscillating electric fields; application to vibrational spectroscopies and nonlinear optical properties* (Presses Universitaires de Namur, Belgium, 2002).
47. O. Quinet, B. Champagne, and B. Kirtman, J. Comp. Chem. **22**, 1920 (2001), erratum : **23**, 1495 (2002).
48. M. W. Schmidt, K. K. Baldridge, J. A. Boatz, S. T. Elbert, M. S. Gordon, J. H. Jensen, S. Koseki, N. Matsunaga, K. A. Nguyen, S. J. Su, T. L. Windus, M. Dupuis, and J. A. Montgomery, J. Comp. Chem. **14**, 1347 (1993).
49. P. J. Davis and P. Rabinowitz, in *Numerical Integration* (Blaisdell Publishing Company, London, 1967), p. 166.
50. D. M. Bishop, *Group Theory and Chemistry* (Dover, New York, 1993).
51. D. M. Bishop, Adv. Chem. Phys. **104**, 1 (1998).
52. D. M. Bishop, Adv. Quantum Chem. **25**, 1 (1994).
53. D. M. Bishop, in *Pauling's Legacy: Modern Modelling of the Chemical Bond*, edited by Z. B. Maksic and W. J. Orville-Thomas (Elsevier, Amsterdam, 1999), Vol. 6.

Table 1. Theoretical Hartree-Fock conformation of the optimized ground state for the 6-31G and 6-31G* basis sets

Molecule	basis set	
	6-31G	6-31G*
$Ph-CH_3$	non-planar (C_s)	non-planar (C_s)
$Ph-CHO$	planar (C_s)	planar (C_s)
$Ph-OH$	planar (C_s)	planar (C_s)
$Ph-NH_2$	planar (C_{2v})	non-planar (C_s)
$Ph-NO_2$	planar (C_{2v})	planar (C_{2v})

Table 2. HF/6-31G hyperpolarizabilities (in a.u.) of $Ph-CH_3$. The optical frequency is $\hbar\omega = 0.042823$ a.u. $= 1.16527$ eV or, equivalently, $\lambda = 1064$ nm. For each contribution or total, the fraction due to electrical anharmonicity is given in parentheses. Vibrational frequencies are in cm^{-1}.

ν	$\beta_{//}(0;0,0)$	$\beta_K(-\omega;\omega,0)$	$\beta_{//}(-2\omega;\omega,\omega)$
23	-0.452 (1.000)	-0.621 (1.000)	-0.904 (1.000)
376	0.223 (1.000)	0.229 (1.000)	0.244 (1.000)
Total A''	0.262 (1.000)	0.121 (1.000)	-0.107 (1.000)
% of $[\beta]^I$	9.5	4.4	-3.7
1583	-3.115 (0.122)	-3.188 (0.122)	-3.356 (0.121)
3185	3.814 (-0.100)	4.031 (-0.099)	4.463 (-0.100)
Total A'	2.480 (0.438)	2.651 (0.411)	2.987 (0.372)
% of $[\beta]^I$	90.5	95.6	103.7
$[\beta]^I$	2.742 (0.491)	2.772 (0.437)	2.880 (0.349)
β^e	2.303	2.245	2.610
$[\beta]^I/\beta^e$	1.19	1.23	1.10

Table 3.   HF/6-31G* hyperpolarizabilities (in a.u.) of Ph-CH$_3$. The optical frequency is $\hbar\omega = 0.042823$ a.u. $= 1.16527$ eV or, equivalently, $\lambda = 1064$ nm. For each contribution or total, the fraction due to electrical anharmonicity is given in parentheses. Vibrational frequencies are in cm^{-1}.

ν	$\beta_{//}(0;0,0)$	$\beta_K(-\omega;\omega,0)$	$\beta_{//}(-2\omega;\omega,\omega)$
17	0.097 (1.000)	-0.118 (1.000)	-0.503 (1.000)
1097	0.207 (1.000)	0.213 (1.000)	0.229 (1.000)
1342	0.226 (1.000)	0.235 (1.000)	0.247 (1.000)
Total A''	0.830 (1.000)	0.646 (1.000)	0.321 (1.000)
% of $[\beta]^I$	25.9	19.8	9.5
1567	-3.776 (0.092)	-3.875 (0.091)	-4.106 (0.089)
3201	4.534 (-0.078)	4.839 (-0.077)	5.470 (-0.076)
Total A'	2.378 (0.469)	2.611 (0.430)	3.071 (0.371)
% of $[\beta]^I$	74.1	80.2	90.5
$[\beta]^I$	3.208 (0.607)	3.257 (0.543)	3.392 (0.430)
β^e	3.084	3.013	3.290
$[\beta]^I/\beta^e$	1.04	1.08	1.03

Table 4.   HF/6-31G hyperpolarizabilities (in a.u.) of Ph-CHO. The optical frequency is $\hbar\omega = 0.042823$ a.u. $= 1.16527$ eVor, equivalently, $\lambda = 1064$ nm. For each contribution or total, the fraction due to electrical anharmonicity is given in parentheses. Vibrational frequencies are in cm^{-1}.

ν	$\beta_{//}(0;0,0)$	$\beta_K(-\omega;\omega,0)$	$\beta_{//}(-2\omega;\omega,\omega)$
869	-0.518 (1.000)	-0.534 (1.000)	-0.564 (1.000)
1157	0.385 (1.000)	0.452 (1.000)	0.609 (1.000)
1198	-0.673 (1.000)	-0.679 (1.000)	-0.680 (1.000)
Total A''	-1.588 (1.000)	-1.423 (1.000)	-0.938 (1.000)
% of $[\beta]^I$	32.3	30.0	23.1
1392	-0.851 (1.001)	-0.908 (1.001)	-1.091 (1.000)
1892	-0.940 (0.633)	-0.996 (0.610)	-1.121 (0.548)
Total A'	-3.330 (1.069)	-3.323 (1.090)	-3.120 (1.164)
% of $[\beta]^I$	67.7	70.0	76.9
$[\beta]^I$	-4.918 (1.047)	-4.746 (1.063)	-4.057 (1.126)
β^e	56.487	62.979	80.385
$[\beta]^I/\beta^e$	-0.0871	-0.0754	-0.0505

Table 5.   HF/6-31G* hyperpolarizabilities (in a.u.) of Ph-CHO. The optical frequency is $\hbar\omega = 0.042823$ a.u. $= 1.16527$ eV or, equivalently, $\lambda = 1064$ nm. For each contribution or total, the fraction due to electrical anharmonicity is given in parentheses. Vibrational frequencies are in cm^{-1}.

ν	$\beta_{//}(0;0,0)$	$\beta_K(-\omega;\omega,0)$	$\beta_{//}(-2\omega;\omega,\omega)$
841	-0.502 (1.000)	-0.519 (1.000)	-0.557 (1.000)
1134	-0.555 (1.000)	-0.553 (1.000)	-0.538 (1.000)
1152	0.517 (1.000)	0.590 (1.000)	0.762 (1.000)
Total A''	-1.762 (1.000)	-1.644 (1.000)	-1.283 (1.000)
% of $[\beta]^I$	36.1	34.4	30.0
1355	-0.703 (1.002)	-0.745 (1.002)	-0.873 (1.002)
2004	-0.793 (0.465)	-0.839 (0.435)	-0.945 (0.351)
Total A'	-3.119 (0.992)	-3.129 (1.000)	-2.997 (1.031)
% of $[\beta]^I$	63.9	65.6	70.0
$[\beta]^I$	-4.881 (0.995)	-4.772 (1.000)	-4.280 (1.022)
β^e	47.027	52.335	66.608
$[\beta]^I/\beta^e$	-0.104	-0.0912	-0.0643

Table 6.   HF/6-31G hyperpolarizabilities (in a.u.) of Ph-OH. The optical frequency is $\hbar\omega = 0.042823$ a.u. $= 1.16527$ eV or, equivalently, $\lambda = 1064$ nm. For each contribution or total, the fraction due to electrical anharmonicity is given in parentheses. Vibrational frequencies are in cm^{-1}.

ν	$\beta_{//}(0;0,0)$	$\beta_K(-\omega;\omega,0)$	$\beta_{//}(-2\omega;\omega,\omega)$
323	-0.134 (1.000)	-0.114 (1.000)	-0.107 (1.000)
1137	-0.200 (1.000)	-0.204 (1.000)	-0.216 (1.000)
1169	-0.105 (1.000)	-0.106 (1.000)	-0.111 (1.000)
Total A''	-0.619 (1.000)	-0.614 (1.000)	-0.658 (1.000)
% of $[\beta]^I$	56.8	54.3	51.1
1386	-0.440 (0.832)	-0.473 (0.830)	-0.547 (0.825)
3406	0.409 (0.377)	0.421 (0.384)	0.450 (0.399)
4048	0.435 (-0.382)	0.436 (-0.387)	0.450 (-0.402)
Total A'	-0.470 (1.406)	-0.517 (1.349)	-0.630 (1.259)
% of $[\beta]^I$	43.2	45.7	48.9
$[\beta]^I$	-1.089 (1.175)	-1.132 (1.160)	-1.288 (1.127)
β^e	-38.031	-38.605	-40.393
$[\beta]^I/\beta^e$	0.0286	0.0293	0.0312

Table 7. HF/6-31G* hyperpolarizabilities (in a.u.) of Ph-OH. The optical frequency is $\hbar\omega = 0.042823$ a.u. $= 1.16527$ eV or, equivalently, $\lambda = 1064$ nm. For each contribution or total, the fraction due to electrical anharmonicity is given in parentheses. Vibrational frequencies are in cm^{-1}.

ν	$\beta_{//}(0;0,0)$		$\beta_K(-\omega;\omega,0)$		$\beta_{//}(-2\omega;\omega,\omega)$	
320	-0.567	(1.000)	-0.560	(1.000)	-0.589	(1.000)
1089	-0.329	(1.000)	-0.337	(1.000)	-0.358	(1.000)
Total A''	-0.924	(1.000)	-0.925	(1.000)	-0.982	(1.000)
% of $[\beta]^I$	67.5		65.2		61.9	
1377	-0.502	(0.353)	-0.542	(0.351)	-0.624	(0.346)
4118	0.401	(-0.532)	0.399	(-0.542)	0.404	(-0.565)
Total A'	-0.445	(1.631)	-0.494	(1.547)	-0.603	(1.418)
% of $[\beta]^I$	32.5		34.8		38.1	
$[\beta]^I$	-1.369	(1.205)	-1.419	(1.190)	-1.586	(1.159)
β^e	-20.469		-20.690		-21.508	
$[\beta]^I/\beta^e$	0.0559		0.0686		0.0737	

Table 8. HF/6-31G hyperpolarizabilities (in a.u.) of Ph-NH$_2$. The optical frequency is $\hbar\omega = 0.042823$ a.u. $= 1.16527$ eV or, equivalently, $\lambda = 1064$ nm. For each contribution or total, the fraction due to electrical anharmonicity is given in parentheses. Vibrational frequencies are in cm^{-1}.

ν	$\beta_{//}(0;0,0)$		$\beta_{\mathrm{K}}(-\omega;\omega,0)$		$\beta_{//}(-2\omega;\omega,\omega)$	
1808	-0.684	(0.784)	-0.713	(0.779)	-0.762	(0.758)
3872	-0.746	(0.293)	-0.800	(0.275)	-0.967	(0.228)
Total A_1	-1.285	(1.223)	-1.306	(1.232)	-1.271	(1.263)
% of $[\beta]^{\mathrm{I}}$	24.0		24.5		24.4	
344	-2.130	(1.000)	-2.170	(1.000)	-2.299	(1.000)
1131	0.269	(1.000)	0.281	(1.000)	0.339	(1.000)
Total A_2	-2.153	(1.000)	-2.177	(1.000)	-2.186	(1.000)
% of $[\beta]^{\mathrm{I}}$	40.3		40.8		42.0	
382	-3.215	(1.000)	-3.342	(1.000)	-3.686	(1.000)
1163	1.389	(1.000)	1.432	(1.000)	1.553	(1.000)
Total B_1	-1.565	(1.000)	-1.531	(1.000)	-1.452	(1.000)
% of $[\beta]^{\mathrm{I}}$	29.3		28.7		27.9	
416	-0.210	(1.000)	-0.211	(1.000)	-0.219	(1.000)
1388	-0.198	(1.000)	-0.199	(1.000)	-0.266	(1.000)
Total B_2	-0.346	(1.000)	-0.319	(1.000)	-0.301	(1.000)
% of $[\beta]^{\mathrm{I}}$	6.5		6.0		5.8	
$[\beta]^{\mathrm{I}}$	-5.349	(1.054)	-5.333	(1.057)	-5.210	(1.064)
β^{e}	131.204		136.134		149.957	
$[\beta]^{\mathrm{I}}/\beta^{\mathrm{e}}$	-0.0408		-0.0392		-0.0347	

Table 9. HF/6-31G* hyperpolarizabilities (in a.u.) of Ph-NH$_2$. The optical frequency is $\hbar\omega = 0.042823$ a.u. $= 1.16527$ eV or, equivalently, $\lambda = 1064$ nm. For each contribution or total, the fraction due to electrical anharmonicity is given in parentheses. Vibrational frequencies are in cm^{-1}.

ν	$\beta_{//}(0;0,0)$		$\beta_K(-\omega;\omega,0)$		$\beta_{//}(-2\omega;\omega,\omega)$	
250	-1.880	(1.000)	-1.956	(1.000)	-2.160	(1.000)
3886	-0.267	(1.000)	-0.286	(1.000)	-0.328	(1.000)
Total A''	-2.441	(1.000)	-2.525	(1.000)	-2.736	(1.000)
% of $[\beta]^I$	-344.7		-323.9		-260.9	
723	1.644	(0.060)	1.714	(0.060)	1.929	(0.061)
892	1.014	(0.046)	1.058	(0.049)	1.210	(0.058)
Total A'	3.149	(-0.043)	3.304	(-0.036)	3.785	(-0.005)
% of $[\beta]^I$	444.7		423.9		360.9	
$[\beta]^I$	0.708	(-3.639)	0.779	(-3.391)	1.049	(-2.626)
β^e	37.385		39.021		43.539	
$[\beta]^I/\beta^e$	0.0189		0.0120		0.0241	

Table 10. HF/6-31G hyperpolarizabilities (in a.u.) of Ph-NO$_2$/6-31G. The optical frequency is $\hbar\omega = 0.042823$ a.u. $= 1.16527$ eV or, equivalently, $\lambda = 1064$ nm. For each contribution or total, the fraction due to electrical anharmonicity is given in parentheses. Vibrational frequencies are in cm^{-1}.

ν	$\beta_{//}(0;0,0)$		$\beta_{\rm K}(-\omega;\omega,0)$		$\beta_{//}(-2\omega;\omega,\omega)$	
440	-2.649	(0.052)	-2.825	(0.046)	-3.325	(0.026)
1482	4.367	(0.534)	4.962	(0.554)	6.728	(0.600)
Total A_1	0.644	(2.796)	1.404	(1.762)	3.958	(1.196)
% of $[\beta]^{\rm I}$	-24.5		-82.8		255.6	
59	1.685	(1.000)	1.900	(1.000)	2.481	(1.000)
1173	-0.493	(1.000)	-0.519	(1.000)	-0.571	(1.000)
Total A_2	1.091	(1.000)	1.304	(1.000)	1.939	(1.000)
% of $[\beta]^{\rm I}$	-41.6		-77.0		125.2	
771	0.576	(1.000)	0.691	(1.000)	1.024	(1.000)
1201	-0.922	(1.000)	-0.925	(1.000)	-0.897	(1.000)
Total B_1	-0.308	(1.000)	-0.030	(1.000)	0.859	(1.000)
% of $[\beta]^{\rm I}$	11.7		1.8		55.4	
1398	-1.336	(1.000)	-1.441	(1.000)	-1.818	(1.000)
1579	-1.116	(1.000)	-1.262	(1.000)	-1.609	(1.000)
Total B_2	-4.049	(1.000)	-4.372	(1.000)	-5.207	(1.000)
% of $[\beta]^{\rm I}$	154.4		258.0		-336.2	
$[\beta]^{\rm I}$	-2.622	(0.559)	-1.694	(0.368)	1.549	(1.502)
$\beta^{\rm e}$	98.193		110.331		142.440	
$[\beta]^{\rm I}/\beta^{\rm e}$	-0.0267		-0.0154		0.0109	

Table 11. HF/6-31G* hyperpolarizabilities (in a.u.) of Ph-NO$_2$. The optical frequency is $\hbar\omega = 0.042823$ a.u. $= 1.16527$ eV or, equivalently, $\lambda = 1064$ nm. For each contribution or total, the fraction due to electrical anharmonicity is given in parentheses. Vibrational frequencies are in cm^{-1}.

ν	$\beta_{//}(0;0,0)$		$\beta_K(-\omega;\omega,0)$		$\beta_{//}(-2\omega;\omega,\omega)$	
435	-1.741	(0.057)	-1.841	(0.050)	-2.109	(0.030)
1640	1.723	(0.969)	1.941	(0.970)	2.550	(0.973)
Total A_1	-0.465	(-3.754)	-0.079	(-28.126)	1.253	(2.966)
% of $[\beta]^I$	12.3		2.4		-73.5	
56	1.296	(1.000)	1.435	(1.000)	1.798	(1.000)
1119	-0.386	(1.000)	-0.405	(1.000)	-0.444	(1.000)
Total A_2	0.796	(1.000)	0.927	(1.000)	1.297	(1.000)
% of $[\beta]^I$	-21.1		-27.8		-76.1	
745	0.209	(1.000)	0.252	(1.000)	0.384	(1.000)
1084	-0.257	(1.000)	-0.260	(1.000)	-0.234	(1.000)
1141	-1.022	(1.000)	-1.037	(1.000)	-1.044	(1.000)
Total B_1	-0.962	(1.000)	-0.833	(1.000)	-0.377	(1.000)
% of $[\beta]^I$	25.5		25.0		22.1	
1356	-0.807	(1.000)	-0.853	(1.000)	-1.024	(1.000)
1868	-1.005	(1.000)	-1.099	(1.000)	-1.318	(1.000)
Total B_2	-3.143	(1.000)	-3.344	(1.000)	-3.878	(1.000)
% of $[\beta]^I$	83.3		100.5		227.5	
$[\beta]^I$	-3.774	(0.415)	-3.329	(0.312)	-1.705	(-0.445)
β^e	46.682		53.489		71.982	
$[\beta]^I/\beta^e$	-0.0808		-0.0622		-0.0243	

CHAPTER 3

POLARIZABILITY AND HYPERPOLARIZABILITY IN SMALL SILICON CLUSTERS

Claude Pouchan, Daisy Y. Zhang, Didier Bégué

Laboratoire de Chimie Structurale, UMR 5624 FR2606 "IPREM" Université de Pau et de Pays de l'Adour, F-64000 Pau, France
E-mail: claude.pouchan@univ-pau.fr

A brief review of the research effort that has been focused on studying the structure and static dipole polarizablity (α) in small silicon clusters is given, followed by a report of our recent calculations on the first hyperpolarizability (β) values in Si_3, Si_9 and Si_{12}, as well as the α values for Si_n, with n = 3 – 10. Our data show that, unlike the α values, theoretically predicted β values are, as expected, extremely method-sensitive, basis-set-sensitive, and furthermore, show great variation among the structural isomers, which, in contrast, all have very close mean α values. Among the various conformational isomers, the estimated mean β values are the greatest in the most oblate-shaped isomers; and among the silicon clusters of different sizes, the mean β values show trend of Si_3 >Si_9 >Si_{12}, which is consistent with the understanding that first hyperpolarizability decreases as the shape of the molecule becomes more spherical, as well as the fact that the flat Si_3 molecule obviously has the most oblate shape compared to the other two isomers.

1. Introduction

The silicon clusters have received, and continue to receive enormous research attention ever since the mid-80s.[1] The intense interest in silicon clusters was initiated by the fact that silicon crystal had been the most important semiconducting material in microelectronics industry, and clusters, as intermediates that bridge between the atom and the macroscopic crystal, provide a unique opportunity to help discover and predict new properties of the bulk as novel materials. However, with the development in nanostructure technology, minimum device features are rapidly approaching the size of atomic clusters. Consequently, the properties of clusters are no longer simply pieces of data that are used for projecting new properties of the bulk solid, but silicon clusters are themselves potential new materials.[2,3] Extensive experimental work on semiconductor atomic clusters took place with the development of laser vaporization techniques[4-9] and molecular beam technique.[10] These techniques allow measurements on the unstable clusters to be performed in an isolated, collision-free environment. Due to the instability of the clusters, direct measurement of the specific arrangement of the atoms in a cluster is never possible. Therefore, the structures of clusters can never been identified directly, but can only be derived indirectly from various properties measured in the experiments. Such experiments include photoelectron spectroscopy,[11-17] photodissociation[1, 18-21] and collision-induced dissociation experiments,[22-24] ion mobility measurements,[25-30] ionization potential measurements,[31,32] Raman (42) and IR spectroscopy,[31,34] and polarizability measurements.[35-38] However, knowing the structure is the base for understanding as well as predicting any physical and chemical properties in any system. Therefore, in the study of clusters, computational chemistry plays a vital role in filling the gap for effective research to take place, namely, from observing properties → establishing the structure → predicting new properties. By computing a physical properties with a certain structure as a candidate, theoretical values are to be compared with the experimental values. If the experimental procedure as well as the theoretical method and the level of calculations have been tested and shown to give credible values of such a property, the structure of the cluster can then be established as the model candidate used in the calculation.

Therefore, developing appropriate theoretical methods for accurately predicting reliable values of various properties for a system has become an ever important research area, especially in clusters science. Enormous effort has been devoted to creating and testing new theoretical methods in order to establish the structures of the silicon clusters.[39-89] Presently, the geometries for small silicon clusters, Si_n, with $n \leq 10$ have been firmly established by both *ab initio*[39-47] and density functional theory (DFT) methods.[48-56] Medium-size clusters have also been investigated with first-principles techniques,[57-59] empirical and *ab initio* based tight-binding models,[60-64] and force-field theory methods.[65,66]

The established structures for Si_n show that they are not only nothing close to the structure in the silicon crystal lattice, but also very different from each other. Each takes a unique structure, driven by the energy gained by overlapping as much as possible the extra electron density on the surface of the clusters, which results in the reconstruction of the structures in order to tight up those "dangling bonds" on the surface. In the structures of small silicon clusters, the silicon atom is often seen to be hypercoordinated, and therefore, the situation of bonding is expected to be very different from either in carbon clusters or in the silicon diamond. Therefore, understanding the nature of chemical bonding in small silicon clusters is another interest research topic.

Herein, we report a piece of theoretical study of the polarizability (α) and hyperpolarizability (β) of small silicon clusters, Si_n, with $n \leq 12$.

Why polarizability and hyperpolarizability? Because polarizability is a fundamental electronic property of a system and hyperpolarizability is an extremely useful property in nanostructure devices. For a cluster, these value serve as indicators for its structure, such that, as mentioned earlier, accurate prediction of the α or β values by theoretical methods provides means to deduce the structure which cannot be directly measured experimentally. But more importantly, the value of polarizability of the clusters is a unique quantity – its relates, on one hand, macroscopically to the dielectric constant of the bulk, and on the other hand, microscopically to the molecular orbitals and chemical bonding.

The polarizability α_{ij} is defined as the second derivative of the total electronic energy, or the first derivative of the dipole moment, with respect to the electric field component

$$\alpha_{ij} = -\partial^2 E / \partial F_i \partial F_j = \partial \mu_i / \partial F_j \ (i,j = x, y, z)$$

where E, F, and μ_i, are the total energy of the cluster, the external electric field, and component of the cluster electric dipole, respectively. The normally measured polarizability in an experiment is the average polarizability which equals the invariant trace $\langle \alpha \rangle = (1/3) \ (\alpha_{xx} + \alpha_{yy} + \alpha_{zz})$. For bulk silicon, the average polarizability per atom value, $\overline{\alpha}$, relates to the static dielectric constant of the bulk solid, ε, by the Clausius-Mossotti relation,

$$\overline{\alpha} = \frac{3}{4\pi} \left(\frac{\varepsilon - 1}{\varepsilon + 2} \right) v$$

where v is the volume per Si atom in the unit cell. The bulk atomic polarizability value has been calculated by Jackson to be 3.64 Å³/atom,[90] by taking the value for ε = 11.8 and v = 19.5[91]

In contrast to the enormous effort in establishing the structures, reports on the polarizability of silicon clusters are much scarce. The first polarizability calculations were carried out by Rantala et al in 1990, where they used an empirical tight-binding method,[92] but results show serious overestimation of these values. Later in 1996, Rubio et al reported their studies by using an ab initio pseudopotential plane-wave technique,[93,94] continued with reports by Chelikowsky and coworkers by using a pseudopotential based finite-difference method,[95,96] Jackson by the first-principle methods,[97,98] and Maroulis et al by both *ab initio* and DFT methods but with large flexible basis sets of Gaussian-type functions.[99] Recent work in this area has been mainly focused on testing the performance of various DFT methods.[90, 100-106]

Certain variations in the values predicted by the various methods do exist, however, all predicted $\overline{\alpha}$ values are shown to be larger for the clusters than for the bulk. Even though this finding is opposite to the experimental values measured by Schafer and coworkers,[37] which shows the $\overline{\alpha}$ values to be smaller than the bulk and to approach the bulk limit from below as the cluster size increases, an explanation for the relatively high polarizability of small clusters has been offered such that the extra electron density on the surface of the clusters forms relatively weaker and therefore, more polarizable bonds than in the crystalline system.

The correlation between chemical bonding and the magnitude of polarizability is well reflected in the following sum-over-states (SOS) expression of α_{ii},[107-109]

$$\alpha_{ii} \propto \sum_{l,k} \left| \langle k|\vec{i}|l \rangle \right|^2 \Big/ E_l - E_k \quad i=x,y,z$$

where α_{ii} is shown to be proportional to the square of the matrix element, $|<k|\,\vec{i}\,|l>|^2$, and inversely proportional to the energy gap between the bonding orbital, k, and the antibonding orbital, l, between which the transition takes place. According to this expression, the presence of the dangling-bonds in the clusters contributes to a lower energy gap between the bonding and antibonding orbitals, i.e. the HOMO-LUMO gap, and consequently increases the α_{ii} values.

Furthermore correlations between geometry, HOMO-LUMO gap, and the polarizability values have been investigated by Jackson,[97] Deng,[102] Pouchan,[105] and Zhang.[106] In all reports, no straightforward correlation has been observed between the size of the HOMO-LUMO gap and the magnitude of polarizability. In contrast, there have been observed correlations between the geometry and polarizability. As clusters with a more prolate shape consistently give higher theoretical polarizability than more compact structures, quantities such as the normalized moment of inertia I/I_0[102] and the standard deviation of the bonding distances[105] have been used to measure the "degree of prolation", and quantitative relationships between these quantities and the size of polarizability have been been proposed.

2. Computational methodology

Polarizability (α) values listed in Table 1 under "B3LYP" were carried out in Gaussian 98 suite of program.[110] by applying the B3LYP method,[111,112] and utilizing the SV0 basis set.[99] The SV0 basis set consists of 5s4p2d, with all the coefficients optimized for calculating the α values. Detailed construction of the SV0 basis set and its performance in calculating polarizability of silicon clusters have been described and reported in a recent paper by Maroulis et al.[99] Geometry optimization were carried out at the same level of theory using 6-311G* basis set.[113,114] Data listed under VWN and BLYP are polarizability values calculated by using the Amsterdam Density Functional (ADF) program [115] at LDA/VWN (Local density approximation/Vosko-Wilk-Nusair[116]) and GGA (General gradient approximation)/BLYP levels of theory. The basis set employed, for both geometry optimization and polarizability calculations, was the even-tempered basis set, ET-QZ3p, which consists of 8s6p3d1f and is of quadruple ζ quality and with 3 polarization functions included for the Si atoms.

For calculating the first hyperpolarizability, β, values for Si_3, a very extended basis set, 9s7p6d5f1g, which is also developed by Maroulis was used with the HF, MP2, B3LYP and B3PW91 methods. These four calculations were carried out by using the Gaussian 98 suite of programs.[110] Additional β values for Si_3 were also computed by using the ADF program at LDA/VWN and GGA/BLYP levels of theory with various basis sets. A total of 8 basis sets were used to test their performance in predicting the β values. These basis sets include DZ (double-ζ), TZ2P (triplet-ζ with 2 polarization functions), ET-pVQZ (quadruple-ζ quality, 6s4p2d1f even-tempered basis), ET-QZ+5P (quadruple-ζ quality, with 5 polarization functions and 1 set of diffuse s, p Slater-type orbitals (STO) added, 8s6p3d2f even-tempered basis), ET-QZ3P (quadruple-z quality, with 3 polarization functions included, 6s4p2d1f even-tempered basis), and with 1, 2, and 3 sets of diffuse s, p, d, and f STO added to the ET-QZ3P-1Diffuse, ET-QZ3P-2Diffuse, and ET-QZ3P-3Diffuse basis sets, respectively.

3. Results and discussion

3.1. Polarizability, α, in Si_n with $n = 3$ to 10

The structures at the optimized geometries for the Si_n clusters are shown in Figure 1, together with the point group for each of the Si_n molecules. These clusters all possess fairly high symmetry at the optimized structure. Theoretical studies carried out earlier at various levels of theory have predicted all these structures to be the global minima,[47] and available experimental results have confirmed these theoretical predictions.[89]

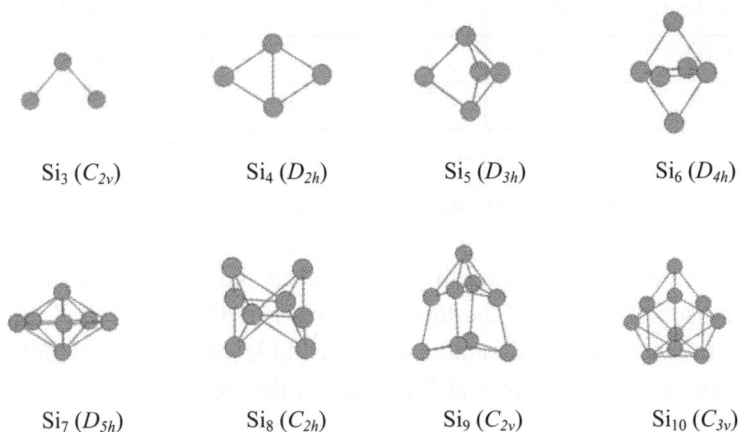

Si$_3$ (C_{2v})   Si$_4$ (D_{2h})   Si$_5$ (D_{3h})   Si$_6$ (D_{4h})

Si$_7$ (D_{5h})   Si$_8$ (C_{2h})   Si$_9$ (C_{2v})   Si$_{10}$ (C_{3v})

Fig. 1 Structures for Si$_n$ (n=3 – 10)

The mean dipole polarizability per atom, α, for all the Si_n molecules, with $n = 3 - 10$, are summarized in Table 1. All three DFT methods predict exactly the same trends in the α values. Among the three methods, GGA/BLYP predicts the largest α values, while LDA/VWN predicts the smallest ones, and the B3LYP method, values in between. The differences in the α values remain more or less constant throughout the Si_n series. For example, almost all BLYP values are higher than the LDA/VWN values by a constant of around 1.7 au/atom; and the B3LYP values are all higher than the LDA/VWN ones by between 0.1 −1.4 au/atom.

Table 1. DFT results for the polarizability per atom, $\overline{\alpha}$, for Si_n (n=3-10).

Si_n (sym)	$\overline{\alpha}$ (a.u.)			literature
Si_3 (C_{2v})	36.3	34.9	37.3	$35.5^a,35.5^b,35.4^c$
Si_4 (D_{2h})	34.7	34.1	35.7	$34.3^a,34.3^b,34.4^c$
Si_5 (D_{3h})	33.7	32.4	34.0	$32.9^a,32.5^b,33..2^c$
Si_6 (D_{4h})	30.9	30.3	31.8	$29.8^a,30.1^b,27.2^c$
Si_7 (D_{5h})	30.3	29.5	31.3	$30.5^a,29.5^b,30.1^c$
Si_8 (C_{2h})	31.7	30.5	32.3	$30.5^b,31.4^c$
Si_9 (C_{2v})	30.6	39.6	31.4	$29.6^b,29.6^c$
Si_{10} (C_{3v})	30.3	29.1	30.7	$29.1^b,28.7^c$

[a]ref. 98   [b]ref. 94   [c]ref. 103

$\overline{\alpha}$ values are listed in atomic units as $(\alpha_{xx} + \alpha_{yy} + \alpha_{zz})\,e^2 a_0^2 E_h^{-1}/3$

The α values from the literature are collected and included in Table 1 for comparison. The first of the three numbers listed under "lit. α" for Si_3 − Si_7 were the α values calculated at CCSD(T) level of theory with SV0 as the basis set by Maroulis et al.[99] Except for the case of Si_7, these *ab initio* values are essentially the same as those reported by Chelikowsky and co-workers,[94] who have carried out the calculations by using the ab initio pseudo-potential based finite-difference method. Except for the sizable differences in the α values calculated for Si_7, which could be due to the difference in the geometries which were optimized at different levels of theory, both sets of the *ab initio* values are in extremely good agreement with the LDA results. The differences in the α values between the two methods are by less than 0.07 au/atom for all Si_n except for n=3 and n=4, for which cases the differences are just only 0.3 au/atom. The last set of α values listed under "lit. α" were reported by Bazterra et al[103] and were calculated by the DFT-B3PW91 methods with Sadlej's basis set. Except for the Si_6 case, most of the α values from the B3PW91 calculations lie in between those of VWN and B3LYP.

3.2. First hyperpolarizability, β, in Si_n with $n = 3$, 9, and 12

Dissociative recombination of monohydride ions has been reviewed recently.[2,32] The process which involves an electronic excitation of the ionic core has not been further investigated since the last review.[32] An imaging experiment with OH^+ in CRYRING led to discovery of a very small kinetic energy release

Theoretical research in predicting the hyperpolarizabilities is only at its beginning. Maroulis and Pouchan have pioneered in such calculations for the first hyperpolarizablity of Si atom,[117] and the second hyperpolarizability of Si_4.[118] Other work in this area include calculations of the β values for hydrogen-terminated silicon clusters.[119-122]

The first hyperpolarizability values have been calculated for Si_3 by using both *ab initio* as well as DFT methods and with various basis sets. The results are summarized in Table 2.

The structure used for these calculations was optimized at CCSD(T)/cc-pVQZ level of theory. It contains the C_{2v} symmetry with the bond length being 2.187 Å, and the bond angle 78.1°.

The $\overline{\beta}$ values are calculated as

$$\overline{\beta} = (\beta_{zxx} + \beta_{xzx} + \beta_{xxz} + \beta_{zyy} + \beta_{yzy} + \beta_{yyz} + 3\beta_{zzz})/5 .$$

In a C_{2v} structure, $\beta_{zxx} = \beta_{xzx} = \beta_{xxz}$, $\beta_{zyy} = \beta_{yzy} = \beta_{yyz}$
Therefore, $\overline{\beta} = 3(\beta_{zxx} + \beta_{zyy} + \beta_{zzz})/5$
The β_{zxx}, β_{zyy}, and β_{zzz} values are listed under β_{ijk}.

Table 2. *ab-initio* and DFT results for the first hyperpolarizability (β) values for Si_3 at various levels of theory.

method		β_{ijk} (a.u.)[a]	$\overline{\beta}$ (a.u.)
HF		-99.9, 300.1, 48.9	149.5
MP2	(9s7p6d5f1g)	-153.4,264.6,23.3	80.7
B3LYP		-149.1,282.5,41.5	104.9
B3PW91		-132.4,266.9,39.2	104.2
DZ		-305.1,409.9,27.0	79.0
		-321.4,430.2,26.7	81.3
TZ2P		-234.9,461.9,23.1	150.0
		-266.5,521.9,26.6	169.2
ET-pVQZ		-270.6,416.3,-14.7	78.6
		-214.2,382.9,-8.5	96.1
ET-QZ3P		-164.1,339.8,48.4	134.5
		-191.6,424.4,54.3	172.3
ET-QZ3P-1Diffuse		-165.8,273.7,37.5	87.2
		-199.9,323.1,40.4	98.2
ET-QZ3P-2Diffuse		-158.5,252.0,37.7	78.8
		-188.1,294.9,40.0	88.1
ET-QZ3P-3Diffuse		-133.4,262.9,44.4	104.4
		-184.1,312.5,47.5	105.5
ET-QZ+5P		-156.9,270.4,38.5	91.2
		-195.6,327.9,39.3	103.0

[a]β_{ijk} values are listed as β_{zzz}, β_{yyz} ($=\beta_{zyy} = \beta_{yzy}$), β_{xxz} ($=\beta_{zxx} = \beta_{xzx}$). All β values are listed in atomic unit as β $e^3a_0^3E_h^{-2}$.

Data in the first 4 rows are the β values calculated with a very large basis set, as it involves 9s7p6d5f1g basis functions. Four methods have been used to carry out the calculations with this basis set, namely, the *ab initio* methods at HF and MP2 level, and the DFT methods with B3LYP and B3PW91.

Results show that the $\overline{\beta}$ values vary in a sizable range, with the MP2 method predicting the smallest $\overline{\beta}$ value (80.7 au), while the HF method the largest value (149.5 au). The difference between these two $\overline{\beta}$ values is 25% of the averaged $\overline{\beta}$ value, which is enormous. Both DFT methods, however, predict essentially the same values of around 104 au, and both of the two values lie about one-third the way between the two extreme values, closer to the MP2 values. In comparing with the B3LYP values, the HF $\overline{\beta}$ values is overestimated, largely due to the underestimated β_{zzz} value to be less negative, as well as the over estimated β_{yyz} value. In contrast, the smaller MP2 value is due to the overestimated of β_{zzz} to be more negative, the underestimation of both β_{yyz} and β_{xxz}.

The method-dependence of the $\overline{\beta}$ values shown here is consistent with a recent study by Maroulis and Pouchan on the assessment of the theoretical methods on the second hyperpolarizability (γ) of Si_4[118] ($\overline{\beta}$ = 0 due to the D_{2h} symmetry of the Si_4 structure), where the HF method is shown to consistently predict the smallest γ values of around 8.7 x 10^4 au, the MP2 method, the largest values of around 1.1 x 10^5 au. The CCSD(T) values are in between the two, and lies slightly lower than the MP2 values. Also reported in this paper, is the γ values calculated by the B3LYP method at various basis sets. The DFT method was shown to predict γ values very close to the CCSD(T) values, and therefore, was recommended as a ideal method for such calculations, due to the accuracy of the results as well as the much lower cost in computing time than methods such as CCSD(T).

The rest of the data in Table 2 are the β values calculated by DFT methods by using the ADF program. The first and second row in each entry for a specific basis set are values by, respectively, the LDA/VWN and GGA/BLYP methods.

Again, the $\overline{\beta}$ values span a large range, from 78.8 au to 172.3 au. The GGA $\overline{\beta}$ values are shown to be larger than the LDA values in all cases, with each β component, β_{ijk} to be larger in their absolute values. Diffuse functions are shown to greatly improve the $\overline{\beta}$ values. For example, with each diffuse function added to the ET-QZ3P basis set, the difference between the GGA and LDA values decreases from 38 au with no diffuse function, to 11, 10, and 1 au, with respectively, 1, 2, and 3 diffuse functions included. In addition, the ET-QZ3P-3Diffuse value is essentially the same as the B3LYP/9s7p6d5f1g value. By comparisons between each component of the β_{ijk} values, the LDA method seems to give values closer to the B3LYP/9s7p6d5f1g values than the GGA method.

We have also performed calculations on the first hyperpolarizability on the next smallest silicon cluster after Si_3 that has a non-zero β value, namely, Si_9. Due to the size of the cluster, we have carried out calculations at GGA/B3LYP/ET-QZ3P level of theory, noting that such level probably predicts, as in Si_3, the upper limit of the β values.

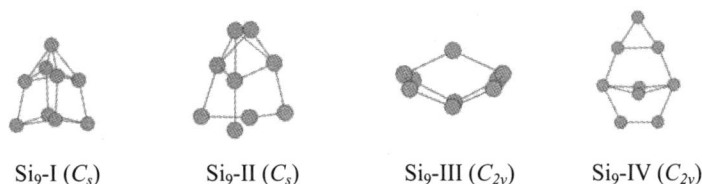

Si$_9$-I (C_s)   Si$_9$-II (C_s)   Si$_9$-III (C_{2v})   Si$_9$-IV (C_{2v})

Fig. 2 Structures for four low-energy isomers of Si$_9$

Both the $\overline{\alpha}$ and $\overline{\beta}$ values predicted at GGA/BLYP/ET-QZ3P level are listed in Table 3.

Table 3. Calculation results for $\overline{\alpha}$ and β for four Si$_9$ isomers at BLYP/ET-QZ3P level of theory.

isomers	$\alpha_{xx}\ \alpha_{yy}\ \alpha_{zz}$ (a.u.)	$\overline{\alpha}/9$ (a.u.)	$\overline{\beta}$ (a.u.)
Si$_9$-I (Cs)	285.9,286.8,265.6	31.0	-54.1
Si$_9$-II (Cs)	239.5,272.0,366.4	32.5	-41.6
Si$_9$-III (C$_{2v}$)	373.9,382.4,210.4	35.8	-141.5
Si$_9$-IV (C$_{3v}$)	201.3,311.9,447.58	35.6	-34.1

The initial structures were taken from ref.103, and relaxed by geometry optimization at GGA/BLYP/ET-QZ3P level. All these structures are stationary points on the potential energy surface. Structure I has the lowest energy, but the other three isomers, II – IV, all have low energies close to I. Among the four structures, I is the most spherical, which is reflected by close values of α_{ii} (i = x, y, z). Structures II and IV are both prolate, and III is oblate. Unlike the α values, which are all fairly close to each other, the β values vary dramatically. The oblate structure gives the most negative value of 141.4 au, and the prolate structures give the least negative values. These results seem to suggest that the β values are much more sensitive to the shape of the cluster than the α values.

In order to confirm this observation, we have also performed parallel calculations on 5 structural isomers of Si_{12}.

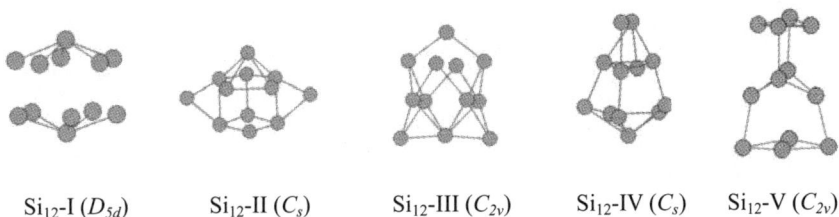

Si_{12}-I (D_{5d}) Si_{12}-II (C_s) Si_{12}-III (C_{2v}) Si_{12}-IV (C_s) Si_{12}-V (C_{2v})

Fig. 3 Structures for five low-energy isomers of Si_{12}.

Like in the Si_9 case, the initial structures were taken from ref.54, and then optimized at GGA/BLYP/ET-QZ3P level, at which level of theory both the α and β values were calculated. Again, these structures are all local minima on the potential energy surface, and the electronic energies are all very close to each other. Both the α and β values are listed in Table 4.

Table 4. Calculation results for $\overline{\alpha}$ and $\overline{\beta}$ for four Si_{21} isomers at BLYP/ET-QZ3P level of theory.

isomers	$\alpha_{xx}\ \alpha_{yy}\ \alpha_{zz}$ (a.u.)	$\overline{\alpha}/9$ (a.u.)	$\overline{\beta}$ (a.u.)
Si_{12}-I (D_{5d})	384.4,384.7,347.6	31.0	-50.0
Si_{12}-II (Cs)	324.6,381.5,444.6	32.0	27.0
Si_{12}-III (C_{2v})	310.9,388.0,436.5	31.6	39.2
Si_{12}-IV (Cs)	413.8,416.0,309.7	31.6	-64.3
Si_{12}-V (C_{2v})	341.7,342.0,533.1	33.8	11.0

Similar behavior of $\overline{\alpha}$ and $\overline{\beta}$ were observed, with the $\overline{\alpha}$ values remain close approximity of each other, while the $\overline{\beta}$ values jump all over the map. Furthermore, the $\overline{\beta}$ values show opposite signs among the isomers. The most oblate shape, structure IV, gives the largest absolute $\overline{\beta}$ value of 64.3 au, and the most prolate shape of V gives the smallest absolute value of 11.0 au, and most spherical structure, I, has the $\overline{\beta}$ value (−50.0 au) lying in between the two extremes.

4. Conclusions

Our results show that, while the static dipole polarizability of small silicon clusters can be fairly accurately predicted by a large range of theoretical methods at reasonable size of basis sets, first polarizability calculations are, however, much more demanding. In spite of the very large basis sets used, β values are shown to vary a great deal among the various methods. Diffuse functions are shown to be essential in improving the first hyperpolarizability calculations.

In contrast to the static dipole polarizability, which are shown not very sensitive to the geometry of the structure of a certain cluster, i.e. spherical vs. prolate or oblate, first hyperpolarizability, $\overline{\beta}$ values, are extremely sensitive to the overall structure of the cluster. Our results show that, among the various conformational isomers, the estimated mean β values are the greatest in the most oblate-shaped isomers; and among the silicon clusters of different sizes, the mean β values show trend of Si$_3$ >Si$_9$ >Si$_{12}$, with $\overline{\beta}$ being 172 a.u. for Si$_3$, between −34 and 142 a.u. for Si$_9$, and between −64 and 39 a.u. for Si$_{12}$. Even though each β value reported for Si$_9$ and Si$_{12}$ itself might be much larger than the true first hyperpolarizability, owing to the lack of the diffuse functions included at the level of theory where the calculations were carried out, the trends in these β values are consistent with the understanding that first hyperpolarizability decreases as the shape of the molecule becomes more spherical, as well as the fact that the flat Si$_3$ molecule obviously has the most oblate shape compared to the other two isomers.

Acknowledgments

DYZ gratefully acknowledges the generous support from the Fulbright Organization and Couseil Regional d'Aquitaine, and the hospitality of the people at Laboratoire de Chimie Structurale at Université de Pau et de Pays de l'Adour, during her stay for her sabbatical year at the institute. This work is part of the COST project D26.

References

1. J. R. Heath, Y. Liu, S. C. O'Brien, Q. L. Zhang, R. F. Curl, F. K. Tittel, and R. E. Smalley, *J. Chem. Phys.* **83**, 5520 (1985).
2. M. F. Jarrold, *Science* **252**, 1085 (1991).
3. W. L. Brown, R. R. Freeman, K. Raghavachari, and M. Schluter, *Science* **235**, 860 (1987).
4. T. G. Dietz, M. A. Duncan, D. E. Powers, and R. E. Smalley, *J. Chem. Phys.* **74**, 6511 (1981).
5. M. D. Morse, J. B. Hopkins, P. R. R. Langridge-Smith, and R. E. Smalley, *J. Chem. Phys.* **79**, 5316 (1983).
6. J. L. Gole, J. H. English, and V. E. Bondybey, *J. Phys. Chem.* **86**, 2560 (1982).
7. V. E. Bondybey, and J. H. English, J. Chem. Phys. **80**, 568 (1984).
8. D. J. Trevor, R. L. Whetten, D. M. Cox, and A. Kaldor, J. Am. Chem. Soc. **107**, 5181 (1985).
9. S. J. Riley, E. K. Parks, G. C. Nieman, L. G. Pobo, and S. Wexler, *J. Chem. Phys.* **80**, 1360 (1984).

10. J. A. Becker, *Angew. Chem. Int. Ed. Engl.* **36**, 1390 (1997).
11. O. Cheshnovsky, S. H. Yang, C. L. Pettiette, M. J. Craycraft, Y. Liu, and R. E. Smalley, *Chem. Phys. Lett.* **138**, 119 (1987).
12. T. N. Kitsoupoulos, C; J. Chick, A. Weaver, and D. M. Neumark, *J. Chem. Phys.* **93**, 6908 (1990).
13. T. N. Kitsoupoulos, C; J. Chick, A. Weaver, and D. M. Neumark, *J. Chem. Phys.* **95**, 1441 (1991).
14. C. S. Xu, T. R. Taylor, G. R. Burton, and D. M. Neumark, *J. Chem. Phys.* **108**, 1395 (1998).
15. T. N. Kitsopoulos, C. J. Chick, A. Weaver, and D. M. Neumark, *J. Chem. Phys.* **93**, 6108 (1990).
16. C. C. Arnold and D. M. Neumark, *J. Chem. Phys.* **99**, 3353 (1993).
17. G. Meloni, M. J. Ferguson, S. M. Sheehan, and D. M. Neumark, *Chem. Phys. Lett.* **399**, 389 (2004).
18. Q. L. Zhang, Y. Liu, R. F. Curl, F. K. Tittel, and R. E. Smalley, *J. Chem. Phys.* **88**, 1670 (1988).
19. L. A. Bloomfield, R. R. Freeman, and W. L. Brown, *Phys. Rev. Lett.* **54**, 2246 (1985).
20. L. A. Bloomfield, M. E. Guesie, R. R. Freeman, and W. L. Brown, *Chem. Phys. Lett.* **121**, 33 (1985).
21. K. D. Rinnen and M. L. Mandich, *Phys. Rev. Lett.* **69**, 1823 (1992).
22. W. Begemann, K. H. Meiwas-Broer, and H. O. Lutz, *Phys. Rev. Lett.* **73**, 2248 (1986).
23. M. F. Jarrold, and E. C. Honea, *J. Phys. Chem.* **95**, 9181 (1991).
24. J. M. Hunter, J. L. Fye, M. F. Jarrold, and J. E. Bower, *Phys. Rev. Lett.* **73**, 2063 (1994).
25. M. F. Jarrold, and J. E. Bower, *J. Chem. Phys.* **96**, 9180 (1992).
26. M. F. Jarrold, and V. A. Constant, *Phys. Rev. Lett.* **67**, 2994 (1991).
27. M. F. Jarrold, *J. Phys. Chem.* **99**, 11 (1995).
28. M. F. Jarrold, Y. Ijiri, and U. Ray, *J. Chem. Phys.* **94**, 3607 (1991).
29. K. M. Ho, A. A. Shvartsburg. B. C. Pan, Z. Y. Lu, C; Z. Wang, J. G. Wacker, J. L. Fye, and M. F. Jarrold, *Nature*, **392**, 582 (1998).
30. R. R. Hudgins, M. Imai, M. F. Jarrold, and P. Dugourd, *J. Chem. Phys.* **111**, 7865 (1999).
31. D. J. Trevor, D. M. Cox, K. C. Reichmann, R. O. Brichman, and A. Kaldor, *J. Phys. Chem.* **91**, 2598 (1987).
32. M. F. Jarrold, J. E. Bower, and K. Creegan, *J. Chem. Phys.* **90**, 3615 (1989).
33. S. Li, R. Z. Van Zee, W. Weltner, Jr., and K. Raghavachari, *Chem. Phys. Lett.* **243**, 275 (1995).
34. E. C. Honea, A. Ogura, C. A. Murray, K. Raghavachari, W. O. Sprenger, M. F. Jarrold, and W. L. Brown, *Nature*, **366**, 42 (1993).
35. J. A. Becker, Angew. Chem. Int. Ed. Egnl. **36**, 1390 (1997).

36. S. Vijayalakshmi, M. A. George, and H. Grebel, Appl. Phys. Lett. **70**, 708 (1997).
37. R. Schafer, S. Schlecht, J. Woenckhaus, and J. A. Becker, Phys. Rev. Lett. **76**, 471 (1996).
38. J. A. Becker, S. Schlecht, R. Schafer, J. Woenckhaus, and F. Hensel, *Mater. Sci. Eng.*, **A217/218**, 1 (1996).
39. K. Raghavachari and V. Logovinsky, *Phys. Rev. Lett.* **55**, 2853 (1985).
40. K. Raghavachari, *J. Chem. Phys.* **84**, 5672 (1986).
41. K. Raghavachari and C. M. Rohlfing, J. Chem. Phys. **89**, 2219 (1988).
42. K. Raghavachari, *Z. Phys.* **D12**, 61 (1989).
43. C. M. Rohlfing and K. Raghavachari, *J. Chem. Phys.* **96**, 2114 (1992).
44. N. Binnggeli and J. R. Chelikowsky, *Phys. Rev.* **B50**, 11764 (1994).
45. P. Ballone, W. Andreoni, R. Car, and M. Parrinello, *Phys. Rev. Lett.* **60**, 271 (1988).
46. C. H. Patterson and R. P. Messmer, *Phys. Rev.* **B42**, 7520 (1990).
47. I. Vasiliev, S. Ogut, and J. R. Chelikowsky, *Phys. Rev. Lett.* **78**, 4805 (1997).
48. R. Fournier, S. B. Sinnot, and A. E. DePristo, *J. Chem. Phys.* **97**, 4149 (1992).
49. O. F. Sankey, D. J. Nikewshi, D. A. Drabold, and J. D. Dow, *Phys. Rev.* **B41**, 127150 (1990).
50. W. Andreoni, R. Car, and M. Parrinello, *Phys. Rev.* **B41**, 10243 (1990).
51. N. Binggeli and J. R. Chelikowsky, **Phys. Rev. B50**, 11746 (1994).
52. N. Binggeli, J. L. Martins and J. R. Chelikowsky, *Phys. Rev. Lett.* **68**, 2956 (1992).
53. P. W. Andreoni, *Z. Phys.* **D19**, 31 (1991).
54. M. V. Ramakrishna and A. Bahel, *J. Chem. Phys.* **104**, 9833 (1996).
55. J. R. Chelikowsky and N. Binggeli, *Mater. Sci. Forum*, **232**, 87 (1996).
56. U. Rothlisberger, W. Andreoni, and P. Giannozzi, *J. Chem. Phys.* **92**, 1248 (1992).
57. M. R. Pederson, K. Jackson, D. V. Porezag, Z. Hajnal, and Th. Frauenheim, Phys. *Rev.* **B54**, 2863 (1996).
58. E. Kaxiras and K. Jackson, *Phys. Rev. Lett.* **71**, 727 (1993).
59. J. C. Grossman and L. Mitas, *Phys. Rev.* **B52**, 16735 (1995).
60. I. Kwon, R. Biswas, C. Z. Wang, K. M. Ho, and C. M. Soukoulis, **Phys. Rev. B49**, 7242 (1994).
61. M. Menon and K. R. Subbaswamy, *Phys. Rev.* **B47**, 12754 (1993).
62. P. Ordejon, D. Lebedenko, and M. Menon, *Phys. Rev.* **B50**, 5645 (1994).
63. D. Tomanek and M. A. Schluter, *Phys. Rev.* **B36**, 1208 (1987).
64. O. F. Sankey, D. J. Niklewski, D. A. Drabold, and J. D. Dow, *Phys. Rev.* **B41**, 12750 (1990).
65. J. R. Chelikowsky, J. C. Phillips, M. Kamal, and M. Strauss, *Phys. Rev. Lett.* **62**, 292 (1989).
66. J. R. Chelikowsky and J. C. Phillips, *Phys. Rev.* **B41**, 5735 (1990).
67. J. Harris, *Phys. Rev.* **B31**, 1770 (1985).
68. J. Song, S. E. Ulloa, and D. A. Drabold, *Phys. Rev.* **B53**, 8042 (1996).
69. Y. Kanemitsu, K. Suzuki, Y. Nakayoshi, and Y. Masumoto, *Phys. Rev.* **B46**, 1992, 3916.

70. M. S. Hybertsen and M. Needels, *Phys. Rev.* **B48**, 1993, 4608.
71. J. C. Phillips, *J. Chem. Phys.* **87**, 1712 (1987).
72. L. Mitas, J. C. Grossman, I. Stich, and J. Tobik, *Phys. Rev. Lett.* **84**, 1479 (2000).
73. J. C. Grossman and L. Mitas, *Phys. Rev. Lett.* **74**, 1323 (1995).
74. J. R. Chelikowsky and J. C. Phillips. *Phys. Rev. Lett.* **63**, 1653 (1989).
75. B. Liu, Z. Y. Lu, B. C. Pan, C. Z. Wang, and K. M. Ho, *J. Chem. Phys.* **109**, 9401 (1998).
76. S. Wie, R. N. Barnett, and U. Landman, *Phys. Rev.* **B55**, 7935 (1997).
77. A. A. Shvartsbury, B. Liu, M. Jarrold, and K. M. Ho, *J. Chem. Phys.* **112**, 4517 (2000).
78. B. X. Li and P. L. Cao, *Phys. Rev.* **B62**, 15788 (2000).
79. T. Fruuenheim, F. Weich, T. Kohler, S. Uhlmann, D. Porezag, and G. Seifert, *Phys. Rev.* **B52**, 11492 (1995).
80. D. J. Wales, *Phys. Rev.* **A49**, 2195 (1994).
81. Z. Y. Lu, C. Z. Wang, and K. M. Ho, *Phys. Rev.* **B61**, 2329 (2000).
82. B. X. Li and P. L. Cao, *Phys. Rev.* **A62**, 23201 (2000).
83. B. X. Li and P. L. Cao, J. Phys. Condens. Matter **13**, 1 (2000).
84. Y. Luo, J. Zhao, and G. H. Wang, *Phys. Rev.* **B60**, 10703 (1999).
85. G. Pacchioni and J. Koutechky, *J. Chem. Phys.* **84**, 3301 (1986).
86. A. A. Shvartsburg, M. F. Jarrold, B. Liu, Z. Y. Lu, C. Z. Wang, and K. M. Ho, *Phys. Rev. Lett.* **81**, 4616 (1998).
87. J. Zhao, X. Chen, Q. Sun, F. Liu, and G. Wang, *Phys. Lett.* **A198**, 243 (1995).
88. K. Fuke, K. Tsukamotó, F. Misaizu, and M. Sanekata, *J. Chem. Phys.* **99**, 7807 (1993).
89. E. C. Honea, A. Ogura, D. R. Peale, C. Félix, C. A. Murray, K. Raghavachari, W. O. Sprenger, M. F. Jarrold, and W. L. Brown, *J. Chem. Phys.* **110**, 12161 (1999).
90. K. Jackson, M. Pederson, C.-Z. Wang, and K.-M. Ho, *Phys. Rev.* **A59**, 3685 (1999).
91. Y.-M Juan, E. Kaxiras, and R. G. Gordon, *Phys. Rev.* **B51**, 9521 (1995).
92. T. T. Rantala, M. I. Stockman, D. A. Jelski, and T. F. George, *J. Chem. Phys.* **93**, 7427 (1990).
93. A. Rubio, J. A. Alonso, X. Blase, L. C. Balbas, and S. G. Louie, *Phys. Rev. Lett.* **77**, 247 (1996).
94. A. Rubio, J. A. Alonso, X. Blase, L. C. Balbas, and S. G. Louie, *Phys. Rev. Lett.* **77**, 5442 (1996).
95. I. Vasiliev, S. Ogut, and J. R. Chelikowsky, *Phys. Rev. Lett.* **78**, 4805 (1997).
96. I. Vasiliev, S. Ogut, and J. R. Chelikowsky, *Phys. Rev. Lett.* **82**, 1919 (1999).
97. K. Jackson, M. R. Pederson, D. Porezag, Z. Hajnal, and R. Frauenheim, *Phys. Rev.* **B55**, 2549 (1997).
98. K. Jackson, *Phys. Stat. Sol.* **B217**, 293 (2000).
99. G. Maroulis, D. Begué, and C. Pouchan, *J. Chem. Phys.* **119**, 794 (2003).

100. A. Sieck, D. Porezag, Th. Frauenheim, M. R. Pederson, and K. Jackson, *Phys. Rev.* **A56**, 4890 (1997).
101. F. Torrens, *Physica* **E13**, 67 (2000).
102. K. Deng, J. and C. T. Yang, Chem. Phys. Rev. **A61**, 025201 (2000).
103. V. E. Bazterra, M. C. Caputo, M. B. Ferraro, and P. Fuentealba, *J. Chem. Phys.* **117**, 11158 (2002).
104. B. K. Panda, S. Mukherjee, and S. N. Behera, *Phys. Rev.* **B63**, 45404 (2001).
105. C. Pouchan, D. Bégué, and D. Y. Zhang, *J. Chem. Phys.* **121**, 4628 (2004).
106. D. Y. Zhang, D. Bégué, and C. Pouchan, *Chem. Phys. Lett.* **398**, 283 (2004).
107. M. Brieger, A. Renn, A. Sodeik, and A. Hese, *J. Chem. Phys.* **75**, 1 (1983).
108. M. Brieger, *J. Chem. Phys.* **89** 275 (1986).
109. D. Bégué, M. Merawa, and C. Pouchan, *Phys. Rev.* **A57** 2470 (1998).
110. M. J. Frisch, G. W. Trucks, H. B. Schlegel, G. E. Scuseria, M. A. Robb, J. R. Cheeseman, V. G. Zakrzewski, J. A. Montgomery Jr., R. E. Stratmann, J. C. Burant, S. Dapprich, J. M. Millam, A. D. Daniels, K. N. Kudin, M. C. Strain, O. Farkas, J. Tomasi, V. Barone, M. Cossi, R. Cammi, B. Mennucci, C. Pomelli, C. Adamo, S. Clifford, J. Ochterski, G. A. Petersson, P. Y. Ayala, Q. Cui, K. Morokuma, D. K. Malick, A. D. Rabuck, K. Raghavachari, J. B. Foresman, J. Cioslowski, J. V. Ortiz, B. B. Stefanov, G. Liu, A. Liashenko, P. Piskorz, I. Komaromi, R. Gomperts, R. L. Martin, D. J. Fox, T. Keith, M. A. Al-Laham, C. Y. Peng, A. Nanayakkara, C. Gonzalez, M. Challacombe, P. M. W. Gill, B. Jonhson, W. Chen, M. W. Wong, J. L. Andres, C. Gonzalez, M. Head-Gordon, E. S. Replogle, and J. A. Pople, Gaussian 98, Revision A.6, (Gaussion Inc., Pittsburgh, PA, 1998).
111. A. D. Becke, *J. Chem. Phys.* **98**, 5648 (1993).
112. C. Lee, W. Yang, and R. G. Parr, *Phys. Rev.* **B37**, 785 (1998).
113. J. A. Pople, *J. Chem. Phys.* **72**, 650 (1980).
114. T. Clark, J. Chandrasekhar, and P. v. R. Schleyer, *J. Comp. Chem.* **4**, 294 (1983).
115. F. M. Bickelhaupt, *J. Comput. Chem.* **22**, 931 (2001).
116. S. H. Vosko, L. Wilk, and M. Nusair, *Can. J. Phys.* **58**, 1200 (1980).
117. G. Maroulis and C. Pouchan, J. *Phys. B: At. Mol. Opt. Phys.* **36** 2011 (2003).
118. G. Maroulis and C. Pouchan, *Phys. Chem. Chem. Phys.* **5**, 1992 (2003).
119. B. Kirtman and M. Hasan, *J. Chem. Phys.* **96**, 470 (1992).
120. E. A. Perpete, J. -M. Andre, and B. Champagne, *J. Chem. Phys.* **109**, 4624 (1998).
121. Y. Mochizuki and H. Agren, *Chem. Phys. Lett.* **336**, 451 (2001).
122. B. Jansik, B. Schimmelpfennig, P. Norman, Y. Mochizuki, Y. Luo, and H. Agren, *J. Phys. Chem.* **A106**, 395 (2002).

CHAPTER 4

THEORETICAL CALCULATIONS OF THE STATIC DIPOLE POLARIZABILITY OF ATOMS AND SMALL ATOMIC CLUSTERS

Patricio Fuentealba

Departmento de Física, Facultad de Ciencias
Universidad de Chile, Santiago, Chile
e-mail: pfuentea@macul.ciencias.uchile.cl

The calculations of the static dipole polarizability of atomic clusters will be discussed. The theoretical methodologies to do the calculations will be briefly analyzed. First, the most important aspects of the calculations on atoms, where there are more information, will be presented. Then, representative results for some families of clusters will be shown and discussed. The technical difficulties will be analyzed and it will be shown that in many cases there are significative deviations between the theoretical calculations and the experimental values.

1. Introduction

The static electric dipole polarizability is a measure of the distortion of the electronic density under the effect of an external static electric field. Since the majority of the atomic interactions occurring in nature are governed by electric forces is no surprising that the dipole polarizability is an important quantity in a variety of phenomenon [1,2]. Specially in studies of intermolecular forces [1], electron scattering [3] and in the last time in studies of clusters [4]. The present article is devoted to the last point, the dipole polarizability of atomic clusters.

In general, for atomic clusters, the chemical bonding is profoundly changed in the formation of a cluster whereas the solid–state-like properties like the dielectric constant are of no significance in a cluster of few atoms. The dipole polarizability, however, represents a suitable molecular property that can take on the role of the dielectric constant and can be measured without touching the cluster [5]. The dipole polarizability comprises the information of the bonding character in the cluster, and on the other hand, through the Classius–Mosotti relation it yields information about the dielectric constant of the solid–like macroscopic particle formed with the cluster. In general, the experimental measurement of any property of neutral clusters is a difficult task and the dipole polarizabilities are one of the few properties which are amenable of an experimental measurement. In presence of an electric field the neutral clusters are deflected in linear order due to their dipole polarizabilities [6], allowing their measurement. Hence, the polarizabilities can be one of the most important piece of information about the nature of the bonding and the geometrical structure of the neutral clusters. To fulfill this goal it is necessary to have a reliable theoretical model to understand the principal factors influencing in the polarizabilities. The jellium model has been successfully applied to study the polarizability of large clusters approaching the bulk properties. For small clusters, however, it is expected more bonding and geometrical effects than the incorporated in the jellium model. Therefore, ab initio calculations are necessary, and the experience gained in calculating the dipole polarizabilities of atoms and molecules is useful. The importance of polarizability data to rationalize experimental observations has been recently stressed [5], justifying the effort of looking for reliable theoretical models to calculate it and make valuable suggestions for further experiments. The basic ideas of the research on atomic clusters have been reviewed from both, the experimental [4] and the theoretical side [7].

The atomic clusters are experimentally difficult to detect, and not too much about them is known. They are usually produced in gas phase by laser vaporization, and one of the few cluster property that can be measured with relative easy and accuracy is the dipole polarizability. To study the electronic structure of clusters one of the primordial characteristic necessary to know is the geometry, and experimentally it is not possible to determine it. Therefore, one of the first goal of the theoretical calculation on atomic clusters is the determination of the geometry of each cluster. However, when the number of atoms in the cluster increases the situation is really complex because there are many isomers separated by a narrow energy gap. It is the calculation of some property, like the dipole polarizability, and its correspondence with the experimental data the only way to postulate the isomer of low energy. Hence, the calibration and comparison of the different theoretical methodologies for calculating the dipole polarizability of atomic clusters is an important issue. From the theoretical point of view, to obtain the most stable isomer and its geometry is already a complicated issue. Reliable geometries are obtained only taking into account the correlation effects, and the posteriori calculation of the dipole polarizability needs a basis set flexible enough with diffuse functions. Therefore,this article will concentrate only on the small clusters where calculations of this type are possible.

2. Theory

The principal definitions will be briefly reviewed following the work by Buckingham [1]. The hamiltonian of an neutral electronic system in weak interaction with an external electric field is

$$H = H^{(0)} - \mu_\alpha F_\alpha - \frac{1}{3}\Theta_{\alpha\beta}F_{\alpha\beta} - \dots \tag{1}$$

where H^0 is the hamiltonian of the free system, μ_α and $\Theta_{\alpha\beta}$ are the dipole moment and quadrupole moment operators, respectively. F_α and $F_{\alpha\beta}$ are the electric field and the field gradient at the origin. The subscripts α, β are the x, y, z components of the respective vector or tensor. The Einstein convention for the summation has been used: a repeated subscript denotes a summation over all three cartesian components. Hence, taking into account only the effects produced by the electric field, the energy of the perturbed system is given by the series

$$E = E^{(0)} - \mu_\alpha^{(0)}F_\alpha - \frac{1}{2}\alpha_{\alpha\beta}F_\alpha F_\beta - \frac{1}{6}\beta_{\alpha\beta\gamma}F_\alpha F_\beta F_\gamma - \frac{1}{24}\gamma_{\alpha\beta\gamma\delta}F_\alpha F_\beta F_\delta F_\delta \dots \tag{2}$$

where $E(0)$ and $\mu_\alpha^{(0)}$ are the energy and dipole moment of the isolated system, respectively. $\alpha_{\alpha\beta}$ are the components of the static dipole polarizability tensor, the object of interest in this work. $\beta_{\alpha\beta\gamma}$ and $\gamma_{\alpha\beta\gamma\delta}$ are the components of the first and second hyperpolarizability tensors, respectively. The dipole polarizability describes the linear distortion of the electronic cloud due to the external electric field and it has, therefore, units of volume. Using the Hellmann-Feynman theorem the series expansion for the total dipole moment can be found

$$\mu_\alpha = -\frac{\partial E}{\partial F_\alpha} = \mu_\alpha^{(0)} + \alpha_{\alpha,\beta}F_\beta + \frac{1}{2}\beta_{\alpha\beta\gamma}F_\beta F_\gamma + \frac{1}{24}\gamma_{\alpha\beta\gamma\delta}F_\beta F_\gamma F_\delta \tag{3}$$

It is important to mention that only the first, nonvanishing moment is independent of the origin of coordinates. Hence, for a charged system the dipole moment depends on the choice of the origin (it will be zero at the center of charge) and so also the dipole polarizability depends on the origin of coordinates for a charged system.

Using again the Hellman–Feynman theorem another two important expressions for the dipole polarizability can be found

$$\alpha_{\alpha,\beta} = -\frac{\partial \mu_\alpha}{\partial F_\beta} \tag{4}$$

and

$$\alpha_{\alpha,\beta} = -\frac{\partial^2 E}{\partial F_\alpha \partial F_\beta} \tag{5}$$

It is important to remember that the Hellmann–Feynman theorem is valid for the exact wave function or one of the Hartree-Fock type. However, in the practice it has been shown that the use of the last two equations to calculate the dipole polarizability produces accurate results for any wave function better than Hartree-Fock. The series expansion of Eqs.(2) and (3) have, of course, their perturbation theory counterpart. Hence, the dipole polarizability is proportional to the energy correction at second order of perturbation theory when the electric field is the perturbation, and considering a cartesian coordinate system the components of the polarizability tensor are

$$\alpha_{ij} = 2 \sum_{k \neq 0} \frac{< 0|p_i|k >< k|p_j|0 >}{E_k - E_0} \qquad i,j = x, y, z \tag{6}$$

where p_i is one of the component of the dipole operator and E_k is the energy of the system in the quantum state k. Another related expression can be obtained using the linear response function $\chi(\vec{r_1}, \vec{r_2})$ [8]

$$\alpha_{ij} = 2 \int \int \chi(\vec{r_1}, \vec{r_2}) p_i p_j d\vec{r_1} d\vec{r_2} \tag{7}$$

Rarely the nine components of the dipole polarizability tensor are necessary. The number of constants required to specify the tensor depends on the symmetry group of the system. Hence, for a system of symmetry type C_1 one needs 6 components and for a symmetry type T_d only one. Usually, the experiments yield the average dipole polarizability

$$\alpha = \frac{1}{3}(\alpha_{xx} + \alpha_{yy} + \alpha_{zz}) \tag{8}$$

Much more difficult are the measurements of the anisotropy

$$\Delta\alpha = \sqrt{\frac{1}{2}[(\alpha_{xx} - \alpha_{yy})^2 + (\alpha_{xx} - \alpha_{zz})^2 + (\alpha_{yy} - \alpha_{zz})^2]} \qquad (9)$$

Eq.(7) is also interesting because it permits to relate the dipole polarizability to some chemical reactivity indices defined in density functional theory [9]. Berkowitz and Parr [10] derived an interesting relationship among the linear response function and the softness kernel, $s(\vec{r_1}, \vec{r_2})$, the softness function, $s(\vec{r})$, and the global softness, S

$$\chi(\vec{r_1}, \vec{r_2}) = -s(\vec{r_1}, \vec{r_2}) + \frac{s(\vec{r_1})s(\vec{r_2})}{S} \qquad (10)$$

all these softness quantities should be interpreted as a measure of the reactivity of a system. Its relation to the polarizability has been long ago postulated. The first numerical evidence of a proportionality was given by Politzer [11], and then using Eq.(7) numerical as well as theoretical evidences have been put forward by several authors [12-15]. This equation is important in the chemical reactivity theory because it relates empirical chemical concepts like softness and its inverse the hardness, which are not susceptible of an experimental measurement, with the dipole polarizability, a well defined quantum observable and experimentally measurable.

The principal formal equations have been already presented. Now, the question is how in practical terms to calculate the dipole polarizability. There have been in the past many attempts to solve the coupled Hartree–Fock equations associated to the perturbation series of Eq.(6) [16], and also the respective coupled Kohn–Sham equations [8]. However, the equations are numerically complicated for systems without spherical symmetry. Nowadays by far the most used methods are based on Eqs.(4) and (5). This technique, known as the finite field method [17] was a long time used on its numerical version [18]. It consists in doing two calculations, one for the isolated system and the other for the same system in presence of an electric field F. The energy difference, in the limit of electric field going to zero, is proportional to the dipole polarizability:

$$\alpha = \lim_{F \to 0} \frac{2\Delta E}{F^2} \qquad (11)$$

It has, however, numerical problems: the choice of the magnitude of the electric field is critical. It should be sufficiently large to produce an energy difference greater than the numerical errors, but small enough to reproduce the limit of zero field. Usually, fields of around 0.001 a.u. were used. This type of calculations were put forward in a brilliant way by Bishop and collaborators [19,20]. By now, most of the currently used softwares can compute the second derivative of Eq. (6) analytically avoiding the complication on the choice of the magnitude of the field. The accuracy of the calculations depend strongly on two factors. First, the basis set used. It is necessary the use of diffuse functions to take into account the electron cloud distortion due to the electric field. Second, the electron correlation effects which can be determinant in the evaluation of the dipole polarizability.

3. Results and Discussion

For light atoms the static electric dipole polarizabilities are known with reasonable accuracy [21] , and almost all the traditional methods of the quantum chemistry and atomic physics have been applied in calculating them. Without being exhaustive we mention the works of Werner and Meyer [18], Bishop et al (see, for instance, Ref. 19,20), Sadlej and collaborators (see for instance Ref. 23), Maroulis and Thakkar (see, for instance, Ref. 25), and Das and Thakkar [24] Those works show, besides good final numbers, that the incorporation of the electronic correlation is an important and difficult task, and that the basis set must be very carefully chosen. However, for heavy atoms the situation is more complicated. There is no experimental values for many atoms and the calculations are more difficult. In some cases, as it will be show later, the use of pseudopotentials simplify a lot the computations, but in systems like the transition metal atoms the large influence of the dynamical correlation makes the calculations very unreliable. On the other hand, density functional methods are being in the last time widely used in many fields of atomic, molecular physics, and quantum chemistry. Specially, for the determination of cluster structures in system with more than ten atoms they are the only workable possibility. Therefore it is important to study and understand its features and weaknesses.

In order to understand the reliability of the theoretical calculations on clusters it is useful to study first the applications to atoms.

3.1. *Atoms*

In this subsection only the main points of the study of the static dipole polarizability of atoms will be presented. The discussion will be focus on the points which are important for the study of the static dipole polarizability of atomic clusters, the matter of the next subsection. A more exhaustive discussion of the dipole polarizability of atoms is presented in another chapter of this book. Let us start with the first row atoms Li–Ne because the principal effects in the polarizability are well understood and there are reliable values for comparison. Various different approximations to the exchange and correlation functionals have been tested for the calculation of the dipole polarizability [26]. In Table 1 some of the obtained results are presented in comparison with the numerical restricted Hartree–Fock results [28] and CEPA (Coupled Electron Pairs Approximations) [18] values.

Table I. Atomic dipole polarizabilities. Mean values (in au.)

	Li	Be	B	C	N	O	F	Ne
HF	168.0	45.54	22.02	11.89	7.140	4.797	3.288	2.367
B88	149.4	45.13	25.72	14.49	8.821	6.454	4.488	3.221
PW91	163.1	43.69	20.48	11.15	6.750	4.529	3.129	2.265
LYP	154.0	43.80	20.37	11.07	6.701	4.601	3.188	2.309
BPW91	142.9	42.68	23.43	13.35	8.224	5.912	4.165	3.018
BLYP	137.7	43.18	23.67	13.49	8.308	6.070	4.275	3.095
B3PW91	146.3	42.85	22.47	12.67	7.790	5.536	3.889	2.820
LB94	156.0	37.73	21.88	10.32	6.967	4.678	3.278	2.554
HF[a]	170.1	45.62	22.02	12.03	7.358	4.734	3.284	2.377
PNO–CEPA[b]	164.5	37.84	20.47	11.84	7.430	5.412	3.759	2.676

[a] values from Ref. [28]
[b] values from Ref. [18]

There are some results using only an exchange functional which should be compared with the Hartree–Fock values, other ones use only a correlation functional together with the Hartree–Fock exchange and also the most popular hybrid functionals which should be compared with the CEPA results. The main conclusion of this work [26] was that the right asymptotic behavior of the exchange potential is essential to obtain reliable values as it is numerically demonstrated with the functional LB98 which is the only one in Table 1 with the right asymptotic behavior. However, it was also demonstrated that this effect it is important in atoms but not in molecules where the bonding effects predominate. For instance, using the hybrid functional B3PW91 they obtained the values of 11.68 a.u. , 10.59 a.u. and 8.617 a.u. for the average dipole polarizability of N_2, O_2 and F_2 molecules, respectively. For the same molecules, MR–CI calculations yields[27] 11.80 a.u., 10.27 a.u. and 8.22 a.u., respectively. Hence, we are in the strange position that the density functional methods are not able to deal with a property of the, in principle, more easier systems like the atoms but they perform well for the most complicated systems like molecules.

Table II. The dipole polarizabilities of the group 1 elements from K to Cs. Mean values (in au.)

	K	Rb	Cs
Pseudopotential[a]	290.7	319.1	395.1
Pseudopotential[b]	292	316	404
Pseudopotential[c]	293	321	402
all–electron[d]	291.12	316.17	396.02
Expt.[e]	293 ± 6	319 ± 6	401 ±0.6

[a] Relativistic energy consistent nine valence electron pseudopotential at the CCSD(T) level. Values from Ref. [29]
[b] Relativistic four component energy adjusted one valence electron pseudopotential. Values from Ref. [30]
[c] Relativistic one valence electron semiempirical pseudopotential. Values from Ref. [31]
[d] All–electron CCSD(T) at the scalar relativistic Douglas–Kroll level. Values from Ref.[29]
[e] experimental values from Refs.[32,33]

On going to heavier atoms an accurate all–electron calculation is more difficult and the use of pseudopotentials seems to be unavoidable. Special attention has been paid to the dipole polarizability of the heavier alkaline metal atoms. Recently, an exhaustive study of the polarizability of the heavier alkaline metal atoms including Fr and the E119 atom has been presented [29]. All–electron calculations at the coupled–cluster level within the Douglas–Kroll relativistic formalism have been done, and compared with pseudopotential calculations. On Table 2, can be seen that for the heavy alkali atoms the pseudopotential results reproduce fairly good the more exhaustive all–electron ones.

The experimental and theoretical values of the atomic dipole polarizability is, for some atoms, still a controversial matter and for others completely unknown. For instance, very few is known about the dipole polarizability of the transition metal atoms. They are from the experimental side unknown and the theoretical calculations are very difficult because of the great electron correlation effects. For others, in principle, simple atoms there are controversies between the experimental and the theoretical values. For Sodium atom, exhaustive theoretical calculations including relativistic effects recommended a value of 164.5 ± 0.5 a.u. [34] which is higher than the recent experimental value [35] of 162.8 ±0.8. An even worse discrepancy has been found for the Aluminium atom. Recent exhaustive theoretical calculations [36] estimate the dipole polarizability of Aluminium atom between 57.6 and 58.4 a.u. with an uncertainty of 1 a.u.. This value is far away from the experimental one of 45.89 ± 2.02 a.u. [37]. This fact is specially important because in some experimental works the dipole polarizability of atomic clusters is measured with respect to the informed experimental value of Aluminium atom.

3.2. *Atomic Clusters*

The study of atomic clusters is an active area of research. The atomic clusters are particles composed of a countable number of atoms, starting with the diatomic molecule to a vaguely defined upper bound of several hundred thousand atoms. The size of many of them is in the range of the nanometer and, therefore, they are specially important in the area of the nanoelectronics [38]. The clusters present properties which are not easily related to the bulk. This might be expected, since for them, the surface effects are very important. Even for a cluster with 1000 atoms, about a quarter of them lie on the surface. Hence, the clusters are important in order to understand the evolution of properties from the atoms to the solid. From the formal point of view, it is interesting to note that clusters with different number of atoms can present different type of bonds. For instance, the magnesium dimer, Mg_2, presents a bond of the van der Waals type, the larger clusters, say, Mg_{10} present covalent bonds, and when the cluster starts to grow at some number of atoms the bond should resemble the one of the bulk, a metallic bond.

Since the experimental measurements of the dipole polarizabilities have been mainly concentrated in alkali metal and silicon clusters the theoretical efforts have been also concentrated mainly on these families of clusters. Therefore, the dipole polarizabilities of the alkali metal clusters will be discussed and then, the dipole polarizabilities of the silicon clusters to conclude with other studies on various families of clusters.

On Table 3 one can see the various values for the dipole polarizability of the first elements of the family of lithium atom clusters.

Table 3. Comparison of calculated and experimental dipole polarizabilities (in a.u.) for Lithium clusters.

	B3LYP [a]	PW91 [b]	B3LYP [c]	MP2 [c]	Exp. values [b]
Li	142.3	151.2	143.7	168.5	164.0
Li_2	199.2	221.4	198.5	202.8	221.1
Li_3	320.1	340.1	330.5	236.2	233.6
Li_4	355.5	394.1	352.4	343.7	327.2
Li_5	452.0	465.6	450.3	430.8	427.5
Li_6	526.0	502.1	491.5	507.1	359.2
Li_7	531.0	562.2	531.9	412.1	539.0
Li_8	581.1	599.2	596.0	609.2	559.3

[a] values from Ref. [39] basis set: D95+diffuses
[b] values from Ref. [40] basis set: 6–31G*
[c] values from Ref. [41] basis set: 6–311+G(d)

The most recently theoretical calculations[39-41] together with the experimental values [40] are displayed. Comparing first the theoretical values among themselves one observes significative differences. They are a clear indication of the importance of both, correlation and basis set. It seems that a basis set of the type 6-31G* is too small to bring reliable values and it is the explanation for the large deviation of the results labelled PW91 with respect to the other two density functional based methods labelled B3LYP. Between the last two calculations all the differences are only attributable to the basis set. They are, however, smaller than 5% with exception of Li_6. The other theoretical values displayed in Table 3 corresponds to MP2 calculations. The differences with respect to the DFT based values are not so significative with the exception of Li_3 and Li_7 where the differences deserve a clear explanation, especially for Li_3 a small system where very exhaustive calculations are possible. Li_3 presents two C_{2v} isomers energetically very close. One is a obtuse triangle and the other one is an acute triangle, but the electronic state is different, $^2B_1$ and $^2B_2$, respectively. However, exhaustive coupled cluster calculations [42] on both isomers with the aug-cc-pvtz basis sets yields a value of 329.2 a.u. closer to the DFT based values and an MP2 calculations with the same basis set (6–311+G(d)) yields a value of 310.4 a.u.. Hence, the result informed in Ref. 40 seems to be wrong. For Li_4 there are also exhaustive theoretical calculations [43]. At the MP2 level with an specially constructed basis set of the 7s5p2d type they obtained a value of 381.5 a.u., and at the MP4 level with a bigger basis set of the 15s7p1d type they obtained a value of 380.1 a.u.. Both values are considerable bigger than the ones displayed in Table 3 with the exception of the PW91 calculations where surely some error compensation explain the agreement. The preceding comparison among different theoretical calculations shows clearly that the issue is not resolved, there are difference of more of ten percent among them. Therefore, it will be desirable to obtain more accurate calculations for the family of lithium clusters. Going on to the comparison with the experimental values the situation is even worse. The differences between the DFT based values and the experimental one for the atom were already explained and they do not need more explanation. However, for the polyatomic systems there is a systematic deviation and in almost all the clusters the theoretical values are greater than the experimental ones. This is very difficult to explain, for all the effects presents in the experiment and not in the calculations, i. e. temperature, vibrational contributions, volume dilatation and others, can only raise the theoretical values making the deviation bigger. It is also interesting to note another qualitative difference. The experimental values increase going from Li to Li_5, goes through a minimum at Li_6 to increase again for the other clusters. This effect is not present in any of the theoretical results. The MP2 values are the only one with a minimum but at Li_7, and the B3LYP calculations of the second column presents a depletion at Li_7. Hence, all the discussed discrepancies show that the study of the evolution of the lithium cluster dipole polarizability is, by far, not completed.

Let us go now to the study of the dipole polarizability of sodium clusters. Here, there are more theoretical and experimental values. The first experimental measurements of the dipole polarizability of sodium clusters were done twenty years ago[6]. A new set of experimental data has been recently reported [40]. In Table 4 some selected theoretical calculations are compared with the experimental values

Table 4. Comparison of calculated and experimental dipole polarizabilities (in a.u.) for Sodium clusters.

	PW91 [a]	B3LYP [b]	MP2 [b]	Exp. values [a]
Na	162.6	143.8	166.9	159.4
Na_2	256.4	230.9	255.4	265.2
Na_3	429.1	397.8	431.7	444.7
Na_4	521.7	482.8	508.6	565.5
Na_5	644.4	595.8	635.3	629.9
Na_6	700.4	651.1	699.3	754.3
Na_7	727.4	694.6	655.5	808.2
Na_8	788.1	748.0	797.7	900.9

[a] values from Ref. [39] basis set: D95+diffuses
[b] values from Ref. [40] basis set: 6–31G*
[c] values from Ref. [41] basis set: 6–311+G(d)

Unfortunately, the experimental values informed by Knight et al[6] are only displayed in a curve, and it is not possible to have the absolute values.

However, qualitatively one finds that the values are slightly greater than the new experimental ones. For the bigger members of the series there is also other experimental values to compare with. Tikhonov et al [48] have informed values of 854.9 ± 23.6 a.u. and 955.5 ± 21.6 a.u. for the dipole polarizability of Na_7 and Na_8, respectively, which compare well with the values of 808.2 a.u. and 900.9 a.u. informed in Table 4 for Na_7 and Na_8, respectively. There are also other theoretical values [44-45] which are not displayed because the trends are well characterized by the data on Table 4. They were selected to follow the same kind of calculations already discussed for the Lithium clusters. There is a great computational complication in going from lithium to sodium clusters. The number of electrons increases considerably and so the basis set. For instance, the number of primitive functions necessary to calculate the cluster of Na_8 is almost the same as the number of functions necessary to calculate Li_{29}. Therefore to keep the same accuracy in the calculations is a serious task. A natural way of retaining the accuracy level is the use of pseudopotentials within the framework of DFT calculations. This route has been followed by Kronik et al [46] to calculate the dipole polarizability of clusters up to Na_{20}. Unfortunately, they only informed a curve without a Table with the absolute values. Hence, it is not possible to compare with the other calculations. Looking again to the Table 4, one can see that the deviations among the different theoretical calculations are smaller than for lithium clusters and the values are, in general, below the experimental ones. Therefore, it is very reasonable to think that the temperature effects which are not considered in the theoretical calculations could raise the values improving the comparison with the experimental ones. The point has been further discussed by Kronik et al [46]. Although the sodium clusters are from a computational point of view more complicated than lithium clusters the agreement between theory and experiment is better. It is also interesting to compare the dipole polarizability of lithium and sodium clusters. For the small members of the series the values are similar, but starting up n=3 or n=4 the sodium clusters present significant greater values. Note that, according to the experimental values, already on Li_3 there is almost no variation with respect to Li_2. The static dipole polarizability of small mixed sodium–lithium clusters is also known. They have been experimentally measured [49] and theoretical calculated [50]. The dipole polarizability of the hydrogenated lithium clusters has been also studied [56]. They are important as prototypes for the study of a possible metal–insulator transition and also as a model for a metallic cluster on the surface of an ionic insulator[57,58]. On this study [56] has been found that the polarizability shows a clear decrease when more hydrogen atoms are added to a given lithium cluster, and they predict the same behavior in the dipole polarizability of other ionic mixed clusters such as Na_nCl_m.

The dipole polarizability of potassium clusters has been measured by Knight et al [6] . However, to our knowledge, there is no theoretical calculations to compare with. The experimental values show for potassium clusters the same trends already observed for sodium clusters. It is also interesting to observe that the polarizabilities for corresponding clusters of sodium and potassium, normalized to their respective atomic polarizabilities, are nearly identical. Whereas the polarizabilities of lithium clusters are clearly smaller. For the heavier alkali metal, rubidium and cesium clusters, the dipole polarizabilities are not known.

The other family of clusters which is well studied is the one of silicon clusters. For the small members of the series the calculated dipole polarizabilities are displayed in Table 5. It is encouraging to note that there are seven different theoretical calculations and all of them differ for no more than a few percent. The most exhaustive and computationally demanding CCSD(T) calculations do not differ for more than one percent with the results of the hybrid functionals B3PW91 or B3LYP. The only exception is Si_6 where the difference is due to the use of different isomers. The density functional calculations yields a C_{2v} isomer as the most stable one, whereas the CCSD(T) calculation uses a D_{4h} isomer. It is interesting to note that even when the isomers are different the polarizabilities are similar. It is also interesting to mention that the results for silicon clusters validate the use of pseudopotentials [46,51] for calculating the dipole polarizability. The point is important in going to the clusters with more atoms where an all–electron calculation is prohibitive. The density functional calculations with the local density approximation (LDA) are slightly lower than the hybrid ones, but they are still in a fairly agreement. The greater deviation is presented for the Si_2 molecule. However, it is known that the silicon dimer is an especially challenging case for the theoretical methods. The ground state is a triplet ($^3\Sigma_g$) with various low lying excited states. This fact produces immediately two practical problems. The system is very difficult to describe with only one Slater determinant and the self–consistent procedure has difficulties in converging to the right state. For instance, a CCSD(T) calculation with a 6-311+G* basis set yields a binding energy of 2.66 eV [51] which is 20% lower than the experimental value of 3.24 eV., and the dipole polarizability is of 83.8 a.u. which compare fairly well with the LDA value [46]. However, repeating the calculations with the Sadlej's basis set of the type 7s5p2d leads

Table 5. Comparison of theoretical dipole polarizabilities (in a.u.) for Silicon clusters.

	B3PW91 [a]	B3LYP [b]	MP2 [c]	MP4 [c]	CCSD(T) [c]	LDA [d]	LDA [e]
Si	39.3					39.7	41.3
Si_2	96.5					52.9	84.9
Si_3	106.1	109.	106.7	105.6	106.7	105.	106.
Si_4	137.4	139.	138.0	137.9	137.3	137.	137.
Si_5	166.0	168.	166.2	165.7	164.5	163.	162.
Si_6	184.2	185.	180.1	180.1	178.8	183.	181.
Si_7	210.2	212.	217.7	217.1	213.6	206.	206.
Si_8	250.5	253.				245.	244.

[a] values from Ref. [50], all–electron, basis set:7s5p2d
[b] values from Ref. [51], all–electron, basis set: 5s4p2d
[c] values from Ref. [52], all–electron, basis set: 5s4p2d
[d] values from Ref. [53], all–electron, basis set: 6s5p4d
[e] values from Ref. [54], pseudopotential, cartesian grid

to a dipole polarizability of 94.1 a.u.. Neglecting the triple excitations, a CCSD calculation yields a dipole polarizability of 89.6 a.u.. Note that the values range from 52.9 to 96.5 a.u.. It is clear that the silicon dimer still constitutes an open problem for the theoretical determination of the dipole polarizability. One should not expect similar problems for the remaining clusters because the existence of various low lying excited states is an inherent characteristic of the dimer. Unfortunately, for these small members of the family of silicon clusters there are no experimental values to compare with. The dipole polarizability for silicon clusters between nine and 120 atoms have been measured [59], and there are also some theoretical results at the LDA level [60-62]. The three cited theoretical works present polarizabilities in a fairly agreement among them but with significant deviations with respect to the experimental values. The theoretical values show relatively small variations for the clusters in the range of ten to twenty atoms with all values clearly larger than the bulk limit. On the contrary, the experimental values show large variations and the values, in average, are below the bulk limit. Just like in the case of lithium metal clusters the experimental values are lower than the theoretical ones. Hence all the effects not considered in the theoretical calculations can only raise the values and there is no explanation for the discrepancy. In order to elucidate the possible effect of considering a mixture of various isomers in the experimental measurements the cluster of Si_9 has been taken as a test case. Using a modified genetic algorithm code fourteen isomers have been found [51] and the dipole polarizability has been calculated using the hybrid functional B3PW91 and the Sadlej basis set of 7s5p2d functions. The dipole polarizabilities range from 286. a.u. to 326 a.u.. The values are significatively larger than the experimental one of 176 ± 49 a.u.. However, attention should be paid to the fact that the experimental measurements has been determined by calibration with Al_N clusters [4], which in turn has been determined by calibration with the dipole polarizability of aluminium atom, and it has been already documented [36] the large discrepancy between theory and experiment in relation to the dipole polarizability of aluminium atom. Using the theoretical value instead of the experimental one the measured dipole polarizability of the silicon clusters would augment in almost 20% improving something the concordance with the theoretical calculations. However, the deviations are still significative and not well understood.

The dipole polarizability of some transition metal atomic clusters has been also studied. In particular the copper clusters has received in the last years considerable attention. On Table 6 various theoretical calculations of the dipole polarizability of copper clusters are displayed. There are four different methodologies employed to calculate the polarizability of the copper clusters. Three of them used pseudopotentials and only one is an all–electron calculation. All of them use some density functional method. It is however to remark that only the Stuttgart pseudopotential include the relativistic effects. After the discussion of the atomic polarizabilities it should be no a surprise that for the atoms the different calculations present significative deviations. For copper atom the influence of the relativistic effects in the polarizability has been estimated to reduce it in around 7.0 a.u.[71] which is almost the difference between the value reported in the second column and the other ones. Very recently, for Cu_2 the relativistic effects on the dipole polarizability have been estimated to reduce it also in around 7.0

Table 6. Comparison of theoretical dipole polarizabilities (in a.u.) for copper clusters.

	B3LYP[a]	B88P86[b]	B3PW91[c]	B3LYP[d]
Cu	39.4	47.0	52.6	
Cu_2	73.8	78.5	77.6	76.6
Cu_3	129.1	130.1	137.1	131.6
Cu_4	148.4	151.5	153.9	153.8
Cu_5	196.9	192.1	196.5	195.6
Cu_6	213.4	217.6	221.2	221.4
Cu_7	233.3	233.1	242.9	241.7
Cu_8	263.4	256.8	267.6	272.9

[a] values from Ref. [39], Stuttgart pseudopotential,augmented basis set
[b] values from Ref. [68], all–electron,augmented basis set
[c] values from Ref. [69], Los Alamos pseudopotential, DZ basis set
[d] values from Ref. [70], Los Alamos pseudopotential, augmented DZ basis set

a.u. [72]. Nothing is known about the relativistic effect in the dipole polarizability of larger systems. The geometry of all clusters in the different works are similar exception made of Cu_8 where one work found the tetrahedron structure as the ground state whereas the other ones found a C_{2v} structure. It is interesting to note that even in this case the polarizability values are similar demonstrating once again that different geometries do not affect significatively the mean polarizability as long as the structures are compact. Looking at Table 6 one can observe that the values of the last two columns are very similar. They are both calculated using the same pseudopotential and only slightly different functionals and basis sets. It is gratifying to see that using one functional or another the results are almost the same. Both

set of values are, however, systematically greater than the values informed
on the other two columns. This could be possible due to the different basis
sets. The calculations informed in the first two columns use bigger basis
set including diffuse functions which seems to be important in the calculation of the dipole polarizability. Unfortunately, a recently experimental
measurement of the dipole polarizability of copper clusters start up Cu_9
[73] and nothing is known about the polarizability of the small members of
the family. However, the theoretical calculations of Refs. (68,70) have also
computed the polarizability of Cu_9 (in Ref. (70) the values Cu_n, $n \leq 13$
are presented). The comparison with the experimental values is astonished.
Refs. 68 and 70 calculate a value of 295.2 a.u. and 309.9 a.u. , respectively,
whereas the experimental value is of 984. a.u., more than three times bigger.
The same enormous difference occurs for the clusters with 10 to 13 atoms.
There is no explanation for those differences.

There are also other families of clusters for which the dipole polarizability has been calculated and/or experimentally measured. Polarizabilities of
Germanium clusters, Ge_n, up to n=10, and Ga_nAs_m ($n + m \leq 8$) have
been calculated [55] using a pseudopotential method and the local density
approximation. There is also a recently study using an hybrid functional
[63]. While it was shown that the LDA approximation works well for the
dipole polarizability of silicon clusters unfortunately in the case of Germanium clusters the values with the more sophisticated hybrid functional are
significatively larger. The mixed Ge–Te clusters has been also studied [64].
The polarizabilities were found to be very close to the bulk value with a
nonmonotonic variation as a function of the cluster size. The dipole polarizability of carbon clusters has been also theoretically studied [65] using
an hybrid density functional and large basis set. Here, it still controversial
which isomers are the ground state. A linear geometry is the preferred one
for the clusters with an odd number of atoms. For the clusters with an even
number of atoms the cyclic and linear isomers are almost isoenergetics and
it is impossible to predict which one is the most stable. The theoretical
calculations predicted [65] that, in particular, a jet formed by the two isomers of C_6, cyclic and planar, will split up in the presence of an electric
field. Unfortunately, the experimental values are not known. In a study
of the stability of As_n (n=4,20,28,32,36,60) cage structures [66] the dipole
polarizability has been calculated. In particular the As_{60} which presents a
fullerene-like structure has a polarizability four or five times greater than
the fullerene. For the phosphorous family of clusters only the dipole polarizability of P_4 is experimentally and theoretically known [67] with a good

agreement among them.

4. Concluding Remarks and Outlook

From the theoretical point of view the formal equations and models to calculate and understand the dipole polarizability are well known and clear. There is also a practical methodology to calculate it. The finite field methods are well documented and there is enough experience in atoms and molecules which is very useful in the calculation of the dipole polarizability of atomic clusters. One knows the practical difficulties in doing reliable calculations. The basis set should be large and with diffuse functions. A systematic study of the basis set convergence in the calculated dipole polarizability has been done for atoms and small molecules and clusters, but not for a family of clusters like the ones discussed in this work. The correlations effects are also very important. Specially in clusters they are not only important for the direct calculation of the dipole polarizability, but also for the determination of the geometry and the most stable isomer. Furthermore, the most practical density functional methods to include the correlations effects have, until now, not been validated for the calculation of the dipole polarizability of atomic clusters. It is necessary a large number of comparisons with methods of first principle beyond Hartree–Fock and desirable for a family of clusters. However, the main difficult to understand the dipole polarizability behavior in atomic clusters is the incredible large deviations between the experimental values and the theoretical ones for some family of clusters. As long as this differences are not explained the advances in the theoretical calculations will not be conclusive. Since the research on atomic clusters is a relative new branch of science it is not surprise that there are so many open problems. However, the recent theoretical advances together with the always more powerful computers permit us to be optimistic about the future of this branch of the atomic cluster science.

References

1. A. D. Buckingham in *Adv. in Chemical Physics* **12**, 107 (1967).
2. D. M. Bishop *Rev. Mod. Phys.* **62**, 343 (1990).
3. F. A. Gianturco and A. Jain, *Phys. Rep.*, **143** 347 (1986).
4. W. A. de Heer, *Rev. Mod. Phys.* **65**, 611 (1993).
5. J. A. Becker, *Angew. Chem. Int. Ed.* **36**, 1390 (1997).
6. W. D. Knight, K. Clemenger, W. A. de Heer and W. A. Saunders, *Phys. Rev. B* **31**, 2539 (1985).
7. M. Brack *Rev. Mod. Phys.* **65**, 677 (1993).

8. G. D. Mahan and K. R. Subbaswamy, *Local Density Theory of Polarizability* Plenum Press (1990).
9. R. G. Parr and W. Yang, *Density Functional Theory of Atoms and Molecules* Oxford Press (1989).
10. M. Berkowitz and R. G. Parr, *J. Chem. Phys.* **88**, 2554 (1988).
11. P. Politzer, *J. Chem. Phys.* **86**, 1072 (1987).
12. P. Fuentealba and O. Reyes, *Theochem* **282**, 65 (1993).
13. T. K. Ghanty and S. Ghosh, *J. Phys. Chem.* **97**, 4951(1993).
14. A. Vela and J. L. Gazquez, *J. Am. Chem. Soc.* **112**, 1490 (1990).
15. Y. Simon and P. Fuentealba, *J. Phys. Chem.* A **102**, 2029 (1998).
16. A. Dalgarno, *Adv. Phys.* **11**, 281 (1962).
17. H. D. Cohen and C. C. J. Roothaan, *J. Chem. Phys.* **43**, S341 (1965).
18. H. J. Werner and W. Meyer, *Mol. Phys.* **31**, 855 (19767).
19. D. M. Bishop, M. Chaillet, C. Larrieu and C. Pouchan, *Phys. Rev.* **A31**, 2785 (1985).
20. G. Maroulis and D. M. Bishop, *J. Phys.* **B18** 3653 (1985).
21. T. M. Miller and B. Bederson, *Adv. At. Mol. Phys.* **13**, 13 (1976).
22. J. Pipin and D. M. Bishop *Phys. Rev.* **45A**, 2736 (1992).
23. V. Kello, A. Sadlej and K. Faegri *Phys. Rev.* **47A**, 1715 (1993).
24. A. K. Das and A. J. Thakkar, *J. Phys.* B **31**, 2215 (1998)
25. G. Maroulis and A. J. Thakkar, *J. Phys.* B **22**, 2439 (1989).
26. P. Fuentealba and Y. Simon–Manso, *J. Phys. Chem.* A **101**, 4231 (1997).
27. D. Spelsberg and W. Meyer *J. Chem. Phys.* **101**, 1282 (1994).
28. J. Stiehler and J. Hinze *J. Phys.* B **28**, 4055 (1995).
29. I. S. Lim, P. Schwerdtfeger, B. Metz and H. Stoll, *J. Chem. Phys.* **122**, 104103 (2005).
30. M. Dolg *Theor. Chim. Acta* **93**, 141 (1996).
31. P. Fuentealba *J. Phys.* B **15**, L555 (1982).
32. R. W. Molof, H. L. Schwartz, T. M. Miller and B. Bederson, *Phys. Rev.* **A10**, 1131 (1974).
33. J. Amini and H. Gould *Phys. Rev. Lett.* **91**, 153001 (2003).
34. G. Maroulis *Chem. Phys. Lett.* **334**, 207 (2001).
35. C. R. Ekstrom, J. Schiemdmayer, M. S. Chapman, T. D. Hammond and D. E. Pritchard *Phys. Rev.* **A51**, 3883 (1995).
36. P. Fuentealba *Chem. Phys. Lett.* **397**, 459 (2004).
37. P. Milani, I. Moullet and W. A. de Heer, *Phys. Rev.* A **42**, 5150 (1990).
38. W. Ekardt, *Metal Clusters* Wiley, U.S.A., (1999).
39. P. Fuentealba, L. Padilla and O. Reyes, *J. Compt. Methods in Sciences and Engineering* **4**, 589 (2004).
40. D. Rayane, A. R. Allouche, E. Benichou, R. Antoine, M. Aubert, Ph. Dugourd, M. Broyer, C. Ristoti, F. Chandezon, B. Hubert and C. Guet *Eur. Phys. J.* D **9**, 243 (1999).
41. K. Chandrakumar, T. Ghanty and S. Ghosh, *J. Phys. Chem.* A **108**, 6661 (2004).
42. P. Fuentealba, unpublished results.
43. G. Maroulis and D. Xenides, *J. Phys. Chem.* A **103**, 4590 (1999).

44. P. Calaminici, K. Jug and A. Koester, *J. Chem. Phys.* **111**, 4613 (1999).
45. J. Guan, M. E. Casida, A. M. Koester and D. Salahub *Phys. Rev. B* **52**, 2184 (1995).
46. L. Kronik, I. Vasiliev, M. Jain and J. R. Chelikowsky *J. Chem. Phys.* **115**, 4322 (2001).
47. J. L. Martins, J. Buttet and R. Car *Phys. Rev. B* **31**, 1804 (1085).
48. G. Tikhonov, V. Kasperovich, K. Wong and V. Kresin *Phys. Rev. A* **64**, 63202 (2001).
49. R. Antoine, D. Rayane, A. Allouche, M. Aubert–Frecon, E. Benichou, F. Dalby, Ph. Dugourd and M. Broyer *J. Chem. Phys.* **110**, 5568 (1999).
50. M. D. Deshpande, D. G. Kanhere, I. Vasiliev and R. M. Martin, *Phys. Rev. A* **65**, 033202 (2002).
51. V. E. Bazterra, M. C. Caputo, M. B. Ferraro and P. Fuentealba *J. Chem. Phys.* **117**, 11158 (2002).
52. C. Pouchan, D. Begue and D. Zhang, *J. Chem. Phys.* **121**, 4628 (2004).
53. G. Maroulis, D. Begue and C. Pouchan, *J. Chem. Phys.* **119**, 794 (2003).
54. A. Sieck, D. Porezag, Th. Frauenheim, M. P. Pederson and K. Jackson, *Phys. Rev. A* **56**, 4890 (1997).
55. I. Vasiliev, S. Ogut and J. Chelikowsky, *Phys. Rev. Lett.* **78**, 4805 (1997).
56. P. Fuentealba and O. Reyes, *J. Phys. Chem. A* **103**, 1376 (1999).
57. G. Rajagopal, R. Barnett and U. Landman, *Phys. Rev. Lett.* **67**, 727 (1991).
58. P. Fuentealba and A. Savin *J. Phys. Chem. A* **105**,11531 (2001).
59. R. Schaefer, S. Schlecht, J. Woenckhaus and J. Becker, *Phys. Rev. Lett.* **76**, 471 (1996).
60. K. Jackson, M. Pederson, C. Wang and K. Ho, *Phys. Rev. A* **59**, 3685 (1999).
61. K. Deng, J. Yang and C. Chan, *Phys. Rev. A* **61**, 25201 (2000).
62. D. Zhang, D. Begue and C. Pouchan, *Chem. Phys. Lett.* **398**, 283 (2004).
63. J. Wang, M. Yang, G. Wang and J. Zhao, *Chem. Phys. Lett.* **367**, 448 (2003).
64. R. Natarajan and S. Ogut, *Phys. Rev. B* **67**, 235326 (2003).
65. P. Fuentealba, *Phys. Rev. A* **58**, 4232 (1998)
66. T. Baruah, M. R. Pederson, R. R. Zope and M. R. Beltran, *Chem. Phys. Lett.* **387**, 476 (2004)
67. U. Hohm, A. Loose, G. Maroulis and D. Xenides, *Phys. Rev. A* **61**, 053202 (2000).
68. P. Calaminici, A. Koester and A. Vela, *J. Chem. Phys.* **113**, 2199 (2000).
69. P. Jaque and A. Toro, *J. Chem. Phys.* **117**, 3208 (2002).
70. Z. Cao, Y. Wang, J. Zhu, W. Wu and Q. Zhang, *J. Phys. Chem. B* **106**, 9649 (2002).
71. P. Fuentealba, H. Stoll, L. v. Szentpály, P. Schwerdtfeger y H. Preuss. *J. Phys. B* **16** L323 (1983).
72. T. Saue and H. Jensen, *J. Chem. Phys.* **118**, 522 (2003).
73. M. B. Knickelbein, *J. Chem. Phys.* **120**, 10450 (2004).

CHAPTER 5

ELONGATION METHOD FOR POLYMERS AND ITS APPLICATION TO NONLINEAR OPTICS

Feng Long Gu,[a,*] Akira Imamura,[b] and Yuriko Aoki [a, c]

a) Group, PRESTO, Japan Science and Technology Agency (JST), Kawaguchi Center Building, Honcho 4-1-8, Kawaguchi, Saitama, 332-0012, Japan

b) Department of Mathematics, Faculty of Engineering, Hiroshima Kokusai Gakuin University, Nakano 6-20-1, Aki-ku, Hiroshima-city, 739-0321, Japan

c) Department of Molecular and Material Sciences, Interdisciplinary Graduate School of Engineering Sciences, Kyushu University, 6-1 Kasuga-Park, Fukuoka, 816-8580, Japan

Contents

* Email: gu@cube.kyushu-u.ac.jp

1. Introduction

Quantum chemistry is an important tool for studying a wide range of problems in physical chemistry and chemical physics.[1] With the recent remarkable developments of computational methods and powerful computers, such as supercomputers and workstations with fast CPU and large memory, it becomes possible to solve complex problems in chemistry. Quantum chemistry calculation has already become a popular and powerful tool to investigate the molecular structures and other chemically related properties.[2] There exist many quantum chemistry program packages for different specific purposes. Among them, the Gaussian program (its latest version G03[3]) from GAUSSIAN company, GAMESS,[4] and CRYSTAL,[5] are the most popular to many users ranging from beginners to highly trained theoreticians. Like an NMR spectrometer to an experimentalist, a computer equipped with quantum chemistry programs is an essential "experimental" tool to elucidate electronic, optical, and magnetic properties of molecules, polymers, and solids.

Ab initio quantum chemistry calculations, also known as first principles quantum-chemical calculations, are typically based on canonical molecular orbitals (CMOs) that extend over the whole system, in which the matrix of the Lagrange multipliers is diagonal and the diagonal elements are called orbital energies. One of the advantages of the CMO based calculations is that the canonical solutions of the Hartree-Fock equations are convenient for the physical interpretation of the Lagrange multipliers. That is, the ionization energy within the "frozen MO" approximation is simply given as the negative orbital energy (known as Koopmans' theorem [6]), and the electron affinity of a neutral molecule is given as the orbital energy of the corresponding anion. The CMOs may be considered as a set of orbitals for carrying out the variational calculation. The matrix of the Lagrange multipliers is diagonal only in the CMO basis. The total energy, however, depends only on the total wave function, which is a Slater determinant in terms of the occupied MOs. The total wave function is unaltered under any kind of unitary transformation among the occupied MOs. Thus, after having obtained the CMOs, other sets of MOs for different purposes can be generated by linear combination of the original CMOs. Localized MOs and hybridized MOs are commonly used.

The reason for using other sets of MOs instead of CMOs is the two main disadvantages of the CMO based treatment: (a) there is no direct relation to most traditional chemical concepts and (b) it is not possible to use the locality or nearsightedness of the underlying electronic structure. Consequently, such calculations based on CMOs have to be performed restrictedly to the whole system and the number of two-electron integrals can become very huge so that computer capacities are easily reached the limit. Due to these limitations, CMO based quantum chemistry calculations confront a big challenge for large systems. Although methodologies for small molecules can be very much sophisticated and the computational results sometimes can be used to calibrate experimental measurements, CMO based quantum chemical approach is difficult to apply to large systems at the *ab initio* level.

If a large system has a translational periodicity, one can utilize the crystal orbital (CO) method by taking into account the periodic boundary condition. Along this line, remarkable progress has been made for calculating the electronic structure of polymers,[7] surfaces and crystals[8] with complete periodicity on the Born-von Kàrmàn periodic boundary conditions.[9] The characteristic features of the periodic systems, such as the band structure and the density of states, can be determined by the CO method. For quasi-one-dimensional polymers, program packages from Erlangen[10] and Namur[11] groups are most known. While for two- and three-dimensional systems, CRYSTAL[5] is one of the most widely used program packages for the bulk electronic properties by *ab initio* calculations.

However, interesting chemical and physical properties in electron conductivity, lattice defects, end effects, etc. depend strongly upon the aperiodic nature. For example, random arrangements of various amino acids play an essential role in catalytic activity of enzymes. For such problems, some quantum chemistry approaches have been proposed to investigate the effects of defects, as well as end effects. One straightforward way is the cluster approximation. In this approach, the interesting properties are obtained by extrapolating the properties of a series of small clusters into infinity. The problem in this approach is that the extrapolation is not always reliable and stable. More seriously, some properties even cannot be extrapolated. Besides these difficulties, the cluster approximation is still inappropriate because the size of the systems extending in three dimensions is increasing too fast. Therefore, computational chemistry shows its own vainness for large aperiodic systems. Conventional quantum chemistry approach is very difficult to be applied to large aperiodic systems, such as random polymers, adsorbed surfaces, and defective crystals.

Due to the above-mentioned problems in conventional quantum chemistry calculations, considerable efforts have been devoted in recent years to exploit locality in order to develop computational methods for large systems. Approaches for this purpose demand calculation procedures in a linear scaling with the number of electrons. They are so-called order N, e.g., $O(N)$, procedure. These methods partition the whole system into fragments associated with localized molecular orbitals (LMOs), instead of CMOs. In principle, the localized orbitals may be atomic or fragment orbitals that are determined *a priori* or they may be orbitals that are optimized during the calculation.

In the early 1990s, Imamura *et al* developed an approach, called elongation method,[12] based on the linear scaling requirement. The elongation method, by its name, was originally proposed for calculating electronic states of large quasi-one dimensional periodic and non-periodic polymers. The elongation idea comes from the experimental polymerization/copolymerization mechanism. By the elongation method, any periodic or random polymer can be theoretically synthesized in

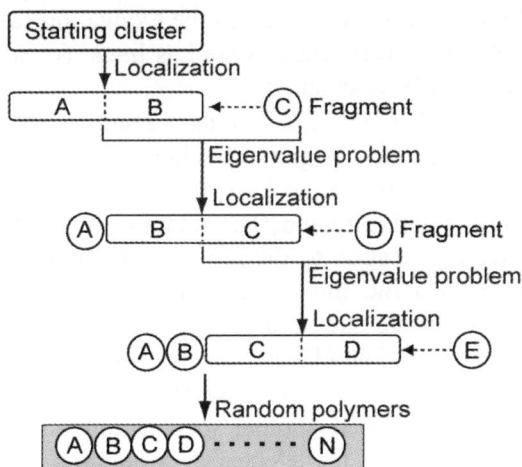

Figure 1. Schematic illustration for the elongation method.

analog with the polymerization experiments. The polymer chain is built up by stepwise adding a monomer unit to a starting cluster. It is general in the sense that the adding monomers during the elongation steps are arbitrary, which means that a random polymer can be synthesized. If and only if individual adding monomers are always the same, a periodic polymer is constructed. Figure 1 presents a schematic illustration of the elongation method.

In the elongation procedure, the delocalized CMOs of a starting cluster are first localized into frozen and active regions in specified parts of the molecule. Next, a monomer is attacking to this cluster to the active region, and the eigenvalue problem is solved by disregarding the LMOs which have no or very weak interaction with the attacking monomer. By repeating this procedure, the length of the polymer chain is increased step by step to any desired length. The obvious advantage of the approach lies in the fact that one can avoid solving very large secular equations for large aperiodic systems. The LMO representation, by which the elongation method is working on, allows one to freeze one part of the system far away from the polymer chain propagation site. The frozen part is disregarded in the elongation SCF procedure. This reduces the number of variational degrees of freedom, and most importantly the number of two-electron integrals that have to be evaluated in the system can be drastically reduced. Thus, an important reduction in computation time can be achieved. The elongation method together with a cut-off technique achieves a major reduction without significant loss of accuracy. The CPU time consumed in the elongation calculations shows a large increase in the efficiency as well as linear (or sub-linear) scaling for the elongation SCF calculations.

A simple scheme of 2×2 unitary rotations was first adopted to obtain the LMOs for the elongation method. [12] In this scheme, a pair of CMOs is successively rotated to form one LMO in the frozen region and another one in the active region. The rotations are continued iteratively until no further localization to the previous iteration is obtained. Using this localization scheme the elongation method has been applied to determine the electronic structure of periodic and aperiodic polymers at both the Hartree-Fock [13] and DFT [14] levels of theory. The convergence of this 2×2 scheme is slow especially for covalently bonded systems. Imamura *et al* [15] proposed another localization scheme based on the stationary conditions of the electronic structures for the elongation method. In this scheme, the frozen orbitals, called as stationary orbitals can be extracted from the CMOs and thus the active orbitals are just the reminders of the whole CMOs excluding the stationary orbitals. While, the stationary orbitals are not unique during the elongation SCF process and the use of this scheme is not straightforward.

As mentioned, the convergence of the 2×2 scheme is slow, especially for delocalized covalently bonded systems. For large basis sets the computation times can become excessive. Furthermore, an initial division of the CMOs into two distinct subsets belonging either to the frozen or the active region is required before performing the localization rotations. Since CMOs are delocalized over the whole system without clear division to frozen or active region, thus poor division in CMOs will yield poor localization. It is strongly desired to develop an improved localization scheme, both reliable and efficient. A new localization scheme has been developed by Gu *et al* [16] based on the density matrix approach on the regional localized molecular orbitals (RLMOs). It has been demonstrated that RLMO scheme works well even for highly delocalized systems.

The elongation method is designed for studying electronic structures for periodic and aperiodic polymers. One application of the elongation method to an aperiodic system is to study a local perturbation in an extended periodic system, which includes (a) a polymer interacting with a small molecule or including an abnormal bonding (one-dimensional); [17] (b) a surface having a defect structure or a local adsorption (two-dimensional); [13] and (c) a crystal including an impurity atom or a lattice defect (three-dimensional). The elongation treatment for these problems includes a procedure which is essentially similar to that for the corresponding orbital or the interaction frontier orbital. [18]

Nonlinear optics (NLO), or more general photonics, has been a subject of much theoretical and experimental interest for several decades since the advent of lasers, which provide high-intensity coherent source light. NLO materials have high potential utilization in optical and electro-optical devices. Some NLO phenomena are useful to, for example, alter the frequency of light and to amplify one source of light with another; second-order harmonic generation (SHG) can be used to increase the capacity of stored information on optical disks. Materials with potential utilization in optical and electro-optical devices are of most interest. [19] Organic π-conjugated molecules, polymers, and molecular crystals are commonly considered for NLO candidates since their properties can be easily tailored by chemical modification. In this regard, molecular design can be helpful to experimentalists who wish to synthesize NLO materials with desired properties. Experimentalists seek and synthesize new efficient NLO materials, while for theoretical studies, the major aim is to find the relationships between NLO properties and chemical structures, and thus predict new NLO materials with fast NLO responses. Such an understanding provides a useful guide to experimentalists for seeking NLO materials with specific properties. Theoretical studies on NLO by means of quantum chemical calculations have been achieved to tailor molecular structures for better NLO properties.

Quantum chemical calculations have been proven to be a reliable and practical tool for investigating the NLO properties of small and medium-size chemical species. Although considerable efforts have been directed at linear scaling methods for large molecules or molecular clusters, such calculations can still be formidable, especially at the *ab initio* level. For large systems with translational symmetry, one can use the CO method based on Born-von Kàrmàn periodic boundary conditions. This method has been utilized to determine various electronic structure properties of polymers and crystals. NLO properties require special treatment because the scalar interaction potential is non-periodic and unbounded no matter how weak the applied field[7b]. Recently, a time-dependent coupled-perturbed Hartree-Fock CO method has been developed for calculating the dipole moment and (hyper)polarizabilities of polymers. [20] This method has now been applied successfully to determine static[21] and dynamic [22] properties of several polymers containing small and moderate size unit cells, although electron correlation has not yet be incorporated. Electron correlation at the MP2 level for determining the static polymeric polarizabilities has been investigated by Otto *et al.* [23] In the latter connection, it has been shown that density functional theory, with conventional exchange-correlation potentials, does not satisfactorily describe the response of such systems to an electric field. New developments in current-density functional theory, which appear promising, may eventually remedy this situation. However, at the time-dependent coupled-perturbed Hartree-Fock level, polymers containing large unit cells are computationally burdensome. Besides, there are many special features of CO calculations that complicate these calculations such as the pseudo-linear dependence of basis sets, band crossing, and numerical instability in transforming the density matrix from k-space to direct space. For this reason, it remains of interest to consider the complementary finite oligomer approach wherein results for increasingly large oligomers are extrapolated to the infinite chain limit. The finite oligomer method has been widely used to determine the (hyper)polarizabilities of polymers[24] and it has also been applied to crystals.[25] Advantages of the oligomer approach include the fact that standard quantum chemistry codes can be utilized without modification and non-periodic perturbations are readily treated. The major disadvantage is that convergence with respect to increasing the number of unit cells can be slow, making the extrapolation difficult or even impossible. Although improved extrapolation techniques have been developed by Champagne *et al,*[26] it is desirable to find other more efficient ways for finite oligomer calculations on longer chains so that the extrapolation based on them becomes reliable.

There are many different approaches to deal with large systems based on the so-called $O(N)$ methods. The first step is to reduce the computational efforts for the most time consuming two-electron integral calculations. The two-electron integral computational effort will become linearly scaling with the system size by adopting screening technique followed by a very fast multipole moment method.[27] As the Fock matrix has been constructed with a computational effort linearly scaling with system size, the second step is to speed up the diagonalization procedure of the Fock matrix to gain the total linear scaling. This can be done by reformulating the SCF equations in terms of a minimization of an energy functional which depends directly on the density matrix.[28] Along this way the Hartee-Fock method is possible to achieve the linear scaling with the system size. Recently, the linear scaling electron correlation methods have also been proposed.[29]

There exist some other ways to obtain the linear scaling for large systems. One of these may be generically referred to as the "combination of fragments" approach. A particular example is the local space approximation method of Kirtman and co-workers.[30] Another approach is the above-mentioned elongation method proposed by Imamura and co-workers. The elongation method has been applied to NLO field by using a finite filed treatment in the presence of an electric field. Since the interaction potential due to an applied electric field is extremely non-local, it is of interest to see whether or not the elongation method is reliable for NLO property calculations. In order to assess the situation, a finite field treatment has been formulated for the elongation method (abbreviated as elongation-FF).[31] The validity and efficiency of this elongation-FF method has been checked for some model polymers and also applied to determine static (hyper)polarizabilities of long-chain oligomers. This novel elongation-FF method has been demonstrated to be a powerful tool to study the NLO properties of long-chain donor and/or acceptor molecules with various substitution patterns and block copolymers.[32] The polyacetylene and poly(paraphenylene) chains with terminal donor and/or acceptor substituents and more general substitution patterns on poly(para-phenylene-ethynylene) chains as well as a π-conjugated block copolymer of pyrrole and silole containing a quantum well have been investigated by using the elongation-FF method.

Organic molecular crystals are promising nonlinear materials. Two molecular crystals are investigated by the elongation-FF method. One is the single crystal 3-Methyl-4-nitropyridine-1-oxide ($C_6H_6O_3N_2$ or POM) and the other one is 2-Methyl-4-nitroaniline ($C_7H_8O_2N_2$ or MNA) with monoclinic structure. The experimental data evidence that POM is a promising NLO material for a pico- and femtosecond optics in the near infrared. Its non-linear optical coefficient is 15 times higher than that of $K_2H_2PO_4$ or KDP, and with small phase matching hall width (5 angular min /cm) and dispersion (0.5 angular min A).[33] For MNA, due to its crystal orientation, it has two nonlinear coefficients. The d_{31} and d_{11} coefficients are respectively 5.8 and 40 times larger than those of $LiNbO_3$. In order to evaluate the nonlinear susceptibilities of the molecular crystals, the supermolecule approach has been adapted for larger and larger clusters to mimic the solid properties. In the supermolecule approach the 2D- or 3D-cluster properties are estimated from the 1D clusters extended along three crystal axes. The large 1D-clusters are required to reliably extrapolate to the infinite limit and this is a big suffer for the conventional supermolecule calculations. At this point, the elongation method is the most suitable because very large clusters can be treated without so much computational cost as in the conventional approach.

In section 2, the elongation method with three different localization schemes is presented. Section 3 is the applications of the elongation method to model polymers and to the modeling of chemiabsorption on surface. In section 4, the linear scaling is obtained for the elongation method implemented with the cutoff technique. The application of the elongation method to NLO will be given with some tests for its validity in section 5. Section 6 is the summary and future prospects of the elongation method.

2. The elongation method

The elongation method is described in this section by first introducing three different localization schemes, and then the elongation Hartree-Fock self-consistent Field (SCF) procedures are presented.

2.1. Localization schemes for the elongation method

The elongation method begins with the CMOs, φ^{CMO}, of an oligomer containing an appropriate number (N) of monomer units. The CMOs of this oligomer (called starting cluster) are assumed known from any conventional quantum chemistry calculation. The CMOs are delocalized over the whole space of the starting cluster,

$$\varphi^{CMO} = C_{AO}^{CMO} \chi^{AO} \tag{1}$$

where vector χ^{AO} groups the atomic orbitals (AOs) of the starting cluster, and C_{AO}^{CMO} is the transformation coefficients from AOs to CMOs obtained by solving the Hartree-Fock-Roothaan equations in the conventional quantum chemistry approach. This starting cluster is then divided into two regions, frozen region and active region, assuming that an attacking monomer is going to interact with the active region of this starting cluster. The eigenvalue problem is solved on the basis of the active LMOs and the CMOs of the attacking monomer instead of the whole AOs in the conventional quantum chemistry calculations. As shown in Fig. 1, the elongation method takes the CMOs of a starting cluster followed by a localization procedure to generate two sets of LMOs, in such a way that one set of the LMOs is located in one end of the starting cluster (region A or frozen region) which is remote away from the interactive point, while the rest of the LMOs are located in the active region and ready to interact with the attacking monomer.

As the frozen and active regions are defined, the CMOs of the starting cluster are split into two sets. One of them is localized into the frozen region while the other set is localized into the active region. The required localization scheme for the elongation method should be efficient and reliable. There are three different localization schemes adopted in the elongation method. In the next, three different localization schemes are described and their merits and disadvantages are discussed and compared.

2.1.1. 2×2 *unitary localization scheme*

The 2×2 unitary transformation localization scheme was first adopted to localize CMOs into LMOs for the elongation method. This scheme is similar to the treatment proposed by Edminston and Rüdenberg.[34] Before performing the 2×2 rotations, it is required to divide CMOs into two sets, one set for frozen region and the other for active region. A pair of CMOs, one from the frozen region and the other one from the active region, is transformed by a 2×2 rotation to form one LMO in the frozen region and another one in the active region. The rotations are repeated to all CMOs respectively in occupied space and in virtual space until no further localization is gained.

For any CMO of the starting cluster, one can initially assign it to frozen region (region A) by checking the quantity,

$$\omega = \sum_{r,s}^{on\,A} C_{ri} S_{rs} C_{si} \qquad (2)$$

where S is the overlap matrix and summations over r and s are within the frozen region. One can assign the CMOs associated with largest ω values to A as the CMOs in the frozen region. The remaining CMOs are assigned to the active region.

Assuming φ_i^{CMO} is the CMO from the frozen region and φ_j^{CMO} is from the active region (region B), the transformed LMOs, ϕ_i^{LMO} and ϕ_j^{LMO}, are obtained by the following 2×2 rotation,

$$\begin{pmatrix} \phi_i^{LMO} \\ \phi_j^{LMO} \end{pmatrix} = \begin{pmatrix} \sin\theta & \cos\theta \\ -\cos\theta & \sin\theta \end{pmatrix} \begin{pmatrix} \varphi_i^{CMO} \\ \varphi_j^{CMO} \end{pmatrix} \qquad (3)$$

Combining Eqs.(1) and (3), LMOs are expanded in the AO basis set as follows,

$$
\begin{cases}
\phi_i^{LMO} = \left(\sum_r^{on\,A} + \sum_r^{on\,B} \right)(\sin\theta\, C_{ri}^{CMO} + \cos\theta\, C_{ri}^{CMO})\chi_r = \phi_i^{LMO}(A) + \phi_i^{LMO}(B) \\
\phi_j^{LMO} = \left(\sum_s^{on\,A} + \sum_s^{on\,B} \right)(-\cos\theta\, C_{si}^{CMO} + \sin\theta\, C_{si}^{CMO})\chi_s = \phi_j^{LMO}(A) + \phi_j^{LMO}(B)
\end{cases}
$$

$$(4)$$

where $\phi_j^{LMO}(A)$ $(\phi_j^{LMO}(B))$ is the jth LMO with the summation over only to the AOs in region A (B). The rotation angle θ in Eq. (4) is determined by maximizing the degree of the localization defined by,

$$
Q_{ij} = \left\langle \phi_i^{LMO}(A) \middle| \phi_i^{LMO}(A) \right\rangle + \left\langle \phi_j^{LMO}(B) \middle| \phi_j^{LMO}(B) \right\rangle \qquad (5)
$$

Insert Eqs. (4) into (5), Q_{ij} can be expressed as a trigonometric function of θ, that is,

$$
Q_{ij} = \alpha_{ij} \sin^2\theta + \beta_{ij} \cos^2\theta + \gamma_{ij} \sin 2\theta \qquad (6)
$$

in which

$$
\begin{cases}
\alpha_{ij} = \sum_{r,s}^{on\,A} C_{ri} S_{rs} C_{si} + \sum_{r,s}^{on\,B} C_{rj} S_{rs} C_{sj} \\
\beta_{ij} = \sum_{r,s}^{on\,A} C_{rj} S_{rs} C_{sj} + \sum_{r,s}^{on\,B} C_{ri} S_{rs} C_{si} \\
\gamma_{ij} = \sum_{r,s}^{on\,A} C_{ri} S_{rs} C_{sj} - \sum_{r,s}^{on\,B} C_{ri} S_{rs} C_{sj}
\end{cases}
\qquad (7)
$$

It is straightforward to determine θ with the maximum value of Q_{ij}. The θ value is given as

$$\theta = \frac{\pi}{4} - \frac{\varpi}{2} \tag{8}$$

With

$$\varpi = \tan^{-1} \frac{\beta_{ij} - \alpha_{ij}}{2\gamma_{ij}} \tag{9}$$

The occupied and vacant CMOs are separately rotated by the above-mentioned transformation in order to keep the density matrix invariant. A schematic illustration of the LMOs obtained from CMOs by this 2×2 rotation scheme is depicted in Fig. 2. The frozen LMOs are localized in A region. The active LMOs, however, are localized mostly in the B region, which are ready to interact with the attacking monomer.

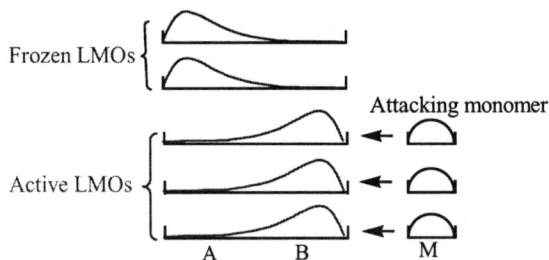

Figure 2. Schematic illustration of LMOs in frozen and active regions.

It should be mentioned that this 2×2 unitary localization scheme is efficient and reliable for non-bonded systems, such as water molecule chain. It could be intractable for covalently bonded system, such as polyacetylene and polyethylene, etc.

For covalently bonded or π-conjugated systems, however, it is not trivial to assign CMOs into frozen and active regions. Therefore, *prior to* the 2×2 unitary localization, one can first transfer the original AO basis CMOs into hybridized AOs (HAO) basis by the procedure of Del Re. [35] HAOs can be obtained by the principle of maximum overlapping between bonds. This procedure is illustrated in Fig. 3 as following with an example of a tetrahedral bonded carbon atom in polyethylene.

As shown in Fig. 3, the central carbon atom, a, is bonded with four neighboring atoms, b, c, d, and e. The overlap matrix between atom a and its neighbors, S_{ax} ($x=b$, c, d, or e), can be used to determine the HAOs by the following eigenvalue equation,

$$S_{ax} S_{ax}^{\dagger} U_{ax} = \lambda_{ax}^{2} U_{ax} \tag{10}$$

The eigenvector of U_a corresponding to the largest λ_{ax}^{2} value is chosen as the hybridized AO. The selected U_a, however, are not necessarily unitary matrix. The symmetric orthogonalization procedure of P-O. Löwdin[36] is adopted to construct the orthonormal HAOs. After this procedure is performed to all atoms, the full transformation matrix, U, from AO to HAO is built by sum over all individual atom matrices. The full matrix is unitary and block diagonal with each atom having its own submatrix linked to the neighbors. By using the HAOs, Eq. (1) becomes

$$\varphi_i^{\text{CMO}} = \sum_m (\sum_r C_{ri}^{\text{CMO}} U_{rm}) \chi_m^{\text{AO}} \tag{11}$$

and, the 2×2 unitary localization scheme is then performed on the HAO basis instead of AO basis.

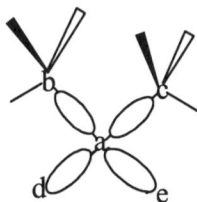

Figure 3. Illustration of construction for HAOs in polyethylene.

The 2×2 unitary localization with HAO basis is more efficient especially for the covalently π-conjugated systems. The reason for this is that the boundary between region A and region B is well defined by a node point among the HAOs, which are already localized between bonds of two atoms.

2.1.2 *Stationary space analysis scheme*

Maekawa and Imamura[37] proposed another localization scheme for the elongation method based on the stationary conditions of the electronic structures. The stationary orbitals can be extracted from the CMOs. The remaining MOs define the active space.

Let's assume that the CMOs of the starting cluster (A), $\varphi_i^{(A)}$, and the elongated cluster (the starting cluster and attacking monomer, A+M), $\psi_i^{(A+M)}$, are known from any conventional quantum chemistry calculation. For the elongation calculations, we should decide which orbital in $\varphi^{(A)}$ set is stationary, unaltered by the interaction between A and M. The stationary orbitals can be determined by the following two conditions:

(i) the orthonormality condition

$$\left\langle \varphi_i^{(A),a} \middle| \varphi_j^{(A),b} \right\rangle = \delta_{ab}\delta_{ij}$$

$$\text{(12a)}$$

$$\left\langle \varphi_i^{(A),a} \middle| \psi_j^{(A+M),b} \right\rangle = 0 \tag{12b}$$

and

(ii) the variational condition

$$\left\langle \varphi_i^{(A),occ} \middle| \hat{F}^{(A+M)} \middle| \varphi_j^{(A),vac} \right\rangle = 0$$

$$\text{(13a)}$$

$$\left\langle \varphi_i^{(A),occ} \middle| \hat{F}^{(A+M)} \middle| \psi_j^{(A+M),vac} \right\rangle = \left\langle \varphi_j^{(A),vac} \middle| \hat{F}^{(A+M)} \middle| \psi_i^{(A+M),occ} \right\rangle = 0 \tag{13b}$$

where a, b = occ (occupied) or vac (vacant) orbitals, $\hat{F}^{(A+M)}$ is the Fock operator of the elongated cluster. A two-step procedure was adopted to get the stationary orbitals.

The first step is to form two overlap matrices between the occupied and vacant CMOs of the starting cluster, $\varphi_i^{(A),occ}$ and $\varphi_i^{(A),vac}$, and the AOs of the attacking monomer M, $\chi_r^{(M)}$, that is, $S_{ir}^{occ} = \left\langle \varphi_i^{(A),occ} \middle| \chi_r^{(M)} \right\rangle$ and $S_{ir}^{vac} = \left\langle \varphi_i^{(A),vac} \middle| \chi_r^{(M)} \right\rangle$. It should be mentioned that those matrices are not square but rectangular since the dimension of $\varphi_i^{(A),occ}$ or $\varphi_i^{(A),vac}$ is different from that of $\chi_r^{(M)}$. However, the square matrices, $S_1 = S^{occ} S^{occ\dagger}$ and $S_2 = S^{vac} S^{vac\dagger}$, can be formed and diagonalized, separately.

$$S_1 \phi^{occ} = \varepsilon_1 \phi^{occ}$$
$$(14a)$$
$$S_2 \phi^{vac} = \varepsilon_2 \phi^{vac} \tag{14b}$$

The eigenvectors from Eqs. (14a) and (14b) are used for the next step.

The second and the last step of this treatment is to construct the rectangular Fock matrices as follows,

$$F^{occ} = \left\langle \phi^{occ} \middle| \hat{F}^{(A+M)} \middle| \phi^{vac} \oplus \chi' \right\rangle$$
$$(15a)$$
$$F^{vac} = \left\langle \phi^{vac} \middle| \hat{F}^{(A+M)} \middle| \phi^{occ} \oplus \chi' \right\rangle \tag{15b}$$

where χ' is the AOs of attacking monomer M corresponding to the non-zero eigenvalues of the overlap matrices from Eqs. (14a) and (14b). Similar to the first step, the square matrices, $F_1 = F^{occ} F^{occ\dagger}$ and $F_2 = F^{vac} F^{vac\dagger}$, are formed and diagonalized separately. The stationary orbitals correspond to the zero eigenvalues (threshold is set as 10^{-8}). All the remaining orbitals together with those from the attacking monomer will span the active space in the next elongation step.

2.1.3. Regional localized molecular orbitals (RLMOs) scheme

Gu *et al* [16] developed another localization scheme for the elongation method based on the regional molecular orbitals. This new localization scheme is more efficient and accurate than the other two schemes outlined above. Some model systems ranging from non-bonding molecule chains to highly delocalized polymers are taken to test this new scheme and it is found that for all cases the elongation error per unit cell is satisfactorily small if a starting cluster is properly large. In this subsection, the detailed scheme based on the regional molecular orbitals will be presented.

The density matrix in an atomic orbital (AO) basis is defined as:

$$D^{AO} = C_{AO}^{CMO} d \, C_{AO}^{CMO\dagger} \qquad (16)$$

where d is the diagonal occupation number matrix and C_{AO}^{CMO} is the transformation matrix from AOs to CMOs as defined in Eq. (1). For a restricted Hartree-Fock wavefunction the occupation number is 2 for the doubly-occupied spatial orbitals or 0 for the unoccupied orbitals. The orthonormal condition is given as,

$$C_{AO}^{CMO\dagger} S^{AO} C_{AO}^{CMO} = 1 \qquad (17)$$

where S^{AO} is the overlap matrix in AO basis and 1 is the unity matrix. From Eqs. (16) and (17), one can easily obtain the idempotence relation in a non-orthogonal AO basis as

$$D^{AO} S^{AO} D^{AO} = 2D^{AO}. \qquad (18)$$

The density matrix can be transferred from a non-orthogonal AO to an orthogonal AO (OAO) basis by adopting the Löwdin's symmetric orthogonalization procedure[36] in order to keep the new basis least distorted from the original AO. The transformation matrix,

$$X = V\varepsilon^{1/2}V^\dagger = X^\dagger \tag{19}$$

from non-orthogonal AO to OAO basis is obtained by diagonalizing the AO basis overlap matrix, $S^{AO}V = \varepsilon V$, where V and ε are the eigenvectors and eigenvalues of S^{AO}, respectively. Thus, the density matrix in the OAO basis is transferred from AO basis as

$$D^{OAO} = XD^{AO}X^\dagger \tag{20}$$

by using Eqs.(16), (20), and the fact that $X^\dagger X = XX^\dagger = S^{AO}$ one can easily verify that

$$D^{OAO}D^{OAO} = 2D^{OAO} \tag{21}$$

This is the idempotence relation of the density matrix in an OAO basis. As a consequence, the eigenvalues of D^{OAO} must be either 2 or 0. Therefore, the eigenvectors of D^{OAO} are well separated into occupied or vacant subspaces.

Now, one can partition the starting cluster into two regions, region A (frozen region) and region B (active region) and localize the CMOs into these two regions. Region B is the one with atoms adjacent to the interactive end of the cluster whereas region A is at the opposite end far away from the interactive center. It should be mentioned here that by using the density matrix, the partition of the starting cluster into two regions are unique since the AOs belonging either to A or B region are well defined. This is different from the other two localization schemes, whereas the partition is not unique. As one has already seen that the division of CMOs into two regions is not so straightforward and poor selection leads to poor localization.

The desired RLMOs can be obtained in two steps. In the first step, a regional orbital (RO) space is constructed by separately diagonalizing the D^{OAO}(A) and D^{OAO}(B), where D^{OAO}(A) and D^{OAO}(B) are the sub-blocks of D^{OAO} with AOs belonging to A and B regions, respectively. These eigenvectors span the RO space. The second step is to perform a unitary transformation between the occupied and unoccupied blocks of D^{RO} in such a way as to preserve the localization as much as possible. The procedure is similar to the construction of the

Feng Long Gu, Akira Imamura and Yuriko Aoki

natural bond orbitals (NBOs) but it is generalized to localized "regional orbitals" rather then localized "bond orbitals".

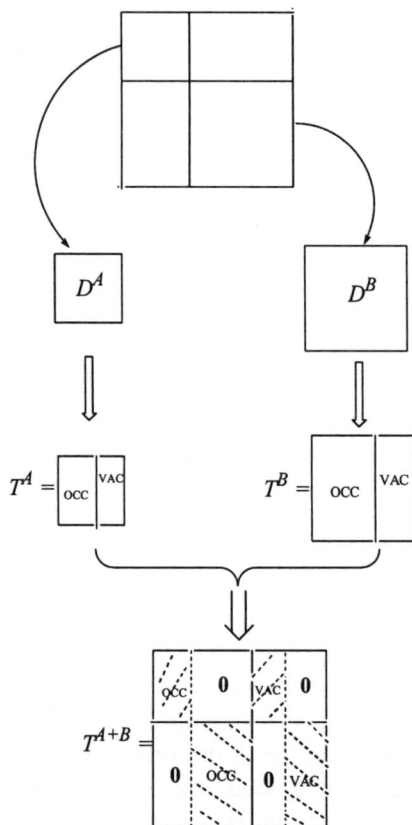

Figure 4. Construction of the transformation matrix from OAO basis to RAO basis.

In the first step, the transformation from OAOs to ROs is given by the direct sum of T^A and T^B, *e.g.*

$$T = T^{A} \oplus T^{B} \tag{22}$$

where T^{A} and T^{B} are the eigenvectors of D^{OAO} (A) and D^{OAO} (B), respectively. Figure 4 shows the matrices of D^{OAO} (A) and D^{OAO} (B) and the schematic construction of the T matrix. The corresponding eigenvalues are divided into three sets corresponding to ROs that are approximately doubly-occupied (the value is close to 2), singly-occupied (close to 1) and empty (close to 0). The singly-occupied orbitals in A and B can be used to construct hybrid orbitals to from covalent bonding/antibonding pairs. Alternatively, one can re-combine the singly-occupied orbitals from A and B regions to build new ROs approximately doubly-occupied or approximately empty. This is equivalent to figuratively transfer an electron from each singly-occupied orbital of A to the corresponding singly-occupied orbital of B to create an ionic pair $(A^{+}B^{-})$. If region A is more electronegative than region B, one can simply create the ionic pair $(A^{-}B^{+})$. For the covalently bonded systems, the results from either choice will be the same or with very tiny difference. For a non-bonded system, such as water chain, there are only two sets, either doubly-occupied or empty. The resulting RO density matrix is

$$D^{RO} = T^{t} D^{OAO} T \tag{23}$$

and by using Eqs. (16) and (20) the transformation coefficients from ROs to CMOs may be written as

$$C_{RO}^{CMO} = T^{t} X C_{AO}^{CMO}. \tag{24}$$

Using Eq.(21) and the unitary condition $TT^{t} = T^{t}T = 1$, one can verify that

$$D^{RO} D^{RO} = 2D^{RO} \tag{25}$$

Except for orthogonalization tails the ROs given above are completely localized to region A or region B. However, they are not completely occupied or unoccupied. Thus, the second, and final, step is to carry out a unitary transformation between the occupied and unoccupied blocks of

D^{RO} to keep the localization as much as possible. This is done using the same Jacobi procedure that is employed in ref. [38] to convert NBOs into localized molecular orbitals. If U is the transformation that diagonalizes D^{RO} , *i.e.*

$$D^{RLMO} = U^{\dagger} D^{RO} U \tag{26}$$

then the only non-zero elements of D^{RLMO} are equal to 2 (*cf.* Eq.(25)). The unitary transformation from CMO to RLMO is given as:

$$C^{CMO}_{RLMO} = U^{\dagger} T^{\dagger} X C^{CMO}_{AO} \tag{27}$$

Finally, the original AOs basis RLMOs is given by,

$$C^{RLMO}_{AO} = X^{-1} T U \tag{28}$$

After the well-localized RLMOs are obtained by the above procedure, the elongation method can proceed, which will be described in details below.

2.2. The elongation SCF procedures

For a given suitably large starting cluster, the interaction between the frozen region and the attacking monomer is minimized by using LMOs. The elongation Hartree-Fock equation is solved self-consistently in the localized orbital space of the interactive region, or more precisely, the working space consists of LMOs of the active region and CMOs of the attacking monomer. This solution yields a set of CMOs in the reduced space which can be localized again into a new frozen region and a new active region. The whole procedure is repeated until the desired length is reached. The important feature of the elongation method is that the Hartree-Fock equation is solved only for the interactive region instead of the whole system. As the system enlarges, the size of the interactive region is almost the same as that of the starting cluster and the CPU required in the elongation SCF is more or less constant.

Assuming that the CMOs of a starting cluster are localized by one of the above-described localization schemes into A_1 and B_1 regions, we can start the elongation SCF procedure. A_1 is the frozen region which is defined by atoms far away from the propagation center. B_1 is the active region composed of the remaining atoms. It should be mentioned that the LMOs in region A_1 (B_1) contain some tailings to the B_1 (A_1) region. This is due to the consequence that the localization is not perfect and also the fact that the interaction between A_1 and B_1 can be only minimized but not neglected. That is why the definition of the regions extends to all atomic orbitals (AOs).

Thus, the resulting orthonormal LMOs is expressed as

$$\varphi_i^{\text{LMO}}(X) = \sum_{j=1}^{N} \sum_{m} L_{mi}^{(j)}(X) \chi_m^{(j)} \tag{29}$$

where the superscript j is the index for the jth fragment and $L_{mi}^{(j)}(X)$ ($X=$ A_1 or B_1) is the LMO coefficient for the mth AO in the jth fragment of the ith LMO localized in region X. The localization schemes outlined in the above subsections are the ways to find the unitary transformation matrix between $L^{(j)}(X)$ and $C^{(j)}(X)$, that is

$$L_{mi}^{(j)} = \sum_{k} C_{mk}^{(j)} U_{ki} \tag{30}$$

in which U is the unitary transformation from the CMOs to LMOs.

After the LMOs are obtained for the regions A_1 and B_1 of the starting cluster, in the first elongation step, we keep the LMOs in the A_1 region frozen during the elongation SCF procedure. Our working space is defined by the LMOs assigned to B_1 region together with CMOs of a monomer M_1, e.g.

$$L^{(j)\prime}(B_1 + M_1) = L^{(j)}(B_1) \oplus C \ (M_1) \tag{31}$$

$L_i^{(j)\prime}(B_1 + M_1)$ is a mixed basis consisting of LMOs from region B_1 and CMOs from the attacking monomer, M_1. In practice, the CMOs of M_1 are just a set of initial orbitals. The LMO basis Fock matrix, or more precisely the LMO-CMO basis Fock matrix, can be transformed from AO basis Fock matrix:

$$F_{ij}^{LMO\text{-}CMO}(B_1 + M_1) =$$

$$\sum_{k=1}^{N}\sum_{l=1}^{N}\sum_{\mu}^{\mu_k}\sum_{\nu}^{\nu_l} L_{\mu i}^{(k)}{}'(B_1 + M_1)F_{\mu\nu}^{AO}(A_1 + B_1 + M_1)L_{\nu j}^{(l)}{}'(B_1 + M_1) \qquad (32)$$

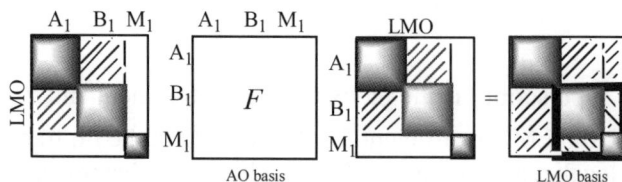

Figure 5. LMO basis Fock matrix constructed from AO basis.

The schematic illustration of the LMO-CMO basis Fock matrix is presented in Fig. 5. The AO basis Fock matrix is for the whole system while the LMO-CMO basis Fock matrix is restricted in the interactive region. The LMO-CMO basis overlap matrix can be obtained by the similar transformation as in Eq. (32). The Hartree-Fock equation of the interactive region (the square area with bold borders in the right-hand side of Fig. 5), for the first elongation step is the $B_1 + M_1$ region, then becomes:

$$F^{LMO\text{-}CMO}(B_1 + M_1)U(B_1 + M_1) =$$
$$S^{LMO\text{-}CMO}(B_1 + M_1)U(B_1 + M_1)\varepsilon(B_1 + M_1) \qquad (33)$$

It can be seen from Eq. (33) that the dimension of the HF equation in the elongation SCF procedure is reduced to the size of the interactive region $(B_1 + M_1)$ instead of the whole space $(A_1 + B_1 + M_1)$. This is the unique feature of the elongation method and it is one of the main sources of the CPU time saving. For the elongation method, another time saving is the employment of the cutoff technique, see below in section 4. It should also be noted that the contribution of the frozen orbitals is accounted for from the AO basis Fock matrix (Eq.(32)) through the terms in the total density matrix.

One thing should be mentioned that neglecting the Fock matrix elements between the A_1 and B_1 regions has little influence on the electron density distribution and the total energy. This can be attributed to the fact that the neglected elements are all within occupied or vacant space. The elements between A_1 and B_1 regions are naturally vanished if they are between occupied and vacant spaces. Therefore, we can diminish the dimension of the matrix, and the elongation SCF can be performed in the interactive region instead of the whole system.

After solving Eq. (33), the CMOs of the B_1+M_1 region is given by the overall transformation from LMOs,

$$C_{\mu i}^{(j)}(B_1 + M_1) = \sum_p L_{\mu p}^{(j)\prime}(B_1 + M_1)U_{pi}(B_1 + M_1) \tag{34}$$

The total density matrix is constructed as

$$D_{\mu\nu}^{AO} = \sum_j^{occ}\sum_i L_{\mu i}^{(j)}(A_1)dL_{\nu i}^{(j)}(A_1) + \sum_j^{occ}\sum_i C_{\mu i}^{(j)}(B_1 + M_1)dC_{\nu i}^{(j)}(B_1 + M_1) \tag{35}$$

where d is the occupancy number. The total AO basis Fock matrix is then constructed as the usual way,

$$F_{\mu\nu}^{AO}(A_1 + B_1 + M_1) = H_{\mu\nu}^{core} + \sum_{\lambda\sigma} D_{\lambda\sigma}^{AO}[(\mu\nu|\sigma\lambda) - \frac{1}{2}(\mu\lambda|\sigma\mu)] \tag{36}$$

The AO basis Fock matrix obtained from Eq. (36) is taken and transferred to form the LMO basis Fock matrix (Eq. 32) and then the elongation SCF is repeated. The elongation SCF is converged as the difference of the density matrix or the total energy between the previous iteration and the current is smaller than a threshold, normally set as 10^{-6} for the density or 10^{-8} a.u. for the total energy.

After the first elongation SCF converged, the next elongation step is prepared by localizing $C(B_1 + M_1)$ into two regions, a new frozen region A_2 and a new active region B_2. The new active region B_2 is ready to interact with a new attacking monomer M_2. With this fashion, the elongation procedure is continued until the desired length of a polymer is reached.

3. Applications of the elongation method

We have performed some sample calculations for testing the validity and efficiency of the elongation method. The results are summarized in the following subsections. The comparison with the conventional HF calculations is given. The performance of the elongation calculations for different systems is presented. The elongation calculations were performed by employing three different localization schemes and it has been demonstrated that the RLMO scheme is the most reliable and efficient one.

3.1 Model systems to test the elongation method

In ref. 12, the idea of the elongation method was tested on hypothetical polymer systems consisting of regularly and irregularly arranged hydrogen molecules, trans-zigzag polyethylene, and isotactic, syndiotactic, and atactic polypropylenes. Although the calculations were performed at the extended Hückel level, the idea of the elongation method is at least the same as that at the *ab initio* level except that the overlap matrix is no longer unity. The 2×2 unitary localization procedure was employed in ref. 12.

For a hydrogen chain with the inter-molecule distance being 1.0 Å and the intra-molecule distance 0.7 Å, the starting cluster consists of 15 H_2 molecules with the frozen region containing one hydrogen molecule, and attacking monomer (one hydrogen molecule) was added to the cluster one by one. It is found that the total energies of the hydrogen chains elongated from 15 units up to 100 units can be reproduced with the deviation from the conventional results less than 2×10^{-4} eV.

It is much more interesting to test the applicability of the elongation method with a more realistic polymer. The next sample is polyethylene. Two elongation

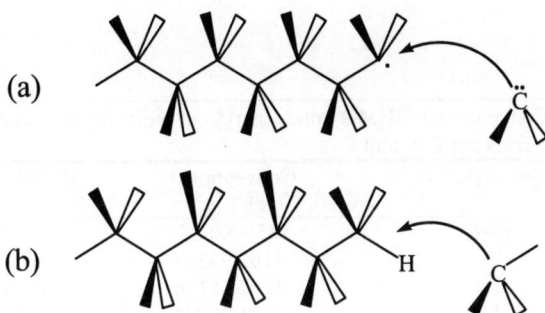

Figure 6. The elongation propagation in (a) open-shell model and (b) closed-shell model.

processes arc illustrated in Fig. 6a and 6b, the first is for open-shell model and the second is for closed-shell model. These are the elongation processes for the theoretical synthesis of polyethylene. In the open-shell case, see Fig. 6a, the elongation process starts from the CMOs of the starting cluster, octanyl radical $[CH_3-(CH_2)_6-CH_2]\cdot$, which is obtained from a conventional calculation. A methylene unit is attacking this starting cluster one after another. This simulates the experimental propagation to form $[CH_3-(CH_2)_{N-1}-CH_2]\cdot$. In the final elongation step, the radical polymer was terminated by CH_3 to obtain the polyethylene $[CH_3-(CH_2)_N-CH_3]$. The closed-shell model is illustrated in Fig. 6b. The starting cluster is an octanane molecule $[CH_3-(CH_2)_6-CH_3]$. The elongation process is the same as the open-shell model except the terminal hydrogen has to be removed during the elongation propagation. For more details of the closed-shell model, please see below in section 3.2.

In practical calculations for polyethylene, for obtaining better LMOs, the hybridized AOs were used as a basis set instead of AOs in the 2×2 unitary localization scheme. The dimension of the eigenvalue problem is kept constant (always containing eight carbon atoms) throughout the elongation steps since one $CH_2\cdot$ unit is frozen as another $CH_2\cdot$ attacking unit is added to the clusters. As an open-shell model was adopted in the elongation process, the cluster was elongated starting from the octanyl radical up to $[CH_3-(CH_2)_{25}-CH_2]\cdot$ followed by a final step with CH_3 to make the neutral polyethylene $[CH_3-(CH_2)_{26}-CH_3]$. The

Table 1. Conventional EHMO total energy of polyethylene and the elongation errors per CH_2 unit.

$[CH_3\text{-}(CH_2)_N\text{-}CH_2]\cdot$	Conventional (eV)	$\Delta E \ (10^{-5} \ eV)$
N=6	-850.32685	
9	-1162.94567	0.07
10	-1267.15176	0.08
11	-1371.35783	8.00
15	-1688.18208	7.59
25	-2830.24266	7.19
$[CH_3\text{-}(CH_2)_{26}\text{-}CH_3]$	-2953.04209	4.43

conventional EHMO energies and the elongation error per CH_2 unit are summarized in Table 1.

From Table 1, one can see that the first and second elongation steps reproduce quite well the conventional energies. The elongation error, $\Delta E = E^{elg} - E^{cvl}$, is defined as the energy difference between the elongation and conventional calculations. The elongation error per CH_2 unit of the starting cluster is obvious zero since in the starting cluster there is no approximation introduced. For n=9 and 10, the elongation results have the error per unit less than 1×10^{-6} eV, while for the further elongation steps, the error per unit is increased to 8×10^{-5} eV and then steadily reduced to the final step as 4.43×10^{-5} eV. The error is due to the poor localization quality for polyethylene. The error will be reduced if a new localization scheme is employed (see below).

3.2 Elongation with the RLMO localization scheme

It has been observed that for the 2×2 localization scheme the convergence to LMOs is very slow, especially for delocalized covalently bonded systems such as polyacetylene and cationic cyanines. For large basis sets beyond minimum basis set, the computational time of the 2×2 scheme can become excessive. Furthermore, in the 2×2 scheme we must initially assign the CMOs either to the frozen or the active region to make series of unitary rotations. But this initial division is not always reliable and a poor selection will yield poor localization or even never convergent during the 2×2 localization.

3.2.1. *Comparison of the RLMO and 2×2 localization schemes*

In the case of covalently bonded polymers, the starting cluster and growing terminal can be treated in two different ways; either with or without capping atom(s) (normally hydrogen atom(s)) in the initial Hartree-Fock calculation. If there is no capping atom(s) employed, then the starting cluster (and growing terminal) will be an open-shell system (Fig. 6a); otherwise it will be a closed-shell system (Fig. 6b). In this work we utilize the closed-shell model for the elongation method. When a closed-shell oligomer is attacked by a monomer, the reaction corresponds to "substituting" the terminal capping atom(s) by the monomer. This substitution requires deleting the proton from the Hamiltonian as well as a procedure for replacing the terminal capping atom's AO coefficient(s) in the CMOs of A and B. For this purpose we use AOs on the atom of the monomer that participates in the new bond. The choice of corresponding atomic orbitals is not important because the coefficients are very small for the CMOs of A and the resulting CMOs of B+M are self-consistently corrected during the elongation SCF procedure. Fig. 7 illustrates the procedure how to remove the terminal atom(s). For simplicity, we suppose hydrogen atom is the terminal atom to be removed. The AO corresponding to the terminal hydrogen is first removed from the basis. The $N \times N$ matrix of the LMOs is becoming a rectangular one with dimension of $(N-1) \times N$. This makes the LMOs linear dependent and the orthonormalization condition is not fulfilled. The linear dependence in the LMOs should be removed and the remaining LMOs be re-orthonormalized. This can be done by diagonalizing the linear dependent LMOs after one AO removed from the basis, that is

$$U^{\dagger}(L^{\dagger}SL)U = \varepsilon \qquad (37)$$

where S is the overlap matrix, L is the LMOs after one AO removed from the basis, U and ε are the eigenvectors and eigenvalues of $L^{\dagger}SL$, respectively. The eigenvectors corresponding to zero eigenvalue are deleted and the re-orthonormalized LMOs are obtained as $L' = LU$.

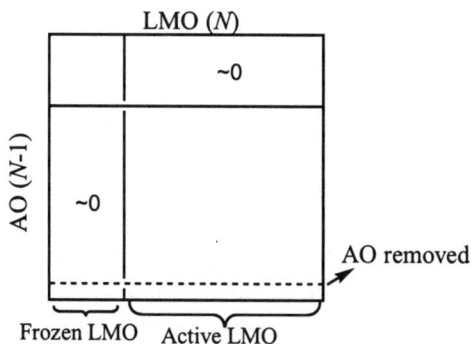

Figure 7. Illustration of the terminal atom removal.

For polyethylene, if one uses the 2×2 localization scheme, one has to work on the HAOs instead of AOs. It is expected that the elongation error will be reduced if the RLMO localization scheme is employed in the elongation calculations. Table 2 summarizes the elongation errors by employing the 2×2 and RLMO schemes, where the elongations calculations with RLMO scheme are performed at the *ab initio* level with STO-3G[39] and 6-31G[40] basis sets. Table 2 clearly shows that the RLMO localization scheme improves the elongation performance. The elongation error of the polyethylene chain with 28 carbon atoms is 0.31×10^{-5} eV for STO-3G basis set and to 3.38×10^{-5} eV for 6-31G basis set. The elongation errors for both STO-3G and 6-31G are smaller than that by the 2×2 scheme with the aid of hibridization orbitals at EHMO level.

Next, we test the reliability and efficiency of the RLMO localization scheme by two strongly delocalized and covalently bonded systems, polyglycine and cyanines, the performance of the regional LMO (RLMO) scheme will be examined as well.

Table 2. Elongation errors per CH_2 unit at different levels of theory for polyethylene, energies in 10^{-5} eV.

$[CH_3\text{-}(CH_2)_N\text{-}$ $CH_2]\cdot$	EHMO a)	HF/STO-3G b)	HF/6-31G b)
N=9	0.07	0.56	1.13
10	0.08	0.53	2.11
11	8.00	0.51	2.00
15	7.59	0.43	2.36
25	7.19	0.32	3.03
$[CH_3\text{-}(CH_2)_{26}\text{-}$ $CH_3]$	4.43	0.31	3.38

a) 2×2 unitary localization scheme employed.
b) RLMO localization scheme employed.

3.2.2. *Polyglycine and cationic cyanine chains*

The geometry structures of polyglycine and cyanines are depicted in Figure 8. The convetional total energies and the elongation errors for different size starting clusters (N_{st}) are presented in Table 3. For polyglycine N_{st} ranges from 4 to 8, where N_{st} counts the number of residues ($-CO-NH-CH_2-$) in the starting cluster. This allows us to check the effects of N_{st} on the elongation error. For the case N_{st}=4 (totally four residues with one residue in the frozen region and three in the active region) the elongation errors referred to the conventional energies are within a maximum value of 7.4×10^{-6} a.u. at a total of 20 residues. The elongation error per residue, $\Delta E(N) - \Delta E(N-1)$, against N is plotted in Fig. 9 for different N_{st}. For any given N, the curve with the smallest N_{st} is always on the top, showing that the elongation error is the maximum for the smallest starting cluster as one might expect. However, the difference between starting clusters becomes smaller as the number of residues increases and, in all instances, the error monotonically approaches a value of roughly 6.0×10^{-7} a.u. This indicates that the elongation error is not accumulated and it saturates to a small asymptotic limit as more and more elongation procedures performed.

Feng Long Gu, Akira Imamura and Yuriko Aoki

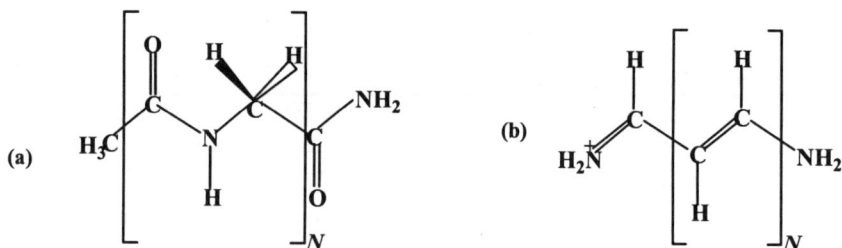

Figure 8. Geometry structures of (a) polyglycine and (b) cyanines.

Table 3. RHF/STO-3G total energy of polyglycine obtained from conventional calculations and the energy difference $\Delta E = E^{elg} - E^{cvl}$ for different size starting clusters (N_{st}).

	E(total, in a.u.)		$\Delta E = E^{elg} - E^{cvl}$ (in 10^{-6} a.u.)				
N	Conventional	$N_{st}=4$	5	6	7	8	
4	-856.14997927	0.00					
5	-1060.26460708	0.13	0.00				
6	-1264.37928877	0.38	0.16	0.00			
7	-1468.49397305	0.72	0.43	0.16	0.00		
8	-1672.60867708	1.11	0.79	0.44	0.17	0.00	
9	-1876.72338190	1.55	1.20	0.82	0.45	0.17	
10	-2080.83809595	2.02	1.66	1.24	0.83	0.46	
11	-2284.95281027	2.51	2.13	1.70	1.25	0.84	
12	-2489.06752962	3.02	2.63	2.18	1.72	1.26	
13	-2693.18224908	3.54	3.15	2.68	2.20	1.73	
14	-2897.29697156	4.07	3.68	3.20	2.71	2.21	
15	-3101.41169409	4.61	4.21	3.73	3.23	2.72	
16	-3305.52641858	5.16	4.76	4.27	3.76	3.24	
17	-3509.64114310	5.72	5.31	4.82	4.30	3.78	
18	-3713.75586896	6.28	5.87	5.37	4.85	4.32	
19	-3917.87059484	6.84	6.43	5.93	5.41	4.87	
20	-4121.98532166	7.41	6.99	6.49	5.96	5.42	

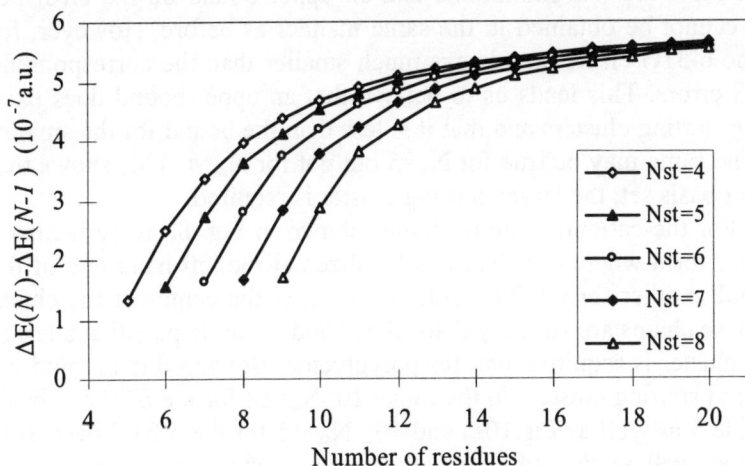

Figure 9. The elongation error with respect to the size of starting cluster for polyglycine at HF/STO-3G level.

Table 4. RHF/6-31G total energy of polyglycine obtained from conventional calculations and the energy difference $\Delta E = E^{elg} - E^{cvl}$ for different size of the starting clusters (N_{st}).

	E(total, in a.u.)	$\Delta E = E^{elg} - E^{cvl}$ (in 10^{-6} a.u.)				
N	Conventional	N_{st}=4	5	6	7	8
4	-866.98679700	0.00				
5	-1073.69571468	0.33	0.00			
6	-1280.40474591	0.84	0.39	0.00		
7	-1487.11377297	1.57	0.95	0.41	0.00	
8	-1693.82284353	2.42	1.74	0.99	0.42	0.00
9	-1900.53191219	3.38	2.63	1.82	1.01	0.42
10	-2107.24100130	4.41	1.48	0.29	0.10	0.02
11	-2313.95008940	5.50	1.94	0.43	0.16	0.03
12	-2520.65918886	6.62	2.45	0.58	0.24	0.06

It is also of interest to see how the RLMO scheme is affected by the choice of basis set. In Table 4 RHF/6-31G results from conventional and elongation calculations are given for polyglycine chains containing up to 12 residues. Using the 6-31G basis set the curves analogous to those in Fig.9 are non-monotonic and an upper bound on the error per residue cannot be obtained in the same manner as before. However, for $N_{st} \geq 6$ the 6-31G errors are always much smaller than the corresponding STO-3G errors. This leads us to believe that an upper bound does exist for these starting clusters and that it is less than the bound for the smaller basis. The same may be true for $N_{st} = 5$ but not for $N_{st} = 4$. This shows that for larger basis set, the larger starting cluster is required.

For the cationic cyanine chains shown in Fig. 8b there is also a resonance form where the charge is localized at the left hand end of the chain and another form with a soliton defect at the center of the chain. Thus, these chains are strongly delocalized and we anticipate that a larger starting cluster is required than for polyglycine. Bearing this in mind we considered starting clusters in the range $10 \leq N_{st} \leq 18$ for the STO-3G basis (see Table 5 as well as Fig.10a) and $10 \leq N_{st} \leq 15$ for the 6-31G basis (see Table 6 as well as Fig.10b). For all starting clusters, the frozen region contains 13 atoms: $H_2N^+ = CH - CH = CH - CH = CH -$ and the remainders are in the active region. Given a cyanine of fixed length the error per $- CH = CH -$ unit, not unexpectedly, decreases as the size of the starting cluster increases. Therefore, for a particular starting cluster the error is always largest for the first elongation step. Using the 6-31G basis this error decreases monotonically as the chain is lengthened. Although the differences between the various starting clusters also decrease, one cannot tell whether the differences eventually become negligible or not. With the STO-3G basis there is an initial decrease in the error per unit as N increases but, then, for longer chains the error per unit goes through a minimum and increases before ultimately saturating. It is evident in this case that the value for large N depends significantly on N_{st}. For either basis set, however, the important point is that, once again, the error does not get compounded as the chain is elongated; instead it levels off to a fairly small value (the order of 10^{-4} a.u. for $N_{st} = 10$). This is small compared to the energy per $-CH = CH-$ unit of about 77 a.u. It should also be noted that the limiting error per unit falls off by an order of magnitude when N_{st} is increased from 10 to 15. We judge that the results are satisfactory for $N_{st} \geq 10$. One can imagine, for example, if $N_{st} = 4$ for cyanines the limiting error per residue is on the order of 10^{-2} a.u.

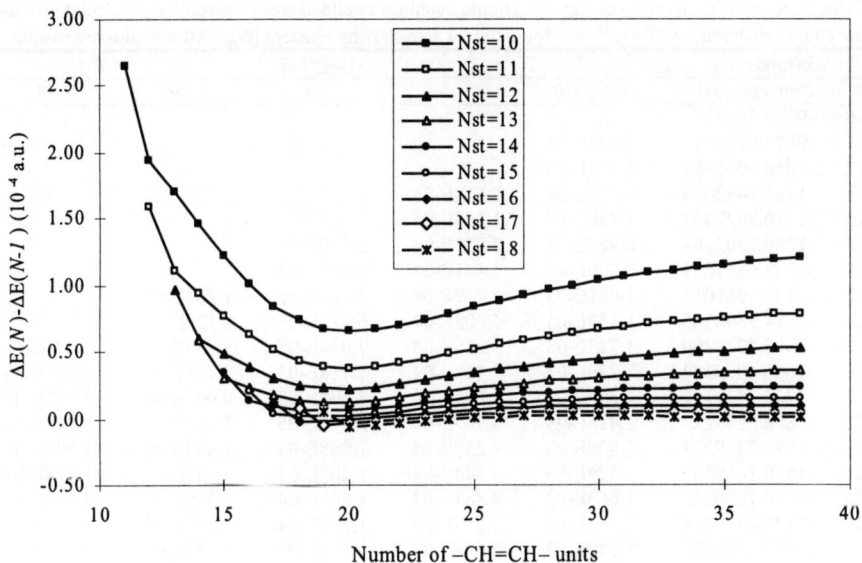

Figure 10(a). Elongation error as a function of N_{st} versus the number of unit cells for cyanines at STO-3G level.

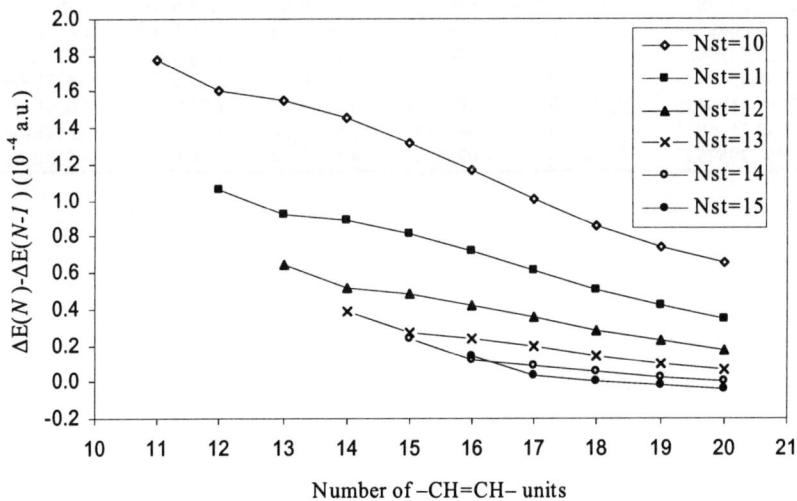

Figure 10(b). Elongation error as a function of N_{st} versus the number of unit cells for cyanines at 6-31G level.

Table 5. RHF/STO-3G total energy of cationic cyanines obtained from conventional calculations and the energy differences $\Delta E = E^{elg} - E^{cvl}$ for different size starting clusters (N_{st}). All energies are in a.u.

	E(total)	$\Delta E = E^{elg} - E^{cvl}$				
N	Conventional	$N_{st}=10$	12	14	16	18
10	-907.0294875					
11	-982.9678631	2.643E-04				
12	-1058.9060880	4.585E-04				
13	-1134.8442013	6.290E-04	9.618E-05			
14	-1210.7822352	7.748E-04	1.563E-04			
15	-1286.7202161	8.963E-04	2.052E-04	3.490E-05		
16	-1362.6581659	9.970E-04	2.441E-04	4.839E-05		
17	-1438.5961016	1.081E-03	2.749E-04	5.785E-05	1.264E-05	
18	-1514.5340365	1.155E-03	3.002E-04	6.476E-05	1.126E-05	
19	-1590.4719800	1.223E-03	3.229E-04	7.074E-05	9.739E-06	4.713E-06
20	-1666.4099384	1.289E-03	3.452E-04	7.718E-05	8.761E-06	-1.085E-06
21	-1742.3479152	1.356E-03	3.687E-04	8.504E-05	9.002E-06	-5.577E-06
22	-1818.2859122	1.426E-03	3.945E-04	9.486E-05	1.084E-05	-8.878E-06
23	-1894.2239294	1.500E-03	4.230E-04	1.068E-04	1.435E-05	-1.093E-05
24	-1970.1619663	1.579E-03	4.544E-04	1.207E-04	1.941E-05	-1.178E-05
25	-2046.1000215	1.663E-03	4.887E-04	1.365E-04	2.579E-05	-1.160E-05
26	-2122.0380934	1.751E-03	5.255E-04	1.538E-04	3.319E-05	-1.064E-05
27	-2197.9761805	1.843E-03	5.648E-04	1.725E-04	4.135E-05	-9.121E-06
28	-2273.9142810	1.940E-03	6.062E-04	1.922E-04	5.003E-05	-7.280E-06
29	-2349.8523935	2.040E-03	6.495E-04	2.127E-04	5.903E-05	-5.311E-06
30	-2425.7905166	2.143E-03	6.944E-04	2.339E-04	6.820E-05	-3.360E-06
31	-2501.7286489	2.250E-03	7.408E-04	2.556E-04	7.744E-05	-1.538E-06
32	-2577.6667893	2.359E-03	7.886E-04	2.778E-04	8.670E-05	8.390E-08
33	-2653.6049368	2.471E-03	8.374E-04	3.004E-04	9.590E-05	1.461E-06
34	-2729.5430905	2.584E-03	8.874E-04	3.233E-04	1.050E-04	2.574E-06
35	-2805.4812496	2.700E-03	9.382E-04	3.465E-04	1.141E-04	3.423E-06
36	-2881.4194135	2.817E-03	9.899E-04	3.700E-04	1.231E-04	4.026E-06
37	-2957.3575815	2.936E-03	1.042E-03	3.937E-04	1.320E-04	4.394E-06
38	-3033.2957532	3.056E-03	1.095E-03	4.176E-04	1.409E-04	4.553E-06

Table 6. RHF/6-31G total energy of cationic cyanines obtained from conventional calculations and the energy difference $\Delta E = E^{elg} - E^{cvl}$ for different size starting clusters (N_{st}). All energies are in a.u.

	E(total)			$\Delta E = E^{elg} - E^{cvl}$			
N	Conventional	$N_{st}=10$	11	12	13	14	15
10	-917.9884528						
11	-994.8440339	1.773E-04					
12	-1071.6994528	3.373E-04	1.064E-04				
13	-1148.5547423	4.926E-04	1.989E-04	6.432E-05			
14	-1225.4099290	6.382E-04	2.874E-04	1.160E-04	3.899E-05		
15	-1302.2650349	7.702E-04	3.687E-04	1.640E-04	6.589E-05	2.361E-05	
16	-1379.1200784	8.865E-04	4.402E-04	2.065E-04	8.954E-05	3.555E-05	1.423E-05
17	-1455.9750747	9.870E-04	5.012E-04	2.420E-04	1.088E-04	4.466E-05	1.726E-05
18	-1532.8300368	1.073E-03	5.519E-04	2.705E-04	1.233E-04	5.046E-05	1.796E-05
19	-1609.6849756	1.147E-03	5.938E-04	2.928E-04	1.334E-04	5.315E-05	1.631E-05
20	-1686.5398999	1.212E-03	6.287E-04	3.101E-04	1.401E-04	5.339E-05	1.283E-05

Thus far we have been using the energy as the criterion for the accuracy of our elongation treatment. It is of interest to examine other properties as well. So, for the cationic cyanines, Mulliken atomic charges computed by the conventional and elongation methods were compared. Figure 11 is a plot of the charge difference for the case $N=20$ obtained using the RHF/6-31G basis. The sizes of the starting cluster for the elongation calculations are $N_{st}=10$ and $N_{st}=15$. From Fig. 11, one can see that the magnitude of the maximum charge difference and the range of atoms over which the larger differences occur, both decrease as N_{st} increases. For $N_{st}=10$ the maximum charge difference is slightly larger than 0.01e, which we regard as borderline accuracy – i.e. this is the smallest starting cluster that one should use. The largest errors are in the vicinity of the border between regions A and B. As noted in Sec. 2.3, in order to have all ROs approximately doubly-occupied or approximately empty, we figuratively transfer an electron from each singly-occupied orbital of region A to the corresponding singly-occupied orbital of region B, which is probably why the largest errors occur near the border. In contrast, for water chains no electron transfer is needed in the localization procedure and the atomic charges in the conventional and elongation calculations are almost identical (the maximum difference is less than 10^{-4}e).

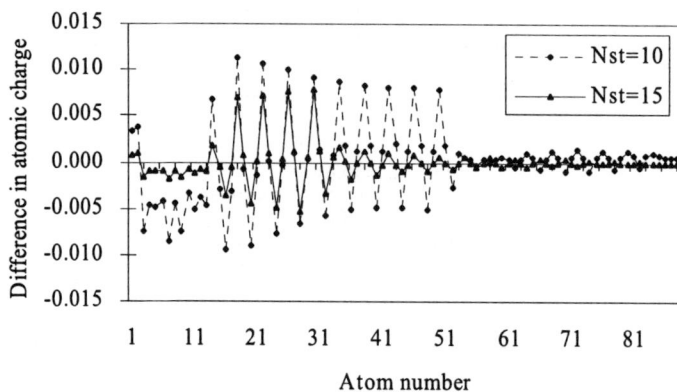

Figure 11. Difference in charge distribution between the elongation and conventional calculations for two different sizes of starting cluster of cyanines.

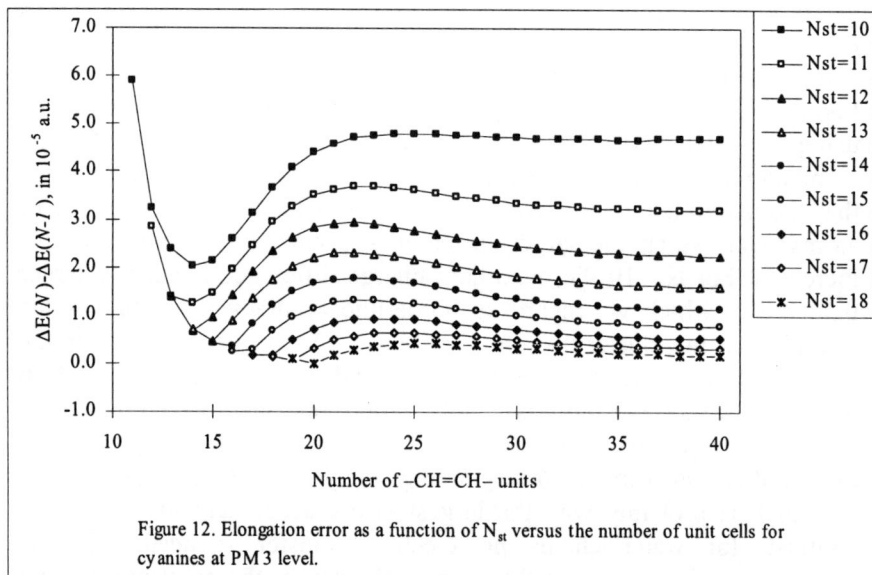

Figure 12. Elongation error as a function of N_{st} versus the number of unit cells for cyanines at PM3 level.

At this point a comment about the negative sign of some ΔE values in the last column of Table 5 is in order. In the limit of perfect localization the elongation method is variational, which means that a negative sign for ΔE cannot occur. However, the CMOs for B $\oplus$ M are obtained with the frozen LMOs of region A. Thus, the orthogonality constraint between orbitals of the two different regions can be slightly violated and, in that event, the conditions for the variation theorem are no longer fulfilled. In the semiempirical treatment (see below), on the other hand, this never happens because of the strong orthogonality approximation.

For comparison purposes a set of calculations was carried out using the PM3 semiempirical Hamiltonian. As might have been expected, the results for the energy error per –CH=CH– unit, shown in Fig.12, are similar to those obtained using the STO-3G minimum basis set in an *ab initio* treatment (see Fig.10a). Both sets of curves exhibit a minimum at some intermediate chain length. However, in contrast with the *ab initio* case, the semiempirical curves do not increase monotonically for larger N but go through a small maximum and converge to the long chain limit from above rather than below.

3.3 Modeling chemisorption on surface

There are two conventional ways to study the local perturbation in periodic systems. One is the cluster or embedded cluster model based on the molecular orbital method. In this approach, the system is modeled by a supermolecule embedded in the bulk surroundings (molecular approach). The other approach is the supercell model based on the crystal orbital method, where a large unit cell is considered to include a local perturbation with the translation symmetry (crystal approach). The elongation method, however, can be employed to investigate the local perturbation of periodic systems.

The elongation method was applied to the locally perturbed system of two-dimensional periodic surface[15]. The frozen and active orbitals are obtained by the stationary space analysis scheme. The perturbed system was chosen as a carbon monoxide (CO) being adsorbed vertically on the (001) surface of magnesium oxide (MgO) crystal. The Pisani's basis sets constructed with nine independent functions (three s and six p) were adopted for Mg^{2+} (8-61G) and O^{2-} (8-51G),[41] while for CO molecule, the Pople's STO-3G basis set was employed.

First, we calculated the electronic structures of the perfect MgO crystal surfaces which consist of up to five layers and CO molecules arranged periodically with various surface coverages. The (001) surface of MgO crystal forms a square lattice and the value of 2.1056 Å for the bulk MgO crystal is used as the distance between nearest-neighboring Mg^{2+} and O^{2-} ions in the surfaces. The bond distance of CO molecule is fixed as 1.15 Å. It was found that a system with three-layered slab MgO on the (001) surface and a coverage with a ratio of 1/8 between CO and Mg^{2+} are required to study the local CO adsorption on the (001) surface of MgO crystal.

The adsorption of CO on the MgO surface is modeled by putting the CO molecule on the top layer of the (001) surface in such a way that the CO molecule vertically approaching to the Mg^{2+} ion with various distances (d) between C atom and Mg^{2+} ion. The energy minimum is located by changing the value of d with an increment of 1.0 Å. The total energies obtained by the conventional supercell method and from the elongation method are listed in Table 7. It can be seen from Table 7, the agreements between the elongation and supercell calculations are quite satisfactory at the various distances and it is confirmed that the crystal elongation approach is sufficiently reliable and applicable. The CPU times by both methods indicate that our approach becomes increasingly advantageous when the system becomes much larger. Especially, as for the system of a three-layered slab model with the low coverage of 1/8 which can be expected to be the most realistic model for the local CO adsorption on the MgO (001) surface, our approach shows obviously great advantage in computational time. From Table 7, one can obviously see that there is an energy minimum around the d value of 2.3 Å.

Table 7. Total energies (in a.u.) and the CPU ratio (conventional supercell vs elongation method) of the CO molecule adsorption on the MgO (001) surface with different *d* values.

d(Å)	Conventional	Elongation	ΔE (10^{-5} a.u.)	CPU ratio
6.0	-6702.710855	-6702.710855	0.00	2.00
5.0	-6702.711030	-6702.711028	0.20	2.03
4.0	-6702.711648	-6702.711645	0.30	1.99
3.0	-6702.715639	-6702.715627	1.20	1.76
2.0	-6702.713544	-6702.713532	1.20	1.53

The elongation method has been applied to the surface adsorption. The frozen and active orbitals were determined by diagonalizing SS^+ and FF^+ matrices corresponding to the orthonormality and the variational conditions, respectively. Further, the interaction range by the aperiodic effects in the perturbed system can be determined self-consistently in the elongation procedures. The crystal elongation method can be expected to contribute the new development in the surface science as well as the advanced cluster approach. Along this direction, new development is in progress in our laboratory.

4. Toward Linear Scaling

Due to the difficulties in the conventional quantum chemistry calculations, much efforts have been devoted to develop so-called $O(N)$ methods for electronic structure calculations. $O(N)$ method is so attractive to quantum chemists because the computational cost scales linearly with the number of atoms (N) in the system, unlike the $\propto N^4$ in the conventional treatment. If $O(N)$ method is employed, large systems can be treated even at high levels of quantum theory. To achieve the linear scaling, one has to avoid solving the eigenvalue problem for the whole system, but only in the small and local space. This is the common feature for different $O(N)$ methods. One way, based on the fragmentation techniques,[42] is to divide the entire system into smaller molecular fragments, each characterized by localized quantities. By solving fragment Hartree-Fock (HF) or Kohn-Sham (KS) equations the whole system electronic structures as well as the total energy are determined by some special treatments. The accuracy of those methods is dependent upon the size of the fragments and the fragment division. Besides the fragmentation techniques, some other methods have been developed towards the linear scaling for every step of HF or KS equations. Some of these approaches have origin in condensed-matter physics and are based on sparse matrix algebra.[43] The sparse matrix elements beyond a certain localization radius are simply neglected. By enlarging the localization radius one can control the computation error in a systematic way, similar to quantum-chemistry based approaches where the size of the fragment plays the same role.

The existence of $O(N)$ methods is connected with the nearsightedness approximation. [44] Namely, a given static property defined in a certain observation region does not "see" the change in the external potential at a given point in the space if and only if this point is far away from the observation region. The elongation method exactly follows this approximation. In analogy with the polymerization/copolymerization mechanism, the elongation method "synthesizes" the electronic structure of a polymer chain by stepwise adding a monomer unit to a starting oligomer. Every step is followed by molecular orbital localization procedure. The LMOs that are far away from the chain propagation center are kept frozen during the whole elongation process. Thus, they do not change when we modify the system external potential by adding a new monomer unit. This picture of interactions is in agreement with traditional chemistry concepts that the properties of bonds between given atoms are determined by these atoms and their nearest neighbors.

The elongation method allows us to diminish the dimension of the variational space that remains practically constant. Furthermore, the elongation method may also "diminish" the size of the system by the so-called cutoff technique. The cutoff technique substantially reduces the number of two-electron integrals that have to be evaluated in the SCF process. In this section the cutoff technique designed for the elongation method will be presented, and the test calculations for polyglycine are performed by employing the cutoff technique in the elongation procedure.

4.1 Cutoff technique for the elongation method

Starting cluster   Attacking molecule

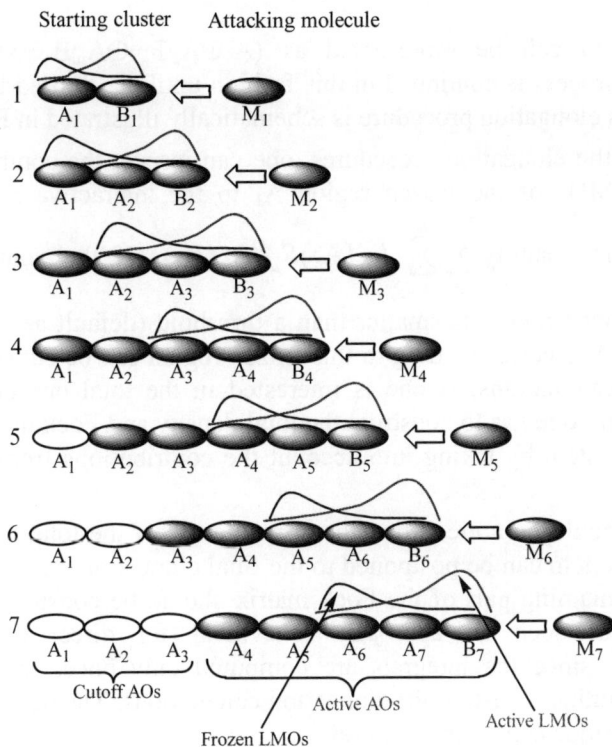

Figure 13. LMOs in the elongation processes.

The elongation calculation starts from a conventional calculation performed on a properly large starting cluster followed by a localization procedure in such a way that the CMOs of the starting cluster are localized into A_1 and B_1 regions. Region A_1 consists of atoms that are most remote from the chain-propagation center. B_1 region consists of the remaining atoms. In the first elongation step, LMOs from A_1 region are kept frozen while those assigned to B_1 region together with CMOs of the attacking monomer M_1 is the interactive region, where the elongation SCF is working on. After the elongation SCF converged, the CMOs of the interactive region, localized into a new frozen region, A_2, and a new active region, B_2, which are ready to interact with a new attacking monomer M_2. As more and more elongation steps are performed, the frozen region becomes larger and larger whereas the interactive region is more or less kept the same size. In general, for the nth elongation step the whole system can be represented as $(A_1 \oplus A_2 \oplus \cdots \oplus A_n | B_n \oplus M_n)$. The elongation process is continued in this fashion until the desired length is gained. This elongation procedure is schematically illustrated in Fig. 13.

During the elongation procedures, one can inspect the contributions from the LMOs of the frozen region A_1 to the interactive region by checking the quantity $\sum\limits_{\mu}^{A_1} \sum\limits_{\nu}^{B_n} L_{\mu i}(A_1) S_{\mu\nu}^{AO} L_{\nu j}(B_n)$. If the contribution from the frozen region is smaller than a threshold (default as 10^{-9}), the AOs and LMOs corresponding to this frozen region can be cut off in the elongation calculations. If one is interested in the total energy of the whole system, one has to construct the total density and Fock matrices of the whole system by taking into account the contributions from the cut off part.

It should be noticed that the computation of the total energy of the whole system can be postponed to the final elongation step. In such a case, the remaining part of the Fock matrix should be constructed. The advantage of such an approach is evident for direct type SCF calculations, since the integrals are computed only once, and such a situation is rather a custom for elongation calculations. The details of the cut-off technique can be found in ref. [45].

4.2 Polyglycine as an example

Polyglycine in C5 conformation, the geometrical structure is depicted in Fig. 8a, is chosen as an example to test the accuracy and reliability of the elongation cutoff technique. The polyglycine chains are built by the elongation method containing from 20 to 150 glycine units. In every elongation step, the cluster is terminated at both ends by converting $-CH_2-X$ to $-CH_3$. The minimal STO-3G basis sets was employed in the calculations. We chose the minimal basis set since the main objective is simply to demonstrate the scaling properties and the accuracy of the elongation method. The method, however, is general and can be applied to any basis sets. The cutoff calculation starts up if the interaction between the cutoff fragment and the active region is less than 10^{-9}. The starting cluster contains 20 glycine units with three different partitions of the polymer, e.g. the size of the frozen region is taken as 4, 8, and 10 glycine units. The second-order SCF (SOSCF) technique for orbital optimization[46] is utilized for both conventional and elongation calculations to speed up the SCF convergence.

Table 8 collects the total energies of C5 polyglycine clusters obtained by the conventional and elongation methods, as well as the elongation errors, $\Delta E = E^{elg} - E^{cvl}$. The conventional energy is utilized as a reference to compare with the elongation results and furthermore with the elongation cutoff energies. From Table 8, it is found that the elongation energies reproduce the exact results very well. The absolute value of ΔE is no more than 5.0×10^{-7} a.u. It turns out that all these points correspond to normal elongation calculations.

Table 8. Conventional RHF/STO-3G total energies (a.u.) for C5 conformer of polyglicine and the energy differences $\Delta E = E^{elg} - E^{cvl}$ with $N_{st}=20$. All energies are in a.u. In the elongation calculations, $N_{frozen}=4$ (10) means four (ten) units are frozen in the starting cluster.

	E(total)	$\Delta E = E^{elg} - E^{cvl}$	
N	Conventional	$N_{frozen}=4$	$N_{frozen}=10$
40	-8204.27989963	-4.83E-07	1.06E-06
60	-12286.57450490	-4.43E-07	-1.54E-07
80	-16368.86911866	-4.08E-07	-1.10E-07
100	-20451.16374085	-4.19E-07	-0.60E-07
120	-24533.45837184	-4.20E-07	-0.13E-07
140	-28615.75301868	-4.22E-07	-0.21E-07

The cutoff step starts from the cluster with $N=32$. It can be seen from Table 8, the cutoff calculations reproduce the conventional results very well with almost constant deviation (around 4.2×10^{-7} a.u). It is found that the cutoff threshold is a key parameter for the cutoff energies. When it is set to 10^{-5}, the error will be of the order 10^{-6} a.u.

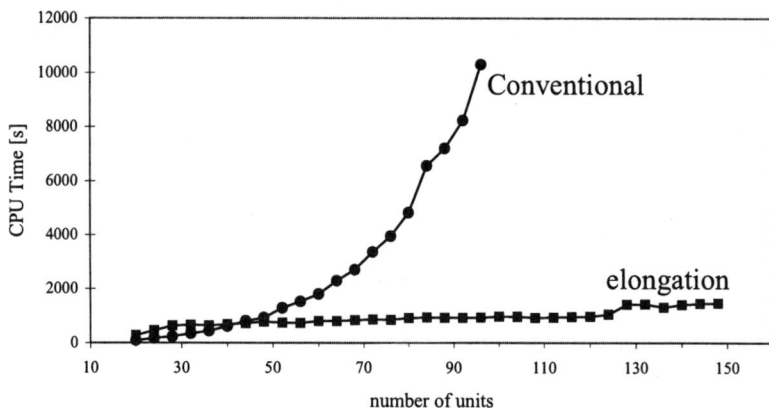

Figure 14. CPU time against the number of units for the conventional and the elongation calculations.

It is interesting to compare the CPU time in the elongation and the conventional calculations. In Figure 14, CPU time for the conventional (with filled circles) and the elongation (with filled squares) calculations as a function of the system size (number of glycine units) is depicted. The elongation calculations require additional CPU time in the localization and the "hydrogen removal" procedures. Thus, it is not surprised that the CPU time consumed by the elongation calculations is slightly more than the conventional treatment for small clusters. While from $N=44$, the elongation CPU time is less than the conventional one since the conventional CPU time is increased as a few powers of N. The curve of the elongation CPU time is very flat and the gap between these two curves is getting wider and wider as the system size increased. The CPU time for conventional calculations quickly goes to "infinity" while the elongation one seems to be saturated. Some small variations are connected with the different number of iterations in the SCF process.

It is obvious that the conventional calculation is performed for each system without referring to the previous calculations, whereas the elongation calculations are relied on the previous steps.

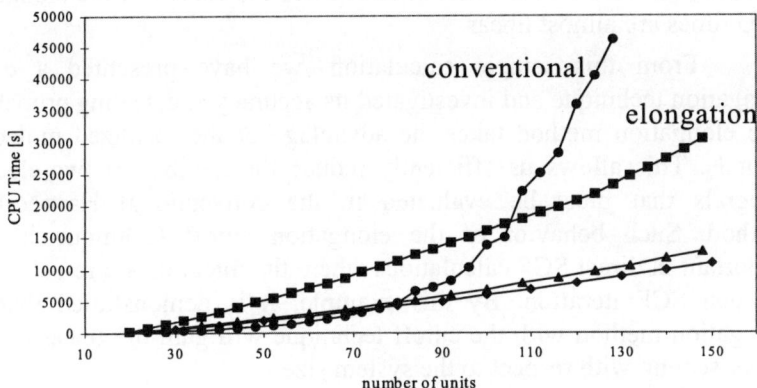

Figure 15. CPU time for the conventional calculations versus the total elongation CPU time. The curves with filled triangle and diamond are the elongation CPU times with respectively eigth and ten unit cells frozen in each step.

Therefore, we should compute the overall elongation CPU time as a summation of all the elongation calculations as $\tau_N = \sum_{i=0}^{N} \tau_i^{\text{elg}}$. This total elongation CPU time is plotted in Fig. 15 together with the conventional step CPU time. One can observe that almost linear dependence for the elongation calculations whereas the conventional CPU is rapidly increased with respect to the number of the unit cells. The linear correlation between the exact and estimated CPU times is higher than 99%. However, the slope of this line is a bit too large, so the elongation method can not compete with the conventional one until N=108. This is due to so many intermediate elongation steps. To highlight the cutoff technique, we have performed some other elongation calculations with eight (filled triangles) or ten (filled diamonds) glycine units frozen and another eight or ten units added in each elongation step. These results are also displayed in Figure 15. In the first case, the first cutoff step was performed for N=36, while in the second case for N=40. Now, the advantage of the cutoff elongation calculations is seen much earlier. When the polyglycine is enlarged by eight units, the overall elongation time is lower than the conventional one after 80 units. Increasing the size of the building block to ten glycine units slightly reduces the overall CPU time. In both cases the curves of the elongation CPU times are almost linear.

From this sample calculation, we have presented a cutoff elongation technique and investigated its accuracy and scaling properties. The elongation method takes the advantage of the localized molecular orbitals. This allows us efficiently reduce the number of two-electron integrals that must be evaluated in the conventional Hartree-Fock method. Such behavior of the elongation cutoff technique is very important in direct SCF calculations where the integrals are re-calculated in each SCF iteration. By this example, it is demonstrated that the elongation method with the cutoff technique will gain the linear or sub-linear scaling with respect to the system size.

5. Application of the elongation method to nonlinear optics

In this section, the elongation method in the presence of an electric field is formulated and the reliability of the elongation finite-field (elongation-FF) calculation is checked by some sample calculations. As it has been already seen in section 4, the elongation method together with the cutoff technique will be linear scaling with respect to the system size. This feature is important for the oligomer approach wherein the conventional treatment is difficult to deal with large systems. Our goal is to check the reliability of the elongation method in the presence of an external electric field. The novel finite-field approach is adopted for the elongation method. The elongation-FF method is then applied to evaluate the static (hyper)polarizabilities of some polymers as well as the molecular crystals of POM and MNA. The comparison is made between the elongation-FF and the conventional results as well as the experimental measurements if they are available.

5.1 Elongation method in the presence of an electric field

The elongation method for unperturbed systems is described in details in section 2. In this section, we give the appropriate equations for the elongation method when a system is perturbed by a uniform static electric field.

The perturbation due to a static homogeneous electric field, E, is:

$$H' = -e\vec{E} \cdot \vec{r} \tag{38}$$

For the starting cluster, the field-dependent CMOs are obtained by solving the HF equation

$$F(\vec{E})C(\vec{E}) = SC(\vec{E})\varepsilon(\vec{E}) \tag{39}$$

The AO basis Fock matrix is constructed in the presence of electric field as follows:

$$F(\vec{E}) = H + H' + D(\vec{E})[2J - K] \tag{40}$$

where H is the core Hamiltonian matrix, $D(E)$ is the field-dependent density matrix, J and K are the Coulomb and exchange supermatrices, respectively.

After the CMOs of the starting cluster are obtained, one can localize these CMOs into frozen and active regions. Then, the system is elongated following the same procedures as the unperturbed case but using the field-dependent Fock matrix from Eq.(40).

The total energy of the elongated system in the presence of the electric field, $W(E)$, is evaluated in the conventional HF formula,

$$W(\vec{E}) = \frac{1}{2} Tr\{D(\vec{E})[H^{core}(\vec{E}) + F^{core}(\vec{E})]\} + E_{N-N} \tag{41}$$

where for example, the field-dependent density matrix of the first elongation step is given by Eq. (35) in which the coefficients are field-dependent.

In the presence of the electric field, the total energy can be expressed as

$$W(\vec{E}) = W(0) - \mu_i E_i - \frac{1}{2!}\alpha_{ij}E_i E_j - \frac{1}{3!}\beta_{ijk}E_i E_j E_k - \frac{1}{4!}\gamma_{ijkl}E_i E_j E_k E_l + ... \tag{42}$$

where, i, j, k, and l are indices for Cartesian axes and the usual summation convention is used for repeated indices. The (hyper)polarizability tensors can be evaluated by numerical differentiation of the field dependent energy as

$$
\alpha_{ij} = -\left(\frac{\partial^2 W(\vec{E})}{\partial E_i \partial E_j} \right)_{\vec{E}=0} \tag{43a}
$$

$$
\beta_{ijk} = -\left(\frac{\partial^3 W(\vec{E})}{\partial E_i \partial E_j \partial E_k} \right)_{\vec{E}=0} \tag{43b}
$$

$$
\gamma_{ijkl} = -\left(\frac{\partial^4 W(\vec{E})}{\partial E_i \partial E_j \partial E_k \partial E_l} \right)_{\vec{E}=0} \tag{43c}
$$

In practice, a numerical differential treatment is adopted to evaluate the static (hyper)polarizabilities. For diagonal components of the (hyper)polarizabilities, total energies obtained with electric field magnitudes of ±0.001 and ±0.0005 a.u. together with the unperturbed total energy are required to evaluate the (hyper)polarizabilities, those are,

$$
\alpha = \frac{1}{E^2} \left\{ \frac{5}{2} W(0) - \frac{4}{3}[W(+E) + W(-E)] + \frac{1}{12}[W(+2E) + W(-2E)] \right\} \tag{44a}
$$

$$
\beta = \frac{1}{E^3} \left\{ [W(+E) - W(-E)] - \frac{1}{2}[W(+2E) - W(-2E)] \right\} \tag{44b}
$$

$$
\gamma = \frac{1}{E^4} \left\{ -6W(0) + 4[W(+E) + W(-E)] - [W(+2E) + W(-2E)] \right\} \tag{44c}
$$

where E is the magnitude of the applied electric field.

5.2 Nonlinear optical properties of polymers

5.2.1. *Reliability of the elongation-FF method*

In order to test the validity of the elongation method in presence of external electric field, semi-empirical PM3 (hyper)polarizability calculations were carried out for model systems: a molecular hydrogen chain, a water chain, a polyacetylene (PA) chain, and a PA with one-end substituted by an NO_2 group (PA-NO_2). For the hydrogen chain the intra- and inter-molecular distances are set to be 0.7 and 1.0 Å, respectively. The geometry of the water chain is taken from ref.[47] and the optimized geometry from ref.[48], is adopted for PA, while the geometry of PA-NO_2 is the one optimized at the HF/6-311G** level.[49] The semi-empirical elongation method was implemented and linked to the MOPAC2000 program package,[50] and the 2×2 unitary localization scheme was employed to obtain the LMOs for the elongation-FF calculations.

First, we consider the behavior of the elongation method as a function of the size of the starting cluster. Table 9 lists the relative errors (in percentage) of the (hyper)polarizabilities between the elongation and conventional PM3 calculations, where the elongation results are obtained for the first elongation step. The relative percentage error is defined as $(P^{elg} - P^{cvl})/P^{cvl} \times 100\%$, in which $P = \alpha$, β, or γ. From Table 9 one can see that $N_{st}=2$ for the elongation calculations of hydrogen chain and water chain do not provide satisfactory values for the longitudinal (hyper)polarizabilities as compared to the conventional results. The relative errors of the hydrogen chain are 5.81% for α and 82.3% for γ; and for water chain the errors are 1.17% for α, 1.71% for β, and 11.0% for γ. We find also for the hydrogen chain with $N_{st}=2$, the elongation error of the heat of formation is about 3%. This test reveals that a starting cluster with $N_{st}=2$ is too small for the elongation-FF calculations even for these two simple model systems. However, when six unit cells are included in the starting cluster of the hydrogen chain (i.e. $N_{st}=6$), the elongation reproduces almost the same α value as the conventional one, and the relative error in γ is less than 2%. By increasing the size of the starting cluster for the hydrogen chain to $N_{st}=10$, the error in γ is significantly reduced to only 0.03%. For water chain, α and β from the elongation calculation with $N_{st}=10$ reaches high accuracy, while the error in γ is double than that of the hydrogen chain. If ten more unit cells were added to the starting cluster of the water chain, we obtain almost the identical values of α, β, and γ to the conventional results (the error in γ is only 0.01%). This shows that $N_{st}=20$ for the water chain elongation (hyper)polarizability calculations is pretty sufficient.

Table 9. Relative errors (%) of (hyper)polarizabilities between the elongation and conventional PM3 calculations. The elongation results are for the first elongation step.

		$N(N_{st})=3(2)$	7(6)	11(10)	16(15)	21(20)
H_2 chain	α	5.81	0.00	0.00	-	-
	γ	82.3	1.99	0.03	-	-
H_2O	α	1.17	-	0.00	-	0.00
chain	β	1.71	-	0.00	-	0.00
	γ	11.0	-	0.06	-	0.01
PA	α	-	-	0.15	0.00	0.00
	γ	-	-	15.8	0.10	0.04
PA-NO_2	α	-	-	-	-	0.00
	β	-	-	-	-	0.01
	γ	-	-	-	-	0.00

It is much more interesting to examine the convergence behavior of a covalently bonded organic system, such as PA. In Table 9, we also list the results for PA oligomers elongated from a starting cluster with $N_{st}=10$, 15 and 20. From Table 9, one can see that the convergence of PA' α and γ with respect to N_{st} is slower than those of the hydrogen and water chain. For PA with $N_{st}=10$, α and γ differ by 0.15% and 16%, respectively from the conventional results, whereas for hydrogen chain and water chain with the same N_{st} value, the accuracy in γ is only 0.03% and 0.06%. This result is not surprising in view of the greater electron delocalization along PA chains. As N_{st} increased to 15, the elongation results for PA are much improved; α is essentially converged, and γ differs by an error of 0.1%, which is similar to what was found for the water cluster with $N_{st}=10$. For PA with $N_{st}=20$, the error in γ is reduced to 0.04%. This shows that $N_{st}=20$ for PA elongation-FF calculation is big enough. It is also interesting to test the convergence of the covalently bonded system with β, so we substitute one terminal H atom in PA by NO_2, that is PA-NO_2 as an example to test the convergence in the elongation-FF calculations. Due to the central symmetry destroyed in PA, PA-NO_2 has non-zero β. With $N_{st}=20$, the elongation results of PA-NO_2 are almost the same as the conventional ones with only a very small error (0.01%) in β.

In the above we focused on the behavior of the elongation-FF method with varying the size of the starting cluster. Now we perform the elongation process till six steps while keeping the size of the starting cluster fixed. As the convergence of the elongation-FF calculation against N_{st} summarized in Table 9, we have shown that for the hydrogen chain, $N_{st}=10$ is sufficiently enough, whereas for the water chain, PA, and PA-NO$_2$, the starting clusters with $N_{st}=20$ are needed for the accuracy in the elongation-FF calculations. Table 10 lists the elongation results for the above four model systems elongated to six elongation steps. For the comparison purpose, the conventional results are listed in the row just below the corresponding elongation values. The results obtained for the starting cluster should be identical for conventional and the elongation methods since there is no approximation introduced at this stage. For subsequent elongation steps, however, this is not necessarily the case.

From Table 10, we see that the elongation results for these four systems agree perfectly with the conventional ones. All the elongation-FF calculations produce almost the same α per unit cell ($\Delta\alpha=\alpha_N-\alpha_{N-1}$) as those of the conventional. The largest difference between them is in the $N+5$ elongation step for PA, which gives 0.07% in $\Delta\alpha$. As expected the hyperpolarizabilities per unit cell, $\Delta\beta$ and $\Delta\gamma$, are particularly not as well converged as the corresponding $\Delta\alpha$. The $\Delta\gamma$ values obtained from the elongation method differ from those of the conventional calculation by less than 0.5%. There exists one exception for the water chain with a difference of about 2% obtained in the $N+2$ elongation step. The reason for this is due to the localization quality of the 2×2 localization scheme. The same calculation was repeated by employing the RLMO localization scheme and it is found that this difference has become negligible (less than 0.02%). One important thing can be seen from Table 10, the difference between the conventional and the elongation values remains almost constant, which means the elongation error is not cumulative. Although evaluation of β requires one order lower than that of γ in the numerical differentiation, the elongation calculations of β are not so accurate as those of α and γ's. For the water chain, β is still accurate and the difference in $\Delta\beta$ is lower than 0.02%. For PA-NO$_2$, however, the difference in $\Delta\beta$ is fairly big, with the maximum difference of 4%. If we define β per unit cell in an alternative way, e.g. $\Delta\beta=\beta/N$, the elongation (vs. conventional) $\Delta\beta$'s are 3746 (3746), 3634 (3634), 3524 (3523), 3418 (3416), 3315 (3313), 3216 (3213) a.u. for $N+1$ to $N+6$, respectively. The difference in $\Delta\beta$ becomes very small since the elongation error is averaged over all unit cells.

Table 10. Comparison between the elongation and the conventional PM3 calculations for the static (hyper)polarizabilities per unit cell of hydrogen chain, water chain, PA, and PA-NO$_2$. The size of the starting cluster for Hydrogen chain is 10 while for the other systems are 20. The first line is the results obtained from the elongation calculations, while the second line is from the conventional ones.

		$N+1$	$N+2$	$N+3$	$N+4$	$N+5$	$N+6$
$(H_2)_n$	$\Delta\alpha$(au)	24.179	24.321	24.427	24.506	24.566	24.614
		24.179	24.323	24.429	24.509	24.570	24.618
	$\Delta\gamma(10^3$au)	46 672	48 576	49 971	51 088	51 960	52 661
		46 541	48 497	49 964	51 090	51 978	52 672
$(H_2O)_n$	$\Delta\alpha$(au)	9.713	9.713	9.714	9.714	9.714	9.714
		9.713	9.713	9.714	9.714	9.714	9.714
	$\Delta\beta$(au)	-24.027	-24.014	-24.002	-24.004	-24.001	-23.994
		-24.025	-24.016	-24.008	-24.002	-23.995	-23.988
	$\Delta\gamma(10^3$au)	4630	4767	4780	4778	4803	4802
		4639	4859	4749	4745	4757	4753
PA	$\Delta\alpha$(au)	190.86	191.36	191.80	192.12	192.60	192.72
		190.86	191.34	191.79	192.16	192.46	192.73
	$\Delta\gamma(10^3$au)	16402	16828	17165	17567	17696	17944
		16464	16857	17197	17490	17742	17961
PA-NO$_2$	$\Delta\alpha$(au)	190.8	191.2	191.5	191.8	192.0	192.3
		190.8	191.1	191.4	191.8	191.9	192.2
	$\Delta\beta$(au)	1481	1279	1111	969	847	744
		1481	1277	1101	951	823	714
	$\Delta\gamma(10^3$au)	17122	17411	17654	17864	18045	18202
		17124	17403	17645	17855	18038	18197

It should be mentioned that although not shown, the total energies and the heat of formations between the elongation and conventional methods always differ by less than 10^{-6} eV in every case. Although the difference in β is not as small as those in α and γ, and there is a small initial increase in the error upon increasing N, the important point is that the elongation error does not accumulate as the elongation is further processed. These results demonstrate the reliability of the elongation-FF method for (hyper)polarizability calculations for different systems.

In the elongation method, there is no translational symmetry implied, and the attacking monomer can be of any nature. Therefore, the elongation method can be applied not only to periodic stereoregular polymers but also to donor-acceptor chains, block copolymers as well as polymers with a random distribution of monomers. In ref. 32, the elongation method has been applied to investigate the terminal donor/acceptor substitution effects on the NLO responses for polyacetylene, poly(paraphenylene), and poly(*para*-phenylene-ethynylene) chains. It has been demonstrated that the elongation method is a powerful tool to study NLO properties for a periodic polymer, wherein the translational symmetry is destroyed by end substitutions. In the following, we present the NLO properties of a block copolymer studied by the elongation-FF approach.

5.2.2. NLO properties of block copolymer with a quantum well structure

Since the early work of Esaki and Tsu [51] in 1969, quantum well electronic and optical devices have matured into commercial products and have the promise of being more useful in the future. Quantum well structure is a heterostructure of alternating semiconductor layers with different band gaps exhibited novel properties. Such quantum well structures have been found to exhibit large optical nonlinearity. The potential applications in NLO materials are, such as optical bistable devices, all-optic directional couplers, and degenerate four-wave mixers.[52] For a review of quantum well devices, we refer the reader to ref. [53].

In this sub-section, we apply the elongation-FF method to determine the (hyper)polarizabilities of a quantum well structure, π-conjugated polyheterol system as shown in Figure 16a. One can see that in this block polymer, each unit cell contains ten five-member rings with five all-trans pyrrole rings followed by five all-trans silole rings. In the elongation calculations, the starting cluster is consisted of one such unit cell. The polymer chain is elongated by adding one ring for each step, up to a chain containing 25 rings.

In order to characterize the electronic structure of polyheterol chain, the local density of states (LDOS) calculation was carried out following the method as described in details in ref. [54],

$$\rho_\mu^{j_\mu}(E) = \sum_i \delta(E - E_i) \sum_\nu \sum_{j_\nu}^{all} C_{\mu i}^{(j_\mu)} S_{\mu\nu}^{(j_\mu, j_\nu)} C_{\nu i}^{(j_\nu)} \ . \tag{45}$$

(a)

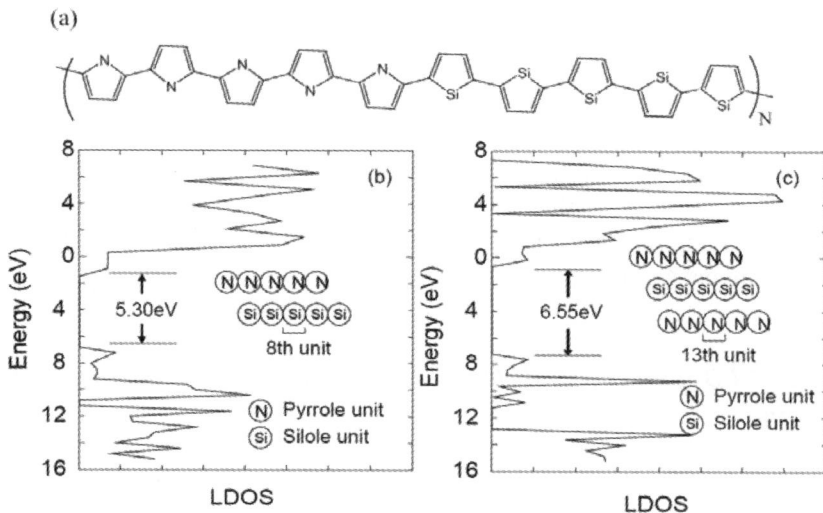

Figure 16. (a) Geometry structure of polyhetrol and LDOS of (b) silole ring and (c) pyrrole ring.

The LDOS of the 8th ring of polyheterol chain containing 10 rings, and the 13th ring of polyheterol chain containing 15 rings are obtained. The 8th ring is the central silole ring of the first silole block, while the 13th ring is at the center of the second pyrrole block. The LDOS together with the energy gaps are displayed in Figure 16b and 16c. It can be seen that the band gap of the silole block is 5.30 eV, whereas the band gap of the pyrrole block is 6.55 eV. The difference between these two band gaps indicates a quantum well structure in polyheterol.

Our calculated values of α, β, and γ are plotted versus ring number (M) in Figure 17a-c, respectively. The (hyper)polarizabilities of pure pyrrole and silole chains are plotted on the same figures for comparison. From Figure 17c, we see that the increment in γ due to adding a pyrrole blocks (corresponding to M=11-15 and 21-25 of the polyheterol chain) is enhanced by a small amount due to copolymerization, while the increment due to the silole blocks (corresponding to M=16-20 of the polyheterol chain) decreases more substantially. The same is true for α, but to a lesser extent. Due to the centrosymmetry in polypyrrole and polysilole, the β values are zero. While for polyheterol chain, the copolymerization destroys the centrosymmetry that leads to a roughly constant negative value over each pyrrole block and a roughly constant positive value over each silole block. The magnitude of β in either case is fairly small.

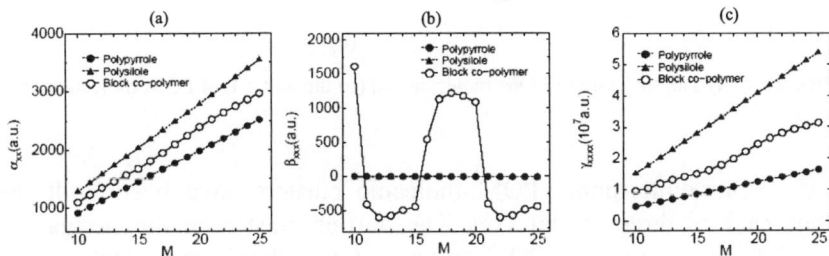

Figure 17. Plots of the (hyper)polarizabilities of polyhetrol against number of rings.

5.3 Nonlinear susceptibilities of molecular crystals POM and MNA

In this section, the static polarizability, first and second hyperpolarizabilities of POM and MNA molecular crystals are evaluated by the above mentioned elongation-FF method at the semi-empirical AM1 level.[55] As these two molecular crystals present important second-order nonlinear susceptibilities, they have already been the subject of many theoretical investigations.[56,57] The POM crystal parameters are taken from ref. [58] with $a = 20.890$Å, $b = 6.094$Å, and $c = 5.135$Å. The POM crystal is orthorhombic with space group $P_{2_1 2_1 2_1}$. The POM molecule and the unit cell of POM crystal are depicted in Fig. 18a and 18b, respectively. The unit cell of the POM crystal consists of four POM molecules, distributed in such a way that the only non-zero component of $\chi^{(2)}$, the macroscopic analog of β, is $\chi^{(2)}_{abc}$. Because of weak dipole-dipole interactions between POM molecules, the molecules adopt a non-parallel stacking. The interaction between the neighboring POM molecules is quite small in the POM crystal.

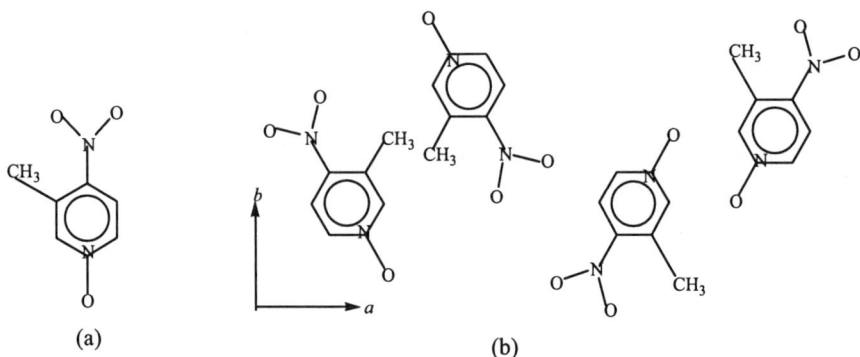

(a)   (b)

Figure 18. (a) The structure of POM molecule and (b) the unit cell of POM molecular crystal.

One-dimensional POM molecule clusters have been built up along each of three crystal axes. The starting cluster for the elongation calculations contains two POM unit cells (eight POM molecules), one in the frozen region and the other one in the active region. The RLMO localization scheme was employed in the elongation-FF calculations for POM clusters. In the first elongation step, one POM unit cell is added to the starting cluster from the active region. Since the CMOs of the starting cluster have been well localized, the Hartree-Fock equation is solved only for the interactive region, which consists of the active region of the starting cluster and the attacking monomer. The frozen part of the starting cluster is excluded from the current elongation SCF procedure. Its contribution to the total density matrix, however, is accounted for when the total energy is calculated. These elongation steps can be repeated until the required size of the POM cluster reached. Then, the (hyper)polarizabilities are obtained by numerically differentiating the field-dependent total energies for each elongation step as formulated in Eq. (44).

The crystal packing effect on the (hyper)polarizabilities is characterized by a packing ratio R defined as

$$R_P = \frac{P^{cluster}}{NP^{unit}} \quad ,(P = \alpha, \beta, \gamma) \qquad (46)$$

where N is the number of unit cells in the molecular cluster. To determine the crystal NLO responses, the molecular packing effect is taken into account through the packing ratio. The effective (hyper)polarizabilities of the crystal unit cell are thus estimated by using the multiplicative scheme as

$$P^{eff} = P^{cell} \times R_P^a(\infty) \times R_P^b(\infty) \times R_P^c(\infty) \quad ,(P = \alpha, \beta, \gamma) \qquad (47)$$

where $R_P^a(\infty)$, $R_P^b(\infty)$, and $R_P^c(\infty)$ are the packing ratios in the infinity limit for the (hyper)polarizabilities, P, along axes a, b, and c, respectively.

5.3.1. *Elongation-FF vs conventional calculations for POM clusters*

The advantage of the elongation method is that one can treat very large systems without paying for too much computational costs. The reliability of the elongation method for calculating the total energy and its derivatives with respect to the external electric fields have been demonstrated in Section 5.2. For the POM crystal, we check the reliability of the NLO calculations by comparing the elongation NLO results with the conventional ones from ref. [59]. For the purpose of comparison, the AM1 static α, β as well as the crystal packing ratio calculated from the conventional and the elongation methods are listed in Table 11. The POM clusters contain six unit cells along three different axes. For the elongation calculation, it starts from a starting cluster with two unit cells and elongated step by step till the POM cluster containing six units. From Table 11, it can be seen that the difference between the conventional and the elongation α or β is always small for POM clusters, except α_{bb} (and β_{abc}) along b axis with the discrepancy of 3.8% (and 2.9%) from the conventional value. This difference is most probably due to the numerical instability and the different approaches used for the calculation of the NLO responses. In ref. 59, it was using the TDHF/AM1 approach while in the present work the elongation-FF approach is adopted.

Table 11. Comparison of the static (hyper)polarizability tensors between the conventional and the elongation calculations at the AM1 level. The crystal packing ratio is in the parenthesis. The systems contain six unit cells of the POM molecular crystal.

	Along *a* axis		Along *b* axis		Along *c* axis	
	conventional	elongation	conventional	elongation	conventional	elongation
α_{aa}	1969(1.039)	1962(1.038)	1810(0.967)	1815(0.968)	1827(0.964)	1839(0.966)
α_{bb}	2390(0.955)	2388(0.960)	2846(1.151)	2738(1.110)	2027(0.873)	2060(0.882)
α_{cc}	1527(0.964)	1525(0.973)	1383(0.820)	1392(0.835)	1768(1.116)	1717(1.088)
β_{abc}	-4808(0.983)	-	-5286(1.081)	-	-3909(0.799)	-
		4799(0.984)		5120(1.050)		3897(0.799)

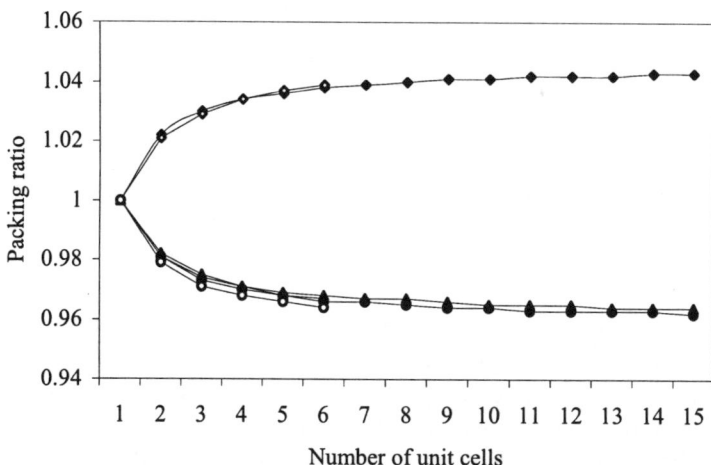

Figure 19. Crystal packing ratios of POM clusters built along *a* axis. The curves with closed (open) diamonds, triagles, and circles are from the elongation (conventional) calculations.

The crystal packing ratios of α_{aa}, α_{bb}, and α_{cc} along *a* axes obtained from the elongation and the conventional calculations are presented in Fig. 19. The curves with closed diamond, triangle, and circle are obtained from the elongation calculations for the packing ratios of α_{aa}, α_{bb}, and α_{cc}, respectively, whereas the curves with open diamond, triangle, and circle are obtained from the conventional calculations for the packing ratios of α_{aa}, α_{bb}, and α_{cc}, respectively. One can see that the conventional (with unit cell number smaller than six) and the elongation packing ratios are in a nice agreement. The elongation results are extended to larger clusters beyond six unit cells till N=15 because the elongation calculation is cheaper and able to treat large systems. The packing ratios are seemingly not convergent with respect to the number of unit cell until the point with six units. Consequently, for the conventional calculation an extrapolation for the packing ratio is needed to estimate the infinity value. From Fig. 19, however, one can see that the curves of the packing ratios are almost flat after N=10 and thus the extrapolation to infinity from this point is more reliable, or even no extrapolation is needed for the POM cluster containing 15 unit cells.

The slow convergence of packing ratio R – especially for the nonlinear responses – as a function of the number of stacked unit cells shows the performance of the elongation method for addressing the NLO properties of molecular crystals.

Table 12. AM1 Diagonal components of the static polarizability tensors and crystal packing ratios (R). The estimated R values from the multiplicative scheme are presented as well.

Cluster	α_{aa}	R	α_{bb}	R	α_{cc}	R
1a	315	1.000	411	1.000	263	1.000
2a	644	1.022	807	0.982	516	0.981
2b	618	0.981	859	1.045	492	0.935
2c	623	0.989	749	0.911	546	1.038
3a	973	1.030	1202	0.975	768	0.973
3b	918	0.971	1320	1.071	718	0.910
3c	929	0.983	1079	0.875	835	1.058
2a×2b	1277	1.013	1673	1.018	954	0.907
Estimated R		1.003		1.026		0.917
3a×2b	1941	1.027	2483	1.007	1416	0.897
Estimated R		1.010		1.019		0.910
3a×3b	2904	1.024	3853	1.042	2045	0.864
Estimated R		1.000		1.044		0.885

Before we start to study the macroscopic NLO responses of the POM crystal, it is interesting to test the validity of Eq. 47, that is, the multiplication scheme. Table 12 and 13 collect the AM1 static polarizability and hyperpolarizabilities of POM clusters and the corresponding packing ratios extended in one and two crystal axes. The calculated R values are generally in good agreement with those obtained by the multiplicative scheme. From Table 12, one can see that, for example, $R(\alpha_{aa})$=1.024 for a $3a \times 3b$ POM cluster (containing 9 unit cells or 36 POM molecules). The estimated value by the multiplicative scheme is that $R^{3a} \times R^{3b}$=1.030×0.971=1.000 with an error of 2.4% from the calculated one. The difference in R for the other polarizability components from direct calculation and the multiplicative scheme is less than 2%. For β_{abc}, the R values calculated and estimated are 1.018 and 1.012, respectively. The other calculated R values of β_{abc} for the $2a \times 2b$ and $3a \times 2b$ POM clusters are still in good agreement with the estimated ones. While for the second hyperpolarizability, the agreement is not so good as those of α and β. For γ_{aaaa}, the calculated and estimated R values of the $3a \times 3b$ POM cluster are 1.982 and 1.756, respectively, with the difference of 11% between them. For γ_{bbbb}, however, the agreement is quite good. The calculated R value is 1.376 while the estimated one is 1.379. The worst case is for γ_{cccc}, the error between the calculated (0.592) and estimated (0.688) R values is about 16%. This requires further investigation for the validity of the multiplicative scheme for second hyperpolarizability whether the higher order numerical differentiation is not a possible source of this large difference.

Table 13. AM1 Diagonal components of the static hyperpolarizability tensors and crystal packing ratios (R). The estimated R values from the multiplicative scheme are presented as well.

Cluster	β_{abc}	R	γ_{aaaa}	R	γ_{bbbb}	R	γ_{cccc}	R
1a	-813	1.000	18389	1.000	133199	1.000	9605	1.000
2a	-1610	0.990	43821	1.192	244726	0.919	18661	0.971
2b	-1643	1.010	47761	1.299	358558	1.346	15034	0.783
2c	-1429	0.879	36349	0.988	173010	0.649	25998	1.353
3a	-2407	0.987	69517	1.260	355912	0.891	27697	0.961
3b	-2501	1.025	76876	1.394	618643	1.548	20642	0.716
3c	-2047	0.839	53616	0.972	217252	0.544	44234	1.535
$2a \times 2b$	-3266	1.004	116783	1.588	643200	1.207	29199	0.760
Estimated R		1.000		1.548		1.237		0.760
$3a \times 2b$	-4891	1.003	200009	1.813	921613	1.153	44805	0.777
Estimated R		0.997		1.637		1.199		0.752
$3a \times 3b$	-7452	1.018	328015	1.982	1649620	1.376	51208	0.592
Estimated R		1.012		1.756		1.379		0.688

5.3.2. *Macroscopic NLO responses of POM crystal*

Table 14. Diagonal components of the static polarizability tensors and crystal packing ratios (R) for POM clusters elongated along three crystal axes till 15 stacked unit cells. The results for the unit cell are also listed. The values in parenthesis are for accounting for electron correlation effects.

N	α_{aa}	R	α_{bb}	R	α_{cc}	R
1	315	1.000	411	1.000	263	1.000
Along a axis						
5	1632	1.036	1992	0.969	1273	0.968
10	3280	1.041	3968	0.965	2535	0.964
15	4929	1.043	5944	0.964	3796	0.962
Along b axis						
5	1517	0.963	2262	1.101	1168	0.888
10	3008	0.955	4648	1.131	2290	0.871
15	4498	0.952	7046	1.143	3411	0.865
Along c axis						
5	1536	0.975	1734	0.844	1422	1.081
10	3046	0.967	3361	0.818	2906	1.105
15	4554	0.964	4985	0.809	4397	1.115

	a axis	b axis	c axis
α^{eff}	302	366	244
$\chi^{(1)}$	0.86(1.17)	1.04(1.35)	0.70(0.97)
n	1.36(1.47)	1.43(1.53)	1.30(1.40)
Exp n	1.66	1.83	1.63

For POM clusters extending along one of the crystal axes, the static diagonal components of the polarizability tensors (α_{aa}, α_{bb}, and α_{cc}) and their corresponding crystal packing ratios are calculated by the elongation-FF method at the AM1 level and the results are reported in Table 14. In Fig. 20 the crystal packing ratios versus the number of the stacked unit cells along three different crystal axes are also plotted. One can see from Fig. 20a-c that the packing ratios are greater than unity for $R(\alpha_{aa})$ along a axes, $R(\alpha_{bb})$ along b axes, and $R(\alpha_{cc})$ along c axes, whereas, the other packing ratios are smaller than 1.0. It is clear from the figure that the packing ratios of the polarizability tensors are almost convergent for POM cluster containing $N=15$ unit cells. Thus the infinite one-dimensional crystal packing ratios can be simply approximated by the values for $N=15$ without loosing the accuracy.

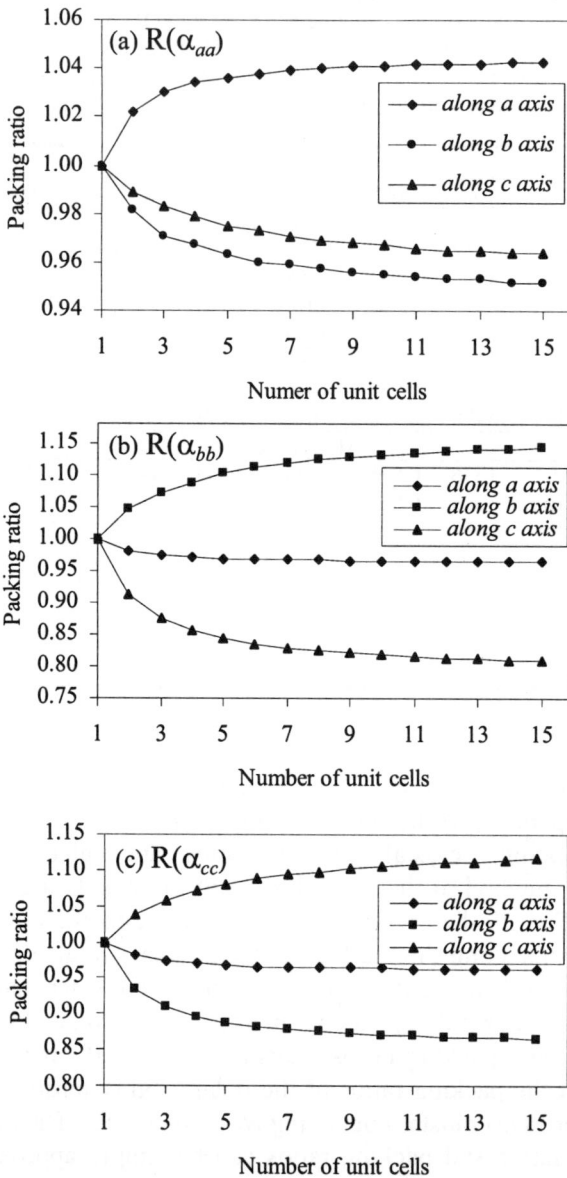

Figure 20. The packing ratios of the polarizability tensors of POM clusters built along three crystal axes.

The effective polarizabilities (α^{eff}) are evaluated by considering the packing ratios in three axes using the multiplicative scheme from Eq. 47. And then the linear susceptibility is obtained as

$$\chi_{\xi\xi}^{(1)} = \frac{\alpha_{\xi\xi}^{\text{eff}}}{\varepsilon_0 V^{\text{cell}}} \tag{48}$$

Where V^{cell} is the unit cell volume of POM crystal, ($V^{\text{cell}}=a\times b\times c=20.890\times6.094\times5.135=653.7$ $\text{Å}^3=653.7\times10^{-30}$ m^3) and $\varepsilon_0=8.8542\times10^{-12}$ $\text{C}^2\text{J}^{-1}\text{m}^{-1}$. The conversion factor of polarizability from a.u. to S.I is 1 a.u. of $\alpha = 1.6488\times10^{-41}$ C^2 m^2J^{-1}. The refractive indices are thus given as

$$n_\xi = \sqrt{\chi_{\xi\xi}^{(1)} + 1} \tag{49}$$

for $\zeta=a$, b, or c. The effective polarizabilities (α^{eff}), the linear susceptibilities ($\chi^{(1)}$) and the refractive indices (n) are evaluated from Eqs. 47-49, respectively, and also reported in Table 14 together with the experimental refractive indices. The experimental results were obtained at a frequency $\omega=1064$ nm in ref. [60].

From Table 14, one can see that the calculated refractive indices are underestimated compared to the experimental results. There are two main factors we should consider. First one is the frequency dispersion effect, but this effect is not so prominent for linear polarizability. The other one is the electron correlation, which is not included in our AM1 calculations though it is somehow accounted for through semi-empirical parameterization procedures. The electron correlation effect can be estimated by performing Hartree-Fock and MP2 calculations and taken by the ratio between the HF and MP2 polarizabilities. If one uses the correlation factors obtained from ref. 59 (1.36, 1.30 and 1.38 for α_{aa}, α_{bb} and α_{cc}, respectively), and the values of refractive indices are becoming $n_a=1.47$, $n_b=1.53$, and $n_c=1.40$ which are comparable to the experimental values of $n_a=1.66$, $n_b=1.83$, and $n_c=1.63$ measured at $\omega=1064$ nm.

The first hyperpolarizability and their packing ratios of the POM clusters are listed in Table 15 as a function of the stacked unit cells. The packing ratios for β_{abc} along three different axes are actually convergent when $N=15$ and thus the infinite ratios are obtained without any extrapolation help.

Table 15. Static β_{abc} tensor and crystal packing ratios (R) for POM clusters elongated along three crystal axes as a function of the number (N) of stacked unit cells.

N	Along a axis		Along b axis		Along c axis	
	β_{abc}	R	β_{abc}	R	β_{abc}	R
1	-813	1.000	-813	1.000	-813	1.000
2	-1610	0.990	-1643	1.010	-1429	0.879
3	-2407	0.987	-2501	1.025	-2047	0.839
4	-3204	0.985	-3370	1.036	-2664	0.819
5	-4002	0.985	-4244	1.044	-3281	0.807
6	-4799	0.984	-5120	1.050	-3897	0.799
7	-5596	0.983	-5999	1.054	-4513	0.793
8	-6393	0.983	-6878	1.058	-5129	0.789
9	-7190	0.983	-7759	1.060	-5744	0.785
10	-7989	0.983	-8639	1.063	-6360	0.782
11	-8784	0.982	-9520	1.065	-6975	0.780
12	-9581	0.982	-10402	1.066	-7590	0.778
13	-10378	0.982	-11283	1.068	-8206	0.776
14	-11175	0.982	-12165	1.069	-8821	0.775
15	-11972	0.982	-13047	1.070	-9437	0.774
∞		*0.981*		*1.084*		*0.768*

One can evaluate the value of the effective β_{abc} from Eq. (47) with the packing ratios extrapolated to the infinity limit as also listed in Table 15. The effective β_{abc} is obtained as 663 a.u. and 5% smaller than the result of ref. 59. This is due to the extrapolated ratio was overestimated for β_{abc} along the b axis. The second-order nonlinear susceptibility is given by

$$\chi^{(2)}_{abc} = \frac{\beta^{eff}_{abc}}{\varepsilon_0 V^{cell}}$$ (50)

with the amount of 3.67 pm/V. The conversion factor from a.u. to S.I is that 1 a.u. of $\beta = 3.2064 \times 10^{-53}$ C^3 $m^3 J^{-2}$.

The result obtained from the semi-empirical level is lack of the electron correlation effects. This effect can be taken into account by an approximate treatment from ref. 59, introducing a correlation factor. This factor is taken as 0.95 determined in ref. 59. Thus the second-order susceptibility $\chi^{(2)}_{abc}$ becomes 3.5 pm/V by using the value of 663 a.u. for β^{eff}_{abc}. This value is too small compared to the experimental measurement for SHG response (12 pm/V) because the local field factors and the frequency dispersion effects are not taken into account in our current study. In a similar way, frequency dependent factor (F^{ω}) can be estimated as the ratio between the dynamic and static (hyper)polarizability values calculated using the conventional approach for the unit cell. For a wavelength of 1064 nm, F^{ω}=1.63. This leads to the estimated frequency-dependent result of $\chi^{(2)}_{abc}$ = 5.7 pm/V. This is a bit closer to the experimental measurement but still 50% underestimated.

Table 16 summarizes the second hyperpolarizability and their packing ratios for the POM clusters as a function of the stacked unit cells. Unlike the packing ratio for β_{abc}, the packing ratios for γ are seemingly reaching the convergence when N=15, however an extrapolation procedure is somehow needed for obtaining their infinite values. The extrapolated packing ratios of γ are listed in Table 16 as well. The effective γ_{aaaa}, γ_{bbbb}, and γ_{cccc} are evaluated by using Eq. (47) with the extrapolated packing ratios listed in Table 16. Then one can calculated the third-order nonlinear susceptibility as following

$$\chi^{(3)}_{\xi\xi\xi\xi} = \frac{\gamma^{eff}_{\xi\xi\xi\xi}}{\varepsilon_0 V^{cell}} \tag{51}$$

The conversion factor for the second hyperpolarizability from a.u. to S.I is that 1 a.u. of γ= 6.2354×10^{-65} C^4 m^{4}J^{-3}. The third-order nonlinear diagonal susceptibilities are respectively for a, b, and c directions as 378, 951, and 122 pm^2/V^2. The most prominent component for the third-order nonlinear susceptibilities is along b direction.

Table 16. Diagonal components of the static second-order hyperpolarizability tensors and crystal packing rations (R) for POM clusters elongated along three crystal axes as a function of the number (N) of stacked unit cells. The unit for γ is in a.u. and for $\chi^{(3)}$ is pm^2/V^2.

N	γ_{aaaa}	R	γ_{bbbb}	R	γ_{cccc}	R
1	18389	1.000	133199	1.000	9605	1.000
			Along a axis			
2	43821	1.192	244726	0.919	18661	0.971
3	69517	1.260	355912	0.891	27697	0.961
4	95271	1.295	466996	0.877	36728	0.956
5	121047	1.317	578038	0.868	45757	0.953
6	146833	1.331	689058	0.862	54784	0.951
7	172626	1.341	800065	0.858	63811	0.949
8	198421	1.349	911065	0.855	72838	0.948
9	224219	1.355	1022059	0.853	81864	0.947
10	250019	1.360	1133050	0.851	90890	0.946
11	275820	1.364	1244037	0.949	99915	0.946
12	301622	1.367	1355023	0.848	108941	0.945
13	327424	1.370	1466007	0.847	117966	0.945
14	353228	1.372	1576990	0.846	126992	0.944
15	379031	1.374	1687972	0.845	136017	0.944
∞		*1.404*		*0.849*		*0.940*
			Along b axis			
2	47761	1.299	358558	1.346	15034	0.783
3	76876	1.394	618643	1.548	20642	0.716
4	105589	1.435	897992	1.685	26212	0.682
5	134050	1.458	1188388	1.784	31757	0.661
6	162361	1.472	1485480	1.859	37285	0.647
7	190579	1.481	1786857	1.916	42804	0.637
8	218738	1.487	2091097	1.962	48317	0.629
9	246861	1.492	2397319	2.000	53824	0.623
10	274956	1.495	2704958	2.031	59329	0.618
11	303030	1.498	3013637	2.057	64831	0.614
12	331087	1.500	3323099	2.079	70332	0.610
13	359133	1.502	3633160	2.098	75830	0.607
14	387168	1.504	3943690	2.115	81328	0.605
15	415196	1.505	4254592	2.129	86825	0.603
∞		*1.531*		*2.319*		*0.577*
			Along c axis			
2	36349	0.988	173010	0.649	25998	1.353
3	53616	0.972	217252	0.544	44234	1.535
4	70398	0.957	261491	0.491	63529	1.654
5	86936	0.946	305522	0.459	83447	1.738
6	103341	0.937	349382	0.437	103753	1.800
7	119665	0.930	393116	0.422	124314	1.849
8	135939	0.924	436759	0.410	145048	1.888
9	152178	0.919	480334	0.401	165904	1.919
10	168393	0.916	523859	0.393	186849	1.945
11	184589	0.913	567344	0.387	207859	1.967
12	200771	0.910	610798	0.382	228920	1.986
13	216943	0.907	654228	0.378	250019	2.002
14	233107	0.905	697637	0.374	271149	2.016
15	249264	0.904	741030	0.371	292302	2.029
∞		*0.887*		*0.337*		*2.179*
$R^a \times R^b \times R^c$		1.907		0.663		1.182
γ^{eff}	35068		88311		11353	
$\chi^{(3)}$	378		951		122	

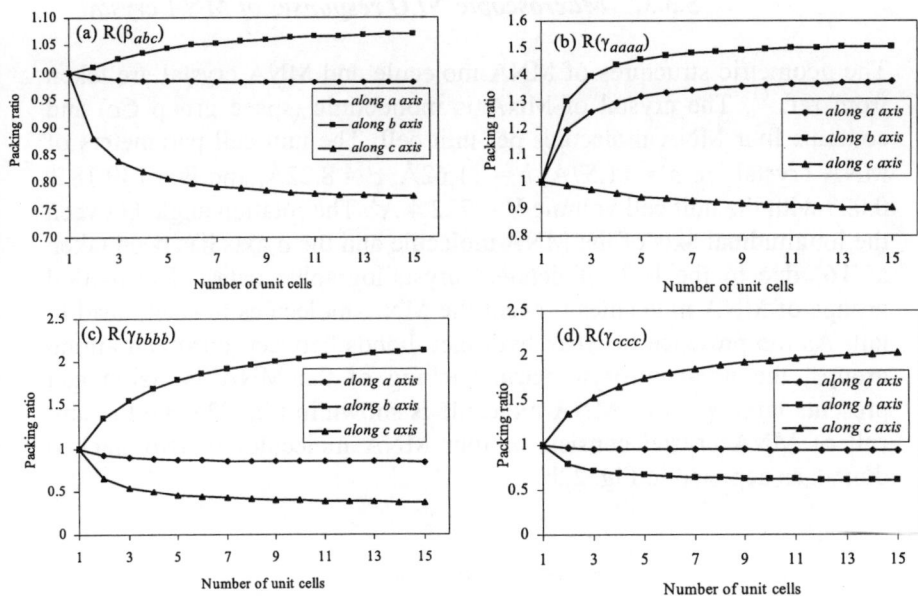

Figure 21. The packing ratios of the first and second hyperpolarizability tensors along three axes.

In Figure 21, the packing ratios of the first and second hyperpolarizaqbilities are plotted against the number of stacked POM units. Unlike to the polarizability and the first hyperpolarizability, the second hyperpolarizability tensors, γ_{bbbb} and γ_{cccc} have strong crystal packing effects with the packing ratios reaching the value over 2.1. The effective γ_{aaaa} , γ_{bbbb} , and γ_{cccc} are obtained as 35, 88, and 11×10^3 a.u., respectively.

From Fig. 21, we find that the crystal packing effects when extending the cluster along the *b* axis are the most prominent. For the first hyperpolarizability tensor component, β_{abc}, see Fig. 21a, packing along the *b* axis leads to an increase by about 8% whereas packing along the *a* and *c* axis is associated with a decrease by 2 and 23%, respectively. For the second hyperpolarizabilities, see Fig. 21b-d, the crystal packing effects are stronger than for the polarizability (Fig. 20) and first hyperpolarizability (Fig. 21a). POM clusters packing along *b* direction leads to a large increase by 120%, whereas it decreases by 64% when POM clusters packing along *c* and decreases 16% along *a* for the dominant γ_{bbbb} component.

5.3.3.   *Macroscopic NLO responses of MNA crystal*

The geometric structures of MNA molecule and MNA crystal are taken from ref. [61]. The crystal of MNA is monoclinic (space group Cc) and contains four MNA molecules per unit cell. The unit cell parameters of MNA crystal are a = 11.57Å, b = 11.62Å, c = 8.22Å, and β = 139.18 ± 0.02° with the unit cell volume V = 722.4 Å^3. The rotation angle between the longitudinal axis of the MNA molecule and the a axis has been taken as 16° due to the lack of detailed crystallographic data.   The methyl groups of MNA molecules prevent the MNA molecules to stack head to tail. As the presence of weak hydrogen bonds between nitro and amino groups, the non-centrosymmetric packing of the MNA is stable and presents large $\chi^{(2)}$. The MNA molecule is shown in Fig. 22a and the unit cell of MNA crystal consists of four MNA molecules or two stacked dimers as depicted in Fig. 22b.

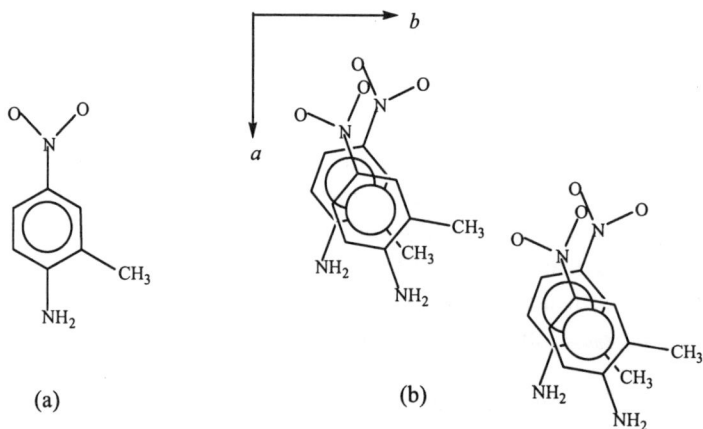

Figure 22. (a) MNA molecule and (b) the unit cell structures of MNA crystal.

AM1 static polarizability tensor α_{aa} and the crystal packing ratios along three crystal axes are calculated by the elongation-FF method and summarized in Table 17. The MNA clusters are built along a and b axes by stepwise adding MNA dimers along these two directions, respectively. For MNA clusters along b axis, they are built up by adding two dimers (a unit cell) for each elongation step. From Table 17, one can see that the packing ratios along three crystal axes are nearly convergent when MNA cluster containing 15 dimers, e.g. 30 MNA molecules in the MNA cluster.

Table 17. AM1 static α tensors and crystal packing ratios for MNA clusters elongated along three crystal axes as a function of the number (N) of stacked dimers.

N	Along a axis		Along b axis		Along c axis	
	α_{aa}	R	α_{aa}	R	α_{aa}	R
1	229	1.000	229	1.000	229	1.000
2	563	1.229	450	0.983	420	0.917
3	934	1.360			608	0.885
4	1322	1.443	868	0.948	795	0.868
5	1718	1.500			982	0.858
6	2119	1.542	1282	0.933	1169	0.851
7	2521	1.573			1355	0.845
8	2925	1.597	1694	0.925	1542	0.842
9	3330	1.616			1729	0.839
10	3736	1.631	2107	0.920	1915	0.836
11	4142	1.644			2102	0.834
12	4548	1.655	2518	0.916	2288	0.833
13	4954	1.664			2475	0.831
14	5361	1.672	2930	0.914	2661	0.830
15	5768	1.679			2847	0.829

Similar to the treatment of POM crystal, the effective polarizability tensor is calculated by using Eq. 47 together with the packing ratios listed in Table 17. The effective α_{aa} is thus obtained as 291 a.u. The linear susceptibility is then deduced from the effective polarizability as 0.75 with the unit cell volume of 722.4 Å^3 for the MNA crystal. The refractive index along a axis is then given (Eq. 49) as 1.32. This can be compared to the experimental measured refractive index of 2.0±0.1,[62] and the value of 1.68 deduced by Morita *et al.* [63] from fitting the Sellmeir expression. The reasons for this discrepancy are due to the lack of the electron correlation and no frequency dispersion effect taken into account.

The AM1 static first hyperpolarizability tensor component β_{aaa} and the packing ratios against the stacked MNA molecules are reported in Table 18. The elongation-FF calculation shows its advantage since the convergent of the packing ratio along a axis is very slow. Thus, more units are needed to reach the R convergence against the stacked dimers. In our elongation-FF calculations, MNA clusters are built till $N=15$

dimers, totally containing 30 MNA molecules in the cluster. The packing effect along a axis is very strong, the ratio value is more than 5.3 with $N=15$. In Fig. 23, we plot the packing ratio with respect to the number of stacked dimers for MNA clusters built up along axis a. It can be seen that the R values are still not reached the convergence but from this point, an extrapolation to infinity is more reliable than that at $N=8$ point which was done in the conventional calculations. The extrapolated packing ratios are also listed in Table 18. From these values, one can evaluate the effective β_{aaa} and finally the second-order susceptibility of MNA crystal. The calculated effective β_{aaa} is equal to 3736 a.u. and yields a value of 37.5 pm/V for $\chi_{aaa}^{(2)}$. Comparing to the value of 32.6 pm/V obtained in ref. [64], we find that the packing ratio of β_{aaa} along a axis is underestimated in their work due to fewer MNA stacked molecules were included in their calculations and thus the extrapolation to infinity is not so reliable. The extrapolated R value in their work was 5.1 as eight dimers were considered in their work.

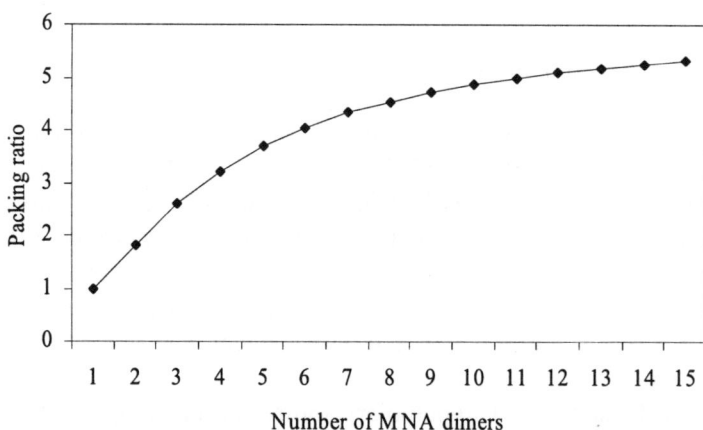

Figure 23. Packing ratio of β_{aaa} along a axis versus the number of stacked MNA dimers.

Table 18. AM1 static β tensors and crystal packing ratios for MNA clusters elongated along three crystal axes as a function of the number (N) of stacked dimers. The unit for β is in a.u. and for $\chi^{(2)}$ is in pm/V.

	Along a axis		Along b axis		Along c axis	
N	β_{aaa}	R	β_{aaa}	R	β_{aaa}	R
1	1730	1.000	1730	1.000	1730	1.000
2	6304	1.822	3220	0.931	2571	0.743
3	13544	2.610			3412	0.657
4	22323	3.226	5632	0.814	4248	0.614
5	31945	3.693			5080	0.587
6	42032	4.049	7982	0.769	5910	0.569
7	52387	4.326			6739	0.556
8	62902	4.545	10311	0.745	7566	0.547
9	73519	4.722			8393	0.539
10	84204	4.867	12632	0.730	9220	0.533
11	94934	4.989			10046	0.528
12	105698	5.091	14948	0.720	10872	0.524
13	116485	5.179			11697	0.520
14	127291	5.256	17262	0.713	12523	0.517
15	138111	5.322			13348	0.514
∞		*6.100*		*0.701*		*0.505*
β^{eff}	3736					
$\chi^{(2)}$	37.5					

The theoretical $\chi^{(2)}_{aaa}$ (37.5 pm/V) of MNA underestimates the experimental value of 300 pm/V by one order of magnitude because two main corrections, the electron correlation and the frequency dispersion, are not included in our calculations. For the missing electron correlation, one can estimate it by multiplying a ratio factor obtained from HF and MP2 hyperpolarizability calculations. The frequency dispersion effect can then be roughly estimated by the ratio factor between SHG and static β. As estimated in ref. 49, combining these two corrections can bring a factor of four to the theoretical $\chi^{(2)}_{aaa}$ value. This leads to a roughly estimated value of $\chi^{(2)}_{aaa}$ as 150 which is still 50% underestimated with respect to the experimental measurement.

For the second hyperpolarizability, one can see from Table 19 that the packing effect is much stronger along a axis, while along the other two axes are almost the same as those of the first hyperpolarizability. The packing ratio for γ_{aaaa} along a axis is reaching

10 when the MNA cluster containing 15 dimers. The extrapolation is needed for the infinity packing ratios and the extrapolated R values are also listed in the table. The effective γ_{aaaa} is given as 318511 a.u. It is more than 3 times larger than that of one MNA molecule due to strong packing effect along a direction. From Eq. 51, one can obtain the third-order nonlinear susceptibility as 3105 pm^2/V^2.

Table 19. AM1 static γ tensors and crystal packing ratios for MNA clusters elongated along three crystal axes as a function of the number (N) of stacked dimers. The unit for γ is in a.u. and for $\chi^{(3)}$ is in pm^2/V^2.

N	Along a axis		Along b axis		Along c axis	
	γ_{aaaa}	R	γ_{aaaa}	R	γ_{aaaa}	R
1	102580	1.000	102580	1.000	102580	1.000
2	550128	2.681	187916	0.916	145330	0.708
3	1267457	4.119			189015	0.614
4	2177854	5.308	320026	0.780	232627	0.567
5	3206631	6.252			275759	0.538
6	4305592	6.996	448868	0.729	319298	0.519
7	5446549	7.585			362343	0.505
8	6612157	8.057	575499	0.701	404951	0.493
9	7793598	8.442			447792	0.485
10	8985866	8.760	702741	0.685	490574	0.478
11	10185409	9.027			533648	0.473
12	11389726	9.253	829692	0.674	575790	0.468
13	12597888	9.447			617933	0.463
14	13809189	9.616	957400	0.667	660773	0.460
15	15021418	9.762			702683	0.457
∞		11.500		0.600		0.450
γ^{eff}	318511					
$\chi^{(3)}$	3105					

5.3.4. *Ab initio elongation-FF results*

Table 20. RHF/STO-3G Diagonal components of the static (hyper)polarizability tensors and crystal packing ratios (R) for POM clusters elongated along *a* axis as a function of the number (*N*) of stacked unit cells. Cutoff starts from *N*=4.

N	α_{aa}	R	β_{abc}	R	γ_{aaaa}	R
1	212	1.000	-384	1.000	3276	1.000
2	431	1.017	-762	0.992	9546	1.457
3	649	1.020	-1141	0.990	15224	1.549
4	868	1.024	-1519	0.989	20942	1.598
5	1087	1.025	-1897	0.988	26638	1.626
6	1306	1.027	-2275	0.987	32340	1.645
7	1525	1.028	-2653	0.987	38043	1.659
∞		*1.031*		*0.985*		*1.688*
∞(AM1)		*1.043*		*0.981*		*1.404*

The elongation-FF method is extended to the *ab initio* level. For the *ab initio* case due to huge computational costs, one has to use the cut-off technique designed for the elongation method, which is described in section 4. The elongation method is to reduce the dimension of the working space by using the LMO basis. Besides this, as the frozen region is far away from the interactive region and thus the interaction is very small, this frozen part can be cut off by the cutoff technique built in the elongation program package. In such a way, we substantially reduce the number of two-electron integrals so that very large systems can be treated by the elongation method, whereas for the conventional method it is burdensome or even impossible.

The POM clusters are built along *a* axis. The starting cluster contains two POM unit cells (eight POM molecules) with one in frozen and the other one in active region. The HF/STO-3G elongation calculations were performed. The POM clusters extended along *a* axis till seven unit cells. By checking the interactions between the frozen unit and the attacking unit, we can start the cutoff procedure after *N*=3. That means from this point, the elongation calculations can be carried out by discarding the most remote frozen unit. The total energy, however, is evaluated by adding the contributions from the cutoff unit after the elongation SCF is converged. By this cutoff technique, the POM cluster containing four units is actually carried out by only considering three unit cells. For the other POM clusters with *N*>4, the next remote unit is cut off as similar to the case for *N*=4. So the elongation calculations for the POM clusters with $N{\geq}4$, the working dimension contains always three POM unit cells since one more unit is cut off as the cluster elongated one more step.

Table 20 displays the (hyper)polarizability results of the POM clusters up to seven units obtained at the HF/STO-3G level by the elongation-FF method. The cutoff starts from the cluster containing four unit cells. One can see from Table 20 that the packing ratios are nearly convergent for α_{aa} and β_{abc}, while for γ_{aaaa} the convergence is slow and thus an extrapolation is needed for infinity limit. Comparing the AM1 and HF/STO-3G results (see Table 20 together with Figs. 20a, 21a and 21b), one can find that the packing ratios for α_{aa} and β_{abc} are almost the same determined from these two levels. For POM cluster with $N{=}7$, the AM1 (HF/STO-3G) packing ratios for α_{aa} and β_{abc} along a axis are 1.04 (1.03) and 0.98 (0.99), respectively, whereas the packing ratio obtained at the AM1 level is 1.40. It is 17% lower than that (1.69) obtained at the HF/STO-3G level. As the minimum basis set is too small to determine (hyper)polarizabilities, *ab initio* elongation calculations at a large basis set are highly demanded.

6. Summary and future prospects

In this chapter, the elongation method for determining electronic structures of large periodic and aperiodic systems has been described. Unlike to the conventional treatment, the elongation method is working on the localized molecular orbital basis instead of canonical molecular basis. Thus, the localization quality is an essential for the elongation method. Three localization schemes have been developed for the elongation method. The newly proposed localization scheme based on the density matrix method is more efficient and reliable even for strong delocalized systems, such as cationic cyanine chains. The elongation SCF procedure is performed only in the interactive region, not the whole space. This is the one source of the time saving. Another CPU time saving is gained by introducing a so-called cutoff technique. For further time savings the cutoff 2e-integrals can be approximated by a fast multipole expansion.

We have applied the elongation method to determine the NLO properties of polymers and molecule crystals by means of finite-field scheme. The validity of the elongation-FF method has been well tested by comparing the elongation (hyper)polarizabilities with the conventional ones. It is found that the elongation-FF can produce NLO results in good agreement with the conventional ones. Since the elongation method reduces the working space of the underlying system, it allows us to investigate large systems, where by the conventional method they are not easy to deal with.

As the usefulness and the reliability of the elongation-FF method are demonstrated both at the semi-empirical and *ab initio* levels, we plan to implement the CPHF or TDHF procedures into the elongation method to determine the frequency-dependent (hyper)polarizabilities. If this is done, the frequency dispersion effect can be directly taken into account and compared to the experiments. To account for the electron correlation effects, the MP2 scheme with TDHF method is currently under investigation in our group. Alternatively, the elongation DFT method as proposed in ref. 14 is considered for accounting for the electron correlations.

Acknowledgments

The authors greatly acknowledged the financial support from the Research and Development Applying Advanced Computational Science and Technology of the Japan Science and Technology Agency (ACT-JST), a Grant-in-Aid for Scientific Research from the Ministry of Education, Culture, Sports, Science, and Technology (MEXT) of Japan, the Alexander von Humboldt Foundation, the Fond des Deutsches Krebsforschungszentrum, and Group, PRESTO, Research Development Corporation of Japan. We thank our co-authors for the elongation method and its application to many different areas. The elongation method has been benefited from long-standing collaborations with Professors Sándor Suhai, János Ladik, and Peter Otto, also from fruitful collaborations with Professors David. M. Bishop, Benoît Champagne, and Bernard Kirtman both on the developing the elongation method and its application to the field of nonlinear optics. We also want to thank Dr. Kazunari Naka for his continuous help in the computer technique support.

REFERENCES

1. (a) A. Szabo and N. S. Ostlund, Modern Quantum Chemistry (McGraw-Hill, New York, 1989); (b) F. Jensen, Introduction to Computational Chemitry (John Wiley & Sons, Chichester, 2003); (c) R. McWeeny and B. T. Sutcliffe, Methods of Molecular Quantum Mechanics, 2nd ed. (Academic, New York, 1976).
2. W. J. Hehre, L. Radom, P. V. R. Schleyer, and J. A. Pople, Ab initio Molecular Orbital Theory (John Wiley & Sons, Chichester, 1985).
3. Gaussian 03, Revision A.11.4, M. J. Frisch, et al, Gaussian, Inc., Wallingford CT, 2004.
4. GAMESS/Version 14, Jan. 2003 (R2) from Iowa State University, M.W. Schmidt, K.K. Baldridge, J.A. Boatz, S.T. Elbert, M.S. Gordon, J.H. Jensen, S. Koseki, N. Matsunaga, K.A. Nguyen, S.J. Su, T.L. Windus, together with M. Dupuis, J.A. Montgomery, J. Comput. Chem. 14, 1347 (1993).
5. CRYSTAL03, V.R. Saunders, R. Dovesi, C. Roetti, R. Orlando, C.M. Zicovich-Wilson, N.M. Harrison, K. Doll, B. Civalleri, I.J. Bush, Ph. D'Arco, M. Llunell.
6. T. A. Koopman, Physica, 1,104 (1933).
7. (a) J. Ladik, Acta Phys. Hung. 18,173 and 185 (1965); (b) J. Ladik, Quantum Chemistry of Polymers as Solids (New York: Plenum, 1987); (c) J. M. André, L. Gouverneur, and G. Leroy, Int. J. Quantum Chem. 1, 427 and 451 (1967); (d) J. M. André, D. H. Mosley, B. Champagne, J. Delhalle, J. G. Fripiat, J. L. Brédas, D. J. Vanderveken, and D. P. Vercauteren, in Methods and Techniques in Computational Chemistry: METECC-94, edited by E. Clementi (STEF, Cagliari, 1993), Vol. B, Chap. 10, p. 423.
8. C. Pisani, R. Dovesi, and C. Roetti, Lecture Notes in Chemistry, Vol. 48, Springer Verlag, Heidelberg (1988).
9. M. Born and T. von Kàrmàn, Z. Phys. 13, 297 (1912); Z. Phys. 14,15 (1913).
10. P. Otto, Integral and HF CO Program Package, Institute for Theoretical Chemistry, Friedrich-Alexander University, Erlangen.
11. (a) J. G. Fripiat, B. Champagne, F. Castet, J. M. André, T. D. Poulsen, D. H. Mosley, J. L. Brédas, V. Bodart, J. Delhalle, and D. P. Vercauteren, LCAO AB Initio Band Structure Calculations for Polymers (PLH2001); (b) D. Jacquemin, B. Champagne, J. M. André, E. Deumens, and Y. Öhrn, J. Comput. Chem. 23,1430 (2002).
12. A. Imamura, Y. Aoki, and K. Maekawa, J. Chem. Phys. 95, 5419 (1991).
13. (a) Maekawa and A. Imamura, Int. J. Quantum Chem. 47, 449 (1993); (b) Kurihara, Y. Aoki, and A. Imamura, J. Chem. Phys. 107, 3569 (1997).
14. (a) Y. Aoki, S. Suhai, and A. Imamura, J. Chem. Phys. 101, 10808 (1994); (b) Y. Aoki, S. Suhai, and A.

Imamura, Int. J. Quantum Chem. 52, 267 (1994); (c) M. Mitani, Y. Aoki, and A. Imamura, Int. J. Quantum Chem. 54, 167 (1995).

15. M. Mitani and A. Imamura, J. Chem. Phys. 103, 663 (1995).
16. F. L. Gu, Y. Aoki, J. Korchowiec, A. Imamura, and B. Kirtman, J. Chem. Phys. 121, 10385 (2004).
17. (a) M. Mitani and A. Imamura, J. Chem. Phys. 98, 7086 (1993); (b) M. Mitani, Y. Aoki, and A. Imamura, J. Chem. Phys. 100, 2346 (1994).
18. (a) A. T. Amos and G. G. Hall, Proc. R. Soc. London, Ser. A263, 483 (1961); (b) K. Fukui, N. Koga, and H. Fujimoto, J. Am. Chem. Soc. 103, 196 (1981); (c) H. Fujimoto, N. Koga, and K. Fukui, J. Am. Chem. Soc. 103, 7452 (1981).
19. (a) Handbook of Advanced Electronic and Photonic Materials, edited by H. S. Nalwa (Academic, San Diego, 2000), Vol. 9; (b) Organic Materials for Nonlinear Optics, edited by D. S. Chemla and J. Zyss (Academic, New York, 1987), Vols. I and II; (c) P. N. Prasad and D. J. Williams, Introduction to Nonlinear Optical Effects in Molecules and Polymers (Wiley, New York, 1991); (d) C. E. Dykstra, S.-Y. Liu, and D. J. Malik, Adv. Chem. Phys. 75, 37 (1989); (e) A. A. Hasanein, Adv. Chem. Phys. 85, 415 (1993); (f) Special Issue, Chem. Rev. 94, 1 (1994); (g) Special Issue, J. Phys. Chem. 104, 4671 (2000); (h) D. M. Bishop, Adv. Quantum Chem. 25, 1 (1994); (i) Nonlinear Optical Materials, ACS Symposium Series 628, edited by S. P.Karna and A. T. Yeates (American Chemical Society, Washington, D.C.,1996); (j) B. Kirtman and B. Champagne, Int. J. Rev. Phys. Chem. 16, 389 (1997).
20. B. Kirtman, F. L. Gu, and D. M. Bishop, J. Chem. Phys. 113, 1294 (2000).
21. (a) D. M. Bishop, F. L. Gu, and B. Kirtman, J. Chem. Phys. 114, 7633 (2001) ; (b) B. Kirtman, B. Champagne, F. L. Gu, and D. M. Bishop, Int. J. Quantum Chem. (Special Issue for Löwdin), 90, 709 (2002); (c) B. Champagne, D. Jacquemin, F. L. Gu, Y. Aoki, B. Kirtman, and D. M. Bishop, Chem. Phys. Lett. 373, 539 (2003) .
22. (a) F. L. Gu, D. M. Bishop, and B. Kirtman, J. Chem. Phys. 115, 10548 (2001) ; (b) F. L. Gu, Y. Aoki, and D. M. Bishop, J. Chem. Phys. 117, 385 (2002).
23. P. Otto, A. Martinez, A. Czaja, and J. Ladik, J. Chem. Phys. 117, 1908 (2002).
24. See for example, B. Champagne and B. Kirtman, in Handbook of Advanced Electronic and Photonic Materials, edited by H. S. Nalwa (Academic, San Diego, 2000), Vol. 9, Chapt. 2, P63.
25. D. M. Bishop and F. L. Gu, Chem. Phys. Lett. 317, 322 (2000).
26. (a) B. Champagne, D. Jacquemin, J. M. André, and B. Kirtman, J. Phys. Chem. A101, 3158 (1997); (b) E. J. Weniger and B. Kirtman, in T. E. Simas, G. Avdelas, and J. Vigo-Aguiar special Issue "Numerical Mwthods in Physics, Chemistry and Engineering" Computers and Mathematics with Applications, 45, 189 (2003).
27. (a) L. Greengard, The rapid evaluation of potential fields in particle systems. The MIT Press, 1987; (b) H. G. Petersen, D. Soelvason, J. W. Perram, and E.R. Smith, J. Chem. Phys. 101, 8870 (1994).
28. J. M. Millan and G. E. Scuseria, J. Chem. Phys. 105, 2726 (1996).
29. (a) H.-J. Werner, F. R. Manby, and P. J. Knowles, J. Chem. Phys. 118, 8149 (2003); (b) P. Pulay, Chem. Phys. Lett. 100, 151 (1983); (c) M. Schütz, G. Hetzer, and H.-J. Werner, J. Chem. Phys. 111, 5691 (1999); (d) G. Hetzer, M. Schütz, H. Stoll, and H.-J. Werner, J. Chem. Phys. 113, 9443 (2000); (e) C. Hampel and H.-J. Werner, J. Chem. Phys. 104, 6286 (1996); (f) M. Schütz and H.-J. Werner, J. Chem. Phys. 114, 661 (2001); (g) M. Schütz, Phys. Chem. Chem. Phys. 4, 3941 (2002).
30. (a) B. Kirtman, Topics in Current Chemistry, Localization and Correlation –A Tribute to Ede Kapuy, P. Surjan, ed., Springer-Verlag, Vol. 203, p. 147; (b) B. Kirtman, J. Chem. Phys. 86, 1059 (1982); (c) B. Kirtman and C. de Melo, J. Chem. Phys. 75, 4592, (1981).
31. F. L. Gu, Y. Aoki, A. Imamura, D. M. Bishop, and B. Kirtman, Mol. Phys. 101, 1487 (2003).

32. S. Ohnishi, F. L. Gu, K. Naka, A. Imamura, B. Kirtman, and Y. Aoki, J. Phys. Chem. A108, 8478 (2004).
33. (a) S. Boomadevi, H. P. Mittal, and R. Dhansekaran, J. Cryst. Growth, 261, 55 (2004); (b) R. Glaser and G. S. Chen, Chem. Mater. 9, 28 (1997).
34. C. Edmiston and K. Rüdenberg, Rev. Mod. Phys. 35, 457 (1963).
35. G. Del Re, Theor. Chim. Acta (Berl.), 1, 188 (1963).
36. P-O Löwdin, J. Chem. Phys. 18, 365 (1949).
37. K. Maekawa and A. Imamura, J. Chem. Phys. 98, 534 (1992).
38. A. E. Reed and F. Weinhold, J. Chem. Phys. 83, 1736 (1985).
39. W. J. Hehre, R. F. Stewart, and J. A. Pople, J. Chem. Phys. 51, 2657 (1967).
40. (a) R. Ditchfield, W. J. Hehre, and J. A. Pople, J. Chem. Phys. 54, 724 (1971); (b) W. J. Hehre, R. Ditchfield, and J. A. Pople, J. Chem. Phys. 56, 2257 (1972).
41. M. Causà, R. Dovesi, C. Pisani, and C. Roetti, Phys. Rev. B33, 1308 (1986).
42. (a) W. Yang, Phys. Rev. Lett. 66, 447 (1991); (b) W. Yang, Phys. Rev. A44, 7823 (1991); (c) W. Yang and T.-S. Lee, J. Chem. Phys. 103, 5674 (1995); (d) K. Kitaura, T. Sawai, T. Asada, T. Nakano, and M. Uebayasi, Chem. Phys. Lett. 312, 319 (1999); (e) K. Kitaura, E. Ikeo, T. Asada, T. Nakano, and M. Uebayasi, Chem. Phys. Lett. 313, 701, (1999); (f) D. G. Fedorov, R.M. Olson, K. Kitaura, M. S. Gordon, and S. Koseki, J. Comput. Chem. 25, 872(2004); (g) D. G. Fedorov and K. Kitaura, J. Chem. Phys. 120, 6832 (2004); (h) Y. Komeiji, Y. Inadomi, and T. Nakano, Comput. Biol. & Chem. 28 (2004) 155; (i) J. J. P. Stewart, Int. J. Quantum Chem. 58, 133 (1996); (j) S. Yokojima, X. J. Wang, D.H. Zhou, and G. H. Chen, J. Chem. Phys. 111, 10444 (1999); (k) K. Babu and S. R. Gadre, J. Comput. Chem. 24, 484 (2003).
43. (a) A.D. Daniels, J. M. Millam, and G. E. Scuseria, J. Chem. Phys. 107, 425 (1997); (b) M. Challacombe, J. Chem. Phys. 110, 2332 (1999); (c) M.Challacombe and E. Schwegler, J. Chem. Phys. 104, 4685 (1996); M. Challacombe and E. Schwegler, J. Chem. Phys. 106, 5528 (1997); (d) E. Schwegler and M. Challacombe, J. Chem. Phys. 111, 6223 (1999); (e) D.G. Hanhere, A. Dhavele, E. V. Ludeña, and V. Karasiev, Phys. Rev. A 62 , 065201 (2000); (f) B. Saravanan, Y. Shao, R. Baer, P. N. Ross, and M. Head-Gordon, J. Comput. Chem. 24, 618 (2003); (g) C. H. Choi, K. Ruedenberg, and M. S. Gordon, J. Comput. Chem. 22, 1481 (2001); (h) X. P. Li, R. W. Nunes, and D. Vanderbilt, Phys. Rev. B 47, 10891 (1993); (i) M. S. Daw, Phys. Rev. B 47, 10985 (1993); (j) P. Ordejón, D. A. Drabold, M. P. Grumbach, and R. M. Martin, Phys. Rev. B 48, 14646 (1993); (k) P. Ordejón, E. Artach, and J. M. Soler, Phys. Rev. B 53, 10441 (1996); (l) S. Goedecker and L. Colombo, Phys. Rev. Lett. 73, 122 (1994); (m) F. Mauri, G. Galli, and R. Car, Phys. Rev. B 47, 9973 (1993); (n) G. Galli, Phys. Stat. Sol. b217, 231 (2000).
44. W. Kohn, Phys. Rev. Lett. 76, 3168 (1996).
45. (a) J. Korchowiec, F. L. Gu, A. Imamura, B. Kirtman, and Y. Aoki, Int. J. Quantum Chem. 102, 785 (2005); (b) J. Korchowiec, F. L. Gu, and Y. Aoki, ICCMSE2004- conference proceedings-submitted.
46. G. Chaban, M. W. Schmidt, and M.S. Gordon, Theor. Chem. Acc. 97, 88 (1997).
47. F. L. Gu, P. Otto, and J. Ladik, J. molec. Model. 3, 182 (1997).
48. E. A. Perpéte and B. Champagne, J. molec. Struct.: THEOCHEM, 487, 39 (1999).
49. (a) G. A. Petersona, A. Bennett, T. G. Tensfeldt, M. A. Al-Laham, W. A. Shirley and J. Mantzaris, J. Chem. Phys. 89, 2193 (1988); (b) G. A. Petersson and M. A. Al-Laham, J. Chem. Phys. 94, 6081 (1991);
50. MOPAC2000, © Fujitsu Limited, 1999; J.J.P. Stewart, Quantum Chemistry Program 7 Exchange, 455.
51. (a) L. Esaki and R. Tsu, IBM Internal Research Report RC 2418; Mar. 26, 1969; (b) L. Esaki, and R. Tsu, IBM J. Res. DeV. 1970, 14, 61.
52. (a) H. M. Gibbs, Optical Bistability: Controlling Light with Light; Academic Press, New York, 1985; (b) D. A. B. Miller, D. S. Chemla, D. J. Eilenberger, P. W. Smith, A. C. Gossard, and W. Wiegmann, Appl. Phys. Lett.

42, 925 (1983); (c) R. Jin, C. L. Chuang, H. M. Gibbs, S. W. Koch, J. N. Polky, and G. A. Pubanz, Appl. Phys. Lett. 53, 1791 (1988).

53. B. R. Nag, Physics of Quantum Well Devices; Kluwer Academic, Boston, 2000.

54. Y. Aoki and A. Imamura, J. Chem. Phys. 97, 8432 (1992).

55. M.J.S. Dewar, E.G. Zoebisch, E.F. Healy, and J.J.P. Stewart, J. Am. Chem. Soc. 107, 3902 (1985).

56. (a) F. Castet and B. Champagne, J. Phys. Chem. A105, 1366 (2001); (b) M. Guillaume, E. Botek, B. Champagne, F. Castet, and L. Ducasse, Int. J. Quantum Chem. 90, 1378 (2002); (c) F. L. Gu, B. Champagne, and Y. Aoki, Lecture Series on Computer and Computational Sciences, 1, 779 (2004); (d) F. Castet, L. Ducasse, M. Guillaume, E. Botek, and B. Champagne, Lecture Series on Computer and Computational Sciences, 1, 783 (2004).

57. (a) S. Di Bella, M.A. Ratner, and T.J. Marks, J. Am. Chem. Soc. 114, 5842 (1992); (b) B. Champagne, E. A. Perpète, T. Legrand, D. Jacquemin, and J.-M. André, J. Chem. Soc. Fraday Trans. 94, 1547 (1998); (c) X. L. Zhu, X. Z. You, Y. Zhong, Z. Yu, and S. L. Guo, Chem. Phys. 253, 241 (2000); (d) X. L. Zhu, X. Z. You, and Y. Zhong, Chem. Phys. 253,287 (2000).

58. F. Baert, P. Schweiss, G. Heger, and M. More, J. Molec. Struct. 178, 29 (1988).

59. M. Guillaume, E. Botek, B. Champagne, F. Castet, and L. Ducasse, J.Chem. Phys. 121, 7390 (2004).

60. J. Zyss, D. S. Chemla, and J. F. Nicoud, J. Chem. Phys. 74, 4800 (1981).

61. G. F. Lipscomb, A. F. Garito, and R. S. Narang, J. Chem. Phys. 75, 1509 (1981).

62. B. F. Levine, C. G. Bethea, C. D. Thurmond, R. T. Lynch, and J. L. Bernsteain, J. Appl. Phys. 50, 2523 (1979).

63. R. Morita, N. Ogasawara, S. Umagaki, and R. Ito, Jpn. J. Appl. Phys. 26, 1711 (1986).

64. F. Castet and B. Champagne, J. Chem. Phys. 105, 1366 (2001).

CHAPTER 6

RESPONSES OF MOLECULAR VIBRATIONS TO INTERMOLECULAR ELECTROSTATIC INTERACTIONS AND THEIR EFFECTS ON VIBRATIONAL SPECTROSCOPIC FEATURES

Hajime Torii

Department of Chemistry, School of Education, Shizuoka University
836 Ohya, Shizuoka 422-8529, Japan
E-mail: torii@ed.shizuoka.ac.jp

Various types of the responses of molecular vibrations to inter-molecular electrostatic interactions are discussed. Based on the expansion formula of the potential energy with respect to the electric field and vibrational coordinates, it is shown that electrical property derivatives (dipole derivatives, *etc.*) are basic quantities for describing those responses. It is also shown that different kinds of responses are derived from the same expansion formula depending on the nature of the electric field operating on molecules. Electric field-induced structural displacements, vibrational frequency shifts induced by those structural displacements, modulation of intermolecular electric fields and polarization properties due to liquid dynamics, vibrational polarizability and hyperpolarizabilities, and transition dipole coupling are discussed from this viewpoint. Theoretical methods for analyzing the nature of the vibrational modes involved in these phenomena are presented. Their effects on the vibrational spectroscopic features are discussed in detail with some typical examples.

1. Introduction

Molecular vibrations in condensed phases are affected by intermolecular interactions. Accordingly, we can see various effects of intermolecular interactions on vibrational spectroscopic features, i.e., vibrational frequency shifts, vibrational transition intensity modulations, and vibrational band shapes. The meaning of this is twofold. On the one hand, we can obtain information on the intermolecular interactions from the analyses of the vibrational band profiles, but on the other hand, to interpret the vibrational band profiles correctly, we should know the precise mechanisms that the intermolecular interactions affect the vibrations.

It has long been recognized that, in many cases, electrostatic interactions play an important role in the effects on molecular vibrations. For example, the sensitivity of the amide I band profiles of polypeptides and proteins to their secondary structures[1–6] has been explained by the transition dipole coupling (TDC) mechanism,[7–14] which is a mechanism of electrostatic coupling between the vibrations of different peptide groups. Recently, the effects of electrostatic interactions between peptide groups and solvent water molecules on the vibrational frequencies of individual peptide groups have also been studied extensively.[15–22] It has been known that there are similar effects on the vibrations of dipolar and hydrogen-bonding molecules in liquids. We can observe those effects as the noncoincidence effect (NCE) of the polarized Raman spectra in the frequency domain,[23–45] and the ultrafast decay of transient absorption anisotropy in the time domain.[46,47]

In all of the above examples, important quantities are the derivatives of the electrical properties (dipoles, polarizabilities, *etc.*) with respect to the molecular vibrational coordinates, which are generally called *electrical property derivatives*.[48] It is well known that the infrared (IR) and Raman intensities are proportional to the square of the dipole derivatives and polarizability derivatives, respectively. However, these derivative quantities are also important in intermolecular interactions, because they describe the responses of molecular vibrations to external electric field in general, not limited to the electric field of radiation. For example, dipole derivatives appear not only in the formula for the IR intensities but also in those for the electric field-induced changes in the molecular structures,[49–51] the polarizabilities of molecules related to those structural changes (called *vibrational polarizabilities*),[52–54] and the TDC.[7–14,25–27]

In the present chapter, we first discuss how various formulas containing electrical property derivatives are related to each other. Those relations between formulas will help understand various forms of the responses of molecular vibrations to intermolecular electrostatic interactions in a consistent way. We then show, with typical examples, some explicit forms of the responses of molecular vibrations to intermolecular electrostatic interactions. We discuss in detail what kind of molecular motions are likely to be involved in those responses, as well as how significant are their effects on the vibrational spectroscopic features.

2. Basic Formulas

Since dipole moment and polarizability are expressed as the first and second derivatives of potential energy with respect to electric field E ($\mu_j = -\partial V/\partial E_j$ and $\alpha_{ij} = -\partial^2 V/\partial E_i \partial E_j$, where i and j (=1, 2, 3) stand for the x, y and z directions), we have dipole derivatives and polarizability derivatives (as well as higher-order terms) in the expansion of the potential energy with respect to the electric field and vibrational coordinates. The expansion formula is given as

$$V = V_0 + \frac{1}{2}\sum_p k_p q_p^2 + \dots$$

$$- \sum_j \left[\mu_j + \sum_p \frac{\partial \mu_j}{\partial q_p} q_p + \frac{1}{2}\sum_{p,r} \frac{\partial^2 \mu_j}{\partial q_p \partial q_r} q_p q_r + \dots \right] E_j$$

$$- \frac{1}{2}\sum_{i,j} \left[\alpha_{ij} + \sum_p \frac{\partial \alpha_{ij}}{\partial q_p} q_p + \frac{1}{2}\sum_{p,r} \frac{\partial^2 \alpha_{ij}}{\partial q_p \partial q_r} q_p q_r + \dots \right] E_i E_j + \dots \quad (1)$$

where q_p is the pth (mass-weighted) normal mode of a given molecule and k_p is its quadratic force constant. As noted in Sec. 1, E_j may be the electric field of radiation or the external electric field operating on a given molecule due to intermolecular interactions. Accordingly, we have the following three cases depending on the type of electric field taken as E_j.[48]

Case 1: When E_j is the electric field of radiation, the term containing $\partial\mu_j/\partial q_p$ gives rise to absorption or emission of light accompanied by a vibrational transition by one quantum, and the term containing $\partial\alpha_{ij}/\partial q_p$ gives rise to Raman scattering. The formulas for these optical processes are well known, so we do not give them explicitly here.

Case 2: When E_j is static or slowly varying as compared with the time scale of q_p, the position of the potential energy minimum is displaced from the original one at $q_p = 0$.[49-51] To the first order in E_j and neglecting higher-order derivatives of dipole moment, the displacement is expressed as

$$\delta q_p = \frac{1}{k_p}\sum_j \frac{\partial\mu_j}{\partial q_p}E_j \qquad (2)$$

This displacement δq_p gives rise to a shift of the quadratic force constant δk_p when the vibrational mode q_p is (mechanically) anharmonic. Neglecting the cross terms of the mechanical anharmonicity, we have

$$\delta k_p = f_p\,\delta q_p$$
$$= \frac{f_p}{k_p}\sum_j \frac{\partial\mu_j}{\partial q_p}E_j \qquad (3)$$

where f_p is the cubic force constant of the mode q_p. Because of this effect, the vibrational frequencies of the modes with large cubic (relative to quadratic) force constants and large dipole derivatives are likely to be perturbed significantly by the intermolecular electrostatic interactions. This is one of the main factors of the phenomenon called the vibrational Stark effect.[50,55-58] An additional effect of the electric field on δk_p arises from the terms containing the dipole second derivatives (called electrical anharmonicity).[11,59] Neglecting the cross terms ($\partial^2\mu_j/\partial q_p\partial q_r$ with $p \neq r$) and combining with Eq. 3, we have

$$\delta k_p = \sum_j \left(\frac{f_p}{k_p}\frac{\partial\mu_j}{\partial q_p} - \frac{\partial^2\mu_j}{\partial q_p^2} \right)E_j \qquad (4)$$

In the example shown in Sec. 3.1 (the amide I mode of N-methylacetamide), the effect of the first term is larger, and is partially cancelled by that of the second term. It is expected that the situation is similar in many cases.

Another effect induced by the displacement δq_p is its contribution to the molecular polarizability, which is called vibrational polarizability.[52-54] Because δq_p gives rise to an induced dipole moment $\delta \mu_i$ as

$$\delta \mu_i = \sum_p \frac{\partial \mu_i}{\partial q_p} \delta q_p = \sum_j \sum_p \frac{1}{k_p} \frac{\partial \mu_i}{\partial q_p} \frac{\partial \mu_j}{\partial q_p} E_j \tag{5}$$

the formula for the vibrational polarizability (in the static limit) is given as[52-54]

$$\alpha_{ij}^{(v)} = \sum_p \frac{1}{k_p} \frac{\partial \mu_i}{\partial q_p} \frac{\partial \mu_j}{\partial q_p} \tag{6}$$

As a result, the vibrational modes with large dipole derivatives and small quadratic force constants have some contribution to the molecular polarizability.

Case 3: When the electric field (on the mth molecule, now denoted as E_m and its component as $E_{j,m}$) is modulated by *intramolecular* vibrations of different molecules (n), the vibrations of those molecules are directly coupled with each other. Within the dipole approximation, E_m is expressed as

$$E_m = \sum_{n(\neq m)} T_{mn} \left(\mu_n + \sum_r \frac{\partial \mu_n}{\partial q_{r,n}} q_{r,n} + ... \right) \tag{7}$$

where μ_n is the dipole moment of the nth molecule, $q_{r,n}$ is its rth normal mode, and T_{mn} is the dipole interaction tensor between the mth and nth molecules, which is given as

$$T_{mn} = \frac{3 r_{mn} r_{mn} - r_{mn}^2 I}{r_{mn}^5} \tag{8}$$

where $r_{mn} = r_m - r_n$ is the distance vector between the two molecules, r_{mn} is the length of this vector, and I is a 3×3 unit tensor. Adding suffix m in Eq. 1 to indicate that the quantities in this equation are those for the

mth molecule, substituting the electric field there with Eq. 7, and taking the summation over m, we have the term given as

$$V = V_0 + \dots - \sum_{m,n(m>n)} \sum_p \sum_r \frac{\partial \mu_m}{\partial q_{p,m}} T_{mn} \frac{\partial \mu_n}{\partial q_{r,n}} q_{p,m} q_{r,n} + \dots \qquad (9)$$

(Here the summation is limited to $m > n$ to avoid double counting.) This term represents the direct electrostatic coupling of the vibrations of the mth and nth molecules, because it is bilinear with respect to $q_{p,m}$ and $q_{r,n}$. Taking the derivative with respect to $q_{p,m}$ and $q_{r,n}$, we obtain the coupling constant given as

$$\frac{\partial^2 V}{\partial q_{p,m} \partial q_{r,n}} = -\frac{\partial \mu_m}{\partial q_{p,m}} T_{mn} \frac{\partial \mu_n}{\partial q_{r,n}} \qquad (10)$$

This is the formula for the TDC.

It is thus clear that the electrical property derivatives are the basic quantities for describing the responses of molecular vibrations to intermolecular electrostatic interactions. The explicit examples of cases 2 and 3 described above are shown in the following sections to see how the general formulas given above are applicable to real molecular systems.

3. Structural Changes Induced by Electric Field and Their Consequences in Vibrational Properties

As shown in Eqs. 3 and 4, a structural displacement δq_p induced by intermolecular electrostatic interactions combined with the direct effect of the latter gives rise to a shift of the quadratic force constant δk_p of an anharmonic vibrational mode q_p. This effect is large for a mode with a large cubic (relative to quadratic) force constant and large dipole first (and second) derivatives. A natural consequence is that one can see significant frequency shifts for those stretching modes of polar bonds that are usually conceived as strongly anharmonic, such as the O–H and N–H stretching modes. In fact, the frequency shifts of these modes induced by hydrogen bonding amount to a few hundred cm^{-1}. However, this does not mean that the frequency shifts induced by intermolecular electrostatic interactions are seen only in those special cases. For many other vibrational modes also, the magnitudes of the dipole first (and second) derivatives and the cubic force constants are sufficiently large so that we can see significant effects of intermolecular electrostatic interactions.

In the liquid state, the magnitude and the direction of the electric field coming from the neighboring molecules are inhomogeneously distributed over certain ranges. This results in a distribution of the frequency shift δk_p given by Eq. 4 and, hence, gives rise to band broadening. Depending on the form of the distribution of δk_p, the vibrational band may be asymmetric or may even contain subbands. Moreover, if the dynamics of the liquid is sufficiently fast, there is a rapid modulation of δk_p as time evolves. In this case, the inhomogeneously broadened vibrational band is narrowed to a certain extent by the motional narrowing effect.[60]

Therefore, for the vibrational modes with sufficiently large dipole first (and second) derivatives and cubic force constants, both the average frequency position and the band shape are sensitive to the nature of the vibrational mode itself and its environment.

3.1. *Example: the Amide I Mode of N-Methylacetamide*

It is well known that the amide I bands of polypeptides and proteins are sensitive to their secondary structures. As mentioned in Sec. 1, the main origin of this sensitivity has been explained by the TDC mechanism, which is given in Eq. 10. Because of the large magnitude of the dipole derivative, the amide I modes of neighboring peptide groups are strongly coupled with each other due to this mechanism. Since the TDC constants depend on the distances and relative orientations of the peptide groups involved in the coupling as shown in Eq. 10, the vibrational spectroscopic features arising from this coupling are strongly dependent on the structure of the system. In the vibrational exciton picture,[9,10,13,61–67] where the amide I modes are regarded as constituting a separate vibrational subspace (called the amide I subspace[9]), the peptide–peptide vibrational interactions are represented as the off-diagonal terms of the vibrational Hamiltonian or the force constant matrix (F matrix). In this sense, the sensitivity of the amide I band profiles to the secondary structures is mainly an *off-diagonal* effect.

H. Torii

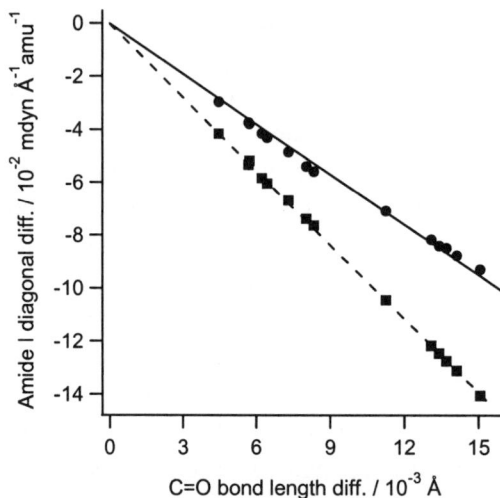

Fig. 1. Relation between the shifts in the diagonal elements of the F matrix in the amide I subspace (δk_I) and the C=O bond length ($\delta s_{C=O}$) of the NMA oligomers from the values of an isolated NMA molecule calculated at the B3LYP/6-31+G(2df,p) level.[21] •: diagonal force constants calculated by the average partial vector method; ■: contribution of the mechanical anharmonicity calculated by the first term of Eq. 11; solid and broken lines: the least-squares fitted lines (constrained to pass the origin) for these points.

As discussed in Sec. 2, however, the large magnitude of the dipole derivative of the amide I mode means that the force constants k_p of this mode of individual peptide groups, which constitute the diagonal terms of the vibrational Hamiltonian or the F matrix in the vibrational exciton picture, are also strongly dependent on the peptide–peptide (and peptide–solvent) interactions. It is important to elucidate this dependence to understand the amide I band profiles correctly. In fact, this point has been extensively studied in recent studies[16,19–21] by ab initio molecular orbital (MO) and density functional (DFT) calculations. N-methyl-acetamide (NMA) has been employed for this purpose as a model compound of the peptide group.

Figure 1 shows in filled circles the relation between the shifts in the diagonal elements of the F matrix in the amide I subspace (denoted as δk_I) and the C=O bond length ($\delta s_{C=O}$) of the NMA oligomers (from dimer to pentamer in the antiparallel configuration[68–70]) from the values of an

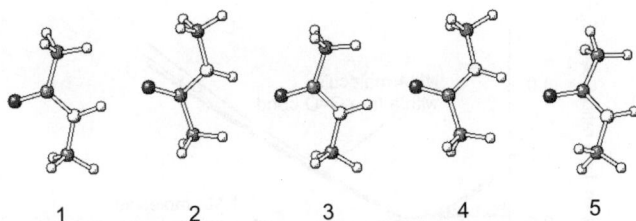

Fig. 2. Structure of the NMA pentamer optimized at the B3LYP/6-31+G(2df,p) level, with the numbering of molecules from the "free C=O" side.

isolated NMA molecule calculated at the B3LYP/6-31+G(2df,p) level.[21] The structure of the NMA pentamer is shown in Fig. 2 as an example, with the numbering of molecules that will be used in the later discussion. The values of δk_I of individual peptide groups in the oligomers are evaluated by the average partial vector (APV) method that has been developed recently.[20] It is clearly seen that δk_I is approximately proportional to $\delta s_{C=O}$. This result suggests that the structural displacement and the mechanical anharmonicity is a major factor for the changes in δk_I as shown in Eq. 3 and the first term of Eq. 4. However, because both the first and second terms in Eq. 4 are controlled by the electric field coming from other molecules in the system, this result alone does not confirm that the first term is indeed the most important factor.

To clarify this point, it is essential to calculate the contribution of the first term separately. Since we have obtained the structural parameters for each molecule in the oligomers from the DFT calculations, we can easily calculate the structural displacements along all the internal coordinates of each molecule, from which it is possible to evaluate the contribution of the mechanical anharmonicity to δk_I. Here it is convenient to adopt the internal-normal mixed coordinate system representation of the mechanical anharmonicity,[20,21] in which the cubic force constants are represented as the derivatives in the second order in the normal coordinate system (q) and the first order in the internal coordinate system (s). We have

$$\delta k_I = \sum_r \frac{\partial^3 V}{\partial q_I^2 \partial s_r}\delta s_r - \frac{\partial^2 \mu}{\partial q_I^2} E \qquad (11)$$

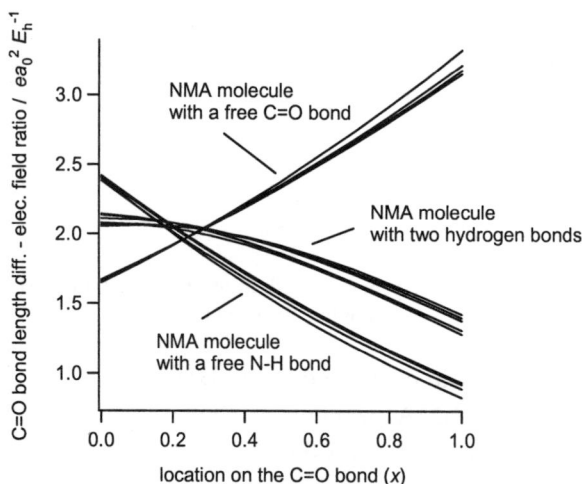

Fig. 3. Plot of the ratio between the C=O bond length change and the electric field ($\delta s_{C=O}/E_{proj(C=O)}$) against the position on the C=O bond (x), where $E_{proj(C=O)}$ is the projection of the electric field evaluated at $r = x\, r_O + (1 - x)\, r_C$ onto the direction of $\partial \mu/\partial s_{C=O}$, calculated at the B3LYP/6-31+G(2df,p) level.[21]

where q_I is the amide I normal mode and δs_r is the displacement along the rth internal coordinate. The contribution of the first term of Eq. 11 is shown in filled squares in Fig. 1. It is seen that it overestimates the value of δk_I. A similar result is also obtained at the HF/6-31+G(2df,p) level.[20] The above result demonstrates that the contribution of the mechanical anharmonicity is indeed dominant, but the partial cancellation by the contribution of the electrical anharmonicity, represented by the second term and mounts to about 20% in the present case, is also important for the correct estimation of the value of δk_I.

To apply Eqs. 2–4 to the cases of real molecular systems, it is necessary to examine where in the molecule we should evaluate the electric field, since these equations are strictly valid for a uniform electric field. For this purpose, the values of $\delta s_{C=O}/E_{proj(C=O)}$ are calculated, where $E_{proj(C=O)}$ is the projection of the electric field onto the direction of the dipole derivative $\partial \mu/\partial s_{C=O}$ of the C=O stretching internal coordinate evaluated at $r = x\, r_O + (1 - x)\, r_C$ on the C=O bond ($0 \leq x \leq 1$), where r_O

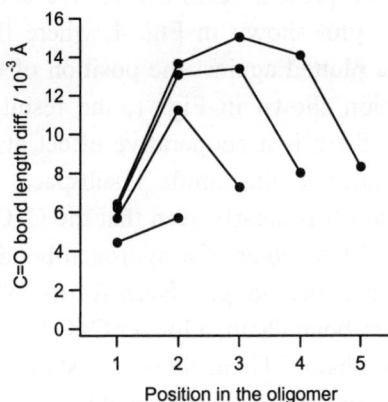

Fig. 4. Changes in the C=O bond lengths of the NMA molecules in the oligomers from that of an isolated NMA molecule calculated at the B3LYP/6-31+G(2df,p) level.[21] The NMA molecules are numbered as shown in Fig. 2.

and r_C are the positions of the carbonyl oxygen and carbon atoms, respectively. Since we have

$$\delta s_{C=O} \cong \frac{1}{k_{C=O}} \frac{\partial \mu}{\partial s_{C=O}} E \tag{12}$$

from Eq. 2 and $(1/k_{C=O}) \times |\partial \mu / \partial s_{C=O}|$ may be regarded as a molecular property of NMA, the ratio $\delta s_{C=O}/E_{proj(C=O)}$ should be approximately constant if $\delta s_{C=O}$ is induced by the intermolecular electric field according to the present scheme. The result is shown in Fig. 3. It is seen that the value of $\delta s_{C=O}/E_{proj(C=O)}$ nearly coincides in the region of $0.2 \leq x \leq 0.3$, with the approximate crossing point at $x \cong 0.23$. This result demonstrates that it is possible to describe the changes in the C=O bond length (and hence those in the values of δk_1 because of the relation shown in Fig. 1) by the formulas of the electric field-induced effects given above as far as the electric fields are evaluated at appropriate positions.

Concerning the C=O bond length and the frequency of the amide I mode of a peptide group, it has been discussed[17,71–79] that there is a cooperative effect, i.e., these quantities depend on the length of the hydrogen-bond chain that the peptide group belongs as well as on the location of the peptide group in the chain. In the case of the NMA oligomers treated in the present section also, the cooperative effect is recognizable from the plot shown in Fig. 4, where the changes in the C=O bond lengths are plotted against the position of the peptide group. (Because of the relation shown in Fig. 1, the result shown in Fig. 4 directly indicates that there is a cooperative effect also in the diagonal elements of the F matrix in the amide I subspace, and hence in the frequency of this mode.) It is clearly seen that the C=O bond length gets longer as we go toward the center of a hydrogen-bond chain, and as the hydrogen-bond chain becomes longer. Even for the NMA molecules on the edges of a hydrogen-bond chain, a longer C=O bond is calculated in a longer hydrogen-bond chain. Then, the result shown in Fig. 3 demonstrates that this cooperative effect is explained by the enhancement of the electric field operating among the NMA molecules in the oligomers.

A further detailed analysis[21] shows that there are three important factors for this enhancement of the intermolecular electric field. One is the variation of the intermolecular distances arising from that of the hydrogen-bond lengths. For example, the distance between molecules 1 and 2 in the NMA pentamer is shorter than that in the NMA dimer by about 2 %, giving rise to the enhancement of the electric field of the molecular dipole (inversely proportional to the cube of the intermolecular distance) by about 6 %. The second factor is the existence of the electric field directly operating between distant molecules, e.g., between molecules 1 and 3 in the trimer. Because of this effect, there is an enhancement of the electric field by 5 to 16 %, except for the cases without distant molecules (e.g., molecule 2 in the trimer), as compared with the simple sum of the electric field from the nearest neighbor molecules. The third factor is the polarization of the intervening molecule(s) between distant molecules. According to the dipole-induced dipole (DID) mechanism,[80,81] the electric field on the mth molecule operating from the nth molecule is enhanced as

$$\Delta E_{m \leftarrow n} = T_{mk}\, \alpha_k \cdots T_{kn}\, \mu_n \qquad (13)$$

where k denotes the intervening molecule(s), α_k is its molecular polarizability, μ_n is the dipole moment of the nth molecule, and T_{mk} is the dipole interaction tensor given in Eq. 8. This effect gives rise to enhancement of the intermolecular electric field by up to about 20%.

This result indicates that the effect of molecular polarization is an important factor for the changes in the structural parameters and, hence, in the vibrational properties of the amide I mode of the peptide group induced by the peptide–peptide interactions. It is expected that this is also the case for many other vibrational modes with large dipole derivatives.

3.2. *Origin of the Intermolecular Electric Field: Charges, Dipoles, and Quadrupoles*

Intermolecular electrostatic interactions are usually represented as those between partial charges on atomic sites or as those between molecular dipoles (and molecular charges for charged species). In some cases, polarizabilities of atoms or molecules are also taken into account. Indeed, the atomic partial charges constitute the lowest-order terms as far as we choose atoms in molecules as the interaction sites, and the same is true for molecular charges and dipoles if we treat each molecule as a whole. It is empirically well recognized that these methods work well in many cases. However, there is no guarantee that they work well in *all* cases, so that it is necessary to carefully consider the way of representing intermolecular electrostatic interactions if any peculiar behavior of intermolecular interactions is encountered.

It has been shown in recent studies[51,82] that, for some well-known molecules containing Cl, Br, and/or S atoms, atomic charges alone are not sufficient for reasonable representation of their intermolecular electrostatic interactions. As examples, the electric fields around CCl_4, HCl, and Cl_2 molecules calculated at the MP3/6-31+G(2df,p) level of the *ab initio* MO method are shown in Fig. 5. It is clearly seen that the electric field around CCl_4 is going out of the Cl atom on the line extended from the C–Cl bond, indicating that the Cl atom looks as if it had a positive charge, in contrast to the expectation from the values of electronegativity of the C and Cl atoms in common electronegativity scales. On the side of the C–Cl bond, the electric field is going toward the Cl atom. In the case of HCl, the electric field around the Cl atom on the line extended from the H–Cl bond is much weaker because of the cancellation with the effect of molecular dipole moment, but the vector

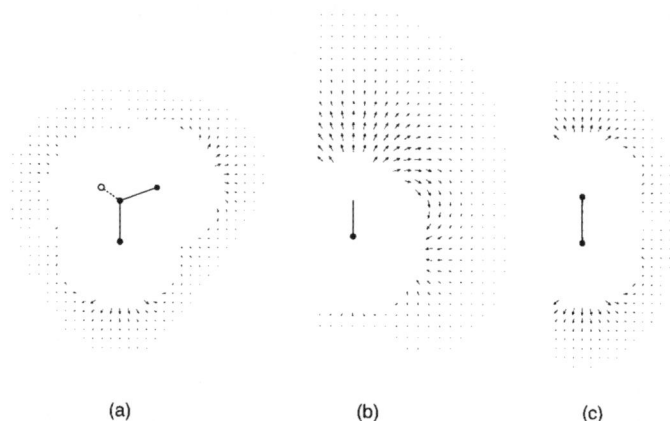

(a)   (b)   (c)

Fig. 5. The electric fields around (a) CCl$_4$, (b) HCl, and (c) Cl$_2$ molecules calculated at the MP3/6-31+G(2df,p) level.[51] The filled circles represent the carbon and chlorine atoms located on the sheet. The open circle in (a) represents two chlorine atoms located above and below the sheet. The hydrogen atom of HCl is located at one end of the thick straight line without a filled circle in (b). The points within 1.6 Å from the hydrogen atom or within 2.8 Å from any other atoms are excluded. In (b) and (c), the electric field on the left side of the molecule is symmetric to the one on the right side, and is omitted in this figure.

of the electric field is also going out of the Cl atom on that line in the vicinity (within ~3.8 Å) of the atom. The same form of electric field is also found around a Cl$_2$ molecule. Since there is only one (for CCl$_4$ and HCl) or zero (for Cl$_2$) adjustable atomic partial charge in the atomic charge model because of the electrical neutrality of these molecules, these forms of electric fields cannot be correctly represented only by atomic partial charges.

Detailed analyses of the forms of these electric fields[51] have shown that, for reasonable representation of these electric fields, it is necessary to take into account the existence of atomic (electrical) quadrupoles on the Cl atoms. The magnitudes of those quadrupoles are in the range of $\Theta(\text{Cl}) = 1.2\text{--}1.6 \ ea_0^2$. In the cases of analogous molecules with different halogen atoms, we have $\Theta(\text{F}) = 0.3\text{--}0.5 \ ea_0^2$ and $\Theta(\text{Br}) = 1.8\text{--}2.2 \ ea_0^2$, so that the value of $\Theta(\text{X})$ gets larger as we go down the periodic table. A similar result is also obtained from the comparison between CO$_2$ and

CS_2.[51] The effect of atomic quadrupoles is not recognizable for the electric field around CO_2, but the effect of $\Theta(S)$ is clearly seen for the electric field around CS_2.

The atomic quadrupoles of those heavy atoms originate from the anisotropy of the electron density.[51] For example, in the case of Cl_2, there are 17 occupied molecular orbitals, out of which the highest seven are the valence orbitals. These orbitals are characterized as linear combinations of the $3s$, $3p_x$, $3p_y$, and $3p_z$ atomic orbitals of the two Cl atoms. There are two occupied molecular orbitals each with the $3s$, $3p_x$, and $3p_y$ characters, taking the z axis as the molecular axis. However, there is only one occupied molecular orbital with the $3p_z$ character, because the other linear combination of $3p_z$ is the σ^* orbital and is therefore unoccupied. As a result, the electron density is substantially higher in the x and y directions than in the z direction around the Cl atoms to the extent equivalent to the atomic quadrupole of $\Theta(Cl)$ = 1.25 ea_0^2. This is considered to be the electronic structural origin of the large atomic quadrupoles of the Cl atoms. Similar situations are also expected for the other heavy atoms discussed above.

For the interaction energies between nonpolar molecules, it is generally expected that the electrostatic interactions involving the Cl and/or other heavy atoms play only a minor role even if we take into account the effect of their atomic quadrupoles. For example, in the case of the CCl_4–CCl_4 interaction energy, the electrostatic part is only a few percent of the total.[83,84] However, the effect of the atomic quadrupoles may be significant for the interactions with polar molecules. As an example, an optimized structure of the acetone–$(CCl_4)_4$ cluster calculated at the HF/6-31+G(2df,p) level[82] is shown in Fig. 6. It is seen that the four CCl_4 molecules are located at almost symmetric positions around the C=O group of acetone, and one C–Cl bond of each CCl_4 molecule points to the carbonyl oxygen with $r(O...Cl)$ in the range of 3.25–3.79 Å. The close contacts between O and Cl atoms are considered to be stabilized by the effect of the atomic quadrupoles of the Cl atoms.

Because of the existence of those atomic quadrupoles, the C=O bond of the acetone molecule is in a rather strong electric field of about 4.6 $\times$ 10^{-3} $E_h e^{-1} a_0^{-1}$ *aside from* the electronic polarization effect. This magnitude of the electric field is equivalent to that generated by an

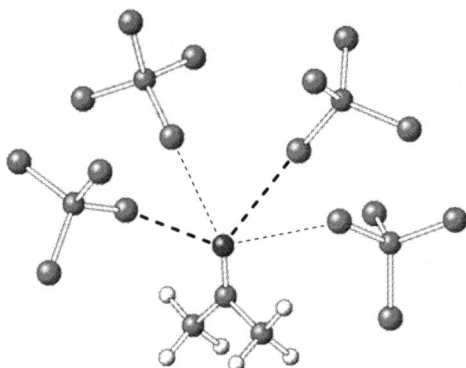

Fig. 6. An optimized structure of the acetone–(CCl₄)₄ cluster calculated at the HF/6-31+G(2df,p) level.[82] The CCl₄ molecules connected by thick broken lines are located on the front side relative to the carbonyl oxygen, and those connected by thin broken lines are located on the other side.

isolated water molecule at about 3 Å from the H atom on the line extended from the O–H bond.[82] Induced by this electric field, the vibrational frequency of the C=O stretching mode of the acetone–(CCl₄)₄ cluster becomes lower by 17.5 cm^{-1} from that of an isolated acetone molecule, in accord with the scheme described above in Secs. 2 and 3.1. This result demonstrates that permanent multipole moments of nonpolar molecules may have a significant effect on the vibrational spectroscopic properties of the molecules interacting with them.

4. Modulation of the Electric Fields in Liquids: Field-Modulating Modes (FMMs)

As noted at the beginning of Sec. 3, the magnitude and the direction of the electric field coming from the neighboring molecules are inhomogeneously distributed and modulated as time evolves in the liquid state. In this relation, it is interesting to see what kind of intermolecular vibrations contribute significantly to the modulation of the electric field. By referring to Eq. 1, within the instantaneous normal mode (INM) picture[85,86] and to the first order in the electric field, the potential energy of a solution may be expanded as

$$V = V_0 + \frac{1}{2}\sum_p k_p q_p^2 + \frac{1}{2}\sum_r \left[K_r Q_r^2 - 2F_r Q_r + ...\right] + ...$$

$$-\sum_j \left[\mu_j^{(0)} + \sum_p \frac{\partial \mu_j}{\partial q_p} q_p + ...\right]\left[E_j^{(0)} + \sum_r \frac{\partial E_j}{\partial Q_r} Q_r + ...\right] + ... \quad (14)$$

where q_p, k_p, and μ_j are now the quantities of the solute, Q_r is the rth (mass-weighted) intermolecular normal mode of the solution, K_r is its quadratic force constant, F_r is the force along this normal mode at the instantaneous configuration of the liquid structure ($Q_r = 0$), and E_j is the electric field on the solute. Here, μ_j at $q_p = 0$ and E_j at $Q_r = 0$ are explicitly denoted with the superscript (0). It is easily noticed that the intermolecular modes with large values of $\partial E_j / \partial Q_r$ induce large changes in E_j.

One problem that should be considered at this point is that there are many intermolecular normal modes Q_r (too many to examine the nature of each), and a large number of them are expected to have large values of $\partial E_j / \partial Q_r$. This problem is related in part to the delocalized nature of the intermolecular normal modes. It is therefore preferable to develop a method for extracting the intermolecular vibrational motions (which may be different from *normal* modes) that are really responsible for the changes in E_j. The picture of the field-modulating modes (FMMs) has been developed for this purpose.[87]

To derive the FMMs, we take the set of normalized mass-weighted translational and librational (rotational) coordinates of the molecules in solution, denoted as $R_{l,m}$ [$1 \le l \le 3$ for translations and $4 \le l \le 6$ for librations ($4 \le l \le 5$ in the case of linear molecules), m denoting the molecule number]. $\partial E_j / \partial Q_r$ may be expanded by $\{R_{l,m}\}_{1 \le l \le 6, 1 \le m \le N}$ as

$$\frac{\partial E_j}{\partial Q_r} = \sum_{m=1}^{N} \sum_{l=1}^{6} \frac{\partial E_j}{\partial R_{l,m}} \frac{\partial R_{l,m}}{\partial Q_r} \quad (15)$$

where N is the number of molecules in the system.

The summation over l and m in Eq. 15 may be regarded as a scalar product of two $6N$-dimensional vectors e_j and q_r defined as $(e_j)_u = \partial E_j / \partial R_{l,m}$ and $(q_r)_u = \partial R_{l,m} / \partial Q_r$ with $u = 6m + l$. Since $\{R_{l,m}\}_{1 \le l \le 6, 1 \le m \le N}$ is a set of normalized mass-weighted coordinates, the vectors q_r are normalized and orthogonal to each other by definition. We consider an ortho-gonal transformation from $\{q_r\}_{1 \le r \le M}$ to another set of $6N$-dimensional vectors $\{s_t\}_{1 \le t \le M}$, where the number of normal modes included in the derivation of FMMs is denoted as M. Since $\{e_j\}_{1 \le j \le 3}$ generates a three-dimensional subspace in the $6N$-dimensional space of intermolecular vibrational motions, we can take $\{s_t\}_{1 \le t \le M}$ such that $e_j \cdot s_t = 0$ ($1 \le j \le 3$; $4 \le t \le M$) is satisfied. We then define a set of vibrational coordinates S_t as $(s_t)_u = \partial R_{l,m}/\partial S_t$ (with $u = 6m + l$). The condition of $e_j \cdot s_t = 0$ ($1 \le j \le 3$; $4 \le t \le M$) means that $\partial E_j/\partial S_t = 0$ for $1 \le j \le 3$ and $4 \le t \le M$, i.e., $\{S_t\}_{4 \le t \le M}$ do not modulate the electric field in any direction. Then, since $\{S_t\}_{1 \le t \le M}$ is related to $\{Q_r\}_{1 \le r \le M}$ by an orthogonal transformation, we may regard $\{S_t\}_{1 \le t \le 3}$ as representing the intermolecular vibrational modes responsible for the modulation of the electric field at the position where the solute molecule is located, i.e., the FMMs.

To obtain the explicit forms of the FMMs, we diagonalize an $M \times M$ matrix H defined as

$$H_{rs} = \sum_{j=1}^{3} \frac{\partial E_j}{\partial Q_r} \frac{\partial E_j}{\partial Q_s} \tag{16}$$

It is easily derived from the algebraic property of the vectors explained in the last paragraph that there are at most three nonzero eigenvalues for H. The rest of the eigenvalues are zero. The eigenvectors for the nonzero eigenvalues represent the FMMs. Each nonzero eigenvalue is equal to the square of the modulation of the electric field by the corresponding FMM.

Concerning the properties of the FMMs thus defined, the following two points are noteworthy. (1) In general, each FMM is not equal to any one of the intermolecular normal modes, but is represented by a linear combination of them. The theoretical derivation described above shows that, if we take appropriate linear combinations of the intermolecular normal modes, it is possible to construct a set of vibrational motions with the property that only (at most) three of them are responsible for the modulation of the electric field on the solute. (2) The dimensionality of the vector space of the FMMs comes from the fact that $\{e_j\}_{1 \le j \le 3}$ generates a three-dimensional space, or in other words, there are three independent elements of $\partial E_j/\partial R_{l,m}$ for each $R_{l,m}$. As a result, if we take only one or two

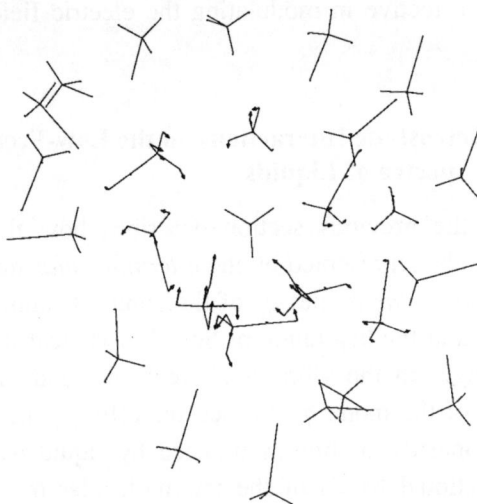

Fig. 7. An example of the vibrational patterns of FMMs calculated for an acetonitrile solution of the nitrate ion.[87] The nitrate ion is located at the center. The atomic motions are shown by arrows.

components of the electric field, the dimension of the vector space of the FMMs decreases accordingly.

As an example, the vibrational pattern of one of the FMMs calculated for one configuration (sampled by MD simulation) of an acetonitrile solution of the nitrate ion $(NO_3^-)^{[87]}$ is shown in Fig. 7. It is seen that the motions of the acetonitrile (solvent) molecules in the first solvation shell (within about 5.5 Å) contribute significantly to the modulation of the electric field on the nitrate ion (solute), although the intermolecular normal modes are extensively delocalized. A further detailed analysis[87] shows that the librational (rotational) motions rather than the translational motions of the acetonitrile molecules are the main origin of the modulation of the electric field. By inspecting the tensor $\partial S_t / \partial Q_r$ $(1 \leq t \leq 3; 1 \leq r \leq M)$ as a function of the frequency of the intermolecular normal modes (Q_r), it is possible to examine the frequency distribution of the normal modes contributing to the FMMs. In the case of the present example, among the modes with large amplitudes of the librations around the x and y axes of acetonitrile (taking the z axis as the molecular axis), those in the 200–400 cm^{-1} region rather than the 0–200 cm^{-1} region tend to be more effective in modulating the electric field of the nitrate ion.

5. Effect of Electrostatic Interactions on the Low-Frequency Vibrational Spectra of Liquids

Equation 14 in the previous section describes how the dynamics of molecules in liquids, represented by the *intermolecular* normal modes Q_r, affect the *intramolecular* modes q_p of the solute. Combined with Eq. 4 and the discussion at the beginning of Sec. 3, it is clear that this effect is seen as the changes in the vibrational frequency and the shape of the vibrational band of the mode q_p. In fact, the effect of the modulation of the electrical properties of liquids induced by liquid dynamics is also seen in the vibrational bands of the intermolecular modes themselves, which appear in the low-frequency region.[88,89] In this case, the modulations of the total dipole moment or polarizability of the system interact directly with the radiation field to give rise to light absorption or scattering.

According to the dipole-induced dipole (DID) mechanism,[80,81,90] the dipole moment μ_m of the mth molecule enhanced by intermolecular electrostatic interactions is given as

$$\mu_m = \mu_m^{\mathrm{M}} + \alpha_m^{\mathrm{M}} \sum_{\substack{n=1 \\ (n \neq m)}}^{N} T_{mn}\mu_n \qquad (17)$$

where the superscript M now stands for the quantities intrinsic to the molecule, i.e., without perturbation by intermolecular electrostatic interactions. The sum of $T_{mn}\mu_n$ over n in the second term represents the electric field coming from the dipole moments of the surrounding molecules. By taking the sum of μ_m over m, we obtain the total dipole moment of the system, the time correlation function of which is related to the low-frequency (far IR) absorption spectra. The polarizability α_m of the mth molecule is obtained as the derivative of μ_m with respect to the electric field, and is expressed as[90]

$$\alpha_m = \alpha_m^{\mathrm{M}} + \sum_{\substack{n=1 \\ (n \neq m)}}^{N} \left(\alpha_m^{\mathrm{M}} T_{mn} \alpha_n + \beta_m^{\mathrm{M}} T_{mn} \mu_n \right) \tag{18}$$

where β_m is the first hyperpolarizability of the mth molecule. The sum of α_m over m is the total polarizability of the system, the time correlation function of which is related to the low-frequency Raman scattering and the optical Kerr effect (OKE). The second part in the parentheses of Eq. 18 represents the changes in the molecular polarizability due to the dipole field $T_{mn}\mu_n$ of the surrounding molecules. The enhancement mechanism including this part is called extended DID (XDID) mechanism.[90]

In the following discussion, we take the case of the low-frequency depolarized Raman spectra. In the time-domain formulation, the explicit formula of the low-frequency depolarized Raman spectra is given as[88–92]

$$R(\omega) = \frac{8\pi c}{h} \left[1 - \exp(-hc\tilde{v}/kT) \right] \, \mathrm{Im} \int_0^{\infty} dt \, \exp(i2\pi c \tilde{v} t) \left[-\frac{\partial}{\partial t} C(t) \right]$$

$$\tag{19}$$

where $C(t)$ is the time correlation function of the anisotropic part of the total polarizability Π, expressed as

$$C(t) = \sum_{\substack{i,j=X,Y,Z \\ (i \neq j)}} \left\langle \Pi_{ij}(t)\Pi_{ij}(0) \right\rangle \tag{20}$$

where X, Y, and Z denote the axes of the laboratory frame, and the bracket stands for the statistical average. In parallel with Eq. 18, Π is divided into two terms as $\Pi = \Pi^{\mathrm{UMP}} + \Pi^{\mathrm{XDID}}$, where

$$\Pi^{\mathrm{UMP}} = \sum_{m=1}^{N} \alpha_m^{\mathrm{M}} \tag{21}$$

is the unperturbed molecular polarizability (UMP) term,[90] and

$$\Pi^{\mathrm{XDID}} = \sum_{\substack{m,n=1 \\ (m \neq n)}}^{N} \left(\alpha_m^{\mathrm{M}} T_{mn} \alpha_n + \beta_m^{\mathrm{M}} T_{mn} \mu_n \right) \tag{22}$$

is the XDID term. As a result, $C(t)$ is divided into three terms as

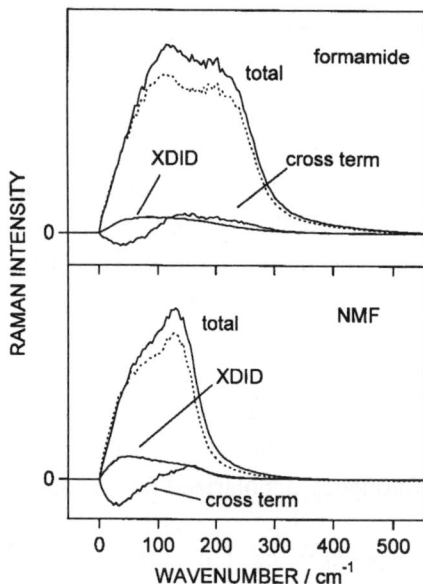

Fig. 8. Calculated Raman spectra of liquid formamide and N-methylformamide (NMF). Solid lines: the total profile, the XDID term, and the cross term calculated with the XDID mechanism; dotted lines: the UMP term. Intensities are given in an arbitrary scale.[90]

$$C(t) = C^{\text{UMP}}(t) + C^{\text{XDID}}(t) + C^{\text{U-XD}}(t) \qquad (23)$$

where the first and second terms are the auto correlation functions of Π^{UMP} and Π^{XDID}, respectively, and the third term is the cross correlation function between Π^{UMP} and Π^{XDID}. $R(\omega)$ is also divided into three terms accordingly:

$$R(\omega) = R^{\text{UMP}}(\omega) + R^{\text{XDID}}(\omega) + R^{\text{U-XD}}(\omega) \qquad (24)$$

As an example, the low-frequency depolarized Raman spectra of liquid formamide and N-methylformamide (NMF)[90] are shown in Fig. 8. It is seen that there are noticeable changes in the Raman intensities, although the overall spectral profiles are not strongly affected by the enhancement due to the XDID mechanism.

Within the scheme of the INM theory,[85,86] an important quantity for calculating Raman spectra is the derivative of α_m in Eq. 18 (or its sum over m, Π) with respect to the intermolecular normal modes Q_r or the translational and librational coordinates $R_{l,n}$ ($1 \le l \le 6$; $1 \le n \le N$). The derivatives with respect to the librational coordinates are given as

$$\frac{\partial \alpha_m}{\partial R_{l,n}} = \delta_{mn} \left[\frac{\partial \alpha_m^{\text{M}}}{\partial R_{l,m}} \left(1 + \sum_{\substack{k=1 \\ (k \ne m)}}^{N} T_{mk} \alpha_k \right) + \frac{\partial \beta_m^{\text{M}}}{\partial R_{l,m}} \sum_{\substack{k=1 \\ (k \ne m)}}^{N} T_{mk} \mu_k \right]$$

$$+ \alpha_m^{\text{M}} \sum_{\substack{k=1 \\ (k \ne m)}}^{N} T_{mk} \frac{\partial \alpha_k}{\partial R_{l,n}} + \beta_m^{\text{M}} \sum_{\substack{k=1 \\ (k \ne m)}}^{N} T_{mk} \frac{\partial \mu_k}{\partial R_{l,n}} \tag{25}$$

This formula indicates that, in some cases, first hyperpolarizability derivatives may also be important for calculating the Raman intensities of the intermolecular modes.

6. How to Recognize the Vibrational Modes with Large Effects of Electrostatic Interactions: Intensity-Carrying Modes (ICMs)

As discussed in Sec. 2, vibrational modes with large dipole derivatives are likely to be affected strongly by intermolecular electrostatic interactions. For small molecules with one or a few polar bonds, it will be rather easy to recognize the nature of the vibrational modes with large dipole derivatives. However, for large nonsymmetrical molecules with delocalized normal modes, such as molecules with conjugated π-electron systems, many normal modes are likely to have non-negligible dipole derivatives, so that it is difficult to characterize the vibrational motions that are really responsible for generating those dipole derivatives. The intensity-carrying mode (ICM) theory[93] is expected to be helpful to solve this problem.

The derivation of the ICMs is quite similar to that of the FMMs described in Sec. 4. The dipole derivatives with respect the intra-molecular normal modes, $\partial \mu_j / \partial q_p$, are expanded with the set of nor-malized mass-weighted cartesian vibrational coordinates $\{x_s\}_{1 \le s \le 3L}$ as

$$\frac{\partial \mu_j}{\partial q_p} = \sum_{s=1}^{3L} \frac{\partial \mu_j}{\partial x_s} \frac{\partial x_s}{\partial q_p} \tag{26}$$

where L is the number of atoms in a given molecule. The summation over s in Eq. (26) may be regarded as a scalar product of two $3L$-dimensional vectors d_j and x_p defined as $(d_j)_s = \partial\mu_j/\partial x_s$ and $(x_p)_s = \partial x_s/\partial q_p$. Note that, by definition, the vectors x_p are normalized and orthogonal to each other. We consider an orthogonal transformation from $\{x_p\}_{1 \leq p \leq V}$ to another set of $3L$-dimensional vectors $\{w_t\}_{1 \leq t \leq V}$, where the number of normal modes included in this derivation is denoted as V. Since $\{d_j\}_{1 \leq j \leq 3}$ generates a three-dimensional subspace in the $3L$-dimensional space of intramolecular vibrational motions, we can take $\{w_t\}_{1 \leq t \leq V}$ such that $d_j \cdot w_t = 0$ $(1 \leq j \leq 3; 4 \leq t \leq V)$ is satisfied. We then define a set of vibrational coordinates w_t as $(w_t)_s = \partial x_s/\partial w_t$. The condition of $d_j \cdot w_t = 0$ $(1 \leq j \leq 3; 4 \leq t \leq V)$ means that $\partial\mu_j/\partial w_t = 0$ for $1 \leq j \leq 3$ and $4 \leq t \leq V$, i.e., $\{w_t\}_{4 \leq t \leq V}$ do not generate dipole derivatives in any direction. Then, since $\{w_t\}_{1 \leq t \leq V}$ is related to $\{q_p\}_{1 \leq p \leq V}$ by an orthogonal transformation, we may regard $\{w_t\}_{1 \leq t \leq 3}$ as representing the intramolecular vibrational modes responsible for generating the dipole derivatives in a given molecule. These modes are called ICMs (or *IR ICMs* to distinguish them from those derived from other electrical property derivatives discussed below).[93] In general, each of these ICMs is not equal to any one of the intramolecular normal modes, but is represented by a linear combination of them.

To obtain the explicit forms of these ICMs, we diagonalize a $V \times V$ matrix M defined as

$$M_{rs} = \sum_{j=1}^{3} \frac{\partial\mu_j}{\partial q_r} \frac{\partial\mu_j}{\partial q_s} \qquad (27)$$

There are at most three nonzero eigenvalues for M, and their eigenvectors represent the ICMs. Each nonzero eigenvalue is equal to the square of the dipole derivatives generated by the corresponding ICM.

The ICMs may also be derived from the expansion of polarizability derivatives $\partial\alpha_{ij}/\partial q_p$ or first hyperpolarizability derivatives $\partial\beta_{ijk}/\partial q_p$. The number of these types of ICMs is (at most) six and ten, respectively, which is equal to the number of independent elements of the derivative for each q_p. In general, the ICMs derived from different electrical property derivatives have different vibrational patterns. However, depending on the electronic structure of the system, they may have quite similar vibrational patterns.

The ICM theory was originally developed to clarify the nature of the vibrational modes with large IR intensities.[94] Since $\{w_t\}_{1 \le t \le V}$ is related to $\{q_p\}_{1 \le p \le V}$ by an orthogonal transformation and $\partial \mu_j / \partial w_t = 0$ holds for $1 \le j \le 3$ and $4 \le t \le V$, we have

$$\sum_{p=1}^{V} \left| \frac{\partial \mu_j}{\partial q_p} \right|^2 = \sum_{t=1}^{V} \left| \frac{\partial \mu_j}{\partial w_t} \right|^2 = \sum_{t=1}^{3} \left| \frac{\partial \mu_j}{\partial w_t} \right|^2 \tag{28}$$

for each j ($1 \le j \le 3$). Then, since the total IR intensity S_{IR} is simply proportional to the sum of the squares of the dipole derivatives of all the normal modes taken into account, we have

$$S_{\text{IR}} = \frac{\pi N_{\text{A}}}{3c^2} \sum_{p=1}^{V} \sum_{j=1}^{3} \left| \frac{\partial \mu_j}{\partial q_p} \right|^2 = \frac{\pi N_{\text{A}}}{3c^2} \sum_{t=1}^{3} \sum_{j=1}^{3} \left| \frac{\partial \mu_j}{\partial w_t} \right|^2 \tag{29}$$

i.e., the total IR intensity is strictly conserved upon the transformation from $\{q_p\}_{1 \le p \le V}$ to $\{w_t\}_{1 \le t \le V}$.

The ICM theory is expected to be useful also for elucidating the nature of the vibrational modes important for the vibrational polarizability $\alpha_{ij}^{(v)}$, given in Eq. 6, and the vibrational hyper-polarizabilities $\beta_{ijk}^{(v)}$ and $\gamma_{ijkl}^{(v)}$, given (in the static limit) as

$$\beta_{ijk}^{(v)} = \sum_p \frac{1}{k_p} \left[\frac{\partial \mu_i}{\partial q_p} \frac{\partial \alpha_{jk}}{\partial q_p} + (j, ik) + (k, ij) \right] \tag{30}$$

$$\gamma_{ijkl}^{(v)} = \sum_p \frac{1}{k_p} \left[\frac{\partial \alpha_{ij}}{\partial q_p} \frac{\partial \alpha_{kl}}{\partial q_p} + (ik, jl) + (il, jk) \right]$$

$$+ \sum_p \frac{1}{k_p} \left[\frac{\partial \mu_i}{\partial q_p} \frac{\partial \beta_{jkl}}{\partial q_p} + (j, ikl) + (k, ijl) + (l, ijk) \right] \tag{31}$$

where (j, ik), *etc.*, denote the terms obtained by permutation of indices, since these formulas contain electrical property derivatives.[52–54,93] Since the ICM theory can concentrate the contribution to the electrical property derivatives into a few-dimensional vibrational subspace, it may be useful for constructing a few-dimensional model (such as the one-dimensional VB-CT model[95–98]). An important difference from the case of the IR (and

(a)

(b)

(c)

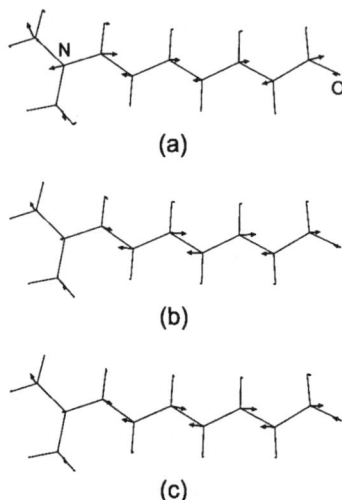

Fig. 9. Vibrational patterns of the primary ICMs of 7-dimethylamino-2,4,6-heptatrienal calculated at the MP2/6-31+G(d,p) level from (a) dipole derivatives, (b) polarizability derivatives, and (c) first hyperpolarizability derivatives.[93] The CH stretching vibraions are eliminated in the derivation of these ICMs.

Raman) intensities is that, in the formulas for $\alpha_{ij}^{(v)}$, $\beta_{ijk}^{(v)}$, and $\gamma_{ijkl}^{(v)}$, we have the reciprocal force constant $1/k_p$ in the summation over p. As a result, they are not strictly conserved upon orthogonal transformation from $\{q_p\}_{1 \le p \le V}$ to $\{w_t\}_{1 \le t \le V}$. The contributions of low-frequency modes may have to be considered separately in some cases.

As an example, we take the case of 7-dimethylamino-2,4,6-heptatrienal [$(CH_3)_2N-(CH=CH-)_3CHO$], which is one of the push–pull type polyene molecules and, hence, has a conjugated π electron system with an intramolecular charge transfer. Calculating the explicit forms of the ICMs of this molecule, it is found[93] that, irrespective of the electrical property derivatives from which the ICMs are derived, there is only one ICM with a large magnitude of electrical property derivative (called *primary ICM*).

The vibrational patterns of those primary ICMs calculated at the MP2/6-31+G(d,p) level are shown in Fig. 9. It is clearly seen that the CN, CC, and CO bonds stretch and contract alternately along the conjugated

Table 1. Vibrational and electronic contribution (in atomic units) to the longest-axis component of the polarizability and hyperpolarizabilities of 7-dimethylamino-2,4,6-heptatrienal calculated at the MP2/6-31+G(d,p) level.[93]

	α_{xx} / au	β_{xxx} / 10^3 au	γ_{xxxx} / 10^5 au	
			α^2	$\mu\beta$
Vibrational component				
1-D model (IR ICM)	43.7	−9.4	6.7	8.1
normal modes, 2000–500 cm^{-1}	49.3	−9.7	8.4	8.3
largest value for a normal mode	13.4	−4.3	4.5	3.7
all normal modes	130.9	−18.3	12.5	15.7
Electronic component	335.7	−10.0	10.9[a]	

[a] This hyperpolarizability cannot be decomposed into α^2 and $\mu\beta$ terms.

chain in all the three ICMs. In other words, these ICMs mainly consist of the bond alternation mode of the conjugated chain. Since the vibrational patterns of these ICMs are similar to each other, it is expected that, to construct a one-dimensional model for the vibrational polarization properties of this molecule, one of these primary ICMs may be (rather arbitrarily) taken as the vibrational coordinate involved in the model.

The calculated polarization properties are summarized in Table 1. For the one-dimensional (1-D) model, the vibrational polarizability and hyperpolarizabilities are calculated as

$$\alpha_{ij}^{(v)} = \frac{1}{k_1} \frac{\partial \mu_i}{\partial w_1} \frac{\partial \mu_j}{\partial w_1} \tag{32}$$

$$\beta_{ijk}^{(v)} = \frac{1}{k_1} \left[\frac{\partial \mu_i}{\partial w_1} \frac{\partial \alpha_{jk}}{\partial w_1} + (j,ik) + (k,ij) \right] \tag{33}$$

$$\gamma_{ijkl}^{(v)} = \frac{1}{k_1} \left[\frac{\partial \alpha_{ij}}{\partial w_1} \frac{\partial \alpha_{kl}}{\partial w_1} + (ik,jl) + (il,jk) \right]$$

$$+ \frac{1}{k_1} \left[\frac{\partial \mu_i}{\partial w_1} \frac{\partial \beta_{jkl}}{\partial w_1} + (j,ikl) + (k,ijl) + (l,ijk) \right] \tag{34}$$

where w_1 and k_1 denote the vibrational mode in the model (the primary IR ICM in the present case) and its quadratic force constant. Only the x-axis components are shown, where the longest axis of the moment of inertia, which is nearly parallel to the direction of the conjugated chain, is taken as the x axis. The α^2 and $\mu\beta$ terms of γ_{xxxx} denote the first and second terms, respectively, of Eqs. 31 and 34.

It is clearly seen that the one-dimensional model (first row) provides good approximations to the total vibrational polarizability and hyper-polarizabilities arising from the modes in the 2000–500 cm^{-1} region (second row). The largest values calculated for one normal mode in this frequency region (third row) are significantly smaller, indicating that the ICM picture provides a better one-dimensional model than the normal mode picture. Comparing the values on the second and fourth rows, however, it is noticed that the vibrational polarizability and hyper-polarizabilities arising from the modes below 500 cm^{-1} are also large. Therefore, it may be said that the ICM picture is useful to obtain insight into the contribution of the vibrational modes in the mid-IR region to the vibrational polarizability and hyperpolarizabilities, but the contribution of the low-frequency modes may have to be considered separately.

7. Electrostatic Vibrational Coupling between Molecules: Transition Dipole Coupling (TDC)

In Sec. 4 and the beginning of Sec. 3, we have discussed the cases where the electric field on a solute molecule is modulated by *intermolecular* vibrational modes Q_r and this modulation affects the vibrational frequency shift δk_p of the *intramolecular* mode q_p of the solute. It has been implicitly assumed in this discussion that the frequencies of q_p and Q_r are significantly different from each other, so that the electric field is *slowly* modulated as compared with the time scale of q_p (case 2 in Sec. 2).

When the electric field is modulated by *intramolecular* vibrational modes of other molecules in the system, we have the situation discussed as case 3 in Sec. 2. In this case, the vibrational modes of different molecules are directly coupled with each other as shown in Eq. 10. The effect of this direct coupling, called transition dipole coupling (TDC),[7-14,25-27] is most clearly seen in the resonant case, where the

Transfer of vibrational excitation
(time domain picture)

ω_m

Delocalization of vib. modes
(freq. domain picture)

ω_n

Off-diagonal coupling

molecule m molecule n

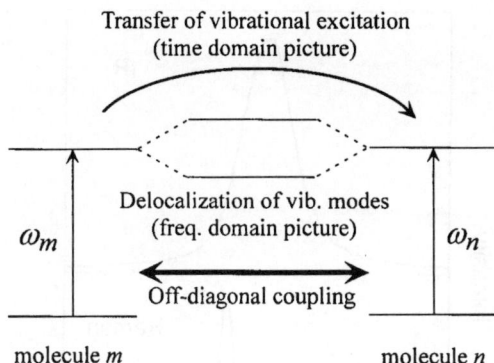

Fig. 10. Scheme of the resonant vibrational coupling between molecules in the frequency- and time-domain pictures.

intrinsic frequencies of the interacting vibrational modes are sufficiently close to each other as compared with the magnitude of the coupling.[27]

In the frequency-domain picture, the resonant vibrational coupling between molecules gives rise to delocalization of vibrational modes. When the vibrations of two molecules are resonantly coupled as shown in Fig. 10, we obtain two delocalized normal modes. One of them is the symmetric (in-phase) linear combination of the vibrations of the two molecules, and the other is the antisymmetric (out-of-phase) counterpart. The sign of the vibrational coupling constant determines which of the two delocalized modes is located on the higher-frequency side. When the vibrations of many molecules are coupled with varying magnitudes in the liquid, the vibrational pattern of each delocalized normal mode is not so simple. However, even in this case, the frequency positions of delocalized normal modes are affected in a similar way by the vibrational patterns and the vibrational coupling constants. The noncoincidence effect (NCE) of vibrational spectra,[23–45] which is discussed below, is a phenomenon that is relevant to this frequency-domain picture of intermolecular resonant vibrational coupling.

In the time-domain picture, a vibrational excitation localized on one or a few molecules cannot be regarded as an eigenstate of the liquid system, and is rapidly transferred to other molecules, as shown also in Fig. 10. When the molecules involved in the coupling are oriented

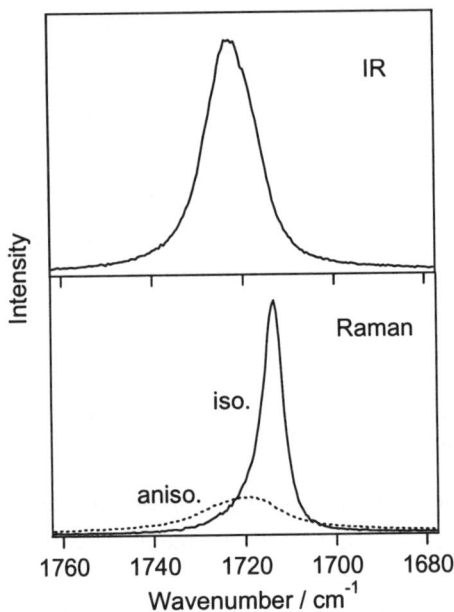

Fig. 11. Calculated IR and polarized Raman spectra in the C=O stretching region of liquid acetone.[99] In the Raman spectrum, the isotropic and anisotropic components are shown in solid and dotted lines, respectively.

randomly in various directions, the polarization of the vibrational excitations initially made will decay rapidly as time evolves. Experimentally, such a phenomenon has been observed for the OH stretching mode of liquid water as the decay of the transient IR absorption anisotropy.[46,47]

As an example of the vibrational modes with large effects of resonant vibrational coupling between molecules, we take the case of the C=O stretching mode of liquid acetone.[26,29,30,35,36,38,39,99] The IR and polarized Raman spectra calculated for this mode[99] are shown in Fig. 11. It is clearly seen that the isotropic component of the Raman spectrum is located at a lower-frequency position than the anisotropic component and the IR band. This phenomenon is called the noncoincidence effect (NCE). The NCE defined as $\tilde{\nu}_{NCE} = \tilde{\nu}_{aniso} - \tilde{\nu}_{iso}$ is positive (ca. 5.5 cm^{-1}) in this case. The magnitude and the sign of the NCE depend on the nature of the vibrational mode and the structure of the liquid. For example, the NCE is large and positive (ca. 40 cm^{-1}) for the O–H stretching mode of liquid methanol,[40,42,45] whereas it is negative (ca. –4 cm^{-1}) for the C–O stretching mode of the same liquid.[41,42,45] This dependence arises from the sensitivity of the TDC constants defined in Eq. 10 to the distances and relative orientations of the molecules involved in the coupling. As a result, by analyzing the spectroscopic features related to the TDC, such as the NCE, we can obtain information on the liquid structures.

As mentioned at the beginning of Sec. 3.1, vibrational coupling due to the TDC mechanism also plays an important role in the spectral profiles of the amide I band of polypeptides and proteins.[7–14] In this case, TDC constants depend on the distances and relative orientations of the peptide groups involved in the coupling. It is well known that the peptide groups in typical secondary structures, such as α-helix and β-sheet, have different relative orientations. As a result, spectral profiles determined by the TDC mechanism are sensitive to the secondary structures.

8. Conclusions

In the present chapter, we have discussed various types of the responses of molecular vibrations to intermolecular electrostatic interactions and their effects on the vibrational spectroscopic features. Electrical property derivatives, which appear in the expansion of the potential energy with respect to the electric field and vibrational coordinates (Eq. 1), are important for describing the responses of molecular vibrations to electric field in general.

When the electric field is static or slowly varying as compared with the time scale of the intramolecular vibrational modes (case 2 in Sec. 2), the molecular structure is displaced as shown in Eq. 2. Those displacements give rise to shifts of the quadratic force constants k_p (and hence the vibrational frequencies) of mechanically anharmonic vibrational modes as shown in Eq. 3. Combined with the effect of the electrical anharmonicity, δk_p is expressed as Eq. 4 or 11. The contributions of the first and second terms in these equations are examined in Sec. 3.1 by taking the case of *N*-methylacetamide as an example. In many cases, electric fields operating between molecules are considered to originate mainly from partial charges on atomic sites or from molecular dipoles (and molecular charges for charged species), but the higher multipoles may also be important, especially for molecules containing Cl, Br, and/or S atoms, as shown in Sec. 3.2. The structural displacements induced by inter-molecular electrostatic interactions also give rise to molecular polarization as shown in Eq. 5, from which the formula for the vibrational polarizability (Eq. 6) is derived. The intensity-carrying mode (ICM) theory discribed in Sec. 6 is useful for characterizing the vibrational motions that are really responsible for this phenomenon.

In the liquid state, electric fields coming from the surrounding molecules are modulated by the liquid dynamics. We can see this effect as the changes in the vibrational band shapes, such as band broadening arising from the distribution of δk_p. To facilitate extraction of the inter-molecular vibrational motions that are really responsible for the modula-tion of the electric field, the field-modulating mode (FMM) picture is developed as shown in Sec. 4. The effect of the modulation of the electrical properties of liquids induced by liquid dynamics is also seen in the vibrational band profiles in the low-frequency region as explained in Sec. 5. The enhancement of molecular dipoles and polarizabilities due to the (extended) dipole-induced dipole (DID) mechanism is important in this respect.

When the electric field operating on a molecule is modulated by *intramolecular* vibrations of different molecules (case 3 in Sec. 2), we have the situation where the vibrations of those different molecules are directly coupled with each other as shown in Eqs. 9 and 10. The effect of this direct coupling, called transition dipole coupling (TDC), is most clearly seen in the resonant case, as the noncoincidence effect (NCE) in the frequency domain, and the decay of transient absorption anisotropy in the time domain, as indications of the delocalization of vibrational excitations.

Summarizing the discussion described above, molecular vibrations respond to intermolecular electrostatic interactions in various ways. It is therefore expected that useful information on the structures, dynamics, and interactions in condensed phases are obtained by analyzing the vibrational spectroscopic features arising from those responses.

Acknowledgments

The studies described in this chapter have been supported by Grant-in-Aids for Scientific Research from the Ministry of Education, Culture, Sports, Science, and Technology (MEXT) and from the Japan Society for the Promotion of Science (JSPS). Part of the calculations was carried out on the computers at the Research Center for Computational Science of the National Institutes of Natural Sciences at Okazaki.

References

1. H. H. Mantsch, H. L. Casal and R. N. Jones, in *Spectroscopy of Biological Systems, Advances in Spectroscopy, Vol. 13*, Ed. R. J. H. Clark and R. E. Hester (Wiley, New York, 1986), p.1.
2. S. Krimm and J. Bandekar, *Adv. Protein Chem.* **38**, 181 (1986).
3. H. Torii and M. Tasumi, in *Infrared Spectroscopy of Biomolecules*, Ed. H. H. Mantsch and D. Chapman (Wiley-Liss, New York, 1996), p. 1.
4. T. Miyazawa and E. R. Blout, *J. Am. Chem. Soc.* **83**, 712 (1961).
5. D. M. Byler and H. Susi, *Biopolymers* **25**, 469 (1986).
6. W. K. Surewitz, H. H. Mantsch and D. Chapman, *Biochemistry* **32**, 389 (1993).
7. S. Krimm and Y. Abe, *Proc. Natl. Acad. Sci. U. S. A.* **69**, 2788 (1972).
8. W. H. Moore and S. Krimm, *Proc. Natl. Acad. Sci. U. S. A.* **72**, 4933 (1975).
9. H. Torii and M. Tasumi, *J. Chem. Phys.* **96**, 3379 (1992).
10. H. Torii and M. Tasumi, *J. Raman Spectrosc.* **29**, 81 (1998).
11. S. Cha, S. Ham and M. Cho, *J. Chem. Phys.* **117**, 740 (2002).
12. Q. Huang and R. Schweitzer-Stenner, *J. Raman Spectrosc.* **35**, 586 (2004).
13. R. Huang, J. Kubelka, W. Barber-Armstrong, R. A. G. D. Silva, S. M. Decatur and T. A. Keiderling, *J. Am. Chem. Soc.* **126**, 2346 (2004).
14. C. M. Cheatum, A. Tokmakoff and J. Knoester, *J. Chem. Phys.* **120**, 8201 (2004).
15. H. Torii, T. Tatsumi and M. Tasumi, *J. Raman Spectrosc.* **29**, 537 (1998).
16. S. Ham, J. H. Kim, H. Lee and M. Cho, *J. Chem. Phys.* **118**, 3491 (2003).
17. N. Kobko and J. J. Dannenberg, *J. Phys. Chem. A* **107**, 6688 (2003).
18. T. M. Watson and J. D. Hirst, *J. Phys. Chem. A* **107**, 6843 (2003).
19. P. Bouř and T. A. Keiderling, *J. Chem. Phys.* **119**, 11253 (2003).
20. H. Torii, *J. Phys. Chem. A* **108**, 7272 (2004).
21. H. Torii, *J. Mol. Struct.* **735/736**, 21 (2005).
22. S. Ham, S. Hahn, C. Lee, T.-K. Kim, K. Kwak and M. Cho, *J. Phys. Chem. B* **108**, 9333 (2004).
23. G. Fini, P. Mirone and B. Fortunato, *J. Chem. Soc. Faraday Trans. 2*, **69**, 1243 (1973).

212   *H. Torii*

24. C. H. Wang and J. McHale, *J. Chem. Phys.* **72**, 4039 (1980).
25. D. E. Logan, *Chem. Phys.* **103**, 215 (1986).
26. H. Torii and M. Tasumi, *J. Chem. Phys.* **99**, 8459 (1993).
27. H. Torii, in *Novel Approaches to the Structure and Dynamics of Liquids: Experiments, Theories and Simulations*, Ed. J. Samios and V. A. Durov (Kluwer, Dordrecht, The Netherlands, 2004), p. 343.
28. M. G. Giorgini, M. Musso and P. Ottaviani, *Mol. Phys.* **99**, 1485 (2001).
29. W. Schindler, P. T. Sharko and J. Jonas, *J. Chem. Phys.* **76**, 3493 (1982).
30. M. G. Giorgini, G. Fini and P. Mirone, *J. Chem. Phys.* **79**, 639 (1983).
31. V. M. Shelley, A. Talintyre, J. Yarwood and R. Buchner, *Faraday Discuss. Chem. Soc.* **85**, 211 (1988).
32. H. D. Thomas and J. Jonas, *J. Chem. Phys.* **90**, 4144 (1989).
33. V. M. Shelley and J. Yarwood, *Chem. Phys.* **137**, 277 (1989).
34. A. Mortensen, O. Faurskov Nielsen, J. Yarwood and V. Shelley, *J. Phys. Chem.* **98**, 5221 (1994).
35. M. Musso, M. G. Giorgini, G. Döge and A. Asenbaum, *Mol. Phys.* **92**, 97 (1997).
36. J. E. Bertie and K. H. Michaelian, *J. Chem. Phys.* **109**, 6764 (1998).
37. R. Schweitzer-Stenner, G. Sieler, N. G. Mirkin and S. Krimm, *J. Phys. Chem. A* **102**, 118 (1998).
38. H. Torii, M. Musso, M. G. Giorgini and G. Döge, *Mol. Phys.* **94**, 821 (1998).
39. H. Torii, *J. Phys. Chem. A* **106**, 3281 (2002).
40. C. Perchard and J. P. Perchard, *J. Raman Spectrosc.* **3**, 277 (1975).
41. T. W. Zerda, H. D. Thomas, M. Bradley and J. Jonas, *J. Chem. Phys.* **86**, 3219 (1987).
42. M. Musso, H. Torii, P. Ottaviani A. Asenbaum, and M. G. Giorgini, *J. Phys. Chem. A* **106**, 10152 (2002).
43. T. Uemura, S. Saito, Y. Mizutani and K. Tominaga, *Mol. Phys.* **103**, 37 (2005).
44. A. Sokolowska and Z. Kecki, *J. Raman Spectrosc.* **24**, 331 (1993).
45. Z. Kecki, A. Sokolowska and J. Yarwood, *J. Mol. Liq.* **81**, 213 (1999).
46. S. Woutersen and H. J. Bakker, *Nature* **402**, 507 (1999).
47. H. Torii, *Chem. Phys. Lett.* **323**, 382 (2000).
48. H. Torii, *Vib. Spectrosc.* **29**, 205 (2002).
49. J. M. Luis, M. Duran and J. L. Andrés, *J. Chem. Phys.* **107**, 1501 (1997).
50. E. S. Park and S. G. Boxer, *J. Phys. Chem. B* **106**, 5800 (2002).
51. H. Torii, *J. Chem. Phys.* **119**, 2192 (2003).
52. D. M. Bishop, *Rev. Mod. Phys.* **62**, 343 (1990).
53. D. M. Bishop, *Adv. Chem. Phys.* **104**, 1 (1998).
54. B. Kirtman, B. Champagne and J. M. Luis, *J. Comput. Chem.* **21**, 1572 (2000).
55. D. M. Bishop, *J. Chem. Phys.* **98**, 3179 (1993).
56. A. Chattopadhyay and S. G. Boxer, *J. Am. Chem. Soc.* **117**, 1449 (1995).
57. E. S. Park, S. S. Andrews, R. B. Hu and S. G. Boxer, *J. Phys. Chem. B* **103**, 9813 (1999).

58. S. S. Andrews and S. G. Boxer, *J. Phys. Chem. A* **106**, 469 (2002).

59. H. Torii and M. Tasumi, in *Computer Aided Innovation of New Materials II*, Ed. M. Doyama, J. Kihara, M. Tanaka and R. Yamamoto (Elsevier, Amsterdam, 1993), p. 1287.

60. R. Kubo, *Adv. Chem. Phys.* **15**, 101 (1969).

61. P. Hamm, M. Lim, W. F. Degrado and R. M. Hochstrasser, *Proc. Natl. Acad. Sci. U. S. A.* **96**, 2036 (1999).

62. S. Woutersen and P. Hamm, *J. Phys. Chem. B* **104**, 11316 (2000).

63. A. Piryatinski, S. Tretiak, V. Chernyak and S. Mukamel, *J. Raman Spectrosc.* **31**, 125 (2000).

64. A. M. Moran, S.-M. Park and S. Mukamel, *J. Chem. Phys.* **118**, 9971 (2003).

65. R. Schweitzer-Stenner, F. Eker, Q. Huang, and K. Griebenow, *J. Am. Chem. Soc.* **123**, 9628 (2001).

66. Y. Mu, D. S. Kosov and G. Stock, *J. Phys. Chem. B* **107**, 5064 (2003).

67. F. Eker, K. Griebenow, X. Cao, L. A. Nafie, and R. Schweitzer-Stenner, *Biochemistry* **43**, 613 (2004).

68. H. Torii, T. Tatsumi, T. Kanazawa and M. Tasumi, *J. Phys. Chem. B* **102**, 309 (1998).

69. T. M. Watson and J. D. Hirst, *J. Phys. Chem. A* **106**, 7858 (2002).

70. S. Ham and M. Cho, *J. Chem. Phys.* **118**, 6915 (2003).

71. N. Kobko, L. Paraskevas, E. del Rio and J. J. Dannenberg, *J. Am. Chem. Soc.* **123**, 4348 (2001).

72. N. Kobko and J. J. Dannenberg, *J. Phys. Chem. A* **107**, 10389 (2003).

73. H. Guo, N. Gresh, B. P. Roques and D. R. Salahub, *J. Phys. Chem. B* **104**, 9746 (2000).

74. Y.-D. Wu and Y.-L. Zhao, *J. Am. Chem. Soc.* **123**, 5313 (2001).

75. Y.-L. Zhao and Y.-D. Wu, *J. Am. Chem. Soc.* **124**, 1570 (2002).

76. W. A. Herrebout, K. Clou and H. O. Desseyn, *J. Phys. Chem. A* **105**, 4865 (2001).

77. H. Guo and M. Karplus, *J. Phys. Chem.* **96**, 7273 (1992).

78. H. Guo and M. Karplus, *J. Phys. Chem.* **98**, 7104 (1994).

79. W.-G. Han, and S. Suhai, *J. Phys. Chem.* **100**, 3942 (1996).

80. T. Keyes, D. Kivelson and J. P. McTague, *J. Chem. Phys.* **55**, 4096 (1971).

81. B. M. Ladanyi and T. Keyes, *Mol. Phys.* **33**, 1063 (1977).

82. H. Torii, *Chem. Phys. Lett.* **365**, 27 (2002).

83. I. R. McDonald, D. G. Bounds and M. L. Klein, *Mol. Phys.* **45**, 521 (1982).

84. H. Torii, *Chem. Phys. Lett.* **393**, 153 (2004).

85. T. Keyes, *J. Phys. Chem. A* **101**, 2921 (1997).

86. R. M. Stratt, *Acc. Chem. Res.* **28**, 201 (1995).

87. H. Torii, *J. Phys. Chem. A* **106**, 1167 (2002).

88. O. Faurskov Nielsen, *Ann. Rep. Prog. Chem., Sect. C, Phys. Chem.* **90**, 3 (1993).

89. O. Faurskov Nielsen, *Ann. Rep. Prog. Chem., Sect. C, Phys. Chem.* **93** 57 (1996).

90. H. Torii, *Chem. Phys. Lett.* **353**, 431 (2002).

91. L. C. Geiger and B. M. Ladanyi, *Chem. Phys. Lett.* **159**, 413 (1989).
92. S. Mukamel, *Principles of Nonlinear Optical Spectroscopy* (Oxford University Press, New York, 1995).
93. H. Torii, *J. Comput. Chem.* **23**, 997 (2002).
94. H. Torii, Y. Ueno, A. Sakamoto and M. Tasumi, *J. Phys. Chem. A* **103**, 5557 (1999).
95. D. Lu, G. Chen, J. W. Perry and W. A. Goddard, III, *J. Am. Chem. Soc.* **116**, 10679 (1994).
96. C. Castiglioni, M. Del Zoppo and G. Zerbi, *Phys. Rev. B* **53**, 13319 (1996).
97. H. Torii and M. Tasumi, *J. Phys. Chem. B* **101**, 466 (1997).
98. M. Cho, *J. Phys. Chem. A* **102**, 703 (1998).
99. H. Torii, M. Musso and M. G. Giorgini, in preparation.

CHAPTER 7

THE (HYPER)POLARIZABILITIES OF LIQUID WATER MODELLED USING COUPLED CLUSTER/MOLECULAR MECHANICS RESPONSE THEORY METHODS

Jacob Kongsted, Anders Osted, Kurt V. Mikkelsen

*Department of Chemistry, H. C. Ørsted Institute, University of Copenhagen,
Universitetsparken 5, DK-2100 Copenhagen Ø, Denmark
E-mail: kongsted@theory.ki.ku.dk (J. Kongsted)
E-mail: osted@theory.ki.ku.dk (A. Osted)
E-mail: kmi@theory.ki.ku.dk (K.V. Mikkelsen)*

Ove Christiansen

*Department of Chemistry, University of Aarhus,
Langelandsgade 140, DK-8000 Aarhus C, Denmark
E-mail: ove@chem.au.dk (O. Christiansen)*

This chapter concerns highly accurate methods for predicting molecular properties of molecules in solution. Our emphasize is on solvent effects or the description of a localized molecular fragment within a larger molecular system, where we solve the problem at hand by considering the effect of a medium, e.g. the solvent or the surrounding molecular system, on the relevant part of the system. Here, we refer to the fact that most chemistry in larger molecular systems may be located at a specific region, while the rest of the molecule is almost unchanged. For example, in the case of a well localized electronic transition in a solute molecule, the solvent molecules may be seen as secondary molecules, since they are not directly involved in the physical process. Here, we will focus on the molecular properties of liquid water. We present and review the theory and applications of two coupled cluster methodologies: (I) the coupled cluster/dielectric continuum model (CC/DC) and (II) the coupled cluster/dielectric continuum model (CC/DC).

1. Introduction

A useful approach for describing a solvated system is through classical simulations using either Molecular Dynamics (MD) or Monte Carlo (MC) methods [1]. While these methods generally perform well for electrical properties of ground state molecules, problems arise when considering excited states. Methods which overcome these problems while still being feasible for larger molecular structures, belong to the mixed quantum/classical models. Here, the total system is divided into at least two parts and the chemical interesting part, is then described using a (high level) quantum mechanical method, whereas the larger part is described using a much coarser level of theory. The dielectric continuum (DC) models [2,3,4,5,6,7,8,9,10,11,12,13,14,15,16,17,18,19,20], which belong to the class of implicit solvation models, treat the solvent as a continuous medium characterized by a (macroscopic) dielectric constant and the solute resides inside a cavity in the dielectric continuum. The DC models include the effects of an average solvent structure around the solute molecule. However, the DC models often suffer from problems due to the complete neglect of the specific solvent structure, which may lead to a poor description of specific intermolecular interactions. Obviously, the concept of a cavity is due to man rather than nature and the treatment of the shape of the cavity, which ranges from spherical to molecular shaped [21], is an issue giving rise to some complications and debate. The advantages of the dielectric models are (i) the drastic reduction in the degrees of freedom as compared to direct inclusion of solvent molecules, (ii) the relatively simple coupling between the quantum mechanical and classical systems and (iii) the fact that these models are not that demanding in computational resources. Thus, solvent effects on molecules feasible for a vacuum quantum mechanical calculation may in general also be described using dielectric continuum models. Furthermore, temperature effects may approximately be included through the temperature dependence of the dielectric constant.

Turning now to the class of explicit solvation models, e.g. models in which the discrete nature of the solvent molecules are kept intact, we find the so-called combined quantum mechanical/molecular mechanical (QM/MM) [22,23,24,25,26,27,28,29,30,31,32,33,34,35,36,37,38,39,40,41,42,43,44,45,46,47] method very promising. In the following we will refer to atoms/molecules belonging to the part of the system treated using quantum mechanics as QM particles and the atoms/molecules belonging to the classically treated part of the system as MM particles. In the QM/MM model, the total configuration of the solute-solvent system is derived using simulations techniques followed by one or more quantum mechanical calculations on the interesting part of the system. The many-body quantum mechanical Hamiltonian is in these quantum calculations modified so that the effects of the surrounding medium are introduced. Normally, this procedure involves the construction of effective one-electron operators, but direct approaches [28,29], which in addition include modifications of the two electron part of the molecular Hamiltonian, have also been considered. The QM/MM method also allows for an accurate description of a smaller part of a larger system, for example a consideration of the active site of an enzyme. In fact, the first formulation of the QM/MM method was on theoretical studies of enzymatic reactions [23].

A special class of QM/MM methods, which is actually rather to be seen as intermediate between the implicit and explicit solvation models, describes the solvent using a Langevin dipole model. This method was originally formulated by Warshel and coworkers [23,42,48,49,50] and has recently been modified by Lipiński et al. [51]. In the Langevin dipole model, a three dimensional grid of rotatable point dipoles is established in the region around the solute. These dipoles represent the molecular dipole moments of the solvent molecules. The permanent part of this is determined from a Langevin type equation, while the induced part is described by assigning to each dipole a point polarizability. The Langevin dipole model has for example successfully been used to study solvent effects on electronic properties such as the $\pi \to \pi^*$ transition in large conjugated molecules [42]. Another intermediate method is the combined reference interaction site model/self-consistent-field method (RISM/SCF) by Ten-no et al. [45,52]. In this method, the solute-solvent structure, e.g. site-site correlation functions, is found by solving the RISM integral equations (using for example vacuum values for the solute partial charges) and the electronic structure of the solute molecule is then determined at the SCF level of theory using a solvated Fock operator, which takes into account the electrostatic potential induced by the solvent molecules. However, since the partial charges assigned to the solute molecule depend on the solute wave function, the RISM and SCF equations are coupled. Thereby, new partial solute charges are determined and the RISM equations are solved again using the updated values for the solute partial charges. This procedure is continued until both the electronic and solvent structures becomes self-consistent. The advantage of using the RISM method to determine the solute-solvent structure is due to the potential reduction in computational time as compared to MD or MC methods. Recently, the RISM method has also been combined with a multiconfigurational self-consistent-field (MCSCF) wave function description [53].

An interesting QM/MM model has been proposed by Sánchez et al. in which the solvent response is described by means of point charges in a way so that they reproduce the average solvent electrostatic potential (ASEP) calculated from MD simulations [34]. The ASEP is then introduced into the coupled Hamiltonian and new solute partial charges are determined which serves as input for a new MD simulation. This procedure is finally terminated upon convergence in the solute charges and solute energy. This ASEP/MD method has been applied to determine lower order electric molecular properties of molecules in solution using either a SCF [34] or a MCSCF [31] descriptions of the solute molecule. For a very recent application of the ASEP/MD to the study of solvent effects on the $n \to \pi^*$ electronic transition in acrolein, we refer to Ref. [54], which also includes a list of references to previous studies using the ASEP/MD method.

An alternative to QM/MM methods would be to consider the super-molecular or semi-continuum approaches [55]. In the former approach, the whole or a larger part of the QM system is treated using quantum mechanics, and, for non-local properties, the relevant property is then evaluated using the differential shell approach [55]. The latter approach includes, in addition to the super-molecular system, a dielectric continuum to describe the long range electrostatic forces. In order to calculate solvent shifts, e.g. changes in the molecular properties as compared to the vacuum reference, basis set superpositions errors (BSSE) must be considered. The QM/MM approach, as presented in this work, avoids this problem naturally. Thus, any super-molecular calculation combined with the differential shell approach is not necessarily a more accurate method than employing the QM/MM scheme.

The QM/MM model may be improved by increasing either the QM level or the parametrization of the MM force field. Our QM/MM approach is based on the notion that the solute molecule is treated using the dynamical correlated coupled cluster approach. Furthermore, while the majority of QM/MM methods make use of effective pair potentials in both the MM/MM and QM/MM couplings, we include an explicit term accounting for the (many-body) polarization of the MM system. As shown later, this leads to a reaction field like Hamiltonian in line with a QM system coupled to a dielectric continuum. Thus, the strengths of our QM/MM method are (i) the use of a highly correlated wave function approach (ii) an explicit account of polarization effects in the MM environment and (iii) the use of response theory to calculate molecular properties. We note that the response theory formalism allows calculations of dynamical molecular properties (as well as excitation energies and transition properties) without the restrictions dictated by finite field (FF) or few-state methods. By introducing the combined coupled cluster/molecular mechanics method we thereby open for a systematic and highly accurate description of solvent effects on molecular properties.

We will focus on the calculation of optical molecular properties represented by the investigations of the linear and non-linear optical (NLO) properties of liquid water. Theoretical predictions of NLO properties of molecules in both vacuum and condensed phases are a very important task and may eventually lead to the tailoring of NLO materials. Accurate calculations of these properties for atoms and small molecules have become feasible either through the use of the FF technique or through the more elaborate use of response theory. In contrast, most experiments of NLO properties are mainly performed on systems in condensed phases. Spectra and (hyper)polaizabilities may be very sensitive to environmental effects and hence, to compare theoretically derived NLO properties with experimental investigations, requires in most cases an accurate description of both the frequency dependence of the molecular property and the surrounding medium. Both of these effects are accounted for in the outlined QM/MM approach.

This chapter is structured in the following way: In the second section we introduce coupled cluster theory for states in vacuum. In the third section we discuss the basics of two solvation models (i) the combined coupled cluster/dielectric continuum (CC/DC) model and (ii) the combined coupled cluster/molecular mechanics (CC/MM) model. Also, we shortly discuss the implementation of these two models as outlined in the DALTON quantum chemistry program package [56]. In the fourth section we outline the derivation of the CC/DC and CC/MM linear, quadratic and cubic response functions. Furthermore, we discuss the implementation of these response functions in DALTON. In the fifth section we present a study of some relevant linear and non-linear electric properties of liquid water as described using the CC/DC and CC/MM models. The final section presents a summary.

2. Coupled Cluster Theory for States in Vacuum

Our focus is on a coupled cluster (CC) description of the solute molecule and therefore we begin with a short introduction to coupled cluster theory for the electronic structure of molecules in vacuum. The second quantization (SQ) representation of the molecular electronic Hamiltonian is given as [57]

$$\hat{H} = \sum_{pq} h_{pq}\hat{E}_{pq} + \frac{1}{2}\sum_{pqrs} g_{pqrs}\hat{e}_{pqrs} + h_{nuc}. \tag{1}$$

where

$$h_{pq} = \int \phi_p^*(\vec{r})\left(-\frac{1}{2}\nabla^2 - \sum_{m=1}^{M}\frac{Z_m}{|\vec{r}-\vec{R}_m|}\right)\phi_q(\vec{r})d\vec{r} \tag{2}$$

$$g_{pqrs} = \int\int \frac{\phi_p^*(\vec{r_1})\phi_r^*(\vec{r_2})\phi_q(\vec{r_1})\phi_s(\vec{r_2})}{|\vec{r_1}-\vec{r_2}|}d\vec{r_1}d\vec{r_2} . \tag{3}$$

The one-electron excitation operator, $\hat{E}_{pq}$, is defined as

$$\hat{E}_{pq} = \sum_{\sigma}\hat{a}_{p\sigma}^{\dagger}\hat{a}_{q\sigma} . \tag{4}$$

Similarly, the two-electron excitation operator, $\hat{e}_{pqrs}$, is defined as

$$\hat{e}_{pqrs} = \sum_{\sigma\tau}\hat{a}_{p\sigma}^{\dagger}\hat{a}_{r\tau}^{\dagger}\hat{a}_{s\tau}\hat{a}_{q\sigma} , \tag{5}$$

where the fermion creation, $\hat{a}_{p\sigma}^{\dagger}$, and annihilation, $\hat{a}_{p\sigma}$, operators act on the electron in the p'th orbital with the projected spin σ. These operators fulfill the usual electronic anticommutator relations [57], $[\hat{a}_{p\sigma}^{\dagger}, \hat{a}_{q\sigma}]_+ = \delta_{pq}$. The set $\{|\phi_p(\vec{r})\rangle\}$ represents the molecular orbitals. The last term in eq.(1) is the nuclear repulsion energy.

Turning now to the coupled cluster wave function, we note that this wave function ansatz was first introduced in nuclear physics in the early sixties [58,59] and later applied in quantum chemistry by Čížek et al. in the sixties and early seventies [60,61,62]. In the eighties, further development of coupled cluster electronic structure theory were considered by several groups including contributions from the groups of Bartlett [63,64], Pople [65,66] and Schaefer [67,68]. After the general acceptance of coupled cluster theory in the late eighties, this area has been the subject of much research in terms of both formal developments, implementational strategies and numerous applications all around the world. Today, the coupled cluster wave function is unequivocally referred to as *the* choice of *ab-initio* electronic structure model for high accuracy calculations. For more references and historical as well as conceptual and numerical aspects of coupled cluster theory, we refer for example to the text by Gauss [69].

The CC wave function is defined through the following exponential ansatz [60,70]

$$|CC\rangle = \exp(\hat{T})|HF\rangle \tag{6}$$

with the cluster operator, $\hat{T}$, written as

$$\hat{T} = \hat{T}_1 + \hat{T}_2 + \hat{T}_3 + \ldots + \hat{T}_n = \sum_{i=1}^{n} \sum_{\mu_i} t_{\mu_i} \hat{\tau}_{\mu_i} . \tag{7}$$

The t_{μ_i} parameters are the excitation amplitudes and $\hat{\tau}_{\mu_i}$ are the corresponding i-electron excitation operators, which simultaneously excite i-electrons from occupied to unoccupied orbitals. The one-electron excitation operator (spin-coupled to a singlet) can for example be represented as $\hat{E}_{ai}$. The state, $|HF\rangle$, is the Hartree-Fock reference wave function. When the cluster expansion in eq.(7) includes all possible electronic excitations in the molecular system, the parameterization of the CC wave function possesses the flexibility to represent the full configuration interaction (FCI) wave function.

In contrast to most other quantum chemical wave function approaches, the coupled cluster wave function is treated using a non-variational procedure. Thus, the excitation amplitudes defining the cluster operator are not found using a standard variational procedure on the energy functional but rather using a projection technique. Inserting the wave function ansatz into the time-independent Schrödinger equation and projecting onto the reference state, we have for the coupled cluster energy

$$E_{CC} = \langle HF| \exp(-\hat{T})\hat{H}\exp(\hat{T})|HF\rangle \tag{8}$$

where $\hat{H}$ is the electronic many-body Hamiltonian. In a similar way, we may project onto the set of excited Slater determinants

$$e_{\mu_i} = \langle \mu_i| \exp(-\hat{T})\hat{H}\exp(\hat{T})|HF\rangle = 0 \tag{9}$$

where $\langle \mu_i| = \langle HF|\hat{\tau}_{\mu_i}^\dagger$, $\langle \mu_i|\nu_j\rangle = \delta_{\mu,\nu}\delta_{i,j}$, defining the amplitude equations. In eqs.(8 - 9) we have, before the projection is carried out, pre-multiplied from the left with the operator $\exp(-\hat{T})$, leading to the so-called linked coupled cluster equations [57]. These equations are often written in terms of the similarity-transformed Hamiltonian, $\tilde{H}^T \equiv \exp(-\hat{T})\hat{H}\exp(\hat{T})$. The idea in expressing eqs.(8 - 9) in terms of $\tilde{H}^T$ is that a Baker-Campbell-Hausdorff (BCH) expansion may be carried out which, for many purposes, leads to computationally convenient expressions. Also, the linked version of coupled cluster theory represents a convenient way in which approximations in the amplitude equations, e.g. eq.(9), may be introduced and still preserving the size-extensivity of the coupled cluster model (see the following discussion in section 2.1).

Concerning the evaluation of molecular properties, we consider a variational formulation of the quantum mechanical method. This is simply due to the fact that the Hellman-Feynman theorem, in its standard form, is not valid for non-variational wave functions [57]. It is possible to introduce a variational Lagrangian as [71,72,73]

$$L_{CC}(t,\bar{t}) = E_{CC}(t) + \sum_{i,\mu_i} \bar{t}_{\mu_i} e_{\mu_i} = E_{CC}(t) + \bar{t}e(t) \qquad (10)$$

where the vector $\bar{t}$ contains the Lagrangian multipliers and the vector $e(t)$ consists of the amplitude equations (eq.(9)). We require that the Lagrangian is simultaneously stationary with respect to t and $\bar{t}$ and we obtain

$$\frac{\partial L_{CC}(t,\bar{t})}{\partial \bar{t}_{\mu_i}} = e_{\mu_i}(t) = \langle \mu_i | \exp(-\hat{T})\hat{H}\exp(\hat{T})|HF\rangle = 0 \qquad (11)$$

$$\frac{\partial L_{CC}(t,\bar{t})}{\partial t_{\nu_j}} = \frac{\partial E_{CC}(t)}{\partial t_{\nu_j}} + \sum_{i,\mu_i} \bar{t}_{\mu_i} \frac{\partial e_{\mu_i}(t)}{\partial t_{\nu_j}}$$

$$= \eta_{\nu_j} + \sum_{i,\mu_i} \bar{t}_{\mu_i} A_{\mu_i \nu_j}$$

$$= \langle \Lambda | [\hat{H}, \hat{\tau}_{\nu_j}] | CC \rangle = 0, \qquad (12)$$

with $\langle \Lambda |$ defined as

$$\langle \Lambda | = (\langle HF| + \sum_{i,\mu_i} \bar{t}_{\mu_i} \langle \mu_i |) \exp(-\hat{T}) . \qquad (13)$$

In eq.(12) we have furthermore introduced the η vector with elements given as

$$\eta_{\nu_j} = \langle HF | \left[\hat{H}, \hat{\tau}_{\nu_j} \right] | CC \rangle \qquad (14)$$

together with the non-symmetric Jacobian

$$A_{\mu_i \nu_j} = \langle \mu_i | \exp(-\hat{T}) \left[\hat{H}, \hat{\tau}_{\nu_j} \right] | CC \rangle . \qquad (15)$$

When these equations are fulfilled, the variational Lagrangian gives the CC energy. By using this Lagrangian technique, we may show that in CC theory, an expectation value for a real operator is evaluated according to the asymmetric expression

$$\langle \hat{X} \rangle = \langle \Lambda | \hat{X} | CC \rangle. \tag{16}$$

If, on the other hand, imaginary operators are considered, the proper expression for the expectation value would be

$$\langle \hat{X} \rangle = \frac{1}{2} \left(\langle \Lambda | \hat{X} | CC \rangle + \langle \Lambda | \hat{X} | CC \rangle^* \right) \tag{17}$$

which yields manifestly real properties. This also follows from requiring the energy to be real.

2.1. Approximate Coupled Cluster Models

As discussed previously, the CC wave function possesses the flexibility to represent the FCI wave function. Truncating the expansion leads to a series of approximate CC models. For example, truncating the cluster operator after the second term, i.e. $\hat{T} = \hat{T}_1 + \hat{T}_2$, defines the coupled cluster singles and doubles (CCSD) [63] model. Furthermore, truncation of the cluster operators after inclusion of specific excitations allows for a systematic increase in accuracy using a hierarchy of CC methods, CCS,CC2,CCSD,CC3 [74,75] etc. (or SCF,MP2,CCSD,CCSD(T) [76] if only ground state static properties are studied [77]). Here, we will discuss two such approximate CC wave functions, the CCSD and CC2 truncations.

First, including all singles and doubles excitations, i.e.

$$\hat{T} = \hat{T}_1 + \hat{T}_2 \tag{18}$$

where $\hat{T}_1 = \sum_{\mu_1} t_{\mu_1} \hat{\tau}_{\mu_1}$ and $\hat{T}_2 = \sum_{\mu_2} t_{\mu_2} \hat{\tau}_{\mu_2}$ contain the single and double excitation operators, respectively, leads to the CCSD [63] model. Defining $\hat{T}_1$ transformed operators as

$$\tilde{O} = \exp{(-\hat{T}_1)} \hat{O} \exp{(\hat{T}_1)}, \tag{19}$$

the CCSD amplitude equations may be written as [78]

$$e_{\mu_1} = \langle \mu_1 | \tilde{H} + \left[\tilde{H}, \hat{T}_2 \right] HF \rangle = 0 \tag{20}$$

and

$$e_{\mu_2} = \langle \mu_2 | \tilde{H} + \left[\tilde{H}, \hat{T}_2 \right] + \frac{1}{2} \left[\left[\tilde{H}, \hat{T}_2 \right], \hat{T}_2 \right] | HF \rangle = 0. \tag{21}$$

The structure of the η^{CCSD} vector and the $\mathbf{A}^{CCSD}$ matrix become [79]

$$\eta^{CCSD} = \left(\eta_{\nu_1} \ \eta_{\nu_2} \right) = \left(\langle HF| \left[\tilde{H}, \hat{\tau}_{\nu_1} \right] |HF\rangle \ \langle HF| \left[\tilde{H}, \hat{\tau}_{\nu_2} \right] |HF\rangle \right) \tag{22}$$

and

$$\mathbf{A}^{CCSD} = \begin{bmatrix} \langle \mu_1| \left[\tilde{H} + \left[\tilde{H}, \hat{T}_2 \right], \hat{\tau}_{\nu_1} \right] |HF\rangle & \langle \mu_1| \left[\tilde{H}, \hat{\tau}_{\nu_2} \right] |HF\rangle \\ \langle \mu_2| \left[\tilde{H} + \left[\tilde{H}, \hat{T}_2 \right], \hat{\tau}_{\nu_1} \right] |HF\rangle & \langle \mu_2| \left[\tilde{H} + \left[\tilde{H}, \hat{T}_2 \right], \hat{\tau}_{\nu_2} \right] |HF\rangle \end{bmatrix} \tag{23}$$

respectively.

A very useful approximation to the CCSD truncation is the second-order approximate coupled cluster singles and doubles model (CC2). This model is defined using arguments based on perturbation theory. The Hamiltonian is written as $\hat{H} = \hat{F} + \hat{U}$, where $\hat{F}$ is the Fock operator and $\hat{U}$ the fluctuation potential, e.g. the difference between the instantaneous electron-electron repulsion and the Fock potential. Introducing this partitioning into the CCSD optimization equations, the CC2 model is defined as follows [80,81]: The singles equations are retained in their original form, but the doubles equations are approximated to be correct to first order in the fluctuation potential with singles treated as zeroth order parameters. The reason for treating the singles as zeroth order parameters is due to their importance in describing the response to external perturbations and as approximate orbital relaxation parameters. The CC2 amplitude equations accordingly become [79,80,81]

$$e_{\mu_1} = \langle \mu_1| \tilde{H} + \left[\tilde{H}, \hat{T}_2 \right] |HF\rangle = 0 \tag{24}$$

and

$$e_{\mu_2} = \langle \mu_2| \tilde{H} + \left[\hat{F}, \hat{T}_2 \right] |HF\rangle = 0 . \tag{25}$$

In the CC2 case the vector η retains the same structure as in eq.(22) [79]. However, the expression for the $\mathbf{A}^{CC2}$ matrix becomes [79]

$$\mathbf{A}^{CC2} = \begin{bmatrix} \langle \mu_1| \left[\tilde{H} + \left[\tilde{H}, \hat{T}_2 \right], \hat{\tau}_{\nu_1} \right] |HF\rangle & \langle \mu_1| \left[\tilde{H}, \hat{\tau}_{\nu_2} \right] |HF\rangle \\ \langle \mu_2| \left[\tilde{H}, \hat{\tau}_{\nu_1} \right] |HF\rangle & \langle \mu_2| \left[\hat{F}, \hat{\tau}_{\nu_2} \right] |HF\rangle \end{bmatrix} . \tag{26}$$

We note that (in vacuum) we obtain a major simplification since $\langle \mu_2 | \left[\hat{F}, \hat{\tau}_{\nu_2} \right] | HF \rangle = \delta_{\mu_i \nu_j} \omega_{\mu_2}$ where ω_{μ_2} is the sum or the difference between the orbital energies ($\epsilon_a + \epsilon_b - \epsilon_i - \epsilon_j$ for an i, j to a, b double excitation). The advantage of the CC2 model is clearly attributed to the computational scaling which is only N^5 as compared to the N^6 scaling of the CCSD model (N is the number of basis functions). In addition, special partitioning techniques may be employed giving very efficient CC2 implementations [82].

3. Solvent Models

In this section we present two different models for describing a molecular system including environmental effects. The methods considered are all within the area of mixed classical-quantum approaches. The models provide the means of describing generally a molecule in a molecular environment and not just a solvated system. This is of importance when modeling for example enzymatic processes.

The first method treats the solvent as a homogeneous dielectric medium. Furthermore, the molecular cavity is assumed to be spherical. In this approach all intrinsic structural anisotropies of the solvent are neglected. The second approach is of the combined quantum mechanics/molecular mechanics type. Here, the discrete nature of the solvent molecules is kept intact, which means that the latter model includes structural effects of the solvent. Both the dielectric continuum and the combined quantum mechanics/molecular mechanics models may be implemented for different quantum descriptions of the solute molecule.

3.1. *The Quantum Mechanical/Dielectric Continuum Model*

This section presents a basic introduction to the spherical cavity dielectric continuum solvent model as described in details in Refs. [3,6,19]. The molecular charge distribution $\rho(\vec{r}\,')$ of the solute molecule (assumed to be located inside the cavity with radius a) gives rise to an electric field $\tilde{\mathbf{E}}(\vec{r})$ at the position $\vec{r}$. Using arguments from classical electrostatics we obtain

$$\tilde{\mathbf{E}}(\vec{r}) = -\vec{\nabla} \int d\vec{r}\,' \frac{\rho(\vec{r}\,')}{|\vec{r} - \vec{r}\,'|} . \tag{27}$$

This field induces a polarization in the surrounding medium. Correspondingly an additional interaction term is included in the molecular Hamiltonian for the solute molecule. The polarization vector, $\mathbf{P}(\vec{r})$, is related to the total electric field in the medium, $\mathbf{E}(\vec{r})$, through the macroscopic susceptibilities, $\chi^{(n)}$, as

$$\mathbf{P}(\vec{r}) = \chi^{(1)}\mathbf{E}(\vec{r}) + \chi^{(2)}\mathbf{E}(\vec{r})\mathbf{E}(\vec{r}) + \chi^{(3)}\mathbf{E}(\vec{r})\mathbf{E}(\vec{r})\mathbf{E}(\vec{r}) + \ldots \quad (28)$$

where the numerical coefficients in the expansion have been absorbed into the susceptibilities. The total electric field in the medium contains the contribution in eq.(27) together with a contribution due to the polarization potential from the induced charges in the medium. Normally, this expansion is truncated after the linear term. For isotropic systems, the second term in eq.(28) vanishes identically due to symmetry reasons, e.g. following Böttcher [83] a reversal in the direction of the electric field leads to a reversal in the direction of the polarization. At the microscopic level this simply reflects the fact that the orientational average of the first hyperpolarizability vanishes identically for achiral molecules. The polarization vector in eq.(28) is more generally referred to as the dipole polarization and is the averaged microscopic dipole moment times an appropriate density. For molecules which do not possess a permanent dipole moment, e.g. centro symmetrical molecules like for example benzene, it may be necessary to consider also the quadrupole polarization, $\mathbf{Q}(\vec{r})$. An extension of the formalism to include the effect of the quadrupole polarization (in addition to the dipole polarizability) has recently been developed by Jeon and Kim [84,85,86]. However, for polar solvents such as water, the first non–vanishing and dominating multipole moment is the dipolar term. The susceptibilities introduced in eq.(28) are generally tensors, but for isotropic systems scalars may be used, i.e. the average value of the tensor quantity, $\chi^{(n)}$. However, the idea in the dielectric continuum model is that the effects of the solvent should be included in a very simple and intuitive manner. Therefore, a scalar first order susceptibility is also often used to describes the solvent in the case of solutions. We note that the scalar first order susceptibility is directly related to the dielectric constant of the solvent through

$$\epsilon^{(1)} = 1 + 4\pi\chi^{(1)} . \quad (29)$$

Including higher order susceptibilities in eq.(29) will lead to intensity dependent dielectric constants, and thereby to an intensity dependence of the refractive index. In this chapter we will only consider the linear approximation to eq.(28), and since we are mainly interested in either liquid water or water solutions, we will not consider the effect of the quadrupole polarization. This means that from now on we define the dielectric constant of the solvent as $\epsilon \equiv \epsilon^{(1)}$.

Returning to eq.(27), a multipole expansion of the solute charge distribution with the origin of the expansion at the center of the cavity leads to a total energy [3,6]

$$E^{eq} = E_{vac} + \sum_{lm} g_l(\epsilon) \langle M_{lm}^\dagger \rangle \langle M_{lm} \rangle \tag{30}$$

where the symbol *eq* means equilibrium, i.e. we have assumed that the charge distribution and the solvent is in equilibrium, a point which we will return to later. In eq.(30) the terms $\langle M_{lm} \rangle$ are the expectation values of the solute charge multipole moments. These are written in terms of the spherical polynomials, $S_{lm}(\vec{r}\,')$, as

$$\langle M_{lm} \rangle = \int_V d\vec{r}\,' \rho(\vec{r}\,')\, S_{lm}(\vec{r}\,') \tag{31}$$

$$S_{lm}(\vec{r}\,') = \sqrt{\frac{4\pi}{(2l+1)}}\,(\vec{r}\,')^l\, Y_{lm}(\Omega), \tag{32}$$

where $Y_{lm}(\Omega)$ are the spherical harmonics. The function $g_l(\epsilon)$, is the reaction field factor which depends on the spherical cavity dimension (a), the order of the multipole expansion (l) and the dielectric constant (ϵ),

$$g_l(\epsilon) = -\frac{a^{-(2l+1)}(l+1)(\epsilon-1)}{2\,[l+\epsilon(l+1)]}. \tag{33}$$

Concerning a quantum mechanical description of the solute molecule, the charge multipole density moments are expressed in terms of the real Hermitian operators, $\hat{T}_{lm}$,

$$\hat{T}_{lm} = \hat{T}_{lm}^n - T_{lm}^e\,. \tag{34}$$

The first contribution is due to the solute nuclei and is given as

$$T_{lm}^n = \sum_K Z_K t_{lm}(\vec{R}_K) \tag{35}$$

while the latter term, the electronic contribution, is described through the one-electron operator

$$\hat{T}_{lm}^e = \sum_{pq} t_{pq}^{lm} \hat{E}_{pq}. \tag{36}$$

The one-electron integrals, $t_{pq}^{lm} = \langle \phi_p | t_{lm}(\vec{r}) | \phi_q \rangle$, and the $t_{lm}(\vec{R}_K)$ elements are found as linear combinations of the spherical polynomials in eq.(32) defined as

$$t_{l,0}(\vec{r}) = S_{l,0}(\vec{r}),$$

$$t_{l,m}(\vec{r}) = \frac{1}{\sqrt{2}}\left[S_{l,m}(\vec{r}) + S_{l,-m}(\vec{r})\right]$$

$$t_{l,-m}(\vec{r}) = \frac{1}{\sqrt{2}i}\left[S_{l,m}(\vec{r}) - S_{l,-m}(\vec{r})\right]. \tag{37}$$

3.2. The Coupled Cluster/Dielectric Continuum Model

In order to derive the coupled cluster/dielectric continuum (CC/DC) model, we follow Ref. [19]. Here a variational CC/DC Lagrangian is introduced as

$$L_{CC/DC}\left(\mathbf{t},\bar{\mathbf{t}}\right) = L_{CC}^{vac}\left(\mathbf{t},\bar{\mathbf{t}}\right) + \sum_{lm} g_l(\epsilon)\langle\Lambda|\hat{T}_{lm}|CC\rangle^2. \tag{38}$$

The term $L_{CC}^{vac}\left(\mathbf{t},\bar{\mathbf{t}}\right)$ is the vacuum Lagrangian but evaluated using the condensed phase wave functions. Applying the variational condition to this Lagrangian we obtain

$$\frac{\partial\left(L_{CC/DC}\left(\mathbf{t},\bar{\mathbf{t}}\right)\right)}{\partial\bar{t}_{\mu_i}} = \langle\mu_i|\exp\left(-\hat{T}\right)\left(\hat{H} + \hat{T}^g(\epsilon)\right)\exp\left(\hat{T}\right)|HF\rangle = 0 \tag{39}$$

$$\frac{\partial\left(L_{CC/DC}\left(\mathbf{t},\bar{\mathbf{t}}\right)\right)}{\partial t_{\mu_i}} = \langle\Lambda|\left[\left(\hat{H} + \hat{T}^g(\epsilon)\right),\hat{\tau}_{\mu_i}\right]|CC\rangle = 0 \tag{40}$$

where the effective one-electron operator $\hat{T}^g(\epsilon)$ has been defined as

$$\hat{T}^g(\epsilon) = -2\sum_{lm} g_l(\epsilon)\langle\Lambda|\hat{T}_{lm}|CC\rangle\hat{T}_{lm}^e. \tag{41}$$

Note, that compared to the previous described vacuum case, the $\mathbf{t}$ and $\bar{\mathbf{t}}$ equations are coupled due to the presence of the interaction operator $\hat{T}^g(\epsilon)$ which contains an expectation value, and thereby has a simultaneous $\mathbf{t}$ and $\bar{\mathbf{t}}$ dependence.

3.3. The Combined Quantum Mechanics/Molecular Mechanics model

In this section we consider the general equations describing the combined quantum mechanics/molecular mechanics model including polarization interactions. As in the previous section, the equations are formulated using the SQ representation. In the end of this section we consider the derivation of the specific combined coupled cluster/molecular mechanics method.

The basic idea in the QM/MM approach is to divide the system of interest into several parts which are treated at different levels of theory e.g. one part is described by quantum mechanics (QM) and the other part is described by molecular mechanics (MM). In the QM part the electrons and the nuclei are treated at positions $\vec{r}_j$ and $\vec{R}_m$, respectively. In the MM part the particles are represented through effective charges positioned at the atomic sites, $\vec{R}_s$, and induced dipole moments, here located at the center-of-mass of each MM molecule $(\vec{R}_a)$. Alternatively, distributed induced dipole moments may be included. The total energy is decomposed into three contributions [23,24,25,29,30]

$$E = E_{QM} + E_{QM/MM} + E_{MM} \qquad (42)$$

where E_{QM} is the usual quantum mechanical energy described by the Born-Oppenheimer many-body vacuum Hamiltonian, $\hat{H}_{QM}$ (eq.(1)). The term, $E_{QM/MM}$, represents the interaction between the QM and MM systems and E_{MM} describes the classically treated part of the total system, which is represented by molecular mechanics. In some situations, the total system (QM + MM) may be embedded in a DC which, in order to describe the boundary between the MM and DC interface, introduces further terms in eq.(42), see for example Refs. [47,87].

The separability in eq.(42) implies a similar separability of the Hamiltonian. In a mean-field approximation, the QM/MM Hamiltonian is divided into three contributions

$$\hat{H}_{QM/MM} = \hat{H}^{el} + \hat{H}^{vdw} + \hat{H}^{pol} . \qquad (43)$$

In the following we will derive expressions for the three operators considered in eq.(43). The first term, $\hat{H}^{el}$, represents the electrostatic interactions between the electrons and the nuclei in the QM system and the partial charges in the MM system. Thus, we define $\hat{H}^{el}$ as

$$\hat{H}^{el} = -\sum_{s=1}^{S} \hat{N}_s + E_{S,N}^{el,nuc} \qquad (44)$$

with the electronic contribution given as

$$\hat{N}_s = \sum_{pq} \langle \phi_p | \frac{q_s}{|\vec{R}_s - \vec{r}|} | \phi_q \rangle \hat{E}_{pq} . \qquad (45)$$

The interactions of the MM partial charges and the QM nuclei are denoted by $E_{S,N}^{el,nuc}$. The index s runs over all the sites in the MM system. These sites are usually chosen to be located at the MM atoms and q_s is the charge at site s. The position vector for site s is labeled $\vec{R}_s$.

For the calculation of molecular properties it is usually very important to include explicit polarization effects in the interaction between the QM and MM systems. The QM system is naturally polarized by the effective charges in the MM part of the system, but this only accounts for the so-called static reaction potential [88]. In order to go beyond this approximation it is necessary to consider the response reaction potential, i.e. the response of the classically treated molecules due to the perturbation from both the molecules treated using QM and MM theory. This is somehow similar to the description of the induced polarization in the dielectric continuum case. However, the dielectric continuum model describes the solvent by a macroscopic parameter. In the case of a discrete representation of the solvent molecules a microscopic model must be used. Hence, we assign a molecular polarizability to each MM molecule. This polarizability could be either isotropic or anisotropic. These polarizabilities give rise to induced dipole moments. Below, we will consider the energy contributions due to the induced dipole moments.

The part of the interaction energy between all molecules in the system which depends directly on the induced dipole moments (μ_a^{ind}) is written as

$$E[\mu_a^{ind}] = - \sum_{a=1}^{A} \mu_a^{ind} \left(\mathbf{E}^s(\vec{R}_a) + \langle \hat{\mathbf{R}} \mathbf{r}_a \rangle + \mathbf{E}^n(\vec{R}_a) \right)$$

$$- \frac{1}{2} \sum_{a,a'(a \neq a')}^{A} \mu_a^{ind} \mathbf{T}_{aa'} \mu_{a'}^{ind} \tag{46}$$

where $\mathbf{T}_{aa'}$ is the dipole interaction tensor,

$$\mathbf{T}_{aa'} = \frac{1}{|\vec{R}_a - \vec{R}_{a'}|^3} \left[\frac{3(\vec{R}_a - \vec{R}_{a'})(\vec{R}_a - \vec{R}_{a'})^T}{|\vec{R}_a - \vec{R}_{a'}|^2} - \mathbf{1} \right] \tag{47}$$

and the index a refers to the center-of-mass of each MM molecule. Here, $\mathbf{E}^n(\vec{R}_a)$ is the electric field due to the QM nuclei at the center-of-mass a of each MM molecule. The term $\mathbf{E}^s(\vec{R}_a)$ is the electric field due to the partial charges situated at the other MM molecules. Furthermore, the operator

$\hat{\mathbf{R}}\mathbf{r}_a$ is the QM electronic electric field operator

$$\hat{\mathbf{R}}\mathbf{r}_a = \sum_{pq} \langle \phi_p | \frac{\vec{r} - \vec{R}_a}{|\vec{r} - \vec{R}_a|^3} | \phi_q \rangle \hat{E}_{pq} \ . \tag{48}$$

The induced dipole moment at the center-of-mass a of each MM molecule, μ_a^{ind}, can be related to the total electric field at this specific site, $\mathbf{E}_a^{total} = \langle \hat{\mathbf{R}}\mathbf{r}_a \rangle + \mathbf{E}^n(\vec{R}_a) + \mathbf{E}^s(\vec{R}_a) + \mathbf{E}^{ind}(\vec{R}_a)$, and in a linear approximation we have

$$\mu_a^{ind} = \alpha_a \mathbf{E}_a^{total} \tag{49}$$

where α_a is the polarizability tensor at the center-of-mass of each MM molecule. Note that this linear approximation is (microscopically) similar to the (macroscopical) linear approximation obtained from eq.(28) in the case of a dielectric continuum description of the solvent. The individual components in the expression for $\mathbf{E}_a^{total}$ are given as (i) the vector $\mathbf{E}^n(\vec{R}_a)$, which is the electric field due to the QM nuclei

$$\mathbf{E}^n(\vec{R}_a) = \sum_{n=1}^{N} \frac{Z_n(\vec{R}_a - \vec{R}_n)}{|\vec{R}_a - \vec{R}_n|^3} \tag{50}$$

where the nuclear charge is denoted by Z_n. (ii) The field $\mathbf{E}^s(\vec{R}_a)$ is the electric field at the center-of-mass a of each MM molecule due to the partial MM charges

$$\mathbf{E}^s(\vec{R}_a) = \sum_{s \notin a} \frac{q_s(\vec{R}_a - \vec{R}_s)}{|\vec{R}_a - \vec{R}_s|^3} \ , \tag{51}$$

where the summation $s \notin a$ is restricted to sites not belonging to the molecule with the center-of-mass a. (iii) The field due to the induced dipole moments, $\mathbf{E}^{ind}(\vec{R}_a)$, written in terms of the dipole tensor, $\mathbf{T}_{aa'}$,

$$\mathbf{E}^{ind}(\vec{R}_a) = \sum_{a' \neq a} \mathbf{T}_{aa'} \mu_{a'}^{ind} \tag{52}$$

In addition to eq.(46) we also have a self energy, E_{self}, of the classical induced dipole moments [89]

$$
\begin{aligned}
E_{self} &= \sum_{a=1}^{A} \int_{0}^{\mu_{a}^{ind}} \mathbf{E}_{a}^{total'} d\mu_{a}^{ind'} = \sum_{a=1}^{A} \int_{0}^{\mu_{a}^{ind}} \mu_{a}^{ind'} \alpha_{a}^{-1} d\mu_{a}^{ind'} \\
&= \frac{1}{2} \sum_{a=1}^{A} \mu_{a}^{ind} \alpha_{a}^{-1} \mu_{a}^{ind} \\
&= \frac{1}{2} \sum_{a=1}^{A} \mu_{a}^{ind} \left(\mathbf{E}^{s}(\vec{R}_{a}) + \langle \hat{\mathbf{R}} \mathbf{r}_{a} \rangle + \mathbf{E}^{n}(\vec{R}_{a}) \right) \\
&\quad + \frac{1}{2} \sum_{a,a'(a \neq a')}^{A} \mu_{a}^{ind} \mathbf{T}_{aa'} \mu_{a'}^{ind} \; .
\end{aligned}
\tag{53}
$$

Addition of the energy contributions in eq.(46) and eq.(53) gives

$$
\begin{aligned}
E[\mu_{a}^{ind}] + E_{self} &= -\frac{1}{2} \sum_{a=1}^{A} \mu_{a}^{ind} \left(\langle \hat{\mathbf{R}} \mathbf{r}_{a} \rangle + \mathbf{E}^{n}(\vec{R}_{a}) \right) \\
&\quad - \frac{1}{2} \sum_{a=1}^{A} \mu_{a}^{ind} \mathbf{E}^{s}(\vec{R}_{a}) \; .
\end{aligned}
\tag{54}
$$

The first term in eq.(54) is here referred to as the polarization energy for the QM solute in the solvent due to the MM polarizabilities. For the QM part of the system it can be described in terms of the effective operator

$$
\hat{H}^{pol} = -\frac{1}{2} \sum_{a=1}^{A} \mu_{a}^{ind} \cdot (\hat{\mathbf{R}} \mathbf{r}_{a} + \mathbf{E}^{n}(\vec{R}_{a})) \; .
\tag{55}
$$

The second term in eq.(54) only involves particles and induced moments in the MM system. Correspondingly, this last term, $-\frac{1}{2} \sum_{a=1}^{A} \mu_{a}^{ind} \mathbf{E}^{s}(\vec{R}_{a})$, is included in the MM interaction energy discussed later. However, this term is still explicitly accounted for in the optimization of the QM wave function.

The mean-field method completely neglects the coupling of instantaneous fluctuations in the charge distribution between the QM molecular system and the classically treated molecules. This means that specific intermolecular interactions beyond the electrostatic ones are not included in the coupling between the two systems. These include for example the dispersion energy. For the classical molecules close to the QM system short range contributions to the interaction energy cannot normally be neglected.

Therefore, the effect of short range interactions are introduced using a parameterized potential which includes an attractive part (mainly due to dispersion) and a repulsive component, which models the short range repulsion. Different choices of potentials can be made (see for example Ref. [90]), but the far most common form is the 6-12 Lennard-Jones potential. The energy contributions due to short range interactions are here collectively referred to as the van der Waals contributions and is thus modeled with a potential of the form

$$\hat{H}^{vdw} = \sum_{a=1}^{A} \sum_{m:center} \left[\frac{A_{ma}}{|\vec{R}_m - \vec{R}_a|^{12}} - \frac{B_{ma}}{|\vec{R}_m - \vec{R}_a|^{6}} \right] . \tag{56}$$

In eq.(56) the index a (m) refers to the center-of-mass of each MM (QM) molecule. Since this potential is independent of the electronic degrees of freedom, the van der Walls term is actually considered as a constant in connection to solving the electronic Schrödinger equation. This simplifies matter greatly, but may on the other hand also be one of the crucial assumptions of the model.

In a mean-field and linear approximation the induced dipole moments are related to the wave function through eq.(49). Accordingly, we may introduce the polarizability and the electric fields in place of the induced moments by inserting this equation once into eq.(55). Taking the expectation value of eq.(43) and inserting eq.(49) (one time) we arrive at the following expression for the QM/MM energy [29,91]

$$E_{QM/MM} = E^{vdw} + E^{el} + E^{pol}$$
$$= E^{vdw} + E_{S,N}^{el,nuc} - \sum_{s=1}^{S} \langle \hat{N}_s \rangle$$
$$- \frac{1}{2} \sum_{a=1}^{A} \langle \hat{R}\hat{r}_a \rangle^T \alpha_a \left\{ \langle \hat{R}\hat{r}_a \rangle + \mathbf{O}_a^{ns}(\vec{R}_a) \right\} + O_{ind}^{ns} . \tag{57}$$

Here the vector $\mathbf{O}_a^{ns}(\vec{R}_a)$ and the energy term O_{ind}^{ns} are defined as

$$\mathbf{O}_a^{ns}(\vec{R}_a) = 2\mathbf{E}^n(\vec{R}_a) + \mathbf{E}^s(\vec{R}_a) + \mathbf{E}^{ind}(\vec{R}_a) \tag{58}$$

and

$$O_{ind}^{ns} = -\frac{1}{2} \sum_{a=1}^{A} \left[\left(\mathbf{E}^{n'}(\vec{R}_a) \right)^T \alpha_a \left\{ \mathbf{E}^n(\vec{R}_a) + \mathbf{E}^s(\vec{R}_a) + \mathbf{E}^{ind}(\vec{R}_a) \right\} \right] . \tag{59}$$

For the optimization of the QM system coupled to the discrete and polarizable environment this is the important energy contribution along with the usual vacuum type expression $\langle \hat{H}_{QM} \rangle$ and the term in eq.(60), (discussed below) which depends on the induced dipole moments.

For the total MM energy, we decompose this energy into an intramolecular term, E_{MM}^{intra}, and an intermolecular contribution, $E_{MM/MM}$. The intermolecular MM energy is given as

$$
E_{MM/MM}[|\Psi\rangle] = \frac{1}{2} \sum_{s,s'(s \neq s')}^{S} \frac{q_s q_{s'}}{|\vec{R}_s - \vec{R}_{s'}|} - \frac{1}{2} \sum_{a=1}^{A} \mu_a^{ind} \mathbf{E}^s(\vec{R}_a)
$$
$$
+ E_{MM/MM}^{vdw} \tag{60}
$$

where the term $E_{MM/MM}^{vdw}$ is the van der Waals MM/MM energy. The second term is the MM part of the energy in eq.(54). In the optimization of the wave function we explicitly consider the effect of the induced dipole moment term in eq.(60). The intramolecular MM energy is usually described using a standard MM potential, and will not be considered further.

3.4. *The Combined Coupled Cluster/Molecular Mechanics Method*

In the previous subsections we have described the QM/MM energy contributions and CC theory for molecules in vacuum. Here, we wish to combine these two descriptions. We solve this nontrivial problem by using the concept of a variational Lagrangian.

In order to derive the optimization conditions for the CC/MM wave function, we extend the vacuum CC Lagrangian by augmenting it with the interaction term in a way similar to the approach where the CC method is coupled to a dielectric continuum [19]. This is done by adding the vacuum Lagrangian and the CC/MM interaction energy now written in terms of CC expectation values (eq.(16)). The CC/MM Lagrangian accordingly becomes [29]

$$L_{CC/MM}(t, \bar{t}) = \langle \hat{H}_{QM} \rangle - \sum_{s=1}^{S} \langle \hat{N}_s \rangle$$

$$- \frac{1}{2} \sum_{a=1}^{A} \langle \hat{\mathbf{R}\mathbf{r}_a} \rangle^T \alpha_a \left[\langle \hat{\mathbf{R}\mathbf{r}_a} \rangle + \mathbf{O}_a^{ns}(\vec{R}_a) \right]$$

$$+ E^{vdw} + E_{S,N}^{el,nuc} + O_{ind}^{ns} + E_{MM} . \qquad (61)$$

We emphasize that this Lagrangian is non-linear in both the t and $\bar{t}$ parameters. Similar to the vacuum case we require this Lagrangian to be stationary with respect to both the t and $\bar{t}$ parameters. By introducing the one-electron interaction operator, $\hat{T}^g$, as

$$\hat{T}^g = - \sum_{s=1}^{S} \hat{N}_s - \sum_{a=1}^{A} \left[\langle \Lambda | \hat{\mathbf{R}\mathbf{r}_a} | CC \rangle + \frac{1}{2} \mathbf{E}_a^{ns}(\vec{R}_a) \right]^T \alpha_a \hat{\mathbf{R}\mathbf{r}_a} \qquad (62)$$

where

$$\mathbf{E}_a^{ns}(\vec{R}_a) = 2\mathbf{E}^n(\vec{R}_a) + 2\mathbf{E}^s(\vec{R}_a) + \mathbf{E}^{ind}(\vec{R}_a) \qquad (63)$$

we obtain the optimization conditions for the CC/MM wave function

$$\frac{\partial L_{CC/MM}(t, \bar{t})}{\partial \bar{t}_{\mu_i}} = \langle \mu_i | \exp(-\hat{T}) \left[\hat{H}_{QM} + \hat{T}^g \right] \exp(\hat{T}) | HF \rangle = 0 \quad (64)$$

$$\frac{\partial L_{CC/MM}(t, \bar{t})}{\partial t_{\nu_i}} = \langle \Lambda | [\hat{H}_{QM} + \hat{T}^g, \hat{\tau}_{\nu_i}] | CC \rangle = 0 . \qquad (65)$$

Note, that since the $\hat{T}^g$ operator depends on both the t and $\bar{t}$ parameters we find that eq.(64) and eq.(65) are coupled. Clearly, this represents an additional complication compared to the corresponding optimization conditions for a molecule in vacuum. The CC/MM method is non-variational and the CC/MM total energy is generally not bounded from below by the exact total energy. However, since the lack of a lower bound has not shown to be a problem for neither vacuum CC nor CC/DC calculations, we do not expect this to be a problem in the case of a CC/MM wave function.

Before we consider some implementational aspects of the outlined solvent models, we note that the optimization conditions of the CC/DC and CC/MM models have similar forms. Actually, these two sets of equations (eqs.(39 - 40) and eqs.(64 - 65)) only differ in the definition of the interaction operators, eqs.(41, 62) leading to the conclusion that the implementation of the CC/DC and CC/MM models may be performed analogously. However, we should pay attention to the fact that the two solvent models are very different physically. The CC/MM model accounts explicitly for the discrete nature of the solvent while all structural effects are treated in an averaged way using the CC/DC model.

Finally, in order to summarize the different intermolecular energy components and how they are accounted for in the CC/DC and CC/MM models we outline the following:

- Electrostatic interactions arising from the static multipole moments of the solvent molecules are included at the point charge level in the CC/MM model but are not considered in the CC/DC model.

- Polarization interactions (or simply the induction energy) can be included explicitly in both the CC/MM and CC/DC models. In both cases the response potential is treated using a linear approximation. However, in the CC/MM model the polarization effects are treated directly at the microscopic level while the CC/DC model makes use of a macroscopic parameter, the dielectric constant, in order to introduce this energy component.

- Exchange interactions and effects of charge transfer are not accounted for in neither the CC/MM nor the CC/DC model.

- The dispersion interactions between the QM molecule and the solvent is not included directly in neither the CC/MM nor the CC/DC model. However, the CC/MM model includes a parametrized potential which in an averaged way introduces the effect of dispersion. Ideally, this potential is obtained by fitting the interaction energy corrected for the electrostatic component (including the induction energy) to an analytical expression. Hence, this component also contains contributions from charge penetration effects, exchange effects, higher order electrostatics and higher order couplings between the induction, dispersion and exchange contributions [92].

- Due to the Pauli principle, the electrons in the QM part of the system should be exposed to a significant non-Coulombic repulsion upon penetration into regions of space formally occupied by the solvent electrons. This highly repulsive component is, like the dispersion, included in the CC/MM model but only in an averaged way. Actually, this effect is not described as an electronic operator but rather as a potential independent of the QM electrons. In the CC/DC model this term is completely missing, which means that the QM electrons (potentially) may penetrate into the dielectric medium and thereby cause unphysical interactions. Dielectric continuum models which simulate the effect of the Pauli repulsion have been addressed in for example Refs. [8,18,93,94,95,96], while QM/MM repulsive operators have been considered in Refs. [43,46,97,98].

3.5. *Implementational Aspects of the CC/DC and the CC/MM Models*

In this section we will shortly discuss some general implementational aspects of the CC/DC and CC/MM models. Since the optimization conditions for the CC/DC and CC/MM models only differ in the definition of the $\hat{T}^g$ operator, we may discuss the two different solvation models simultaneously. The following discussion is based on Ref. [91].

As noted in the previous sections, the CC/DC and CC/MM optimization equations for t and $\bar{t}$ are coupled due to the introduction of the $\hat{T}^g$ operator. For their solution we adopted an approach where we, for a given $\hat{T}^g$ operator, solve them as if they were completely decoupled and then perform some outer iterations on the T^g operator. Furthermore, when solving the t and $\bar{t}$ equations iteratively through a set of outer iterations, we only to perform a few iterations in the t equations and then use this set of t parameters in the solution of the $\bar{t}$ equations. Also, in the solution of the $\bar{t}$ equations we only perform a few iterations. With these sets of (un)converged t and $\bar{t}$ parameters we now update the $\hat{T}^g$ operator and repeat the solution of the t and $\bar{t}$ parameters until convergence is obtained in both the t and the $\bar{t}$ parameters, and also with respect to the outer iterations. For the case of liquid water, test calculations leads to the conclusion that the total number of iterations is minimized if the maximum number of iterations in the solution of the t and $\bar{t}$ equations are 3 and 4, respectively.

4. Response Theory and Molecular Properties for Solvated Molecules

This section will be concerned with the theory of response functions. First, we will briefly address the CC response functions for molecular systems in vacuum, and next we will consider the evaluation of CC response functions including environmental effects. A detailed derivation of CC response functions for molecules in vacuum may be found in Ref. [71], which also includes references to previous work on this subject. After a discussion concerning some implementational aspects of the CC/DC and CC/MM models, we will consider the connection between response functions and molecular properties. Finally, we will discuss how to relate the calculated molecular properties to macroscopic properties.

We note that the issue of response functions coupled to a surrounding medium has been considered by other authors. Thus, Mikkelsen et al. have presented linear [99,100] and non-linear [101,102] response functions for a molecular system treated at the MCSCF level of theory combined with a (spherical cavity) dielectric continuum description of the solvent. Furthermore, Yu et al. considered solvent effects on the frequency dependent first hyperpolarizability using the time-dependent Hartree-Fock description for the solute and an Onsager approximation [83,103] for the solvent [10]. In addition, linear response functions have been considered in combination with PCM using either SCF [104,105] or DFT [106,107,108] or MCSCF [109] descriptions of the solute molecule. Also, non-linear response have been considered using the coupled perturbed Hartree-Fock method [110] or the time-dependent Hartree-Fock model [111]. Very recently, linear response functions derived within the second order polarization propagator approximation (SOPPA) [112,113,114] combined with a (spherical cavity) dielectric continuum description of the solvent, have been presented [115]. Also, linear and non-linear response functions for molecules in combination with inhomogeneous solvation has been considered by Jørgensen et al. [116,117,118]. Concerning solvent models at the CC level, Christiansen et al. have presented linear response functions combined with a (spherical cavity) dielectric continuum description of the solvent [119].

Going beyond a dielectric continuum description of the solvent, linear and non-linear response functions have been addressed by Poulsen et al. using MCSCF combined with a discrete and polarizable solvent model [120,121]. Furthermore, Jensen et al. have, using a similar representation of the solvent, combined DFT and MM, and thereby derived both linear and non-linear response functions [38,122]

4.1. Theoretical Aspects

4.1.1. Coupled Cluster Response Theory for States in Vacuum

The derivation of response functions is closely connected to fundamental principles in time-dependent perturbation theory [71,123,124,125,126,127,128]. Here, we will first define an appropriate quasienergy followed by the identification of the response functions as simple derivatives.

In the following we consider a molecular system in vacuum described by a time-dependent Hamiltonian, $\hat{H}$,

$$\hat{H} = \hat{H}_{QM} + \hat{V}^t , \qquad (66)$$

where $\hat{H}_{QM}$ is the vacuum unperturbed molecular Born-Oppenheimer Hamiltonian and $\hat{V}^t$ is a time-dependent perturbation. The term, $\hat{V}^t$, may be written as a Fourier expansion

$$\hat{V}^t = \sum_{k=-N}^{N} \exp(-i\omega_k t)\hat{V}^{\omega_k}$$

$$= \sum_{k=-N}^{N} \exp(-i\omega_k t) \sum_y \epsilon_y(\omega_k)\hat{Y} . \qquad (67)$$

It is required that $\epsilon_y(\omega_k) = \epsilon_k(-\omega_k)^*$ and that the operators $\hat{Y}$ are Hermitian. This form of the perturbation Hamiltonian is very general as it allows for the inclusion of both static and dynamical perturbations. For example, if an external (classical) homogeneous and periodic dynamical electric field is applied to the molecule, the interaction Hamiltonian becomes [129]

$$\hat{V}^t = -\mathbf{E}_0\boldsymbol{\mu}\left\{\exp(-i\omega t) + \exp(i\omega t)\right\} . \qquad (68)$$

Here, $\mathbf{E}_0$ is the field strength and $\boldsymbol{\mu}$ is the electric dipole moment. Eq.(68) is referred to as the electric dipole approximation, where it is assumed that the size of the molecule is negligible compared to the spatial variations in an external field. We emphasize that the interaction operators are not restricted to a homogeneous field. In the case of an electric field which varies in space, the interaction operator includes electric quadrupole and potentially higher order multipole expansions of the electric field [130].

Before considering the CC response functions, a comment concerning the exact response functions is appropriate. We write the (exact) time-dependent wave function in phase-isolated form as [129,71]

$$|\bar{O}(t)\rangle = \exp(-iF(t))|\tilde{O}(t)\rangle \qquad (69)$$

where $F(t)$ is a real function of time. The time-dependent wave function, $|\tilde{O}(t)\rangle$, is normalized to unity and furthermore $|\tilde{O}(t)\rangle$ reduces to the time-independent wave function, $|O\rangle$, in the limit of no perturbation. Also, we assume that before the perturbation was switched on, the time-independent wave function, $|O\rangle$, is the solution to the time-independent Schrödinger equation and that $|O\rangle$ is normalized to unity.

The time-evolution of the system is described by the time-dependent Schrödinger equation

$$\hat{H}|\bar{O}(t)\rangle = i\frac{\partial}{\partial t}|\bar{O}(t)\rangle. \tag{70}$$

Inserting the ansatz for the time-dependent wave function (eq.(69)) into the time-dependent Schrödinger equation we obtain

$$\exp(-iF(t))\left[\hat{H} - i\partial_t - \dot{F}(t)\right]|\tilde{O}(t)\rangle = 0 \tag{71}$$

or in phase isolated form

$$\left[\hat{H} - i\partial_t - \dot{F}(t)\right]|\tilde{O}(t)\rangle = 0 , \tag{72}$$

where a dot indicates a time-derivative and $\partial_t = \frac{\partial}{\partial t}$. Projecting eq.(72) onto the time-dependent phase isolated wave function, we obtain an equation for $\dot{F}(t)$

$$\dot{F}(t) = \langle\tilde{O}(t)|\left[\hat{H} - i\partial_t\right]|\tilde{O}(t)\rangle . \tag{73}$$

If the perturbation is absent eq.(73) reduces to

$$\dot{F}(t) = E_0 , \tag{74}$$

i.e. $\dot{F}(t)$ reduces to the stationary energy. The quantity

$$Q(t) \equiv \dot{F}(t) = \left\langle\tilde{O}(t)|\left[\hat{H} - i\partial_t\right]|\tilde{O}(t)\right\rangle \tag{75}$$

is denoted the time-dependent quasienergy [71].

For any wave function that satisfies the time-dependent variation principle [131]

$$\delta\langle\tilde{O}(t)|\left[\hat{H} - i\partial_t\right]|\tilde{O}(t)\rangle + i\partial_t\langle\tilde{O}(t)|\delta\tilde{O}(t)\rangle = 0 \tag{76}$$

where δ represents a first order variation in the wave function, we obtain, by differentiating the time-dependent quasienergy with respect to a perturbation strength parameter, the time-dependent Hellmann-Feynman theorem

$$\frac{dQ}{d\epsilon} = \frac{d\left\langle \tilde{O}(t) \left| \left[\hat{H} - i\partial_t\right] \right| \tilde{O}(t) \right\rangle}{d\epsilon}$$
$$= \left\langle \tilde{O}(t) \left| \frac{\partial \hat{H}}{\partial \epsilon} \right| \tilde{O}(t) \right\rangle - i\partial_t \left\langle \tilde{O}(t) \left| \frac{d\tilde{O}(t)}{d\epsilon} \right. \right\rangle . \tag{77}$$

For the perturbation in eq.(67) we obtain

$$\frac{\partial \hat{H}}{\partial \epsilon_x(\omega)} = \hat{X}\exp(-i\omega t) , \tag{78}$$

and eq.(77) may be written as [123,132]

$$\left\langle \tilde{O}(t) | \hat{X} | \tilde{O}(t) \right\rangle \exp(-i\omega t) = \frac{dQ(t)}{d\epsilon_x(\omega)} + i\partial_t \left\langle \tilde{O}(t) \left| \frac{d\tilde{O}(t)}{d\epsilon_x(\omega)} \right. \right\rangle . \tag{79}$$

Assuming a common period T for the perturbations in eq.(67) and denoting the time-average of an arbitrary periodic function of time, $f(t)$, over a period by

$$\{f(t)\}_T = \frac{1}{T}\int_{-T/2}^{T/2} f(t)dt \tag{80}$$

we then obtain the following important relations for the time-averaged quasienergy [71,132]

$$\delta\{Q(t)\}_T = 0 \tag{81}$$
$$\frac{d\{Q(t)\}_T}{d\epsilon_x(\omega)} = \{\langle \tilde{O}(t)|\hat{X}|\tilde{O}(t)\rangle \exp(-i\omega t)\}_T \tag{82}$$

where δ represents a first order variation and where we have used that the time-average of a time-differentiated periodic function is zero. Eq.(81) is the time-averaged variational condition and eq.(82) represents the time-averaged Hellmann-Feynman theorem.

The time evolution of the expectation value of a time-independent Hermitian operator $\hat{X}$ can be expanded as [71]

$$\langle \bar{O}(t)|X|\bar{O}(t)\rangle = \langle \tilde{O}(t)|X|\tilde{O}(t)\rangle = \langle 0|X|0\rangle$$

$$+ \sum_{k_1} \exp(-i\omega_{k_1}t) \sum_y \langle\langle X,Y\rangle\rangle_{\omega_{k_1}} \epsilon_y(\omega_{k_1})$$

$$+ \frac{1}{2} \sum_{k_1,k_2} \exp(-i(\omega_{k_1}+\omega_{k_2})t) \sum_{y,z} \langle\langle X,Y,Z\rangle\rangle_{\omega_{k_1},\omega_{k_2}} \epsilon_y(\omega_{k_1})\epsilon_z(\omega_{k_2})$$

$$+ \frac{1}{3!} \sum_{k_1,k_2,k_3} \exp(-i(\omega_{k_1}+\omega_{k_2}+\omega_{k_3})t)$$

$$\cdot \sum_{y,z,u} \langle\langle X,Y,Z,U\rangle\rangle_{\omega_{k_1},\omega_{k_2},\omega_{k_3}} \epsilon_y(\omega_{k_1})\epsilon_z(\omega_{k_2})\epsilon_u(\omega_{k_3}) + O(4) \qquad (83)$$

where $O(4)$ indicates higher-order terms. The linear response function $\langle\langle X,Y\rangle\rangle_{\omega_{k_1}}$, quadratic response function $\langle\langle X,Y,Z\rangle\rangle_{\omega_{k_1},\omega_{k_2}}$ and so on determine the expansion coefficients of the Fourier components. Inspection of the complex conjugate of eq.(83) together with the fact that the expectation value of a Hermitian operator is real, leads to the following symmetry condition for the response functions

$$\langle\langle X,Y,Z,...\rangle\rangle_{\omega_{k_1},\omega_{k_2},...} = (\langle\langle X,Y,Z,...\rangle\rangle_{-\omega_{k_1},-\omega_{k_2},...})^*. \qquad (84)$$

These symmetry relations are of course fulfilled in exact theory, but in the approximate theories they will not automatically be fulfilled. By the same token, the quasienergy is real for exact theory, but this is not guaranteed in approximate theories.

Inserting the expansion for the expectation value of the Hermitian operator, $\hat{X}$, eq.(83) in eq.(82) we obtain

$$\frac{d\{Q(t)\}_T}{d\epsilon_x(\omega)} = \langle 0|\hat{X}|0\rangle\delta(\omega) + \sum_y \sum_{k_1} \langle\langle X,Y\rangle\rangle_{\omega_{k_1}} \epsilon_y(\omega_{k_1})\delta(\omega+\omega_{k_1})$$

$$+ \frac{1}{2} \sum_{y,z} \sum_{k_1,k_2} \langle\langle X,Y,Z\rangle\rangle_{\omega_{k_1},\omega_{k_2}} \epsilon_y(\omega_{k_1})\epsilon_z(\omega_{k_2})\delta(\omega+\omega_{k_1}+\omega_{k_2})$$

$$+ \quad ... \qquad (85)$$

Using eq.(85) we finally obtain the response functions as derivatives of eq.(85) i.e. as derivatives of the quasienergy [71,126]

$$\langle \hat{X} \rangle = \frac{d\{Q(t)\}_T}{d\epsilon_x(0)} \tag{86}$$

$$\langle\langle X, Y \rangle\rangle_{\omega_{k_1}} = \frac{d^2\{Q(t)\}_T}{d\epsilon_x(\omega_0)d\epsilon_y(\omega_{k_1})} \tag{87}$$

$$\langle\langle X, Y, Z \rangle\rangle_{\omega_{k_1}, \omega_{k_2}} = \frac{d^3\{Q(t)\}_T}{d\epsilon_x(\omega_0)d\epsilon_y(\omega_{k_1})d\epsilon_z(\omega_{k_2})} \tag{88}$$

$$\langle\langle X, Y, Z... \rangle\rangle_{\omega_{k_1}, \omega_{k_2}...} = \frac{d^{n+1}\{Q(t)\}_T}{d\epsilon_x(\omega_0)d\epsilon_y(\omega_{k_1})d\epsilon_z(\omega_{k_2})...} \tag{89}$$

where

$$\omega_0 = -\sum_{i=1}^{n}\omega_{k_i} \tag{90}$$

which also follows from eq.(85). Note, that even though not explicit stated in the above equations, the response functions are defined as derivatives of the quasienergy taken at zero field-strength.

Let us now concentrate on the coupled cluster parametrization. The time-dependent CC wave function is parametrized as

$$|\overline{CC}(t)\rangle = \exp(-iF(t))|\widetilde{CC}(t)\rangle = \exp(-iF(t))\exp(\hat{T}(t))|HF\rangle \tag{91}$$

where $F(t)$ is a time-dependent phase and $|HF\rangle$ denotes the time-independent and optimized Hartree-Fock (HF) wave function.

The cluster operator is now time-dependent

$$\hat{T}(t) = \sum_{i=1}^{n}\hat{T}_i(t) = \sum_{i,\mu_i}\hat{\tau}_{\mu_i}t_{\mu_i}(t) \tag{92}$$

where the time dependence enters through the amplitudes, $t_{\mu_i}(t)$. The time-independent operator, $\hat{\tau}_{\mu_i}$, is an i-fold electronic excitation operator. We assume that the excitation operators commute and that the excitation operators satisfy the killer condition $\langle HF|\hat{\tau}_{\mu_i} = 0$. By insertion of eq.(91) into the time-dependent Schrödinger equation

$$\left(\hat{H} - i\frac{\partial}{\partial t}\right)|\overline{CC}(t)\rangle = 0 \tag{93}$$

we obtain, after pre-multiplying with $\exp(iF(t))\exp(-\hat{T}(t))$,

$$\exp(-\hat{T}(t))\hat{H}\exp(\hat{T}(t))|HF\rangle = \left(\dot{F}(t) + i\sum_{i,\mu}\hat{\tau}_{\mu_i}\dot{t}(t)_{\mu_i}\right)|HF\rangle. \tag{94}$$

Projection of eq.(94) onto the state $\langle HF|$ gives an equation for the time-dependent CC quasienergy, $Q(t)$,

$$Q(t) = \dot{F}(t) = \langle HF|\hat{H}\exp(\hat{T}(t))|HF\rangle = \langle HF|\hat{H}|\widetilde{CC}(t)\rangle . \quad (95)$$

Due to the non-variational nature of the CC wave function, we do not define the CC response functions as derivatives of the time-average of $Q(t)$, but rather we construct, as in the time-independent case, a CC quasienergy Lagrangian, $L_{CC}(t)$. As discussed in Refs. [71,75] we consider

$$L_{CC}(t) = Q(t) + \sum_{\mu_i,i} \bar{t}_{\mu_i}(t)(e_{\mu_i}(t) - i\partial t_{\mu_i}(t)/\partial t)$$

$$= \langle\tilde{\Lambda}| \left(\hat{H} - i\frac{\partial}{\partial t}\right) |\widetilde{CC}\rangle \quad (96)$$

where the set of time-dependent parameters $\bar{t}_{\mu_i}(t)$ are the Lagrangian multipliers and $\langle\tilde{\Lambda}|$ is defined as

$$\langle\tilde{\Lambda}| = (\langle HF| + \sum_{i,\mu_i} \bar{t}_{\mu_i}(t)\langle\mu_i|)\exp(-\hat{T}(t)) . \quad (97)$$

In addition to the introduction of the state $\langle\tilde{\Lambda}|$ we have also extended the definition of the CC amplitude equation (eq.(9)) to the time-dependent case

$$e_{\mu_i}(t) - i\partial t_{\mu_i}(t)/\partial t = \langle\mu_i|\exp(-\hat{T}(t))\hat{H}\exp(\hat{T}(t))|HF\rangle - \frac{i\partial t_{\mu_i}(t)}{\partial t} . \quad (98)$$

Expanding this Lagrangian in orders of the perturbation and performing the time-average we may, as shown in detail in Ref.[71], obtain the response functions as derivatives of the time-averaged CC quasienergy $\{L(t)\}_T$, while the stationary condition $\delta\{L(t)\}_T = 0$, which is equivalent to eq.(81), gives us the response equations. In particular, the linear, quadratic and cubic response functions are given as

$$\langle\langle\hat{X},\hat{Y}\rangle\rangle_{\omega_y} = \frac{1}{2}C^{\pm\omega}\frac{d^2\{L_{CC}(t)\}_T}{d\epsilon_x(\omega_x)d\epsilon_y(\omega_y)} \quad (99)$$

$$\langle\langle\hat{X},\hat{Y},\hat{Z}\rangle\rangle_{\omega_y,\omega_z} = \frac{1}{2}C^{\pm\omega}\frac{d^3\{L_{CC}(t)\}_T}{d\epsilon_x(\omega_x)d\epsilon_y(\omega_y)d\epsilon_z(\omega_z)} \quad (100)$$

$$\langle\langle\hat{X},\hat{Y},\hat{Z},\hat{U}\rangle\rangle_{\omega_y,\omega_z,\omega_u} = \frac{1}{2}C^{\pm\omega}\frac{d^4\{L_{CC}(t)\}_T}{d\epsilon_x(\omega_x)d\epsilon_y(\omega_y)d\epsilon_z(\omega_z)d\epsilon_u(\omega_u)} ,(101)$$

respectively, where the derivatives are taken at zero field strength and the sum of the frequencies equals zero, e.g. $\omega_x = -\omega_y$ for linear response. The operator $C^{\pm\omega}$ is defined as

$$C^{\pm\omega} f^{XY}(\omega_x, \omega_y) = f^{XY}(\omega_x, \omega_y) + \left(f^{XY}(-\omega_x, -\omega_y)\right)^* \qquad (102)$$

for linear response and so forth for higher order response. This operator ensures that the approximate response function possesses the correct symmetry with respect to a sign change in the frequency variable followed by a complex conjugation of the response function. In the next sections we will derive explicit expressions for the CC response functions and CC response equations for a molecular system coupled to an environment. For derivations and implementations of the vacuum CC response functions following the strategy outlined above, see for example Refs. [71,74,75,79,78,133,134].

The response functions described in this section are directly related to (frequency dependent) molecular properties for example the dipole-(hyper)polarizabilities. Molecular properties derived from the linear response function are called second order properties and so forth for the higher order properties. We will return to this issue later.

4.1.2. Coupled Cluster/Dielectric Continuum or Molecular Mechanics Response Theory

In this section we briefly discuss the CC solvent Lagrangians. First, we consider the CC/DC case and second we describe the CC/MM case. Finally, we may, treating the two solvent approaches in a unified way, derive the appropriate response functions.

Concerning the CC/DC model, we augment, as described in Ref. [119], the vacuum Lagrangian in eq.(96) with the multipole expanded interaction term. However, for the calculation of dynamical properties the problem of equilibrium versus nonequilibrium solvent configuration must be considered. In the CC/DC model this problem is solved by treating the total polarization vector as a sum of two contributions : the optical (fast) and inertial (slow) polarization. Thereby, an optical and an inertial susceptibility are considered. These are related to the solvent optical (ϵ_{op}) and static (ϵ_{st}) dielectric constants as $\epsilon_{op} = 1 + 4\pi\chi_{op}$ and $\epsilon_{st} = \epsilon_{op} + 4\pi\chi_{in}$. Furthermore, it is assumed that the optical polarization is always in equilibrium with the solute, while this is not the case for the inertial polarization. The appropriate energy expression for the solute-solvent system has been derived previously (see for example Ref. [119] and references therein). Therefore, we simply give the appropriate nonequilibrium CC/DC Lagrangian as [119]

$$^{neq}L_{CC/DC}(t) = Q(t) + \sum_{\mu,i} \bar{t}_{\mu_i}(t)(e_{\mu_i}(t) - i\partial t_{\mu_i}(t)/\partial t)$$

$$+ \sum_{lm} g_l(\epsilon_{op})\langle\tilde{\Lambda}|\hat{T}_{lm}|\widetilde{CC}\rangle^2$$

$$+ \sum_{lm} g_l(\epsilon_{st}, \epsilon_{op}) \left[2\langle\tilde{\Lambda}|\hat{T}_{lm}|\widetilde{CC}\rangle - \langle\Lambda|\hat{T}_{lm}|CC\rangle \right] \langle\Lambda|\hat{T}_{lm}|CC\rangle$$

$$(103)$$

where the function $g_l(\epsilon_{st}, \epsilon_{op})$ is defined by

$$g_l(\epsilon_{st}, \epsilon_{op}) = g_l(\epsilon_{st}) - g_l(\epsilon_{op}) .$$ $$(104)$$

In the equilibrium case this Lagrangian reduces to

$$^{eq}L_{CC/DC}(t) = Q(t) + \sum_{\mu,i} \bar{t}_{\mu_i}(t)(e_{\mu_i}(t) - i\partial t_{\mu_i}(t)/\partial t)$$

$$+ \sum_{lm} g_l(\epsilon)\langle\tilde{\Lambda}|\hat{T}_{lm}|\widetilde{CC}\rangle^2 .$$ $$(105)$$

Concerning the response functions in the CC/MM approach, we construct the corresponding time dependent CC/MM energy functional. As discussed in Ref. [29] we consider

$$L_{CC/MM}(t) = Q(t) + \sum_{\mu,i} \bar{t}_{\mu_i}(t)(e_{\mu_i}(t) - i\partial t_{\mu_i}(t)/\partial t)$$

$$- \sum_{s=1}^{S} \langle\tilde{\Lambda}|\hat{N}_s|\widetilde{CC}\rangle$$

$$- \frac{1}{2} \sum_{a=1}^{A} \langle\tilde{\Lambda}|\mathbf{R}\hat{\mathbf{r}}_a|\widetilde{CC}\rangle\alpha_a \left\{ \langle\tilde{\Lambda}|\mathbf{R}\hat{\mathbf{r}}_a|\widetilde{CC}\rangle + \mathbf{O}_a^{ns}(\vec{R}_a) \right\}$$

$$+ E^{vdw} + E_{S,N}^{el,nuc} + O_{ind}^{ns} + E_{MM} ,$$ $$(106)$$

i.e. we augment the vacuum Lagrangian with the time-dependent expression for the interaction energy term. In the construction of eq.(106) we assume that the induced dipole moments entering the terms $\mathbf{O}_a^{ns}(\vec{R}_a)$ and O_{ind}^{ns} are fixed, i.e. they are maintained at the values found in the optimization of the wave function. We note that this does not imply that the effect of changes in the induced moments from changes in the wave function due to the perturbation is neglected. This is simply due to the fact that we have in the time-dependent quasienergy (eq.(106)) introduced the equation $\mu_a^{ind} = \alpha_a \cdot \mathbf{E}_a^{total}$ explicitly. However, the total electric field depends on the induced moments of the other molecules treated using MM and these induced dipole moments are kept fixed.

According to the $2n + 1$ rule [71], we only need the first order wave function responses in order to calculate up to third order properties. Thus, for discussing the second and third order properties, for example frequency dependent (first) hyperpolarizabilities, we only need to solve the first order response equations [135]

$$(\mathbf{A} - \omega_y \mathbf{1}) t^Y(\omega_y) + \boldsymbol{\xi}^Y + \bar{t}^Y(\omega_y) \mathbf{J}(\omega_y) = 0 \qquad (107)$$

$$\bar{t}^Y(\omega_y)(\mathbf{A} + \omega_y \mathbf{1}) + \boldsymbol{\eta}^Y + \mathbf{F} t^Y(\omega_y) = 0 \qquad (108)$$

which is found using the stationary condition on the time-average of the CC/DC or CC/MM Lagrangians. The explicit expressions for the matrices and vectors in eqs.(107 - 108) are given in Table 1. However, to solve for fourth order properties, for example the second hyperpolarizability, we also need to consider the second order response equations

$$(\mathbf{A} - (\omega_y + \omega_z)\mathbf{1}) t^{YZ}(\omega_y, \omega_z) + P\left(Y(\omega_y), Z(\omega_z)\right) \left[\mathbf{A}^Y t^Z(\omega_z) \right.$$
$$+ \frac{1}{2} \mathbf{B} t^Y(\omega_y) t^Z(\omega_z) + \bar{t}^Y(\omega_y) \mathbf{K} t^Z(\omega_z) + \frac{1}{2} \bar{t}^{YZ}(\omega_y, \omega_z) \mathbf{J} \right] = 0 \quad (109)$$

$$\bar{t}^{YZ}(\omega_y, \omega_z)(\mathbf{A} + (\omega_y + \omega_z)\mathbf{1}) + P\left(Y(\omega_y), Z(\omega_z)\right) \left[\mathbf{F}^Y t^Z(\omega_z) \right.$$
$$+ \frac{1}{2} \mathbf{G} t^Y(\omega_y) t^Z(\omega_z) + \frac{1}{2} \mathbf{F} t^{YZ}(\omega_y, \omega_z)$$
$$+ \bar{t}^Y(\omega_y) \left(\mathbf{A}^Z + \mathbf{B} t^Z(\omega_z) \right) + \frac{1}{2} \bar{t}^Y(\omega_y) \bar{t}^Z(\omega_z) \mathbf{K} \right] = 0 , \qquad (110)$$

where we have defined the matrices entering eqs.(109 - 110) in Tables 1 and 2. In order to derive the equations for the response functions we (i) expand the the CC/DC or CC/MM Lagrangians in order of the perturbation (ii) collect orders of the expanded CC/DC or CC/MM Lagrangians (iii) perform the time-average according to eq.(80) and (iv) perform the differentiation of the time-averaged quasienergy according the eqs.(99 - 101). Accordingly, we obtain from the CC/DC Lagrangian in eq.(103) and the CC/MM Lagrangian in eq.(106) the linear response functions [135]

$$\langle\langle \hat{X}, \hat{Y} \rangle\rangle_{\omega_y} = \frac{1}{2} C^{\pm\omega} P\left(X(\omega_x), Y(\omega_y)\right) \left[\boldsymbol{\eta}^X t^Y(\omega_y) + \frac{1}{2} \mathbf{F} t^X(\omega_x) t^Y(\omega_y) \right.$$
$$- \frac{1}{2} \bar{t}^X(\omega_x) \bar{t}^Y(\omega_y) \mathbf{J} \right] , \qquad (111)$$

the quadratic response functions [136]

$$\langle\langle \hat{X}, \hat{Y}, \hat{Z} \rangle\rangle_{\omega_y,\omega_z} = \frac{1}{2} C^{\pm\omega} P\left(X(\omega_x), Y(\omega_y), Z(\omega_z)\right) \left[\left(\frac{1}{2}\mathbf{F}^X + \frac{1}{6}\mathbf{G}t^X(\omega_x) \right) t^Y(\omega_y) \right.$$
$$\left. + \bar{t}^X(\omega_x)\left(\mathbf{A}^Y + \frac{1}{2}\mathbf{B}t^Y(\omega_y) \right) + \frac{1}{2}\bar{t}^X(\omega_x)\bar{t}^Y(\omega_y)\mathbf{K} \right] t^Z(\omega_z)$$

$$(11$$

as well as the cubic response function

$$\langle\langle \hat{X}, \hat{Y}, \hat{Z}, \hat{U} \rangle\rangle_{\omega_y,\omega_z,\omega_u} = \frac{1}{2} C^{\pm\omega} P\left(X(\omega_x), Y(\omega_y), Z(\omega_z), U(\omega_u)\right)$$
$$\left[\frac{1}{8}\mathbf{F}t^{XY}(\omega_x,\omega_y)t^{ZU}(\omega_z,\omega_u) + \frac{1}{2}\mathbf{F}^X t^Y(\omega_y)t^{ZU}(\omega_z,\omega_u) \right.$$
$$+ \frac{1}{2}\bar{t}^X(\omega_x)\left(\mathbf{A}^Y + \mathbf{B}t^Y(\omega_y) \right) t^{ZU}(\omega_z,\omega_u)$$
$$+ \frac{1}{4}\mathbf{G}t^X(\omega_x)t^Y(\omega_y)t^{ZU}(\omega_z,\omega_u)$$
$$+ \left(\frac{1}{6}\mathbf{G}^X + \frac{1}{24}\mathbf{H}t^X(\omega_x) \right) t^Y(\omega_y)t^Z(\omega_z)t^U(\omega_u)$$
$$+ \bar{t}^X(\omega_x)\left(\frac{1}{2}\mathbf{B}^Y + \frac{1}{6}\mathbf{C}t^Y(\omega_y) \right) t^Z(\omega_z)t^U(\omega_u)$$
$$+ \frac{1}{4}\bar{t}^X(\omega_x)\bar{t}^Y(\omega_y)\left(\mathbf{K}t^{ZU}(\omega_z,\omega_u) + \mathbf{L}t^Z(\omega_z)t^U(\omega_u) \right)$$
$$\left. - \frac{1}{8}\bar{t}^{XY}(\omega_x,\omega_y)\bar{t}^{ZU}(\omega_z,\omega_u)\mathbf{J} \right] .$$

$$(113$$

In these expressions we have used the definitions of the vectors and matrices given in Table 1 and Table 2 and introduced the symmetrizer defined as (for linear response)

$$P\left(X(\omega_x), Y(\omega_y)\right) g^{XY}(\omega_x,\omega_y) = g^{XY}(\omega_x,\omega_y) + g^{YX}(\omega_y,\omega_x) . \quad (114)$$

In Table 1 and Table 2 the effective one-electron operators, which describe the direct solvent effects, have been introduced in an unified way such that the expressions for the matrix elements are valid for both the CC/DC and CC/MM case. However, the appropriate Lagrangian should of course be used, i.e. in the first column of Tables 1 and 2 $L_{CC/MM}$ should be replaced by $L_{CC/DC}$ in the dielectric continuum case.

Thus in the CC/MM case these operators are defined as

$$\hat{T}^g = -\sum_{s=1}^{S} \hat{N}_s - \sum_{a=1}^{A} \left[\langle \Lambda | \hat{\mathbf{R}\mathbf{r}_a} | CC \rangle + \frac{1}{2}\mathbf{E}_a^{ns}(\vec{R}_a) \right]^T \alpha_a \hat{\mathbf{R}\mathbf{r}_a} \quad (115)$$

where the vector $\mathbf{E}_a^{ns}(\vec{R}_a)$ is defined as

$$\mathbf{E}_a^{ns}(\vec{R}_a) = 2\mathbf{E}^n(\vec{R}_a) + 2\mathbf{E}^s(\vec{R}_a) + \mathbf{E}^{ind}(\vec{R}_a) \ . \tag{116}$$

and

$$\hat{T}^{g\nu_j} = -\sum_{a=1}^{A}\langle\Lambda|\left[\hat{\mathbf{R}\mathbf{r}}_a, \hat{\tau}_{\nu_j}\right]|CC\rangle\boldsymbol{\alpha}_a\hat{\mathbf{R}\mathbf{r}}_a \tag{117}$$

$$\hat{T}^{g\nu_j\sigma_k} = -\sum_{a=1}^{A}\langle\Lambda|\left[\left[\hat{\mathbf{R}\mathbf{r}}_a, \hat{\tau}_{\nu_j}\right], \hat{\tau}_{\sigma_k}\right]|CC\rangle\boldsymbol{\alpha}_a\hat{\mathbf{R}\mathbf{r}}_a \tag{118}$$

$$\hat{T}^{g\nu_j\sigma_k\delta_k} = -\sum_{a=1}^{A}\langle\Lambda|\left[\left[\left[\hat{\mathbf{R}\mathbf{r}}_a, \hat{\tau}_{\nu_j}\right], \hat{\tau}_{\sigma_k}\right], \hat{\tau}_{\delta_k}\right]|CC\rangle\boldsymbol{\alpha}_a\hat{\mathbf{R}\mathbf{r}}_a \tag{119}$$

$$^{\nu_j}\hat{T}^{g} = -\sum_{a=1}^{A}\langle\bar{\nu}_j|\hat{\mathbf{R}\mathbf{r}}_a|CC\rangle\boldsymbol{\alpha}_a\hat{\mathbf{R}\mathbf{r}}_a \tag{120}$$

$$^{\nu_j}\hat{T}^{g\sigma_k} = -\sum_{a=1}^{A}\langle\bar{\nu}_j|\left[\hat{\mathbf{R}\mathbf{r}}_a, \hat{\tau}_{\sigma_k}\right]|CC\rangle\boldsymbol{\alpha}_a\hat{\mathbf{R}\mathbf{r}}_a \ . \tag{121}$$

In the CC/DC model the $\hat{T}^g$ operator is defined as

$$\hat{T}^g(\epsilon) = -2\sum_{lm} g_l(\epsilon)\langle\Lambda|\hat{T}_{lm}|CC\rangle\hat{T}_{lm}^e \tag{122}$$

while the other operators are defined as

$$\hat{T}^{g\nu_j} = -2\sum_{lm} g_l(\epsilon_{op})\langle\Lambda|\left[\hat{T}_{lm}, \hat{\tau}_{\nu_j}\right]|CC\rangle\hat{T}_{lm}^e \tag{123}$$

$$\hat{T}^{g\nu_j\sigma_k} = -2\sum_{lm} g_l(\epsilon_{op})\langle\Lambda|\left[\left[\hat{T}_{lm}, \hat{\tau}_{\nu_j}\right], \hat{\tau}_{\sigma_k}\right]|CC\rangle\hat{T}_{lm}^e \tag{124}$$

$$\hat{T}^{g\nu_j\sigma_k\delta_l} = -2\sum_{lm} g_l(\epsilon_{op})\langle\Lambda|\left[\left[\left[\hat{T}_{lm}, \hat{\tau}_{\nu_j}\right], \hat{\tau}_{\sigma_k}\right], \hat{\tau}_{\delta_l}\right]|CC\rangle\hat{T}_{lm}^e \tag{125}$$

$$^{\nu_j}\hat{T}^{g} = -2\sum_{lm} g_l(\epsilon_{op})\langle\bar{\nu}_j|\hat{T}_{lm}|CC\rangle\hat{T}_{lm}^e \tag{126}$$

$$^{\nu_j}\hat{T}^{g\sigma_k} = -2\sum_{lm} g_l(\epsilon_{op})\langle\bar{\nu}_j|\left[\hat{T}_{lm}, \hat{\tau}_{\sigma_k}\right]|CC\rangle\hat{T}_{lm}^e \tag{127}$$

In Table 1 we have listed the vectors and matrices relevant for the linear and quadratic CC/DC and CC/MM response functions, while the additional matrices relevant for the cubic CC/DC and CC/MM response functions have been listed in Table 2. From Tables 1 and 2 we observe that the perturbation dependent quantities have no direct solvent contributions while the **J**, **K** and **L** matrices have no vacuum part. However, the **A**, **F**, **B**, **G**, **C** and **H** matrices have contributions from both the vacuum and the solvent part of the CC quasienergy. From eqs.(107 - 110) we note that the pure solvent matrices (**J**, **K** and **L**) introduce a coupling between the **t** and $\bar{\mathbf{t}}$ response equations. However, in vacuum these matrices vanish and the response equations decouple. Moreover, in the situation where the solvent is described as a nonpolarizable discrete medium, e.g. only point charges and Lennard-Jones parameters are used to define the classical molecules, the response equations also decouple. Furthermore, in this case the $\hat{T}^g$ operator only includes the point charge perturbation. However, this perturbation may also be included directly in the one-electron part of the (vacuum) Hamiltonian, and the procedure for determining the response functions is thereby seen to be the same as in the vacuum case, but with modified one-electron integrals. However, including explicit polarization effects leads to a somehow more complicated solution scheme for the response equations.

4.2. *Implementational Aspects of the CC/DC and CC/MM Response Functions*

The calculation of linear and higher order response properties together with their residues, can be reduced into a sequence of linear algebraic steps. For a detailed discussion of linear, non-linear properties and excitation energies for the vacuum CCS, CC2 and CCSD response functions, we refer to Refs. [79,78,133,134]. In this section we will discuss some general implementational aspects of the CC/DC and CC/MM linear and higher order response functions. Concerning the CC2 model, we note that the implementation follows from the CCSD implementation. However, the special partitioning of the Hamiltonian is to be remembered. Thus, as described in Ref. [137], when carrying out the substitution $\hat{H} \to \hat{H} + \hat{T}^g(\epsilon)$ in the optimization conditions for the wave function, implies within the CC2 partitioning scheme that $\hat{F} \to \hat{F} + \hat{T}^g(\epsilon)$. The following discussion is based on Refs. [135,136,138]

4.2.1. *Linear response*

In the CC/DC and CC/MM models all the interaction contributions are described using effective one-electron operators. As observed in the previous section concerning the linear response function, we only need linear transformations of trial vectors with the derived coupling or solvent contributions to the **A**, **F** and **J** matrices defined in Table 1. Thus, we need the following additional transformations $^{solvent}\rho = {}^{solvent}\mathbf{A}t^C$, $^{solvent}\sigma = \bar{\mathbf{t}}^{B\,solvent}\mathbf{A}$,

Table 1. Matrices and vectors for linear and quadratic response functions.

Quantity	Derivative expression	Vacuum Contribution	Solvent contribution
$A_{\mu_i\nu_j}$	$\dfrac{\partial^2\{L_{CC/MM}^{(2)}\}^T}{\partial t_{\mu_i}^{(1)}(\omega_x)\partial t_{\nu_j}^{(1)}(\omega_y)}$	$\langle\bar{\mu}_i|\left[\hat{H}_{QM},\hat{\tau}_{\nu_j}\right]|CC\rangle$	$\langle\bar{\mu}_i|\left[\hat{T}^g,\hat{\tau}_{\nu_j}\right]|CC\rangle + \langle\bar{\mu}_i|\hat{T}^{g\nu_j}|CC\rangle$
$F_{\mu_i\nu_j}$	$\dfrac{\partial^2\{L_{CC/MM}^{(2)}\}^T}{\partial t_{\mu_i}^{(1)}(\omega_x)\partial t_{\nu_j}^{(1)}(\omega_y)}$	$\langle\Lambda|\left[\left[\hat{H}_{QM},\hat{\tau}_{\mu_i}\right],\hat{\tau}_{\nu_j}\right]|CC\rangle$	$\langle\Lambda|\left[\left[\hat{T}^g,\hat{\tau}_{\mu_i}\right],\hat{\tau}_{\nu_j}\right]|CC\rangle$ $+\tfrac{1}{2}P^{\mu_i\nu_j}\langle\Lambda|\left[\hat{T}^{g\mu_i},\hat{\tau}_{\nu_j}\right]|CC\rangle$
$\xi_{\mu_i}^Y$	$\dfrac{\partial^2\{L_{CC/MM}^{(2)}\}^T}{\partial\epsilon_g(\omega_y)\partial t_{\mu_i}^{(1)}(\omega_y)}$	$\langle\bar{\mu}_i|\hat{Y}|CC\rangle$	
$\eta_{\mu_i}^Y$	$\dfrac{\partial^2\{L_{CC/MM}^{(2)}\}^T}{\partial\epsilon_g(\omega_y)\partial t_{\mu_i}^{(1)}(\omega_y)}$	$\langle\Lambda|\left[\hat{Y},\hat{\tau}_{\nu_i}\right]|CC\rangle$	
$J_{\mu_i\nu_j}$	$\dfrac{\partial^2\{L_{CC/MM}^{(2)}\}^T}{\partial t_{\mu_i}^{(1)}(\omega_x)\partial t_{\nu_j}^{(1)}(\omega_y)}$		$\tfrac{1}{2}P^{\mu_i\nu_j}\langle\bar{\mu}_i|\nu\hat{T}^g|CC\rangle$
$B_{\mu_i\nu_j\sigma_k}$	$\dfrac{\partial^3\{L_{CC/MM}^{(3)}\}^T}{\partial t_{\mu_i}^{(1)}(\omega_x)\partial t_{\nu_j}^{(1)}(\omega_y)\partial t_{\sigma_k}^{(1)}(\omega_z)}$	$\langle\bar{\mu}_i|\left[\left[\hat{H}_{QM},\hat{\tau}_{\nu_j}\right],\hat{\tau}_{\sigma_k}\right]|CC\rangle$	$\langle\bar{\mu}_i|\left[\left[\hat{T}^g,\hat{\tau}_{\nu_j}\right],\hat{\tau}_{\sigma_k}\right]|CC\rangle$ $+P^{\nu_j\sigma_k}\langle\bar{\mu}_i|\left[\hat{T}^{g\nu_j},\hat{\tau}_{\sigma_k}\right]|CC\rangle$
$G_{\mu_i\nu_j\sigma_k}$	$\dfrac{\partial^3\{L_{CC/MM}^{(3)}\}^T}{\partial t_{\mu_i}^{(1)}(\omega_x)\partial t_{\nu_j}^{(1)}(\omega_y)\partial t_{\sigma_k}^{(1)}(\omega_z)}$	$\langle\Lambda|\left[\left[\left[\hat{H}_{QM},\hat{\tau}_{\mu_i}\right],\hat{\tau}_{\nu_j}\right],\hat{\tau}_{\sigma_k}\right]|CC\rangle$	$\langle\Lambda|\left[\left[\left[\hat{T}^g,\hat{\tau}_{\mu_i}\right],\hat{\tau}_{\nu_j}\right],\hat{\tau}_{\sigma_k}\right]|CC\rangle$ $+\tfrac{1}{2}P^{\mu_i\nu_j\sigma_k}\langle\Lambda|\left[\left[\hat{T}^{g\mu_i},\hat{\tau}_{\nu_j}\right],\hat{\tau}_{\sigma_k}\right]|CC\rangle$
$F_{\mu_i\nu_j}^X$	$\dfrac{\partial^3\{L_{CC/MM}^{(3)}\}^T}{\partial\epsilon_x(\omega_x)\partial t_{\mu_i}^{(1)}(\omega_y)\partial t_{\nu_j}^{(1)}(\omega_z)}$	$\langle\Lambda|\left[\left[\hat{X},\hat{\tau}_{\mu_i}\right],\hat{\tau}_{\nu_j}\right]|CC\rangle$	
$A_{\mu_i\nu_j}^X$	$\dfrac{\partial^3\{L_{CC/MM}^{(3)}\}^T}{\partial\epsilon_x(\omega_x)\partial t_{\mu_i}^{(1)}(\omega_y)\partial t_{\nu_j}^{(1)}(\omega_z)}$	$\langle\bar{\mu}_i|\left[\hat{X},\hat{\tau}_{\nu_j}\right]|CC\rangle$	
$K_{\mu_i\nu_j\sigma_k}$	$\dfrac{\partial^3\{L_{CC/MM}^{(3)}\}^T}{\partial t_{\mu_i}^{(1)}(\omega_x)\partial t_{\nu_j}^{(1)}(\omega_y)\partial t_{\sigma_k}^{(1)}(\omega_z)}$		$P^{\mu_i\nu_j}\langle\bar{\mu}_i|\left[\nu_j\hat{T}^g,\hat{\tau}_{\sigma_k}\right]|CC\rangle$

Table 2. Additional matrices used for the cubic and higher order CC/MM response function.

Quantity	Derivative expression	Vacuum Contribution	Solvent contribution
$C_{\mu_i \nu_j \sigma_k \delta_l}$	$\dfrac{\partial^4 \{L_{CC/MM}^{(4)}\}^T}{\partial t_{\mu_i}^{(1)}(\omega_x)\partial t_{\nu_j}^{(1)}(\omega_y)\partial t_{\sigma_k}^{(1)}(\omega_z) t_{\delta_k}^{(1)}(\omega_u)}$	$\langle \bar{\mu}_i | [[[\hat{H}_{QM}, \hat{\tau}_{\nu_j}], \hat{\tau}_{\sigma_k}], \hat{\tau}_{\delta_l}] |CC\rangle$	$\langle \bar{\mu}_i | [[[\hat{T}^g, \hat{\tau}_{\nu_j}], \hat{\tau}_{\sigma_k}], \hat{\tau}_{\delta_l}] |CC\rangle$ $+ \frac{1}{2} P^{\nu_j \sigma_k \delta_l} \left(\langle \bar{\mu}_i | [[\hat{T}^{g\nu_i}, \hat{\tau}_{\sigma_k}], \hat{\tau}_{\delta_l}] + [\hat{T}^{g\nu_j \sigma_k}, \hat{\tau}_{\delta_l}] |CC\rangle \right)$ $+ \frac{1}{6} P^{\nu_j \sigma_k \delta_l} \langle \bar{\mu}_i | \hat{T}^{g\nu_j \sigma_k \delta_l} |CC\rangle$
$H_{\mu_i \nu_j \sigma_k \delta_l}$	$\dfrac{\partial^3 \{L_{CC/MM}^{(4)}\}^T}{\partial t_{\mu_i}^{(1)}(\omega_y)\partial t_{\nu_j}^{(1)}(\omega_z)\partial t_{\sigma_k}^{(1)}(\omega_u)}$	$\langle \Lambda | [[[[\hat{H}_{QM}, \hat{\tau}_{\mu_i}], \hat{\tau}_{\nu_j}], \hat{\tau}_{\sigma_k}], \hat{\tau}_{\delta_l}] |CC\rangle$	$\langle \Lambda | [[[[\hat{T}^g, \hat{\tau}_{\mu_i}], \hat{\tau}_{\nu_j}], \hat{\tau}_{\sigma_k}], \hat{\tau}_{\delta_l}] |CC\rangle$ $+ \frac{1}{6} P^{\mu_i \nu_j \sigma_k \delta_l} \langle \Lambda | [[[\hat{T}^{g\mu_i}, \hat{\tau}_{\nu_j}], \hat{\tau}_{\sigma_k}], \hat{\tau}_{\delta_l}] |CC\rangle$ $+ \frac{1}{8} P^{\mu_i \nu_j \sigma_k \delta_l} \langle \Lambda | [[\hat{T}^{g\mu_i \nu_j}, \hat{\tau}_{\sigma_k}], \hat{\tau}_{\delta_l}] |CC\rangle$
$G^X_{\mu_i \nu_j \sigma_k}$	$\dfrac{\partial^4 \{L_{CC/MM}^{(4)}\}^T}{\partial \epsilon_x (\omega_x)\partial t_{\mu_i}^{(1)}(\omega_y)\partial t_{\nu_j}^{(1)}(\omega_z)\partial t_{\sigma_k}^{(1)}(\omega_u)}$	$\langle \Lambda | [[[\hat{X}, \hat{\tau}_{\mu_i}], \hat{\tau}_{\nu_j}], \hat{\tau}_{\sigma_k}] |CC\rangle$	
$B^X_{\mu_i \nu_j \sigma_k}$	$\dfrac{\partial^4 \{L_{CC/MM}^{(4)}\}^T}{\partial \epsilon_x (\omega_x)\partial t_{\mu_i}^{(1)}(\omega_y)\partial t_{\nu_j}^{(1)}(\omega_z)\partial t_{\sigma_k}^{(1)}(\omega_u)}$	$\langle \bar{\mu}_i | [[\hat{X}, \hat{\tau}_{\nu_j}], \hat{\tau}_{\sigma_k}] |CC\rangle$	
$L_{\mu_i \nu_j \sigma_k \delta_l}$	$\dfrac{\partial^4 \{L_{CC/MM}^{(4)}\}^T}{\partial t_{\mu_i}^{(1)}(\omega_x)\partial t_{\nu_j}^{(1)}(\omega_y)\partial t_{\sigma_k}^{(1)}(\omega_z)\partial t_{\delta_k}^{(1)}(\omega_u)}$		$\frac{1}{2} P^{\mu_i \nu_j} P^{\sigma_k \delta_l} \left(\langle \bar{\mu}_i | [[\nu_j \hat{T}^g, \hat{\tau}_{\sigma_k}], \hat{\tau}_{\delta_l}] |CC\rangle \right.$ $\left. + \langle \bar{\mu}_i | [\nu_j \hat{T}^{g\sigma_k}, \hat{\tau}_{\delta_l}] |CC\rangle \right)$

$\zeta = \bar{t}^B \mathbf{J}$ and $^{solvent}\gamma = {}^{solvent}\mathbf{F}t^C$ where t^C and $\bar{t}^B$ are a right and a left trial vectors, respectively. Here, we just consider the transformations of the Jacobian.

For the first transformation we obtain

$$
\begin{aligned}
{}^{solvent}\rho_{\mu_i} &= \sum_{\nu_j, j} \langle \mu_i | \exp(-\hat{T}) \left[\hat{T}^g, \hat{\tau}_{\nu_j} \right] | CC \rangle t^C_{\nu_j} \\
&+ \sum_{\nu_j, j} \langle \mu_i | \exp(-\hat{T}) \hat{T}^{g\nu_j} | CC \rangle t^C_{\nu_j} \\
&= \langle \mu_i | \exp(-\hat{T}) \left[\hat{T}^g, \hat{C} \right] | CC \rangle + \langle \mu_i | \exp(-\hat{T}) \hat{T}^{gC} | CC \rangle
\end{aligned}
$$

(128)

where $\hat{C} = \sum_{\nu_j, j} \hat{\tau}_{\nu_j} t^C_{\nu_j}$ and

$$
\begin{aligned}
\hat{T}^{gC} &= -2 \sum_{lm} g_l(\epsilon_{op}) \langle \Lambda | [\hat{T}_{lm}, \hat{C}] | CC \rangle \hat{T}^e_{lm} \\
&= -2 \sum_{lm} g_l(\epsilon_{op}) \sum_{j, \nu_j} \eta^{\hat{T}_{lm}}_{\nu_j} t^C_{\nu_j} \hat{T}^e_{lm}
\end{aligned}
$$

(129)

for the CC/DC case and

$$
\begin{aligned}
\hat{T}^{gC} &= -\sum_{a=1}^{A} \langle \Lambda | [[\hat{\mathbf{R}}\mathbf{r}_a, \hat{C}]] | CC \rangle \alpha_a \hat{\mathbf{R}}\mathbf{r}_a \\
&= -\sum_{a=1}^{A} \sum_{j, \nu_j} \eta^{\hat{\mathbf{R}}\mathbf{r}_a}_{\nu_j} t^C_{\nu_j} \alpha_a \hat{\mathbf{R}}\mathbf{r}_a
\end{aligned}
$$

(130)

for the CC/MM case. We have used the definition of the η^X vector given in Table 1. The first term in eq.(128) is calculated as a standard one-electron Hamiltonian contribution [78]. The last contribution is, in the CC/MM case, equivalent to the calculation of a ξ^X vector with an operator constructed by taking the sum of dot products of $\eta^{\hat{\mathbf{R}}\mathbf{r}_a}$ vectors and the **C** trial vector and multiplying this number and the polarizability on the $\hat{\mathbf{R}}\mathbf{r}_a$ operator. In the CC/DC case we simply sum over the angular quantum numbers l and m and multiply with the $g_l(\epsilon_{op})$ function instead of the polarizability.

The left transformation of the Jacobian matrix is given as

$$
\begin{aligned}
^{solvent}\sigma_{\nu_j} &= \sum_{\mu_i,i} \bar{t}^B_{\mu_i} \langle \mu_i | \exp(-\hat{T}) \left[\hat{T}^g, \hat{\tau}_{\nu_j} \right] |CC\rangle \\
&+ \sum_{\mu_i,i} \bar{t}^B_{\mu_i} \langle \mu_i | \exp(-\hat{T}) \hat{T}^{g\nu_j} |CC\rangle \\
&= \langle HF| \sum_{\mu_i,i} \bar{t}^B_{\mu_i} \hat{\tau}^\dagger_{\mu_i} \exp(-\hat{T}) \left[\hat{T}^g, \hat{\tau}_{\nu_j} \right] |CC\rangle \\
&+ \langle HF| \sum_{\mu_i,i} \bar{t}^B_{\mu_i} \hat{\tau}^\dagger_{\mu_i} \exp(-\hat{T}) \hat{T}^{g\nu_j} |CC\rangle \\
&= \langle \bar{B}| \left[\hat{T}^g, \hat{\tau}_{\nu_j} \right] |CC\rangle + \langle \bar{B}| \hat{T}^{g\nu_j} |CC\rangle
\end{aligned}
\tag{131}
$$

where $\langle \bar{B}| = \left(\langle HF| \sum_{\mu_i,i} \bar{t}^B_{\mu_i} \hat{\tau}^\dagger_{\mu_i} \right) \exp(-\hat{T})$. The first term in eq.(131) is calculated as a standard one-electron Hamiltonian contribution [79]. For the last term in eq.(131) we have

$$
\begin{aligned}
\langle \bar{B}| \hat{T}^{g\nu_j} |CC\rangle &= -\sum_{a=1}^A \langle \Lambda| \left[\hat{\mathbf{R}r_a}, \hat{\tau}_{\nu_j} \right] |CC\rangle \alpha_a \langle \bar{B}| \hat{\mathbf{R}r_a} |CC\rangle \\
&= -\sum_{a=1}^A \eta^{\hat{\mathbf{R}r}_a}_{\nu_j} \alpha_a \sum_{\mu_i,i} \bar{t}^B_{\mu_i} \xi^{\hat{\mathbf{R}r}_a}_{\mu_i} .
\end{aligned}
\tag{132}
$$

The $\xi^{\hat{\mathbf{R}r}_a}_{\mu_i}$ vector elements can be evaluated using the vacuum CC linear response theory. Thereby, the operator $-\sum_a \sum_{\mu_i,i} \bar{t}^B_{\mu_i} \xi^{\hat{\mathbf{R}r}_a}_{\mu_i} \alpha_a \hat{\mathbf{R}r_a}$ may be constructed. Performing a η-transformation of this operator we finally arrive at the result in equation eq.(132). The corresponding contribution in the QM/DC case is obtained by performing the substitutions: $\sum_{a=1}^A \rightarrow 2\sum_{lm}$, $\alpha_a \rightarrow g_l(\epsilon_{op})$, $\hat{\mathbf{R}r_a} \rightarrow \hat{T}_{lm}$.
Finally, the linear transformation of the $\mathbf{F}$ and $\mathbf{J}$ matrices can be performed in the same way as for the $\mathbf{A}$ matrix [119,135].

4.2.2. Quadratic Response

In order to calculate properties described by the quadratic response function, we need in addition to consider linear transformations of the $\mathbf{K}$, $\mathbf{B}$ and $\mathbf{G}$ matrices. First we consider the calculation of the $\mathbf{K}$ matrix. The $\mathbf{K}$ matrix is linearly transformed with two response vectors from the left. Using the expanded expression from Table 1 we obtain for the dielectric continuum case

$$(\bar{t}^A \bar{t}^B \mathbf{K})_{\sigma_k} = \sum_{i,\mu_i} \sum_{j,\nu_j} -2 \sum_{lm} g_l(\epsilon_{op}) \left(\bar{t}^A_{\mu_i} \langle \bar{\mu}_i | [\hat{T}_{lm}, \hat{\tau}_{\sigma_k}] | CC \rangle \bar{t}^B_{\nu_j} \langle \bar{\nu}_j | \hat{T}_{lm} | CC \rangle \right.$$
$$\left. + \bar{t}^B_{\nu_j} \langle \bar{\nu}_j | [\hat{T}_{lm}, \hat{\tau}_{\sigma_k}] | CC \rangle \bar{t}^A_{\mu_i} \langle \bar{\mu}_i | \hat{T}_{lm} | CC \rangle \right)$$
$$= -2 \sum_{lm} g_l(\epsilon_{op}) \left(\langle \bar{A} | [\hat{T}_{lm}, \hat{\tau}_{\sigma_k}] | CC \rangle \langle \bar{B} | \hat{T}_{lm} | CC \rangle \right.$$
$$\left. + \langle \bar{B} | [\hat{T}_{lm}, \hat{\tau}_{\sigma_k}] | CC \rangle \langle \bar{A} | \hat{T}_{lm} | CC \rangle \right) . \tag{133}$$

By defining the operator $^B \hat{T}^g$ as

$$^B \hat{T}^g = -2 \sum_{lm} g_l(\epsilon_{op}) \langle \bar{B} | \hat{T}_{lm} | CC \rangle \hat{T}^e_{lm} \tag{134}$$

we arrive at the following expression

$$(\bar{t}^A \bar{t}^B \mathbf{K})_{\sigma_k} = \langle \bar{A} | [^B \hat{T}^g, \hat{\tau}_{\sigma_k}] | CC \rangle + \langle \bar{B} | [^A \hat{T}^g, \hat{\tau}_{\sigma_k}] | CC \rangle$$
$$= {}' \eta^{B \hat{T}^g}_{\sigma_k} + {}' \eta^{A \hat{T}^g}_{\sigma_k} . \tag{135}$$

Here, the $'$ indicates that the HF part of the η vector has been omitted. The same expression for the linear transformed $\mathbf{K}$ matrix is found in the CC/MM case, but the specific definition of the $^B \hat{T}^g$ operator is of course different. Thereby, we in CC/MM theory obtain

$$^B \hat{T}^g = - \sum_{a=1}^{A} \langle \bar{B} | \hat{\mathbf{R}} \mathbf{r}_a | CC \rangle \boldsymbol{\alpha}_a \hat{\mathbf{R}} \mathbf{r}_a . \tag{136}$$

This is the appropriate effective operator to be considered in eq.(135) using the CC/MM approach. The $^B \hat{T}^g$ operator may be expressed as a contraction of a $\boldsymbol{\xi}$ vector and a left trial vector.

For the $\mathbf{B}$ matrix we note that the contribution due to the $^B \hat{T}^g$ operator simply follows the one-electron part of the vacuum Hamiltonian. For the additional contributions, we proceed as for the $\mathbf{K}$ matrix. Expanding the expression in Table 1 we obtain

$$^{solvent}(\mathbf{B} \mathbf{t}^A \mathbf{t}^C)_{\sigma_k} = \langle \bar{\sigma}_k | [\hat{T}^{gA}, \hat{C}] | CC \rangle + \langle \bar{\sigma}_k | [\hat{T}^{gC}, \hat{A}] | CC \rangle$$
$$+ \langle \bar{\sigma}_k | T^{gAC} | CC \rangle$$
$$= A^{\hat{T}^{gA}}_{\sigma_k C} + A^{\hat{T}^{gC}}_{\sigma_k A} + \xi^{\hat{T}^{gAC}}_{\sigma_k} \tag{137}$$

where the additional effective operator has been defined as

$$\hat{T}^{gAC} = -2\sum_{lm} g_l(\epsilon_{op})\langle\Lambda|[[\hat{T}_{lm},\hat{A}],\hat{C}]|CC\rangle\hat{T}^e_{lm}$$

$$= -2\sum_{lm} g_l(\epsilon_{op})\sum_{i,\mu_i}\sum_{j,\nu_j} F^{\hat{T}_{lm}}_{\mu_i\nu_j} t^A_{\mu_i} t^C_{\nu_j}\hat{T}^e_{lm} \qquad (138)$$

for the CC/DC case and

$$\hat{T}^{gAC} = -\sum_{a=1}^{A}\langle\Lambda|[[\hat{\mathbf{R}}\mathbf{r}_a,\hat{A}],\hat{C}]|CC\rangle\alpha_a\hat{\mathbf{R}}\mathbf{r}_a$$

$$= -\sum_{a=1}^{A}\sum_{i,\mu_i}\sum_{j,\nu_j} F^{\hat{\mathbf{R}}\mathbf{r}_a}_{\mu_i\nu_j} t^A_{\mu_i} t^C_{\nu_j}\alpha_a\hat{\mathbf{R}}\mathbf{r}_a \qquad (139)$$

in the CC/MM case.

Proceeding to the **G** matrix we note that the first contribution, $\langle\Lambda|[[[\hat{T}^g,\hat{\tau}_{\mu_i}],\hat{\tau}_{\nu_j}],\hat{\tau}_{\sigma_k}]|CC\rangle$, vanishes since $\hat{T}^g$ is a one electron operator [57] (see the discussion concerning the cubic response function in the next section). For the last contribution to the **G** matrix in Table 1 we have

$$^{solvent}(\mathbf{G}\mathbf{t}^B\mathbf{t}^C)_{\sigma_k} = \langle\Lambda|[[\hat{T}^{gB},\hat{C}],\hat{\tau}_{\sigma_k}]|CC\rangle + \langle\Lambda|[[\hat{T}^{gC},\hat{\tau}_{\sigma_k}],\hat{B}]|CC\rangle$$

$$+ \langle\Lambda|[[\hat{T}^{g\sigma_k},\hat{C}],\hat{B}]|CC\rangle$$

$$= F^{\hat{T}^{gB}}_{C\sigma_k} + F^{\hat{T}^{gC}}_{\sigma_k B} + \eta^{\hat{T}^{gCB}}_{\sigma_k} \qquad (140)$$

where the appropriate effective operators have been defined above.

For all the solvent modifications we first construct the appropriate effective operators by looping over the quantum number lm (the CC/DC) or the center of mass number a (the CC/MM case). Thereafter, we perform the specific transformation and then finally dotting the linear transformed vector onto the appropriate response vector we calculate the solvent contribution to the given property.

4.2.3. Cubic Response

In order to implement the cubic response function we would, according to the 2n+1 rule for the **t** parameters and the 2n+2 rule for the multipliers ($\bar{\mathbf{t}}$) [71], need to solve for the first and second order **t** parameters and first order $\bar{\mathbf{t}}$ parameters. However, as seen from the expressions for the CC/DC and CC/MM linear, quadratic and cubic response functions, the 2n+2 rule is actually weakened to a 2n+1 rule upon introducing a coupling to the solvent. Thus, for the cubic CC/DC or CC/MM response function, we need in addition to solve for the second order $\bar{\mathbf{t}}$ parameters. Also, we emphasize that in the CC/MM (including explicit polarization) and CC/DC models the second order $\mathbf{t}$ and $\bar{\mathbf{t}}$ response vectors are coupled.

According to the expression for the cubic CC/DC or CC/MM response function, eq.(113), additional linear transformations of the $\mathbf{H}$, $\mathbf{C}$ and $\mathbf{L}$ matrices, as defined in Table 2 need to be derived. However, the expression for the $\mathbf{H}$ and $\mathbf{C}$ matrix elements in Table 2 may, in the same way as for the $\mathbf{G}$ matrix described in the previous section, be simplified in the following way. Consider a BCH expansion of $\exp(-\hat{T})\hat{A}\exp(\hat{T})$, where $\hat{A}$ is a general second quantization operator

$$\exp(-\hat{T})\hat{A}\exp(\hat{T}) = \hat{A} + \left[\hat{A},\hat{T}\right] + \frac{1}{2}\left[\left[\hat{A},\hat{T}\right],\hat{T}\right]$$
$$+ \frac{1}{6}\left[\left[\left[\hat{A},\hat{T}\right],\hat{T}\right],\hat{T}\right]$$
$$+ \frac{1}{24}\left[\left[\left[\left[\hat{A},\hat{T}\right],\hat{T}\right],\hat{T}\right],\hat{T}\right]$$
$$+ \dots \tag{141}$$

where $\hat{T} = \sum_{\mu_i i} t_{\mu_i}\hat{\tau}_{\mu_i}$. It can be shown [57], that this expansion ends after five terms if $\hat{A}$ is a two-electron operator and after three terms if $\hat{A}$ is a one-electron operator. Since $\hat{T}^g$ and $\hat{T}^{g\nu}$ are effective one-electron operators the expression for the $\mathbf{C}$ matrix elements reduces to

$$C_{\mu_i\nu_j\sigma_k\delta_l} = \frac{\partial^4\{L_{\text{CC/MM}}^{(4)}\}_T}{\partial\bar{t}_{\mu_i}^{(1)}(\omega_x)\partial t_{\nu_j}^{(1)}(\omega_y)\partial t_{\sigma_k}^{(1)}(\omega_z)t_{\delta_k}^{(1)}(\omega_u)}$$
$$= \langle\bar{\mu}_i|\left[\left[\left[\hat{H}_{QM},\hat{\tau}_{\nu_j}\right],\hat{\tau}_{\sigma_k}\right],\hat{\tau}_{\delta_l}\right]|CC\rangle$$
$$+ P^{\nu_j\sigma_k\delta_l}\langle\bar{\mu}_i|\frac{1}{2}\left[\left[\hat{T}^{g\nu_j},\hat{\tau}_{\sigma_k}\right],\hat{\tau}_{\delta_l}\right] + \frac{1}{2}\left[\hat{T}^{g\nu_j\sigma_k},\hat{\tau}_{\delta_l}\right] + \frac{1}{6}\hat{T}^{g\nu_j\sigma_k\delta_l}|CC\rangle \tag{142}$$

and for the $\mathbf{H}$ matrix we find

$$H_{\mu_i\nu_j\sigma_k\delta_l} = \frac{\partial^4\{L_{\text{CC/MM}}^{(4)}\}_T}{\partial t_{\mu_i}^{(1)}(\omega_x)\partial t_{\nu_j}^{(1)}(\omega_y)\partial t_{\sigma_k}^{(1)}(\omega_z)t_{\delta_k}^{(1)}(\omega_u)}$$
$$= \langle\Lambda|\left[\left[\left[\hat{H}_{QM},\hat{\tau}_{\mu_i}\right],\hat{\tau}_{\nu_j}\right],\hat{\tau}_{\sigma_k}\right],\hat{\tau}_{\delta_l}\right]|CC\rangle$$
$$+ \frac{1}{8}P^{\mu_i\nu_j\sigma_k\delta_l}\langle\Lambda|\left[\left[\hat{T}^{g\mu_i\nu_j},\hat{\tau}_{\sigma_k}\right],\hat{\tau}_{\delta_l}\right]|CC\rangle . \tag{143}$$

Furthermore, the $\mathbf{G}^X$ matrix vanishes if $\hat{X}$ is a one-electron operator and no higher derivatives than $\mathbf{H}$ do exist [71].

Formally, frequency dependent fourth order properties are found from the cubic response function. However, if one of the frequencies is equal to zero, which is the case for example for electric field induced second harmonic generation (EFISHG), frequency dependent fourth order properties may be found from FF calculations on third order properties, i.e.

$$\gamma_{ijkl}(-\omega_i;\omega_j,\omega_k,0) = \frac{d\beta_{ijk}(-\omega_i,\omega_j,\omega_k)}{d\epsilon_l} \qquad (144)$$

where ϵ_k is the field strength parameter and $\omega_i = \omega_j + \omega_k$.

As discussed in this section, the introduction of a reaction-field methodology in CC theory leads to a coupling of the two sets of t and $\bar{t}$ response equations. However, as discussed previously, the two sets of response equations actually decouple in the CC/MM case *if* the classical molecules are defined *only* through point charges and Lennard-Jones parameters, and the expression for the response functions are the same as in vacuum but with modified one-electron integrals. This means that restricting the calculation of fourth order properties to the case where at least one frequency is equal to zero we may, for the CC/DC and CC/MM (including polarization) models, use our analytical implementation of the quadratic CC/DC and CC/MM response functions [136] combined with the finite field technique. In addition, we may in the CC/MM case where explicit polarization effects are excluded, use the analytical implemented vacuum cubic response function [134], but with modified one-electron integrals, to calculate fourth order properties. However, one major drawback is that for the CC/DC model only equilibrium results may be obtained when using the finite field approach.

4.3. *Molecular Properties*

4.3.1. *General Considerations*

In this section we shortly describe the connection between response functions and molecular properties. As in section 4.1.1 we will rely on the semi-classical description for the interaction between a molecule and the electromagnetic radiation, i.e. the field is described classically while the electronic structure of the molecule is considered using quantum mechanics. We note that this semi-classical approximation neglects the presence of the molecule on the electrostatic field, i.e. the electrostatic field may be described in terms of the usual scalar, $\phi(r,t)$ and vector, $\mathbf{A}(r,t)$ potentials. In classical electrodynamics, the interaction between a charged particle and an electromagnetic field is described by introducing the vector and scalar potentials into the Hamiltonian. Expansion of the Hamiltonian, and using the Coulomb gauge, leaves us with a Hamiltonian, which has a linear and a quadratic dependence on the vector potential. The vector potential is expanded around some molecular origin

$$\mathbf{A}(r,t) = \mathbf{A}^0(t) + \{\nabla\mathbf{A}(t)\}^0 \cdot \mathbf{r} + \cdots \tag{145}$$

to a given order. Truncating the expansion after the first term leaves us with the electric dipole approximation, i.e. we assume that the vector potential is homogeneous over the molecular system. However, truncation after the second term and neglecting the quadratic dependence of the vector potential (the weak-field approximation [139]) introduces two additional contributions (i) the magnetic dipole interaction and (ii) the electric quadrupole interaction. In this approximation the explicit expression for the multipolar Hamiltonian becomes [140]

$$\hat{H} = \hat{H}_0 - \boldsymbol{\mu} \cdot \mathbf{E}^0(t) - \mathbf{m} \cdot \mathbf{B}^0(t) - \mathbf{Q}\{\nabla\mathbf{E}(t)\}^0 \tag{146}$$

where ^{0} indicates that the quantities are evaluated in the center of the expansion of the vector potential and where $\hat{H}_0$ is the unperturbed molecular Hamiltonian.

Let us first discuss the electric dipole approximation more carefully, i.e. the vector potential is truncated after the first term which is equivalent to $\hat{H} = \hat{H}_0 - \boldsymbol{\mu} \cdot \mathbf{E}(t)$ (in this discussion we drop the superscript 0 on the electric field). We observe that this approximation is justified for the case where $\frac{r}{\lambda} \ll 1$ (r is the molecular dimension and λ is the wavelength). For radiation with $\lambda \simeq 380 - 720$ nm (or equivalently frequencies in the interval of 0.06 - 0.12 au), i.e. visible light, the electric dipole approximation is expected to be valid for molecules of dimension up to a few nanometers. However, if the considered electronic transition is dipole forbidden, higher order terms may be significant.

Choosing $\hat{X} = \boldsymbol{\mu}$, e.g. the electric dipole operator, in the expression for the time evolution of the expectation value of a time-independent Hermitian operator, eq.(83), allows for an identification of the response functions in eq.(83) as the molecular dipole polarizabilities [129], i.e.

$$\alpha_{xy}(-\omega;\omega) = -\langle\langle\mu_x, \mu_y\rangle\rangle_\omega \tag{147}$$

$$\beta_{xyz}(-\omega_1 - \omega_2; \omega_1, \omega_2) = -\langle\langle\mu_x, \mu_y, \mu_z\rangle\rangle_{\omega_1\omega_2} \tag{148}$$

$$\gamma_{xyzu}(-\omega_1 - \omega_2 - \omega_3; \omega_1, \omega_2, \omega_3) = -\langle\langle\mu_x, \mu_y, \mu_z, \mu_u\rangle\rangle_{\omega_1\omega_2\omega_3} \tag{149}$$

where $xyzu$ refer to the molecular axes. The quantities α_{ij}, β_{ijk} and γ_{ijkl} are the (permanent) electric dipole-dipole polarizability, the electric dipole-dipole-dipole first hyperpolarizability and the electric dipole-dipole-dipole-dipole second hyperpolarizability, respectively, of the unperturbed molecule.

The molecular properties in eqs.(147 - 149) describe a variety of physical processes. For example, the frequency dependent dipole-dipole polarizability is primarily responsible for the refractive index, while the second harmonic generation (SHG) dipole-dipole-dipole first hyperpolarizability, e.g. eq.(148) with $\omega_1 = \omega_2$, is responsible for the temperature dependent part of the electric field induced second harmonic generation (EFISHG) experiment. Correspondingly, the EFISHG dipole-dipole-dipole-dipole second hyperpolarizability determines the temperature independent part of the EFISHG experiment. Finally, third harmonic generation (THG) is described by the THG dipole-dipole-dipole-dipole second hyperpolarizability, e.g. eq.(149) with $\omega_1 = \omega_2 = \omega_3$.

Concerning coupled cluster theory we note, as pointed out in Refs. [119], that the specific coupling of the $\mathbf{t}$ and $\bar{\mathbf{t}}$ responses together with the fact that the CC/DC and CC/MM Lagrangians are non-linear in the $\bar{\mathbf{t}}$ parameters causes some complications in the determination of transition properties. However, an approximate decoupling of the $\mathbf{t}$ and $\bar{\mathbf{t}}$ parameters in the eigenvalue equation makes the calculation of excitation energies straightforward. Thus, as discussed in Ref. [119], we proceed with a slightly additional simplification: the contribution from the $\mathbf{J}$ matrix is neglected in the determination of transition properties. In this way we obtain the same formal expression for transition properties as in the vacuum case, but now we have included the QM/DC or QM/MM contributions. Thereby, the electronic excitation energies, ω_f, are found as the eigenvalues of the coupled cluster Jacobian [71,141],

$$\mathbf{A}\mathbf{R}^f = \omega_f \mathbf{R}^f \tag{150}$$

$$\mathbf{L}^f \mathbf{A} = \mathbf{L}^f \omega_f \tag{151}$$

where $\mathbf{R}^f(\mathbf{L}^f)$ denotes a right(left) eigenvector of the Jacobian matrix. Furthermore, from the residues of the linear CC/DC and CC/MM response functions we obtain the transition strength

$$S_{XY}^{of} = \frac{1}{2}(T_{0f}^X T_{f0}^Y + (T_{0f}^Y T_{f0}^X)^*). \tag{152}$$

The "left and right transition moments" of exact theory ($T_{f0}^X = \langle f|\hat{X}|0\rangle$ and $T_{0f}^X = \langle 0|\hat{X}|f\rangle$) are in CC response theory replaced by

$$T_{f0}^X = \mathbf{L}^f \boldsymbol{\xi}^X \tag{153}$$

$$T_{0f}^X = \boldsymbol{\eta}^X \mathbf{R}^f + \bar{\mathbf{M}}^f(\omega_f)\boldsymbol{\xi}^X \tag{154}$$

where the vector $\bar{\mathbf{M}}^f(\omega_f)$ is defined through the equation

$$\bar{\mathbf{M}}^f(\omega_f)(\omega_f \mathbf{1} + \mathbf{A}) + \mathbf{F}\mathbf{R}^f = 0 . \tag{155}$$

Finally we note that the residues of the quadratic and cubic response functions allow for a determination of a variety of molecular properties such as two-photon and three-photon transition properties and excited state first and second order properties (see for example Ref. [129] and references therein).

5. Electric Properties of Molecules in Condensed Phases.

In this section we present and discuss some applications of the CC/MM and CC/DC response formalism. We consider liquid water where we will present electronic excitation energies [135] together with electric dipole-(hyper)polarizabilities [135,136,138]. Water is by far the most interesting liquid and solvent to investigate due to its presence in biological systems and unique properties in terms of hydrogen bonding, dipole moment and higher order electric molecular properties. Generally, it has also been the molecule of choice when benchmarking solvent models.

5.1. *Electronic Electric Response Properties of Liquid Water*

In this section we will consider the lowest electronic excitation energies together with some electric response properties of liquid water. All CC/MM results in this section will refer to an averaged water configuration. The average configuration of water is based upon MD simulations and the generation of the cluster has been described in detail in for example Ref. [91]. The total water cluster used in the combined quantum mechanics/molecular mechanics calculations contains 128 water molecules and one of these molecules is treated using coupled cluster theory, while the other 127 water molecules represent the MM part of the system. The results termed "nonpol" denote calculations that neglect the explicit MM polarization, whereas the "pol" results include the explicit MM polarization. All the calculations are performed using the intramolecular water geometry $R(OH) = 0.9572$ Å and $< (HOH) = 104.49°$ taken from Ref. [142]. The necessary force field parameters are all taken from Ref. [89], i.e. $\bar{\alpha} = 9.718$ a.u., $q_H = 0.3345$ and $q_O = -0.669$ for the (isotropic) polarizability and the partial charges, respectively, and $A_{ma} = 2.083 \cdot 10^6$ a.u. and $B_{ma} = 45.41$ a.u. for the Lennard-Jones parameters. Using the averaged geometry we only have to perform *one* CC/MM calculation, but we are in turn only left with a single number for each property and not a statistical distribution. We note that the use of just one solvent structure may be a crude approximation and that sampling over different solvent configurations can be very important. However, we try to solve this problem by defining an averaged structure as discussed above, even though a proper sampling over different configurations clearly would represent an improvement [143]. We note that the use of averaged structures (or mean field models) are widely used. This includes, for example, methods based on the calculation of the averaged solvent electrostatic potential from molecular mechanics data [34,144] or methods which average over configurations in order to calculate mean electric field components [145,146] and for the calculation of rate constants for S_N2 reactions [142]. Also, methods which use a lowest energy configuration based on MD has been presented [147]. The issue of statistically converged molecular properties of liquid water modelled by CC/MM is currently being investigated.

In Figure 1 we show the water molecule treated using quantum mechanics together with the first solvation shell as found in the averaged water structure. From Figure 1 we observe that the water molecule treated using quantum mechanics performs four hydrogen bonds. Also, the structure of the water cluster approximately possesses a tetrahedral structure. Both observations are in line with experimental observations.

When including results using the (nonequilibrium) dielectric continuum model, a single water molecule is enclosed in a spherical cavity surrounded by a dielectric continuum. The radius of the cavity is 4.0 au. In the continuum calculations we have used $\epsilon_{st} = 78.54$ and $\epsilon_{op} = 1.778$ [3,100] for the

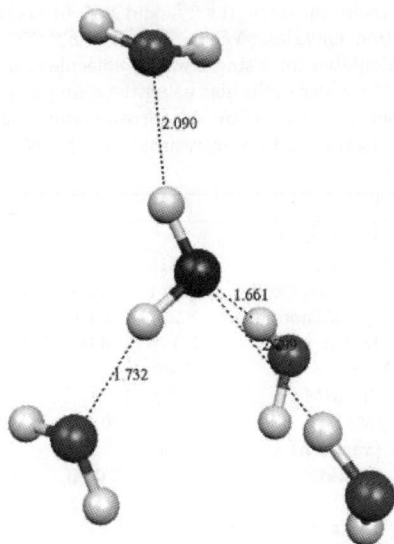

Fig. 1.   Water including the first solvation shell. Distances in Ångstrøm

dielectric constants of water. If not stated differently, we will assume that the water molecule described using quantum mechanics is placed in the xz plane with the oxygen in the origin and hydrogens at positive z-values. The molecule has a constrained C_{2v} geometry and thereby has a two-fold axis aligned in the z-direction.

5.1.1. *Electronic Excitation Energies*

In Table 3 we present excitation energies and oscillator strengths calculated for a single water molecule in a cluster of 127 water molecules using the d-aug-cc-pVTZ basis set [148]. The oscillator strengths have been calculated using the length gauge formulation Also shown are the corresponding gas phase results and results employing the nonequilibrium dielectric continuum model (Noneq.Cont.) [100,119]. The CC(/MM) results are from Ref. [135].

Experimentally, data of liquid water gives the first maximum of the one-photon absorption spectrum to be ∼ 8.2 eV and another maximum is found to be ∼ 9.9 eV. [149,150,151]. We note that the band around 9.9 eV is rather

Table 3. Excitation energies E_{ex} (eV), length gauge oscillator strengths $^n f_r$ and shift in electronic excitation energies, $\Delta E = E_{ex}^{\text{liquid}} - E_{ex}^{\text{vacuum}}$, (in eV) calculated for a single water molecule in a cluster of 127 water molecules using the d-aug-cc-pVTZ basis set. Also shown are the corresponding vacuum and dielectric continuum results. CC(/MM) results are from

Model	E_{ex}	$^n f_r$	ΔE
$^1A_1(\tilde{X}) \rightarrow 1^1B_1$			
CCSD(Vacuum)	7.616	0.051	—
CCSD(Noneq.Cont.)	7.711	0.067	0.095
CCSD/MM(nonpol)	8.237	0.072	0.621
CCSD/MM(pol)	8.176	0.079	0.560
HF/MM	9.49		0.84
MCSCF/MM	8.62		0.77
DFT/MM	8.25	0.075	0.53
Exp. (vacuum)	7.4		—
Exp. (liquid)	8.2	0.06	0.8
$^1A_1(\tilde{X}) \rightarrow 1^1A_2$			
CCSD(Vacuum)	9.368	—	—
CCSD(Noneq.Cont.)	9.752	—	0.384
CCSD/MM(nonpol)	10.021	0.002	0.653
CCSD/MM(pol)	9.965	0.006	0.597
HF/MM [121]	11.3		1.0
MCSCF/MM [121]	10.5		0.9
DFT/MM [122]	10.18	0.003	0.65
$^1A_1(\tilde{X}) \rightarrow 2^1A_1$			
CCSD(Vacuum)	9.881	0.058	—
CCSD(Noneq.Cont.)	9.852	0.029	-0.029
CCSD/MM(nonpol)	10.664	0.094	0.783
CCSD/MM(pol)	10.556	0.113	0.675
DFT/MM	10.27	0.112	0.59
Exp. (vacuum)	9.7		—
Exp. (liquid)	9.9		0.2

diffuse. The corresponding gas phase results are around 7.4 eV and 9.7 eV, respectively [150,152,153]. Thus, the solvent effects are approximately 0.8 eV and 0.2 eV, respectively, from a naive interpretation, but the origin and meaning of the band maxima may of course be rather different in vacuum and solution.

From Table 3 we find that both CC/MM models predict a blue shift of the lowest excitations in accordance with the experimental results. Furthermore, the MM polarization accounts for as much as up to 16 % on the reported shifts in the excitation energies. Neglecting the MM polarization overestimates the excitation energies when compared to the description including polarization effects. From the results concerning the oscillator strength, we find that the first and the third electronic excitations reported in Table 3 are to be compared with the experimental values ~ 8.2 eV and 9.9 eV, respectively. However, while the absolute value and the shift in the first excitation energy compare well with experiment, the absolute value and the shift in the third excitation energy is overestimated compared to experiment. The experimental oscillator strength for the first excitation is about 0.06 [149] which also compares well with the results reported in Table 3. For the results employing the nonequilibrium dielectric continuum model, we find that the solvent effect on the first excitation is underestimated as compared to the CC/MM description and to experiment. Also, the shift of the third excitation has the wrong sign when compared to the CC/MM and experiment. Thus, the continuum model cannot reproduce the experimental data. This has also been concluded in Ref. [154].

We note that the HF/MM, the combined multiconfigurational self-consistent-field/molecular mechanics (MCSCF/MM) and the DFT/MM results in Table 3 are obtained using the same averaged water structure as used in the CC/MM calculations. As seen from Table 3, the HF/MM and MCSCF/MM methods overestimate the excitation energies as compared to CCSD/MM. However, for the first excitation the shift upon solvation compare reasonable well with CCSD/MM. The DFT/MM approach combined with the statistical averaging of (model) orbital potentials (SAOP) exchange-correlation potential of Baerends et al. [155,156,157], compare well with the CCSD/MM results reported in Table 3. However, as discussed in Ref. [122], the SAOP calculations belongs to the more elaborate results of Ref. [122], and using functionals such as BLYP [158,159] the first excitation in Table 3 is predicted to be 6.49 eV with a solvent shift of 0.25 eV, e.g. worse than the HF/MM results.

5.1.2. *The Electric Dipole-Dipole Polarizability of Liquid Water*

The electric dipole-dipole polarizability may be calculated from the linear response function, and using the outlined theoretical solvent models, environmental effects may be included using either a dielectric continuum or a discrete representation of the solvent. In Table 4 we report the molecular electric dipole-dipole polarizability for water in gas and liquid phases and compare the result with experimental data. The experimental liquid value in Table 4 have been derived using the Lorenz-Lorentz [83] equation

$$\alpha(-\omega;\omega) = \frac{3M_w}{4\pi N_a \rho} \frac{n(\omega)^2 - 1}{n(\omega)^2 + 2} \qquad (156)$$

where M_w is the molecular weight, N_a Avogadro's number, ρ the density and $n(\omega)$ the refractive index of liquid water [160].

We note that eq.(156) is only valid in the high frequency region and furthermore, the condensed phase polarizabilities derived using this equation are rather to be considered as "vacuum" polarizabilities since the effect of the solvent is included in the local field and not in the derived properties. However, geometrical effects due to the presence of the solvent is still included. The theoretical properties presented in Table 4 is solute properties, which mean that the experimental results should include the reaction field effect before comparison. The reaction field effect may be *estimated* by considering the Onsager approach to solvation [83,103]. This leads to a reaction field correction factor, F_R^ω, for water at a frequency of 0.0856 au of approximately 1.06 leading to 10.61 au for the liquid phase polarizability. This is indicated in parenthesis in Table 4. We note that this correction to the experimental results is solely based on the Onsager approximation. However, as shown in Ref. [163], the Onsager approximation for the linear polarizability for water compares relatively good with the more extended

---

Table 4. Experimental and theoretical values for the electric dipole-dipole polarizabilities. The calculations have been performed using the d-aug-cc-pVTZ basis set. The terms ω is the frequency, α_{ii} the polarizability tensor elements, $\langle\alpha\rangle_{ZZ}$ the isotropic polarizability and $\Delta\langle\alpha\rangle_{ZZ}$ the anisotropy. Results are in au.

Model	ω	α_{xx}	α_{yy}	α_{zz}	$\langle\alpha\rangle_{ZZ}$	$\Delta\langle\alpha\rangle_{ZZ}$
Experimental (gas phase) [161]	0.0000	—	—	—	9.83	—
Experimental (gas phase) [162]	0.0880	9.55	10.31	9.91	9.92	0.67
Experimental (liquid phase) [160]	0.0856	—	—	—	10.01(10.61)	—
CCSD	0.0000	9.36	9.98	9.61	9.65	0.54
CCSD	0.0856	9.70	10.17	9.86	9.91	0.59
CCSD/DC(noneq)	0.0856	10.84	10.43	10.40	10.56	0.60
CCSD/MM(nonpol)	0.0856	9.53	9.80	9.66	9.66	0.61
CCSD/MM(pol) [135]	0.0856	9.96	10.39	10.38	10.24	1.14
MCSCF/MM [121]	0.0856	9.71	10.0	10.0	9.93	—
DFT/MM [122]	0.0856	9.99	9.89	10.10	9.99	—

calculations ($L_{max} = 10$).

At the MP2 level ZPVA corrections to the static polarizability (for water in vacuum) have been calculated to be around 0.29 au [161]. Adding this contribution to the (vacuum) electronic component in Table 4, we obtain a value of 9.94 au which compares good with the experimental data. We note that the experimental static dipole-dipole polarizability is found from a quadratic extrapolation of refractive index data and therefore excludes most of the pure vibrational contributions. The effect of triples excitations on the polarizability has been considered in Ref. [164], and are found to be small for water. Assuming that the electronic component and the ZPVA contribution have the same frequency dependence we obtain $\Delta\alpha^{ZPVA} \simeq$ 0.3 au at a frequency of 0.0856 au, which should be added to the pure theoretical electronic dynamical contributions given in Table 4 before comparison with the dynamical experimental data. We note that for the liquid case this is of course only approximate, since the vibrational contributions (as well as the molecular geometry) will change upon solvation. For the anisotropy (invariant) a static ZPVA contribution around 0.24 au [161] has been reported which leads to a total theoretical anisotropy around 0.78 au ($\omega = 0.0856$ au) which compares relatively good with the experimental data in Table 4.

From the condensed phase results in Table 4 we find that excluding the correction factor, F_R^ω, the CCSD model tends to overestimate the polarizabilities as compared to the experimental observations. However, including this factor leaves an "experimental" polarizability of 10.61 au which, when correcting for the ZPVA contribution, compares much better with the theoretical predicted polarizabilities. Specially, the CCSD/MM(pol) model (10.54 au) is seen to be in very good agreement with this "experimental" result. In fact, we find from Table 4 that the CC/MM(pol) and CC/DC results compare somehow better with this "experimental" polarizability than the DFT and MCSCF results. We note that this comparison between "experimental" and theoretical properties is only approximate as indicated by the many approximations considered above.

Experimentally, we observe that for a frequency of 0.0856 au an increase in the polarizability is observed upon solvation. For the theoretical data in Table 4 we also observe this increase in polarizability, except in the case where the solvent is modeled as a discrete nonpolarizable medium. We note that the MCSCF/MM and DFT/MM results in Table 4 use the same averaged water cluster as the CC/MM calculations and are hence directly comparable. Furthermore, for the DFT/MM results in Ref. [122], we note that a large spread in values are found upon using different functionals. For example, for the isotropic polarizability at a frequency of 0.0856 au, values in the interval 9.75 - 11.85 au, i.e. a spread $\sim$ 2 au, is reported. Specially, the popular BLYP [158,159] exchange-correlation functional is found to give rather high values for the polarizabilities.

5.1.3. *Non-linear Optical Response Properties of Liquid Water*

At the microscopic level, the properties governing the non-linear processes are, to the lowest order, the frequency dependent first and the second hyperpolarizabilities. These properties may be derived from the quadratic and cubic response functions, respectively. Thus, using the CC/DC or CC/MM models outlined in this chapter, solvent effects on these microscopic properties may be calculated. Experimentally, the quantity measured in the EFISHG experiment is the macroscopic third order non-linear susceptibility, $\chi^{(3)}(-2\omega; \omega, \omega, 0)$, and this quantity is related to the microscopic effective second order hyperpolarizability, which contains a contribution from both the first and second hyperpolarizabilities

$$\bar{\gamma}(-2\omega; \omega, \omega, 0) = \gamma_{||}(-2\omega; \omega, \omega, 0) + \frac{\mu \beta_{||}(-2\omega; \omega, \omega)}{3kT} , \qquad (157)$$

where $\gamma_{||}(-2\omega; \omega, \omega, 0)$ is the scalar components of the fourth order hyperpolarizability tensor, and the dipolar contribution involves the vector component in the direction of the dipole moment of the first hyperpolarizability. Both terms include vibrational effects. We have calculated β and γ using CC response theory including solvent effects. For the CC/MM(pol) and CC/DC cases we have, in order to calculate γ, performed finite field calculations on β, while we have used the analytical implementation in the CC/MM(nonpol) case.

In Figure 2 we report the dispersion of $\gamma_{||}(-2\omega; \omega, \omega, 0)$ for water calculated using CCSD, CCSD/MM(nonpol) and CC/MM(pol) combined with the d-aug-cc-pVTZ basis set [148]. The results are from Ref. [138]. The parameters defining the CC/MM model are the same as for the linear polarizability. As seen from Figure 2, the same trends as for the linear polarizability are found upon solvation, i.e. treating the solvent as a polarizable environment

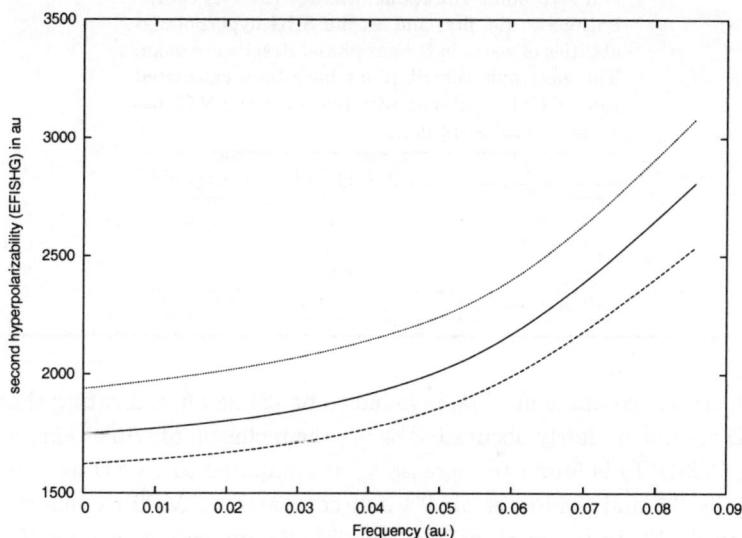

Fig. 2. The frequency dependent second hyperpolarizability, $\gamma_{||}(-2\omega;\omega,\omega,0)$, for water calculated using CCSD combined with the d-aug-cc-pVTZ basis set. (——) : CCSD Vacuum; (- - - - -) : CCSD/MM(nonpol); (·····) : CCSD/MM(pol).

leads to an increase in the property as compared to the vacuum calculation, whereas using a nonpolarizable description of the solvent the property is decreased as compared to the vacuum calculation.

In order to compare the calculated vacuum properties to experimental data, we have in Table 5 collected the CCSD results of the SHG $\beta_{||}$ and $\gamma_{||}$ at the frequency of 0.0656 au. Also, we have listed literature values for the ZPVA and PV contributions. As seen from Table 5, the dominating contribution is the electronic part of the hyperpolarizabilities. The ZPVA value for $\gamma_{||}$ in Table 5 equal to 51 au is for the *static case*. Assuming a dispersion exactly equal to the electronic counterpart we obtain a ZPVA contribution of 69 au, i.e. an increase of around 35 %. However, in any case the theoretical hyperpolarizabilities compare excellently with the experimental data indicating the high accuracy of the CCSD method.

The effects of triples excitations on $\beta_{||}$ have been considered in Ref. [164] using the CC3 model [74]. At a frequency of 0.0656 au, a best estimate of

Table 5.   Electronic (elec.), pure vibrational (PV)
and zero-point vibrational average (ZPVA) contributions to the first and second SHG hyperpolarizabilities of water in the gas phase. Results are in au.
The electronic contributions have been calculated
using CCSD combined with the d-aug-cc-pVTZ basis set. Results are in au.

	$\beta^{gas}_{\parallel\ \omega=0.0656}$	$\gamma^{gas}_{\parallel\ \omega=0.0656}$
elec. [136,138]	-21.73	2355
PV [165]	-0.21	-10.9
ZPVA [166]	-0.95	51
Total	-22.89	2395
Experimental [167]	-22.0 ± 0.9	2310 ± 120

the electronic component of $\beta_{\parallel}$ is found to be -21.28 au, indicating that our CCSD results are fairly accurate. For $\gamma_{\parallel}$, the inclusion of triples excitations using CCSD(T) is found to increase $\gamma_{\parallel}$ as compared to CCSD as noted by Maroulis [168] and Sekino et al. [169] and compared to MP2 as reported by Reis et al. [166]. In the work of Maroulis [168], the increase is around 10 % as compared to CCSD. However, the CCSD(T) method only allows for calculations of the static second hyperpolarizability (the frequency was modeled using SCF dispersion).

Turning now to the condensed phase polarizabilities, we have in Table 6 collected theoretical and the experimental results for the SHG $\beta_{\parallel}$ and $\gamma_{\parallel}$ for liquid water. We emphasize that the theoretical results include only the electronic contributions. Furthermore, the experimental results are derived from measurements of the macroscopic third order non-linear susceptibility and in order to derive the corresponding microscopic properties effective field factors must be considered. Thereby, the theoretical results in Table 6 are solute properties while, in order to extract the experimental results, both the cavity and reaction field effects have been considered.

Table 6.   Electronic contribution to the first and second SHG hyperpolarizabilities of water in the liquid phase. Results are in au. The electronic contributions have been calculated using CCSD combined with the d-aug-cc-pVTZ basis set. Results are in au.

	CCSD/DC	CCSD/MM(nonpol)	CCSD/MM(pol)	Exp.
$\beta^{liquid}_{\parallel\ \omega=0.0428}$ (Ref. [136])	4.51	-0.16	12.21	11.5 [170]
$\gamma^{liquid}_{\parallel\ \omega=0.0428}$ (Ref. [138])	—	1790	2164	2134 [170]

This obviously means that a direct comparison between the experimental and theoretical hyperpolarizabilities is not straightforward and the experimental numbers merely act as qualitative directions e.g. in terms of signs and magnitudes. In Table 6, the frequency is 0.0428 au because the experiment was performed at this frequency [170]. The first important thing to note from Table 6 is that the experimentally established sign change in the first hyperpolarizability is reproduced using both the CCSD/DC and CCSD/MM(pol) models, but is not found using the CCSD/MM(nonpol) model. Second, we find that for $\beta_{||}$ the CCSD/DC model cannot capture enough solvent effects as compared to the CCSD/MM(pol) model and experimental data. Indeed, the magnitude of $\beta_{||}$ using the CCSD/DC model is underestimated as compared to the CCSD/MM(pol) result, which clearly gives result in line with the experimental values.

In passing we note that using the simple Onsager model, i.e.

$$\beta_{||\ \omega=0.0428}^{solute} = \beta_{||\ \omega=0.0428}^{vacuum} \left(F_R^\omega\right)^2 F_R^{2\omega} \tag{158}$$

predicts $\beta_{||\ \omega=0.0428}^{solute} \simeq$ -22.9 au which obviously has the wrong sign as compared to experimental data.

In Ref. [171] we discussed an approach where the effects of the solvent polarization was modelled implicitly, e.g. by neglecting the explicit polarizability but increasing the magnitude of the partial point charges situated at the nuclei of the molecules treated using MM theory. The charges were in this approach found by requiring that the interaction energy between the QM and MM molecules, e.g. $E_{QM/MM}$, was the same in the situation where the solvent polarization was modelled either implicitly or explicitly. For lower order molecular properties this approach was found to be fairly accurate. In Ref. [171] we used the aug-cc-pVTZ basis set [172]. Repeating this procedure using the d-aug-cc-pVTZ basis set [148] we find the appropriate MM charges to be $q_O = -0.8198$ and $q_H = 0.4099$. Using now this set of partial MM point charges in calculating the first hyperpolarizability we arrive at $\beta_{||} = 3.62$ au. Thus, treating the solvent polarization implicitly reveals the right sign of $\beta_{||}$. However, as compared to both experiment and the case where the solvent polarization is treated explicitly, the magnitude of $\beta_{||}$ is underestimated. Thus, in the calculation of higher order response properties we find it of crucial importance to model the solvent polarization explicitly.

For $\gamma_{||}$ we have not listed the value of the result obtained using the

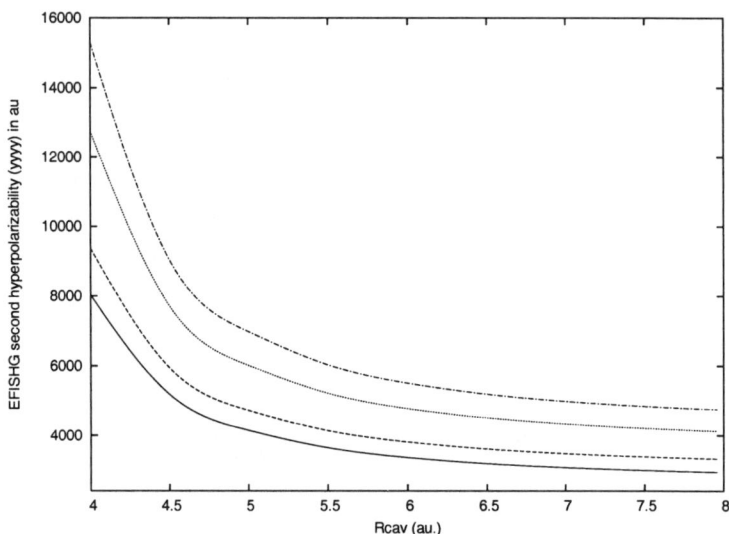

Fig. 3. EFISHG Second hyperpolarizability (γ_{yyyy}) as a function of the cavity radius (R_{cav}) for water calculated at the CC2 and CCSD level of theory. The basis set used is the d-aug-cc-pVTZ basis set. (——) : CCSD ($\omega = 0.0000$); (- - - - -) : CCSD ($\omega = 0.0430$); ($\cdots\cdots$) : CC2 ($\omega = 0.0000$); (—·—·—) : CC2 ($\omega = 0.0430$)

CCSD/DC model. This is due to the fact that in the condensed phase cases, we obtain $\gamma_{||}$ by performing finite field calculations on β, which imposes the restriction that only equilibrium results may be derived. This assumption may represent a potential problem concerning the calculation of dynamical properties. The dielectric continuum model, may also suffer from a much more complex problem. This problem has to do with the treatment of the cavity. In order to investigate the effect of the cavity radius on the second hyperpolarizability, we have in Figure 3 plotted γ_{yyyy}^{EFISHG} as a function of cavity radius (Rcav). The results are from Ref. [138]. The γ_{yyyy} component has been evaluated at the CC2 and CCSD level of theory and for two frequencies ($\omega = 0.0000, 0.0430$ au). From Figure 3 we find that the cavity radius has a significant effect on γ_{yyyy}, and we also find that choosing a cavity radius greater than 4.0 au would improve upon the results, as compared to CC/MM.

The cavity radius of 4.0 au is obtained as the distance from the center-of-mass to the hydrogen atom plus the van der Waals radius of the hydrogen atom. Another method would be to add to the distance from the center-of-mass to the hydrogen atom 1.2 times the van der Waals radius for hydrogen (see for example Ref. [173]). This would lead to a cavity radius of 4.5 au. However, if a CC/DC calculation is performed using a cavity radius equal to 4.5 au, the first hyperpolarizability is not obtained in qualitative agreement with the experimental results, i.e. no sign change in $\beta_{||}$ is observed. Thus, as the third order properties are not correctly described, nothing conclusive can be said about the higher order properties.

Obviously, part of the problems concerning the DC model may be directly related to both the treatment of the shape of the cavity and the missing short range repulsion between the solute molecule and the solvent. Specially when employing large basis sets, the treatment of short range repulsion effects may be mandatory and methods which directly introduce the effect of the Pauli repulsion into the optimization of the wave function might remove part of the artifacts otherwise introduced [8,18,93,94,95,96].

We note that solvent effects on β or γ for water have recently been considered in several papers including the work of Gubskaya et al. [146] (β and γ, charge perturbation at the MP2 and MP4 levels of theory), Sylvester-Hvid et al. [101,102] (β and γ, respectively, MCSCF combined with the dielectric continuum model), Mikkelsen et al. [55] (β, HF semi-continuum calculations including the first solvation shell surrounded by the dielectric continuum and MCSCF dielectric continuum), Poulsen et al. [120] (β, MCSCF combined with molecular mechanics), Jensen et al. [174] (β and γ, DFT combined with molecular mechanics), Macak et al. [163] (β and γ, SCF theory combined with semi-classical [175,176,177] and self-consistent-reaction-field models), Reis et al. [145] (γ, Lorentz correction [83], Onsager reaction field model and a discrete local-field theory) and Reis et al. [166] (β and γ, MP2 based local-field method). Finally, Sylvester-Hvid et al. have very recently extended the before mentioned MCSCF calculations to include, in addition to the dielectric continuum, a first solvation shell of water molecules [178].

Finally, we summarize our best results for $\beta_{||}$ and $\gamma_{||}$ in both gas and liquid phases in Table 7 at the frequency of 0.0656 au. From this table we clearly find that solvent effects on the first hyperpolarizability are significant, while solvent effects on the second hyperpolarizability are more modest. However, as compared to the vibrational contributions for the sec-

Table 7.　Electronic contribution to the first and second SHG hyperpolarizabilities of water in gas and liquid phases. Results are in au. The electronic contributions have been calculated using CCSD combined with the d-aug-cc-pVTZ basis set.

Model	$\beta_{\parallel}$ $\omega=0.0656$	$\gamma_{\parallel}$ $\omega=0.0656$
CCSD	-21.73	2355
CCSD/MM(pol)	12.89	2583
Solvent effects	34.62 ($\simeq -160\%$)	228 ($\simeq 10\%$)

ond hyperpolarizability in Table 5, which are around 40 au, solvent effects are still more important than vibrational effects for the second hyperpolarizability, at least in the dynamical case where the pure vibrational effects usually are rather small.

6. Conclusions

We have explored solvent effects on molecular properties using two different theoretical methods, e.g. the combined coupled cluster/dielectric continuum (CC/DC) and the combined coupled cluster/molecular mechanics (CC/MM) methods. These two models differ significantly in the way solute-solvent interactions are described. In the CC/DC model, all structural properties of the solvent molecules are neglected and the statistical averaging over solvent configurations is implicitly included. In contrast, the CC/MM model directly accounts for the solvent structural anisotropies. We have presented theoretical and implementational aspects of both models, including the extensions required, as compared to the gas-phase calculations, to incorporate the effects of the solvent into the response equations. Using this response theory formalism, we have presented solvent effects on properties described by the linear, quadratic and cubic response functions.

The presented applications of the CC/DC and CC/MM models include linear and non-linear optical properties of liquid water. Generally, we find good agreement between the theoretical predicted quantities and the experimental data as indicated by the results for the electronic excitation energies and (hyper)polarizabilities. For example, we reproduce the experimentally measured solvent induced sign change in the second harmonic generation first hyperpolarizability. The CC/MM model includes the possibility of including solvent polarization effects on the calculated properties. This, in combination with coupled cluster electronic structure theory, is an unique feature, and for water we show that the effect of the solvent polarization is generally significant and cannot be neglected. For example, neglecting the solvent polarization on the second harmonic generation first hyperpolarizability leads to an underestimation in the magnitude of this property, and no solvent induced sign change is observed. Clearly, the CC/MM results including polarization effects are of high quality, and have already been used as benchmarks for other electronic structure methods in combination with molecular mechanics (see e.g. Refs. [38,122]). The CC/MM method therefore extends the well known high accuracy of CC methods for molecules in the gas phase to molecules in the condensed phase, allowing for high accuracy *ab-initio* predictions of molecular properties for molecules in condensed phases.

We emphasize that one of the strengths of the presented methods is, that certain molecular quantities (such as for example molecular dipole moments of solvated molecules), which are not directly experimental accessible, may be evaluated using high level electronic structure theory combined with a classical description of the solvent. Also, using the hierarchy of CC/MM models, specific intermolecular interactions may be included/excluded and the CC/MM model therefore allows for a systematic study of the importance of these specific intermolecular interactions on varies molecular properties. For example, we may choose to include the solvent polarization either directly (through the use of explicit polarizabilities) or indirectly (through the use of increased partial charges). Thus, the theoretical framework presented in this chapter may be used to reveal and study important aspects of solvation, which are not directly accessible from experimental observations.

Acknowledgments

J.K. thanks the Danish Center for Scientific Computing. O.C. acknowledge support by the Danish Center for Scientific Computing, and the Danish National Science Research Council (SNF). K.V.M. thanks Statens Naturvidenskabelige Forskningsråd, Statens Tekniske Videnskabelige Forskningsråd, the Danish Center for Scientific Computing and the EU-network NANOQUANT for support.

References

1. M.P. Allen and D.J. Tildesley. *Computer Simulation of Liquids.* Clarendon Press: Oxford, 1987.

2. J.G. Kirkwood. *J.Chem.Phys.*, **2**, 351, 1934.

3. K.V. Mikkelsen, H. Ågren, H.J.Aa. Jensen, and T. Helgaker. *J.Chem.Phys.*, **89**, 3086, 1988.

4. M.D. Newton. *J.Chem.Phys.*, **58**, 5833, 1973.

5. R. Contreras and A. Aizman. *Int.J.Quant.Chem.*, **27**, 293, 1985.

6. K.V. Mikkelsen, E. Dalgaard, and P. Swanstrøm. *J.Phys.Chem.*, **91**, 3081, 1987.

7. G. Karlström. *J.Phys.Chem.*, **92**, 1315, 1988.

8. G. Karlström. *J.Phys.Chem.*, **93**, 4952, 1989.

9. S.D. Bella, T.J Marks, and M.A. Ratner. *J.Am.Chem.Soc.*, **116**, 4440, 1994.

10. J. Yu and M.C. Zerner. *J.Chem.Phys.*, **100**, 7487, 1994.

11. J.G Ánguán. *Chem.Phys.Lett.*, **241**, 51, 1995.

12. C. Cramer and D. Truhlar. *Chem. Rev.*, **99**, 2161, 1999.

13. O. Tapia and O. Goscinski. *Mol. Phys.*, **29**, 1653, 1975.

14. S. Miertuš, E. Scrocco, and J. Tomasi. *Chem.Phys.*, **55**, 117, 1981.

15. J.G Ánguán. *J. Math. Chem.*, **10**, 93, 1992.

16. D. Rinaldi and J.-L. Rivail. *Theor. Chim. Acta.*, **32**, 57, 1973.

17. D. Rinaldi, M.F. Ruiz-Lopez, and J.-L. Rivail. *J. Chem. Phys.*, **78**, 834, 1983.

18. J.H. McCreery, R.E. Christoffersen, and G.G. Hall. *J.Am.Cham.Soc.*, **98**, 7191, 1976.

19. O. Christiansen and K.V. Mikkelsen. *J.Chem.Phys*, **110**, 1365, 1999.

20. A. Klamt and G. Schüürmann. *J.Chem.Soc.Perkin Trans. 2*, page 799, 1993.

21. J. Tomasi and M. Persico. *Chem.Rev.*, **94**, 2027, 1994.

22. J.O. Noell and K. Morokuma. *Chem. Phys. Lett.*, **36**, 465, 1975.

23. A. Warshel and M. Levitt. *J.Mol.Biol.*, **103**, 227, 1976.

24. U.C. Singh and P.A. Kollman. *J.Comput.Chem.*, **7**, 718, 1986.

25. M.J. Field, P.A. Bash, and M. Karplus. *J.Comput.Chem.*, **11**, 700, 1990.

26. M.A. Thompson. *J.Phys.Chem.*, **100**, 14492, 1996.

27. R.P. Muller and A. Warshel. *J.Phys.Chem.*, **99**, 17516, 1995.

28. B.T. Thole and P.Th. van Duijnen. *Chemical Physics*, **71**, 211, 1982.

29. J. Kongsted, A. Osted, K.V. Mikkelsen, and O. Christiansen. *Mol.Phys*, **100**, 1813, 2002.

30. J. Gao and X. Xia. *Science*, **258**, 631, 1992.

31. M.E. Martín, M.L. Sánchez, F.J.O. del Valle, and M.A. Aguilar. *J.Chem.Phys.*, **113**, 6308, 2000.

32. Q. Cui and M. Karplus. *J.Chem.Phys.*, **112**, 1133, 2000.

33. T.D. Poulsen, J. Kongsted, A. Osted, P.R. Ogilby, and K.V. Mikkelsen. *J.Chem.Phys*, **115**, 2393, 2001.

34. M.L. Sánchez, M.A. Aguilar, and F.J.O. del Valle. *J.Comput.Chem.*, **18**, 313, 1997.

35. M.E. Martín, M.L. Sánchez, M.A. Aguilar, and F.J.O. del Valle. *J.Mol.Struct.(Theochem)*, **537**, 213, 2001.

36. R.V. Stanton, D.S. Hartsough, and K.M. Merz. *J.Phys.Chem.*, **97**, 11868, 1993.

37. I. Tuñón, M.T.C. Martins-Costa, C. Millot, M.F. Ruiz-López, and J.L. Ri-

vail. *J.Comput.Chem.*, **17**, 19, 1996.

38. L. Jensen, P.Th. van Duijnen, and J.G. Snijders. *J.Chem.Phys*, **118**, 514, 2003.

39. J. Sauer and M. Sierka. *J.Comput. Chem.*, **21**(16), 1470, 2000.

40. M. Brandle, J. Sauer, R. Dovesi, and N.M. Harrison. *J.Chem.Phys.*, **109**, 10379, 1998.

41. M. Sierka and J. Sauer. *J.Chem.Phys.*, **112**, 6983, 2000.

42. V. Luzhkov and A. Warshel. *J.Am.Chem.Soc.*, **113**, 4491, 1991.

43. P.N. Day, J.H. Jensen, M.S. Gordon, S.P. Webb, W.J. Stevens, M. Krauss, D. Garmer, H. Basch, and D. Cohen. *J.Chem.Phys.*, **105**, 1968, 1996.

44. M. Svensson, S. Humbel, R.D.J. Froese, T. Matsubara, S. Sieber, and K. Morokuma. *J.Phys.Chem.*, **100**, 19357, 1996.

45. S. Ten-no, F. Hirata, and S. Kato. *Chem. Phys. Lett.*, **214**, 391, 1993.

46. S. Chalmet and M.F. Ruiz-López. *Chem. Phys. Lett.*, **329**, 154, 2000.

47. N.W. Moriarty and G. Karlström. *J.Chem.Phys.*, **106**, 6470, 1997.

48. S.T. Russell and A. Warshel. *J.Mol.Biol.*, **185**, 389, 1985.

49. V. Luzhkov and A. Warshel. *J.Comput.Chem.*, **13**, 199, 1992.

50. F.S. Lee, Z.T. Chu, and A. Warshel. *J.Comput.Chem.*, **14**, 161, 1993.

51. J. Bartkowiak W. Lipiński. *J.Phys.Chem.A*, **101**, 2159, 1997.

52. S. Ten-no, F. Hirata, and S. Kato. *J.Chem.Phys.*, **100**, 7443, 1994.

53. H. Sata, F. Hirata, and S. Kato. *J.Chem.Phys.*, **105**, 1546, 1996.

54. M.E. Martín, A.M. Losa, Fdez.-Galván, and M.A. Aguilar. *J.Chem.Phys.*, **121**, 3710, 2004.

55. K.V. Mikkelsen, Y. Luo, H. Ågren, and P. Jørgensen. *J.Chem.Phys.*, **102**, 9362, 1995.

56. T. Helgaker, H.J.Aa. Jensen, P. Jørgensen, J. Olsen, K. Ruud, H. Ågren, A.A. Auer, K.L. Bak, V. Bakken, O. Christiansen, S. Coriani, P. Dahle, E.K. Dalskov, T. Enevoldsen, B. Fernandez, C. Hättig, K. Hald, A. Halkier, H. Heiberg, H. Hettema, D. Jonsson, S. Kirpekar, R. Kobayashi, H. Koch, K.V. Mikkelsen, P. Norman, M. J. Packer, T.B. Pedersen, T.A. Ruden, A. Sanchez, T. Saue, S.P.A. Sauer, B. Schimmelpfennig, K.O. Sylvester-Hvid, P.R. Taylor, and O. Vahtras. *"Dalton, an ab initio electronic structure program, Release 1.2 (2001)"*,. see http://www.kjemi.uio.no/software/dalton/dalton.html.

57. T. Helgaker, P. Jørgensen, and J. Olsen. *Molecular Electronic Structure Theory.* Wiley, 2000.

58. F. Coester. *Nucl. Phys.*, **7**, 421, 1958.

59. F. Coester and H. Kümmel. *Nucl. Phys.*, **17**, 477, 1960.

60. J. Čížek. *J.Chem.Phys.*, **45**, 4256, 1966.

61. J. Paldus, J. Čížek, and I. Shavitt. *Phys. Rev. A.*, **5**, 50, 1972.

62. J. Čížek and J. Paldus. *Int. J. Quant. Chem.*, **5**, 359, 1971.

63. G.D. Purvis and R.J. Bartlett. *J.Chem.Phys.*, **76**, 1910, 1982.

64. R.J. Bartlett and G.D. Purvis. *Int. J. Quant. Chem.*, **14**, 561, 1978.

65. J.A. Pople, R. Krishnan, H.B. Schlegel, and J.S. Binkley. *Int. J. Quant. Chem.*, **14**, 545, 1978.

66. J.A. Pople, M. Head-Gordon, and K. Raghavachari. *J. Chem. Phys.*, **87**,

J. Kongsted et al.

5968, 1987.
67. G.E. Scuseria, A.C. Scheiner, T.J. Lee, J.E. Rice, and H.F. Schaefer. *J. Chem. Phys.*, **86**, 2881, 1987.
68. A.C. Scheiner, G.E Scuseria, J.E. Rice, T.J. Lee, and H.F. Schaefer. *J. Chem. Phys.*, **87**, 5361, 1987.
69. J. Gauss. *Encyclopedia of Computational Chemistry*. Wiley, New York, 1998. page 615 - 636.
70. T.J. Lee and G.E. Scuseria. *Quantum Mechanical Electronic Structure Calculations with Chemical Accuracy*. Kluwer Academic: Dordrecht, 1995.
71. O. Christiansen, P. Jørgensen, and C. Hättig. *Int.J.Quant.Chem.*, **68**, 1, 1998.
72. J. Arponen. *Ann. Phys.*, **151**, 311, 1983.
73. T. Helgaker and P. Jørgensen. *Theor.Chim.Acta.*, **75**, 111, 1989.
74. O. Christiansen, H. Koch, and P. Jørgensen. *Chem.Phys.Lett.*, **243**, 409, 1995.
75. O. Christiansen, H. Koch, and P. Jørgensen. *J.Chem.Phys.*, **103**, 7429, 1995.
76. K. Raghavachari, G.W. Trucks, J.A. Pople, and M. Head-Gordon. *Chem.Phys.Lett.*, **157**, 479, 1989.
77. F. Jensen. *Introduction to Computational Chemistry*. Wiley, 1999.
78. O. Christiansen, H. Koch, A. Halkier, P. Jørgensen, T. Helgaker, and A. Sánchez de Meras. *J.Chem.Phys.*, **105**, 6921, 1996.
79. O. Christiansen, A. Halkier, H. Koch, P. Jørgensen, and T. Helgaker. *J.Chem.Phys.*, **108**, 2801, 1998.
80. O. Christiansen, H. Koch, and P. Jørgensen. *Chem.Phys.Lett.*, **243**, 409, 1995.
81. O. Christiansen, H. Koch, P. Jørgensen, and T. Helgaker. *Chem.Phys.Lett.*, **263**, 530, 1996.
82. C. Hättig and F. Weigend. *J.Chem.Phys.*, **113**, 5154, 2000.
83. C. Böttcher. *Theory of Electrc Polarization*, volume 1. Elsevier Scientific, Amsterdam, 2nd edition, 1973.
84. J. Jeon and H.J. Kim. *J.Chem.Phys.*, **119**, 8606, 2003.
85. J. Jeon and H.J. Kim. *J.Chem.Phys.*, **119**, 8626, 2003.
86. J. Jeon and H.J. Kim. *J.Phys.Chem. A*, **104**, 9812, 2000.
87. S. Chalmet, D. Rinaldi, and M.F. Ruiz-López. *Int.J.Quant.Chem.*, **84**, 559, 2001.
88. P.Th. DeVries, A.H. van Duijnen and A.H. Juffer. *Int.J.Quant.Chem.: Quant. Chem. Sym.*, **27**, 451, 1993.
89. P. Ahlström, A. Wallqvist, S. Engström, and B. Jönsson. *Mol.Phys.*, **68**, 563, 1989.
90. O. Engkvist, P.O. Åstrand, and G. Karlström. *Chem. Rev.*, **100**, 4087, 2000.
91. J. Kongsted, A. Osted, K.V. Mikkelsen, and O. Christiansen. *J.Phys.Chem A*, **107**, 2578, 2003.
92. A.J. Stone. *Theory of intermolecular forces*. Oxford University Press Inc. New York, 1996.
93. C. Amovilli and B. Mennucci. *J.Phys.Chem B*, **101**, 1051, 1997.
94. A. Bernhardsson, R. Lindh, G. Karlström, and B.O. Roos. *Chem.Phys.Lett.*,

    251, 141, 1996.

95. M. Cossi and V. Barone. *J.Chem.Phys.*, **112**, 2427, 2000.

96. F.M. Floris, J. Tomasi, and J.L. Pascual Ahuir. *J.Comput.Chem.*, **12**, 784, 1991.

97. M. Ben-Nun and T.J. Martínez. *Chem.Phys.Lett.*, **290**, 289, 1998.

98. N.W. Moriarty and G. Karlström. *J.Phys.Chem.*, **100**, 17791, 1996.

99. K.V. Mikkelsen, P. Jørgensen, and H.J.Aa. Jensen. *J.Chem.Phys.*, 1994.

100. K.V. Mikkelsen and K.O. Sylvester-Hvid. *J.Phys.Chem.*, **100**, 9116, 1996.

101. K.O. Sylvester-Hvid, K.V. Mikkelsen, D. Jonsson, P. Norman, and H. Ågren. *J.Chem.Phys.*, **109**, 5576, 1998.

102. K.O. Sylvester-Hvid, K.V. Mikkelsen, D. Jonsson, P. Norman, and H. Ågren. *J.Phys.Chem.A*, **103**, 8375, 1999.

103. L. Onsager. *J.Am.Chem.Soc.*, **58**, 1486, 1936.

104. R. Cammi and J. Tomasi. *Int. J. Quant. Chem.: Quant. Chem. Symp.*, **29**, 465, 1995.

105. R. Cammi and B. Mennucci. *J.Chem.Phys.*, **110**, 9877, 1999.

106. R. Cammi, B. Mennucci, and J. Tomasi. *J.Phys.Chem.A*, **104**, 5631, 2000.

107. B. Mennucci, J. Tomasi, R. Cammi, J.R. Cheeseman, M.J. Frisch, F.J. Devlin, S. Gabrial, and P.J. Stephens. *J.Phys.Chem.A*, **106**, 6102, 2002.

108. Z. Rinkevicius, L. Telyatnyk, O. Vahtras, and K. Ruud. *J.Chem.Phys.*, **121**, 5051, 2004.

109. R. Cammi, L. Frediani, B. Mennucci, and K. Ruud. *J.Chem.Phys.*, **119**, 5818, 2003.

110. R. Cammi, M. Cossi, and J. Tomasi. *J.Chem.Phys.*, **104**, 4611, 1996.

111. R. Cammi, M. Cossi, B. Mennucci, and J. Tomasi. *J.Chem.Phys.*, **105**, 10556, 1996.

112. E.S. Nielsen, P. Jørgensen, and J. Oddershede. *J.Chem.Phys.*, **73**, 6238, 1980.

113. M.J. Packer, E.K. Dalskov, T. Enevoldsen, H.J.Aa. Jensen, and J. Oddershede. *J.Chem.Phys.*, **105**, 5886, 1996.

114. K.L. Bak, H. Koch, J. Oddershede, O. Christiansen, and S.P.A. Sauer. *J.Chem.Phys.*, **112**, 4173, 2000.

115. C.B. Nielsen, S.P.A. Sauer, and K.V. Mikkelsen. *J.Chem.Phys.*, **119**, 3849, 2003.

116. S. Jørgensen, M.A. Ratner, and K.V. Mikkelsen. *J.Chem.Phys.*, **115**, 3792, 2001.

117. S. Jørgensen, M.A. Ratner, and K.V. Mikkelsen. *J.Chem.Phys.*, **115**, 8185, 2001.

118. S. Jørgensen, M.A. Ratner, and K.V. Mikkelsen. *J.Chem.Phys.*, **116**, 10902, 2002.

119. O. Christiansen and K.V. Mikkelsen. *J.Chem.Phys.*, **110**, 8348, 1999.

120. T.D. Poulsen, P.R. Ogilby, and K.V. Mikkelsen. *J.Chem.Phys.*, **115**, 7843, 2001.

121. T.D. Poulsen, P.R. Ogilby, and K.V. Mikkelsen. *J.Chem.Phys.*, **116**, 3730, 2002.

122. L. Jensen, P.Th. van Duijnen, and J.G. Snijders. *J.Chem.Phys*, **119**, 3800,

2003.

123. P.W. Langhoff, S.T. Epstein, and M. Karplus. *Rev.Mod.Phys.*, **44**, 602, 1972.

124. J. Olsen and P. Jørgensen. in *Modern Electronic Structure Theory*, David R. Yarkony, editor, (World Scientific, Singapore). volume 2. chapter 13, pages 857–990. 1995.

125. D.M. Bishop. *Adv. Quant. Chem*, **25**, 2, 1994.

126. K. Sasagane, F. Aiga, and R. Itoh. *J.Chem. Phys.*, **99**, 3738, 1993.

127. J. Oddershede. *Adv. Chem. Phys.*, **69**, 201, 1987.

128. C. Hättig and B. A. Heß. *Chem. Phys. Lett.*, **233**, 359, 1995.

129. J. Olsen and P. Jørgensen. *J.Chem.Phys.*, **82**, 3235, 1985.

130. J. Fiutak. *Can. J. Phys.*, **41**, 12, 1963.

131. J. Frenkel. *Wave-mechanics, Advanced general theory.* Oxford Universiy press, London, 1934.

132. H. Sambe. *Phys.Rev.A.*, **21**, 197, 1973.

133. C. Hättig, O. Christiansen, H. Koch, and P. Jørgensen. *Chem.Phys.Lett.*, **269**, 428, 1997.

134. C. Hättig, O. Christiansen, and P. Jørgensen. *Chem.Phys.Lett.*, **282**, 139, 1998.

135. J. Kongsted, A. Osted, K.V. Mikkelsen, and O. Christiansen. *J.Chem.Phys*, **118**, 1620, 2002.

136. J. Kongsted, A. Osted, K.V. Mikkelsen, and O. Christiansen. *J.Chem.Phys*, **119**, 10519, 2003.

137. A. Osted, J. Kongsted, K.V. Mikkelsen, and O. Christiansen. *Mol. Phys*, **101**, 2055, 2003.

138. J. Kongsted, A. Osted, K.V. Mikkelsen, and O. Christiansen. *J.Chem.Phys*, **120**, 3787, 2004.

139. G.C. Schatz and M.A. Ratner. *Quantum Mechanics in Chemistry.* Prentice Hall, Englewood Cliffs, New Jersey, 1993.

140. L.D. Barron. *Molecular light scattering and optical activity.* Cambridge, Cambridge University Press, 1982.

141. H. Koch and P. Jørgensen. *J.Chem.Phys.*, **93**, 3333, 1990.

142. G.D. Billing and K.V. Mikkelsen. *Chem.Phys.*, **182**, 249, 1994.

143. J. Kongsted, A. Osted, K.V. Mikkelsen, P.O. Åstrand, and O. Christiansen. *J.Chem.Phys*, **121**, 8435, 2004.

144. M.E. Martín, M.A. Aguilar, S. Chalmet, and M. Ruiz-López. *Chem.Phys.Lett.*, **344**, 107, 2001.

145. H. Reis, M.G. Papadopoulos, and D.N. Theodorou. *J.Chem.Phys.*, **114**, 876, 2001.

146. A.V. Gubskaya and P.G. Kusalik. *Mol.Phys.*, **99**, 1107, 2001.

147. L. Jensen, M. Swart, P. Th. van Duijnen, and J.G. Snijders. *J.Chem.Phys*, **117**, 3316, 2002.

148. D.E. Woon and T.H. Dunning. *J.Chem.Phys.*, **100**, 2975, 1994.

149. M.B. Robin. *Higher Excited States of Polyatomic Molecules.* Academic Press; New York, 1985.

150. J.M. Heller, R.N. Jr. Hamm, R.D. Birkhoff, and L.R. Painter. *J.Chem.Phys.*, **60**, 3483, 1974.

151. C.L. Thomsen, D. Madsen, S.R. Keiding, J. Thøgersen, and O. Christiansen. *J.Chem.Phys.*, **110**, 3453, 1999.

152. W. Kerr, G.D. Williams, R.D. Birkhoff, R.N. Hamm, and L.R. Painter. *Phys.Rev.A*, **5**, 2523, 1972.

153. F. Williams, S.P. Varma, and S. Hillenius. *J.Chem.Phys.*, **64**, 1549, 1976.

154. O. Christiansen, T.M. Nymand, and K.V. Mikkelsen. *J.Chem.Phys.*, **113**, 8101, 2000.

155. P.R.T. Schipper, O.V. Gritsenko, S.J.A. van Gisbergen, and E.J. Baerends. *J.Chem.Phys.*, **112**, 1344, 2000.

156. O.V. Gritsenko, P.R.T. Schipper, and E.J. Baerends. *Chem.Phys.Lett.*, **302**, 199, 1999.

157. O.V. Gritsenko, P.R.T. Schipper, and E.J. Baerends. *Int.J.Quant.Chem.*, **76**, 407, 2000.

158. C. Lee, W. Yang, and R.G. Parr. *Physical Review B*, **37**, 785, 1988.

159. A.D. Becke. *Phys. Rev. A*, **38**, 3098, 1988.

160. I. Thormählen, J. Straub, and U. Grigull. *J.Phys.Chem.Ref.Data*, **14**, 933, 1985.

161. A.J. Russell and M.A. Spackman. *Mol.Phys.*, **84**, 1239, 1995.

162. W.F. Murphy. *J.Chem.Phys.*, **67**, 5877, 1977.

163. P. Macak, P. Norman, Y. Luo, and H. Ågren. *J.Chem.Phys.*, **112**, 1868, 2000.

164. O. Christiansen, J. Gauss, and J.F. Stanton. *Chem.Phys.Lett.*, **305**, 147, 1999.

165. D.M. Bishop, B. Kirtman, H.A. Kurtz, and J.E. Rice. *J. Chem. Phys*, **98**, 8024, 1993.

166. H. Reis, S.G. Raptis, and M.G. Papadopoulos. *Chem.Phys.*, **263**, 301, 2001.

167. J.F. Ward and C.K. Miller. *Phys. Rev. A*, **19**, 826, 1979.

168. G. Maroulis. *Chem. Phys. Let.*, **289**, 403, 1998.

169. H. Sekino and R.J. Bartlett. *J.Chem.Phys.*, **98**, 3022, 1993.

170. B.F. Levine and C.G. Bethea. *J.Chem.Phys.*, **65**, 2429, 1976.

171. J. Kongsted, A.Osted, K.V. Mikkelsen, and O. Christiansen. *J.Mol.Structure (Theochem)*, **632**, 207, 2003.

172. R.A. Kendall, T.H. Dunning, and R.J. Harrison. *J.Chem.Phys.*, **96**, 6796, 1992.

173. B. Mennucci, R. Cammi, M. Cossi, and J. Tomasi. *J.Mol.Structure (Theochem)*, **426**, 191, 1998.

174. L. Jensen, P.Th. van Duijnen, and J.G. Snijders. *J.Chem.Phys*, **119**, 12998, 2003.

175. Y. Luo, P. Norman, and H. Ågren. *J.Chem.Phys.*, **109**, 3589, 1998.

176. Y. Luo, P. Norman, P. Macak, and H. Ågren. *J.Chem.Phys.*, **111**, 9853, 1999.

177. P. Norman, P. Macak, Y. Luo, and H. Ågren. *J.Chem.Phys.*, **110**, 7960, 1999.

178. K.O. Sylvester-Hvid, K.V. Mikkelsen, P. Norman, D. Jonsson, and H. Ågren. *J.Phys.Chem.A*, **108**, 8961, 2004.

CHAPTER 8

THE DISCRETE SOLVENT REACTION FIELD MODEL: A QUANTUM MECHANICS/MOLECULAR MECHANICS MODEL FOR CALCULATING NONLINEAR OPTICAL PROPERTIES OF MOLECULES IN CONDENSED PHASE

Lasse Jensen

Northwestern University, Department of Chemistry
2145 Sheridan Road,Evanston, IL 60208-3113, USA
E-mail: l.jensen@chem.northwestern.edu

Piet Th. van Duijnen

Theoretical Chemistry, Materials Science Centre, Rijksuniversiteit Groningen
Nijenborgh 4, 9747 AG Groningen, The Netherlands

In this chapter we review the discrete solvent reaction field (DRF) model as implemented in time-dependent density functional theory (TD-DFT). The DRF model is a polarizable quantum mechanics/ molecular mechanics (QM/MM) model for calculating molecular response properties of molecules in the condensed phase. Using this model effective microscopic properties can be calculated which is related to the macroscopic susceptibilities by Lorentz/Onsager local field factors. The macroscopic susceptibilities can then be compared directly with experimental results. We review some of the applications of the DRF model to calculate the microscopic and macroscopic response properties of molecules in the condensed phase. By carefully validating that DFT calculate accurate response properties in the gas phase we have shown that also in the condensed phase results in good agreement with experiments can be obtained. This is particular true for the refractive index and the pure electronic third-harmonic generation (THG) nonlinear susceptibility. On the other hand the model failed to predict the electric field induced second harmonic generation (EFISH) susceptibility due to a poor handling of the rotational contribution. Future directions to resolve this problem are suggested.

1. Introduction

Understanding the linear and nonlinear optical (NLO) properties of materials is an important research area. Materials exhibiting NLO effects are of great technological importance for use in furture application within electronics and photonics.[1-4] Therefore, accurate predictions of molecular response properties in the condensed phase and in the gas phase are of great interest both from a theoretical and a technological point of view.

The use of quantum chemical methods[5] enables accurate calculations of molecular response properties like the electronic excitations and frequency-dependent (hyper)polarizabilities.[6] However, the accurate prediction of NLO properties of molecules requires large basis sets, high-level electron correlation, frequency-dependence and both electronic and vibrational contributions to be included.[6] Even in the gas phase this is only feasible for small molecules due to a very high computational burden, see Refs. 7–9 for examples where this has been included systematic and therefore good agreement with experiments were found.

A method which has attracted considerable interest, especially within recent years, is Time-Dependent Density Functional Theory (TD-DFT).[10-14] The main reason for this is that (TD-)DFT provides a level of accuracy which in most cases is sufficient at a lower computational requirement than other methods. The use of TD-DFT for calculating molecular response properties in the gas-phase has been shown to be accurate for small and medium size molecules, especially if one uses recently developed density functionals.[15-24]

In recent years applications of TD-DFT to calculate properties of molecules in solution has also been presented.[25-40] Whereas most of these studies have focused on calculating solvatochromic shifts, only a few studies have been devoted to calculating NLO properties of molecules in solution using TD-DFT.[30-33] Since molecular properties like (hyper)polarizabilities are very sensitive to the local environment accurate calculations of these properties could serve as a test for the molecular models used in describing the solvent.

Among the methods for treating solvent effect on molecules are the continuum models,[41-43] the supermolecular model,[44] the frozen density functional approach,[45] *ab initio* molecular dynamics (MD)[46] and the combined quantum mechanical and classical mechanical (QM/MM) models.[47-58] The continuum models have become a standard approach for modeling solvent effects on molecular properties within computational chemistry and are very efficient models. However, in the continuum model the explicit microscopic structure of the solvent are neglected and therefore provides a poor description of the local molecular environment. Also, the results are affected by the choice of the radius and shape of the cavity in which the solute is embedded.[59]

In both the supermolecular models and in *ab initio* MD models all molecules are treated at the same level of theory. This gives a highly accurate description of solvent-solute interaction but due to the high computational demand only a few solvent molecules can be included. A problem of this type of models is that there is no unique way of defining properties of the individual molecules.[60-62] The definition of the molecular properties requires an arbitrary partitioning of the wave function or the electronic charge density among the molecules much in the same way as defining atomic charges. This can be particular problematic when considering response properties.[25,61,63-65]

In the QM/MM methods[47-58] the system is divided into a quantum mechanical part, the solute, and a classical part, the solvent, and the interaction between the two subsystems is described with an effective operator. The solvent molecules are then treated with a classical force field and the method therefore allows for a greater number of solvent molecules to be included. Like in the continuum model the solute is separated from the solvent molecules and the molecular properties of the solute are therefore well defined. The remaining problem is finding an accurate approximate representation of the solvent molecules and the solute-solvent interactions.[66] The discrete representation of the solvent molecules introduces a large number of solvent configurations over which the solute properties must be averaged. This is typically done using Monte Carlo or MD techniques which lead to a large number of quantum mechanical calculations. For this reason the QM/MM method is often employed at a semiempirical level of theory.[57]

The force fields used in the QM/MM methods are typically adopted from fully classical force fields. While this in general is suitable for the solvent-solvent interactions it is not clear how to model the van der Waals interaction between the solute and the solvent.[67] The van der Waals interactions are typically treated as a Lennard-Jones (LJ) potential and the LJ parameters for the quantum atoms are then taken from the classical force field or optimized to the particular QM/MM method[68] for some molecular complexes. However, it is not certain that optimizing the parameters on small complexes will improve the results in a QM/MM simulation[67] of a liquid.

In recent years the classical force fields have been improved in order to also describe the polarization of the molecules.[69-76] The polarization of the classical molecules has also been included in QM/MM studies[29-33, 47, 48, 77-87] and shown that it is important to consider also the polarization of the solvent molecules. Since the inclusion of the solvent polarization leads to an increase in computational time most studies ignore this contribution and use the more simple pair potentials. When the solvent polarization is included it is usually treated using either an isotropic molecular polarizability[78, 82, 84-87] or using distributed atomic polarizabilities[77, 80, 81, 83] according to the Applequist scheme.[88] At short distances the Applequist scheme leads to the so-called "polarizability catastrophy" [88-90] due to the use of a classical description in the bonding region. Thole[90] avoided this problem by introducing smeared out dipoles which mimics the overlapping of the charge distributions at short distances. Thole's model has been shown to be quite succesful in reproducing the molecular polarizability tensor using model atomic polarizability parameters independent of the chemical environment of the atoms.[90-93]

This is the polarization model adopted in the Discrete Solvent Reaction Field (DRF) model[29-33, 48, 54, 94] which is an QM/MM model implemented within wave function theory[48, 54, 94] and, recently, within TD-DFT.[29-33] In the DRF model the permanent electronic charge distribution of the solvent molecules (MM) is modeled by point charges, while distributed atomic polarizabilities are included to model the solvent polarization. The QM/MM interactions are introduced into the Hamiltonian through an effective operator and all interactions are solved self-consistently. An important feature of the model is the inclusion of polarizabilities in the MM part which allows for the solvent molecules to be polarized by the solute and by interactions with other solvent molecules. The advantages of including polarizabilities in the MM part is that all parameters can be obtained from gas phase properties. Furthermore, it is expected that a distributed polarizability approach will give better results than an approach in which one models the molecular polarizability using only a single (anisotropic) polarizability per molecule, especially as the size of the solvent molecule increases.[95] The permanent point charges represent at least the permanent molecular dipole moment, and the distributed atomic polarizabilities the full molecular polarizability tensor. The atomic charges are straightforward obtained using Multipole Derived Charges (MDC)[96] and the distributed polarizabilities by adopting standard parameters or refitting them to match the calculated polarizability tensor.[90-93] This allows for a simple procedure to obtained the solvent model parameters which subsequently can be used in the MD and QM/MM simulations.

The calculation of response properties of molecules in the condensed phase is developed to a much lesser degree than in the gas phase. Comparison with experiments are often done at the microscopic level while the experimental values are extracted from macroscopic quantities usualy using Lorentz local field factors. However, in recent years it has been realized that it is better to calculate the macroscopic quantitites, i.e., the (non-)linear susceptibilites, which can be related to experimental results.[9, 32, 33, 94, 97–102]

The DRF model has been extended to also include the so-called local field factors, i.e. the difference between the macroscopic electric field and the actual electric field felt by the solute.[32, 33, 94] This will enable the calculation of effective microscopic properties which can be related to the macroscopic susceptibilities. The macroscopic susceptibilities can then be compared directly with experimental results. There exist in the literature some other approaches to calculate effective properties and relate these to the macroscopic properties,[9, 97–99, 101–103] which differ from this work in the way the solvent was represented.

In this chapter we will review the DRF model for calculating molecular response properties for molecules in solution. We will limit the review to the recent implementation into TD-DFT[29–33] but most of the theory is general for the DRF model. The chapter is organised as follow: in section 2 the DRF model is presented, in section 3 the relation between microscopic and macroscopic polarization is discussed, in section 4 some applications of the DRF model are presented and in section 5 a short summary and outlook are given.

2. The discrete solvent reaction field model

In the QM/MM method the solvent molecules (MM) are treated with a classical force field and the interactions between the solute and solvent are described with an effective operator. In the QM/MM method the total (effective) Hamiltonian for the system is written as[47–58]

$$\hat{H} = \hat{H}_{\text{QM}} + \hat{H}_{\text{QM/MM}} + \hat{H}_{\text{MM}} \tag{1}$$

where $\hat{H}_{\text{QM}}$ is the quantum mechanical Hamiltonian for the solute, $\hat{H}_{\text{QM/MM}}$, describes the interactions between solute and solvent and $\hat{H}_{\text{MM}}$ describes the solvent-solvent interactions. This separation of the system into two parts is illustrated in Figure 1.

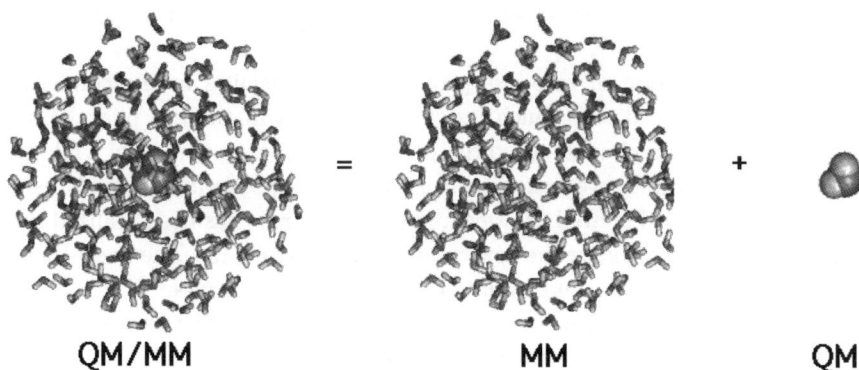

QM/MM   **MM**   **QM**

Fig. 1.   An illustration of the separation of the total system into two subsystem used in the QM/MM model.

Within the Discrete Solvent Reaction Field model the QM/MM operator at a point r_i is given by[29-31]

$$\hat{H}_{\text{QM/MM}} = \sum_i v^{\text{DRF}}(r_i, \omega) = \sum_i v^{\text{el}}(r_i) + \sum_i v^{\text{pol}}(r_i, \omega) \qquad (2)$$

where the first term, v^{el}, is the electrostatic operator and describes the Coulombic interaction between the QM system and the permanent charge distribution of the solvent molecules. The second term, v^{pol}, is the polarization operator and describes the many-body polarization of the solvent molecules, *i.e.* the change in the charge distribution of the solvent molecules due to interaction with the QM part and other solvent molecules.

The charge distribution of the solvent is represented by atomic point charges, hence the electrostatic operator is given by

$$v^{\text{el}}(r_i) = \sum_s \frac{q_s}{R_{si}} = \sum_s q_s T_{si}^{(0)}, \qquad (3)$$

where the zero'th order interaction tensor has been introduced and the index s runs over all atoms of the solvent molecules. In general the interaction tensor to a given order, n, can be written as

$$T_{pq,\alpha_1...\alpha_n}^{(n)} = \nabla_{pq,\alpha_1} ... \nabla_{pq,\alpha_n} \left(\frac{1}{R_{pq}}\right), \qquad (4)$$

where R_{pq} is the distance between the interacting entities.

The many-body polarization term is represented by induced atomic dipoles at the solvent molecules and the polarization operator is given by

$$v^{\text{pol}}(r_i, \omega) = \sum_s \mu_{s,\alpha}^{\text{ind}}(\omega) \frac{R_{si,\alpha}}{R_{si}^3} = -\sum_s \mu_{s,\alpha}^{\text{ind}}(\omega) T_{si,\alpha}^{(1)}. \tag{5}$$

where $R_{si,\alpha}$ is a component of the distance vector and $\mu_s^{\text{ind}}(\omega)$ is the induced dipole at site s. For Greek indices the Einstein summation convention is employed. The induced dipoles are discussed in more detail in the next section.

2.1. The Atomic Induced Dipoles

One of the important features of the DRF model is the use of distributed atomic polarizabilities for describing the many-body polarization and will be described here.

For a collection of atomic polarizabilities in an electric field, assuming linear response, the induced atomic dipole at site s is given by

$$\mu_{s,\alpha}^{\text{ind}}(\omega) = \alpha_{s,\alpha\beta}[F_{s,\beta}^{\text{init}}(\omega) + \sum_{t\neq s} T_{st,\beta\gamma}^{(2)} \mu_{t,\gamma}^{\text{ind}}(\omega)], \tag{6}$$

where $\alpha_{s,\alpha\beta}$ is a component of the atomic polarizability tensor at site s, which for an isotropic atom gives $\alpha_{s,\alpha\beta} = \delta_{\alpha\beta}\alpha_s$, and $T_{st,\alpha\beta}^{(2)}$ is the screened dipole interaction tensor.[29, 90, 91] The screened dipole interaction tensor will be described in the next section. Here we neglect the frequency-dependence of the classical part, i.e. the atomic polarizability is frequency independent, but the model can easily be extended to include also this effect.[92, 93]

$F_{s,\beta}^{\text{init}}(\omega)$ is the initial electric field at site s which arises from both the classical and the quantum mechanical charge distributions. Another important feature of the DRF model is that $F_{s,\beta}^{\text{init}}(\omega)$ has been extended to include the macroscopic electric field.[32] This is important for calculating properties which can be related to experimental results and will be discussed later. The initial field then consist of four terms

$$F_{t,\beta}^{\text{init}}(\omega) = F_{t,\beta}^{\text{QM,el}}(\omega) + \delta_{0,\omega} F_{t,\beta}^{\text{QM,nuc}} + \delta_{0,\omega} F_{t,\beta}^{\text{MM,q}} + F_{t,\beta}^{\text{mac}}(\omega), \tag{7}$$

where $F_{t,\beta}^{\text{QM,el}}(\omega)$ is the field arising from the electronic charge distribution of the QM part,

$$F_{t,\beta}^{\text{QM,el}}(\omega) = -\int \rho(r_i, \omega) \frac{R_{it,\beta}}{R_{it}^3} dr_i = \int \rho(r_i, \omega) T_{it,\beta}^{(1)} dr_i \tag{8}$$

and $F_{t,\beta}^{\text{QM,nuc}}$ is the field arising from the QM nuclei,

$$F_{t,\beta}^{\text{QM,nuc}} = \sum_m \frac{Z_m R_{mt,\beta}}{R_{mt}^3} = -\sum_m Z_m T_{mt,\beta}^{(1)} \tag{9}$$

and $F_{t,\beta}^{\text{MM,q}}$ is the field arising from the point charges at the solvent molecules,

$$F_{t,\beta}^{\text{MM,q}} = \sideset{}{'}\sum_s \frac{q_s R_{st,\beta}}{R_{st}^3} = -\sideset{}{'}\sum_s q_s T_{st,\beta}^{(1)} \tag{10}$$

The prime in Eq 10 indicates that the sum is restricted to sites which do not belong to the same molecule. The last term, $F_{t,\beta}^{\text{mac}}(\omega)$, is the macroscopic electric field

Since the induced dipole in Eq. 6 depends on the induced dipoles at the other sites these equations have to be solved self-consistently. This can be done analytically by rewriting the equations into a $3N \times 3N$ linear matrix equation, with N the number of atoms, as

$$A\mu^{\text{ind}} = F^{\text{init}} \tag{11}$$

and the components of the matrix, $A_{st,\alpha\beta}$, given by

$$A_{st,\alpha\beta} = (\alpha_{s,\alpha\beta}^{-1}\delta_{st} - T_{st,\alpha\beta}^{(2)}). \tag{12}$$

This matrix equation can then be solved for the induced dipoles using standard mathematical tools for solving linear equations. The inverse of the matrix A, the so called relay matrix, is a generalized polarizability matrix which describes the total linear response of the discrete solvent molecules.

2.2. Damping of the induced dipoles

If F^{init} is a uniform external field the polarizability of the classical system can be written as[88]

$$\alpha_{\alpha\beta}^{\text{mol}} = \sum_{p,q}^N B_{pq,\alpha\beta}, \tag{13}$$

where **B** is the relay matrix defined as

$$\mathbf{B} = \mathbf{A}^{-1} = \left(\alpha^{-1} - \mathbf{T}^{(2)}\right)^{-1}. \tag{14}$$

The polarizability parallel, $\alpha_{\parallel}$, and perpendicular, $\alpha_{\perp}$, to the axes connecting two interacting atoms, p and q, are given by Silberstein's equations,[89] which are the exact solutions to Eq. 13,

$$\alpha_{\parallel} = \frac{\alpha_p + \alpha_q + 4\alpha_p\alpha_q/R^3}{1 - 4\alpha_p\alpha_q/R^6}, \tag{15}$$

$$\alpha_{\perp} = \frac{\alpha_p + \alpha_q - 2\alpha_p\alpha_q/R^3}{1 - \alpha_p\alpha_q/R^6}. \tag{16}$$

From Eqs. 15 and 16 it is seen that when R approaches $(4\alpha_p\alpha_q)^{1/6}$, $\alpha_{\parallel}$ goes to infinity and becomes negative for even shorter distances. This problem is generally referred to as the "polarizability catastrophe" and likewise a over polarization will occur in the DRF model if the distance between atoms in the MM part becomes to close.

In order to avoid this "polarizability catastrophe" Thole[90] modified the dipole interaction tensor using smeared-out dipoles. The dipole interaction tensor was first rewritten in terms of a reduced distance $u_{pq,\beta} = R_{pq,\beta}/(\alpha_p\alpha_q)^{1/6}$ as

$$T^{(2)}_{pq,\beta\gamma} = (\alpha_p\alpha_q)^{1/2}t(u_{pq}) = (\alpha_p\alpha_q)^{1/2}\frac{\partial^2\phi(u_{pq})}{\partial u_{pq,\beta}\partial u_{pq,\gamma}} \tag{17}$$

where $\phi(u_{pq})$ is a spherically symmetric potential of some model charge distribution ρ. The screened dipole interaction tensor can be written as

$$T^{(2)}_{pq,\alpha\beta} = \frac{3f^T_{pq}R_{pq,\alpha}R_{pq,\beta}}{R^5_{pq}} - \frac{f^E_{pq}\delta_{\alpha\beta}}{R^3_{pq}}. \tag{18}$$

where the damping functions f^T_{pq} and f^E_{pq} have been introduced. If we consider a exponential decaying charge distribution the screening functions in Eq. 18 are given by[91]

$$f^E_{pg} = 1 - \left[1 + s_{pq} + \frac{1}{2}s^2_{pq}\right]\exp(-s_{pq}) \tag{19}$$

and

$$f^T_{pg} = f^E_{pg} - \frac{1}{6}s^3_{pq}\exp(-s_{pq}), \tag{20}$$

where the term s_{pq} is give by $s_{pq} = aR_{pq}/(\alpha_p\alpha_q)^{1/6}$, with a the screening length, and α_p the atomic polarizability of atom p. The same screening of the dipole interaction tensor is adopted in the DRF model when solving Eq. 6.

The atomic polarizabilities are obtained from a parametrization of the molecular polarizability tensor using Eq. 13.[90–93] This allows for a unique parametrization which often is highly transferable due to the screening of the interaction tensor.[90–93]

2.3. *The QM/MM interaction energy*

The QM/MM interaction energy is in the DRF model given by a sum of three terms,

$$E^{\text{QM/MM}} = E^{\text{elst,el}} + E^{\text{elst,nuc}} + E^{\text{ind}} \tag{21}$$

where the first two terms are the electrostatic interaction between the QM electrons and the classical point charges

$$E^{\text{elst,el}} = -\sum_s q_s \int \rho(r_i) \frac{1}{R_{is}} dr_i \tag{22}$$

and the electrostatic interaction between the QM nuclei and the point charges

$$E^{\text{elst,nuc}} = \sum_s q_s \sum_m \frac{Z_m}{R_{ms}}, \tag{23}$$

respectively.

The last term is the induction energy and is given by[69, 104]

$$E^{\text{ind}} = -\frac{1}{2}\mu^{\text{ind}} F^{\text{QM}}, \tag{24}$$

where F^{QM} is the electric field arising from the QM system, i.e. the field from the QM electrons and nuclei. The induction energy consist of the sum of the energy of the induced dipoles in the electric field and the polarization cost, i.e. the energy needed for creating the induced dipoles.

2.4. *Time-dependent density functional theory*

Within time-dependent density functional theory (TD-DFT) one can study the dynamic response of a system due to an external time-dependent perturbation by considering the time-dependent Kohn-Sham (TD-KS) equations.[10–14] If the peturbation is small we can use response theory to obtain a perturbative solution to the TD-KS equations.

The TD-KS equations including the DRF operator is given by,[30, 31]

$$i\frac{\partial}{\partial t}\phi_i(r,t) = \left[-\frac{1}{2}\nabla^2 + v_{\text{eff}}(r,t)\right]\phi_i(r,t), \tag{25}$$

with the effective potential given by

$$v_{\text{eff}}(r,t) = \int dr' \frac{\rho(r',t)}{|r-r'|} + v^{\text{per}}(t) + v^{\text{DRF}}(r,t) + v_{\text{xc}}(r,t), \tag{26}$$

where $v^{\mathrm{DRF}}(r,t)$ is the operator defined in Eq. 2 and $v^{\mathrm{per}}(t)$ the perturbing field turned on slowly in the distant past. The last term is the time-dependent xc-potential which in the adiabatic approximation is given by

$$v_{\mathrm{xc}}[\rho](r,t) \approx \frac{\delta E_{\mathrm{xc}}[\rho_t]}{\delta \rho_t(r)} = v_{\mathrm{xc}}[\rho_t](r). \qquad (27)$$

The time-dependent electronic density is given by

$$\rho(r,t) = \sum_i^{\mathrm{occ}} n_i |\phi_i(r,t)|^2, \qquad (28)$$

where n_i is the occupation number of orbital i.

2.5. The Frequency-dependent (Hyper)Polarizability

The total dipole moment of a molecule in the presence of a time-dependent electric field, $F_\beta = F_\beta^0 + F_\beta^\omega \cos(\omega t)$, can be expanded as a Taylor series in the applied electric field[105]

$$
\begin{aligned}
\mu_\alpha = {} & \mu_\alpha^0 + \alpha_{\alpha\beta}(0;0)F_\beta^0 + \alpha_{\alpha\beta}(-\omega;\omega)F_\beta^\omega \cos(\omega t) \\
& + \frac{1}{2}\beta_{\alpha\beta\gamma}(0;0,0)F_\beta^0 F_\gamma^0 + \frac{1}{4}\beta_{\alpha\beta\gamma}(0;\omega,-\omega)F_\beta^\omega F_\gamma^\omega \\
& + \beta_{\alpha\beta\gamma}(-\omega;0,\omega)F_\beta^0 F_\gamma^\omega \cos(\omega t) + \frac{1}{4}\beta_{\alpha\beta\gamma}(-2\omega;\omega,\omega)F_\beta^\omega F_\gamma^\omega \cos(2\omega t) \\
& + \cdots
\end{aligned}
\qquad (29)
$$

where $\alpha_{\alpha\beta}$ is the molecular polarizability and $\beta_{\alpha\beta\gamma}$ is the molecular first hyperpolarizability with (α,β,γ) designating Cartesian coordinates. The total dipole moment can be obtained from the trace of the dipole moment matrix, H^α, and the density matrix in the presence of the electric field, $P(F)$,

$$\mu_\alpha = -Tr[H^\alpha P(F)]. \qquad (30)$$

We can expand the density matrix in a Taylor series,

$$P = P^0 + P^\beta F_\beta + \frac{1}{2!}P^{\beta\gamma}F_\beta F_\gamma + \cdots, \qquad (31)$$

where P^0 is the unperturbed density matrix, P^β the linear response and $P^{\beta\gamma}$ the quadratic response. Inserting this expansion into Eq. 30 and comparing with Eq. 29 allows us to identify the dipole moment, the frequency-dependent polarizability, and the frequency-dependent first hyperpolarizabillity as

$$\mu_\alpha = -Tr[H^\alpha P^0] \qquad (32)$$

$$\alpha_{\alpha\beta}(-\omega;\omega) = -Tr[H^\alpha P^\beta(\omega)], \tag{33}$$

$$\beta_{\alpha\beta\gamma}(-\omega_\sigma;\omega_a,\omega_b) = -Tr[H^\alpha P^{\beta\gamma}(\omega_a,\omega_b)], \tag{34}$$

where $\omega_\sigma = \omega_a + \omega_b$. In the following we will present how to obtain the linear and quadratic response properties using the (2n+1) rule.

2.6. Linear response of the density matrix

For linear response properties we need the first-order change in the density to a time-dependent perturbation

$$\delta\rho(r,\omega) = \sum_{s,t} P_{st}(\omega)\phi_s\phi_t^* \tag{35}$$

$$= \sum_{i,a} P_{ia}(\omega)\phi_i\phi_a^* + P_{ai}(\omega)\phi_a\phi_i^*, \tag{36}$$

where P is the first-order density matrix and a,b indicates virtual orbitals , i,j occupied orbitals and s,t indicates general orbitals. By expanding the KS-equations to first-order in the perturbing potential we find that the first-order density matrix is given by

$$P_{st}(\omega) = \frac{\Delta n_{st}}{(\epsilon_s - \epsilon_t) - \omega}\delta v_{st}^{\text{eff}}(\omega), \tag{37}$$

where Δn_{st} is the difference in occupation numbers, i.e. 1 for $st = ai$ and -1 for $st = ia$. The change in the effective potential, $\delta v_{st}^{\text{eff}}$, is dependent on the first order change in the density and is given by

$$\delta v_{st}^{\text{eff}}(\omega) = \delta v_{st}^{\text{per}}(\omega) + \delta v_{st}^{\text{scf}}(\omega)$$

where the self-consistent field, δv^{scf}, denotes terms which depend on the first-order change in the density and is given by

$$\delta v^{\text{scf}}(\omega) = \int dr' \frac{\delta\rho(r',\omega)}{|r - r'|} + v_{\text{xc}}[\delta\rho](r,\omega) + v^{\text{DRF}}[\delta\rho](r,\omega). \tag{39}$$

The contribution from the DRF operator is given by

$$v^{\text{DRF}}[\delta\rho](r_i,\omega) = -\sum_s \mu_{s,\alpha}^{\text{ind}}[\delta\rho](\omega)T_{si,\alpha}^{(1)}, \tag{40}$$

where $\mu_{s,\alpha}^{\text{ind}}[\delta\rho](\omega)$ is the frequency-dependent dipole moments found by solving Eq. 6. It should be noted that for $F_{s,\beta}^{\text{init}}(\omega)$ only the terms depending on the first-order change in the density and the macroscopic electric field is included. The DRF contribution arises from the induced dipoles in the MM part due to the first-order change in the QM charge distribution. Inserting the first order change in the density, Eq. 35, into Eq. 38 allows for the change in the effective potential to be written as

$$\delta v_{st}^{\text{eff}}(\omega) = \delta v_{st}^{\text{per}}(\omega) + \sum_{uv} K_{st,uv} P_{uv}(\omega), \tag{41}$$

where the coupling matrix, K, has been introduced. The coupling matrix will be described in more detail later. Inserting Eq. 41 into Eq. 37 the first-order density matrix can be written as

$$P_{st}(\omega) = \frac{\Delta n_{st}}{(\epsilon_s - \epsilon_t)\omega}[\delta v_{st}^{\text{per}}(\omega) + \sum_{uv} K_{st,uv} P_{uv}(\omega)]. \tag{42}$$

This can be written as a set of coupled linear equations for the first-order density matrix elements using the fact that only elements relating occupied and virtual orbitals are nonzero

$$\sum_{jb}[\delta_{ij}\delta_{ab}(\epsilon_a - \epsilon_i + \omega) + K_{ia,jb}]P_{jb} + \sum_{jb} K_{ia,bj}P_{bj} = -(\delta v_{ia}^{\text{per}}), \tag{43}$$

$$\sum_{jb}[\delta_{ij}\delta_{ab}(\epsilon_a - \epsilon_i - \omega) + K_{ai,bj}]P_{bj} + \sum_{jb} K_{ai,jb}P_{jb} = -(\delta v_{ai}^{\text{per}}) \tag{44}$$

These equation can be written as one matrix equation using the more common notation $X_{jb} = P_{jb}$ and $Y_{jb} = P_{bj}$ as

$$\left[\begin{pmatrix} \mathbf{A} & \mathbf{C} \\ \mathbf{C}^* & \mathbf{A}^* \end{pmatrix} - \omega \begin{pmatrix} -1 & 0 \\ 0 & 1 \end{pmatrix}\right] \begin{pmatrix} X \\ Y \end{pmatrix} = -\begin{pmatrix} \delta v^{\text{per}} \\ \delta v^{\text{per}*} \end{pmatrix}, \tag{45}$$

where the individually matrix elements are defined as

$$A_{ia,jb} = \delta_{ab}\delta_{ij}(\epsilon_a - \epsilon_i) + K_{ia,jb} \tag{46}$$

and

$$C_{ia,jb} = K_{ia,bj}. \tag{47}$$

In TD-DFT the equality $K_{ia.jb} = K_{ia,bj}$ allows for that the equations can be reduced to half the size. This is not the case in TD-HF where this equality is not valid. From the solution of the linear equations in Eq. 45 we have access to the frequency-dependent polarizability or by transforming the left hand side into an eigenvalue equation we can obtain the excitation energies and oscillator strengths.

2.7. The coupling matrix

The coupling matrix describes the linear response of the self-consistent field to changes in the density and consist of three terms.

$$
K_{st,uv} = \frac{\partial v_{st}^{\mathrm{scf}}}{\partial P_{uv}}
$$

$$
= K_{st,uv}^{\mathrm{Coul}} + K_{st,uv}^{\mathrm{xc}} + K_{st,uv}^{\mathrm{DRF}}. \tag{48}
$$

The first term is the Coulomb part given by

$$
K_{st,uv}^{\mathrm{Coul}} = \int \int dr_i dr_j \phi_s^*(r_i)\phi_t(r_i)\frac{1}{|r_i - r_j|}\phi_u(r_j)\phi_v^*(r_j), \tag{49}
$$

the second term is the xc part

$$
K_{st,uv}^{\mathrm{xc}} = \int \int dr_i dr_j \phi_s^*(r_i)\phi_t(r_i)\frac{\delta v_{xc}(r_i,\omega)}{\delta \rho(r_j,\omega)}\phi_u(r_j)\phi_v^*(r_j) \tag{50}
$$

and the last term is the DRF part

$$
K_{st,uv}^{\mathrm{DRF}} = \int dr_i \phi_s^*(r_i)\phi_t(r_i)\frac{\partial v^{\mathrm{DRF}}(r_i,\omega)}{\partial P_{uv}} \tag{51}
$$

$$
= -\int dr_i \phi_s^*(r_i)\phi_t(r_i)\sum_{s,t} B_{st,\alpha\beta}T_{js,\beta}^{(1)}T_{si,\alpha}^{(1)}\frac{\partial \delta\rho(r_j,\omega)}{\partial P_{uv}} \tag{52}
$$

$$
= -\int \int dr_i dr_j \phi_s^*(r_i)\phi_t(r_i)\sum_{s,t} B_{st,\alpha\beta}T_{js,\beta}^{(1)}T_{si,\alpha}^{(1)}\phi_u^*(r_j)\phi_v(r_j) \tag{53}
$$

where B is the relay matrix defined in Eq. 14.

2.8. Quadratic response of the density matrix using the (2n+1) rule

In a manner similar to the linear response a set of equations for the solution of the higher order density response can be constructed.[22-24, 106, 107] However, a more effecient approach is to take advantage of the (2n+1) rule which allows for the quadratic response properties to be rewritten in terms of quantities obtained from the solution of the first order response equations. Within TD-DFT van Gisbergen et al. have shown how this is done for the frequency-dependent first hyperpolarizability[106] in an approach similar to the TD-HF approach of Karna and Dupuis.[108] Here we will present the results obtained by van Gisbergen et al.[106] since the inclusion of the DRF operator will not affect the structure of these equations.

The frequency-dependent first hyperpolarizabilty can be rewritten using the (2n+1) rule as[106]

$$
\begin{aligned}
\beta_{\alpha\beta\gamma}(-\omega_\sigma; \omega_a, \omega_b) &= -Tr[H^\alpha P^{\beta\gamma}(\omega_a, \omega_b)] \\
&= \hat{p} Tr[nU^\alpha(-\omega_\sigma)[G^\beta(\omega_a), U^\gamma(\omega_b)]_-] \\
&\quad + Tr[g_{xc}(\omega_a, \omega_b) P^\alpha(-\omega_\sigma) P^\beta(\omega_a) P^\gamma(\omega_b)] \quad (54)
\end{aligned}
$$

where $[G^\beta(\omega_a), U^\gamma(\omega_b)]_-$ denotes the commutator of $G^\beta(\omega_a)$ and $U^\gamma(\omega_b)$, and $\hat{p}$ the sum of all permutations of $(\alpha, -\omega_\sigma)$, $(\beta, -\omega_a)$, and $(\gamma, -\omega_b)$. The $G^\alpha(\omega_a)$ matrix is the first-order KS matrix[106] and is given by

$$
G^\alpha_{pg}(\omega) = \int dr \phi_p(r) \left[v^\alpha_{per} + v^{Coul}[\delta\rho^\alpha(\omega)] + v_{xc}(r)[\delta\rho^\alpha(\omega)](r) + \right.
$$
$$
\left. v^{DRF}[\delta\rho^\alpha(\omega)](r) \right] \phi_q(r) \quad (55)
$$

which is identical to the effective potential matrix, $\delta v^{eff}(\omega_a)$, in Eq. 38 due to the α-component of the perturbation. The first-order transformation matrix, $U^\alpha(\omega)$, is given by

$$
U^\alpha_{pq}(\omega) = \frac{G^\alpha_{pg}(\omega)}{\epsilon_q^{(0)} - \epsilon_p^{(0)} - \omega} \quad (56)
$$

and is only nonzero for the occupied-virtual block.[106] The last term in Eq. 54 is an additional term in the DFT expression which is not present in the TDHF case and is given by

$$
Tr[g_{xc}(\omega_a, \omega_b) P^\alpha(-\omega_\sigma) P^\beta(\omega_a) P^\gamma(\omega_b)] =
$$
$$
\int\int\int dr dr' dr'' g_{xc}(r, r', r'', \omega_a, \omega_b) \delta\rho^\alpha(r, -\omega_\sigma) \delta\rho^\beta(r', \omega_a) \delta\rho^\gamma(r'', \omega_b) (57)
$$

where the xc kernel, g_{xc}, has been introduced,

$$
g_{xc}(r, r', r'', \omega_a, \omega_b) = \left. \frac{\delta^2 v_{xc}(r)}{\delta\rho(r', \omega_a)\delta\rho(r'', \omega_b)} \right|_{\rho^{(0)}} . \quad (58)
$$

Usually the adiabatic approximation is invoked for this kernel,

$$
g_{xc}(r, r', r'', \omega_a, \omega_b) \simeq g_{xc}(r, r', r'', 0, 0). \quad (59)
$$

2.9. Implementation into ADF

The DRF model has been implemented into a local version of the Amsterdam Density Functional (ADF) program package.[109,110] In ADF the KS equations are solved by numerical integration which means that the effective KS-operator has to be evaluated in each integration point. Since the numerical integration grid is chosen on the basis of the quantum part alone care must be taken when evaluating the DRF operator if the integration points are close to a classical atom. In order to avoid numerical instabilities we introduce a damping of the operator at small distances which is modeled by modifying the distance R_{ij} to obtain a scaled distance S_{ij},[93]

$$S_{ij} = v_{ij}R_{ij} = f(R_{ij}) \ , \tag{60}$$

where v_{ij} is a scaling factor and $f(R_{ij})$ an appropriately chosen function of R_{ij}. Furthermore, each component of R_{ij} is also scaled by v_{ij}, so the reduced distance becomes,

$$S_{ij} = \sqrt{S_{ij,\alpha}S_{ij,\alpha}} = v_{ij}\sqrt{R_{ij,\alpha}R_{ij,\alpha}} = v_{ij}R_{ij} \ , \tag{61}$$

consistent with the definition in Eq. 60. The damped operator can thus be obtained by modifying the interaction tensors in Eqs. 3 and 5,

$$T^{(n)}_{ij,\alpha_1\ldots\alpha_n} = \nabla_{\alpha_1}\ldots\nabla_{\alpha_n}\left(\frac{1}{S_{ij}}\right) \ , \tag{62}$$

which is equivalent to replacing R_{ij} by S_{ij} and $R_{ij,\alpha}$ by $S_{ij,\alpha}$ in the regular formulae for the interaction tensors. The particular form of the scaling function employed here is[93]

$$f(r_{pq}) = \frac{r_{pq}}{erf(r_{pq})}, \tag{63}$$

which was obtained by considering the interaction between two Gaussian charge distributions with unity exponents. The damping, although not optimized to treat this, mimics the short range repulsion due to the overlapping charge densities of the QM part and the MM part. The damping is dependent on the width of the gaussian charge distribution, which in this work was taken to be unit (a.u.). However, both a slightly smaller width[111] and a slightly larger width[112] have been suggested.

The extension to the TD-DFT part has been implemented in the RE-SPONSE module of ADF.[106, 113, 114] In the RESPONSE module the functional derivative of the xc-potential in Eq. 27 and 58 is restricted to the Adiabatic LDA (ALDA) xc-potential. The linear equations for the first-order density matrix in Eq. 37 is solved using an efficient iterative algorithm,[113] and, for that reason, the DRF response operator, Eq. 40, is calculated by solving a set of linear equations like in Eq. 6.

3. Calculating macroscopic and microscopic properties with a QM/MM model

Using the DRF model we can calculate and distinguish between different solvent effects on the microscopic response properties.[32, 33] The first effect is the interactions between the solute and the solvent (pure solvent effect) and is calculated with the macroscopic electric field *not included* in Eq. 7. The microscopic properties calculated in this way will be denoted "solute properties". If the macroscopic electric field is *included* in Eq. 7 we will denote the calculated properties as "effective". The effective properties include, in addition to the pure solvent effect , the influence of the dipole moments induced in the solvent by the macroscopic electric field. Finally, by combining the effective properties with the macroscopic local field factors (Lorentz/Onsager) we can obtain the macroscopic susceptibilities.

3.1. *The macroscopic polarization*

The macroscopic polarization of a material in the presence of a macroscopic electric field, F^{mac}, is expressed as a power series in the field strength as[1,115]

$$\mathcal{P}_I(t) = \mathcal{P}_I^0 + \chi_{IJ}^{(1)} F_J^{\text{mac}}(t) + \chi_{IJK}^{(2)} F_J^{\text{mac}}(t) F_K^{\text{mac}}(t) \tag{64}$$

$$+ \chi_{IJKL}^{(3)} F_J^{\text{mac}}(t) F_K^{\text{mac}}(t) F_L^{\text{mac}}(t) + \cdots \tag{65}$$

where $\mathcal{P}^0$ is the permanent polarization, $\chi^{(1)}$ the linear optical susceptibility, $\chi^{(2)}$ the second-order nonlinear optical susceptibility, and $\chi^{(3)}$ the third-order nonlinear optical susceptibility. The subscripts $I, J, K, L, ...$ denotes space-fixed axes and the Einstein summation convention is used for repeated subscripts. If we consider the macroscopic field to be a superposition of a static and an optical component,

$$F_J^{\text{mac}}(t) = F_{0,J}^{\text{mac}} + F_{\omega,J}^{\text{mac}} \cos(\omega t), \tag{66}$$

the macroscopic polarization can be expressed as[1,115]

$$\mathcal{P}_I(t) = \mathcal{P}_I^0 + \mathcal{P}_I^\omega \cos(\omega t) + \mathcal{P}_I^{2\omega} \cos(2\omega t) + \mathcal{P}_I^{3\omega} \cos(3\omega t) + \cdots . \tag{67}$$

The Fourier amplitudes of the polarization are then given in terms of the frequency-dependent susceptibilities as[1,115]

$$\mathcal{P}_I^{\omega_s} = \delta_{\omega_s,0} \mathcal{P}_I^0 + \chi_{IJ}^{(1)}(-\omega_s;\omega_s) F_{\omega_s,J}^{\text{mac}}$$

$$+ K(-\omega_s;\omega_a,\omega_b) \chi_{IJK}^{(2)}(-\omega_s;\omega_a,\omega_b) F_{\omega_a,J}^{\text{mac}} F_{\omega_b,K}^{\text{mac}} \tag{68}$$

$$+ K(-\omega_s;\omega_a,\omega_b,\omega_c) \chi_{IJKL}^{(3)}(-\omega_s;\omega_a,\omega_b,\omega_c) F_{\omega_a,J}^{\text{mac}} F_{\omega_b,K}^{\text{mac}} F_{\omega_c,L}^{\text{mac}} \tag{69}$$

$$+ \cdots , \tag{70}$$

where the output frequency is given as the sum of input frequencies $\omega_s = \sum_a \omega_a$. The numerical coefficients $K(-\omega_s;\omega_a,\cdots)$ arise from the Fourier expansion of the electric field and polarization and ensures that all susceptibilities of the same order have the same static limit. A tabulation of the coefficients can be found in Ref. 116, 117. The frequency-dependent susceptibilities can then be found from Eq. 69 by differentiation which gives the linear susceptibility

$$\chi_{IJ}^{(1)}(-\omega_s;\omega_s) = \left. \frac{\partial \mathcal{P}_I^{\omega_s}}{\partial F_{\omega_s,J}^{\text{mac}}} \right|_{F^{\text{mac}}=0}, \tag{71}$$

the second-order nonlinear susceptibility

$$\chi_{IJK}^{(2)}(-\omega_s;\omega_a,\omega_b) = K^{-1}(-\omega_s;\omega_a,\omega_b) \left. \frac{\partial^2 \mathcal{P}_I^{\omega_s}}{\partial F_{\omega_a,J}^{\text{mac}} \partial F_{\omega_b,K}^{\text{mac}}} \right|_{F^{\text{mac}}=0}, \tag{72}$$

and the third-order nonlinear susceptibility

$$\chi_{IJKL}^{(3)}(-\omega_s;\omega_a,\omega_b,\omega_c) = K^{-1}(-\omega_s;\omega_a,\omega_b,\omega_c) \left. \frac{\partial^3 \mathcal{P}_I^{\omega_s}}{\partial F_{\omega_a,J}^{\text{mac}} \partial F_{\omega_b,K}^{\text{mac}} \partial F_{\omega_c,L}^{\text{mac}}} \right|_{F^{\text{mac}}=0}. \tag{73}$$

Each of the frequency-dependent susceptibilities corresponds to different physical processes,[1,115] e.g $\chi^{(1)}(-\omega;\omega)$ governs the refractive index, $\chi^{(2)}(-2\omega;\omega,\omega)$ the second harmonic generation (SHG), $\chi^{(3)}(-3\omega;\omega,\omega,\omega)$ the third harmonic generation (THG) and, $\chi^{(3)}(-\omega;\omega,\omega,-\omega)$ the degenerate four-wave mixing (DFWN) or the intensity-dependence of the refractive index.

3.2. The microscopic polarization

Similarly the microscopic polarization (dipole moment) can be expanded in terms oscillating at different frequencies as[97,105]

$$\mu_\alpha(t) = \mu_\alpha^0 + \mu_\alpha^\omega \cos(\omega t) + \mu_\alpha^{2\omega} \cos(2\omega t) + \mu_\alpha^{3\omega} \cos(3\omega t) + \cdots. \tag{74}$$

The microscopic dipole moment is then given by the Taylor expansion[97,105] in Eq. 29 were it is expanded in terms of the actual total electric field, $F_{\omega_b,\gamma}^{\text{tot}}$, felt by the molecule. In the condensed phase the actual electric field felt by the molecule is different from the macroscopic electric field. Therefore, in order to express the macroscopic properties in terms of the microscopic properties we need to relate the actual electric field at a molecule to the macroscopic electric field.

3.3. The local electric field

Lorentz[118] derived a simple relation between the internal electric field, the macroscopic electric field and the macroscopic polarization of the system, and due to its simplicity Lorentz local field theory is still used.[1, 104, 115, 119] The idea is that only close to the molecule we need to consider explicitly the field from nearby molecules. Therefore, the total system is separated into a macroscopic region far from the molecule and a microscopic region close to the molecule. The molecules in the macroscopic region can then be described by the average macroscopic properties. This separation is illustrated in Figure 2.

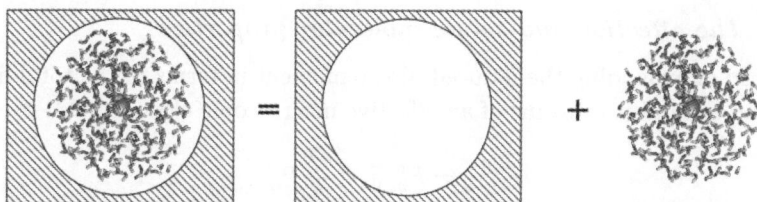

Fig. 2.   Separation of the total system into a macroscopic region far from the central molecule and a microscopic region close to the molecule. In the microscopic region the actual electric field felt by the molecule should be considered in detail.

Inside a macroscopically small, but microscopically large, virtual cavity V we subtract the contribution from the macroscopic electric field and replace it by the correct discrete local field,

$$F_{\omega_s, \alpha}^{\text{tot}} = F_{\omega_s, \alpha}^{\text{mac}} - F_{\omega_s, \alpha}^{\text{pol}} + F_{\omega_s, \alpha}^{\text{disc}}(\Omega), \qquad (75)$$

where $F_{\omega_s, \alpha}^{\text{pol}}$ is the macroscopic electric field in the cavity V and $F_{\omega_s, \alpha}^{\text{disc}}(\Omega)$ is the discrete electric field in the cavity V which depends on the local configuration, Ω, of the molecules inside the cavity. Since we are not allowing the macroscopic region to adjust to the presence of the cavity the polarization remains homogeneous.[104] This approach neglects that a static electric field tends to orient molecules with a permanent dipole[104, 120] and therefore a correction due to Onsager[120] is often used for static electric fields. For a spherical cavity the macroscopic field is simply given in terms of the macroscopic polarization[104, 118] and the total electric field can be written as

$$F_{\omega_s, \alpha}^{\text{tot}} = F_{\omega_s, \alpha}^{\text{mac}} + \frac{4\pi}{3} \mathcal{P}_{\alpha}^{\omega_s} + F_{\omega_s, \alpha}^{\text{ind}}(\Omega) + F_{\alpha}^{\text{perm}}(\Omega) \qquad (76)$$

where we have split the discrete electric field, F^{disc}, into two different contributions, F^{ind} and F^{perm}. The first term arises from the interactions of the macroscopic electric field with the other molecules in the cavity, i.e.

accounts for the induced polarization of the surrounding molecules due to the electric field. The second term accounts for the interactions between the molecules when there is no electric field present, i.e. arises from the permanent charge distribution of the surrounding molecules. The last two terms depend strongly on the local configuration of the molecules in the cavity and are inherently microscopic in nature and it is therefore better to treat these fields explicitly within the microscopic model used.

3.4. *The effective and solute molecular properties*

Instead of expanding the induced dipole moment in terms of the total field we can expand it in terms of an effective macroscopic electric field

$$F^{\mathrm{eff}}_{\omega_s,\alpha} = F^{\mathrm{mac}}_{\omega_s,\alpha} + \frac{4\pi}{3}\mathcal{P}^{\omega_s}_{\alpha}. \tag{77}$$

This expansion defines the so-called effective properties.[32,97] These effective properties could be compared with experimental results corrected for differences between the total field and the macroscopic electric field by the Lorentz/Onsager local field method.[97] They are calculated in the DRF model when the macroscopic electric field is included in Eq. 7.

Since we have separated the discrete field into the two contributions mentioned above we can also choose to expand the induced dipole moment in terms of the field arising directly from the macroscopic electric field,

$$F^{\mathrm{sol}}_{\omega_s,\alpha} = F^{\mathrm{mac}}_{\omega_s,\alpha} + \frac{4\pi}{3}\mathcal{P}^{\omega_s}_{\alpha} + F^{\mathrm{ind}}_{\omega_s,\alpha}(\Omega), \tag{78}$$

where the field arising from the interactions between the molecules when there is no macroscopic field is incorporated into the properties. This gives an expansion which defines the so-called solute properties[32,97] The solute properties are calculated in the DRF model when the macroscopic electric field is not included in Eq. 7.

The difference between the solute and effective properties comes from the inclusion of the macroscopic electric field in Eq. 7 in the latter case. The difference can also be estimated by calculating the dipoles induced by the macroscopic electric field only. They generate an electric field which, in general, counteracts the induced field. The relation between the macroscopic and actual fields can be expressed in terms of a field factor. This allows us to estimate α^{eff} as

$$\overline{\alpha}^{\mathrm{eff}} \sim \mathcal{L}^{\mathrm{eff}}(R_{\mathrm{com}})\overline{\alpha}^{\mathrm{sol}}, \tag{79}$$

and γ^{eff} as

$$\overline{\gamma}^{\text{eff}} \sim \mathcal{L}^{\text{eff}}(R_{\text{com}})^3 \overline{\gamma}^{\text{sol}}. \tag{80}$$

For applying these expressions one has to decide on where the polarizabilities are centered. Here we took the center of mass, R_{com}, which for small molecules is a natural choice, but for larger molecules this is not without problems.

3.5. Relating the macroscopic and the microscopic polarization

The macroscopic polarization is related to the average microscopic dipole moment per molecule by[115, 119]

$$\mathcal{P}_I^{\omega_s} = N_d \langle \mu_\alpha^{\omega_s} \rangle_I \tag{81}$$

where N_d is the number density and the brackets, $\langle \rangle$, denote orientational averaging and relate the molecule-fixed axes to the space-fixed axes.[119, 121] Inserting the expansion of the dipole moment in terms of the effective macroscopic field we can express the macroscopic polarization in terms of the effective (hyper)polarizabilities as,

$$\begin{aligned}
\mathcal{P}_I^{\omega_s} &= N_d \delta_{\omega_s,0} \langle \mu_\alpha^{0,eff} \rangle_I + N_d \langle \alpha_{\alpha\beta}^{\text{eff}}(-\omega_s;\omega_s) \rangle_{IJ} F_{\omega_s,J}^{\text{eff}} \\
&+ \frac{1}{2} K(-\omega_s;\omega_a,\omega_b) N_d \langle \beta_{\alpha\beta\gamma}^{\text{eff}}(-\omega_s;\omega_a,\omega_b) \rangle_{IJK} F_{\omega_a,J}^{\text{eff}} F_{\omega_b,K}^{\text{eff}} \\
&+ \frac{1}{6} K(-\omega_s;\omega_a,\omega_b,\omega_c) N_d \langle \gamma_{\alpha\beta\gamma\delta}^{\text{eff}}(-\omega_s;\omega_a,\omega_b,\omega_c) \rangle_{IJKL} F_{\omega_a,J}^{\text{eff}} F_{\omega_b,K}^{\text{eff}} F_{\omega_c,L}^{\text{eff}} \\
&+ \cdots.
\end{aligned} \tag{82}$$

Since the effective field is macroscopic we have taken it outside the averaging and the macroscopic polarization can be expressed in terms of orientational averages of the effective (hyper)polarizabilities. This is exactly the reason why the total electric field was split into an effective macroscopic part and a microscopic part which was incorporated into the (hyper)polarizabities by expanding the dipole moment in terms of the effective field.

3.6. Local Field Factors

Comparing Eq. 82 with Eq. 69 we see that we have to consider derivatives of the effective electric field with respect to the macroscopic electric field.

This is usually done by introducing the so-called local field factors which relates the macroscopic field to the effective field.

$$F_{\omega_a,J}^{\text{eff}} = \left. \frac{\partial F_{\omega_a,J}^{\text{eff}}}{\partial F_{\omega_a,J}^{\text{mac}}} \right|_{F^{\text{mac}}=0} \times F_{\omega_a,J}^{\text{mac}} = \mathcal{L}_{\omega_a} F_{\omega_a,J}^{\text{mac}} \tag{83}$$

Using the definition of the effective electric field, Eq. 77, we obtain a local field factor

$$\mathcal{L}_{\omega_a} = 1 + \frac{4\pi}{3}\chi^{(1)}(-\omega_a;\omega_a) = 1 + \frac{\epsilon^{(1)}(\omega_a) - 1}{3} = \frac{\epsilon^{(1)}(\omega_a) + 2}{3} \tag{84}$$

where $\epsilon^{(1)}(\omega_a)$ is the optical dielectric constant at frequency ω_a. The is the Lorentz form of the local field factors. However, as mentioned above this does not account for the fact that for a static macroscopic field the molecules tend to orient. Onsager[120] analysed this problem and suggested the following form for the local field factor,

$$\mathcal{L}_0 = 1 + \frac{n^{(1)}(0)^2(\epsilon^{(1)}(0) - 1)}{2\epsilon^{(1)}(0) + n^{(1)}(0)^2} = \frac{\epsilon^{(1)}(0)(2 + n^{(1)}(0)^2)}{2\epsilon^{(1)}(0) + n^{(1)}(0)^2} \tag{85}$$

to be used for static electric fields, where $\epsilon^{(1)}(0)$ is the dielectic contant and $n^{(1)}(0)$ is the refractive index at zero frequency. We see that the Onsager field factor reduces to the Lorentz factor for optical fields by using the relation $n^{(1)}(\omega_a)^2 = \epsilon^{(1)}(\omega_a)$.

3.7. Orientational averaging

In order to relate the molecule-fixed axes to the space-fixed axes we need also to consider molecular rotations (or orientations). The orientational averaging and thermal averaging of the dipole moment will be done using classical theory and is given by[121]

$$\langle \mu_\alpha^{\omega_s} \rangle_I =$$
$$\frac{\int_0^{2\pi} \int_0^\pi [\mu_\alpha^{\text{eff}}\Phi_{\alpha I} + \alpha_{\alpha\beta}^{\text{eff}} F_\beta^{\text{eff}}\Phi_{\alpha I}\Phi_{\beta J} + \cdots] exp(-\Delta E/k_b T) sin\theta d\theta d\phi}{\int_0^{2\pi} \int_0^\pi exp(-\Delta E/k_b T) sin\theta d\theta d\phi} \tag{86}$$

where $\Phi_{\alpha I}$ is the cosine of the angle between the molecular axis α and the laboratory axis I. The angular dependent part of the energy in the presence of the electric field is given by,

$$\Delta E = -\mu_\alpha^{\text{sol}}\Phi_{\alpha I}F_{0,I}^{\text{eff}} + \cdots . \tag{87}$$

We note that it is only the solute properties which are responsible for the change in the energy due to the electric field. If we expand the exponential and only terms of the order $(kT)^{-1}$ are retained we get

$$exp(-\Delta E/k_b T) = 1 + \mu_\alpha^{sol} \Phi_{\alpha I} F_{0,I}^{eff}/k_b T + \cdots . \tag{88}$$

By combining the definitions of the susceptibilities in Eqs. 71, 72, and 73 with the expression for the macroscopic polarization in terms of the effective (hyper)polarizabilities we can obtain a link between the macroscopic and the microscopic properties.

3.8. Refractive index

The macroscopic quantity determining the refractive index is the linear optical polarization due to an optical electric field. By substituting the expression for the effective field, Eq. 77, into Eq. 82 and using the definition of the linear susceptibilitiy, Eq. 71, we obtain

$$\chi_{ZZ}^{(1)}(-\omega;\omega) = N_d \left\langle \alpha_{\alpha\beta}^{eff}(-\omega;\omega) \right\rangle_{ZZ} \left(1 + \frac{4\pi}{3} \chi_{ZZ}^{(1)}(-\omega;\omega) \right) \tag{89}$$

The optical field is considered to be oscillating faster than the permanent dipole moments can be oriented and therefore no rotational contribution exists. The linear susceptibility can then be written in terms of the mean effective polarizability by rewriting Eq. 89 as,

$$\chi_{ZZ}^{(1)}(-\omega_s;\omega_s) = \frac{N_d \overline{\alpha}^{eff}(-\omega_s;\omega_s)}{1 - \frac{4\pi}{3} N_d \overline{\alpha}^{eff}(-\omega_s;\omega_s)} \tag{90}$$

which is the standard expression for the susceptibility corrected for the Lorentz local field,[115, 122] although using the effective rather than the gas phase polarizability. The susceptibility is related to the refractive index or the optical dielectric constant of the system as

$$n^{(1)}(\omega_s) = \sqrt{\epsilon^{(1)}(\omega_s)} = \sqrt{1 + 4\pi \chi_{ZZ}^{(1)}(-\omega_s;\omega_s)}$$
$$= \sqrt{\frac{1 + \frac{8\pi}{3} N_d \overline{\alpha}^{eff}(-\omega_s;\omega_s)}{1 - \frac{4\pi}{3} N_d \overline{\alpha}^{eff}(-\omega_s;\omega_s)}} \tag{91}$$

which is the familar Lorentz-Lorenz or Clausius-Mossotti equation,[104, 118, 122] again with the effective polarizability instead of the gas phase polarizability.

3.9. *Third-Harmonics Generation*

The first nonlinear susceptibility we will consider is the third-order nonlinear susceptibility which arises from three optical electric fields corresponding to the Third-Harmonics Generation (THG) experiments. The THG susceptibility is obtained by substituting Eq. 77 into Eq. 82 and using the definition of the susceptibilitiy, Eq. 73. This gives the THG susceptibility as

$$\chi^{(3)}_{ZZZZ}(-3\omega;\omega,\omega,\omega) = N_d \left\langle \alpha^{\text{eff}}_{\alpha\beta}(-3\omega;3\omega) \right\rangle_{ZZ} \left(\frac{4\pi}{3} \chi^{(3)}_{ZZZZ}(-3\omega;\omega,\omega,\omega) \right)$$
$$+ \frac{1}{6} N_d \left\langle \gamma^{\text{eff}}_{\alpha\beta\gamma\delta}(-3\omega;\omega,\omega,\omega) \right\rangle_{ZZZZ} \mathcal{L}^3_\omega \tag{92}$$

where we see that there is a contribution both from the linear susceptibility and from the third-order nonlinear susceptibility. The isotropic orientation average of the second hyperpolarizability, often refered to as the mean or parallel second hyperpolarizability, is given by[121]

$$\left\langle \gamma^{\text{eff}}_{\alpha\beta\gamma\delta} \right\rangle_{ZZZZ} = \overline{\gamma}^{\text{eff}}_{\parallel} = \frac{1}{15} \sum_{\alpha\beta} (\gamma^{\text{eff}}_{\alpha\alpha\beta\beta} + \gamma^{\text{eff}}_{\alpha\beta\beta\alpha} + \gamma^{\text{eff}}_{\alpha\beta\alpha\beta}) \tag{93}$$

We can then rewrite the THG susceptibility as[32]

$$\chi^{(3)}_{ZZZZ}(-3\omega;\omega,\omega,\omega) = \frac{1}{6} N_d \overline{\gamma}^{\text{eff}}_{\parallel}(-3\omega;\omega,\omega,\omega) \mathcal{L}_{3\omega} \mathcal{L}^3_\omega \tag{94}$$

which is the form for the nonlinear susceptibility well known from standard Lorentz local field theory with $n + 1$ local field corrections, where n is the number of applied fields.

3.10. *Electric Field Induced Second-Harmonic Generation*

The second nonlinear susceptibility we will consider is the third-order nonlinear susceptibility arising from two optical and one static electric field and corresponds to the Electric Field Induced Second-Harmonic Generation (EFISH) experiments. The EFISH susceptibility is obtained by substituting Eq. 77 into Eq. 82 and using the definition of the susceptibilitiy,

Eq. 73 giving

$$
\chi^{(3)}_{ZZZZ}(-2\omega;\omega,\omega,0) =
$$

$$
N_d \left\langle \alpha^{eff}_{\alpha\beta}(-2\omega;2\omega) \right\rangle_{ZZ} \left(\frac{4\pi}{3} \chi^{(3)}_{ZZZZ}(-2\omega;\omega,\omega,0) \right)
$$

$$
+\frac{1}{2} K(-2\omega;\omega,\omega) K^{-1}(-2\omega;\omega,\omega,0) N_d \frac{\partial \left\langle \beta^{eff}_{\alpha\beta\gamma}(-2\omega;\omega,\omega) \right\rangle_{ZZZ}}{\partial F^{mac}_{0,Z}} \mathcal{L}^2_\omega
$$

$$
+\frac{1}{6} N_d \left\langle \gamma^{eff}_{\alpha\beta\gamma\delta}(-2\omega;\omega,\omega,0) \right\rangle_{ZZZZ} \mathcal{L}^2_\omega \mathcal{L}_0 \tag{95}
$$

We see that the EFISH susceptibility consists of three terms. The second term is the rotational contribution given by

$$
\frac{\partial \left\langle \beta^{eff}_{\alpha\beta\gamma}(-2\omega;\omega,\omega) \right\rangle_{ZZZ}}{\partial F^{mac}_{0,Z}} = \mathcal{L}_0 \frac{\overline{\beta}^{eff}_{\parallel}(-2\omega;\omega,\omega)\mu^{sol}_z}{3k_bT} \tag{96}
$$

where the mean hyperpolarizability, $\overline{\beta}_{\parallel}$, in the direction of the dipole moment, here the z axis, is introduced

$$
\overline{\beta}^{eff}_{\parallel} = \frac{1}{5} \sum_\alpha (\beta^{eff}_{z\alpha\alpha} + \beta^{eff}_{\alpha z\alpha} + \beta^{eff}_{\alpha\alpha z}). \tag{97}
$$

The EFISH susceptibilty can be written as,[32]

$$
\chi^{(3)}_{ZZZZ}(-2\omega;\omega,\omega,0) =
$$

$$
\frac{1}{6} N_d \left(\overline{\gamma}^{eff}_{\parallel}(-2\omega;\omega,\omega,0) + \frac{\overline{\beta}^{eff}_{\parallel}(-2\omega;\omega,\omega)\mu^{sol}_z}{3k_bT} \right) \mathcal{L}_{2\omega} \mathcal{L}^2_\omega \mathcal{L}_0
$$

$$
= \frac{1}{6} N_d \Gamma^{eff}_{\parallel}(-2\omega;\omega,\omega,0) \mathcal{L}_{2\omega} \mathcal{L}^2_\omega \mathcal{L}_0 \tag{98}
$$

where it has been used that $K(-2\omega;\omega,\omega) = \frac{1}{2}$ and $K^{-1}(-2\omega;\omega,\omega,0) = \frac{2}{3}$.

4. Selected applications

In this section we will present some of the applications of the DRF model to calculated both microscopic and macroscopic response properties. We will present results for liquid water, liquid acetonitrile and for fullerene C_{60} clusters.

4.1. Response properties of liquid water

The DRF model within TD-DFT was first used to study the microscopic response properties of a water molecule embedded in a small cluster of water molecules.[29-31] We have chosen water not because we consider it an interesting molecule with respect to its NLO properties, although, the experimentally observed sign change in the first hyperpolarizability upon solvation is interesting. The reason is rather that for this particular cluster of water molecules there existed uncorrelated and correlated wave function QM/MM results. This enabled us to assess approximate xc-potentials for calculating molecular response properties in solution. This water cluster corresponds to an average water structure (AWS) obtained from MD simulations and should in an average way represent the local environment around the central water molecule. Although the small water cluster did not provide a realistic model of liquid water it provided a benchmark for the DRF model.

Table 1.   A comparison of the molecular properties of water in the gas phase. All results presented are in a.u. $\overline{\Gamma}_\parallel(-2\omega;\omega,\omega,0)$ in 10^3 a.u. CCSD results are taken from: μ) Ref. 84, α) Ref. 85, β) Ref. 86, and γ) Ref. 87. Exp. results are taken from: μ) Ref. 123, α) Ref. 124, β) Ref. 125, and γ) Ref. 125. DFT results are taken from: μ) Ref. 29, α) Ref. 30, β) Ref. 31, and γ) Ref. 31

	μ	$\overline{\alpha}(-\omega;\omega)$	$\overline{\beta}_\parallel(-2\omega;\omega,\omega)$	$\overline{\gamma}_\parallel(-2\omega;\omega,\omega,0)$	$\overline{\Gamma}_\parallel(-2\omega;\omega,\omega,0)$
Gas phase					
CCSD	0.73	9.52	-19.26	1942	-30.2
DFT	0.71	9.97	-20.42	2021.3	-31.0
Exp.	0.73	9.83	$-19.2\pm5\%$	$1800\pm8\%$	-31.5
AWS					
CCSD/MM	1.07	10.04	12.21	2169	
DFT/DRF	1.04	10.13	8.57(12.77)	2117.6	

In Table 1 we summarize the results for the dipole moment, polarizability, first hyperpolarizability and the second hyperpolarizability of water in the gas phase and for the AWS cluster. The results have been calculated at $\omega = 0.0428$ a.u. ($\lambda = 1064$ nm) and are compared both with experimental and CCSD results in the gas phase and with CCSD/MM results for the AWS cluster. In general we find good agreement between the DFT results and the CCSD results for all properties. If we compare with results obtained from EFISH experiments we see that for water there is an excellent agreement between the calculated and the experimental results for all properties. However, as shown in section 3 the measured quantity in the EFISH experiment is $\overline{\Gamma}_{\|} = \frac{\mu_z \overline{\beta}_{\|}}{3 k_b T} + \overline{\gamma}_{\|}$. Therefore, we also report this value for the different methods in Table 1. Again, we see that there is good agreement between theory and experiment for water. If we compare the results for the AWS again we see a good agreement between the DFT/DRF and the CCSD/MM results. The only exception seems to be the first hyperpolarizability where the DFT/DRF result is smaller than that of CCSD/MM. However, as discussed in Ref. 31 the cause was that at short distances the DRF operator in Eq. 2 is damped. The first hyperpolarizability calculated without the short range damping is also presented in Table 1 (number in paranthesis) and is seen to be in good agreement with the CCSD/MM results. For the other properties only small changes were found by ignoring the damping which indicated that especially the first hyperpolarizability was sensitive to the local molecular environment.

In order to obtain a more realistic model for liquid water we performed[32] a MD simulation on 256 water molecules from which 101 different water structures water, *i.e.*, snapshots of the local molecular environment, was collected. We then performed a DFT/DRF calculation for each of the configurations and averaged the molecular properties obtained from these calculations. We calculated both the solute and effective properties and the results are presented in Table 2. In general we see that the solvent effects on

Table 2.   Frequency-dependent molecular response properties of liquid water taken from Ref. 32. The solute and effective properties are reported as the average value ± the standard error. The standard error is given by $\frac{\sigma}{\sqrt{N}}$ where σ is the standard deviation and N is the number of configurations. All results are in a.u.

	μ	$\overline{\alpha}(-\omega;\omega)$	$\overline{\beta}_{\|}(-2\omega;\omega,\omega)$	$\overline{\gamma}_{\|}(-2\omega;\omega,\omega,0)$
Gas	0.71	9.97	-20.42	2021.3
Solute	1.01±0.01	10.32±0.01	6.82±0.53	2180.52±15.26
Effective	1.01±0.01	9.95±0.01	7.20±0.50	1503.44±10.52

the dipole moment and first hyperpolarizability are large whereas for the polarizability and second hyperpolarizability the solvent effects are only modest. We also see that the solute properties are larger than the effective properties for both the polarizability and for the second hyperpolarizability. For the first hyperpolarizability they are identical within the standard error. In the case of the dipole moment they are trivially identical due to that the macroscopic electric field was not included in the MD simulations. The macroscopic electric field induces dipoles in the solvent which screens the macroscopic electric field. The solute and effective mean polarizability for the different solvent configurations are displayed in Fig. 3 which illustrates the fluctuations due to the different molecular environment. It is therefore

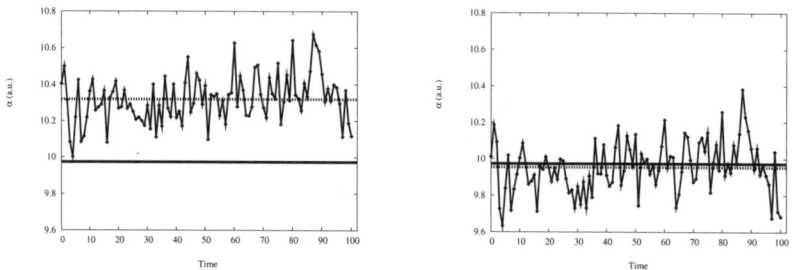

(a) Solute $\bar{\alpha}$. Dashed line (top) is the averaged results and the solid line (bottom) is the gas phase results.

(b) Effective $\bar{\alpha}$. Dashed line (bottom) is the averaged results and solid line (top) is the gas phase results.

Fig. 3.   The mean first polarizability, $\bar{\alpha}(-\omega;\omega)$, at $\omega = 0.0428$ a.u. for water. From Ref. 32

important to account for these fluctuations by sampling over several configurations. This is even more important for the first hyperpolarizability which is extremely sensitive to the local environment.

So far we have only discussed the microscopic properties. However, as mentioned in section 3 we are able to combine the effective properties with Lorentz/Onsager local field factor and obtain the macroscopic susceptibilities directly. This was done in Ref. 32 and compared with experimental results. In Table 3 the refractive index and the third-order nonlinear susceptibility corresponding to THG and EFISH are summarized. Note that the experimental results in Table 3 are converted to match the conventions adopted in this work.[32,105] The more recent value of the reference value for fused silica for the THG experiments[129,130] and the currently accepted value for quartz[131,132] for the EFISH experiment was used, and both values are believed to be more accurate. We see that for the refractive index there is very good agreement between the DFT/DRF results and the experimental data. The DFT/DRT results are slightly larger than the experimental values. This is most likely due to the fact that for water DFT predicts a larger

Table 3.   Macroscopic response properties of liquid water. The frequency is $\omega = 0.0428$ a.u. $\chi^{(3)}$ is in units of 10^{-14} esu. The refractive index $n(\omega)$ is taken from Ref. 126 and for $n(2\omega)$ from Ref. 127 and is at the slightly lower frequency of 0.077 a.u. THG experiment is from Ref. 126 and the EFISH experiment is from Ref. 128 DFT results are from Ref. 32

	$n(\omega)$	$n(2\omega)$	$\chi^{(3)}_{THG}$	$\chi^{(3)}_{EFISH}$
DFT/DRF	1.334	1.342	1.07	4.1
Exp.	1.326	1.333	1.29	10.5

polarizability than CCSD, see Table 1. Also for the THG susceptibility we see that there is a good agreement between DFT/DRF and the experimental result. The DFT/DRF result is slightly lower than the experimental result, but some of the discrepancy can be due to THG susceptibility being estimated from the EFISH effective second hyperpolarizability which shows a smaller dispersion than the THG second hyperpolarizability. Comparing now the EFISH susceptibility calculated using DFT/DRF with the experimental result we see that it is much smaller than the experimental value. Since the rotational contribution, i.e., the term depending on the dipole moment and first hyperpolarizability, is the major contribution it was argued that this term was not modeled accurately.[32] The rotational contribution is not correctly described since in the MD simulation the orientational effect due to the macroscopic electric field is ignored.

4.2. Response properties of liquid Acetronitrile

The DFT/DRF model has also been applied to calculate the microscopic and macroscopic response properties of liquid acetonitrile.[32] Again, acetonitrile was not chosen for its important NLO properties but rather because there exits both theoretical and experimental reference data. Although there has been studies of the NLO properties of acetonitrile using continuum solvation models[9, 133–136] this was the first study using a (polarizable) QM/MM model.

In Table 4 we summarize the results for the dipole moment, polarizability, first hyperpolarizability and the second hyperpolarizability of acetonitrile in the gas phase. The results have been calculated at $\omega = 0.0885$ ($\lambda = 514.5$ nm) and are compared with experimental and CCSD results. In general we find good agreement between the DFT results and the CCSD results for all properties, although not as good as for water. The largest difference of $\sim 25\%$ between the calculated values is in $\overline{\gamma}_{\parallel}$ and $\overline{\beta}$. The DFT

Table 4. A comparison of the molecular properties of acetonitrile in the gas phase. All results presented are in a.u. $\overline{\Gamma}_\|(-2\omega;\omega,\omega,0)$ in 10^3 a.u. CCSD results are taken from Ref. 9. In the case of μ the CCSD(T) results are reported. Exp. results are taken from Ref. 133. $\overline{\alpha}$ from Ref. 137. DFT results are from Ref. 32.

	μ	$\overline{\alpha}(-\omega;\omega)$	$\overline{\beta}_\|(-2\omega;\omega,\omega)$	$\overline{\gamma}_\|(-2\omega;\omega,\omega,0)$	$\overline{\Gamma}_\|(-2\omega;\omega,\omega,0)$
CCSD	1.525	30.23	32.54	5855	233.8
DFT	1.59	31.94	23.99	7317.97	207.9
EXP.	1.542	30.43	26.3	4619	189.4

results for $\overline{\beta}_\|$ is lower than the CCSD results whereas for $\overline{\gamma}_\|$ the opposite is found. If we compare with results obtained from EFISH experiments we see that for $\overline{\beta}_\|$ the DFT is in better agreement with the experiment whereas for $\overline{\gamma}$ it is the CCSD results. Comparing $\overline{\Gamma}_\|$ for acetonitrile the DFT and CCSD results are within 10% and 20%, respectively, of the experimental results.

The microscopic response properties for liquid acetonitrile were calculated in a similar manner as for water by averaging over 101 solvent configurations.[32] In the MD simulation and the subsequently DFT/DRF calculations 128 acetonitrile molecules was included with one molecule treated at the DFT level of theory. This corresponds roughly to a simulations box of a similar size as that used in the water simulation. The calculated solute and effective properties are collected in Table 5.

Table 5. Frequency-dependent molecular response properties of liquid acetonitrile taken from Ref. 32. The solute and effective properties are reported as the average value ± the standard error. The standard error is given by $\frac{\sigma}{\sqrt{N}}$ where σ is the standard deviation and N is the number of configurations. All results are in a.u.

	μ	$\overline{\alpha}(-\omega;\omega)$	$\overline{\beta}_\|(-2\omega;\omega,\omega)$	$\overline{\gamma}_\|(-2\omega;\omega,\omega,0)$
Gas	1.59	31.94	23.99	7317.97
Solute	1.99±5%	34.30±1%	146.82±21%	10662.99±9%
Effective	1.99±5%	32.16±2%	111.80±25%	8136.71±12%

The results for acetonitrile presented in Table 5 show some of the same trends as for water, i.e., strong solvent effects on the dipole moment and first hyperpolarizability and weak solvent effect on the polarizability. However, in contrast to water the solvent effects on the second hyperpolarizability are also strong. The solvent effects on the first hyperpolarizability are particular strong. β^{sol} and β^{eff} are enhanced by around a factor of six and five, respectively, as compared with the gas phase value. Similar large enhancements have been found using continuum models.[9, 133–136] Again, the screening due to the inclusion of the macroscopic electric field in Eq. 7 is evident since the effective properties are lower than the corresponding solute properties. The fluctuation in the solute and effective mean polarizability due to changes in the local molecular environment in the different solvent configurations are displayed in Fig. 4. It is seen that the fluctuations are larger for the efffective polarizability than for the solute polarizability. This is due to the extra induced dipoles in the solvent which enhances the anisotropy of the solvent configurations. Therefore, it is more important to sample over several different configurations when calculating the effective properties.

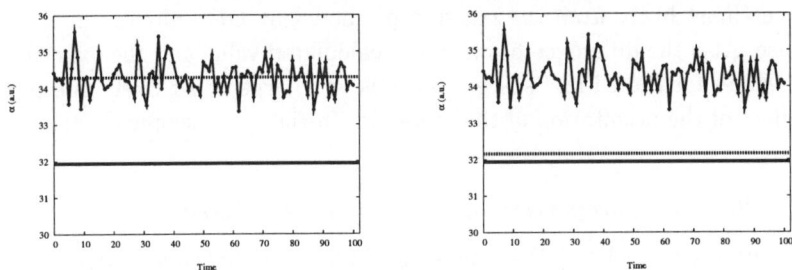

(a) Solute $\overline{\alpha}$. Dashed line (top) is the averaged results and the solid line (bottom) is the gas phase results.

(b) Effective $\overline{\alpha}$. Dashed line (Top) is the averaged results and solid line (bottom) is the gas phase results.

Fig. 4.   The mean first polarizability, $\overline{\alpha}(-\omega;\omega)$, at $\omega = 0.0885$ a.u. for acetronitrile.

The macroscopic response properties of liquid acetonitrile were also calculated.[32] In Table 6 the refractive index and the third-order nonlinear susceptibilities corresponding to THG and EFISH are collected together with experimental results. Also for acetonitrile the experimental data have been converted to match the convention used in this work and corrected for the newer reference values, see the discussion for water.

For the macroscopic properties we see similar trends as for water. The calculated refractive index is in good agreement with the experimental value. The DFT/DRF results are larger than the experimental results,

Table 6.   Macroscopic response properties of liquid acetonitrile. $\chi^{(3)}$ is in units of 10^{-14} esu. For the calculated refractive index the frequency is $\omega = 0.0885$ a.u. The experimental refractive index is taken from Ref. 133 and is reported for the frequency of 0.0856 a.u. The experimental THG result for acetonitrile is measured at $\omega = 0.0239$ a.u.[138] whereas the static calculated value is reported here. The EFISH experiment is from Ref. 134 and both the calculated and experimental results is for $\omega = 0.0885$ a.u. DFT results are from Ref. 32.

	$n(\omega)$	$n(2\omega)$	$\chi^{(3)}_{\text{THG}}$	$\chi^{(3)}_{\text{EFISH}}$
DFT/DRF	1.377	1.434	1.27	35.7
Exp.	1.347	1.384	1.15	13.6

again, most likely due to that DFT predicts a larger polarizability than CCSD. For the THG susceptibility there is good agreement between the calculated and experimental results. The calculated THG susceptibility is in this case larger than the experimental value. Some of the discrepancy comes most likely from the larger γ predicted by DFT already in the gas phase. Also the difference between the calculated value and the experimental values for the EFISH susceptibility is very large, most likely due to the neglect of the orientation of the molecules in the macroscopic electric field.

4.3. Response properties of C_{60} fullerene cluster

Having tested the DFT/DRF model for different liquids and found good results for the refractive index and the THG susceptibility we wanted to test the method on a system for which accurate wave function method were unfeasible. We chose to investigate the response properties of the C_{60} fullerene embedded in a C_{60} cluster.[33] This illustrated both the ability of DFT to treat fairly large systems and at the same time the ability of DFT/DRF to treat a MM part containing several thousand atoms. The fullerene C_{60} is interesting since it belongs to a class of all carbon systems containing a delocalized conjugated π-system. Furthermore, only the polarizability has been measured in the gas phase and the results for the second hyperpolarizability is measured either in solution or for thin films. The C_{60} clusters investigated provides results which can be compared with the thin film results. The largest structure investigated, displayed in Fig. 5, contained 63 C_{60} molecules (3780 atoms) where the central C_{60} molecules was treated using DFT and the rest by DRF. We calculated the frequency-dependent polarizability, second hyperpolarizability, refractive index and both THG and DFWM susceptibilities.[33]

Fig. 5.   Structure of the isotropic $N = 63$ fullerene cluster.

Since we can only calculated the EFISH second hyperpolarizability using the current implementation we estimated the other processes by using dispersion relations.[6] The frequency-dependence of the various $\{\alpha\}$ and $\{\gamma\}$ can be characterized by fitting them to the following dispersion relations[6]

$$\overline{\alpha}(-\omega;\omega) = \overline{\alpha}(0;0) \times (1 + A\omega^2 + \cdots) \tag{99}$$

and

$$\overline{\gamma}(-\omega_s;\omega_1,\omega_2,\omega_3) = \overline{\gamma}(0;0,0,0) \times (1 + A\omega_l^2 + \cdots) \tag{100}$$

where $\omega_s = \omega_1 + \omega_2 + \omega_3$ and $\omega_l^2 = \omega_s^2 + \omega_1^2 + \omega_2^2 + \omega_3^2$. The coefficients A are then transferable between different processes and can therefore be used to estimate the second hyperpolarizability belonging to other processes than the EFISH.[6]

The microscopic results are collected in Table 7 for the isolated C_{60} molecule. The values in Table 7 are in fair agreement with the experiments at both frequencies although we predict slightly larger values. The agreement with recent quantum chemical calculations and results from a frequency-dependent point dipole interaction (PDI) model is also good. The only exception is the CC2 results which are found to be too high compared with the other results. Recently it was shown that for small fullerenes conventional DFT predicts a slightly larger value for the polarizability due to the neglect of extra long-range exchange-correlation effects which could explain the small difference with the experimental results.[139] Also for the second hyperpolarizability there is good agreement between the results reported in Table 7 and results calculated using quantum chemical methods.[22, 140–142]

Table 7. The frequency-dependent mean polarizability, $\bar{\alpha}$ and second hyperpolarizability, $\bar{\gamma}$(EFISH), for C_{60}. Polarizability in a.u. and second hyperpolarizability in 10^3 a.u. DFT results are from Ref. 33, SCF results from Refs. 141, 143, PDI results from Refs. 93,102,144, CC2 results from Ref. 145 and experimental results from Refs. 146,147.

Vacuum		$\bar{\alpha}$	$\bar{\gamma}$
$\omega = 0.0000$	DFT	558.92	97.87
	SCF	506.7	109.2
	PDI	522.8	115.2
	CC2	623.7	
	Exp.	516.2 ± 54	
$\omega = 0.0428$	DFT	571.44	129.24
	SCF	515.6	
	CC2	640.2	
	PDI	527.7	
	Exp.	533 ± 27	

The microscopic response properties for C_{60} in the condensed phase are presented in Table 8. The DFT/DRF results are for the central molecule in the cluster. The results are compared with results obtained using SCRF and the classical PDI model. Also presented in the Table are the effective properties calculated from the relation with the solute properties (DFT/DRF-LF) in Eqs. 79 and 80.

Table 8. The mean polarizability, $\bar{\alpha}$ and second hyperpolarizability, $\bar{\gamma}$, for C_{60} in the condensed phase. Polarizability in a.u. and second hyperpolarizability in 10^3 a.u. DFT results are from Ref. 33, SCRF from Ref. 100 and PDI results from Ref. 102.

Crystal $N = 63$	$\bar{\alpha}^{\text{sol}}$	$\bar{\alpha}^{\text{eff}}$	$\bar{\gamma}^{\text{sol}}$	$\bar{\gamma}^{\text{eff}}$
DFT/DRF	753.84	637.29	298.85	159.09
DFT/DRF-LF	-	633.2	-	177.1
SCRF	632.91	-	278	-
PDI	-	571.8	-	169

Although the methods use for calculating the response properties presented in Table 8 are different in the way they treat the molecular environment and the central molecule the results shows similar trends. All of the methods predict an enhancement of the properties in the condensed phase. Especially the second hyperpolarizability is enhanced as compared with the isolated molecule. We also see that the estimated effective properties, DFT/DRF-LF, is in good agreement with the calculated effective properties, especially for the polarizability. It should therefore be possible to obtain the effective properties directly from the solute properties without the need for an extra set of calculations. However, this should be tested for the liquids as well in order to establish how general the relations is. It is expected that for smaller molecules the approximations in Eqs. 79 and 80 should be better.

In Table 9 we have collected the refractive indices and third-order nonlinear susceptibilities for various processes for C_{60} in the condensed phase. Calculated results using DFT/DRF, PDI and SCRF and experimental results are presented. The experimental results have been converted to match the conventions in the work and corrected for the more accurate reference data.[33] We see that there is very good agreement between the refractive index calculated using DFT/DRF and the results from the PDI model and from experiments.

Table 9. Macroscopic response properties of C_{60} in the condensed phase. Third-order nonlinear susceptibility in 10^{-13} esu. DFT results are from Ref. 33, PDI results from Ref.[102] and SCRF from Ref. 100. The experimental refractive indices are from Refs. 148–154, the THG result is from Ref. 149 and the DFWM result from Ref. 155.

Process	Method	n	$\chi^{(3)}$
STATIC	DFT/DRF	2.2	4.8
	PDI	2.0	3.2
	SCRF	-	2.7
	Exp.	~ 2	
DFWM	DFT/DRF	2.2	5.3
	Exp.	-	36 ± 12
THG	DFT/DRF	2.3	6.3
	Exp.	-	16.1 ± 2.4

The bulk of the experimental studies of $\chi^{(3)}$ for C_{60} films are done in the resonant third-order regime. Only two non-resonant experiments exist for which a comparison with calculated results are valid. This complicates the comparison since the experimental results are for THG and DFWM and differs by a factor of two. However, in the DFWM experiments apart from the electronic contribution, also vibrational contributions are very important. From theory it has been found that for the isolated C_{60} molecule the vibrational contribution to $\gamma(DFWM)$ is of the same order as the electronic contribution.[156] Therefore, it is likely that the same is true in the condensed phase. Our theoretical value for $\chi^{(3)}$ is a factor of two smaller than that from the THG experiment and a factor of five smaller than that from the DFWM experiment. The better agreement with the THG experiments was expected since we have only calculated the pure electronic contribution to $\chi^{(3)}$ and not the vibrational contribution present in the DFWM experiments. The agreement between the calculated results is much better.

5. Summary and Outlook

In this chapter we have reviewed the discrete solvent reaction field (DRF) model as implemented in time-dependent density functional theory (TD-DFT). The DRF model is a polarizable quantum mechanics/ molecular mechanics (QM/MM) model for calculating molecular response properties of molecules in the condensed phase. The classical description of the discrete molecular environment is done using distributed atomic charges and polarizabilities. All of the atomic parameters have been chosen so as to describe molecular gas phase properties of the isolated molecules, i.e., the atomic charges reproduce at least the molecular dipole moment and the atomic polarizabilities reproduce the full molecular polarizability tensor using a modified dipole interaction model. This allows for a straightforward parametrization of the DRF model. The DRF model has been extended to include the so-called local field factors, i.e. the difference between the macroscopic electric field and the actual electric field felt by the solute. This enables the calculations of effective microscopic properties which is related to the macroscopic susceptibilities by Lorentz/Onsager local field factors. The macroscopic susceptibilities can then be compared directly with experimental results.

We have reviewed some of the applications of the DRF model to calculate the microscopic and macroscopic response properties of molecules in the condensed phase. By carefully validating the DFT model for calculating response properties in the gas phase we have shown that also in the condensed phase results in good agreement with experimental results can be obtained. This is particular true for the refractive index and pure electronic third-harmonic generation (THG) nonlinear susceptibility. On the other hand the model failed to predict the electric field induced second harmonic generation (EFISH) susceptibility due to a poor handling of the rotational contribution. It was suggested that this could be improved by treating the orientations of the solvent molecules due to the macroscopic electric field in the MD simulations directly. The alignment of the solvent molecules due to the macroscopic electric field should in this way be included in the configurations for which the response properties are calculated. This would be an important extension and test of the local field approach adopted in this model and future investigation of this should be undertaken.

So far the number of applications of the DRF model to calculate in particular the macroscopic response properties are fairly limited and therefore it is difficult to draw conclusions on the general applicability of the model. Therefore, further application of the model to test different liquids and also molecules in solution should be carried out. Investigation different solvent effects on the solute properties going from non-polar to polar solvent would be valuable to asses local field effects. Furthermore, the extension of the current implementation with other response properties like optical rotation, raman scattering and NMR properties, could provide further tests of the model and, hopefully, extend its usefulness.

320   L. Jensen and P. Th. van Duijnen

References

1. P. N. Prasad and D. J. Williams. *Introduction to nonlinear optical effects in molecules and polymers*. Wiley, New York, 1991.
2. J. Jortner and M. Ratner, editors, *Molecular Electronics*. Blackwell Science, Oxford, 1997.
3. D. R. Kanis, M. A. Ratner, and T. J. Marks. *Chem. Rev.*, **94**, 195–242, 1994.
4. J. L. Brédas, C. Adant, P. Tackx, A. Persoons, and B. M. Pierce. *Chem. Rev.*, **94**, 243–278, 1994.
5. F. Jensen. *Introduction to computational chemistry*. Wiley, 1999.
6. D. M. Bishop. *Adv. Quant. Chem.*, **25**, 1, 1994.
7. C. Hättig and P. Jørgensen. *J. Chem. Phys.*, **109**, 2762, 1998.
8. D. M. Bishop, F. L. Gu, and S. M. Cybulski. *J. Chem. Phys.*, **109**, 8407, 1998.
9. H. Reis, M. G. Papadopoulos, and A. Avramopoulos. *J. Phys. Chem. A*, **107**, 3907, 2003.
10. E. Runge and E. K. U. Gross. *Phys. Rev. Lett.*, **52**, 997, 1984.
11. E. K. U. Gross and W. Kohn. *Adv. Quant. Chem.*, **21**, 255, 1990.
12. R. van Leeuwen. *Int. J. Mod. Phys. B*, **15**, 1969, 2001.
13. M. E. Casida. In D. P. Chong, editor, *Recent Advances in Density-Functional Methods*, page 155. World Scientific, Singapore, 1995.
14. M. E. Casida. In J. M. Seminario, editor, *Recent Developments and Applications of Modern Density Functional Theory*. Elsevier, Amsterdam, 1996.
15. S. J. A. van Gisbergen, V. P. Osinga, O. V. Gritsenko, R. van Leeuwen, J. G. Snijders, and E. J. Baerends. *J. Chem. Phys.*, **105**, 3142, 1996.
16. M. E. Casida, C. Jamorski, M. C. Casida, and D. R. Salahub. *J. Chem. Phys.*, **108**, 4439, 1998.
17. D. J. Tozer and N. C. Handy. *J. Chem. Phys.*, **109**, 10180, 1998.
18. P. R. T. Schipper, O. V. Gritsenko, S. J. A. van Gisbergen, and E. J. Baerends. *J. Chem. Phys.*, **112**, 1344, 2000.
19. M. E. Casida and D. R. Salahub. *J. Chem. Phys.*, **113**, 8918, 2000.
20. M. Grüning, O. V. Gritsenko, S. J. A. van Gisbergen, and E. J. Baerends. *J. Chem. Phys.*, **114**, 652, 2001.
21. M. Grüning, O. V. Gritsenko, S. J. A. van Gisbergen, and E. J. Baerends. *J. Chem. Phys.*, **116**, 9591, 2002.
22. J.-I. Iwata, K. Yabana, and G. F. Bertsch. *J. Chem. Phys.*, **115**, 8773–8783, 2001.
23. H. H. Heinze, F. D. Sala, and A. Görling. *J. Chem. Phys.*, **116**, 9624, 2002.
24. P. Salek, O. Vahtras, T. Helgaker, and H. Ågren. *J. Chem. Phys.*, **117**, 9630, 2002.
25. L. Bernasconi, M. Sprik, and J. Hutter. *J. Chem. Phys.*, **119**, 12417, 2003.
26. D. Sebastiani and U. Rothlisberger. *J. Phys. Chem. B*, **108**, 2807, 2004.
27. M. Sulpizi, P. Carloni, J. Hutter, and U. Rothlisberger. *Phys. Chem. Chem. Phys.*, **5**, 4798, 2003.
28. U. F. Röhrig, I. Frank, J. Hutter, A. Laio, J. VandeVondele, and U. Roth-

lisberger. *ChemPhysChem*, **4**, 1177, 2003.

29. L. Jensen, P. Th. van Duijnen, and J. G. Snijders. *J. Chem. Phys.*, **118**, 514, 2003.

30. L. Jensen, P. Th van Duijnen, and J. G. Snijders. *J. Chem. Phys.*, **119**, 3800, 2003.

31. L. Jensen, P. Th. van Duijnen, and J. G. Snijders. *J. Chem. Phys.*, **119**, 12998, 2003.

32. L. Jensen, M. Swart, and P. Th. van Duijnen. *J. Chem. Phys.*, **122**, 034103, 2005.

33. L. Jensen and P. Th. van Duijnen. *Int. J. Quant. Chem.*, **102**, 612, 2005.

34. R. Cammi, B. Mennucci, and J. Tomasi. *J. Phys. Chem. A*, **104**, 5631, 2000.

35. C. Adamo and V. Barone. *Chem. Phys. Lett.*, **330**, 152, 2000.

36. M. Cossi and V. Barone. *J. Chem. Phys.*, **115**, 4708, 2001.

37. J. Neugebauer, M. J. Louwerse, E. J. Baerends, and T. A. Wesolowski. *J. Chem. Phys.*, **122**, 094115, 2005.

38. S. Fantacci, F. De Angelis, and A. Selloni. *J. Am. Chem. Soc.*, **125**, 4381, 2003.

39. W.-G. Han, T. Lui, F. Himo, A. Toutchkine, D. Bashford, K. M. Hahn, and L. Noodleman. *ChemPhysChem*, **4**, 1084, 2003.

40. T. Liu, W.-G. Han, F. Himo, G. M. Ullman, D. Bashford, A. Toutchkine, K. M. Hahn, and L. Noodleman. *J. Phys. Chem. A*, **108**, 3545, 2004.

41. J. Tomasi and M. Persico. *Chem. Rev.*, **94**, 2027–2094, 1994.

42. C. J. Cramer and D. G. Truhlar. *Chem. Rev.*, **99**, 2161–2200, 1999.

43. M. Orozco and F. J. Luque. *Chem. Rev.*, **100**, 4187–4225, 2000.

44. A. Pullman and B. Pullman. *Quart. Rev. Biophys.*, **7**, 505, 1975.

45. T. Wesolowski and A. Warshel. *J. Phys. Chem.*, **98**, 5183, 1994.

46. R. Car and M. Parrinello. *Phys. Rev. Lett.*, **55**, 2471, 1985.

47. A. Warshel and M. Levitt. *J. Mol. Bio.*, **103**, 227, 1976.

48. B. T. Thole and P. Th. van Duijnen. *Theor. Chim. Acta*, **55**, 307, 1980.

49. U. C. Singh and P. A. Kollman. *J. Comp. Chem.*, **7**, 718, 1986.

50. P. A. Bash, M. J. Field, and M. Karplus. *J. Am. Chem. Soc.*, **109**, 8092, 1987.

51. M. J. Field, P. A. Bash, and M. Karplus. *J. Comp. Chem.*, **11**, 700, 1990.

52. V. Luzhkov and A. Warshel. *J. Comp. Chem.*, **13**, 199, 1992.

53. R. V. Stanton, D. S. Hartsough, and K. M. Merz. *J. Phys. Chem.*, **97**, 11868, 1993.

54. A. H. de Vries, P. Th van Duijnen, A. H. Juffer, J. A. C. Rullmann, J. P. Dijkman, H. Merenga, and B. T. Thole. *J. Comp. Chem.*, **16**, 37, 1995.

55. I. Tuñón, M. T. C. Martins-Costa, C. Millot, M. F. Ruiz-López, and J. L. Rivail. *J. Comp. Chem.*, **17**, 19, 1996.

56. J. Gao. Methods and applications of combined quantum mechanical and molecular mechanical potentials. In K. B. Lipkowitz and D. B. Boyd, editors, *Reviews in Computational Chemistry*, volume 7, pages 119–185. VCH, New York, 1995.

57. J. Gao. *Acc. Chem. Res.*, **29**, 298–305, 1996.

58. J. Gao and M. A. Thompson, editors. *Combined Quantum Mechanical and*

Molecular Mechanical Methods, volume 712. ACS Symposium Series, 1998.

59. V. Barone, M. Cossi, and T. Tomasi. *J. Chem. Phys.*, **107**, 3210, 1997.

60. E. R. Batista, S. S. Xantheas, and H. Jónsson. *J. Chem. Phys.*, **111**, 6011, 1999.

61. L. Jensen, M. Swart, P. Th. van Duijnen, and J. G. Snijders. *J. Chem. Phys.*, **117**, 3316, 2002.

62. L. Delle Site, A. Alavi, and R. M. Lynden-Bell. *Mol. Phys.*, **96**, 1683, 1999.

63. K. V. Mikkelsen, Y. Luo, H. Ågren, and P. Jørgensen. *J. Chem. Phys.*, **102**, 9362–9367, 1995.

64. O. Christiansen, T. M. Nymand, and K. V. Mikkelsen. *J. Chem. Phys.*, **113**, 8101–8112, 2000.

65. J. Kongsted, A. Osted, T. B. Pedersen, K. V. Mikkelsen, and O. Christianse. *J. Phys. Chem. A*, **108**, 8624, 2004.

66. O. Engkvist, P.-O. Åstrand, and G. Karlström. *Chem. Rev.*, **100**, 4087, 2000.

67. Y. Tu and A. Laaksone. *J. Chem. Phys.*, **111**, 7519, 1999.

68. M. Freindorf and J. Gao. *J. Comp. Chem.*, **17**, 386, 1996.

69. J. A. C. Rullmann and P. Th. van Duijnen. *Mol. Phys.*, **63**, 451, 1988.

70. P. Ahlström, A. Wallqvist, S. Engström, and B. Jönsson. *Mol. Phys.*, **68**, 563, 1989.

71. S. Kuwajima and A. Warshel. *J. Phys. Chem.*, **94**, 460, 1990.

72. L. X. Dang. *J. Chem. Phys.*, **97**, 2183, 1992.

73. J.-C. Soetens and C. Milot. *Chem. Phys. Lett.*, **235**, 22, 1995.

74. L. X. Dang and T.-M. Chang. *J. Chem. Phys.*, **106**, 8149, 1997.

75. C. J. Burnham, J. Li, S. S. Xantheas, and M. Leslie. *J. Chem. Phys.*, **110**, 4566, 1999.

76. T. A. Halgren and W. Damm. *Curr. Opin. Struc. Biol.*, **11**, 236, 2001.

77. M. A. Thompson and G. K. Schenter. *J. Phys. Chem.*, **99**, 6374, 1995.

78. G. Jansen, F. Colonna, and J. G. Ángyán. *Int. J. Quant. Chem.*, **58**, 251, 1996.

79. P. N. Day, J. H. Jensen, M. S. Gordon, S. P. Webb, W. J. Stevens, M. Krauss, D. Garmer, H. Basch, and D. Cohen. *J. Chem. Phys.*, **105**, 1968, 1996.

80. J. Gao and M. Freindorf. *J. Phys. Chem. A*, **101**, 3182, 1997.

81. J. Gao. *J. Comp. Chem.*, **18**, 1061, 1997.

82. T. D. Poulsen, J. Kongsted, A. Osted, P. R. Ogilby, and K. V. Mikkelsen. *J. Chem. Phys.*, **115**, 2393, 2001.

83. M. Dupuis, M. Aida, Y. Kawashima, and K. Hirao. *J. Chem. Phys.*, **117**, 1242, 2002.

84. J. Kongsted, A. Osted, K. V. Mikkelsen, and O. Christiansen. *Chem. Phys. Lett.*, **364**, 379, 2002.

85. J. Kongsted, A. Osted, K. V. Mikkelsen, and O. Christiansen. *J. Chem. Phys.*, **118**, 1620, 2003.

86. J. Kongsted, A. Osted, K. V. Mikkelsen, and O. Christiansen. *J. Chem. Phys.*, **119**, 10519, 2003.

87. J. Kongsted, A. Osted, K. V. Mikkelsen, and O. Christiansen. *J. Chem. Phys.*, **120**, 3787, 2004.

88. J. Applequist, J. R. Carl, and K. F. Fung. *J. Am. Chem. Soc.*, **94**, 2952, 1972.
89. L. Silberstein. *Phil. Mag.*, **33**, 521, 1917.
90. B. T. Thole. *Chem. Phys.*, **59**, 341, 1981.
91. P. Th. van Duijnen and M. Swart. *J. Phys. Chem. A*, **102**, 2399, 1998.
92. L. Jensen, P.-O. Åstrand, K. O. Sylvester-Hvid, and K. V. Mikkelsen. *J. Phys. Chem. A*, **104**, 1563, 2000.
93. L. Jensen, P.-O. Åstrand, A. Osted, J. Kongsted, and K. V. Mikkelsen. *J. Chem. Phys.*, **116**, 4001, 2002.
94. H. Reis, M. G. Papadopoulos, C. Hättig, J. G. Ángyán, and R. W. Munn. *J. Chem. Phys.*, **112**, 6161–6172, 2000.
95. F. C. Grozema, R. W. J. Zijlstra, and P. Th. van Duijnen. *Chem. Phys.*, **256**, 217, 1999.
96. M. Swart, P. Th. van Duijnen, and J. G. Snijders. *J. Comp. Chem.*, **22**, 79, 2001.
97. R. Wortmann and D. M. Bishop. *J. Chem. Phys.*, **108**, 1001, 1998.
98. R. Cammi, B. Mennucci, and J. Tomasi. *J. Phys. Chem. A*, **104**, 4690, 2000.
99. R. W. Munn, Y. Luo, P. Macák, and H. Ågren. *J. Chem. Phys.*, **114**, 3105–3108, 2001.
100. Y. Luo, P. Norman, P. Macak, and H. Ågren. *Phys. Rev. B*, **61**, 3060, 2000.
101. H. Reis, M. G. Papadopoulos, and D. N. Theodorou. *J. Chem. Phys.*, **114**, 876, 2001.
102. L. Jensen, P.-O. Åstrand, and K. V Mikkelsen. *J. Phys. Chem. B*, **108**, 8226, 2004.
103. P. Th. van Duijnen, A. H. de Vries, M. Swart, and F. Grozema. *J. Chem. Phys.*, **117**, 8442, 2002.
104. C. J. F. Böttcher. *Theory of Electric Polarization*, volume 1. Elsevier, Amsterdam, 2nd edition, 1973.
105. A. Willetts, J. E. Rice, D. M. Burland, and D. Shelton. *J. Chem. Phys.*, **97**, 7590, 1992.
106. S. J. A. van Gisbergen, J. G. Snijders, and E. J. Baerends. *J. Chem. Phys.*, **109**, 10644, 1998.
107. E. K. U. Gross, J. F. Dobson, and M. Petersilka. In Nalewajski, editor, *Density Functional Theory, Topics in current chemistry*, volume 181. Springer-Verlag Berlin Heidelberg, 1996.
108. S. P. Karna and M. Dupuis. *J. Comp. Chem.*, **12**, 487, 1991.
109. G. te Velde, F. M. Bickelhaupt, E. J. Baerends, C. Fonseca Guerra, S. J. A. van Gisbergen, J. G. Snijders, and T. Ziegler. *J. Comp. Chem.*, **22**, 931, 2001.
110. ADF. http://www.scm.com, 2005.
111. M. Eichinger, P. Tavan, J. Hutter, and M. Parrinello. *J. Chem. Phys.*, **110**, 10452, 1999.
112. H. Takahashi, T. Hori, H. Hashimoto, and T. Nitta. *J. Comp. Chem.*, **22**, 1252, 2001.
113. S. J. A. van Gisbergen, J. G. Snijders, and E. J. Baerends. *Comput. Phys. Commun.*, **118**, 119, 1999.

114. S. J. A. van Gisbergen, J. G. Snijders, and E. J. Baerends. *J. Chem. Phys.*, **103**, 9347, 1995.
115. R. W. Boyd. *Nonlinear Optics*. Academic Press, San Diego, 1992.
116. B. J. Orr and J. F. Ward. *Mol. Phys.*, **20**, 513, 1971.
117. D. P. Shelton and J. E. Rice. *Chem. Rev.*, **94**, 3–29, 1994.
118. H. A. Lorentz. *The theory of electrons*. B. G. Teubner, Leipzig, 1st edition, 1909.
119. P. N. Butcher and D. Cotter. *The elements of nonlinear optics*. Cambridge University Press, Cambridge, 1st edition, 1990.
120. L. Onsager. *J. Am. Chem. Soc.*, **58**, 1486, 1936.
121. D. M. Bishop. *Rev. Mod. Phys.*, **62**, 343, 1990.
122. J. D. Jackson. *Classical Electrodynamics*. John Wiley and Sons, New-York, 2nd edition, 1975.
123. S. L. Shostak, W. L. Ebenstein, and J. S. Muenter. *J. Chem. Phys.*, **94**, 5875, 1991.
124. A. J. Russel and M. A. Spackman. *Mol. Phys.*, **84**, 1239, 1995.
125. P. Kaatz, E. A. Donley, and D. P. Shelton. *J. Chem. Phys.*, **108**, 849, 1998.
126. F. Kajzar and J. Messier. *Phys. Rev. A*, **32**, 2352, 1985.
127. I. Thormahlen, J. Straub, and U. Grigul. *J. Phys. Chem. Ref. Data*, **14**, 933, 1985.
128. B. F. Levine and C. G. Bethea. *J. Chem. Phys.*, **65**, 2429, 1976.
129. U. Gubler and C. Bosshard. *Phys. Rev. B*, **61**, 10702, 2000.
130. C. Bosshard, P. Gubler, U. snd Kaatz, W. Mazerant, and U. Meier. *Phys. Rev. B*, **61**, 10688, 2000.
131. P. Kaatz and D. P. Shelton. *J. Chem. Phys.*, **105**, 3918, 1996.
132. I. Shoji, T. Kondo, and R. Ito. *Optical and Quantum Electronics*, **34**, 797, 2002.
133. M. Stähelin, C. R. Moylan, D. M Burland, A. Willets, J. E. Rice, D. P. Shelton, and E. A. Donley. *J. Chem. Phys.*, **98**, 5595, 1993.
134. A. Willets and J. E. Rice. *J. Chem. Phys.*, **99**, 426, 1993.
135. P. Norman, Y. Luo, and H. Ågren. *J. Chem. Phys.*, **197**, 9535, 1997.
136. Y. Luo, P. Norman, H. Ågren, K. O. Sylvester-Hvid, and K. V. Mikkelsen. *Phys. Rev. E*, **57**, 4778, 1998.
137. G. R. Alms, A. K. Burnham, and W. H. Flygare. *J. Chem. Phys.*, **63**, 3321, 1975.
138. G. R. Meredith, B. Buchalter, and C. Hanzlik. *J. Chem. Phys.*, **78**, 1543, 1983.
139. M. van Faassen, L. Jensen, J. A. Berger, and P. de Boeij. *Chem. Phys. Lett.*, **395**, 274, 2004.
140. S. J. A. van Gisbergen, J. G. Snijders, and E. J. Baerends. *Phys. Rev. Lett.*, **78**, 3097–3100, 1997.
141. D. Jonsson, P. Norman, K. Ruud, and H. Ågren. *J. Chem. Phys.*, **109**, 572–577, 1998.
142. L. Jensen, P. Th. van Duijnen, J. G. Snijders, and D. P. Chong. *Chem. Phys. Lett.*, **359**, 524, 2002.
143. K. Ruud, D. Jonsson, and P. R. Taylor. *J. Chem. Phys.*, **114**, 4331–4332,

2001.

144. J. Kongsted, A. Osted, L. Jensen, P.-O. Åstrand, and K. V. Mikkelsen. *J. Phys. Chem. B*, **105**, 10243, 2001.

145. T. B. Pedersen, A. M. J. Sánchez de Merás, and H. Koch. *J. Chem. Phys.*, **120**, 8887, 2004.

146. R. Antoine, Ph. Dugourd, D. Rayane, E. Benichou, F. Chandezon M. Broyer, and C. Guet. *J. Chem. Phys.*, **110**, 9771–9772, 1999.

147. A. Ballard, K. Bonin, and J. Louderback. *J. Chem. Phys.*, **113**, 5732, 2000.

148. W. Krätschmer, L. D. Lamb, K. Fostiropoulos, and D. R. Huffman. *Nature*, **347**, 354–358, 1990.

149. J. S. Meth, H. Vanherzeele, and Wangm Y. *Chem. Phys. Lett.*, **197**, 26, 1992.

150. Z. H. Kafafi, J. R. Lindle, R. G. S. Pong, F. J. Bartoli, L. J. Lingg, and J. Milliken. *Chem. Phys. Lett.*, **188**, 492–496, 1992.

151. A. Ritcher and J. Sturm. *Appl. Phys. A*, **61**, 163–170, 1995.

152. P. C. Eklund, A. M. Rao, Y. Wang, K. A. Zhou, P.and Wang, J. M. Holden, M.S. Dresselhaus, and G. Dresselhaus. *Thin Solid Films*, **257**, 211–232, 1995.

153. A. F. Hebard, R. C. Haddon, R. M. Fleming, and A. R. Kortan. *Appl. Phys. Lett.*, **59**, 2109–2111, 1991.

154. P. L. Hanse, P. J. Fallon, and W. Krätschmer. *Chem. Phys. Lett.*, **181**, 367, 1991.

155. P. D. Maker and R. W. Terhune. *Phys. Rev.*, **137**, A801, 1965.

156. E. A. Perpéte, B. Champagne, and B. Kirtman. *Phys. Rev. B*, **61**, 13137, 2000.

CHAPTER 9

EXTRAORDINARY FIRST HYPERPOLARIZABILITIES FROM LOOSELY BOUND ELECTRON IN DIPOLE-BOUND ANIONS: $(HF)_N^-$ (N = 2, 3, 4)

Di Wu , Zhi-Ru Li*, Ying Li, Chia-Chung Sun

State Key Laboratory of Theoretical and Computational Chemistry, Institute of Theoretical Chemistry, Jilin University, Changchun 130023, P. R . China

The extraordinary first hyperpolarizabilities ($\beta_0 = 1.5 \times 10^7$, 6.7×10^6 and 4.4×10^6 a.u. for n = 2, 3, 4, respectively), polarizabilities (α) and large dipole moments (μ) for $(HF)_n^-$ (n = 2, 3, 4) anions were got in this work by ab initio calculations at the MP2 level. The β_0 contributions from the anions are divided into two parts. One part is from the neutral core $(HF)_n$ (n = 2, 3, 4), which is very small as about -10^1 a.u. ($\beta_0 = -13.87$, -13.63, and -1.88 a.u., for n = 2, 3, 4, respectively). The other part is from the loosely bound excess electron (itself and its interaction with the neutral core) which contributes $10^7 \sim 10^6$ a.u β_0 value for each dipole-bound anion.

*

Corresponding author, Fax: (+86)431-8945942

E-mail address: lzr@mail.jlu.edu.cn

1. Introduction

In a dipole-bound anion, there are manifold interactions: intra-molecular covalent interaction, intermolecular hydrogen bonded interaction, and both the electrostatic and dispersion interaction between the neutral cluster core and the excess electron. According to extensive theory as well as experimental work [1], it has been established that an excess electrons can be bound by the dipole field of a neutral molecule or cluster system to form 'dipole-bound anion' if the dipole moment of the system is larger than approximately 2.5 D. Small neutral clusters $(HF)_n$ have been studied [2-6]. The $(HF)_n$ clusters with n>2 are not polar due to their most likely cyclic structures [7]. When an excess electron attaching to the $(HF)_n$, a 'zig-zag' structure is preferred to over cyclic one [8].

Recently, the significant works [1, 8-13] from both experimental and theoretical studies on the dipole-bound anions $(HF)_n^-$ (n = 2-4) have dealt with the electron detachment energies and the vibrational frequencies. The dipole-bound anions have loosely bound electrons (excess electrons) with spatially diffusive electron clouds [9]. The average separations between the neutral cores and the excess electrons are large (about 2 Å [14], 5-100 Å [15, 16]). Thus the dipole-bound anions should have extraordinary physical and chemical properties.

The design of novel materials with large nonlinear optical (NLO) properties (i.e., hyperpolarizabilities) is currently of great interest, mainly due to their potential applications [17-19]. As a result the theoretical characterization of species with large NLO properties has been an active area of research [20-25]. In order to understand electronic and optical properties of the dipole-bound anions and to explore the role of the loosely bound electron in the NLO roperties, we performed ab initio calculations of the β_0, α, μ values for the dipole-bound anions.

Our MP2 calculations on $(HF)_n^-$ (n = 2, 3, 4) anions gave extraordinary first hyperpolarizabilities $(\beta_0 \approx 10^7 \sim 10^6$ a.u.), and polarizabilities $(\alpha \approx 10^4$ a.u.), and also large dipole moments $(\mu \approx 10^1$ a.u.).

The loosely bound electron in the anions is responsible for these results. These β_0 values are even larger than those of the organic long-chain conjugated molecules which are considered to have large hyperpolarizabilities β_0 $(1.8 \times 10^5$ a.u. and so on) [26].

2. Calculational methods and numerical results

The optimized structures of the $(HF)_n^-$ at the MP2 level are shown in Fig.1 and Table 1. Our calculations employ basis sets formed by combining a standard valence-type (or aug-cc-pVDZ [27]) set designed to describe the neutral core (molecular cluster) and an extra diffusive set (X) [2, 12, 13] designed to describe the charge distribution of the extra electron (loosely bound electron).

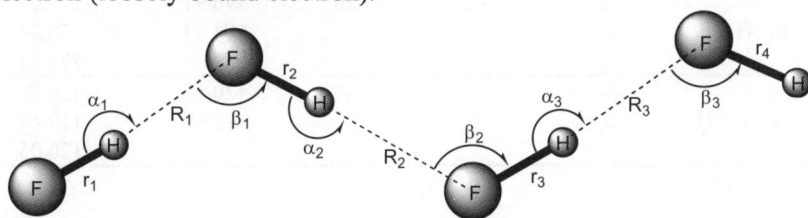

Fig. 1. Geometrical parameters of the HF molecule and dipole-bound anions $(HF)_n^-$ (n = 2, 3, 4).

In the presence of an applied electric field, the energy of a system is a function of the field strength. Hyperpolarizabilities are defined as the coefficients in the Taylor series expansion of the energy in the external electric field [28]. When the external electric field is weak and homogeneous, this expansion becomes

$$E = E^0 - \mu_i F_i - \frac{1}{2}\alpha_{ij}F_iF_j - \frac{1}{6}\beta_{ijk}F_iF_jF_k - \cdots \quad (1)$$

where E^0 is the energy of the unperturbed molecules, F_i is the field at the origin, μ_i, α_{ij}, and β_{ijk} are the components of dipole moment, polarizability and the first hyperpolarizability, respectively. (the repeated Greek subscript implies summation over the Cartesian coordinates x, y, and z)

Table 1. Geometrical parameters (distances in Å, angles in degrees) of HF molecule and dipole-bound anions $(HF)_n^-$ (n = 2,3,4).

		HF	$(HF)_2^-$	$(HF)_3^-$	$(HF)_4^-$
F—H	r_1	0.917	0.933	0.9391	0.9401
	r_2		0.932	0.9469	0.9532
	r_3			0.9419	0.9569
	r_4				0.9516
F...H	R_1		1.788	1.7008	1.6835
	R_2			1.6486	1.5761
	R_3				1.5812
F...H—F	α_1		179.20	176.91	177.04
	α_2			174.93	178.00
	α_3				172.65
H...F—H	β_1		121.40	120.85	120.78
	β_2			121.69	117.44
	β_3				120.05

We computed the total static dipole moment μ, the mean polarizability α, the the mean first hyperpolarizability β_0. They are defined as

$$\mu = (\mu_x^2 + \mu_y^2 + \mu_z^2)^{\frac{1}{2}},$$

$$\alpha = \frac{1}{3}(\alpha_{xx} + \alpha_{yy} + \alpha_{zz}), \tag{2}$$

$$\beta_0 = \frac{1}{5}\sum_i (\beta_{zii} + \beta_{izi} + \beta_{iiz}), i = x, y, z.$$

Firstly, we discuss the dipole moment. Table 2 shows that the large dipole moments, μ (where z is the axis of dipole moment) of the dipole-bound anions $(HF)_n^-$ (n = 2, 3, 4) are 9.9162 (for n = 2), 6.9204 (for n = 3) and 6.2970 a.u. (for n = 4), while the μ of their neutral cores $(HF)_n$ (n = 2, 3, 4) with the same structures of the anions are -1.4552 (for n = 2), -2.3977 (for n = 3) and -3.4854 a.u. (for n = 4). Because for each anion

Table 2. Dipole moments, polarizabilities and first hyperpolarizabilities of the dipole-bound anions $(HF)_n^-$ and the cores $(HF)_n$ (n = 2,3,4) (in a.u.)*

	Basis sets		μ	α_{xx}	α_{zz}	α	β_{xxz}	β_{zzz}	β_0
$(HF)_2^-$	Aug-cc-pVDZ+X	(80)	9.9162	3.63×10^4	3.63×10^4	3.63×10^4	5.22×10^6	1.56×10^7	1.56×10^7
	6-311+G**+X	(72)	9.1195	3.65×10^4	3.65×10^4	3.65×10^4	4.66×10^6	1.39×10^7	1.39×10^7
	6-311+G*+X	(66)	8.8192	3.63×10^4	3.63×10^4	3.63×10^4	4.53×10^6	1.36×10^7	1.36×10^7
$(HF)_2$	Aug-cc-pVDZ+X	(80)	-1.4553	9.69	12.90	10.76	-2.55	-7.81	-13.87
$(HF)_3^-$	Aug-cc-pVDZ+X	(112)	6.9204	3.65×10^4	3.65×10^4	3.65×10^4	2.35×10^6	6.51×10^6	6.73×10^6
	6-311+G**+X	(100)	7.3373	3.63×10^4	3.63×10^4	3.63×10^4	2.87×10^6	8.44×10^6	8.51×10^6
	6-31+G*+X	(79)	6.9843	3.63×10^4	3.63×10^4	3.63×10^4	2.65×10^6	7.83×10^6	7.88×10^6
$(HF)_3$	Aug-cc-pVDZ+X	(112)	-2.3978	12.60	18.92	14.71	-1.66	-19.39	-13.62
$(HF)_4^-$	Aug-cc-pVDZ+X	(144)	6.2970	3.65×10^4	3.65×10^4	3.65×10^4	1.47×10^6	4.47×10^6	4.44×10^6
	6-311+G**+X	(128)	6.5676	3.63×10^4	3.63×10^4	3.63×10^4	1.69×10^6	5.14×10^6	5.11×10^6
	6-31+G*+X	(100)	6.2630	3.63×10^4	3.63×10^4	3.63×10^4	1.52×10^6	4.68×10^6	4.63×10^6
$(HF)_4$	Aug-cc-pVDZ+X	(144)	-3.4854	18.30	28.53	21.71	-1.12	-0.90	-1.88

*Z axis is principal axis of the molecule. 1au = 2.5418 D for μ, 1au = 1.4817 × $10^{-25}cm^3$ for α, 1au = 8.6392 × 10^{-33}esu for β_0.

the excess electron localize at the positive dipole side of the neutral core, the dipole moment directions of the anions opposite to those of their neural cores. The magnitude of the dipole moments for the anions decreases with the molecular number in the anions increasing, while the absolute value of the dipole moments for the neural cores increases with the number of molecules increasing. Then the magnitude of the dipole moments of the anions decreases with the increasing of the absolute value of those of the neural cores.

Secondly, we consider the average polarizabilities (α) of the anions and their neutral cores. For the neutral cores, the α value (10.76 a.u. for n = 2, 14.71 a.u. for n = 3, and 21.71 a.u. for n = 4) increases in rough proportion as 1:1.5:2. For the anions $(HF)_n^-$, the value of α is 3.63×10^4 a.u. for n = 2, 3.65×10^4 a.u. for n = 3 and 3.65×10^4 a.u. for n = 4. It is interesting that the α of the $(HF)_n^-$ (n = 2, 3, 4) are about constant as 3.6×10^4 a.u. which is far larger (about 1000 times) than the α of neutral core $(HF)_n$. The contribution mainly comes from the effect of the diffusive electron cloud of the loosely bound electron but inconsiderable contribution comes from the neutral core.

Thirdly, we focus on the nonlinear optical properties (static first hyperpolarizability, β_0) of the anions $(HF)_n^-$ (n = 2, 3, 4) and the neutral cores $(HF)_n$ (n = 2, 3, 4). For the neutral cores, the β_0 are −13.87 a.u. for $(HF)_2$, -13.63 a.u. for $(HF)_3$ and −1.88 a.u. for $(HF)_4$, which are about -10^1 a.u.. For the dipole-bound anions $(HF)_n^-$ (n=2, 3, 4), the first hyperpolarizabilities β_0 have extraordinary values: $\beta_0 = 1.56 \times 10^7$ a.u. for $(HF)_2^-$, $\beta_0 = 6.73 \times 10^6$ a.u. for $(HF)_3^-$ and $\beta_0 = 4.44 \times 10^6$ a.u. for $(HF)_4^-$. The ratio of β_0 from the anion to the neutral core is $(1.56 \times 10^7 / 1.39 \times 10^1 \approx 10^6$ for n = 2, $6.73 \times 10^6 / 1.36 \times 10^1 \approx 10^5$ for n = 3, $4.44 \times 10^6 / 1.88 \approx 10^6$ for n = 4) about $10^5 \sim 10^6$.

From Fig. 2, the β_0 value of $(HF)_2^-$ are far greater than the very large value of first hyperpolarizability of the long chain conjugated molecule A from reference [26]. For the four atom system $(HF)_2^-$, because it has a loosely bound electron, its β_0 is 1.56×10^7 a.u., but for the molecule A (75 atoms system), its β_0 is 8×10^5 a.u., only about 1% of that of the anion $(HF)_2^-$.

The contribution of the loosely bound electron (Q_{e-}) to nonlinear optical properties β_0 and related properties (μ, α), are shown in Table 3. The results have been obtained by eq (3)

$$Q_{e-} = Q - Q0 \qquad\qquad (3)$$

where Q, $(Q = \mu, \alpha, \beta_0)$ is a physical property value of a dipole-bound anion, Q_{e^-} is the contribution coming from the loosely bound electron and Q_0 is the contribution of the neutral core. For α and β_0, the values of Q_0 are very small (less than 0.1% of α, and 0.0002% of β_0).

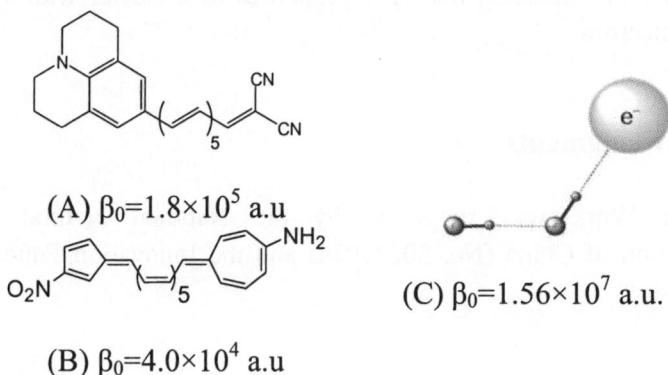

(A) $\beta_0 = 1.8 \times 10^5$ a.u

(B) $\beta_0 = 4.0 \times 10^4$ a.u

(C) $\beta_0 = 1.56 \times 10^7$ a.u.

Fig. 2. The comparison of β_0 between $(HF)_2^-$ and conjugated molecules [26].

Table 3. The contributions (in a.u.) of the loosely bound electron in $(HF)_n^-$ (n = 2, 3, 4)

Q_{e^-}	n = 2	3	4
μ_{e^-}	11.4219	9.3182	9.7824
α_{e^-}	36310.75	36489.50	3648.9
β_{e^-}	1.5×10^7	6.7×10^6	4.4×10^6

It is clear that for the anions $(HF)_n^-$ the contributions to both polarizability and first hyperpolarizability, almost come from the loosely bound electrons and the contributions coming from the neutral cores could be neglected. In enhancing the NLO properties, a loosely bound electron is a considerable factor.

3. Conclusion

In summary, the dipole-bound anions $(HF)_n^-$ ($n = 2$-4) have extraordinary first hyperpolarizabilities and polarizabilities, and large dipole moments, which were found by ab initio theoretical calculations at the MP2 level. The loosely bound electron in the anions is responsible for these results. And it is clear that the loosely bound electron is very important in enhancing the NLO properties of a cluster with a loosely bound electron.

Acknowledgments

This Work was supported by the National Natural Science Foundation of China (No. 20273024) and the Innovation Fund of Jilin University.

References

1. J. Simons, K. D. Jordan, Chem. Rev. 87 (1987) 535.
2. R. Ramaekers, D. M. A. Simith, J. Smets and L. Adamowicz, J. Chem. Phys. 107 (1997) 9475.
3. F. Huisken, M. Kaloudis, A. Kulcke. C. Laush and J. H. Lisy, J. Chem. Phys. 103 (1995) 5366.
4. S. Y. Liu, D. W. Michael. C. E. Dykstra and J. M. Lisy. J. Chem. Phys. 84 (1986) 5032.
5. K. D. Kolenbrandr, C. E. Dy,kstra and J. M. Lisy. J. Chem. Phys. 88 (1988) 5995.
6. C. E. Dykstra, J. Chem. Phys. 94 (1990) 180.
7. T. R. Dyke, B. J. Howard and W. Klemperer, J. Chem. Phys. 56 (1972) 2442.
8. R. Ramaekers, D. M. A. Simith, Y. Elkadi, L. Adamowicz, Chem. Phys. Lett. 277 (1997) 269.
9. J. H. Hendricks, H. L. de Clercq, S. A. Lyapustina and K. H. Bowen, Jr. J. Chem Phys. 107 (1977) 2962.
10. F. Fermi and E. Teller. Phys. Rev. 72 (1947) 399.
11. K. D. Jordan and J.J. Wendoloski. Chem. Phys. 21 (1977) 145.
12. M. Gutowski and P. Skurski, J. Chem. Phys. 107 (1997) 2968.
13. M. Gutowski and P. Skurski, J. Phys. Chem. B. 101 (1997) 9143.
14. T. Tsurusawa and S. Iwata, J. Phys Chem. 103 (1999) 6134.
15. K. D. Jordan and W. Luken, J.Chem. Phys. 64 (1976) 2760.
16. R. N. Barnett, U. Landman, C.L. Cleveland, and J. Jortner, J. Chem. Phys. 88 (1988) 4429.
17. J.Eyss and I. Ledoux, Chem. Rev. 94 (1994) 77.
18. D. M. Burland, Chem. Rev. 94 (1994) 7.
19. D. P. Shelton and J.E. Rice, Chem. Rev. 94 (1994) 3.
20. B.L. Hannond and J.E. Rice, J. Chem. Phys. 97 (1992) 1138.
21. S. R. Marder, L.-T. Cheng, B. G. Tiemann, A. C. Friedli, M. Blan-chard-Desce, J. W. Perry, J. Skindhøj, Science 263 (1994) 511.
22. M. Blanchard-Desce, C. Runser, A. Fort, M. Barzoukas, J,-M Lehn, V. Bloy, V. Alain, Chem. Phys. 199 (1995) 253.
23. M. Blanchard-Desce, V. Alain, P. V. Bedworth, S. R. Marder, A. Fort, C. Runser, M. Barzoukas, S. Lebus, R. Wortmann, Chem. Eur. J. 3 (1997) 1091.
24. K. Bhanuprakash, L.J. Rao, Chem. Phys. Lett. 314 (1999) 282
25. V. Moliner, P. Escribano and E. Peris, New. J. Chem. (1998) 387.
26. V. Alain, L. Thouin, M. B. Desce, U. Gubler, C. Bosshard, P. Günter, J. Muller, A. Fort and M. Barzoukas, Adv. Mater. 11 (1999) 1210.
27. R. A.Kendall, T.H. Duning,Jr., and R. J. Harrison, J. Chem. Phys. 96 (1992) 6796.
28. A.D. Buckingham, Adv. Chem. Phys. 12 (1967) 107.

CHAPTER 10

THIRD-ORDER NONLINEAR OPTICAL PROPERTIES OF OPEN-SHELL AND/OR CHARGED MOLECULAR SYSTEMS

Masayoshi Nakano

Division of Chemical Engineering, Department of Materials Engineering Science, Graduate School of Engineering Science, Osaka University, Toyonaka, Osaka 560-8531, Japan
E-mail: mnaka@cheng.es.osaka-u.ac.jp

This paper explains theoretical and computational investigations of microscopic third-order nonlinear optical (NLO) properties, i.e., the second hyperpolarizabilities (γ), of open-shell and/or charged molecular systems. Contrary to the conventional NLO molecules, which are usually closed-shell neutral π-conjugated systems, the open-shell and/or charged systems turn out to exhibit a unique feature, i.e., sensitive structural dependence of magnitude and sign of γ. From the analysis of virtual excitation processes using perturbation theory, we derive a structure-property relation of γ based on the contribution of symmetric resonance structure with invertible polarization (SRIP). On the basis of this relationship, the ionic radical states of condensed-ring π-conjugated systems tend to exhibit negative sign of γ, which is rare in general. On the other hand, for neutral open-shell systems, the γ values tend to enhance in the intermediate spin multiplicity state. Also, the singlet diradical systems turn out to show a remarkable dependence of γ on the diradical character, i.e., the γ value increases with increase of diradical character, attains the maximum in the intermediate diradcial character region, and then decreases in the large diradical character region. On the basis of these results, the spin multiplicity, diradical charcater and charged state are expected to become control parameters of sign and magnitude of γ for a new class of open-shell NLO systems, i.e., open-shell and/or charged NLO systems.

1. Introduction

For the last two decades, lots of theoretical and experimental studies on the molecular-based third-order nonlinear optical (NLO) properties, which are described by the second hyperpolarizabilities (γ), have been carried out and various guidelines of designing molecules exhibiting large γ values and of controlling their features (sign and magnitudes) have been proposed [1-7]. However, the investigation of NLO properties has been mostly limited to the closed-shell neutral systems though pioneering studies have been carried out to enhance the field-induced charge fluctuations in charged and/or radical states [8-16]. Several studies on these systems aim to design multi-functional materials, for example, combining magnetic and NLO properties [15] or electron conductivity and NLO properties [16]. In this paper, we explain the theoretical and computational studies on the third-order NLO properties of charged and open-shell systems toward the elucidation of structure-property relationships in such systems and a proposition of novel control parameters of third-order NLO properties for such systems.

First, we explain a structure-property relationship of γ on the basis of the perturbational analysis of virtual excitation processes of γ. Namely, the charged radical systems having symmetric resonance structure with invertible polarization (SRIP) [8,12,16] are expected to have unique negative γ, which are the microscopic origin of the self-defocusing effects, in the off-resonant region and to exhibit significant electron-correlation dependences [12-16]. Some crystals composed of these charged radical systems are also predicted to present unique third-order NLO properties (negative γ) as well as high electrical conductivity [15,16]. Further, the open-shell molecules and their clusters involving nitronyl nitroxide radicals, which are important in realizing organic ferromagnetic interaction, belong to the SRIP systems and are predicted to give negative off-resonant γ values.

Second, we provide a viewpoint of the hyperpolarizability of open-shell system based on the chemical bonding nature [13,17,18]. Open-shell systems can be classified according to the strength of electron correlation, i.e., weak-, intermediate- and strong (magnetic)- correlation regimes, which can be exemplified by an equilibrium-, intermediate- and long- bond distance regions of a homogeneous neutral diatomic molecule [17,18]. In previous studies [13], we have predicted a remarkable variation in γ of H_2 model with increasing the bond distance and have suggested a significant enhancement of γ in the intermediate correlation (intermediate bond breaking) regime. This feature is due to the field-induced large fluctuation of intermediate bonding electrons. Such intermediate bond breaking nature in the intermediate correlation regime is, for example, expected to be realized by the increase in the spin multiplicity in open-shell neutral systems [19]. We here consider the spin multiplicity dependence of γ using simple neutral open-shell π-conjugated molecules, i.e., C_5H_7 radicals in the doublet, qualtet and sextet states [19]. In addition, the effect of introducing charge into such open-shell systems on γ is investigated in view of the spin-control of open-shell NLO systems since the control of spin state is often achieved by introducing the charges into systems in molecular magnetism [20].

Third, we consider another type of open-shell singlet NLO system, whose NLO property can be controlled by the diradical character [18,21]. This is regarded as a real system, which directly corresponds to the dissociation model of singlet H_2. As an example, the dependence of γ on the diradical character is investigated using diradical model compounds, *p*-quinodimethane (PQM) models with different both-end carbon-carbon (C-C) bond lengths, by highly correlated *ab initio* molecular orbital (MO) and hybrid density functional theory (DFT) method. The γ value of a real diradical molecule, i.e., 1,4-bis-(imidazole-2-ylidene)-cyclohexa-2,5-diene (BI2Y), which is expected to possess an intermediate diradical character, is also examined. We also consider related benzenoid molecules with different diradical characters to confirm our prediction of dependence of γ on the diradical character observed in H_2 dissociation model. Further, the spin multiplicity effect on γ for such diradical systems is investigated using triplet state models of BI2Y and related molecules.

On the basis of these results, we clarify the features of a new class of open-shell NLO systems, which have novel control parameters, i.e., spin multiplicity, diradical character and charge, of their third-order NLO properties, and are also interesting in view of mutual control of magnetic and NLO properties.

2. Structure-property relationship of γ based on the fourth-order virtual excitation processes

In this section, we explain our structure-property relationship of γ for molecular systems based on the perturbation theory [11,12]. The static γ, which is known to be a good approximation to the off-resonant γ, is considered for convenience, but it can be straightforwardly applied to the dynamic case. The perturbation formula for γ can be approximately partitioned into three types of contributions (I, II and III) as follows [8,11,22,23]:

$$\gamma = \gamma^{I} + \gamma^{II} + \gamma^{III} \tag{1}$$

where

$$\gamma^{I} = \sum_{n=1} \frac{(\mu_{n0})^2 (\mu_{nn})^2}{E_{n0}^3} \tag{2}$$

$$\gamma^{II} = -\sum_{n,m} \frac{(\mu_{n0})^2 (\mu_{m0})^2}{E_{n0} E_{m0}^2} \tag{3}$$

and

$$\gamma^{III} = \gamma^{III-1} + \gamma^{III-2} = \sum_{\substack{m,n=1 \\ (m \neq n)}} \frac{\mu_{0n} \mu_{nn} \mu_{nm} \mu_{m0}}{E_{n0}^2 E_{m0}} + \sum_{\substack{n',m,n=1 \\ (n' \neq m \neq n)}} \frac{\mu_{0n} \mu_{nm} \mu_{mn'} \mu_{n'0}}{E_{n0} E_{m0} E_{n'0}} \tag{4}$$

Here, μ_{n0} is the transition moment between the ground and the nth excited states, μ_{mn} is the transition moment between the mth and the nth excited states, μ_{nn} is the difference of dipole moments between the ground and the nth excited states and E_{n0} is the transition energy given by ($E_n - E_0$). The fourth-order virtual excitation processes (0-i-j-k-0) represented by a series of subscripts in products of transition moments in the numerators in Eqs.(2-4) can be illustrated in Fig. 1(a). The type I(0-n-n-n-0) process, which involves two dipole moment differences between the excited (n) and the ground (0) states (μ_{nn}) on a virtual excitation path, provides positive contributions, while the type II(0-n-0-n-0) process, which involves the ground state (0) in the middle of a virtual excitation path, provides negative contributions. The type III terms can be partitioned into two contributions specified by type III-1(0-

(a) Virtual excitation processes of γ

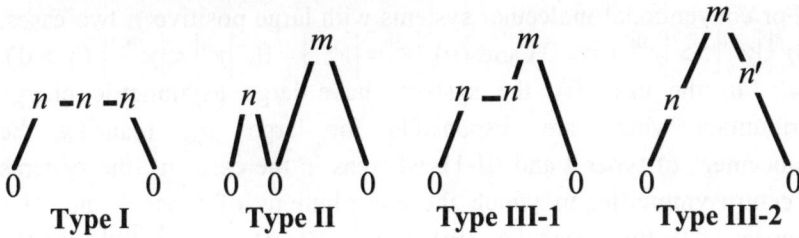

Type I   Type II   Type III-1   Type III-2

(b) Classification of systems with large γ values based on the three-type formula in the three-state $(0, n$ and $m)$ approximation

Symmetry of the system	Main process	Criterion		Sign of γ
Noncentrosymmetric	(I),(III)	large α_0 and μ_{bm}, small E_{n0}		positive
Centrosymmetric	(II), (III-2)	$\alpha_n > 0$	$\dfrac{\alpha_n}{\alpha_0} > \dfrac{E_{n0}}{E_{mn}}$	positive
			$\dfrac{\alpha_n}{\alpha_0} < \dfrac{E_{n0}}{E_{mn}}$	negative
		$\alpha_n < 0$		negative

Fig. 1. Virtual excitation processes of γ (a) and classification of systems with large γ values based on three-type approximate formula (b).

n-n-m-0) and type III-2(0-n-m-n'-0). The type III contributions involve higher excited states (m) as compared to other types. The III-1 process involves only one dipole-moment difference (μ_{nn}) on a virtual excitation path, whereas the type III-2 process has no dipole-moment difference on a virtual excitation path. Although the type III contributions can have a positive or negative value as seen from Eq. (4), total contribution of type III usually becomes positive in sign because the diagonal III-2(0-n-m-n-0) contributions, which are primary contributions, are definitely positive.

Thus, the magnitude and sign of total hyperpolarizability is shown to be closely related to each virtual excitation process.

For conventional molecular systems with large positive γ, two cases, (i) $\left|\gamma^{I}\right|, \left|\gamma^{III}\right| >> \left|\gamma^{II}\right|$ ($\gamma > 0$) and (ii) $\left|\gamma^{I}\right| = \left|\gamma^{III-1}\right| = 0$, $\left|\gamma^{II}\right| < \left|\gamma^{III-2}\right|$ ($\gamma > 0$), exist. In the case (i), the systems have large asymmetric charge distributions which are responsible for large μ_{nn} (causing the enhancement of types I and III-1), whereas in the case (ii), the systems are centrosymmetric, in which the contributions of types I and III-1 disappear. The third case, i.e., (iii) $\left|\gamma^{I}\right| = \left|\gamma^{III-1}\right| = 0$, $\left|\gamma^{II}\right| > \left|\gamma^{III-2}\right|$ ($\gamma < 0$), is interesting because the systems with negative static γ are rare. Such systems are symmetric ($\mu_{nn} = 0$) and exhibit strong dipole transition moment between the ground and the first excited states ($\left|\mu_{0n}\right| > \left|\mu_{nm}\right|$). This indicates that the symmetric systems with large ground-state polarizability (α_0) tend to exhibit negative γ. Figure 1(b) shows the structure-property relationship of γ based on this analysis, which involves the polarizabilities of the ground state (α_0) and the nth excited state (α_n), the excitation energies (E_{n0} and E_{mn}) and dipole moment differences (μ_{nn}) between the nth excited and the ground states [11].

Next, we show several symmetric π-conjugated model systems exhibiting our structure-property relationship. Figure 2 shows the resonance structures mainly contributing to the ground states of (a) charged soliton-like oligomer [23,24], (b) anion radical state of s-indacene [11,12], (c) cation radical state of tetrathiapentalene (TTP) [15,16] and (d) nitronyl nitroxide radical [14]. These resonances between polarized structures with mutually opposite directions can contribute to the stability of the ground-state electronic structures and also correspond to the reduction of the first excitation energy (E_{n0}) and the enhancement of magnitude of transition moment (μ_{n0}) between the ground and the first excited states, leading to the enhancement of α_0. As seen from Fig. 1(b), therefore, a system with a large contribution of resonance structures with invertible polarization (SRIP) tends to exhibit negative γ[11,12]. In fact, high-order electron-correlated *ab initio* MO calculations predict that these SRIP systems exhibit negative static γ according to our structure-property relationship as shown in Sec. 4.

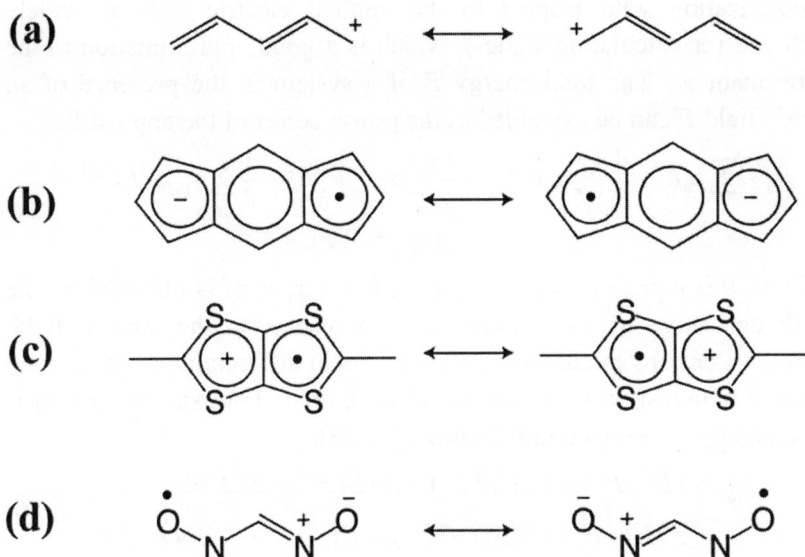

Fig. 2. Examples of symmetric resonance structure with invertible polarization (SRIP): charged-soliton like oligomer (a), anionic radical state of *s*-indacene (b), cationic radical state of tetrachiapentalene (TTP) (c) and nitronyl nitroxide radical (d).

3. Calculation and analysis methods of γ

3.1. *Finite-field calculation*

So far, various theoretical approaches for calculating γ have been presented in the literature. The sum-over-state (SOS) approach [25] based on the time-dependent perturbation theory (TDPT) is useful for elucidating the contribution of transitions between electronic excited states to γ. However, the application of this approach at the *ab initio* MO and DFT level is limited to the relatively small-size molecules since this needs transition properties (transition moments and transition energies) over many excited states, which are very demanding at the first principle calculation level at the present time. Alternatively, the finite-field (FF) approach consisting in numerical or analytical derivatives of total energy or polarization with respect to the applied electric field is widely employed for calculating static γ, which is a good approximation to the off-resonant γ. The total energy E of a system in the presence of an electric field F can be expanded as the power series of the applied field:

$$E = E_0 - \sum_i \mu_0^i F^i - \frac{1}{2}\sum_{ij}\alpha_{ij}F^iF^j - \frac{1}{3}\sum_{ijk}\beta_{ijk}F^iF^jF^k - \frac{1}{4}\sum_{ijkl}\gamma_{ijkl}F^iF^jF^kF^l - \ldots$$

(5)

From this equation, static γ_{ijkl} $(i, j, k, l = x, y, z)$ is obtained by the fourth derivative of total energy E with respect to the applied field. These γ values are calculated by the numerical differentiation of the total energy E with respect to the applied electric field. For example, the γ_{iiii} is expressed by the seven points formula [22,23]:

$$\gamma_{iiii} = \{E(3F^i) - 12E(2F^i) + 39E(F^i) - 56E(0)$$
$$+39E(-F^i) - 12E(-2F^i) + E(-3F^i)\}/\{36(F^i)^4\},$$

(6)

where $E(F^i)$ depicts the total energy in the presence of the electric field F applied in the i direction. In order to avoid numerical errors, several minimum field strengths are used. After numerical differentiations using these fields (mostly, 0.001-0.01 a.u.), we adopt numerically stable γ values for various systems considered in this study. As to the basis sets [26,27] and electron-correlation methods used in this paper are explained in each section. All *ab initio* MO and DFT, e.g., B3LYP [28] and BHandHLYP [29], calculations are performed using the Gaussian 98 program package [29].

3.2. Hyperpolarizability density analysis

In order to elucidate the spatial contributions of electrons to γ, the static second hyperpolarizability (γ) density analysis [13,23,30,31] is applied in the FF approach. This analysis is based on the expansion of charge density function $\rho(r,F)$ in powers of the field F as [32]

$$\rho(r,F) = \rho^{(0)}(r) + \sum_j \rho_j^{(1)}(r)F^j + \frac{1}{2!}\sum_{jk}\rho_{jk}^{(2)}(r)F^jF^k + \frac{1}{3!}\sum_{jkl}\rho_{jkl}^{(3)}(r)F^jF^kF^l + \ldots,$$ (7)

From this equation and the following expansion formula for the dipole moment in a power series of the electric field (in atomic units):

$$\mu^i(F) = \mu_0^i + \sum_j \alpha_{ij} F^j + \sum_{jk} \beta_{ijk}(r) F^j F^k + \sum_{jkl} \gamma_{ijkl} F^j F^k F^l + ..., \quad (8)$$

the static γ can be expressed by [23,30]

$$\gamma_{ijkl} = -\frac{1}{3!} \int r^i \rho_{jkl}^{(3)}(r) d^3 r \quad (9)$$

where

$$\rho_{jkl}^{(3)}(r) = \left. \frac{\partial^3 \rho}{\partial F^j \partial F^k \partial F^l} \right|_{F=0} \quad (10)$$

This third-order derivative of the electron density with respect to the applied electric fields, $\rho_{jkl}^{(3)}(r)$, is referred to as the γ density. In this study, we focus on the longitudinal γ density ($\rho_{iii}^{(3)}(r)$), which is referred to as $\rho^{(3)}(r)$. The positive and negative γ density values multiplied by F^3 correspond, respectively, to the field-induced increase and decrease in the charge density (in proportion to F^3), which induce the third-order dipole moment (third-order polarization) in the direction from positive to negative γ densities. Therefore, the γ density map represents the relative phase and magnitude of change in the third-order charge densities between two spatial points with positive and negative values. The γ densities are calculated for a grid of points using a numerical third-order differentiation of the electron densities calculated by Gaussian 98 [29].

In order to explain the procedure of analysis of the spatial contributions to γ using γ density map, let us consider a pair of localized γ densities ($\rho^{(3)}(r)$) with positive and negative values. The sign of the γ contribution is positive when the direction from positive to negative γ density coincides with the positive direction of the coordinate system. The sign becomes negative in the opposite case. Moreover, the magnitude of the γ contribution associated with this pair of γ densities is proportional to the distance between them.

4. π-Conjugated systems with symmetric resonance structure with invertible polarization (SRIP)

In this section, we examine the static γ values for π-conjugated systems, which are expected to have a large contribution of SRIP. As explained in Sec.2, these systems have a possibility of exhibiting negative static γ and tend to exhibit large electron-correlation dependence. As examples, we consider linear π-conjugated systems with a charged defect (Sec. 4.1), neutral and charged radical states of condensed-ring systems, i.e., tetrathiapentalene, (Sec. 4.2) and nitronyl nitroxide radical (Sec. 4.3).

4.1. *π-Conjugated linear chain with charged defect*

4.1.1. *FF calculations of allyl cation*

Allyl cation is one of the smallest charged polyenes, which is expected to have a large contribution of SRIP in the relatively smaller chain lengths as shown in Fig. 2. In fact, de Melo et al. have reported that the charged soliton-like linear chain exhibits negative longitudinal γ values and significant chain-length dependence as compared to that of neutral ones in the PPP approximation [24]. In our previous paper [23], we have also found that such negative enhancement of γ is limited to relatively short chain-length region: subsequently the sign of γ value reverses and then exhibits a large chain-length dependence toward a large positive value.

The main component of γ (γ_{xxxx}) for allyl cation shown in Fig. 3 is examined. The geometry of this molecule is fully optimized by the *ab initio* HF calculation using 6-31G** basis set. The basis sets used for calculations of γ_{xxxx} are STO-3G, 6-31G, 6-31G**, 6-31+G (ζ_{sp} = 0.0438), 6-31G+d (ζ_d = 0.05), 6-31G+pd (ζ_p = 0.05, ζ_d = 0.05) and 6-31G+pdd (ζ_p = 0.05, ζ_d = 0.2, ζ_d = 0.05). The applied correlation methods are MP2, MP3, MP4DQ (including double(D) and quadruple(Q) excitations), MP4SDQ (including single(S), double(D) and quadruple(Q) excitations), MP4SDTQ (including single(S), double(D), triple(T) and quadruple(Q) excitations), CCSD (including single(S) and double(D) excitations) and CCSD(T) (including single(S), double(D) and perturbative triple(T) excitations). We use 0.01 a.u. as the minimum electric field in the FF method. The γ_{xxxx} value is obtained by the numerical differentiation of total energy E with respect to the field F^x as follows.

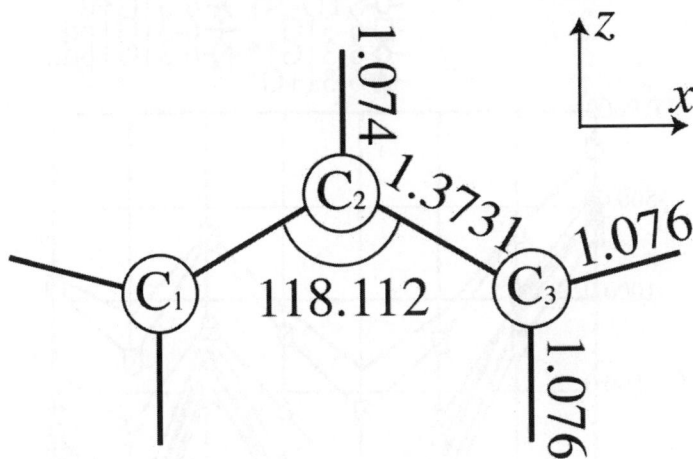

Fig. 3. Structures of allyl cation optimized by the ab initio HF method using 6-31G**. Bond lengths [Å] and bond angles [°] are shown.

    The basis set and electron correlation dependences of γ for allyl cation are shown in Fig. 4. The γ values by all basis sets and correlation methods are negative in sign in contrast to those of neutral allyl molecule [33]. The behaviors of γ_{xxxx} with the correlation effects are qualitatively all similar regardless of their used basis sets. This smaller basis set dependence suggests the unnecessity of diffuse p and d functions in describing cation states. It is however noted that tight d function involved in 6-31G+pdd is turned out to correct the overshot magnitudes of γ by the 6-31+G, 6-31G+d and 6-31G+pd basis sets. On the other hand, the changes of γ with electron correlations are remarkable. The correlation effects contribute to the increase of $|\gamma|$ from MP2 to MP4DQ, while S and T excitation effects in MP4 contribute to the decrease of $|\gamma|$. It is also found that the γ values of CCSD are close to the MP4SDQ values and the CCSD(T) tends to decrease the $|\gamma|$ due to the T excitation effects. Although each order correlation effect is considerably large, the CCSD(T) values are shown to be close to the MP2 values eventually.

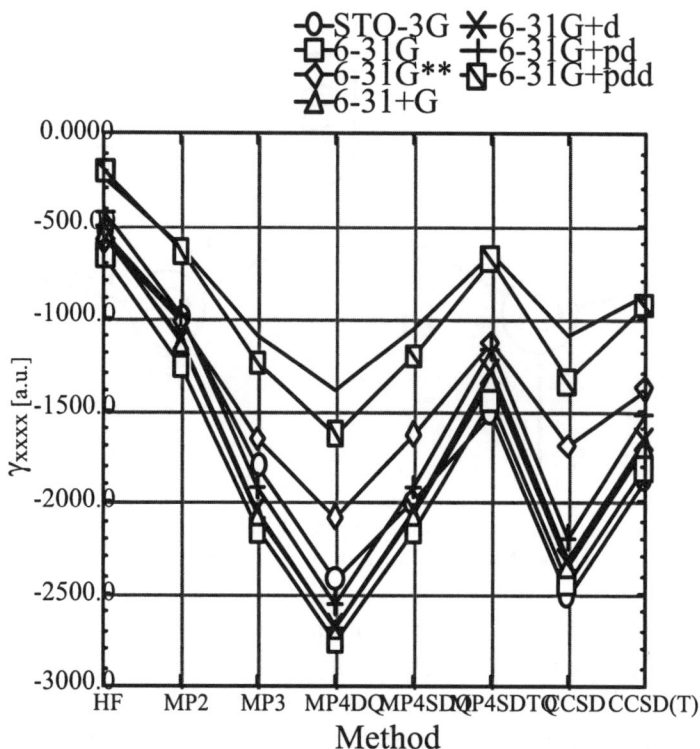

Fig. 4. Basis set and electron correlation dependences of γ_{xxxx} for allyl cation.

4.1.2. Virtual excitation analysis of γ of allyl cation models with various bond alternations

As explained in Sec. 2, the variations in molecular geometry, in particular, the symmetry, could remarkably affect the relative contribution among three types of virtual excitation processes. Indeed, the asymmetric structural change, which provides asymmetric charge distribution, causes type I and III-1 (positive) contributions to γ in contrast to the SRIP systems, which have symmetric charge distribution and tend to give negative γ. For example, de Melo *et al.* [24] has reported that regular polyenes possess positive longitudinal γ values, while small-size charged soliton-like polyenes exhibit large negative γ values. We hear consider the allyl cation models in order to elucidate the relationship between the degree of C-C bond-length alternation and the γ, which is predicted to be negative in the equilibrium geometry as shown in the previous section. Virtual excitation processes are analyzed in the numerical Liouville approach (NLA) [34]. We discuss the variations in the contributions from three types of virtual excitation processes with respect to the bond-length alternation. On the basis of the optimized geometry of allyl cation (Fig. 3), the C-C bond-length alternation, denoted by $\Delta r = l(C_1 - C_2) - l(C_2 - C_3)$, is varied by shifting the middle carbon C_2 along the x axis.

We apply the NLA to the calculation of the off-resonant γ in the third-harmonic generation (THG). The NLA is based on a numerical Fourier transformation of the time series of polarization obtained by using numerical solutions of the quantum Liouville equation. This approach can provide numerically exact on- and off-resonant hyperpolarizabilities including population and phase relaxation effects. To apply the NLA, we first construct a state model that mimics the electronic states of the system and calculate its transition energies and transition moments. These transition properties for excited states are calculated using semiempirical INDO/S with single and double excitation configuration interactions (SDCI) (using all π electrons), the method of which has been proven to reproduce reliable transition properties of organic π-conjugated molecules at a much lower cost than large-scale *ab initio* CI. Figure 5 shows the transition energies and transition moments of the three-state model calculated by the INDO/S-SDCI method for $\Delta r = -1.3 - 1.3$ Å. We investigate the off-resonant THG processes induced by three incident beams with a power density of 100 MW/cm^2 and a frequency of 10000 cm^{-1}, which corresponds to the off-resonant value with respect to the first excited state **1**.

Figure 6(a) shows individual and total types of contributions of real parts of THG γ_{xxxx} (referred to as γ, in this section) of allyl cation with respect to Δr. The ground-state geometry based on the *ab initio* HF calculation corresponds to $\Delta r = 0$. In this case, the total γ is observed to be largely negative. In contrast, the γ increases remarkably with $|\Delta r|$, passes 0, and exhibits a peak in a positive region around $|\Delta r| = 0.8$ Å. For $|\Delta r|$ values larger than 0.8 Å, the γ decreases again. This behavior is found to be similar to that of neutral polyenes end-capped by an acceptor and a donor [35].

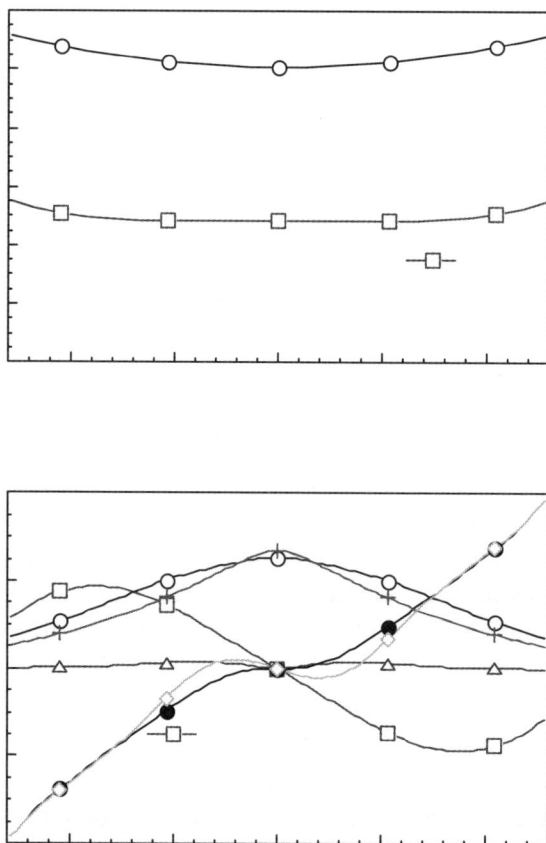

Fig. 5. Variations in transition energies E_{ij} (E_i-E_j) [cm^{-1}] (a), and transition moments μ_{ij} [D] (b) of a three-state model for the shift (in the direction x) of middle carbon atom of allyl cation (Fig. 3). The power density is 100 MW/cm^2 and the frequency of the external fields is 10000 cm^{-1}.

(a)

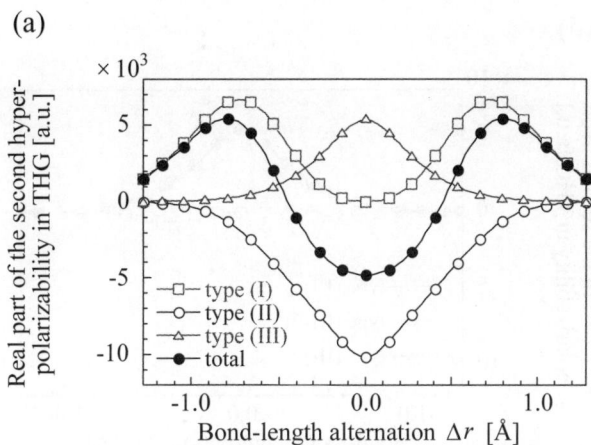

(b)

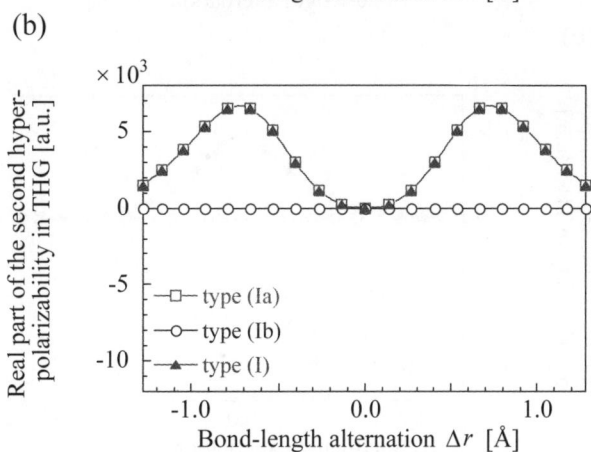

(c)

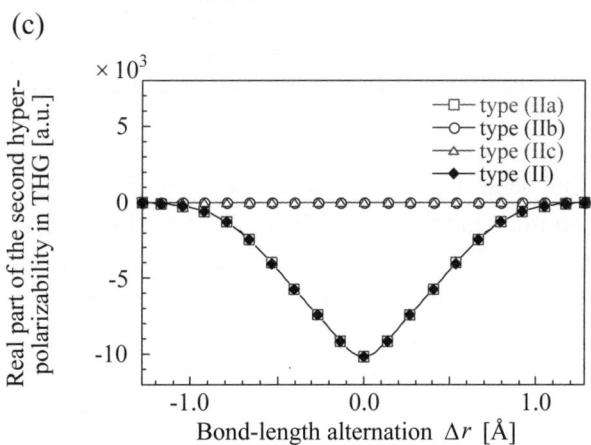

(d)

(e)

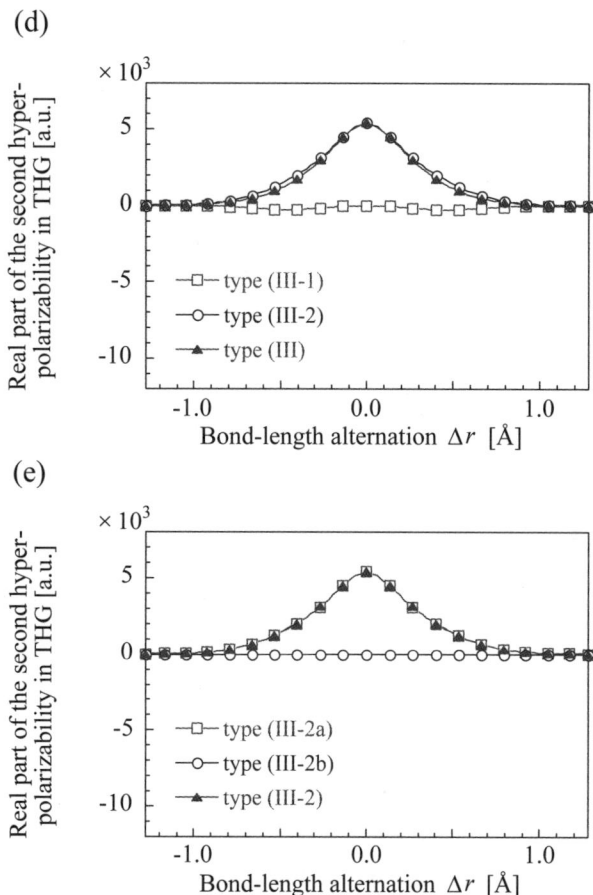

Fig. 6. Real part of the off-resonant γ in THG with respect to the C-C bond-length alternation Δr [Å]: total γ of types I-III contributions (a) and contributions of component types (b)-(e).

As can be seen from Fig. 6(a), the type I process is a main contribution in the region of large $|\Delta r|$ (≥ 0.8 Å), while the type II and III contributions are enhanced around $|\Delta r| = 0$. This is considered to be caused by the fact that the molecule with large $|\Delta r|$ induce large asymmetric charge distributions (in the direction of x axis) which contribute to the type I process predominantly, while the molecule around $|\Delta r| = 0$ has nearly symmetric charge distributions in the direction of x axis, where the type I contribution vanishes. For allyl cation, the |type II| value outweighs the |type III| value, so that the total γ in the region with small Δr turns to be negative.

Figure 6(b) shows the variations in type I, Ia and Ib contributions with respect to Δr. This figure indicates that the type I contribution is almost determined by the type Ia process, which is characterized by a large dipole-moment difference between the **1** and **0** states. This suggests that the excited state **1** is a charge-transfer (CT) state, which is mainly constructed from single excitation configurations.

Variations in each contribution from the type II process Ia - IIc with respect to Δr are shown in Fig. 6(c). Apparently, the type IIa contribution is essential. This contribution takes the largest absolute value near $\Delta r = 0$. As a result, this contribution primarily leads to a large negative γ value of allyl cation in the equilibrium geometry.

The behavior of the type III process is almost determined by that of the type III-2 process, as shown in Fig. 6(d). This contribution consists of the type III-2a and III-2b processes shown in Fig. 6(e). The type III contributions are essentially determined by the type III-2a process, which is constructed from the virtual excitation process through excited states higher than those involved in the type Ia and IIa processes.

From the above analysis, it is concluded that the γ of allyl cation at large Δr is dominated by the type Ia process, while the total γ at small Δr can be qualitatively described by the type IIa and III-2a processes. For allyl cation, the type IIa process provides a larger $|\gamma|$ than the type III-2a process, so that the total γ turns to be negative in the equilibrium geometry.

In summary, the characteristics of hyperpolarizability can be interpreted by the inherent quantum (virtual excitation) processes, which can be modified by the variations in the conformation and charged states. For example, a negative γ for allyl cation in the equilibrium geometry implies that the contributing virtual excitations tend to be limited to the low-lying virtual excitation process (0-1-0-1-0). This strong low-lying excitation process seems to be attributed to the charged defect in allyl cation. Similar tendencies seem to exist in the charged soliton-like polyenes with small chain lengths [23]. In contrast, in the region remote from the equilibrium geometry a large positive contribution to γ is dominated by the virtual excitation process (0-1-1-1-0), which represents the virtual CT in the molecule with asymmetric charge distributions at the ground state.

4.2. *Condensed-ring systems*

The second examples are condensed-ring systems such as pentalene [11,12] and tetrathiapentalene (TTP) [15,16]. The charged radical states of these systems are expected to exhibit a large contribution of SRIP because of the recovery of the aromaticity of each ring based on $(4n + 2)\pi$ rule: for each ring, 6 π - 5 π or 5 π - 6π electron systems for anionic radical state of pentalene and 6 π - 7 π or 7 π - 6π systems for cationic radical state of TTP are realized. Namely, a charge and radical electrons tend to easily fluctuate under external electric fields in the longitudinal direction, leading to SRIP (see Fig. 2(c)). These systems are also interesting from the viewpoint of multi-functional systems because segregated columns (π-π stacking molecular aggregate) composed of condensed-ring molecules are known to exhibit high electric conductivity along the column direction in the incomplete CT states [36]. These columns are generated by the CT between neighboring donor (D) and acceptor (A) columns and partially include charged radical states of monomers in the column.

As examples, we show recent results by Yamada et al. on TTP in the singlet neutral and doublet cationic radical states, which are partially involved in the donor segregated stack as shown in Fig. 7. The geometries of singlet neutral TTP and doublet cationic radical TTP are optimized by the RB3LYP/6-311G** and UB3LYP/6-311G** methods, respectively (see Fig. 8). The longitudinal γ values for both systems are calculated using various electron-correlated *ab initio* MO methods and hybrid DFT (B3LYP and BHandHLYP) methods using 6-31+G* basis sets. The calculated data are given in Table 1. For neutral TTP, all the γ values are positive in sign. The electron correlation at the MP2 (13480 a.u.) and MP4 (14120 a.u.) levels are shown to significantly enhance (by about 2.8 - 3.0 times) the γ at the RHF level (4760 a.u.). The MP2 and MP4 values are shown to well reproduce the γ at the RCCSD(T) level (14100 a.u.), which is the most reliable result in this study.

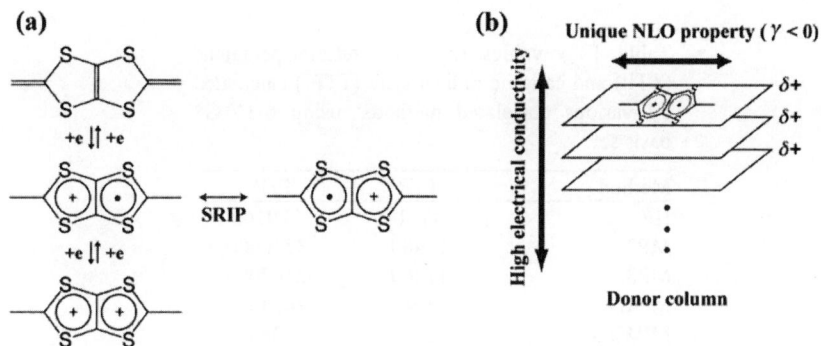

Fig. 7. Two-step redox reaction process for tetrathiapentalene (TTP) (a) and a multi-property aggregate model combining high electrical conductivity and unique third-order optical nonlinearity (b). The mixed-valence stack is composed of partially cationic donor molecules.

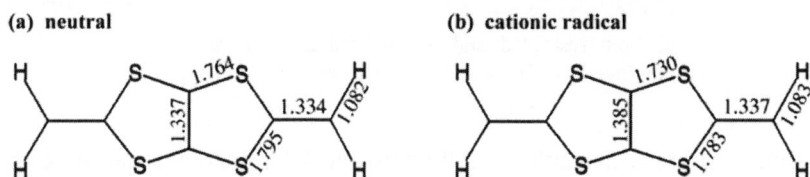

Fig. 8. Geometries of neutral (a) and cationic radical (b) states of tetrathiapentalene (TTP) optimized by the RB3LYP/6-311G** and UB3LYP/6-311G** methods, respectively.

The RB3LYP method ($\gamma = 13750$ a.u.) can also well reproduce the γ at the RCCSD(T) level, while the RBHandHLYP result (9103 a.u.) is shown to be undershot by about 65 % as compared to the RCCSD(T) result. In contrast, for cationic radical TTP, all the methods give negative γ value according to our prediction based on the SRIP. The electron correlation dependence are remarkably large as compared to neutral TTP case. The magnitude of γ at the UMP levels are shown to be about one order overshot than that at the UHF result (-289100 a.u.), which unexpectedly well coincides with γ at the UCCSD(T) level (-289500 a.u.). From the comparison between the UCCSD and UCCSD(T) results, the high-order triple excitation effects, which provide significant positive contributions to γ, are indispensable for obtaining

Masayoshi Nakano

Table 1. γ values [a.u.] for tetrathiapentalene (TTP) and cationic radical state (TTP$^+$) calculated by various correlated methods[a] using 6-31+G* basis set.

Method	TTP	TTP$^+$
HF	4761	-289100
MP2	13480	-3216000
MP3	11260	-2107000
MP4D	13040	-2423000
MP4DQ	12210	-2108000
MP4SDQ	12410	-1008000
MP4SDTQ	14120	-1224000
CCSD	12470	-441100
CCSD(T)	14100	-289500
B3LYP	13750	-131900
BHandHLYP	9104	-408000

a) Spin restricted and unrestricted methods are employed for TTP and TTP$^+$, respectively.

qualitatively correct results. Although the DFT methods can give the same order magnitude as well as a correct sign of γ, the UB3LYP method ($\gamma = -131900$ a.u.) overshoots, while the UBHandHLYP method ($\gamma = -408000$ a.u.) undershoots the γ values at the UCCSD(T) level. However, these results as well as the RBLYP result ($\gamma = -53750$ a.u.) suggest that the hybrid DFT can reproduce the UCCSD(T) result by tuning the weight of HF exchange [20,21]. In fact, in the case of including 30 % of HF exchange and 70% of Becke exchange as well as local (LYP) and non-local (VWN) correlation functionals, the hybrid DFT method gives the longitudinal γ values : 12040 a.u. for neutral TTP and -309400 a.u. for cationic radical TTP. These values well reproduce the UCCSD(T) values, respectively (14100 a.u. for neutral TTP and -289500 a.u. for cationic radical TTP). The longitudinal γ density plots for neutral and cationic radical states of TTP are shown in Fig. 9. For both systems, the primary contributions come from π electron distributions. For neutral TTP shown in Fig. 9(a), the internal π electron contributions are smaller and negative in sign, while outer contributions are larger and positive in sign. Therefore, total γ value becomes positive in sign. In contrast, the γ

density distributions on both sides of five-membered rings of cationic radical TTP provide large negative contributions (see Fig. 9(b)). It is also found that the iso-surface of γ density for cationic radical TTP is one-order larger than that for neutral TTP. This indicates a possibility of not only realizing a negative sign of γ but also a remarkable enhancement of magnitude of γ, which is attractive in view of molecular design of NLO materials.

(a) Neutral   (b) Cationic radical

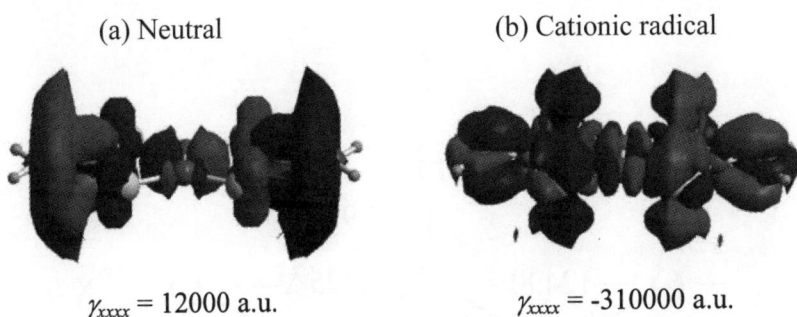

$\gamma_{xxxx} = 12000$ a.u. $\gamma_{xxxx} = -310000$ a.u.

Fig. 9. γ_{xxxx} density plots of neutral (a) and cationic radical (b) states of TTP calculated by the hybrid DFT method including 30 % of HF exchange and 70% of Becke exchange as well as local (LYP) and non-local (VWN) correlation functionals. Red and blue iso-surfaces with the magnitude of 100 a.u. for (a) and 1000 a.u. for (b) show the positive and negative γ desities, respectively.

4.3. *Nitronyl nitroxide radicals*

In previous studies [14], we have investigated the γ for para-nitrophenyl nitronyl nitroxide (2-(4-nitrophenyl)-4,4,5,5-tetramethyl-4,5-dihydro-1H-imidazolyl-1-oxyl-3-oxide), which is referred to as p-NPNN, in the INDO coupled Hartree-Fock (CHF) approximation. The calculated results suggest that the largest component of γ for this system is negative, and is primarily determined by a contribution from the nitronyl nitroxide radical unit. This is in agreement with our prediction based on the SRIP (see Fig. 2(d)). In this section, the basis set and electron-correlation dependences of the γ are investigated by using various electron-correlated *ab initio* MO methods such as spin-unrestricted Møller-Plesset (UMP) perturbation, quadratic configuration interaction (UQCI) and coupled-cluster (UCC) methods.

4.3.1. Calculation model and calculation methods

Figure 10 shows the molecular geometry and a coordinate system of nitronyl nitroxide radical (N^2-oxidoformamidin-N^1-yloxyl radical) optimized by the UMP2 method using the 6-311G** basis set.

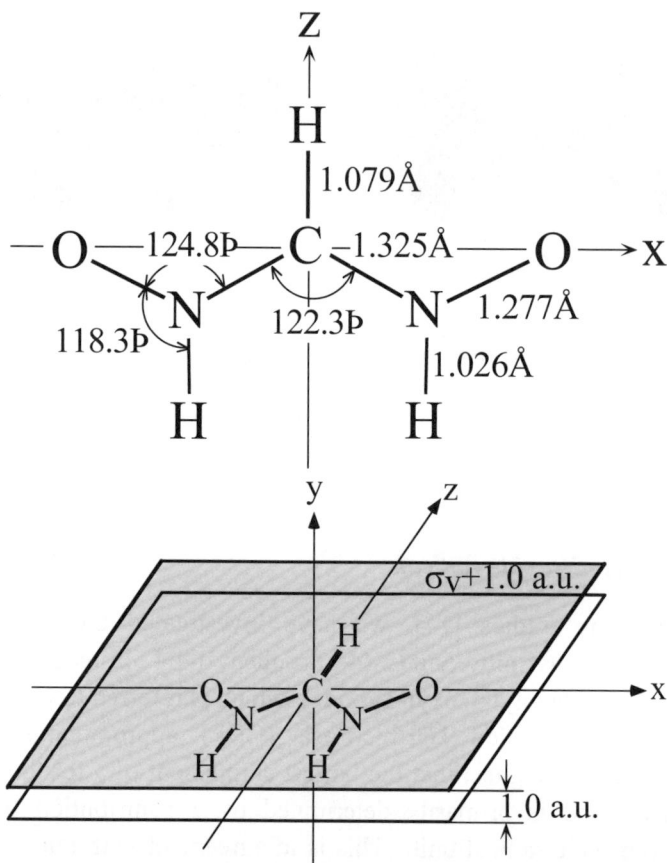

Fig. 10. Molecular geometry and the coordinate system of nitronyl nitroxide radical optimized by UMP2 method using 6-311G** basis set. The plane (σ_V+1 a.u.) is located at 1.0 a.u. above the reflection plane σ_V.

We confine our attention to the γ_{xxxx} (γ) of this system since the SRIP contribution occurs in this direction. The calculations are performed by using the Gaussian 98 program package [28]. We first investigate the basis-set dependence of the γ using the standard (minimal (STO-3G), split-valence (6-31G)) and extended basis sets augmented by diffuse and polarization functions (p and d), which are known to be essential for reproducing semi-quantitative γ for organic π-conjugated systems. The extended basis sets are given in Table 2. These exponents are all determined from the outermost two exponents of 6-31G by an even-tempered method. Secondly, the effects of electron correlations on the γ are examined by the UMP2, UMP3, UMP4DQ, UMP4SDQ, UMP4SDTQ, UCCD(=UQCID), UQCISD, UCCSD and UCCSD(T) methods. Here, the symbols S, D, T and Q denote the inclusion of correlation effects caused by the single, double, triple and quadruple excitations, respectively. The CC method can include these correlation effects to infinite-order. The FF method is applied to the calculation of γ using Eq. (6) with minimum fields of around 0.0015 au. The γ density analysis is also employed to clarify the differences in the spatial contributions of electrons to γ for these results.

Table 2. Exponential Gaussian orbital parameters for polarization and diffuse functions.

	C		N		O	
	p	d	p	d	p	d
6-31G+p	0.0523		0.0582		0.0719	
6-31G+d		0.0523		0.0582		0.0719
6-31G+pd	0.0523	0.0523	0.0582	0.0582	0.0719	0.0719

4.3.2. Basis set and electron-correlation dependences

Figure 11 shows the variations in γ of the nitronyl nitroxide radical for different basis sets and electron-correlation methods. This figure indicates that the STO-3G and 6-31G basis sets cannot provide a sufficient magnitude of γ. In contrast, extended basis sets (6-31G+p and 6-31G+d) are shown to significantly increase the magnitudes of γ. The

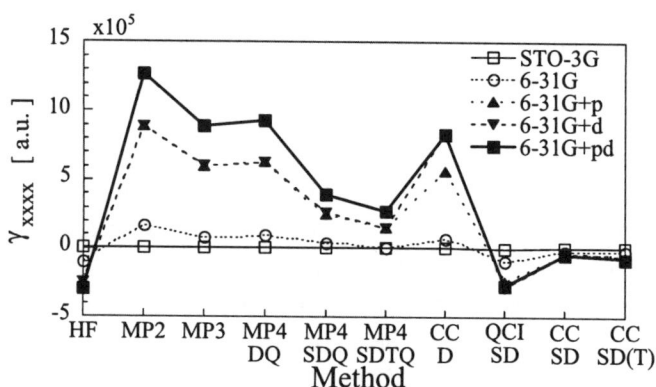

Fig. 11. Variations in γ_{xxxx} of nitronyl nitroxide radical for various basis sets and electron correlation methods.

6-31G+*pd* basis set is found to provide similar electron correlation dependences to those at the 6-31G+*p* and 6-31G+*d* basis sets though the magnitudes of γ at the 6-31G+*pd* are slightly enhanced. Therefore, we consider that the 6-31G+*pd* basis set can provide a qualitative description of the dependence of γ on the various electron correlation effects. This feature is similar to those of other closed-shell conjugated organic systems. This basis-set dependence is also investigated using the γ densities for the results by using 6-31G and 6-31G+*pd* at the UHF level. In the following discussion concerning the electron-correlation dependence of γ, the γ calculated using the most extended basis set, 6-31G+*pd*, are used.

As shown in Fig.11, a remarkable electron correlation dependence of the γ is observed. The UHF method provides a large negative γ, while the D effects by the UMP2 method change the sign and enhance the magnitude of γ by the UHF method. The D effects by the MP3 method decrease γ by the MP2 method. The D and Q effects involved in the UMP4 method are found to provide only a slight positive contribution to the γ. A comparison between the UCCD and UMP4DQ γ values indicates that the correlation effects originating from the higher-order D beyond the fourth-order are not important. The S and T effects involved in the UMP4 method decrease γ value. Particularly, the S effects involved in the UCCSD method remarkably decrease UCCD γ value and then reverse its sign. This feature implies that the contribution from the higher-order S effects beyond the fourth-order are indispensable for a reliable description of the γ for this system. The UQCISD method cannot satisfactorily reproduce γ by the UCCSD method though its approximation level is generally considered to be similar to that of the latter method. However, the sign of γ by the UQCISD method coincides with that by the UCCSD method, and is thus considered to be a good approximation of that by the UCCSD method. The γ by the UCCSD(T) method is shown to be nearly equal to that by the UCCSD method. This feature indicates that γ by the UCCSD method seems to be sufficiently converged.

4.3.3. Analysis of spatial contributions to γ by the density γ

In order to elucidate the relationships among the spatial contributions of electrons, the basis set and electron-correlation dependences of γ, the γ density ($\rho_{xxx}^{(3)}(r)$) plots are investigated. Figure 12 gives contour plots of $\rho_{xxx}^{(3)}(r)$ on the $\sigma_v + 1$ a.u. plane of the nitronyl nitroxide radical shown in Fig. 10. From the plots of $\rho_{xxx}^{(3)}(r)$ by the UHF method (Figs. 11(a) and (b)), diffuse p and d functions are not shown to change the qualitative spatial contributions of electrons to the γ, except for their size differences. As can be seen from all of the $\rho_{xxx}^{(3)}(r)$ plots, the $\rho_{xxx}^{(3)}(r)$ mainly contributing to the γ are distributed along the bond region of the O-N-C-N-O unit. This implies that a field-induced charge fluctuation in the bond region of the O-N-C-N-O unit predominantly determines the features of γ. This is in contrast to the usual case for a neutral closed-shell π-conjugated system, where the contributions to γ from expanded regions remote from the bond regions are dominant.

The tendency of the distribution of the $\rho_{xxx}^{(3)}(r)$ at the UHF and UQCISD levels is found to be dramatically different from that at other levels, and therefore the distribution of $\rho_{xxx}^{(3)}(r)$ can be classified into two types. The first type (type A) includes the $\rho_{xxx}^{(3)}(r)$ by the UHF and UQCISD methods, and the second type (type B) includes the $\rho_{xxx}^{(3)}(r)$ by the UMP2, UMP3, UMP4DQ, UMP4SDQ and UCCD methods. As shown in Fig. 12, γ values of type A are negative, while those of type B

(a)   HF/6-31G
           -102000 a.u.

(b)   HF/6-31G+pd
           -292000 a.u.

(c)   MP2/6-31G+pd
           127000 a.u.

(d)   MP3/6-31G+pd
           88600 a.u.

(e)   MP4DQ/6-31G+pd
           921000 a.u.

(f)   MP4SDQ/6-31G+pd
           390000 a.u.

(g)   CCD/6-31G+pd
           834000 a.u.

(h)   QCISD/6-31G+pd
           -271000 a.u.

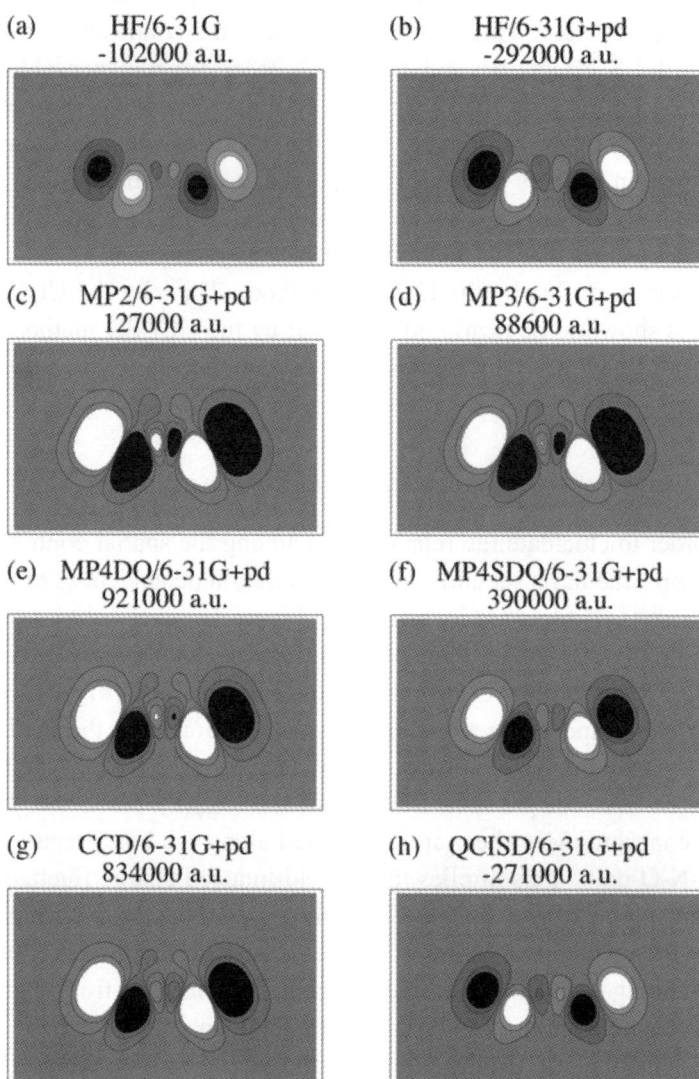

Fig. 12. Contour plots of γ_{xxx} densities on the $\sigma_V + 1$ a.u. plane of nitronyl nitroxide radical (Fig. 10) and γ_{xxx} for various calculation methods. Contours are drawn from -1250 to 1250 a.u. Lighter areas represent the spatial regions with larger $\rho_{xxx}^{(3)}(r)$ values. The white regions correspond to those with $\rho_{xxx}^{(3)}(r)$ larger than 1250 a.u., while the black regions correspond to those with $\rho_{xxx}^{(3)}(r)$ smaller than -1250 a.u.

are positive. A common feature of both types is that contributions from the vicinity of the (α-)carbon atom (in the ONCNO skeleton) are appreciably small, or can be negligible, so that the γ are contributed primarily from the vicinity of the NO groups on the left- and right-hand sides. For type A, it is surprising that there is no difference between the feature of distribution of $\rho_{xxx}^{(3)}(r)$ at the UHF level and that at the UQCISD level. The contributions from the $\rho_{xxx}^{(3)}(r)$ in the vicinity of the N atoms in both NO groups are shown to be positive, while those of the O atoms in both NO groups are shown to be negative. As can be seen from Eq. (9), the contribution from a pair of $\rho_{xxx}^{(3)}(r)$ is more significant when the distance of the pair of $\rho_{xxx}^{(3)}(r)$ is larger. In the present case, since the O-O distance is larger than the N-N distance, the total γ is mainly determined by the contribution from both O atoms, and becomes negative. In contrast, for type B, which mainly include the D or Q effects, the contributions from $\rho_{xxx}^{(3)}(r)$ distributed over the vicinity of the N atoms in both NO groups are shown to be negative, while those of O atoms in both NO groups are shown to be positive, so that the total γ becomes positive. As can be seen from the UMP4DQ and UMP4SDQ results, the S effects tend to decrease $|\rho_{xxx}^{(3)}(r)|$ in the O atom regions in both NO groups, so that γ at the UMP4SDQ level becomes smaller than that at the UMP4DQ level. The distribution of $\rho_{xxx}^{(3)}(r)$ at the UCCD level is shown to provide a positive γ similarly to that at the UMP4DQ level. In contrast, the S effects involved in the higher-order (beyond the fourth-order) perturbation methods are found to reverse the sign of γ at the UCCD level. As a result, an inclusion of the D and Q effects up to infinite-order cannot reproduce the qualitative features of the distributions of $\rho_{xxx}^{(3)}(r)$ at the UQCISD level. It is also noteworthy for this system that the features of $\rho_{xxx}^{(3)}(r)$ at the UHF level (Fig. 12(b)) is very similar to that at the UQCISD level (Fig. 12(h)). This suggests that a cancellation among the S and DQ effects is qualitatively achieved at infinite-order. For this system, an incomplete or unbalanced consideration of electron correlations will lead to incorrect features of the sign and magnitude of the γ.

In summary, the longitudinal components of γ for nitronyl nitroxide radical is shown to be negative in sign and is primarily determined by the field-induced electron fluctuations in the bond region of the O-N-C-N-O unit. Although the feature at the HF level is in qualitatively good agreement with that by the UQCISD methods, this does not immediately imply a good convergence of γ by including low-order UMP electron correlations. Actually, the UMP2, UMP3 and UMP4 methods cannot sufficiently reproduce converged γ, even in the sign. This deficiency, slower convergence, of UMPn results for γ is discussed in relation to the spin contamination effects on γ for open-shell systems in Sec. 5. From a comparison among the UMP4DQ, UMP4SDQ, UCCD and UQCISD $\rho_{xxx}^{(3)}(r)$, a balanced inclusion of electron correlations with S and DQ effects beyond the fourth-order is essential for obtaining at least the correct sign of γ. Judging from the qualitative coincidence of the spatial distributions of $\rho_{xxx}^{(3)}(r)$ at the UHF and UQCISD methods, a cancellation among the S and DQ effects seems to be qualitatively achieved at the infinite-order; thus the INDO CHF results [14] for large nitroxide systems, e.g., p-NPNN, are considered to provide at least a qualitatively correct description of the dominant contribution from its nitronyl nitroxide radical unit. A dominant contribution to γ is found to be determined by the electron fluctuation in the O-N-C-N-O unit, where the SRIP largely contributes to the stability of the nitronyl nitroxide radical.

Considering the sensitive electron fluctuation of the O-N-C-N-O radical unit, the high-order NLO responses would be dramatically changed by adding subtle chemical and physical perturbations. It is further expected from its attractive magnetic property that these systems with nitronyl nitroxide radicals would be one of the feasible candidates of high-order magnetic and optical systems in the future.

5. Spin multiplicity dependence of γ for open-shell π-conjugated systems

In the present and next sections, we introduce some recent investigations of novel NLO systems, i.e., open-shell NLO systems [19]. As is well known, the theoretical and experimental studies for highly-efficient NLO organic systems has during the last three decades mostly focused on closed-shell conjugated compounds [1-7]. Most of structure-property relationships have elucidated that the NLO properties are closely related to the nature and length of the conjugated linker, the strengths of donor and acceptor substituents and the charged states of the system [8,24]. Although several studies have highlighted the potential of open-shell NLO systems [8,9,11,13-16], much less has been achieved yet. In open-shell NLO systems, the spin state constitutes another degree of freedom that could be tuned to match the desired properties or to be used in logic devices.

Open-shell systems can be classified according to the strength of electron correlation, i.e., weak-, intermediate- and strong (magnetic)-correlation regimes, which can be exemplified by an equilibrium-, intermediate- and long- bond distance regions of a homogeneous neutral diatomic molecule [17,18]. We have found the remarkable variation in γ according to increasing the bond distance and suggested the enhancement of γ in the intermediate correlation regime [13]. In addition, the amplitude of the electron correlation is expected to change by modulating the spin and/or the charge of a system. In this section, as the first step toward realizing spin-modulated NLO systems, we focus on the dependency of γ on the spin state (doublet, quartet and sextet). Namely, we investigate the static longitudinal γ values of small-size neutral π-conjugated model, the C_5H_7 radical. Since significant electron correlation dependence and spin contamination effects are predicted for γ of such open-shell system [14-16], we employ the UHF and *post*-UHF methods as well as spin-projected methods. In addition to γ values, the spatial contributions of total, α and β electrons to γ are characterized using the γ density analysis [23,30,31] in order to investigate the spin polarization and electron correlation effects on γ. Second, since such systems involve changes of not only the spin multiplicity but also the charge, the effects of charges on the NLO properties of open-shell systems are investigated. We compare the results of two model compounds with five π-electrons, the neutral radical C_5H_7 and radical cation C_6H_8 in their ground doublet states, and discuss the cooperative effects of spin and charge on γ.

On the basis of these results, the effects of spin and charged states on γ values for open-shell systems are discussed in connection with the proposal of a new class of NLO systems, i.e. open-shell NLO systems, which present the possibility of spin and/or charge control of NLO properties.

5.1. Geometrical structures and calculation methods for neutral radicals, C_5H_7

Figure 13 shows the structures of C_5H_7 radicals in the doublet (a), quartet (b) and sextet (c) states optimized at the UB3LYP level of approximation using 6-311G* basis set. The doublet, quartet, and sextet states are characterized by an excess of α-electrons with respect to β-electrons: one α-electron in excess for the doublet, three for the quartet, and five for the sextet. For each spin multiplicity the lowest energy state has been considered. At the B3LYP/6-311G* level, the corresponding $\langle S^2 \rangle$ values are 0.795, 3.765, and 8.765 for the doublet, quartet, and sextet whereas the exact values are 0.75, 3.75, and 8.75, respectively. The CC bond length alternation is shown to decrease when going from the doublet to the sextet state, where all CC bonds are similar to single bonds. This feature can be understood by the fact that increasing the spin multiplicity corresponds to breaking π bonds.

In the present study, the 6-31G*+pd basis set with p and d exponents of 0.0523 has been employed [26]. For the analysis of electron correlation effects on γ, a succession of methods has been adopted, starting with the UHF scheme. The *post*-UHF methods include the UMPn (n=2-4), UCCSD, UCCSD(T) and UQCISD. In addition, the l-fold spin-projected UMPn methods using the Löwdin type spin projection [37], i.e. PUHF (l=1), PUMP2(l=1) and PUMP3(l=1), have also been applied in order to highlight the effects of spin contamination corrections on γ. Moreover, at the HF and MP2 levels, the corresponding restricted open-shell approaches (ROHF and ROMP2) have been employed while among the DFT schemes, the hybrid B3LYP exchange-correlation functional has been adopted. All calculations have been performed using the Gaussian 98 program package [29].

We confine our attention to the longitudinal components of γ. In order to improve the accuracy on the γ values, a 4-point procedure (equivalent to a 7-point procedure for a non-symmetric case) using field amplitudes of 0.0, 0.0010, 0.0020, and 0.0030 a.u (Eq.(6)) [23] and/or

(a) Doublet state

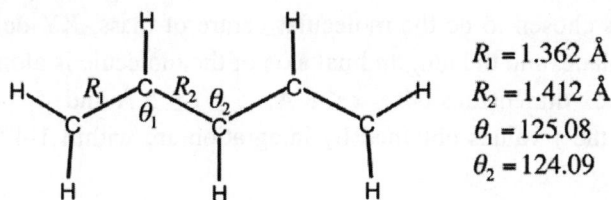

$R_1 = 1.362$ Å
$R_2 = 1.412$ Å
$\theta_1 = 125.08$
$\theta_2 = 124.09$

(b) Quartet state

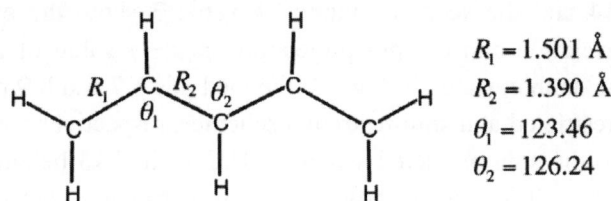

$R_1 = 1.501$ Å
$R_2 = 1.390$ Å
$\theta_1 = 123.46$
$\theta_2 = 126.24$

(c) Sextet state

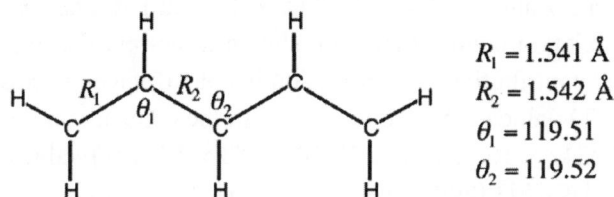

$R_1 = 1.541$ Å
$R_2 = 1.542$ Å
$\theta_1 = 119.51$
$\theta_2 = 119.52$

Fig. 13. Structures of C_5H_7 radicals in doublet (a), quartet (b) and sextet (c) states. The structures are planar and belong to the C_{2v} point group and are optimized by the UB3LYP method using 6-311G* basis set.

the Romberg scheme [38] using field amplitudes of 0.0, 0.0010, 0.0020, and 0.0040 a.u. are adopted. This has enabled to reach an accuracy of 10-100 a.u. on the static longitudinal second hyperpolarizability of the C_5H_7 radical. The γ densities are also calculated for a grid of points using a numerical third-order differentiation of the electron densities (total, α and β) calculated by Gaussian 98. For treating the C_5H_7 radical, the origin is chosen to be the molecular centre of mass, XY defines the molecular plane, and the longitudinal axis of the molecule is along the X-axis. The box dimensions ($-8 \leq x \leq 8$ Å, $-5 \leq y \leq 5$ Å and $-5 \leq z \leq 5$ Å) ensure that the γ values obtained by integration are within 1-4 % of the FF results.

5.2. *Electron correlation effect on γ of the doublet state of C_5H_7 radical*

Figure 14 and the second column of Table 3 show the effects of electron correlation and of spin projection on the γ value of the C_5H_7 radical in the doublet state. In Fig. 14, the methods **1-7** and **8-9** belong to the spin-unrestricted and spin-restricted schemes, respectively, while **10-12** belong to the spin-projected scheme. The method **13** belongs to the DFT scheme. The second-order electron correlation correction is significant and positive: it enhances the γ value by more than 100% [151 $\times 10^2$ a.u. (UHF) and 304 $\times 10^2$ a.u. (UMP2)]. Although at the UMP3 (263 $\times 10^2$ a.u.) and UMP4SDQ (242 $\times 10^2$ a.u.) levels, higher-order electron correlation contributions are shown to correct the overshooting second-order contribution, the correction is not sufficient as compared to the UCCSD(T) value (215 $\times 10^2$ a.u.). From the comparison between the UMP4SDQ (242 $\times 10^2$ a.u.) and UMP4 (258 $\times 10^2$ a.u.) values, and that between the UCCSD (200 $\times 10^2$ a.u.) and UCCSD(T) (215 $\times 10^2$ a.u.), the inclusion of the triple excitations in the fourth-order perturbation treatment brings a small positive contribution to γ.

On the other hand, the spin contamination correction decreases the UHF, UMP2, and UMP3 γ values by 65%, 28%, and 24%, respectively [53 $\times 10^2$ a.u. (PUHF), 220 $\times 10^2$ a.u. (PUMP2) and 201 $\times 10^2$ a.u. (PUMP3)]. Similarly to the non-spin-projected case, the second-order Møller-Plesset correction substantially increases γ whereas at third-order

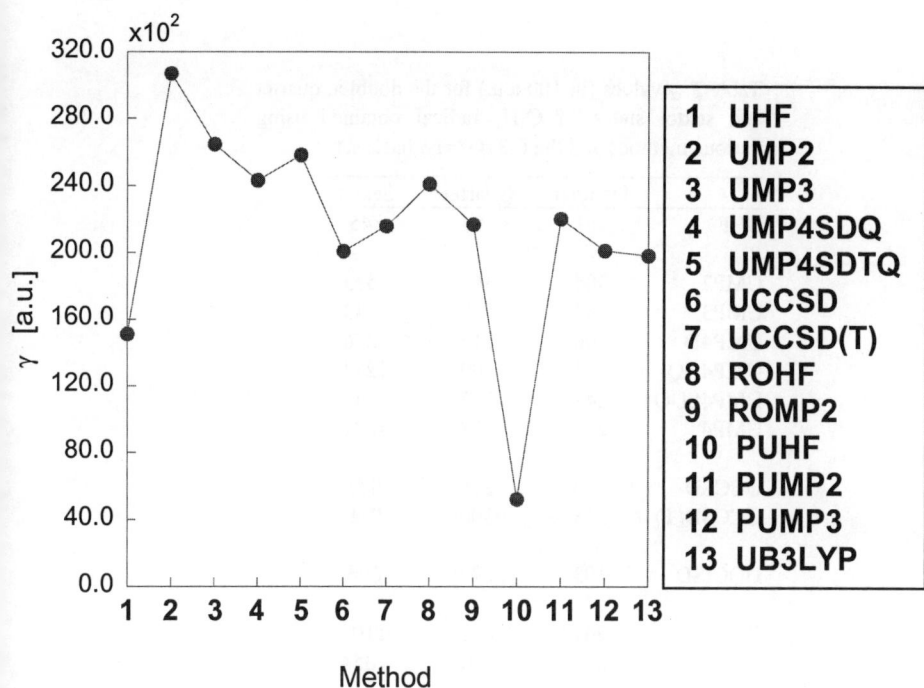

Fig. 14. Electron-correlation dependency of γ [a.u.] for doublet C_5H_7 radical. The UHF, UMP2, UMP3, UMP4SDQ, UMP4SDTQ, UCCSD, UCCSD(T), PUHF, PUMP2, PUMP3 and UB3LYP results using 6-31G*+pd basis sets are shown.

level, this increase is slightly reduced. The low-order spin-projected PUMP2 value is close to the high-order correlated CCSD and CCSD(T) values. This suggests that, in the case of open-shell neutral systems in low spin states, a fast convergence of the γ value with respect to the order of electron correlation requires first the removal of spin contamination. The UQCISD (193×10^2 a.u.) and UB3LYP (198×10^2 a.u.) results are very similar to the best estimates of γ. The restricted open-shell treatments provide γ values which are both in close agreement with the UCCSD(T) results [241×10^2 a.u. (ROHF) and 217×10^2 a.u. (ROMP2)], showing a fast convergence as a function of the inclusion of electron correlation similarly to the unrestricted and projected approaches.

Masayoshi Nakano

Table 3. γ values (in 100 a.u.) for the doublet, quartet and sextet states of C_5H_7 radical obtained using various methods and the 6-31G*+pd basis set.

	Doublet	Quartet	Sextet
UHF	151	184	845
UMP2	304	366	1533
UMP3	263	315	1343
UMP4D	266	315	1356
UMP4DQ	253	300	1284
UMP4SDQ	242	297	1145
UMP4	258	328	1242
UCCSD	200	274	872
UCCSD(T)	215	300	931
UQCISD	193	272	696
ROHF	241	606	1106
ROMP2	217	-242	2354
PUHF	53	147	832
PUMP2	220	333	1539
PUMP3	201	295	1337
B3LYP	198	407	227

Further insights into the spin contamination effects on the γ of the doublet are provided by the expectation value of S^2 listed in Table 4 for HF and MP2 levels of approximations. In particular, at the UHF level the spin contamination is not negligible and it is overcorrected using the PUHF (l=1) scheme. The γ densities as well as their α- and β-electron components have been determined at the UHF, UQCISD, and UB3LYP levels and are shown in Fig. 15. At the UHF level, the β-electron contribution (81×10^2 a.u.) is larger than its α counterpart (76×10^2 a.u.) though the doublet C_5H_7 possesses an excess of α π-electrons.

Table 4. Expectation value of S^2 for the doublet, quartet and sextet states of C_5H_7 radical obtained using various methods and the 6-31G*+pd basis set.

	UHF	UMP2	PUHF	PUMP2	Exact
Doublet	1.171	1.056	0.684	0.728	0.750
Quartet	3.911	3.847	3.746	3.750	3.750
Sextet	8.815	8.770	8.749	8.750	8.750

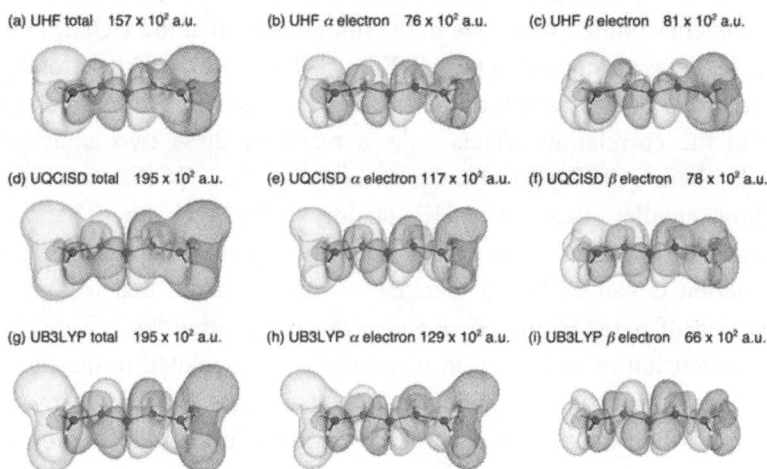

Fig. 15. γ density distributions of the total, α and β electron contributions obtained at the UHF, UQCISD and UB3LYP levels. The yellow and blue meshes represent positive and negative iso-surfaces with ±20 a.u., respectively.

In contrast, their relative contributions are inverted upon inclusion of electron correlation. Indeed, at the UQCISD level, the α-electron contribution is enhanced by 54 % (117×10^2 a.u.) whereas the β-electron contribution is slightly reduced (78×10^2 a.u.). At the UHF and UQCISD levels, the main contributions to γ originate from the π-electron γ densities located at the molecule extremities and the amplitudes are the largest at the correlated level (Figs. 15(a) and 15(d)). At the UHF level (see Figs. 15(b) and 15(c)), the difference between the α- and β-electron contributions comes from the most-contributing end regions where the amplitudes of the β-electron γ densities are larger. The γ density delocalization is also observed for the β-electron contributions in the internal chain region. The larger β-electron γ densities are related to the presence of β-hole in a π-symmetry orbital. Indeed, due to Pauli principle, the β π-electrons are predicted to fluctuate more significantly. This will be referred to as the Pauli effect and is responsible for the larger UHF γ densities of the β electrons in both end regions.

The small reduction of the β-electron γ densities at the UQCISD level compared to the UHF case can be associated with the corresponding reduced spin polarization (Figs. 16(a) and 16(b)). In addition, the β π-electron contribution increases in the middle region at the UQCISD level (Fig. 15(f)) as compared to the UHF level (Fig. 15(c)). This is related to the increase of delocalization of the β π-electrons in the middle region due to the correlation effects. As a result of these two antagonistic effects, the UQCISD β-electron contribution at the UQCISD level is only slightly smaller than its UHF analog. On the other hand, the enhancement of α π-electron contribution upon including electron correlation is caused by an increase of delocalization that leads to the extension of outer region of α π-electron distributions (see Fig. 15(e)). Such extension of outer region is predicted to be related to the fact that significant delocalization in internal chain region is restricted by Pauli principle while the α-spin density is important in the outer region (Fig. 16(b)).

Therefore, correcting the overestimated UHF spin polarization leads to a small reduction of the β π-electron contributions whereas including electron correlation significantly increases the α π-electron contributions so that, at the UQCISD level, the γ contribution per π-electron is similar for the α and β electrons whereas at the UHF level, the β component per π-electron is larger. These tendencies are substantiated by the γ reduction upon using spin-projected methods as well as by the enhancement of γ by the MPn methods. In addition, the UCC and UQCI calculations can involve both electron correlation and spin polarization corrections.

Finally, the applicability of the UB3LYP method to reproduce the γ of doublet state is examined. The spin density at the UB3LYP level closely maps the UQCISD one (Figs. 16(b) and 16(c)) while most of the

(a) Doublet   UHF   (b) Doublet   UQCISD   (c) Doublet   UB3LYP

(d) Quartet   UHF   (e) Quartet   UQCISD   (f) Quartet   UB3LYP

(g) Sextet   UHF   (h) Sextet   UQCISD   (i) Sextet   UB3LYP

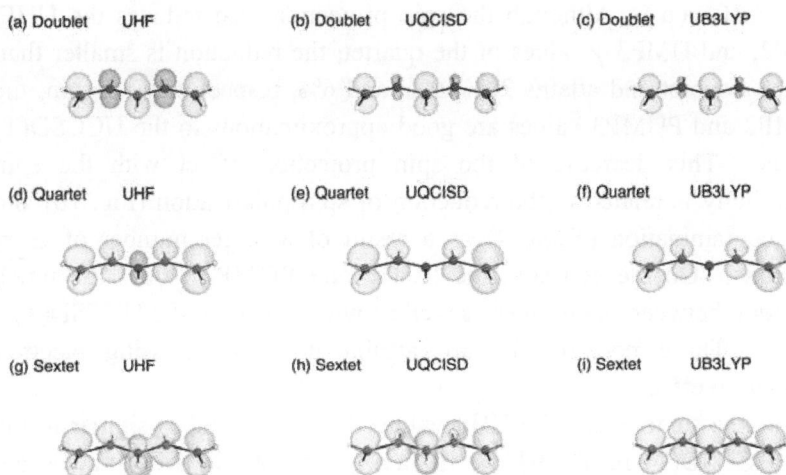

Fig. 16. Spin density distributions for the doublet, quartet and sextet states evaluated at the UHF, UQCISD and UB3LYP levels. The yellow and blue meshes represent α and β spin densities with iso-surface 0.02 a.u., respectively

γ contribution (66%) also comes from the α electrons, which is a bit larger than at the UQCISD level (60%). Although the primary features of γ densities are similar at the UB3LYP and UQCISD levels, the UB3LYP α electron γ density is slightly larger in the outer regions (Fig. 15(h)) whereas for the β electron contribution a decrease is observed at the extremities (Fig. 15(i)).

5.3. *Spin multiplicity effects on γ of C_5H_7 radical*

The spin multiplicity effects on γ are investigated by comparing the doublet, quartet and sextet states of the C_5H_7 radical (Table 3). At the exception of UB3LYP and ROMP2, all methods predict an enhancement of γ with the spin multiplicity, with a larger increase between the quartet and the sextet than between the doublet and the quartet. Similarly to the doublet state, the γ of the quartet is enhanced by almost a factor of 2 when adding second-order electron correlation corrections (UMP2 : 366×10^2 a.u.) to the UHF result (184×10^2 a.u.) while higher-order corrections at the UCCSD(T) level reduce this enhancement by 18 %

$(300 \times 10^2$ a.u.). Although the spin projection also reduces the UHF, UMP2, and UMP3 γ values of the quartet, the reduction is smaller than for the doublet and attains 20%, 9%, and 6%, respectively. Again, the PUMP2 and PUMP3 values are good approximations to the UCCSD(T) results. This decrease of the spin projection effect with the spin multiplicity is related to the reduction of spin polarization (Fig. 16) and spin contamination (Table 4) as a result of a larger number of α π-electrons. On the contrary, the ROHF and ROMP2 values are much different, between themselves as well as with respect to the UCCSD(T) γ value. These poor results are attributed to the missing electron correlation effects.

For the sextet state, the UHF value $(845 \times 10^2$ a.u.) is similar to the best UCCSD(T) result $(931 \times 10^2$ a.u.), whereas the MPn values are strongly overestimated. Contrary to the case of the quartet, the spin projection effect is shown to be negligible as expected (Table 4). Moreover, higher-order electron correlation effects at the UCCSD and UCCSD(T) levels are necessary for correcting the overshot UMPn and PUMPn values. The ROHF and ROMP2 values are larger than their unrestricted (and projected) analogs showing again the impact of the missing electron correlation effects.

Although the UB3LYP method reproduces the UCCSD(T) γ value of the doublet state, it overshoots the γ of the quartet state by 36% while it significantly undershoots (by 76%) the sextet γ value. For the quartet, the deficiency of UB3LYP method seems to originate in the self-interaction error associated with the large CC bond lengths. Indeed, it was pointed out by Mori-Sánchez et al. [39] that approximate exchange-correlation functionals incorrectly describe the polarizability of weakly-interacting molecules with a fractional charge. The alternation between large γ underestimation and large γ overestimation for different spin state, which is obtained when using DFT schemes with usual exchange-correlation functionals, seem to be caused by the drawbacks of exchange-correlation functional related to their shortsightedness [40-42]. More investigation is required to elucidate the origin of these failures.

Fig. 17 compares the UQCISD total, α- and β-electron γ densities of the three spin states. In the quartet state (Fig. 17(d-f)), the total γ density contribution is composed of two large and delocalized positive π-

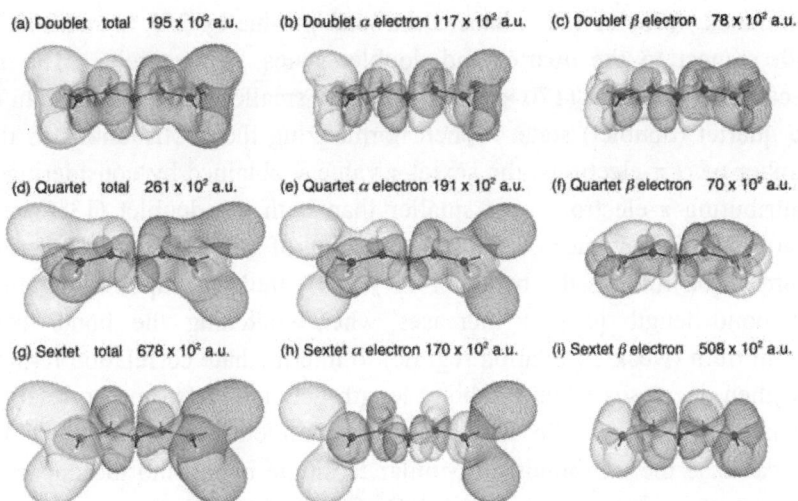

(a) Doublet total 195 x 10^2 a.u.   (b) Doublet α electron 117 x 10^2 a.u.   (c) Doublet β electron 78 x 10^2 a.u.

(d) Quartet total 261 x 10^2 a.u.   (e) Quartet α electron 191 x 10^2 a.u.   (f) Quartet β electron 70 x 10^2 a.u.

(g) Sextet total 678 x 10^2 a.u.   (h) Sextet α electron 170 x 10^2 a.u.   (i) Sextet β electron 508 x 10^2 a.u.

Fig. 17. γ density distributions for the total, α and β electron contributions of the doublet, quartet and sextet states determined at the UQCISD level. The yellow and blue meshes represent positive and negative γ densities with iso-surfaces with ±20 a.u. for doublet and quartet states and ±40 a.u. for sextet state.

electron distributions as well as of two smaller delocalized negative σ-electron distributions. As compared to the doublet (Fig. 17(a-c)), these distributions are enhanced at the level of the end CC bonds due to the β electron contribution (70×10^2 a.u.) as well as in the outer region due to the remarkable extension of the α electron γ density (191×10^2 a.u.). However, when normalizing the contributions to the number of α or β π-electrons, the relative β contribution increases substantially when going from the doublet (39×10^2 a.u. per β π-electron) to the quartet (70×10^2 a.u. per β π-electron) whereas for the α contribution the increase is smaller: from 39×10^2 a.u. per α π-electron in the doublet to 48×10^2 a.u. in the quartet. Such effective enhancements of the α- and β-electron contributions and their delocalized distributions in the quartet state can be explained by considering the intermediate π bond breaking nature of the outer CC bonds ($R_1 = 1.501$ Å) of the quartet state which is associated with larger γ values [13].

In contrast to the quartet state, in the sextet state (Fig. 17(g-i), Table 3), the β electron contribution is so large (508×10^2 a.u.) that it dominates γ (75 %) and enhances the total γ value by 255 % and 360 % with respect to the quartet and doublet states, respectively. The α-electron contribution (170×10^2 a.u) is 11% smaller (45% larger) than in the quartet (doublet) state. When normalizing the contributions to the number of α π-electrons, the sextet γ value – obtained by considering 5 contributing π-electrons – is smaller than both the doublet (13%) and quartet (28%) values. This reduction of the sextet α-electron contribution follows the behavior of γ in H_2 molecule upon elongating the bond length [13]: γ increases when stretching the bond from equilibrium (weak correlation regime) to intermediate correlation regime and then decreases when the bond length gets larger (strong correlation regime). Of course, the analogy is not complete because in the C_5H_7 radical case the CC bonds are similar to single bonds and the unpaired electrons and sites are not identical like in H_2. Actually, the α π-electron γ density (Fig. 17(h)) is particularly extended in the outer region like for the quartet (Fig. 17(e)) and contributes positively to γ. A small localized π-electron feature with negative contributions is located in the internal chain region and could originate in the strong correlation among the 5 α π-electrons (one on each C site) due to Pauli principle. Moreover, σ-electrons bring an additional negative contribution in such a way that the α-electron contribution slightly decreases when going from the quartet to the sextet.

In contrast, as seen from Fig. 17(i), the β electron contribution is larger and positive for the σ-electrons while smaller and negative for the π-electron. In order to elucidate this different hyperpolarization effects in the sextet state, we investigate at the UQCISD level the natural orbitals. Together with their occupation numbers they are displayed in Fig. 18. It is found that in the sextet state the order of the lowest π-orbital **16** and the highest σ-orbital **17** is inverted with respect to the doublet and quartet states in such a way that an unpaired α electron lies in σ-orbital **17** instead of the π-orbital **16**, which is doubly occupied. This feature is understood by the fact that the σ-orbital **17** is composed of localized CC σ-bond distributions with mutually opposite phase while the π-orbital **16** has a delocalized distribution over the entire chain region. As a result, there are six π-electrons (five α and one β) as well as one α electron and a β-hole in a σ-orbital. This feature supports the spin

(a) 16 π $n = 1.922$ (b) 17 σ $n = 1.032$ (c) 18 π $n = 1.013$

(d) 19 π $n = 0.9929$ (e) 20 π $n = 0.9850$ (f) 21 π $n = 0.9829$

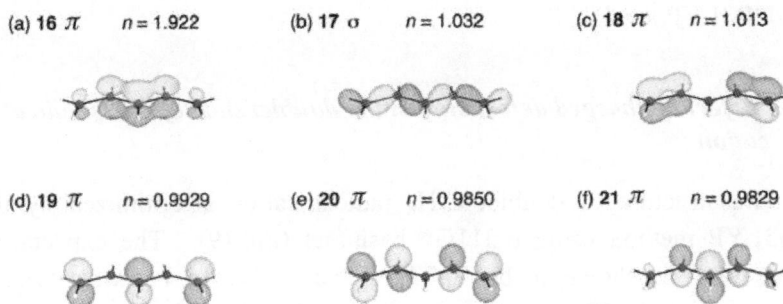

Fig. 18. Natural orbitals and their occupation numbers for the highest **16-21** in sextet state at the UQCISD level. The yellow and blue meshes represent positive and negative iso-surfaces with ± 0.08 a.u., respectively.

distribution of the sextet state at the UHF and UQCISD levels, which presents a partial σ-symmetry character as shown in Figs. 16(g) and 16(h). The presence of a β-hole in the σ-symmetry orbital **17** enables large field-dependent fluctuations of the β-electron density and accounts for the substantial β-electron contribution to γ. On the other hand, the fluctuations of the β-electron in the doubly occupied π-orbital **16** are small because of the stable bonding nature of the orbital. The γ contribution of this π-electron of spin β, which is shown in the middle three C sites region, is negative like for those systems where low-energy virtual excitation processes (type II) dominate the response [11,12]. In analogy to Refs. [13] and [17], the α electrons belong to the strong correlation regime due to the Pauli effects between the five electrons of same spin while the β-electrons belong to intermediate correlation regime. At the UB3LYP level, the σ-orbital of highest energy is below the five singly-occupied π-orbitals so that there is no β-hole in the σ-symmetry orbital. This is confirmed by the UB3LYP spin density of the sextet (Fig. 16(i)). The strong γ underestimation is therefore explained by the remarkable decrease of the β-electron contribution (314 a.u.) due to the non-existence of β σ hole and β π electrons and the smaller α-electron contribution (223×10^2 a.u.) than that (337×10^2 a.u.) in the quartet state at the UB3LYP level, the feature of which is associated with the strong correlation regime for the α π electrons in the sextet state at the UB3LYP level.

5.4. *Effect of charged defect on γ in the doublet state of C_6H_8 radical cation*

The structure of doublet C_6H_8 radical cation is optimized by the UB3LYP method using 6-311G* basis set (Fig.19). The expectation value $<S^2>$ is shown to be 0.779. The 6-31G*+p basis set with p exponent of 0.0523 on C atom is employed. Various levels of approximate results using 6-31G*+p basis set are found to reproduce about 80%-90% of those using 6-31G*+pd basis set [19]. For the analysis of electron correlation effects on γ, the UHF-based schemes are employed and include the UMPn (n=2-4)), the UCCSD, UCCSD(T) and UQCISD methods. In order to elucidate the effects of spin contamination corrections on γ, the PUHF (l=1), PUMP2(l=1) and PUMP3(l=1) are applied. At the HF and MP2 levels, the ROHF and ROMP2 are also employed. Among the DFT schemes, the hybrid B3LYP exchange-correlation functional is adopted. All calculations have been performed using the Gaussian 98 program package [29]. We confine our attention to the longitudinal component (largest component) of the static γ value, which is a good approximation to off-resonant values. The static γ quantities can be obtained by adopting the FF approach (Eq. (6)). The γ density analysis [23,30] is employed in order to characterize the spatial contributions of γ as well as of its α- and β- spin components.

Figure 20 shows the effects of electron correlation and spin projection on the γ value of the C_6H_8 radical cation in the doublet state using 6-31G*+p basis set in comparison with the results on the C_5H_7 radical. The variation in γ of C_6H_8 radical cation for each calculation method except the ROHF method is similar to that of C_5H_7 radical though the γ value at the UCCSD(T) becomes negative in sign (-238 x 10^2 a.u.). The UMP2 correction is shown to enhance γ and change its sign (4387 x 10^2 a.u.) as compared to the UHF value (-2818 x 10^2 a.u.). Although higher-order electron correlation contributions in the UMP scheme cannot really correct for this overshooting, a dominant correction comes from the

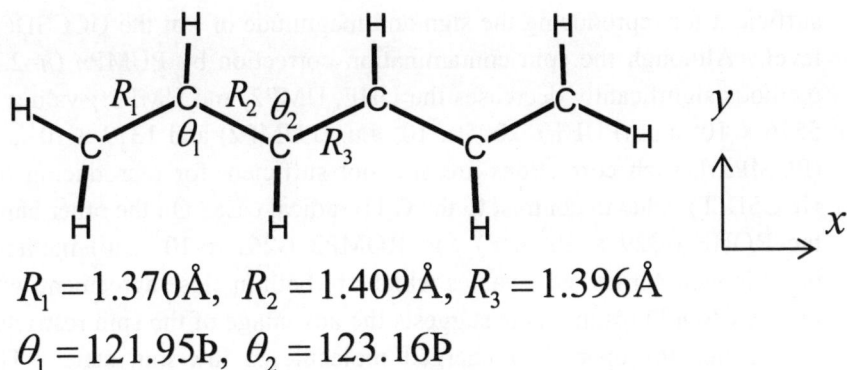

$R_1 = 1.370\text{Å}, \quad R_2 = 1.409\text{Å}, \quad R_3 = 1.396\text{Å}$

$\theta_1 = 121.95°, \quad \theta_2 = 123.16°$

Fig. 19. Structure of C_6H_8 radical cation in doublet state. The structure is planar and belong to the C_{2h} point group and is optimized by the UB3LYP method using 6-311G* basis set. The middle C-C bonds of all models have an angle of 30° with respect to the longitudinal (x) axis.

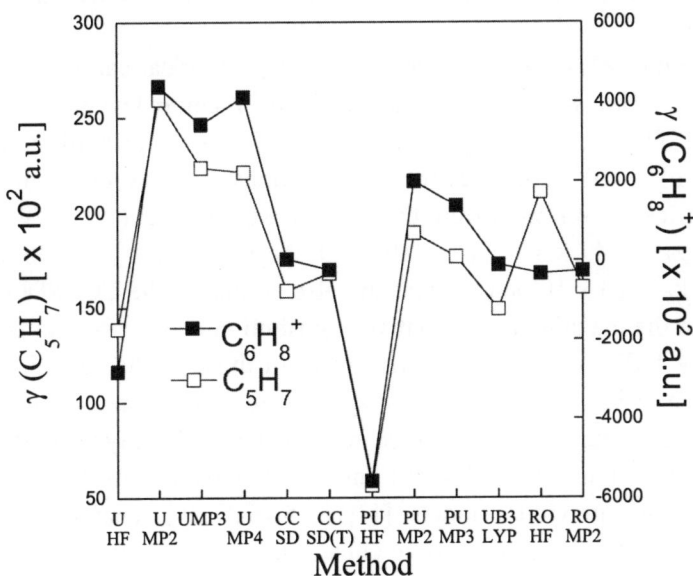

Fig. 20. Electron-correlation dependence of γ [a.u.] for doublet C_6H_8 radical cation and C_5H_7 radical. The UHF, UMP2, UMP3, UMP4SDQ, UMP4SDTQ, UCCSD, UCCSD(T), PUHF, PUMP2, PUMP3, UB3LYP, ROHF and ROMP2 results using 6-31G*+p basis sets are shown.

correlation at the UCCSD level (32 x 10^2 a.u.). Nevertheless, this is not sufficient for reproducing the sign and magnitude of γ at the UCCSD(T) level. Although the spin contamination correction by PUMPn (n=2,3) methods significantly decreases the UHF, UMP2, and UMP3 γ values [-5576 x 10^2 a.u. (PUHF), 2005 x 10^2 a.u. (PUMP2) and 1379 x 10^2 a.u. (PUMP3)], such corrections are still not sufficient for reproducing the UCCSD(T) value in contrast to the C_5H_7 radical case. On the other hand, the ROHF (-329 x 10^2 a.u.) and ROMP2 (-262 x 10^2 a.u.) methods provide well-converged γ values which are both in close agreement with the UCCSD(T) result. This suggests the advantage of the spin restricted approaches for open-shell charged molecule in low-spin state. The UB3LYP (-108 x 10^2 a.u.) result is also negative in sign, while its magnitude is about the half of that at the UCCSD(T) level. It is also noted that the UQCISD result (391 x 10^2 a.u.) is different from the UCCSD result (32 x 10^2 a.u.) and both are positive in sign though both methods significantly reduce the overshot behavior at the UMPn (n=2-4) level.

Figure 21 shows the γ densities of C_6H_8 radical cation at the ROHF level. The sign of γ at the UHF level coincides with that at the UCCSD(T) level, while the feature of α (γ_α = -2811 x 10^2 a.u.) and β (γ_β =14 x 10^2 a.u.) electron contributions to γ at the UHF level is significantly different from those at the ROHF level (γ_α = -79 x 10^2 a.u. and γ_β = -249 x 10^2 a.u.). Although the UB3LYP (γ_α = -46 x 10^2 a.u. and γ_β = -63 x 10^2 a.u.) results are also similar to the UCCSD(T) results, while the magnitude and relative contribution between α and β electrons at the UB3LYP level are different from those at the ROHF level. As shown in Sec. 2, the γ values of symmetric systems are composed of two types of virtual excitation processes, types II and III: type II(III) involves low(high)-lying excited states and has negative(positive) contributions. From our previous studies [11,12,15,16,23,24], charged defects are predicted to cause the relative enhancement of type II contributions and make the γ value negative. In the present case, although the radical (excess a π-electron) tends to increase the α-electron contribution with positive value by enhancing type III contribution similarly to the case of C_5H_7 radical, the effect of charged defect retains the negative γ value. This suggests that the enhancement of type II contribution caused by the

total   -329 x 10^2 a.u. α electron   -79 x 10^2 a.u. β electron   -249 x 10^2 a.u.

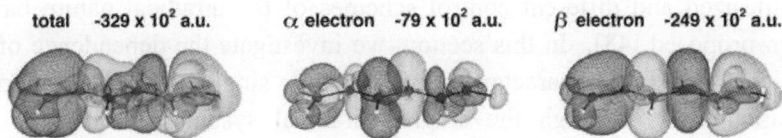

Fig. 21. γ density plots of C_6H_8 radical calculated by the ROHF using 6-31G*+p basis set. The yellow and blue regions indicate positive and negative γ densities with magnitude of 60 a.u, respectively.

introduction of charged defect overcomes the enhancement of type III contribution mainly caused by excess a π-electron. From Fig. 21, it is found that the dominant α-electron contribution with negative values mainly comes from the middle region of π-electron distributions, while the γ densities are significantly reduced in the both end regions, suggesting the increase in the positive (type III) contributions of a π-electrons in both end regions. In contrast, the β π-electrons show primary negative contributions with large magnitude in the both end regions, while smaller positive ones in the middle region. This indicates the negative contributions of β electrons as a whole.

In summary, the introduction of charged defect in radical systems tends to enhance negative contributions to γ. On the basis of the present results, a novel control scheme of third-order NLO properties will be possible by changing charge and spin states for a new class of NLO systems, i.e., open-shell charged conjugated molecules.

6. Diradical character dependence of γ for singlet diradical systems

For open-shell NLO systems, another way to optimize the NLO response consists in considering singlet diradical systems. As shown in the variation in γ of H_2 in the dissociation process [34], the variations in bond breaking nature, i.e., variations in the degree of electron correlation, is known to be well described by the diradical character [20,21], which is an index of the degree of diradical nature and ranges from 0 to 1 as a function of the bond dissociation. In fact, several diradical compounds with various diradical characters have been synthesized and different control schemes of the diradical nature have been proposed [43]. In this section, we investigate the dependence of γ on the diradical character using neutral singlet diradical model compounds. Although the singlet diradical systems are useless as molecular magnets, the singlet diradical systems with intermediate and large diradical characters, i.e., belonging to the intermediate and strong correlation regimes, are expected to exhibit γ values much larger than those with small diradical characters, i.e., belonging to the weak correlation regime. As a result, the diradical NLO systems are expected to be a candidate for tunable NLO systems controlled by a diradical character.

6.1. *Model systems and calculation methods for p-quinodimethane (PQM)*

Figure 22 shows the structure of the p-quinodimethane (PQM) molecule in its singlet ground state. It is described by two resonance forms, i.e., the quinoid and diradical forms. In general, the experimental structures of conjugated diradical systems are well reproduced by the spin-restricted (R) B3LYP and unrestricted (U) B3LYP methods. The optimized structure for the present system with D_{2h} symmetry ($R_1 = 1.351$ Å, $R_2 = 1.460$ Å and $R_3 = 1.346$ Å) at the UB3LYP level using 6-311G* basis set is the same as that at the RB3LYP level. The optimized parameters suggest that PQM presents a large degree of quinoid form instead of diradical nature since R_1 and R_3 are like a C-C double bond length and are smaller than R_2 having a C-C single bond nature. In order to increase the degree of diradical nature, we consider several PQM models with bond-length R_1 changing from 1.35 Å to 1.7 Å under the constraint of $R_2 = R_3 = 1.4$ Å. We first determine the diradical character from UHF calculations. The diradical character y_i related to the HOMO-i and LUMO+i is defined by the weight of the doubly-excited configuration in the MC-SCF theory and is formally expressed in the case of the spin-projected UHF (PUHF) theory as [17,18]

$$y_i = 1 - \frac{2T_i}{1 + T_i^2} \qquad (11)$$

where T_i is the orbital overlap between the corresponding orbital pairs. T_i can also be represented by using the occupation numbers (n_j) of UHF natural orbitals (UNOs):

$$T_i = \frac{n_{\text{HOMO}-i} - n_{\text{LUMO}+i}}{2} \qquad (12)$$

(a) *p*-quinodimethane (PQM) model

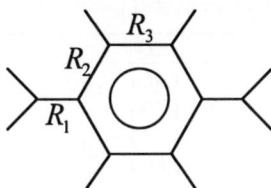

(b) Resonance structure

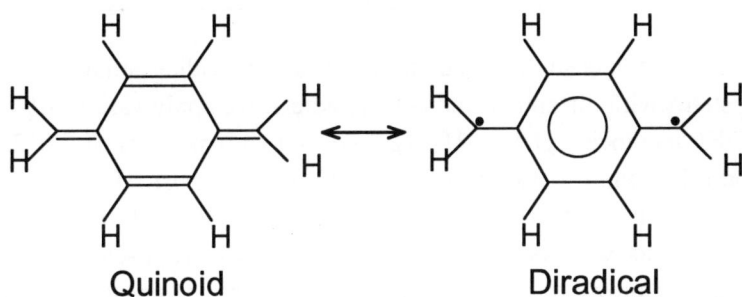

Quinoid   Diradical

Fig. 22. Geometrical parameters of the *p*-quinodimethane (PQM) molecule (a) and its resonance forms (b).

Since the PUHF diradical characters amount to 0 % and 100 % for closed-shell and pure diradical states, respectively, y_i represents the diradical character, i.e., the instability of the chemical bond. Table 5 gives the diradical characters y calculated from Eqs. (11) and (12) using HOMO and LUMO of UNOs for PQM systems with various R_1 values. As expected, it is found that the diradical character y increases when increasing the bond length R_1 starting from the optimized (equilibrium) geometry, which possesses a low y value (0.146), i.e., quinoid like structure.

Table 5. Diradical character y for R_1, R_2 and R_3 in the PQM models shown in Fig. 22 (a). The first row corresponds to the optimized equilibrium geometry with D_{2h} symmetry in the singlet ground state.

R_1	R_2	R_3	
1.351	1.460	1.346	0.1
1.350	1.400	1.400	0.2
1.400	1.400	1.400	0.3
1.450	1.400	1.400	0.4
1.500	1.400	1.400	0.4
1.560	1.400	1.400	0.5
1.600	1.400	1.400	0.6
1.700	1.400	1.400	0.7

In order to choose the calculation methods providing reliable γ values for systems with different diradical characters, we apply various *ab initio* and DFT methods. The 6-31G*+p basis set (p exponents of 0.0523 on carbon (C) atoms) is used [26]. The R(U)HF and *post*-HF methods are employed. The *post*-R(U)HF methods include the R(U)MPn (n=2-4), the R(U)CCSD as well as with R(U)CCSD(T). The RHF-based *post*-HF methods are known to break down in the dissociation (strong correlation) regime since the starting RHF solution is heavily triplet-unstable [17,44,45]. It is known that even higher-order excitation operators, such as SD(T) in RCC treatment [17], are not necessarily adequate for removing the deficient behavior for many electron systems. On the other hand, the UHF and UMPn methods are known to significantly suffer from spin contamination effects and give a small hump in the intermediate bond-dissociation region of the potential energy curve [17] and exhibit incorrect variation in γ in the dissociation process of diatomic molecules [13]. Therefore, the approximate spin projected UHF and UMP2, i.e., APUHF and APUMP2 [46], have also been applied in order to highlight the effects of spin contamination corrections on γ. In this scheme, the lowest spin (LS) UHF-based solutions are projected only by the corresponding highest spin (HS) solutions. The APUHF X energy is given by [46]

$$^{LS}E_{APUHF\,X} = {}^{LS}E_{UHF\,X} + f_{SC}[{}^{LS}E_{UHF\,X} - {}^{HS}E_{UHF\,X}] \qquad (13)$$

where $^{Y}E_{UHF\,X}$ denotes the total energy of the spin state Y determined at the X *post*-HF level based on the UHF solution. The factor f_{SC} for a spin multiplet $(2s+1)$ is the fraction of spin contamination given by

$$f_{SC} = \frac{{}^{LS}\left\langle S^2 \right\rangle_{UHF\,X} - s(s+1)}{{}^{HS}\left\langle S^2 \right\rangle_{UHF\,X} - {}^{LS}\left\langle S^2 \right\rangle_{UHF\,X}} \qquad (14)$$

The AP UHF X methods (X = MPn, CC, etc.) have been successfully applied to the calculations of potential energy surfaces [17], effective exchange integrals in the Heisenberg models for open-shell clusters [47-49] and hyperpolarizabilities of H_2 in the bond breaking region [13]. Among the density functional theory (DFT) schemes, the spin-restricted (R) and unrestricted (U) hybrid B3LYP and BHandHLYP exchange-correlation functionals have been adopted. All calculations have been performed using the Gaussian 98 program package [29]. We confine our attention to the dominant longitudinal component of static γ, which is considered to be a good approximation to off-resonant dynamic γ. The static γ can be obtained by the FF approach [Eq. (6)].

6.2. *Diradical character dependence of γ using PQM models*

Table 6 provides the γ values for PQM models with different diradical characters at the RHF, UHF, RMP2, UMP2, APUMP2, UCCSD(T) and UBHandHLYP levels. The γ values are given in atomic units (a.u.), which can be converted to other units by the relation: 1 a.u. = 5.0366 x 10^{-40} esu.

Table 6. γ [x 10^2 a.u.] for PQM models with different diradical characters y at the RHF, UMF, RMP2, UMP2, APUMP2, UCCSD(T) and UBHandHLYP levels.

y	RHF	UHF	RMP2	UMP2	APUMP2	UCCSD(T)	UBHandHLYP
0.146	$\overline{7.709}$	260.8	273.0	765.4	433.1	232.7	546.9
0.257	$\overline{78.84}$	218.5	477.0	630.5	468.4	459.6	641.1
0.335	$\overline{143.2}$	204.0	774.8	596.9	559.9	628.4	678.0
0.414	$\overline{243.0}$	184.8	1305	544.7	606.9	749.0	688.1
0.491	$\overline{393.6}$	164.6	2249	478.5	575.5	775.4	671.6
0.576	$\overline{671.9}$	141.7	4428	410.1	515.7	737.2	620.6
0.626	$\overline{941.3}$	127.4	7040	363.3	427.5	680.6	572.5
0.731	$\overline{2103}$	99.28	23250	255.1	132.2	488.1	433.7

Figure 23 shows the results calculated by the RHF-based methods as well as the UCCSD(T) result. Among the different levels of approximation, the UCCSD(T) method is considered to provide the most reliable results for the present systems. As expected, it turns out that the γ value at the UCCSD(T) level increases with increasing y in the weak diradical region (y = 0.1-0.5), attains a maximum in the intermediate region ($y \approx 0.5$) and then decreases in the region corresponding to large diradical character ($y > 0.5$). The maximum γ value at the UCCSD(T) level (77500 a.u.) for a y value close to 0.5 is 3.3 times as large as the value at the equilibrium geometry (23300 a.u.) (y = 0.146). The RHF γ value is negative while its magnitude increases strongly with the diradical character. The inclusion of electron correlation at the RMPn (n = 2-4) levels cannot correct this deficient behavior: at the RMP2 and RMP3 levels γ is positive but displays a sharp and overshot increase with y, while at the RMP4 level the evolution as a function of y is correctly

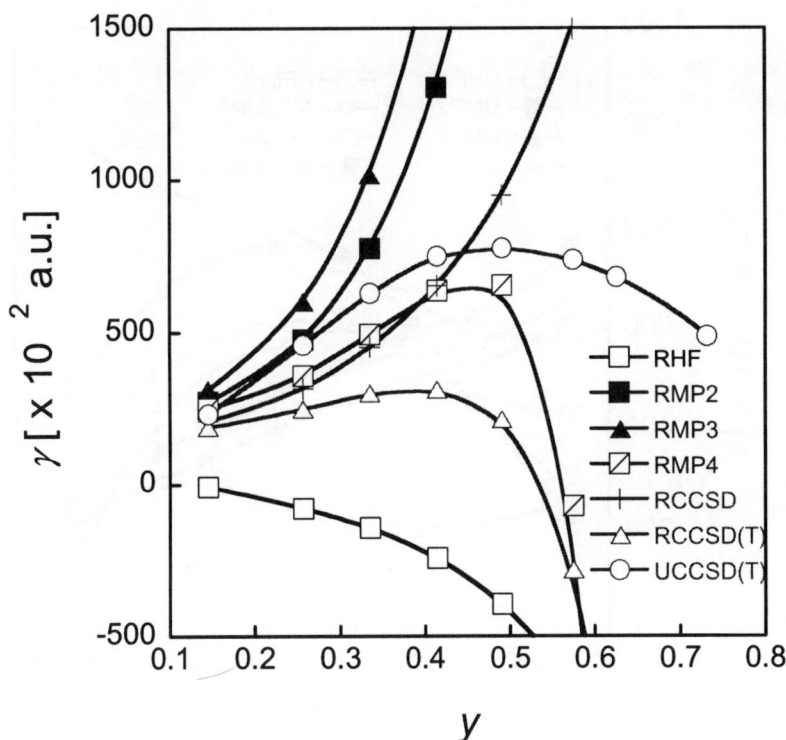

Fig. 23. Diradical character (y) dependence of γ [a.u.] for the PQM models. The RHF, RMP2, RMP3, RMP4, RCCSD ,and RCCSD(T) results as well as the UCCSD(T) results using 6-31G*+p (ζ=0.0523) basis sets are shown.

reproduced up to $y = 0.4$ but deviates towards too negative values for larger y values. Even the RCC methods cannot correct such wrong behavior though the RCCSD and RCCSD(T) methods rather suppress the overshooting behavior of the RMPn results ($n = 2, 3$). Such incorrect behavior is predicted to originate from the triplet instability of the RHF solution in the intermediate and strong correlation regimes [13,17].

Figure 24 shows the results obtained by the UHF-based methods. Although for the equilibrium geometry the UHF method nicely reproduces the UCCSD(T) γ value, the UMPn (n=2-4) methods overshoot γ. At the UHF and UMPn levels γ monotonically decreases

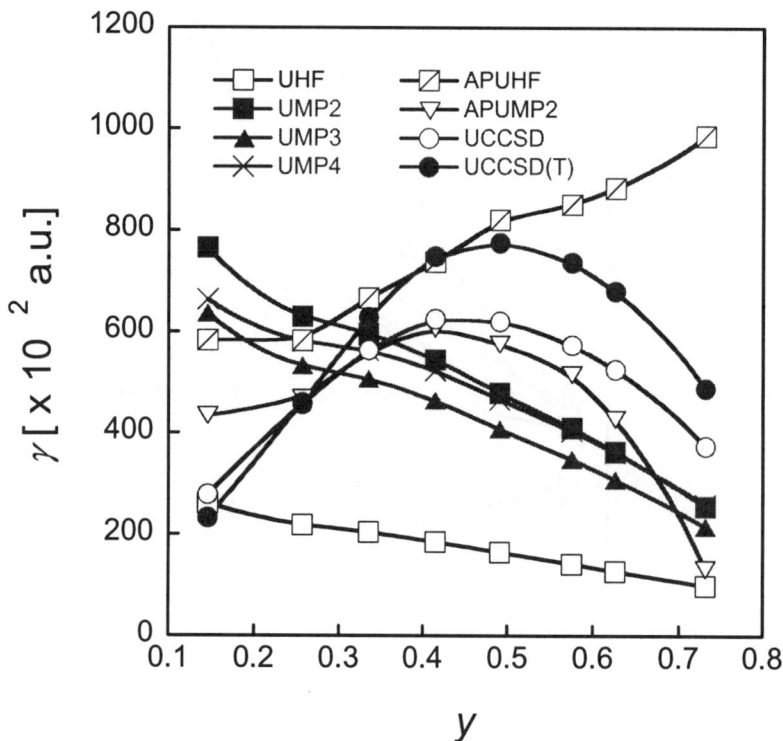

Fig. 24. Diradical character (y) dependence of γ[a.u.] for the PQM models. The UHF, UMP2, UMP3, UMP4, APUHF, APUMP2 UCCSD, and UCCSD(T), results using 6-31G*+p basis sets are shown.

with increasing diradical character, so that for y smaller than 0.3 they are larger than the UCCSD(T) reference value, whereas they are smaller in the intermediate and large diradical character regions.   Adopting the APUHF method, the behavior of γ for y is different: γ increases with increasing diradical character. In contrast, the γ values determined at the APUMP2 level are significantly improved as compared to those at the UHF, UMPn and APUHF levels and qualitatively reproduce the results obtained at the UCCSD level. To obtain a better agreement with the UCCSD(T) results, higher-order correlation methods, i.e., APUMP3 and APUMP4 methods, are predicted to be required. This suggests that not only the non-dynamical correlation involved at the APUHF level but also the dynamical correlation involved at the APUMP*n* levels are indispensable for the description of γ of open-shell singlet systems. Similarly to the case of neutral doublet C_5H_7 radical system [19], obtaining a fast convergence of γ for open-shell singlet systems with respect to the order of electron correlation requires first the removal of spin contamination. The UCCSD method turns out to reproduce qualitatively the variations in γ obtained at the UCCSD(T) level, whereas the triple excitations become important for a quantitative description in the regions with intermediate and large diradical characters ($y > 0.25$).

Figure 25 displays the diradical character dependences of γ evaluated by DFT methods as well as the UCCSD(T) scheme for comparison. The RB3LYP and RBHandHLYP methods undershoot the UCCSD(T) γ value. In addition, although for equilibrium geometry they provide similar γ values, the RB3LYP value increases with y, whereas it decreases using the RBHandHLYP exchange-correlation functional. The UB3LYP results are found to give the same results as the RB3LYP results until $y \approx 0.25$, while an abrupt increase is observed around $y = 0.3$. After the abrupt increase, γ overshoots the UCCSD(T) value and then increases steadily. In contrast, the UBHandHLYP method provides a similar qualitative description of the variation in γ to that at the UCCSD(T) level. Nevertheless, the γ values at the UBHandHLYP level overshoot that at the UCCSD(T) level in the region with small diradical character ($y < 0.3$), while the γ values at the UBHandHLYP level are slightly smaller than that at the UCCSD(T) level in the region with intermediate and large diradical characters ($y > 0.4$). Moreover, the value of y corresponding to the maximum of γ is slightly different at the UBHandHLYP and UCCSD(T) levels of approximation.

Judging from the present results, it is predicted that hybrid DFT methods with exchange-correlation functionals specifically-tuned for reproducing the NLO properties of diradical species could provide satisfactory results in comparison with the more elaborated UCCSD(T) scheme. However, it remains to be assessed whether a given tuned functional can perform efficiently for a large range of diradicals.

The variation in γ for the PQM models with increasing diradical character can be understood from the analogy with the dissociation of H_2.

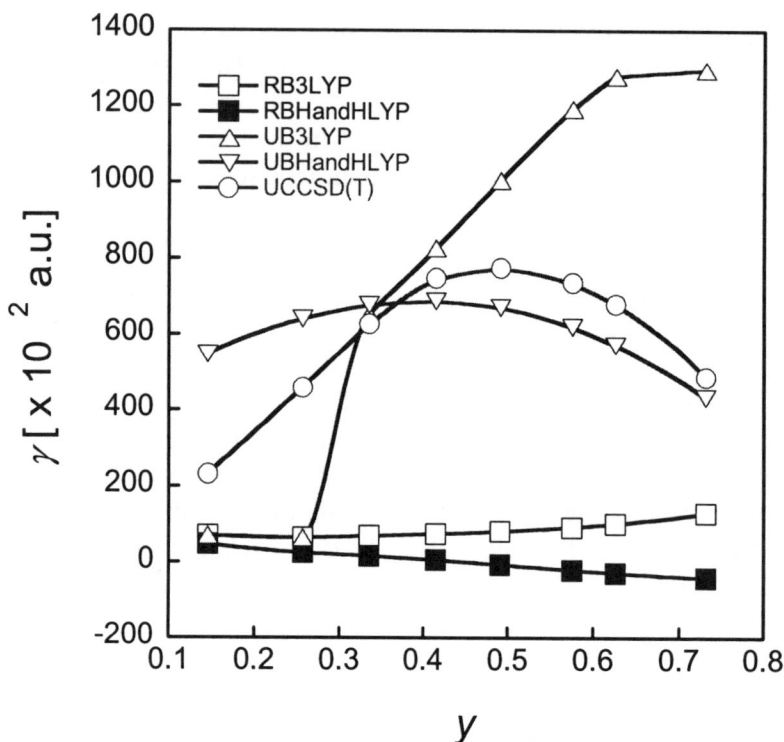

Fig. 25. Diradical character (*y*) dependence of γ [a.u.] for the PQM models. The RB3LYP, RBHandHLYP, UB3LYP and UBHandHLYP results as well as the UCCSD(T) results using 6-31G*+*p* basis sets are shown.

The increase of γ in the intermediate diradical character region is predicted to be caused by the virtual excitation processes with zwitterionic contribution between the radicals on both-end carbon atoms, which corresponds to the intermediate dissociation region for the H_2 molecule [13]. In an analogical way, the intermediately spatial polarized wavefunctions for α and β radical spins on both-end carbon sites in the PQM with intermediate radical character are predicted to contribute to the enhancement of γ through the virtual excitation processes involving the zwitterionic natureas compared to the case in the small diradical character region with a stable bond nature, giving a relatively small polarization. On the other hand, in the large radical character (strong correlation or magnetic) region, the localized spins on both-end sites exhibit less charge polarization to the applied external electric field due to its strong correlation nature, so that the γ value decreases again.

6.3. *Comparison of γ between BI2Y and BI2YN*

Figures 26(a) and 26(b) show optimized structures of 1,4-bis-(imidazole-2-ylidene)-cyclohexa-2,5-diene (BI2Y) and its related molecule (BI2YN) at the UB3LYP/6-31G** level, respectively. It is expected that two resonance forms, i.e., quinoid and diradical forms, contribute to BI2Y (a), while the diradical form little contributes to BI2YN (b) because of the lone pairs of N atoms linked with the middle

(a) BI2Y

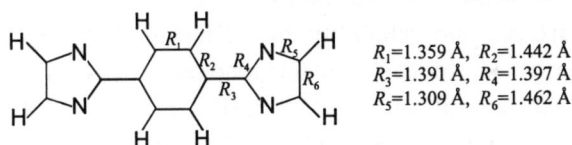

R_1=1.359 Å, R_2=1.442 Å
R_3=1.391 Å, R_4=1.397 Å
R_5=1.309 Å, R_6=1.462 Å

Resonance structures

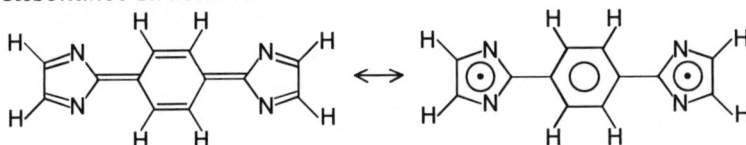

(b) BI2YN

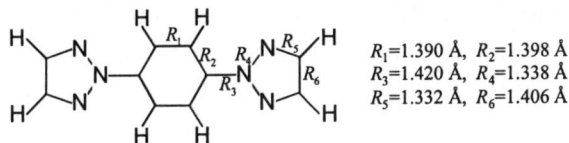

R_1=1.390 Å, R_2=1.398 Å
R_3=1.420 Å, R_4=1.338 Å
R_5=1.332 Å, R_6=1.406 Å

Resonance structure

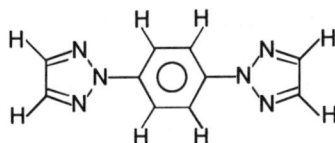

Fig. 26. Geometrical structures of BI2Y (a) and BI2YN (b) optimized by the B3LYP/6-31G**. Resonance structures are also shown.

aromatic benzene ring. In fact, molecules (a) and (b) are turned out to exhibit, respectively, an intermediate (y = 0.423) and very small (y = 0.010) diradical character (y), which is calculated from the UHF natural orbitals (UNOs). The diradical character is an index ranging between 0 (closed-shell) and 1 (complete diradical). We use the 6-31G*+p basis set with p exponent of 0.0523 on carbon (C) atoms and 0.0582 on nitrogen (N) atoms. We confine our attention to the dominant longitudinal components of static γ. The static γ is calculated by the FF approach [Eq. (6)]. From the results for the PQM models (Sec. 6.2), the spin-unrestricted hybrid DFT method (UBHandHLYP) is shown to give reliable γ values for diradical molecules with intermediate diradical characters, while the spin-restricted BHandHLYP (RBHandHLYP) method is shown to give reliable γ values for diradical molecules with small diradical characters or closed-shell molecules, at least for molecules of the size investigated here. Furthermore, using the 6-31G basis set, the reliability of UBHandHLYP to estimate γ of BI2Y is confirmed. Indeed, its value amounts to 174, -40, 939, 690, 484, and 524 x 10^3 a.u at the UHF, PUHF (l = 3), UMP2, PUMP2 (l = 3), UBHandHLYP, and UCCSD(T) levels of approximation, respectively. We therefore apply the R(U)BHandHLYP methods. For comparison, we also apply the UHF and UMP2 methods to the calcuation of γ for BI2Y (a) and the spin-restricted HF (RHF) and RMP2 methods to BI2YN (b).

The γ values at the UHF (2002 x 10^2 a.u.) and UMP2 (9962 x 10^2 a.u.) levels are shown to undershoot and overshoot that (6534 x 10^2 a.u.) at the UBHandHLYP level, respectively. These deficiencies are caused by the spin contamination at the UHF and UMP2 solutions shown in Sec. 5. The situation with respect to spin contamination is more cumbersome for BI2YN (b) although the small diradical character (y = 0.01). Indeed, the γ value (444 x 10^2 a.u.) at the UHF level is close to that (405 x 10^2 a.u.) at the RHF level, the MP2 values are much different [1001 x 10^2 a.u. (UMP2) versus 617 x 10^2 a.u.(RMP2)] while removal of the spin contamination at the PUHF (l = 1) and PUMP2 (l = 1) levels of approximation gives γ values of 47 x 10^2 a.u and 642 x 10^2 a.u. The γ value at the RMP2 [and PUMP2 (l = 1)] level is in good agreement with that (654 x 10^2 a.u.) at the RBHandHLYP level, which is known to well reproduce γ value for relatively small size closed-shell π-conjugated molecule at the CCSD(T) level. The UBHandHLYP solution is found to coincide with RBHandHLYP solution for BI2YN (b) as a result of its very small diradical character ($y = 0.01$). Considering the BHandHLYP results, γ increases by about one order of magnitude by going from BI2YN (b) to BI2Y (a). This trend is in agreement with that observed in p-quinodimethane model (see Sec. 6.2). The enhancement of γ in the intermediate diradical character regime is predicted to originate from the increase of the field-induced electron fluctuation due to the intermediate bond dissociation nature (π-bond breaking in the present case) in such regime. This fluctuation is expected to be reduced both in the stable bond region with small diradical character and in the complete diradical region where the diradical character is close to 1.

6.4. *Spin state dependence of γ for diradical molecules*

We here further investigate the spin state dependence of the static γ values for diradical molecules using 1,4-bis-(imidazole-2-ylidene)-cyclohexa-2,5-diene (BI2Y) in the singlet and triplet spin states. The *ab initio* MO and hybrid DFT methods are employed to obtain γ values. On the basis of the results, we discuss the possibility of the spin-state control of third-order NLO properties of diradical molecular systems.

Molecular geometries for singlet and triplet states of BI2Y are shown in Figure 27. Each geometry is optimized by the UB3LYP/6-31G** method. In the singlet case, the UB3LYP result turns out to coincide with the RB3LYP result. The longitudinal static γ is calculated by the finite-field approach [Eq. (6)]. For the calculation of energy E, we use the 6-31G*+p basis set with p exponent of 0.0523 on carbon (C) atoms and 0.0582 on nitrogen (N) atoms. The diradical character (y) is calculated for singlet BI2Y from the UHF natural orbitals. We apply various spin-unrestricted *ab initio* MO and hybrid DFT, i.e., UBHandHLYP, methods to the calculation of γ values.

The calculated γ values of our model molecules are given in Table 7. Because the electron correlation dependence of γ in diradical systems is known to be remarkable, we first examine the dependence of γ on the calculation methods. The diradical character for singlet BI2Y is shown

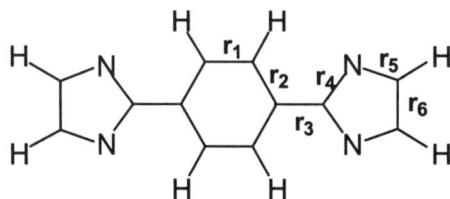

In Å	Singlet	Triplet
r_1	1.359	1.386
r_2	1.442	1.411
r_3	1.391	1.450
r_4	1.397	1.382
r_5	1.309	1.319
r_6	1.462	1.464

Fig. 27. Molecular geometries of singlet and triplet BI2Y optimized, respectively, by RB3LYP and UB3LYP methods using 6-31G** basis set.

Table 7. Calculated $\square$ values [a.u.] for singlet and triplet BI2Y.

Spin state	Method/Basis-set	$\gamma_{\square\tilde{\square}\tilde{\square}\square}$
	UBHandHLYP/6-31G*+p	653400
Singlet	UBHandHLYP/6-31G	484360
	UCCSD(T) /6-31G	524400
	UBHandHLYP/6-31G*+p	211600
Triplet	UBHandHLYP/6-31G	196000
	UCCSD(T) /6-31G	192400

to be an intermediate value ($y = 0.423$). As shown in our previous study [19], the UBHandHLYP method can give reliable γ values for the singlet diradical molecules with intermediate diradical character, at least for molecules of the size investigated here. For the triplet state, the UBHandHLYP method is also expected to give a reliable γ value (211600 a.u.) since the spin contamination is small ($<S^2> = 2.0475$). In fact, for both singlet and triplet states, the γ values calculated by the UBHandHLYP/6-31G method are similar to those by the UCCSD(T)/6-31G method (Table 7). From these results, the UCCSD(T)/6-31G*+p results are expected to be reproduced well by the UBHandHLYP/6-31G*+p results, which are used in the following discussion. On the other hand, the overshot γ values at the UMP2 (289100 a.u.) and spin projected UMP2 [PUMP2(l=1)] (251800 a.u.) levels of approximation are predicted to be caused by the spin contamination effects.

We next investigate the spin state dependence of γ. The γ value for the singlet state calculated by the UBHandHLYP method is about three times as large as that for the triplet state as shown in Table 7. This significant difference of γ value between the singlet and triplet states is predicted to be caused by the difference of the field-induced electron fluctuation effect between the singlet and triplet states. This feature can be understood as follows. The large third-order polarization in the singlet state with intermediate diradical character is caused by the intermediate overlap between a pair of radical orbitals [19]. In the triplet state, which corresponds to the highest spin state for diradical system, such polarization is significantly suppressed because of the non-overlap between a pair of radical orbitals due to the Pauli principle. Also such difference between the singlet and triplet states is predicted to be reduced in the region with large y value for the singlet state because the overlap between diradical distributions becomes negligible in such region for both spin states.

The spin state dependence of γ of diradical molecule with intermediate diradical character, BI2Y, is investigated by the *ab initio* MO and hybrid DFT method. It is found that the calculated γ value for the singlet spin state is about three times as large as that for the triplet spin state. Such remarkable change in γ originate in the intermediate diradical nature of the singlet spin state which leads to larger field-induced electron fluctuations, while the electron fluctuation effect is significantly suppressed for the triplet spin state due to the Pauli principle. In conclusion, the spin state dependence of γ for diradical systems with intermediate diradical character in the singlet state is shown to be remarkable. This feature suggests the possibility of sensitive spin-state control of γ for diradical molecular-based third-order NLO systems.

6.4. *Summary*

In this section, we have shown the enhancement of γ for a real singlet diradical molecule BI2Y (Fig. 27(a)) presenting intermediate diradical character ($y = 0.423$) as compared to that for the related molecule BI2YN

(Fig. 27(b)) having a negligible diradical character ($y = 0.01$). This enhancement feature for BI2Y is predicted to be caused by the large field-induced electron fluctuation in the intermediate π-bond dissociation region. We also show the possibility of drastic change of γ by changing the spin state for such diradical systems using the singlet and triplet BI2Y. These examples indicate that the control of γ value for diradical systems is possible via the control of diradical character by modifying the chemical structures as well as the spin state control of γ.

7. Novel structure-property relationship in γ for open-shell and/or charged systems

In this section, we summarize our two structure-property relationships of γ for open-shell and/or charged systems, and then propose novel molecular systems, which are expected to bear these two structure-property relationships.

As explained in Secs. 2 and 4, on the basis of the analysis of three types of virtual excitation processes, we elucidate that the systems with a large contribution of resonance structure with invertible polarization (SRIP) tend to shown negative γ values as well as large polarizability α (**Rule 1**). Charged radical states of π-conjugated condensed-ring molecules such as cationic radical state of tetrathiapentalene are shown to belong to typical SRIP systems as shown in Sec. 4.2. On the other hand, neutral singlet open-shell systems, i.e., diradical systems, are clarified to exhibit a distinct diradical character dependence of γ: the γ value increases with the increase in the diradical character and attains the maximum in the intermediate and somewhat large diradical character regions and then decreases again in the sufficiently large diradical character region (**Rule 2**). This rule is a part of structure-property relationships for open-shell systems explained in Secs. 5 and 6. By this rule, the singlet diradical systems with intermediate and somewhat large diradical character tend to exhibit larger positive γ value than conventional closed shell NLO systems as shown in Sec. 6. In this section, we propose novel molecular systems having these two features using azulene derivatives.

Azulene is a typical nonalternant aromatic hydrocarbon and its optical properties have been actively studied in view of the application to photonic devices [50,51]. The present model systems shown in Fig. 28(a) are expected to have both SRIP and diradical nature in the non-substituted case, so that the slight chemical modification has a possibility of drastic change in the sign and magnitude of γ by changing the relative contribution of these two factors. The *ab initio* MO and hybrid DFT (BHandHLYP) methods are applied to the calculation of γ for these azulene derivatives in order to clarify the effect of the introduction of donor or acceptor groups into the middle five-membered ring on γ. We consider three types of models: X= H, NH$_2$ and NO$_2$. For X = H case, two types of contributions are expected as shown in Figs. 28(b) and 28(c). Case (b), which involves polarization in the middle azulene structure, generates diradical in the both-end five-membered rings, while case (c) represents a contribution of SRIP composed of the invertible polarization in the left- and right-hand side azulene structures in the lower condensed ring part. As a result, non-substituted system (X = H) is predicted to have two contributions of diradical and SRIP, which causes an enhancement of positive and negative contributions to γ, respectively. If the negative contribution due to the SRIP overcomes the positive contribution due to the intermediate diradcial character, this case (X = H) is predicted to exhibit a negative γ value. The introduction of donor group (X = NH$_2$) is predicted to reduce the polarization in the middle azulene structure and induce the symmetric polarization in the lower condensed-ring part, both of which reduce the diradical character and SRIP contribution (see Fig. 28(d)) and then make the γ positive in sign. In contrast, the introduction of acceptor (NO$_2$) group is predicted to enhance the polarization in the middle azulene structure and to increase the diradical character in the lower both-end five-membered rings (see Fig. 28(e)). This is also predicted to reduce the SRIP contribution shown in Fig. 28(c), so that the γ value will become positive and depend on the diradical character.

The geometries of model molecules shown in Fig. 28 are optimized by the UB3LYP/6-31G** method. Contrary to our speculation, the most stable states for these three models possess asymmetric structures with bond alternations (C_S symmetry). Therefore, the SRIP contributions are

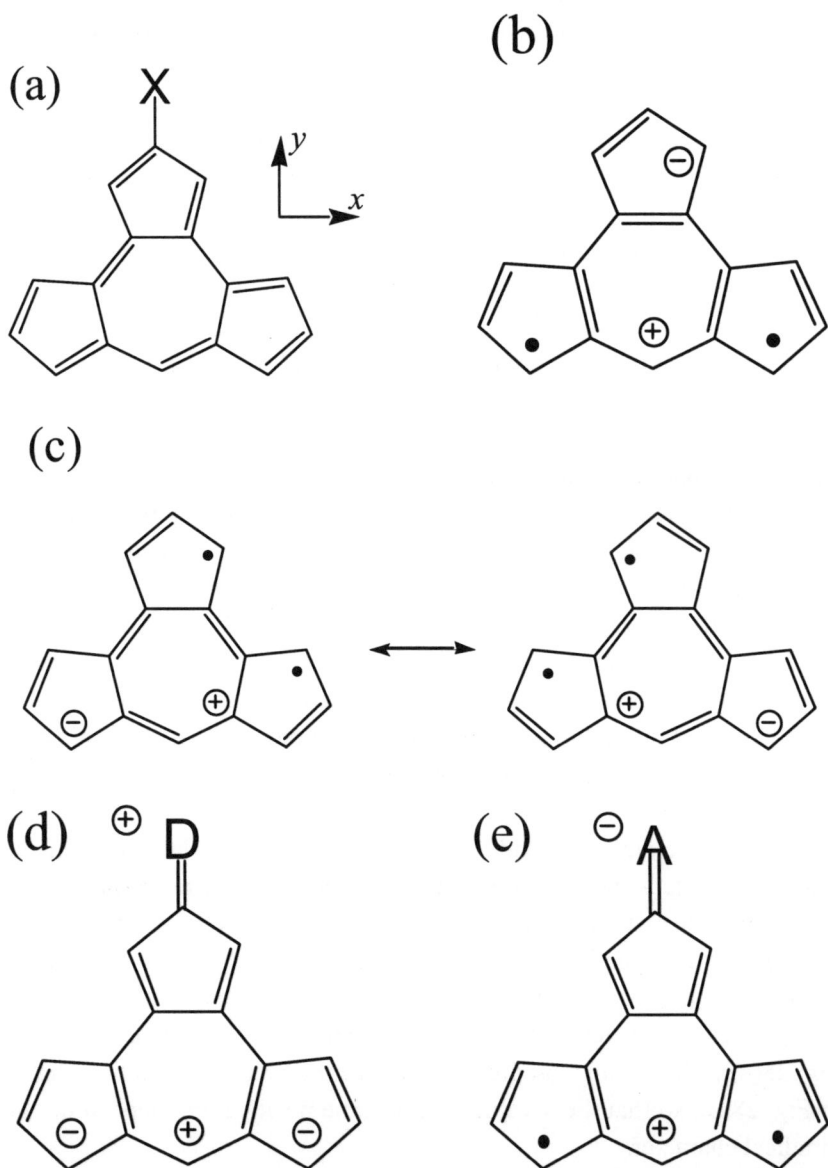

Fig. 28. Azulene derivative models (a) (X = H, NH$_2$, NO$_2$). Resonance forms ((b) and (c)) contribute to non-substituted models, and (d) and (e) show main contributions for donor (D) and acceptor (A) substituted models, respectively.

predicted to decrease for these models. We confine our attention to the static γ_{xxxx} of these models (X = H, NH_2 and NO_2). As shown in Sec. 6.2, the spin-unrestricted hybrid DFT method (UBHandHLYP) is shown to give reliable γ values for diradical molecules with relatively broad range of diradical characters, at least for the size of molecules in this study. Therefore, we employ the UBHandHLYP method using 6-31G*+p basis set with p exponent of 0.0523 on carbon (C) atoms, 0.0582 on nitrogen (N) atoms and 0.0719 on oxygen atoms. The static γ is calculated by the FF approach [Eq. (6)].

The diradical character (y) calculated using the UHF natural orbitals (UNO) and γ values are given in Table 8. The diradical character for model (X = H) is found to be relatively small but slightly larger than that of model (X = NH_2), whereas that for model (X = NO_2) is significantly increased. This feature is in qualitative agreement with our prediction of the variation in diradical character of these models. It turns out that model (X = H) exhibits the largest positive value (1.12 x 10^5 a.u.) though the diradical charcater is slightly larger than that of model (X = NH_2) (γ = 2.60 x 10^4 a.u.). This is due to the additional positive contribution of types I and III-1 in model (X = H) with more asymmetirc charge distribution (dipole moment $\mu_{\square}$ = -0.14 a.u.) as compared to other models [$\mu_{\square}$ = -0.05 a.u. (X = NH_2) and $\mu_{\square}$ = -0.02 a.u. (X = NO_2)]. As a result, the increase in γ for model (X = NO_2) (γ = 6.97 x 10^4 a.u.) can be understood by the increase in diradical character [y = 0.724 (X = NO_2) versus y = 0.384 (X = NH_2)]. It is noted that model (X = NH_2), which hardly has SRIP contribution and has a small diradical character, is shown to provide a similar γ value to that (2.56 x $10^{\square}$ a.u.) of a reference system, anthracene, with a similar π-conjugation length. It is also found that the asymmetric structure (C_S) for model (X = H) is slightly stable by less than 1 kcal/mol as compared to the symmetric structure (C_{2v}), which is predicted to provide a large negative γ value (-3.2 x 10^7 a.u) due to a large SRIP contribution. This suggests that the γ value of model (X = H) shows a remarkable structure dependence including the inversion of sign.

A remarkable structural dependence of γ for donor/acceptor substtituted azulene derivatives (see Fig. 28) is presented based on the SRIP contribution (**Rule 1**) and the diradcial character (**Rule 2**). By changing the donating and accepting intensity of the donor and acceptor

Table 8. γ values for models (X = H, NH$_2$, NO$_2$) calculated by the UBHadnHLYP/6-31G*+p.

X	y	μ_x [a.u.]	γ [a.u.]
H	0.392	-0.14	1.12 x 105
NH2	0.384	-0.05	2.60 x 104
NO2	0.724	-0.02	6.97 x 104

groups (X), we could tune the magnitude of γ for these systems. In comparison with conventional π-conjugated systems with similar π-conjugation lengths, the γ value of acceptor-substituted model (X = NO$_2$) tends to be several times increased. On the other hand, for non-substituted case, a negative contribution due to the SRIP is expected through slight structure changes caused by thermal excitation. This will lead to the change of not only magnitude but also sign of γ. Considering these structure-property relationships, the present model compounds based on azulene structure are expected to be a candidate for novel tunable third-order NLO systems.

Acknowledgments

This work was supported by Grant-in-Aid for Scientific Research (No. 14340184) from Japan Society for the Promotion of Science (JSPS). The author wishes to express his gratitude to Professor Kizashi Yamaguchi for discussions throughout the concept of open-shell NLO systems. The author is deeply grateful to Dr. Mitsutaka Okumura and Dr. Isamu Shigemoto for their cooperation of constructing perturbational analysis based on virtual excitation processes and Dr. Satoru Yamada, Dr. Shinji Kiribayashi for their cooperation of extracting novel structure-property relationship based on the SRIP of third-order NLO systems. Most of works in Secs. 5 and 6 were carried out in collaboration with Dr. Benoît Champagne and Dr. Edith Botek in Namur University, Dr. Kenji Kamada and Dr. Koji Ohta. in AIST, and Dr. Takashi Kubo, Professor

Kazuhiro Nakasuji and Mr. Ryohei Kishi in Osaka University. The author deeply thanks to them for their cooperation, helpful suggestions, and continuous severe discussions. The author also thanks to Professor David Bishop for his useful suggestions and discussions on the hyperpolarizability density analysis.

References

1. D. J. Williams, Ed. *Nonlinear Optical Properties of Organic and Polymeric Materials*; ACS Symposium Series 233 (American Chemical Society, Washington, D.C., 1984).

2. D.S. Chemla and J. Zyss, Eds. *Nonlinear optical properties of organic molecules and crystals*, Vols. 1 and 2 (Academic Press, New York, 1987).

3. N. P. Prasad and D. J. Williams, *Introduction to Nonlinear Optical Effects in Molecules and Polymers* (Wiley, New York, 1991).

4. J. Michl, Ed., *Optical Nonlinearities in Chemistry*, Chem. Rev. **94**, (1994).

5. S.P. Karna and A.T. Yeates, Eds. *Nonlinear Optical Materials. Theory and Modeling*; ACS Symposium Series 628 (American Chemical Society: Washington, D.C., 1996).

6. Ch. Bosshard, K. Sutter, Ph. Prêtre, J. Hulliger, M. Flosheimer, P. Kaatz, and P. Günter, *Organic Nonlinear Optical Materials*, (Gordon and Breach, Basel, 1995).

7. H.S. Nalwa, Ed. *Handbook of Advanced Electronic and Photonic Materials and Devices*, Vol. **9**, *Nonlinear Optical Materials*, (Academic Press, New York, 2001).

8. M. Nakano and K. Yamaguchi, in Trends in Chemical Physics, vol. 5, pp.87-237, (Research trends, Trivandrum, India, 1997).

9. M. Nakano and K. Yamaguchi, *Mechanism of Nonlinear Optical Phenomena for π-Conjugated Systems*, in *"Organometalic Conjugation"* Eds. Akira Nakamura, Norikazu Ueyama and Kizashi Yamaguchi (Koudansha, Springer, 2002).

10. M. Nakano and K. Yamaguchi, in *Advances in Multi-Photon Processes and Spectroscopy*, vol. 15, pp.1-146, (World Scientific, Singapore, 2003).

11. M. Nakano and K. Yamaguchi, Chem. Phys. Lett. **206**, 285 (1993).

12. M.Nakano, S. Kiribayashi, S. Yamada, I. Shigemoto and K. Yamaguchi, Chem. Phys. Lett., **262**, 66 (1996).

13. (a) M. Nakano, H. Nagao and K. Yamaguchi, Phys.Rev.A**55** 1503 (1997). (b) M. Nakano, S. Yamada, K. Yamaguchi, J. Comput. Meth. Sci. Eng. (JCMSE), **4**, 677 (2004). (c) M. Nakano, S. Yamada, R. Kishi, M. Takahata, T. Nitta and K. Yamaguchi, Journal of Nonlinear Optical Physics and Materials, **13**, 411 (2004).

14. (a) M. Nakano, S. Yamada, K. Yamaguchi, Bull. Chem. Soc. Jpn. **71**, 845 (1998). (b) S. Yamada, M. Nakano, K. Yamaguchi, Chem.Phys.Lett. **276**, 375 (1997). (c) S. Yamada, M. Nakano, I. Shigemoto, K. Yamaguchi, Chem. Phys. Lett., **267**, 445

(1997). (d) S. Yamada, M. Nakano and K. Yamaguchi, Int. J. Quant. Chem., **71** 329 (1999). (e) S. Yamada, M. Nakano and K. Yamaguchi, J. Phys. Chem. **103**, 7105 (1999).

15. (a) M. Nakano, S. Yamada and K. Yamaguchi, Chem. Phys. Lett., **311**, 221 (1999). (b) M. Nakano, S. Yamada and K. Yamaguchi, Chem. Phys. Lett. **321**, 491 (2000).

16. (a) M. Nakano, S. Yamada and K. Yamaguchi, J. Phys. Chem.A, **103**,3103 (1999). (b) M.Nakano, S. Yamada and K. Yamaguchi, Mol. Cryst. Liq. Cryst. **337** 369 (1999)

17. S. Yamanaka, M. Okumura, M. Nakano, K. Yamaguchi, J. Mol. Structure, **310**, 205 (1994).

18. K. Yamaguchi, in R. Carbo, M. Klobukowski (Eds.). *Self-Consistent Field. Theory and Applications* (Elsevier, Amsterdam, 1990).

19. (a) M. Nakano, T. Nitta, K. Yamaguchi, B. Champagne and E. Botek, J. Phys. Chem. A 108, 4105 (2004). (b) M. Nakano, B. Champagne, E. Botek, R. Kishi, T. Nitta and K. Yamaguchi, in Advances in Science and technology 42, "Computational Modeling and Simulation of Materials III" Part A P. Vincenzini, A. Lami Eds. pp. 863-870 (Techna Group, Srl, 2004). (c) B. Champagne, E. Botek, M. Nakano, T. Nitta and K. Yamaguchi, J. Chem.Phys. in press.

20. K. Yamaguchi, T. Kawakami, D. Yamaki, Y. Yoshioka, *Theory of Molecular Magnetism*, in *Molecular Magnetism*, K. Ito, M. Kinoshita, Eds. (Kodansha, and Gordon and Breach, 2000,. pp.9-48).

21. K. Yamaguchi, M. Okumura, K. Takada, S. Yamanaka, Int. J. Quantum Chem. Symp., **27**, 501 (1993).

22. M. Nakano, M. Okumura, K. Yamaguchi and T. Fueno, Mol. Cryst. Liq. Cryst. **182A** 1 (1990).

23. M. Nakano, I. Shigemoto, S. Yamada, K. Yamaguchi, J. Chem. Phys. **103**, 4175 (1995).

24. (a) C.P. de Melo and R. Silbey, J. Chem. Phys. **88**, 2558 (1988). (b) C.P. de Melo and R. Silbey, J. Chem. Phys. **88**, 2567 (1988). (c) M. Nakano, M. Okumura, K. Yamaguchi and T. Fueno, Mol. Cryst. Liq. Cryst. **182A**, 1 (1990).

25. J. Orr and J.F. Ward, Mol. Phys. **20**, 513 (1971).

26. G. J. B. Hurst, M. Dupuis and E. Clementi, J. Chem. Phys. **89**, 385 (1988)

27. M. Nakano, H. Fujita, M. Takahata, and K. Yamaguchi, J. Am. Chem. Soc. **124**, 9648 (2002).

28. (a) A. D. Becke, J. Chem. Phys. **98**, 5648 (1993). (b) C. Lee, W. Yang, R. G. Parr, Phys. Rev. B **37**, 785 (1988).

29. M.J. Frisch, G.W. Trucks, H.B. Schlegel, G.E. Scuseria, M.A. Robb, J.R. Cheeseman, V.G. Zakrzewski, J.A. Montgomery, R.E. Stratmann, J.C. Burant, S. Dapprich, J.M. Millam, A.D. Daniels, K.N. Kudin, M.C. Strain, O. Farkas, J. Tomasi, V. Barone, M. Cossi, R. Cammi, B. Mennucci, C. Pomelli, C. Adamo, S. Clifford, J. Ochterski, G.A. Petersson, P.Y. Ayala, Q. Cui, K. Morokuma, D.K. Malick, A.D. Rabuck, K. Raghavachari, J.B. Foresman, J. Cioslowski, J.V. Ortiz,

B.B. Stefanov, G. Liu, A.P. Liashenko, A.P. Piskorz, I. Komaromi, R. Gomperts, R.L. Martin, D.J. Fox, T. Keith, M.A. Al-Laham, C.Y. Peng, A. Nanayakkara, C. Gonzalez, M. Challacombe, P.M.W. Gill, B.G. Johnson, W. Chen, M.W. Wong, J.L. Andres, M. Head-Gordon, E.S. Replogle, and J.A. Pople, GAUSSIAN 98, Revision A11, Gaussian Inc., Pittsburgh, PA, 1998.

30. M. Nakano, K. Yamaguchi and T. Fueno, Chem. Phys. Lett.185, 550 (1991).

31. M. Nakano, H. Fujita, M. Takahata and K. Yamaguchi, Chem. Phys. Lett. 356, 462 (2002).

32. P. Chopra, L. Carlacci, H. King, P. N. Prasad, J. Phys. Chem. 93, 7120 (1989).

33. M. Nakano and K. Yamaguchi, Mol. Cryst. Liq. Cryst. A 255, 139 (1994).

34. (a) M.Nakano and K. Yamaguchi, Phys. Rev. A. 50, 2989 (1994). (b) M.Nakano, K. Yamaguchi, Y.Matuzaki, K.Tanaka and T. Yamabe, J. Chem.Phys.102, 2996 (1995). (c) M.Nakano and K. Yamaguchi, Chem.Phys.Lett. 234, 323 (1995).

35. S. R. Marder, C. B. Gorman, F. Meyers, J. W. Perry, G. Bourhill, J. L. Brédas, B.M. Pierce, Science 265, 632 (1994).

36. J. B. Torrance, Acc. Chem. Res. 12, 79 (1979).

37. P. O. Löwdin, Phys. Rev. 97 1509 (1955).

38. D. Jacquemin, B. Champagne, J. M. André, Int. J. Quantum Chem. 68, 679 (1997).

39. P. Mori-Sánchez, Q. Wu, W. Yang, J. Chem. Phys. 119, 11001 (2003)

40. B. Champagne, E. A. Perpète, S.J.A. van Gisbergen, J. G. Snijders, E. J. Baerends, C. Soubra-Ghaoui, K. A. Robins, B. Kirtman, J. Chem. Phys. 109, 10489 (1999); *erratum*, 110, 11664 (1999).

41. S.J.A. van Gisbergen, P. R. T. Schipper, O. V. Gritsenko, E. J. Baerends, J. G. Snijders, B. Champagne, B. Kirtman, Phys. Rev. Lett. 83, 694 (1999).

42. B. Champagne, E. A. Perpète, D. Jacquemin, S. J. A. van Gisbergen, E. J. Baerends, C. Soubra-Ghaoui, K. A. Robins, B. J. Kirtman, Phys. Chem. A 104, 4755 (2000).

43. (a) D. Scheschkewitz, H. Amii, H. Gomitzka, W. W. Schoeller, D. Bourissou and G. Bertrand, Angew. Chem. Int. Ed. 43, 585 (2004). (b) D. Scheschkewitz, H. Amii, H. Gomitzka, W. W. Schoeller, D. Bourissou and G. Bertrand, Science 295, 1880 (2002). (c) D. R. McMasters and J. Wirz, J. Am. Chem. Soc. 123, 238 (2001). (d) H. Sugiyama, S. Ito and M. Yoshifuji, Angew. Chem. Int. Ed. 42, 3802 (2003). (e) E. Niecke, A. Fuchs, F. Baumeister, M. Nieger and W. W. Schoeller, Angew. Chem. Int. Ed. Engl. 34, 555 (1995).

44. M. Okumura, S. Yamanaka, W. Mori, K. Yamaguchi, J. Mol. Structure, 310, 177 (1994).

45. D. J. Thouless, *The Quantum Mechanics of Many-Body Systems*, (Academic Press, New York 1961).

46. (a) K. Yamaguchi, Y. Takahara, T. Fueno, K. N. Houk, Theor. Chim. Acta, 73, 337 (1988). (b) K. Yamaguchi, M. Okumura, W. Mori, J. Maki, K. Takada, T. Noro, K. Tanaka, Chem. Phys. Lett. 210 201 (1993).

47. K. Yamaguchi, T. Tsunekawa, Y. Toyoda, T. Fueno, Che. Phys. Lett. 143, 371 (1988).

48. K. Yamaguchi, M. Okumura, J. Maki, T. Noro, H. Namimoto, M. Nakano, T. Fueno, K. Nakasuji, Chem. Phys. Lett. **190**, 353 (1992).
49. M. Okumura, S. Yamanaka, W. Mori, K. Yamaguchi, J. Mol. Structure, **310**, 177 (1994).
50. T. Makinoshita, M. Fujitsuka, M. Sasaki, Y. Araki, O. Ito, S. Ito and N. Morita, J. Phys. Chem. A, **108**, 368 (2003).
51. S. Shevyakov, H. Li, R. Muthyala, A. Asato, J. C. Croney, D. M. Jameson and R. S. H. Liu, J. Phys. Chem. A **107**, 3295 (2003).

CHAPTER 11

SEQUENTIAL MONTE CARLO/QUANTUM MECHANICS STUDY OF THE DIPOLE POLARIZABILITY OF ATOMIC LIQUIDS. THE ARGON CASE.

Kaline Coutinho and Sylvio Canuto[*]

Instituto de Física, Universidade de São Paulo, CP 66318, 05315-970, São Paulo, SP, Brazil.

[*] Corresponding author. E-mail canuto@if.usp.br

The sequential QM/MM methodology is suggested to obtain the dipole polarizability of atomic liquids. Using Metropolis Monte Carlo simulation within the NpT ensemble the structures of liquid argon are obtained at T = 91.8 K and p = 1.8 atm. These structures and the calculated density are in very good agreement with the experimental results. After the simulation, the auto-correlation function of the energy is calculated to extract the correlation interval which gives information about the statistical relevance of the structures generated. Seventy statistically relevant configurations, composed of the first solvation shell (14 Ar atoms), are extracted for subsequent quantum mechanical calculations. Using these structures density-functional theory calculations are performed within the B3P86 hybrid functional and the aug-cc-pVDZ basis set to obtain statistically converged values for the dipole polarizability in the liquid case. The result points to a slightly increased value compared to the gas phase result.

1. Introduction

The linear response of an atomic or molecular system to the application of a weak electric field is given by the dipole polarizability [1-3]. This linear, as well as the nonlinear responses, relate to several physical properties and phenomena [4-7]. There are numerous *ab initio* calculations of isolated static atomic and molecular polarizabilities [8-29] including cluster [30] and liquid systems [31]. In recent years there has been an increased interest in describing the effect of a surrounding medium on the electronic properties of a reference atomic or molecular system, including the dipole polarizability and hyperpolarizabilities. Understanding these solvent effects is very important for rationalizing experimental data in general and for planning realistic devices. Some progress has been achieved by treating solvent effects using the so-called continuum models [32-36], where the solvent is described by its macroscopic constants such as the dielectric constant and index of refraction. The solute-solvent interaction is described by the electrostatic moments of the solute and the polarized solvent medium. Sophisticated implementations such as the PCM (polarizable continuum model) [37,38] are now widely available. Solvent effects in dipole polarizabilities and hyperpolarizabilities have been studied using continuum theories [39-43]. Although very successful in some contexts, the statistical nature of the liquid environment is not considered in these models. In addition, applications in the case of neutral closed-shell atoms may be problematic because of the lack of electrostatic moments in the atom. In addition the exchange interaction between the solute and the solvent is not included and in some cases may play an important role. A proper treatment of a liquid system should also consider its statistical nature [44,45]. Indeed, for nonzero temperature there are many possible geometrical arrangements of the molecules (or atoms) of the liquid with equivalent probability. Thus, liquid properties are best described by a statistical distribution [31,46-49], and all properties are statistical averages over an ensemble.

In this paper we use the sequential Monte Carlo quantum mechanics (S-MC/QM) methodology [50-53] to study the dipole polarizability of the liquid argon. In this sequential QM/MM procedure we first generate the structure of the liquid using some sort of computer simulation of liquids. It could be either Molecular Dynamics or Monte

Carlo (MC) simulation. As in most cases the interest lies in the average of a certain property, without any concern for time-sequence, the choice for MC is computationally more convenient. Hence, we use Metropolis MC simulation to generate the structure of the liquid of interest. Next, statistically relevant configurations are separated and submitted to quantum mechanical (QM) calculations. Several QM calculations are made to obtain the ensemble average that is necessary to characterize the statistical nature of the liquid. Statistically converged results can be obtained with an adequate sampling of the configurations [50-52]. This is normally done by calculating the auto-correlation function of the energy [50-52] to obtain the statistical interval of correlation between configurations generated by the simulation. This is one of the greatest advantages of using the sequential procedure. The autocorrelation function is calculated only after the simulation is finished and having the knowledge of the statistical correlation allows an efficient protocol for the subsequent QM calculations. In addition, the portion of the system that will be treated by QM can be selected on the basis of the solvation shells around the solute.

In the present study the dipole polarizability of liquid argon is studied using the S-MC/QM procedure. Explicit argon atoms are used both in the "solute" and in the "solvent" representation. The QM calculations are made using the density-functional theory (DFT) in the hybrid B3P86 exchange-correlation functional [54]. We compare the dipole polarizability in the condensed liquid phase with that for the gas phase. Because of the confinement due to the condensed medium it has been argued that the dipole polarizability should decrease [55,56]. Numerical examples have indeed shown that this is the case for some neutral and anionic systems [55-57]. In special Cl^- [55] and F^- [57] were shown to suffer a strong reduction of the polarizability of ~40 %. *Ar* is a very special system with van der Waals interaction and it is not clear whether the polarizability increases or decreases in the condensed liquid phase. Hence, we address to this problem from the point of view of atomic liquid simulation combined with QM calculations. Special attention is given for a proper representation of the liquid structures and to ensure statistical convergence.

2. Methods

2.1 Monte Carlo simulation

The MC simulation is performed using standard procedures [45] for the Metropolis technique in the *NpT* ensemble. We used the image method combined with the periodic boundary conditions of a cubic box. In our simulation, we used 500 *Ar* atoms at T = 91.8 K and p = 1.8 atm. The atoms interact via the Lennard-Jones potential with parameters, ε = 0.2378 kcal/mol e σ = 3.41Å developed and tested by Maitland and Smith [58]. In the calculation of the pair-wise energy, each atom interacts with all others within a separation that is smaller than the half size of the box. For separations larger than this, we evaluate the long range correction using the pair radial distribution function [45], G(r) ≈ 1. The initial configuration is generated randomly, considering the position of each atom. One MC step is concluded after selecting one atom randomly and trying to translate it in all the Cartesian directions. The maximum allowed displacement of the atoms is auto-adjusted to give an acceptance rate of new configurations around 50%. The simulation was performed with the DICE program [59].

After thermalization, 10.5×10^6 MC steps were performed. The average density was calculated as 1.362 ± 0.015 g/cm³, in very good agreement with the experimental result of 1.365g/cm³ [60] obtained for the same conditions of temperature and pressure. Radial distribution functions and liquid configurations were also generated during the averaging stage in the simulation. After completing the cycle over all 500 atoms a configuration of the liquid is generated and separated. Thus, the total number of configurations generated by the MC simulation is 21×10^3.

QM calculations were performed on these configurations generated by the MC simulation. At this point it is very important to optimize the statistics sampling configurations according to the statistical correlation. As successive configurations are statistically very correlated they will not give important additional information to the average. Therefore, to sample efficiently the configurations for the subsequent QM calculations we calculated the auto-correlation function of the energy [50-52] to obtain the correlation interval. Using an interval of 300 for successive configuration, we obtained a statistical correlation of 13%, which in practice is considered uncorrelated. Then, from the 21×10^3 configurations generated by the MC simulation, we selected 70 statistically uncorrelated configurations, separated by an interval of 300, that represent the entire simulation [61].

2.2 Quantum mechanical calculations

The major interest of this paper is the calculation of the dipole polarizability, α, of *Ar* in the liquid phase, as compared to the value in the gas phase. To obtain that, we performed first-principle calculations, over 70 selected configurations composed of one *Ar* atom surrounded by the first solvation shell (*n Ar* atoms). As the appropriate Boltzmann weights are included in the Metropolis MC the dipole polarizability, or any other property calculated from the MC configurations, is given as a simple average over all the values calculated for each configuration.

An important aspect in the study of dipole polarizability of a reference molecule in solution is related to the separability problem. Of course, the application of an external static homogeneous electric field affects both the reference and the solvent molecules (in the present case, atoms). The use of response function, thus avoiding the external field, minimizes this problem but still has the conceptual problem of distinguishing the solute and the solvent. It is normally assumed that a simple aditivity is a reasonable first approximation, particularly, for weakly interacting systems. The *interaction* polarizability [62] of weakly interactive systems is the subject of Maroulis study in this issue. For a homogeneous liquid this problem is less critical because one can define the dipole polarizability per molecule, provided a sufficiently large number of molecules is included. In practice, however, this is seldom the case, particularly for first-principle calculations. To calculate the polarizability, α, of the liquid *Ar*, we used this procedure. Therefore, as the solute and the solvent are the same entity X, the polarizability of a system X in the liquid phase, $\alpha_{liq}(X)$, is obtained dividing the polarizability of the $(1+n)$ X, $\alpha(X+n\_solv)$ by $n+1$, i.e.

$$\alpha_{liq}(X) = \alpha(X+n\_solv) / (n+1), \qquad (1)$$

where n is the number of molecules selected up to a certain solvation shell. This approximation works better for increasing value of n. One aspect that should be emphasized is that the results given here are statistically converged and therefore spurious errors due to the improper use of statistical distribution is discarded. The QM calculations of the dipole polarizability, α, is made using the Gaussian98 program [63].

3. Results

3.1 Solvation shells

The structures of atomic liquids are analyzed using the radial distribution function (RDF), which defines the solvation shells. In figure 1, the calculated RDF of the liquid *Ar* is shown and in table 1 the maximum positions are compared with the experimental results.

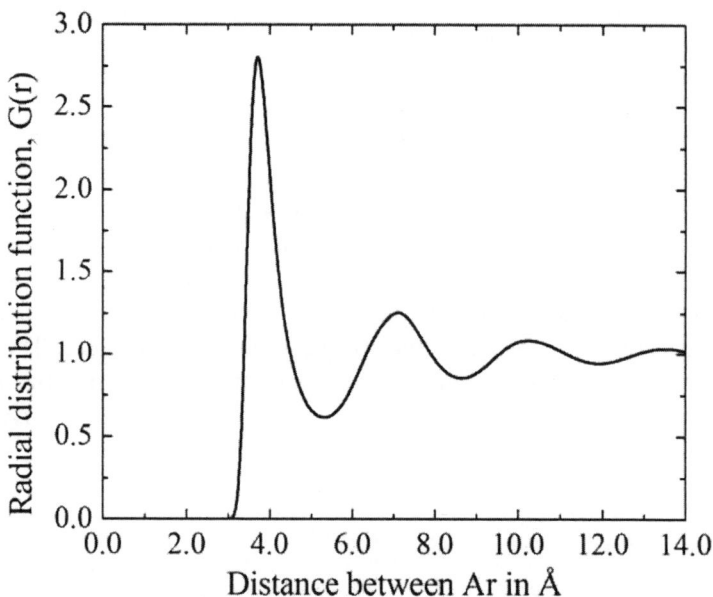

Figure 1: The radial distribution function of the liquid *Ar*.

Four solvation shells are well defined. The integration of these peaks defines the coordination number, or the number of atoms in each solvation shell. The first shell, that starts at 3.20 Å has a maximum at 3.75 Å and ends at 5.35 Å, has in average 13 *Ar* atoms. The second shell, ending on 8.70 Å with a maximum at 7.05 Å, has around 43 *Ar* and the third, which goes to 11.95 Å with a maximum at 10.25 Å, has 92 *Ar*. Therefore, in the first solvation shell there are 13 *Ar* atoms and in total, there are 56 *Ar* atoms up to the second shell and 148 *Ar* atoms up to the third shell.

Table 1: Analysis of the calculated radial distribution function in comparison with the experimental data [60] obtained from X-ray diffraction for liquid Argon at the same thermodynamic condition, $T = 91.8$ K and $p = 1.8$ atm.

Solvation shells	First	Second	Third	Fourth
Experimental	3.79 ± 0.05 Å	6.91 ± 0.05 Å	10.31 ± 0.05 Å	13.67 ± 0.05 Å
This work	3.75 ± 0.10 Å at 5.35 Å 13 atoms	7.05 ± 0.10 Å at 8.70 Å 56 atoms	10.25 ± 0.10 Å at 11.95 Å 148 atoms	13.55 ± 0.10 Å

Experimental density is 1.365 g/cm^3.
Theoretical density is 1.362 ± 0.015 g/cm^3.

The three maxima of the RDF, shown in table 1, are in very good agreement with the experimental results, obtained by Eisenstein and Gingrich [60] using X-ray diffraction in liquid *Ar* in the same condition of temperature and pressure. The results of table 1, together with the calculated density, give a clear indication that the structure of liquid argon is very well described in our MC simulation. Therefore, the configurations generated by the simulation are now used to obtain the average dipole polarizability of liquid argon.

3.2 Polarizability of liquid argon

Before presenting the calculated results for the polarizability in the liquid case it is of interest to discuss first the adequacy of the QM results for the isolated *Ar* atom. There has been a great interest in the dipole polarizability of *Ar* and it has been subjected to several theoretical investigations. The very recent study by Lupinetti and Thakkar [64] gives a good account of previous studies and put forward a recommended value. The true non-relativistic value for isolated *Ar* is suggested as 11.065 ± 0.05 a_0^3. Within this range one finds both recent CCSD(T) [65] as well as early PNO-CEPA [66] calculations. As remarked [64], relativistic corrections should increase the polarizability value by 0.02 ± 0.01 a_0^3. On the experimental side previous values are 11.070 ± 0.007 a_0^3 [67], 11.078 ± 0.010 a_0^3 [68] as well as the result of 11.067 a_0^3, tabulated by Binram [69]. The present theoretical results for the dipole polarizability of the free *Ar* atom, $\alpha_{gas}(Ar)$, estimated using the B3P86 functional are shown in table 2. Only results obtained with basis sets that include diffuse and polarization functions are of interest here. The results on table 2 vary from 5.14 to 11.38 a_0^3 depending on the basis set. Using the augmented correlation-consistent polarized valence double, triple and quadruple-zeta (aug-cc-pVXZ basis, where X=D, T and Q) [70] the results vary from 9.98 and 11.38 a_0^3, what compare reasonably well with the experimental results discussed above. We have then selected the B3P86/aug-cc-pVDZ theoretical model.

Table 2: The calculated values of the dipole polarizability of one *Ar* atom with different levels of quantum mechanical calculations, in comparison with the experimental data.

Method/Bases	$\alpha_{gas}(Ar)$ in a_0^3
B3P86/6-31+G(d)	5.14
B3P86/6-31+G(2d)	7.75
B3P86/6-31+G(2df)	7.75
B3P86/6-31+G(3d)	10.54
B3P86/aug-cc-pVDZ	9.98
B3P86/aug-cc-pVTZ	11.08
B3P86/aug-cc-pVQZ	11.38
Experimental values	11.070 ± 0.007 [a]
	11.078 ± 0.010 [b]
	11.067 [c]

a) Ref. [67],
b) Ref. [68]
c) Ref. [69].

Having discussed the atomic case and before considering the liquid we discuss the dipole polarizability of *Ar-Ar* as a function of the interatomic distance. Figure 2 shows that the dipole polarizability of *Ar-Ar* is larger than the corresponding polarizability at the dissociation limit; i.e. the polarizability of *Ar-Ar* is enhanced with respect to the individual atoms. The increase is systematic but not very large and at the largest polarization the dipole polarizability increases by 2.2 %. Whereas it is clear that the two-body interaction increases the polarizability, in the liquid case the situation is different because of the confining effect that the solvent exerts in the reference system. Hence we now discuss the polarizability in the liquid case.

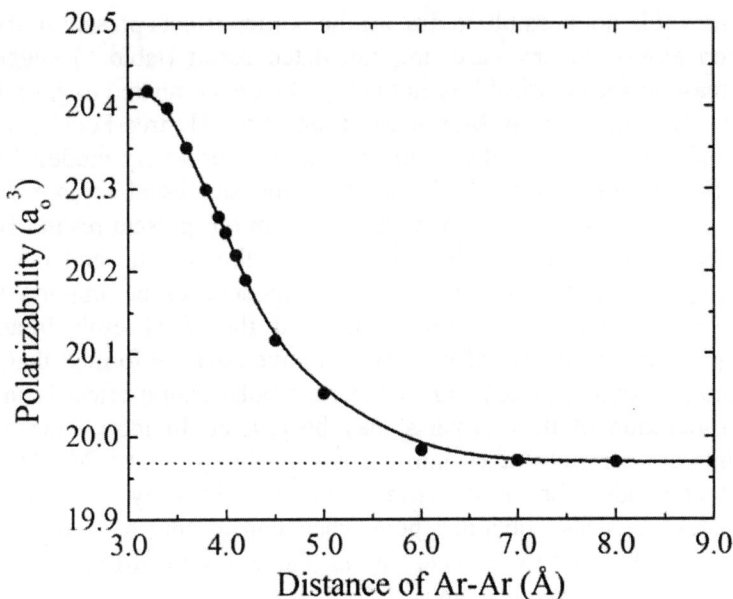

Figure 2: The calculated polarizability of *Ar-Ar* as a function of the interatomic distance.

Finally, table 3 shows the calculated result for the first solvation shell obtained using the liquid structures extracted from the MC simulation. We consider the homogeneous liquid and calculate the polarizability for the entire super-system. To visualize the configuration

Table 3: The calculated values of the dipole polarizability (in a_0^3) of *Ar* using eq. 1 for the first solvation shell. All calculations are performed with the B3P86 functional. Uncertainties are statistical errors obtained over 70 statistically uncorrelated configurations.

Model	$\alpha(Ar+13\_Ar)$	$\alpha_{liq}(Ar)$	$\alpha_{gas}(Ar)$	$\alpha_{sln}/\alpha_{gas}$
Solute(aug-cc-pVDZ) Solvent(aug-cc-pVDZ)	147.14 ± 0.06	10.51 ± 0.01	9.98	105.3%

space that is occupied by the *Ar* atoms, we superimpose in figure 3 all the 70 structures composed by the 14 *Ar* atoms. The *Ar* atoms occupy a spherical volume with radius of 5.35 Å. For a relatively small number of entities it is not possible to discard the border effects present in the outer argon atoms. In any case, the calculated result (table 3) suggests an increase of the polarizability in the liquid case, compared to gas phase, of 5 %. Recent study by Morita and Kato at the Hartree-Fock level using extended basis set for the "solute" *Ar* atom, but a very modest basis set (3-21G) for the rest, the "solvent" *Ar* atoms, suggested a minor *decrease* of ~2 % [55]. This differs qualitatively from our present result. Because of the use of a very modest basis set for the solvent, and the lack of electron correlation effects that are expected to be important, it is difficult to make a direct comparison with their [55] result. In addition, only a limited number of calculations have been performed that cannot guarantee statistical convergence. As the polarization effect is small the consideration of these aspects may be crucial. In fact, some of these points are carefully taken into account in this present S-MC/QM study that concludes for an *increase* in the polarizability. This qualitative conclusion is also obtained here when considering a central *Ar* atom surrounded by the 13 "solvent" atoms and taking the difference $\alpha_{sln}(X) = \alpha(X+n\_solv) - \alpha(0+n\_solv)$.

To demonstrate the statistical convergence, figure 4 shows the average of the polarizability with the increasing number of structures used. The convergence is fast. As it can be easily seen after 40 QM calculations the polarizability has converged. In addition, figure 5 shows the histogram and the normal distribution of calculated values for the dipole polarizability. Note the standard deviation indicating the spread of

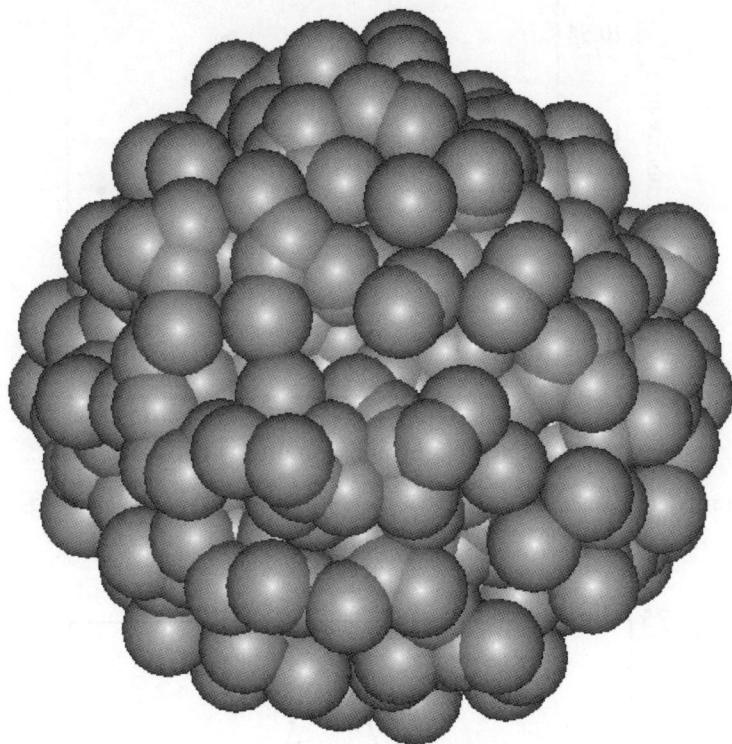

Figure 3: Superposition of the seventy structures of 1+13 *Ar* atoms obtained from the simulation of the liquid *Ar*, showing that the configuration space is a spherical distribution with radius 5.35 Å.

calculated values. Accordingly, 68% of the calculated values are within $10.51 \pm 0.04 \; a_0^3$.

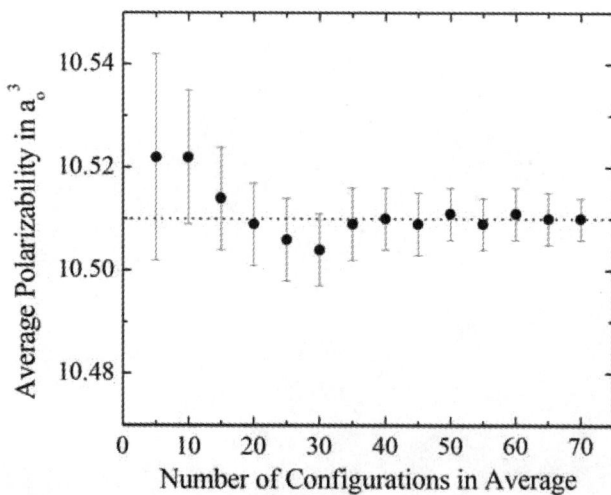

Figure 4: Convergence of the average value of the polarizability of liquid *Ar*. The error bars are the statistical error of the average.

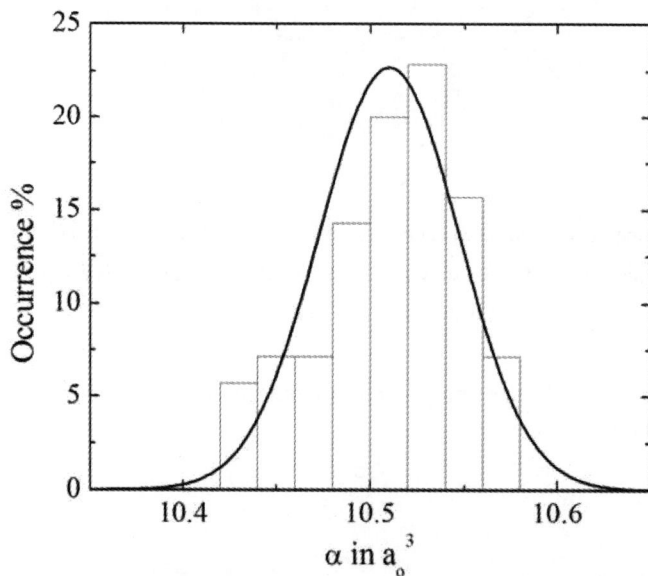

Figure 5: Histogram and normal distribution of calculated values for the dipole polarizability of liquid *Ar*. Standard deviation is 0.037 a_0^3.

4. Concluding remarks

The sequential QM/MM methodology has been suggested as a viable and efficient mean to obtain the dipole polarizability of atomic liquids. Liquid argon is a very challenging system with a small polarization and requires consideration of detailed aspects. Using Monte Carlo simulation the structures of liquid argon were generated. These structures and the calculated density are in very good agreement with the experimental results. The auto-correlation function of the energy derived from the simulation is calculated and analyzed. From this, statistically relevant configurations are extracted for subsequent quantum mechanics calculations. Using the radial distribution function we separated the structures that give the first solvation shell, composed of the central *Ar* atom surrounded by 13 other atoms. Using these structures density-functional theory calculations were performed within the B3P86 hybrid functional and the aug-cc-pVDZ basis set to obtain statistically converged values for the dipole polarizability in the liquid case. The results pointed to a slightly increased value in the liquid case, as compared to the gas phase result.

Acknowledgments

This work has been partially supported by CNPq and FAPESP (Brazil).

418 Coutinho and Canuto

REFERENCES

1. D. M. Bishop, Adv. Quantum Chem. 25, 1 (1994).
2. J. Oddershede, P. Jorgensen and D. L. Yeager, Comp. Phys. Report 2, 33 (1984).
3. D. R. Kanis, M. A. Ratner and T. J. Marks, Chem. Rev. 94, 195 (1994).
4. T. M. Miller and B. Bederson, Adv. Atom. Mol. Phys. 25, 37 (1988).
5. P. N. Prasad and D. J. Williams, Introduction to Nonlinear Optical Effects in Organic Molecules and Polymers, J. Wiley, New York (1991).
6. A. J. Stone, The Theory of Intermolecular Forces, Clarendon Press, Oxford (1996).
7. K. D. Bonin and V. V. Kresin, Electric-Dipole Polarizability of Atoms, Molecules and Clusters, World Scientific, London (1997).
8. S. P. A. Sauer and J. Oddershede, Int. J. Quantum Chem. 50, 317 (1994).
9. B. Kirtman, C. E. Dykstra and B. Champagne, Chem. Phys. Lett.305, 132 (1999).
10. G. Maroulis, Chem. Phys. Lett. 334, 207 (2001).
11. W. A. Parkinson and J. Oddershede, J. Chem. Phys. 94, 7251 (1991)
12. Y. Mochizuki and H. Ågren, Chem. Phys. Lett. 336, 451 (2001)
13. R. J. Doerksen amd A. J. Thakkar, J. Phys. Chem. A 103, 2141 (1999)
14. P. Fuentealba, Chem. Phys. Lett. 397, 459 (2004).
15. T. K. Ghosh, A. K. Das and P. K. Mukherjee, Chem. Phys. Lett. 218, 433 (1994).
16. E. Botek and B. Champagne, Chem. Phys. Lett., 370, 197 (2003)
17. S. Canuto, W. Duch, J. Geertsen, F. Mueller-Plathe, J. Oddershede and G. Scuseria, Chem. Phys. Lett. 147, 435 (1988).
18. S. Canuto, M. A Castro and P. K. Mukherjee, Phys. Rev. A 49, 3315 (1994).
19. G. H. F. Diercksen and A. J. Sadlej, Chem. Phys. Lett. 84, 390 (1981).
20. E. F. Archibong and A. J. Thakkar, Chem. Phys. Lett. 173, 579 (1990).
21. J. Pipin and D. M. Bishop, J. Phys. B: At. Mol. Phys. 25, 17 (1992).
22. P. W. Fowler, P. Jorgensen and J. Olsen, J. Chem. Phys. 93, 7256 (1990).
23. A. K. Das, T. K. Ghosh and P. K. Mukherjee, Phys. Scripta, 50, 354 (1994).
24. D. M. Bishop, Adv, Chem. Phys. 104, 1 (1998).
25. D. M. Bishop, J. Chem. Phys. 100, 6535 (1990).
26. D. M. Bishop, J. M. Luis and B. Kirtman, J. Chem. Phys. 108, 10013 (1998).
27. A. Avramopoulos and M. G. Papadopoulos, Mol. Phys. 100, 821 (2002).
28. O. P. Andrade, A. Aragão, O. A. V. Amaral, T. L. Fonseca and M. A. Castro, Chem. Phys. Lett. 392, 270 (2004).
29. R. Pessoa, M. A. Castro, O. A. V. Amaral and T. L. Fonseca, Chem. Phys. 306, 281 (2004).

30. V. E. Bazterra, M. C. Caputo, M. Ferraro and P. Fuentealba, J. Chem. Phys. 117, 11158 (2002)

31. E. E. Fileti, K. Coutinho, T. Malaspina and S. Canuto, Phys. Rev. 67, 61504 (2003)

32. O. Tapia and O. Goscinski, Molec. Phys. 29, 1653 (1975).

33. J. L. Rivail and D. Rinaldi, Chem. Phys. 18, 233 (1976).

34. J. Tomasi and M. Persico, Chem. Rev. 94, 2027 (1994).

35. M. M. Karelson and M. C. Zerner, J. Phys. Chem. 96, 6949 (1992).

36. J. Tomasi, Theor. Chem. Accounts, 112, 112 (2004).

37. S. Miertus, E. Scrocco and J. Tomasi, J. Chem Phys 1981, 55, 117.

38. M. Cossi, V. Barone, R. Cammi and J. Tomasi, Chem. Phys. Lett. 255, 327 (1996).

39. Y. Luo, H. Ågren, P. Jorgensen and K. V. Mikkelsen, Adv. Quantum Chem. 26, 165 (1995).

40. K. V. Mikkelsen, P. Jorgensen and H. J. Aa. Jensen, J. Chem. Phys. 100, 6597 (1994).

41. R. Cammi, M. Cossi and J. Tomasi, J. Chem. Phys. 104, 4611 (1996).

42. K. V. Mikkelsen, Y. Luo, H. Ågren and P. Jorgensen, J. Chem. Phys. 102, 9362 (1995).

43. A. Willettts and J. E. Rice, J. Chem. Phys. 99, 426 (1993).

44. D. M. Heyes, The Liquid State. Applications of Molecular Simulations, John Wiley, New York (1998)

45. M. P. Allen and D. J. Tildesley, Computer Simulation of Liquids, Clarendon Press, 1987.

46. J. T. Blair, K. Krogh-Jespersen and R. M. Levy, J. Am. Chem. Soc. 111, 6948 (1989).

47. J. Gao, J. Am. Chem. Soc. 116, 9324 (1994).

48. J. Zeng, N. S. Hush and J. R. Reimers, J. Chem. Phys. 99, 1496 (1993).

49. K. Coutinho, S. Canuto and M. C. Zerner, Int. J. Quantum Chem. 65, 885 (1997).

50. K. Coutinho and S. Canuto, Adv. Quantum Chem. 28, 89 (1997).

51. K. Coutinho, S. Canuto and M. C. Zerner, J. Chem. Phys., 112, 9874 (2000).

52. K. Coutinho and S. Canuto, J. Chem. Phys. 113, 9132 (2000).

53. S. Canuto, K. Coutinho and D. Trzresniak, Adv. Quantum Chem. 41, 161 (2002).

54. A. D. Becke, J. Chem. Phys. 98, 5648 (1993); J. P. Perdew, Phys. Rev. B 33, 8822 (1986).

55. A. Morita and S. Kato, J. Chem. Phys. 110, 11987 (1999).

56. P. Th. van Duijnen, A. H. De Vries, M. Swart and F. Grozema, J. Chem. Phys. 117, 8442 (2002).

57. S. Canuto, K. Coutinho and P. K. Mukherjee, Adv. Quantum Chem. 58, 161 (2005).

58. G. C. Maitland and E. B. Smith, Mol. Phys. 22 (1971) 861.

59. K. Coutinho and S. Canuto, DICE: A general Monte Carlo program for liquid simulation, University of São Paulo, 2000.

60. A. Eisenstein and N. S. Gingrich, Phys. Rev. 62 (1942) 261.

61. K. Coutinho, M. J. Oliveira and S. Canuto, Int. J. Quantum Chem. 66, 249 (1998).

62. G. Maroulis, J. Chem. Phys. 113, 1813 (2000).

63. M. J. Frisch, et al. Gaussian 98, Revision A.6, Gaussian, Inc., Pittsburgh PA, 1998.

64. C. Lupinetti and A. J. Thakkar, J. Chem. Phys. 122, 44301 (2005).

65. P. Soldan, E. P. F. Lee and T. G. Wright, Phys. Chem. Chem. Phys. 3, 4661 (2001)

66. E. A. Reinsch and W. Meyer, Phys. Rev. A 14, 915 (1976).

67. U. Hohm and K. Kerl, Mol. Phys. 69, 819 (1990).

68. D. R. Johnston, G. J. Oudemans and R. H. Cole, J. Chem. Phys. 33, 1310 (1960).

69. Kagaku Binran (Japanese), 4th ed., edited by M. Oki et al. Maruzen, Tokyo, 1993.

70. D. E. Woon, T. H. Dunning Jr., J. Chem. Phys. 100, 2975 (1994).

CHAPTER 12

HIGH ORDER POLARIZABILITIES FROM OPTICAL INTERACTION-INDUCED SPECTROSCOPY

Tadeusz Bancewicz*, Yves Le Duff+ and Jean Luc Godet+

* Nonlinear Optics Division, Faculty of Physics,
Adam Mickiewicz University, Umultowska 85, 61-614 Poznań, Poland
E-mail:tbancewi@zon12.physd.amu.edu.pl
+ Laboratoire des Propriétés Optiques des Matériaux et Applications,
Université d'Angers, 2 bd Lavoisier, 49045 Angers, France
E-mail:yleduff@libertysurf.fr; godet@univ-angers.fr

Multipolar polarizability contributions to the collision-induced light scattering spectra of optically isotropic molecules may be put in evidence in the Rayleigh wings and in the vicinities of the ν_1 Raman vibrational lines. Experiments on CH_4 or CF_4 (tetrahedral symmetry) and SF_6 (octahedral symmetry) make possible evaluations of the multipolar polarizabilities provided that isotropic Rayleigh spectra and anisotropic Raman bands are obtained in some relatively large frequency ranges. These evaluations – in particular those based on set inversion analysis – constitute experimental references for quantum chemistry computations of the multipolar polarizabilities. In the case of CH_4, CF_4 and SF_6, the agreement between experimental and *ab initio* values is quite good.

1. Introduction

During collisions, spatial coordinates of molecules change with time and their motions induce temporary changes in the result polarizability of interacting pairs of molecules. These effects are manifested in the dielectric and optical properties of compressed gases[1,2,3,4]. In Cartesian coordinates, the collision-induced (CI) polarizability $\Delta A_{\alpha\beta}$ of the pair of molecules A and B is defined as

$$\Delta A_{\alpha\beta}(AB) = A_{\alpha\beta}^{(tot)}(AB) - A_{\alpha\beta}(A) - A_{\alpha\beta}(B) \qquad (1)$$

where $A_{\alpha\beta}^{(tot)}(AB)$ denotes the total polarizability tensor of the collision complex and $A_{\alpha\beta}(k)$ stands for permanent (intrinsic) polarizability of the

isolated molecule $k = A$ or B. Usually the main contribution to this CI polarizability results from induced purely dipolar dipole-induced dipole (DID)
mechanism. It mainly concerns the anisotropy of $\Delta A_{\alpha\beta}$ where DID contribution is of the first order. Higher order (multipolar) polarizabilities also
play a role in $\Delta A_{\alpha\beta}$[5,6]. However, quantitatively their overall contribution is
usually very weak. Generally, the DID contribution overshadows multipo-
lar ones in most of the collision-induced experiments (virial coefficients of
Kerr constant, dielectric constant, electric-field induced birefringence, light
scattering integrated studies etc). However in frequency resolved light scattering experiments, higher order polarizabilities give intensities of specific
spectral features and their contributions may be separated and studied. In
particular for intermediate frequency shifts, it has been found that higher
order polarizabilities can play a leading role[7,8,9,10,11]. The CI polarizability, Eq.(1), induced by interactions in a molecular pair is a tensor which
have two invariants: its trace $\alpha(t)$ and its anisotropy $\beta(t)$. With the use of
polarization properties of the scattered radiation we are able to measure
the spectral behavior of $\beta(t)$ and $\alpha(t)$ separately studying anisotropic and
isotropic light scattering spectra, respectively. Moreover for the trace, the
DID light scattering mechanism is of the second order and it gives us a
good opportunity to study multipolar polarizabilities.

In this review, we give the theoretical polarizability of a pair of molecules
for two types of molecular symmetry (tetrahedral and octahedral) taking
into account DID as well as multipolar polarizabilities. Then, we discuss
experimental data obtained with molecules for which recent computational
works have been made.

2. Multipolar polarizabilities

2.1. *Pair polarizability*

We consider a system of two interacting molecules A and B in an external field **V**. We calculate, within the first-order (linear in the interaction
tensor $\mathbf{T}_N^{(AB)}$, see e.g.[12]) approximation the excess irreducible spherical
interaction-induced pair polarizability ΔA_{LM}. Eq.(1) in spherical coordinates reads:

$$\Delta A_{LM} = A_{LM}^{(total)} - A_{LM}^{(A)} - A_{LM}^{(B)} \qquad (2)$$

where $A_{LM}^{(total)}$ denotes the total polarizability tensor of the pair and $A_{LM}^{(k)}$
stays for the intrinsic polarizability of species k ($k = A$ or B). The induced

multipole moment ΔQ_{lm} for each molecule can be divided into two parts: the external moment ΔQ_{lm}^{ext} that arises from the external field $\mathbf{V}$ and the internal moment ΔQ_{lm}^{int} that results from the field of the other molecule [13]. Calculating polarizabilities in the first-order approximation we neglect the internal field of a pair. Then the electrostatic part of the interaction energy of a pair of two molecules A and B, perturbed by a static external field $\mathbf{V(r)}$ takes the form [13,14]:

$$U = \sum_{\substack{l_1=1,2,3,\ldots \\ l_2=1,2,3,\ldots}} \left(\frac{2^N}{(2l_1)!\,(2l_2)!}\right)^{1/2} \frac{X_N}{X_{l_2}} \sum_{m_1 m_2,\, n} (-1)^{m_2}\, C(l_1, N, l_2; m_1, n, m_2)$$
$$_{(A)}Q_{l_1 m_1}^{(ext)}\, T_{Nn}^{(AB)}\, _{(B)}Q_{l_2-m_2}^{(ext)} \tag{3}$$

where $N = l_1 + l_2$, $_{(P)}Q_{lm}^{(ext)}$ denotes the m-th component of the l-th order spherical electric multipole moment of the perturbed molecule P, $\mathbf{T}_N^{(AB)}$ is the spherical interaction tensor, $C(a\,b\,c;\alpha\,\beta\,\gamma)$ stands for the Clebsch-Gordan coefficient and $X_{ab\ldots f} = [(2a+1)(2b+1)\ldots(2f+1)]^{\frac{1}{2}}$. As usual $l_i = 1$, $l_i = 2$, $l_i = 3$, $l_i = 4$ stay in Eq.(3) for the dipole, quadrupole, octopole and hexadecapole electric moments, respectively. The multipole moment of the perturbed molecule p, in the first order approximation (neglecting molecular fields [15,17]), becomes [18]:

$$_{(P)}Q_{l_1 m_1}^{(ext)} = {}_{(P)}Q_{l_1 m_1}(0) + \sum_{l_2 n_2} {}_{(P)}\overline{A}_{l_1 l_2}^{m_1 n_2}(0)\, V_{l_2}^{n_2 *}$$
$$+ \frac{1}{2!} \sum_{\substack{l_2,n_2 \\ l_3,n_3}} {}_{(P)}\overline{B}_{l_1 l_2 l_3}^{m_1 n_2 n_3}(0) \times V_{l_2}^{n_2 *} V_{l_3}^{n_3 *} + \ldots \tag{4}$$

where $_{(P)}Q_{l_1 m_1}(0)$ and $_{(P)}\overline{A}_{l_1 l_2}^{m_1 m_2}(0)$ denote, respectively, the permanent multipole moment, the symmetrized spherical reducible $l_1 - l_2$th order multipole polarizability of an unperturbed molecule P and $_{(P)}\overline{B}_{l_1 l_2 l_3}^{m_1 m_2 m_3}(0)$ is its $l_1 - l_2 - l_3$ order reducible multipolar hyperpolarizability. Moreover $\mathbf{V}_k$ stands for the k-order regular solid spherical harmonic. The symbol 0 indicates here the multipole moment and the polarizability of an unperturbed (isolated) molecule. Further on, to simplify notation, we drop the 0 and the bar symbols $\overline{A}$ assuming that we deal with the symmetrized polarizabilities.

Using Eqs.(2) and (4) we calculate the excess reducible dipolar *pair* polarizability tensor as the derivative

$$\Delta A_1^{s_1 \, s_2} = - \left(\frac{\partial^2 U}{\partial V_1^{s_1 *} \partial V_1^{s_2 *}} \right)_{V \to 0} \tag{5}$$

and consequently, taking into account the linear multipolar contributions in Eq.(4), we have

$$
^{(L)}\Delta A_1^{s_1 \, s_2} = - \sum_{l_1 l_2} \left(\frac{2^N}{(2l_1)!(2l_2)!} \right)^{1/2} \frac{X_N}{X_{l_2}} \sum_{m_1 m_2 n} \{ (-1)^{m_2} C(l_1 \, N \, l_2; m_1 \, n \, m_2)
$$
$$
\times \left[{}_{(A)}A_{l_1 \, 1}^{m_1 \, s_1} \, T_{Nn}^{(AB)} \, {}_{(B)}A_{l_2 \, 1}^{-m_2 \, s_2} + {}_{(A)}A_{l_1 \, 1}^{m_1 \, s_2} \, T_{Nn}^{(AB)} \, {}_{(B)}A_{l_2 \, 1}^{-m_2 \, s_1} \right] \} \tag{6}
$$

All polarizability tensors in Eqs.(4) to (6) are reducible under the three-dimensional rotation group. Irreducible tensors may be constructed from them by the standard coupling method:

$$A_l^{m \, s}{}_1 = \sum_{j\phi} C(l \, 1 \, j; m \, s \, \phi) \, A_{j\phi}^{(1,l)} \tag{7}$$

$$^{(L)}\Delta A_{K \, M}^{(1,1)} = \sum_{s_1 s_2} C(1 \, 1 \, K; s_1 \, s_2 \, M) \, {}^{(L)}\Delta A_1^{s_1 \, s_2} \tag{8}$$

Finally, inserting Eq.(6) and Eq.(7) into Eq.(8) and reducing the sum of four Clebsch-Gordan coefficients[19] of Eq.(6), for the excess interaction-induced irreducible spherical multipole pair polarizability tensor we have close $^{(L)}\Delta A_{KM}$ formula [9,20,22]. It is instructive to write this formula in a form of a series

$$^{(L)}\Delta A_{KM} = \sum_{j_1,l_1,j_2,l_2} {}^{(L)}\Delta A_{KM}^{(j_1,l_1,j_2,l_2)} \tag{9}$$

where

$$^{(L)}\Delta A_{KM}^{(j_1,l_1,j_2,l_2)} = \sum_x \mathcal{A}_K^{(j_1,l_1,j_2,l_2)}\{x\} \left\{ \mathbf{T}_N^{(AB)} \otimes \left[{}_{(A)}\mathbf{A}_{j_1}^{(1,l_1)} \otimes {}_{(B)}\mathbf{A}_{j_2}^{(1,l_2)} \right]_{(x)} \right\}_{KM}$$
$$\tag{1}$$

and for the respective components $\mathcal{A}_K^{(j_1,l_1,j_2,l_2)}\{x\}$ we obtain

$$A_K^{(j_1,l_1,j_2,l_2)}\{x\} = \left[(-1)^{j_1+N} + (-1)^{K+j_2}\right] \left(\frac{2^N}{(2l_1)!\,(2l_2)!}\right)^{\frac{1}{2}} X_{j_1\,j_2\,N\,x}$$

$$\times \left\{ \begin{array}{ccc} j_1 & j_2 & x \\ 1 & 1 & K \\ l_1 & l_2 & N \end{array} \right\} \tag{11}$$

For tetrahedral and octahedral molecules these components have been collected in Tables 1 and 2 for subsequent collision-induced multipolar light scattering mechanisms and for isotropic (K=0) and anisotropic (K=2) light scattering parts.

Table 1. Components of the excess multipolar origin isotropic pair polarizability $^{(L)}\Delta A_{00}$

lsm	j_1	l_1	j_2	l_2	$A_0^{(j_1,l_1,j_2,l_2)}$
DQ	0	1	3	2	$\frac{\sqrt{42}}{9}$
DO	0	1	4	3	$\frac{\sqrt{10}}{15}$
QQ	3	2	3	2	$-\frac{\sqrt{165}}{45}$
QO	3	2	4	3	$\frac{\sqrt{\frac{143}{14}}}{45}$
OO	4	3	4	3	$\frac{\sqrt{\frac{13}{35}}}{36}$

Bearing in mind the form of the second derivative of a product of two functions f and g:

$$\frac{\partial^2(fg)}{\partial V_1 \partial V_2} = \frac{\partial f}{\partial V_1}\frac{\partial g}{\partial V_2} + \frac{\partial g}{\partial V_1}\frac{\partial f}{\partial V_2} + f\frac{\partial^2 g}{\partial V_1 \partial V_2} + g\frac{\partial^2 f}{\partial V_1 \partial V_2} \tag{12}$$

we note that first two terms of Eq.(12) give us the pair polarizability of dipole-multipole origin, see Eqs.(6) and (12). The last two terms contribute to the pair polarizability resulting from dipole^2-multipole hyperpolarizability –permanent multipole coupling between molecules A and B of the pair. Here we focus our attention on the latter one. Using Eqs.(3), (5) and (12) and calculating respective derivatives for interaction-induced variation in

Table 2. Components of the excess multipolar origin anisotropic pair polarizability $^{(L)}\Delta A_{2M}$

lsm	j_1	l_1	j_2	l_2	Range of x	$\mathcal{A}_2^{(j_1,l_1,j_2,l_2)}$
DQ	0	1	3	2	3	$-\frac{2\sqrt{7}}{15}$
DO	0	1	4	3	4	$-\sqrt{\frac{11}{105}}$
QQ	3	2	3	2	2–6	$\left\{-\frac{\sqrt{\frac{2}{35}}}{15},0,\frac{\sqrt{\frac{2}{105}}}{3},0,-\frac{\sqrt{\frac{13}{5}}}{3}\right\}$
QO	3	2	4	3	3-7	$\left\{\frac{1}{30\sqrt{105}},-\frac{1}{150\sqrt{7}},-\frac{\sqrt{\frac{11}{42}}}{50},\frac{\sqrt{\frac{13}{10}}}{90},\frac{1}{3\sqrt5}\right\}$
OO	4	3	4	3	4-8	$\left\{\frac{1}{420\sqrt{11}},0,-\frac{\sqrt{\frac{13}{55}}}{180},0,\frac{\sqrt{\frac{17}{5}}}{45}\right\}$

the pair polarizability due to nonlinear molecular polarizabilities we have:

$$^{(NL)}\Delta A_1^{s_1 s_2}{}_1 = -\sum_{l_1 l_2}\left\{\left(\frac{2^N}{(2l_1)!(2l_2)!}\right)^{1/2}\frac{X_N}{X_{l_2}}\sum_{m_1 m_2,n}[(-1)^{m_2}C(l_1,N,l_2;m_1,\right.$$

$$n,m_2)\left.\left(_{(A)}B_{l_1\ 1\ 1}^{m_1 s_1 s_2}T_{Nn}^{(AB)}{}_{(B)}Q_{l_2-s_2}+{}_{(A)}Q_{l_1 s_1}T_{Nn}^{(AB)}{}_{(B)}B_{l_2\ 1\ 1}^{-m_2 s_1 s_2}\right)\right]\right\}\quad(13)$$

Again reducible polarizability and hyperpolarizability tensors $\Delta A_1^{s_1 s_2}{}_1$ and $B_{l_i\ 1\ 1}^{m_i s_1 s_2}$ transform into their irreducible counterparts by standard procedure:

$$^{(NL)}\Delta A_1^{s_1 s_2}{}_1 = \sum_{K,M}C(11K;s_2 s_1 M)\ {}^{(NL)}\Delta A_{KM}\quad(14)$$

$$B_{l_i\ 1\ 1}^{m_i s_1 s_2} = \sum_{a,\xi,J_i,M_i}C(11a;s_2 s_1\xi)\ C(al_i J_i;\xi m_i M_i)\ B_{J_i M_i}[(11)al_i](15)$$

where $B_{J_i M_i}[(11)al_i]$ is the irreducible J_i rank spherical tensor of dipole^2-2**l_i-pole hyperpolarizability of molecule i in a coupling scheme where two dipoles are first connected and, subsequently the 2**l_i-pole multipolar moment.

Applying transformations given by Eq.(14) and Eq.(15) in Eq.(13) and using orthogonality and symmetry relations for Clebsch-Gordan coefficients we easily arrive at the formula $^{(NL)}\Delta A_{KM}$ [23] for the excess contribution to the pair polarizability due to nonlinear polarization of the molecules of a pair. Again we write the *nonlinear* origin pair polarizability as a series:

$$^{(NL)}\Delta A_{KM} = \sum_{j_1,l_1,l_2}{}^{(NL)}\Delta A_{KM}^{(j_1,l_1,l_2)}\quad(16)$$

where

$$^{(NL)}\Delta A_{KM}^{(j_1,l_1,l_2)} = \sum_x \mathcal{B}_K^{(j_1,l_1,l_2)}\{x\}\left[\left\{\mathbf{T}_N^{(AB)} \otimes \left[_{(A)}\mathbf{B}_{j_1}[(11)Kl_1] \otimes _{(B)}\mathbf{Q}_{l_2}\right]_x\right\}_{KM}\right.$$
$$\left. + (-1)^N \left\{\mathbf{T}_N^{(AB)} \otimes \left[_{(B)}\mathbf{B}_{j_1}[(11)Kl_1] \otimes _{(A)}\mathbf{Q}_{l_2}\right]_x\right\}_{KM}\right] \tag{17}$$

and for $\mathcal{B}_K^{(j_1,l_1,l_2)}\{x\}$ components we have:

$$\mathcal{B}_K^{(j_1,l_1,l_2)}\{x\} = (-1)^{(K+l_1+x+1)}\left(\frac{2^N}{(2l_1)!\,(2l_2)!}\right)^{\frac{1}{2}}\begin{Bmatrix} N & l_2 & l_1 \\ j_1 & K & x \end{Bmatrix}\frac{X_{j_1 N}}{X_K} \tag{18}$$

For tetrahedral and octahedral molecules the $\mathcal{B}_K^{(j_1,l_1,l_2)}\{x\}$ components have been collected in Table 3 for subsequent nonlinear origin light scattering mechanisms.

Table 3. Components of the excess nonlinear origin anisotropic pair polarizability $^{(NL)}\Delta A_{2M}$

lsm	j_1	l_1	l_2	Range of x	$\mathcal{B}_2^{(j_1,l_1,l_2)}$
$B_0\,\Omega$	0	2	3	3	$\frac{\sqrt{165}}{450}$
$B_0\,\Phi$	0	2	4	4	$-\frac{\sqrt{\frac{13}{105}}}{60}$
$b_3\,\Omega$	3	1	3	2–6	$\left\{\frac{1}{30\sqrt{7}}, \frac{1}{30}, \frac{\sqrt{\frac{3}{7}}}{10}, \frac{\sqrt{11}}{30}, \frac{\sqrt{\frac{13}{2}}}{15}\right\}$
$b_3\,\Phi$	3	1	4	3–7	$\left\{-\frac{1}{90\sqrt{10}}, -\frac{1}{50\sqrt{6}}, -\frac{\sqrt{11}}{225}, -\frac{\sqrt{\frac{13}{105}}}{15}, -\frac{1}{2\sqrt{210}}\right\}$
$B_4\,\Omega$	4	2	3	3–7	$\left\{\frac{1}{60\sqrt{105}}, \frac{1}{60\sqrt{7}}, \frac{\sqrt{\frac{11}{42}}}{30}, \frac{\sqrt{\frac{13}{10}}}{30}, \frac{1}{6\sqrt{5}}\right\}$
$B_4\,\Phi$	4	2	4	4–8	$\left\{-\frac{1}{60\sqrt{1155}}, -\frac{1}{60\sqrt{105}}, -\frac{\sqrt{\frac{13}{231}}}{60}, -\frac{1}{12\sqrt{105}}, -\frac{\sqrt{\frac{17}{21}}}{60}\right\}$

For the A-B molecular pair the incremental *pair Raman* polarizability tensor $\Delta A_{\alpha\beta}^{(\nu_1)}$ for the normal vibration ν_1 reads:

$$\Delta A_{\alpha\beta}^{(\nu_1)} = \frac{\partial \Delta A_{\alpha\beta}}{\partial Q_A^{\nu_1}}Q_A^{\nu_1} + \frac{\partial \Delta A_{\alpha\beta}}{\partial Q_B^{\nu_1}}Q_B^{\nu_1} \tag{19}$$

where $\Delta A_{\alpha\beta}$ is the *pair Rayleigh* excess polarizability and $Q_p^{\nu_1}$ is the normal coordinate for the mode ν_1 and the molecule p. Moreover for the ν_1

vibration of CF_4 and SF_6 molecules, the Raman polarizability tensor of an isolated molecule is isotropic and, consequently, monomer polarizabilities do not contribute to depolarized Raman scattering. For further details concerning e.g. *pair Raman* multipolar and nonlinear polarizabilities and Raman correlation functions see Ref.[27,40].

2.2. *Theoretical light scattering spectra*

We consider a macroscopically isotropic system composed of $N-$like globular molecules in an active scattering volume V illuminated by laser radiation of angular frequency $\omega_L = 2\pi\nu_L$, polarized linearly in the direction **e**. We analyze the secondary purely collision–induced electromagnetic radiation emitted by the system in response to that perturbation. At a point **R** distant from the center of the sample, the radiation scattered at $\omega_s = 2\pi(\nu_L - \nu)$ is measured on traversal of an analyzer with polarization **n**. Then the quantum-mechanical expression for the pair differential scattering cross section has the form[24]:

$$\frac{\partial^2 \sigma}{\partial\Omega\partial\omega} = k_s^4 \frac{1}{2\pi} \int\limits_{-\infty}^{\infty} dt\, e^{-i\omega t} \underbrace{\langle (\mathbf{n} \cdot \Delta\mathbf{A}(0) \cdot \mathbf{e})(\mathbf{n} \cdot \Delta\mathbf{A}(t) \cdot \mathbf{e}) \rangle}_{F(t)} \tag{20}$$

For the pair double differential cross sections of the, respectively, isotropic and anisotropic collision-induced light scattering we obtain

$$\left(\frac{\partial^2 \sigma}{\partial\Omega\partial\omega}\right)_{iso} = \frac{(\mathbf{e} \cdot \mathbf{n})^2}{3} k_s^4 \frac{1}{2\pi} \int\limits_{-\infty}^{\infty} dt\, e^{-i\omega t} F_{00}(t) \tag{21}$$

$$\left(\frac{\partial^2 \sigma}{\partial\Omega\partial\omega}\right)_{ani} = \frac{3 + (\mathbf{e} \cdot \mathbf{n})^2}{30} k_s^4 \frac{1}{2\pi} \int\limits_{-\infty}^{\infty} dt\, e^{-i\omega t} F_{22}(t) \tag{22}$$

where k_s denotes the wave-vector of scattered light and

$$F_{KK}(t) = \langle \Delta\mathbf{A}_{(k)}(0) \odot \Delta\mathbf{A}_{(k)}(t) \rangle. \tag{23}$$

The correlation function given by equation (23) of scattered light generally deals with a situation when the rotational and translational degrees of freedom are coupled. However, when considering radiation scattered by low density gaseous systems of globular molecules we assume that molecules of the scattering volume are correlated radially but uncorrelated orientationally [6,16,25,26]. In this case when using the form of pair multipolar and nonlinear polarizabilities given respectively by equations (10–11) and (17–18) and the decoupling procedures [19]

$$
(\{\mathbf{P}_a \otimes \mathbf{Q}_b\}_c \odot \{\mathbf{R}_d \otimes \mathbf{S}_e\}_c) = (-1)^{2a+b-d} \sum_g X_c^2 \left\{ \begin{matrix} a & b & c \\ e & d & g \end{matrix} \right\}
$$
$$
\times \left(\{\mathbf{P}_a \otimes \mathbf{R}_d\}_g \odot \{\mathbf{Q}_b \otimes \mathbf{S}_e\}_g \right) \tag{24}
$$

and

$$
\left\{ \{\mathbf{P}_a \otimes \mathbf{Q}_b\}_c \otimes \{\mathbf{R}_d \otimes \mathbf{S}_e\}_f \right\}_k = \sum_{gh} X_{cfgh} \left\{ \begin{matrix} a & b & c \\ d & e & f \\ g & h & k \end{matrix} \right\}
$$
$$
\times \left\{ \{\mathbf{P}_a \otimes \mathbf{R}_d\}_g \otimes \{\mathbf{Q}_b \otimes \mathbf{S}_e\}_h \right\}_k \tag{25}
$$

we are in a position, due to lack of coupling between rotational and translational motion of molecules, to let $g = 0$ in our formula for the correlation function given by equation (23) similar to equation (24). Subsequently we are in a position to let $g = h = 0$ in our following formula for the correlation function of multipolar and nonlinear polarizabilities given by a form similar to equation (25) as a consequence of the uncorrelated orientational motion of molecules A and B.

Averaging isotropically over the initial orientation of molecules A: $\Omega_A(0)$ and B: $\Omega_B(0)$ as well as over the initial orientation $\Omega_{AB}(0)$ of the intermolecular vector we write out our correlation function given by equation (25) as a product of the orientational:

$$
R_j(t) = \langle D_{nn}^j (\delta\Omega(t)) \rangle \tag{26}
$$

and the translational:

$$
S_N(t) = \left\langle D_{00}^N(\delta\Omega_{AB}(t)) R_{AB}(0)^{-(N+1)} R_{AB}(t)^{-(N+1)} \right\rangle \tag{27}
$$

correlation functions. In our equations (26) and (27) $\delta\Omega(t)$ and $\delta\Omega_{AB}(t)$ denote the reorientation angles of the molecule and intermolecular vector, respectively, at the time t.
Then the following relations hold:

$$
\left\langle \mathbf{T}_N^{(AB)}(0) \odot \mathbf{T}_N^{(AB)}(t) \right\rangle = \frac{(2N)!}{2^N} S_N(t) \tag{28}
$$

and

$$\langle \mathbf{A}_j(0) \odot \mathbf{A}_j(t)\rangle = R_j(t)\,(\tilde{\mathbf{A}}_j \odot \tilde{\mathbf{A}}_j) \tag{29}$$

where the tilde denotes the multipolarizability tensor $\mathbf{A}$ in its molecular frame reference system.

When using the form of pair multipolar polarizability given by Eqs.(9–11) and Eqs.(16–18) we note two terms of the pair correlation function:

1. *the square type term*

$$\left\langle \left\{ \mathbf{T}_N(0) \otimes \left[{}_{(A)}\mathbf{C}_{j_1}(0) \otimes {}_{(B)}\mathbf{D}_{j_2}(0) \right]_x \right\}_K \odot \left\{ \mathbf{T}_{N'}(t) \otimes \left[{}_{(A)}\mathbf{F}_{j_3}(t) \otimes {}_{(B)}\mathbf{G}_{j_4}(t) \right] \right. \right.$$

$$\left. \Big\}_K \right\rangle = (-1)^{K+N+j_1+j_2} \delta_{x\,x'}\,\delta_{j_1\,j_3}\,\delta_{j_2\,j_4}\,\delta_{N\,N'}\frac{X_K^2}{X_{N\,j_1\,j_2}^2}\frac{(2N)!}{2^N}\left(\tilde{\mathbf{C}}_{j_1} \odot \tilde{\mathbf{F}}_{j_1}\right)$$

$$\left(\tilde{\mathbf{D}}_{j_2} \odot \tilde{\mathbf{G}}_{j_2}\right) S_N(t)\,R_{j_1}(t)\,R_{j_2}(t)$$

2. *the cross type term*

$$\left\langle \left\{ \mathbf{T}_N(0) \otimes \left[{}_{(A)}\mathbf{C}_{j_1}(0) \otimes {}_{(B)}\mathbf{D}_{j_2}(0) \right]_x \right\}_K \odot (-1)^{N'} \left\{ \mathbf{T}_{N'}(t) \otimes \left[{}_{(B)}\mathbf{F}_{j_3}(t) \otimes \right. \right. \right.$$

$$\left. {}_{(A)}\mathbf{G}_{j_4}(t) \right]_{x'} \Big\}_K \right\rangle = (-1)^{K+x} \delta_{x\,x'}\,\delta_{j_1\,j_4}\,\delta_{j_2\,j_3}\,\delta_{N\,N'}\frac{X_K^2}{X_{N\,j_1\,j_2}^2}\frac{(2N)!}{2^N}$$

$$\times \left(\tilde{\mathbf{C}}_{j_1} \odot \tilde{\mathbf{G}}_{j_1}\right)\left(\tilde{\mathbf{D}}_{j_2} \odot \tilde{\mathbf{F}}_{j_2}\right) S_N(t)\,R_{j_1}(t)\,R_{j_2}(t) \tag{31}$$

Finally we write the light scattering correlation function $F_{KK}(t)$ as a sum of the respective multipolar and/or nonlinear origin contributions $F_{KK}^{(j_1,l_1,j_2,l_2)}(t)$

$$F_{KK}(t) = \sum_{j_1,l_1,j_2,l_2} F_{KK}^{(j_1,l_1,j_2,l_2)}(t) \tag{32}$$

the multipolar origin *square* type correlation function

$$F_{KK}^{(j_1,l_1,j_2,l_2)}(t) = (-1)^{K+N+j_1+j_2}\frac{X_K^2}{X_{N\,j_1\,j_2}^2}\frac{(2N)!}{2^N}\left(\sum_x \left[A_K^{(j_1,l_1,j_2,l_2)}\{x\}\right]^2\right)$$

$$\left(\tilde{\mathbf{A}}_{j_1}^{(1,l_1)} \odot \tilde{\mathbf{A}}_{j_1}^{(1,l_1)}\right)\left(\tilde{\mathbf{A}}_{j_2}^{(1,l_2)} \odot \tilde{\mathbf{A}}_{j_2}^{((1,l_2))}\right) S_N(t)\,R_{j_1}(t)\,R_{j_2}(t)$$

$$\tag{33}$$

the nonlinear origin *square* type correlation function

$$
F_{KK}^{(j_1,l_1,j_2,l_2)}(t) = (-1)^{K+N+j_1+j_2} \frac{X_K^2}{X_{N\,j_1\,j_2}^2} \frac{(2N)!}{2^N} \left(\sum_x \left[\mathcal{B}_K^{(j_1,l_1,N-l_1)}\{x\} \right]^2 \right)
$$
$$
\left(\tilde{\mathbf{B}}_{j_1}[(11)Kl_1] \odot \tilde{\mathbf{B}}_{j_1}[(11)Kl_1] \right) \left(\tilde{\mathbf{Q}}_{j_2} \odot \tilde{\mathbf{Q}}_{j_2} \right) S_N(t)\, R_{j_1}(t)\, R_{j_2}(t)
$$

(34)

the first type multipolar origin–nonlinear origin *cross* type correlation function

$$
F_{KK}^{(j_1,l_1,j_2,l_2)}(t) = (-1)^{K+N+j_1+j_2} \frac{X_K^2}{X_{N\,j_1\,j_2}^2} \frac{(2N)!}{2^N} \left(\sum_x A_K^{(j_1,l_1,j_2,l_2)}\{x\} \right.
$$
$$
\left. \mathcal{B}_K^{(j_1,l_1,N-l_1)}\{x\} \right) \left(\tilde{\mathbf{A}}_{j_1}^{(1,l_1)} \odot \tilde{\mathbf{B}}_{j_1}[(11)Kl_1] \right) \left(\tilde{\mathbf{A}}_{j_2}^{(1,l_2)} \odot \tilde{\mathbf{Q}}_{j_2} \right) S_N(t)\, R_{j_1}(t)\, R_{j_2}(t)
$$

(35)

the second type multipolar origin–nonlinear origin *cross* type correlation function

$$
F_{KK}^{(j_1,l_1,j_2,l_2)}(t) = (-1)^{K} \frac{X_K^2}{X_{N\,j_1\,j_2}^2} \frac{(2N)!}{2^N} \left(\sum_x (-1)^x A_K^{(j_1,l_1,j_2,l_2)}\{x\} \right.
$$
$$
\left. \mathcal{B}_K^{(j_1,l_1,N-l_1)}\{x\} \right) \left(\tilde{\mathbf{A}}_{j_1}^{(1,l_1)} \odot \tilde{\mathbf{Q}}_{j_1} \right) \left(\tilde{\mathbf{A}}_{j_2}^{(1,l_2)} \odot \tilde{\mathbf{B}}_{j_2}[(11)Kl_2] \right) S_N(t)\, R_{j_1}(t)\, R_{j_2}(t)
$$

(36)

Formulas (3–30) apply to a pair of monomers of arbitrary symmetry. Now we consider pairs of molecules of tetrahedral and/or octahedral symmetry. For these symmetries up to $\langle R_{12}^{-12} \rangle$ the excess *nonlinear origin isotropic* pair polarizability vanish: $_{(NL)}\Delta A_0 = 0$. For nonlinear polarizabilities we use for brevity the following notation $\mathbf{b}_3 = \mathbf{b}_3[(11)21]$, $\mathbf{B}_0 = \mathbf{B}_0[(11)22]$, and $\mathbf{B}_4 = \mathbf{B}_4[(11)22]$, Then for multipolar and nonlinear polarizabilities and multipolar moments discussed here the respective scalar tensor products appearing in Eqs. (33–36) are of the form

$$
\left(\tilde{\mathbf{A}}_3^{(1,2)} \odot \tilde{\mathbf{A}}_3^{(1,2)} \right) = 4\,A^2 \quad \left(\tilde{\mathbf{A}}_4^{(1,3)} \odot \tilde{\mathbf{A}}_4^{(1,3)} \right) = 3\,E^2
$$

$$
\left(\tilde{\mathbf{b}}_3 \odot \tilde{\mathbf{b}}_3 \right) = 6\,b_{xyz}^2 \quad \left(\tilde{\mathbf{Q}}_3 \odot \tilde{\mathbf{Q}}_3 \right) = \frac{12}{5}\,\Omega^2 \quad \left(\tilde{\mathbf{b}}_3 \odot \tilde{\mathbf{Q}}_3 \right) = 2\sqrt{\frac{18}{5}}\,b_{xyz}\,\Omega
$$

$$
\left(\tilde{\mathbf{A}}_3^{(1,2)} \odot \tilde{\mathbf{Q}}_3 \right) = 2\sqrt{\frac{12}{5}}\,A\,\Omega \quad \left(\tilde{\mathbf{A}}_3^{(1,2)} \odot \tilde{\mathbf{b}}_3 \right) = 2\sqrt{6}\,A\,b_{xyz} \quad (37)
$$

$$\left(\tilde{\mathbf{B}}_0 \odot \tilde{\mathbf{B}}_0\right) = \frac{6}{5} B_0^2 \quad \left(\tilde{\mathbf{B}}_4 \odot \tilde{\mathbf{B}}_4\right) = \frac{1}{5} (\Delta B)^2 \quad \left(\tilde{\mathbf{Q}}_4 \odot \tilde{\mathbf{Q}}_4\right) = \frac{12}{7} \Phi^2$$

$$\left(\tilde{\mathbf{B}}_4 \odot \tilde{\mathbf{Q}}_4\right) = 2\sqrt{\frac{3}{35}} \Delta B \, \Phi \quad \left(\tilde{\mathbf{A}}_4^{(1,3)} \odot \tilde{\mathbf{Q}}_4\right) = \frac{6\sqrt{7}}{7} E \, \Phi$$

$$\left(\tilde{\mathbf{A}}_4^{(1,3)} \odot \tilde{\mathbf{B}}_4\right) = \sqrt{\frac{3}{5}} E \, \Delta B \tag{38}$$

where $B_0 = B_{z,z,zz} + 2 B_{x,z,xz}$, $\Delta B = 3 B_{z,z,zz} - 4 B_{x,z,xz}$ and b_{xyz} and $B_{\alpha,\beta,\gamma\delta}$ stand for the respective components of the dipole[3] and the dipole[2]–quadrupole hyperpolarizabilities. Moreover A and E stand for the A_{xyz} and $E_{z,z,zz}$ components of the dipole–quadrupole and the dipole–octopole multipolarizability tensors, whereas Ω and Φ denote the octopole and hexadecapole moment of a molecule.

We note that special care must be taken especially when we calculate the correlation functions connected with nonlinear contributions. We illustrate this on the case of the dipole[2]-quadrupole hyperpolarizability–permanent octopole $\mathbf{B}_4 \, \mathbf{T}_5 \, \Omega$ light scattering mechanism. Excess *anisotropic nonlinear* pair polarizability due to the $\mathbf{B}_4 \, \mathbf{T}_5 \, \Omega$ light scattering mechanism has the form

$$_{(PAIR)}\Delta A_{2M} = \sum_x \mathcal{B}_2^{(4,2,3)}\{x\} \left[\left\{\mathbf{T}_5^{(AB)} \otimes \left[_{(A)}\mathbf{B}_4 \otimes_{(B)} \mathbf{Q}_3\right]_{(x)}\right\}_{2M}\right.$$
$$\left. + (-1)^x \left\{\mathbf{T}_5^{(AB)} \otimes \left[_{(A)}\mathbf{Q}_3 \otimes_{(B)} \mathbf{B}_4\right]_{(x)}\right\}_{2M}\right] \tag{39}$$

with the vector $\mathcal{B}_2^{(4,2,3)}\{x\}$ given by the 5th entry of the Table 3. Using Eq.(34) we compute nonlinear *square* correlation function of this mechanism as

$$2 \cdot \frac{5}{7 \cdot 9 \cdot 11} \cdot \frac{10!}{2^5} \cdot \sum_{x=3}^{7} \left(\mathcal{B}_2^{(4,2,3)}\{x\}\right)^2 \cdot \frac{1}{5} \cdot \frac{12}{5} \cdot (\Delta B \, \Omega)^2 = \frac{144}{25} (\Delta B \, \Omega) \tag{40}$$

We note however that, within our model, $\mathbf{B}_4 \, \mathbf{T}_5 \, \Omega$ light scattering mechanism correlates also with $\mathbf{b}_3 \, \mathbf{T}_5 \, \Phi$ nonlinear mechanism as well as with $\mathbf{A} \, \mathbf{T}_5 \, \mathbf{E}$ multipolar mechanism. Correlating $\mathbf{B} \, \mathbf{T}_5 \, \Omega$ with $\mathbf{A} \, \mathbf{T}_5 \, \mathbf{E}$ we obtain

$$4 \cdot \frac{5}{7 \cdot 9 \cdot 11} \cdot \frac{10!}{2^5} \cdot \sum_{x=3}^{7} \left((-1)^x \mathcal{B}_2^{(4,2,3)}\{x\} \cdot \mathcal{A}_2^{(3,2,4,3)}\{x\}\right) \cdot 2\sqrt{\frac{12}{5}} \cdot \sqrt{\frac{3}{5}}$$
$$\times A \, E \, \Delta B \, \Omega = -\frac{576}{7} A \, E \, \Delta B \, \Omega \tag{41}$$

We point out that within our model Eqs. (34) and (35) only the same molecules correlate so since the order of molecules A and B in $\mathbf{B_4 \, T_5 \, \Omega}$ and $\mathbf{A \, T_5 \, E}$ light scattering mechanisms is reverse in our sum of Eq.(41) we have to use the factor $(-1)^x$. Correlating $\mathbf{B_4 \, T_5 \, \Omega}$ light scattering mechanism with the other $\mathbf{b_3 \, T_5 \, \Phi}$ nonlinear mechanisms we obtain

$$4 \cdot \frac{5}{7 \cdot 9 \cdot 11} \cdot \frac{10!}{2^5} \cdot \sum_{x=3}^{7} \left((-1)^x \, \mathcal{B}_2^{(4,2,3)} \{x\} \cdot \mathcal{B}_2^{(3,1,4)} \{x\} \right) \cdot 2\sqrt{\frac{18}{5}} \cdot 2\sqrt{\frac{3}{35}} \cdot \Delta B \, \Omega$$

$$\times b_{xyz} \, \Phi = \frac{96}{7} \Delta B \, \Omega \, b_{xyz} \, \Phi \tag{42}$$

Again due to the reverse order of molecules A and B in both mechanisms considered we have to use the factor $(-1)^x$. Using other multipolar polarizability vectors $\mathcal{A}$ given in Tables 1 and 2 and nonlinear polarizability vectors $\mathcal{B}$ given in Table 3, by Eqs. (33–38) and a symbolic computation program, e.g. *Mathematica* [28] we easily obtain the remaining correlation functions. In Table 4 we collected all multipolar, nonlinear and cross collision–induced light scattering correlation functions $F_{KK}^{j_1,l_1,j_2,l_2}(t)$ resulting from Tables 1–3 and Eqs.(33–36). We consider multipolar, nonlinear and cross CI light scattering functions up to $\langle R_{12}^{-12} \rangle$ ($S_6(t)$). In the next section we use these formulas in our numerical computations and to comparison of the experimental data with theoretical ones. For Raman light scattering correlation functions, see Ref. [27].

3. Comparison of computed multipolar polarizabilities with experimental data

Although multipolar polarizabilities may influence several of the macroscopic properties of dense fluids not many experiments have been achieved to study dipole-quadrupole (DQ), dipole-octopole (DO) or higher order polarizabilities of molecules. To our knowledge, only few experimental techniques have been used to evaluate multipolar polarizabilities and the most accurate measurements of these parameters are based upon the study of light scattering.

At the end of the seventies it has been shown[6,5] that dipole-quadrupole and/or dipole-octopole polarizabilities may influence the Rayleigh wings of molecular gases and experimental values of these polarizabilities were deduced from anisotropic scattering spectra for several optically isotropic molecules. Then, other studies have been achieved concern-

Table 4. The correlation functions $F_{KK}^{(j_1,l_1,j_2,l_2)}(t)$.

Rayleigh CIS	j_1	l_1	j_2	l_2	Isotropic $F_{00}^{(j_1,l_1,j_2,l_2)}(t)$	Depolarized $F_{22}^{(j_1,l_1,j_2,l_2)}(t)$
$(DQ)^2$	0	1	3	2	$\frac{160}{7}\,\alpha^2\,A^2\,R_3(t)\,S_3(t)$	$\frac{192}{7}\,\alpha^2\,E^2\,R_3(t)\,S_3(t)$
$(DO)^2$	0	1	4	3	$\frac{224}{9}\,\alpha^2\,E^2\,R_4(t)\,S_4(t)$	$\frac{220}{9}\,\alpha^2\,E^2\,R_4(t)\,S_4(t)$
$(QQ)^2$	3	2	3	2	$\frac{1408}{189}\,A^4\,(R_3(t))^2\,S_4(t)$	$\frac{125824}{945}\,A^4\,(R_3(t))^2\,S_4(t)$
$(QO)^2$	3	2	4	3	$\frac{416}{21}\,A^2\,E^2\,R_3(t)\,R_4(t)\,S_5(t)$	$\frac{9280}{21}\,A^2\,E^2\,R_3(t)\,R_4(t)\,S_5(t)$
$(OO)^2$	4	3	4	3	$\frac{55}{3}\,E^4\,(R_4(t))^2\,S_6(t)$	$\frac{11330}{21}\,E^4\,(R_4(t))^2\,S_6(t)$
$(B_0\,\Omega)^2$	0	2	-	3	0	$\frac{864}{25}\,B_0^2\,\Omega^2\,R_3(t)\,S_5(t)$
$(B_0\,\Phi)^2$	0	2	-	4	0	$\frac{1584}{35}\,B_0^2\,\Phi^2\,R_4(t)\,S_6(t)$
$QQ{-}b\,\Omega$	3	1	-	3	0	$-\frac{768}{5}\,A^2\,b_{xyz}\,\Omega\,(R_3(t))^2\,S_4(t)$
$(b\,\Omega)^2$	3	1	-	3	0	$\frac{384}{7}\,(b_{xyz})^2\,\Omega^2\,(R_3(t))^2\,S_4(t)$
$QO{-}b\,\Phi$	3	1	-	4	0	$-192\,A\,E\,b_{xyz}\,\Phi\,R_3(t)\,R_4(t)\,S_5(t)$
$(b\,\Phi)^2$	3	1	-	4	0	$\frac{240}{7}\,(b_{xyz})^2\,\Phi^2\,R_3(t)\,R_4(t)\,S_5(t)$
$QO{-}B_4\,\Omega$	4	2	-	3	0	$-\frac{576}{7}\,A\,E\,\Delta B\,\Omega\,R_3(t)\,R_4(t)\,S_5(t)$
$(B_4\,\Omega)^2$	4	2	-	3	0	$\frac{144}{25}\,(\Delta B)^2\,\Omega^2\,R_3(t)\,R_4(t)\,S_5(t)$
$B_4\,\Omega{-}b\,\Phi$	4	2	-	3	0	$\frac{96}{7}\,b_{xyz}\,\Delta B\,\Omega\,\Phi\,R_3(t)\,R_4(t)\,S_5(t)$
$OO{-}B_4\,\Phi$	4	2	-	4	0	$-\frac{1056}{7}\,E^2\,\Delta B\,\Phi\,(R_4(t))^4\,S_6(t)$
$(B_4\,\Phi)^2$	4	2	-	4	0	$\frac{176}{15}\,(\Delta B)^2\,\Phi^2\,(R_4(t))^4\,S_6(t)$

ing a more extended frequency range as well as isotropic scattering intensities [9,10,25,26,29,30,31,32,33,34,35,36,37,38]. On the other hand, the study of interaction induced scattering intensities within vibrational Raman bands has allowed the measurement of the derivative of few multipolar polarizabilities[23,39,40].

A typical experimental set up for the 90 degrees scattering experiment used for these studies has been described in detail elsewhere[27]. In general, the light scattered by the sample illuminated by the beam of a laser is spectrally analyzed with a monochromator. Two kinds of scattering spectra, parallel spectra $(I_{//}(\nu))$ and perpendicular ones $(I_{\perp}(\nu))$ are required to perform a complete analysis of the scattering properties of the sample. They are obtained with a polarization of the laser parallel $(I_{//})$ or perpendicular $(I_{\perp})$ respectively to the scattering plane formed with the axis of the laser and the scattered beam. Then, anisotropic and isotropic intensities, I_{ani} and I_{iso} respectively, are deduced from

$$I_{ani}(\nu) = I_{//}(\nu) \tag{43}$$

$$I_{iso}(\nu) = I_{\perp}(\nu) - \frac{6}{7}I_{//}(\nu) \tag{44}$$

Studies of multipolar polarizabilities generally have been achieved from the scattering intensities generated by independent pairs of interacting molecules. In gaseous samples these intensities are obtained by a virial expansion in the density of the scattering intensities recorded at a series of pressures [27]. Then, these binary experimental spectra are compared with theoretical spectra computed using molecular parameters to be chosen.

Several contributions have to be taken into account in order to compute the theoretical collision-induced light scattering (CILS) spectra. For optically symmetric molecules, the dipole-induced dipole (DID) mechanism and the dipole-multipole mechanisms yield significant contributions to scattering intensities[27]. Other mechanisms exist (short-range and overlap effects, distortion of the molecular frame, influence of the non point-like size of the molecules, etc.) which are not well-known for globular molecules. However in general, the effects of these latter mechanisms are expected to be weak. Scattering intensities from DID and dipole-multipole mechanisms may be computed from Fourier transforms of Eqs.(33–36) and using an intermolecular potential $V(R)$, where R is the intermolecular distance. The DID intensities have two origins: bound and metastable dimers (pairs of molecules trapped in the well of the effective potential) and free dimers (colliding pairs of molecules). Bound dimers contribute mainly to the very low frequency region of the spectra (up to few cm^{-1}). On the contrary, free dimers are responsible for a so-called "translational" spectrum (due to the translational motion of the molecules) in the whole frequency range of the spectrum. This free-dimer DID contribution may be obtained by using classical trajectories of a pair of interacting molecules[3,41]. On the other hand for the dipole-multipole intensities (DQ, DO, etc.), the convolution product of translational and induced rotational stick spectra have to be calculated in a semi-classical way in order to get the associated "rototranslational" spectra[27].

3.1. *Tetrahedral molecules*

Tetrahedral molecules are optically isotropic molecules and in the earlier CILS studies they were considered as similar to rare gas atoms. This is roughly verified if the only integrated interaction induced intensity is considered. However when scattering spectra are investigated tetrahedral molecules show specific features. Here, we will discuss the contributions of multipolar polarizabilities (DQ and DO) to scattering intensities in two frequency regions: in the Rayleigh wings and in a vibrational Raman band.

3.1.1. *CILS in the Rayleigh wings*

Due to its tetrahedral symmetry, the polarizability tensor of the isolated molecule (monomer) is isotropic. The Rayleigh line generated by the monomer takes place in the isotropic spectrum only. However, intensities are observed in the wings of this line for both isotropic and anisotropic spectra. They are of collision-induced origin and for low density gases they are mainly due to dimers (binary interactions). The "Rayleigh" anisotropic binary spectrum results from the anisotropy of the pair polarizability tensor whereas the "Rayleigh" isotropic binary spectrum is connected to the collisional part of the trace of this tensor. According to Eq.(44), the isotropic spectrum may be obtained from the measurement of parallel and perpendicular scattering intensities, but in general, for most of the molecular gases studied up to now, this spectrum is less intense than the anisotropic one. Thus, in the earlier works only anisotropic Rayleigh binary spectra of tetrahedral molecules (CH_4[31,32,34], CF_4[26,33,42], neopentane[25]) have been reported. Two examples of anisotropic binary spectra obtained at 295 K are provided in Fig.1a and Fig.1b, for CH_4 and CF_4 respectively. In Fig.1a, experimental data come from Ref.[32] and from Ref.[11] whereas in Fig.1b they come from Ref.[26].

The isotropic Rayleigh spectra of CH_4 and CF_4 at 295 K are presented in Figs.2a and 2b, respectively. The isotropic Rayleigh spectrum of CH_4 given in Fig.2a has been obtained recently[11]. The one of CF_4 in Fig.2b was the first CILS isotropic binary spectrum to be published for molecular gases[9].

In these four figures, experimental binary intensities are given together with their error bars on an absolute intensity scale (in cm^6) versus frequency shift ν (in cm^{-1}). Moreover, theoretical CH_4 and CF_4 spectral contributions due to DID and dipole-multipole (DQ, DO) mechanisms and their cross terms (QQ, QO, OO) are presented using experimental values of the dipolar polarizability α at $\lambda_L = 514.5\,nm$[43] and the *ab initio* computed values of the dipole-multipole polarizabilities A (DQ) and E (DO)[44] reported in Table 5. For the anisotropic spectra, the nonlinear origin (NL) contributions are calculated by using *ab initio* computed values of the multipoles Ω and Φ and of the hyperpolarizabilities b_{xyz}, $B_{xz,xz}$ and $B_{zz,zz}$ (Ref.[45] for CH_4 and Ref.[46] for CF_4). All theoretical contributions have been computed by using the isotropic parts of the recent CH_4 and CF_4 intermolecular potentials of Palmer and Anchell[47].

Table 5. *Ab initio* computed values of the multipolarizabilities and their bond-length R-derivatives CH_4, CF_4 and SF_6 (Refs. [45,44,46,48])

Polarizability	CH_4	CF_4	SF_6
$|A|$	0.707 $Å^4$	0.972 $Å^4$	
$|A'| = \left|\frac{\partial A}{\partial R}\right|$	-	4.09 $Å^3$	
$|E|$	0.784 $Å^5$	1.154 $Å^5$	4.266 $Å^5$
$|E'| = \left|\frac{\partial E}{\partial R}\right|$	-	5.53 $Å^4$	22.64 $Å^4$

Other potentials are available which modify the values of the scattering intensities of every mechanism. So in a first step, a potential must be adopted. For the anisotropic spectrum, it is worth noting that the total DID contribution (bound and free dimers) is responsible for most of the integrated intensity of the spectrum. Therefore, it is convenient to select potentials for which the anisotropic DID intensity: $M_{0,a}^{DID} =$

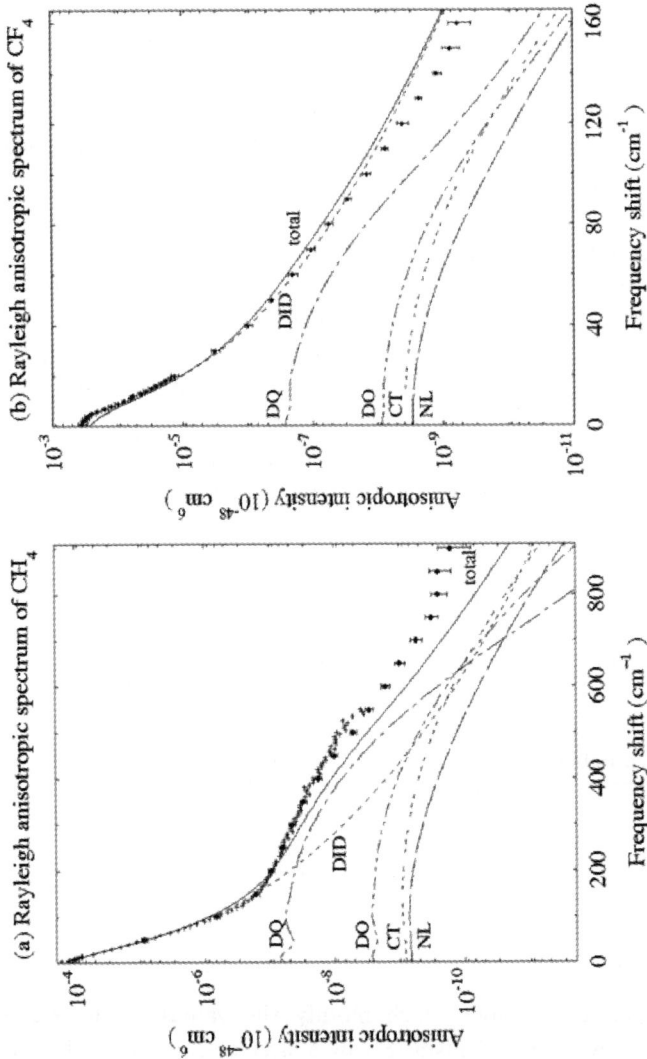

Fig. 1. Two-body anisotropic scattering Stokes spectra (in cm^6) for a) CH$_4$ and b) CF$_4$ at 294.5 K. Experimental data are represented by: a) crosses (+) and full circles (●) from Ref. [32] and Ref. [11], respectively; b) full circles (●) from Ref. [26]. Partial and total theoretical spectra are computed using the CH$_4$ and CF$_4$ potentials of Palmer and Anchell [47]: dipole-induced dipole due to free dimers (DID: - - - -); dipole-quadrupole (DQ: - - -); dipole-octopole (DO: - · · -); double transitions and cross terms QQ+QO+OO (CT: - · · - · · - · · -); nonlinear contribution (NL: ——); total (solid line). Polarizabilities are: a) $\alpha = 2.642\,\text{Å}^3$, $A = 0.707\,\text{Å}^4$ and $E = -0.784\,\text{Å}^5$; b) $\alpha = 2.93\,\text{Å}^3$, $A = 0.972\,\text{Å}^4$ and $E = -1.154\,\text{Å}^5$.

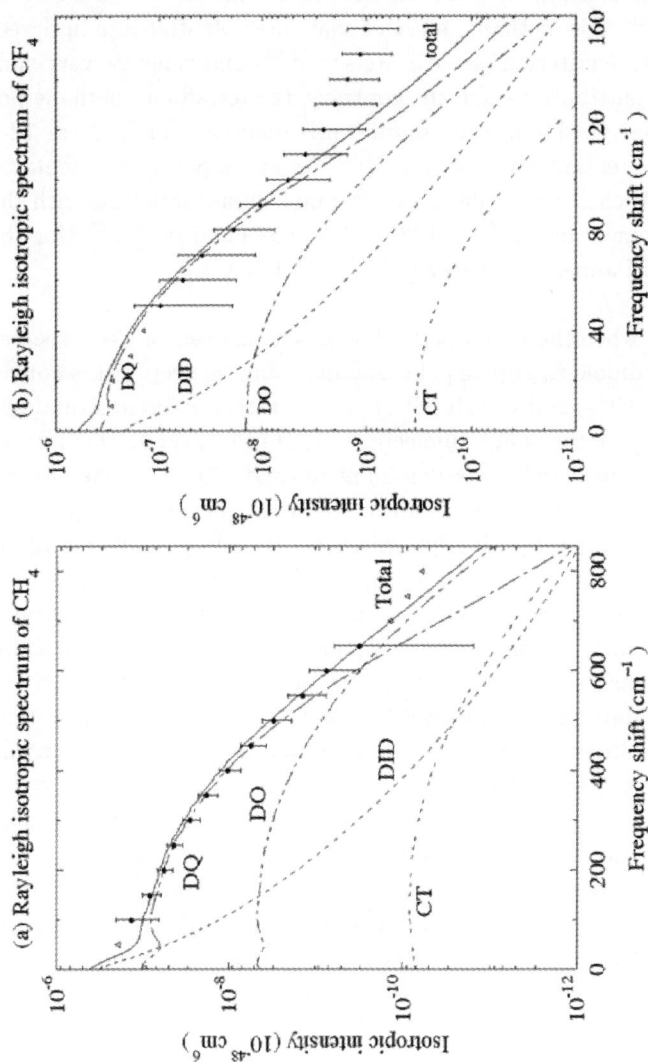

Fig. 2. Two-body isotropic scattering Stokes spectra (in cm⁶) for a) CH₄ and b) CF₄ at 294.5 K. Experimental data are represented by full circles (●) and estimations by triangles (▽): a) from Ref. [11]; b) from Ref. [9]. Partial and total theoretical spectra are computed using the CH₄ and CF₄ potentials of Palmer and Anchell [47]: dipole-induced dipole due to free dimers (DID: - - -); dipole-quadrupole (DQ: - - —); dipole-octopole (DO: - - —); Cross terms QQ+QO+OO (CT: - - - - - -); total (solid line). Polarizabilities are: a) $\alpha = 2.642\,\text{Å}^3$, $A = 0.707\,\text{Å}^4$ and $E = -0.784\,\text{Å}^5$; b) $\alpha = 2.93\,\text{Å}^3$, $A = 0.972\,\text{Å}^4$ and $E = -1.154\,\text{Å}^5$.

$\int_0^\infty \beta_{DID}^2(R) \exp(-U(R)/k_B T) 4\pi R^2\, dR$ is close to the zero-order spectral moment of the anisotropic spectrum $M_{0,a}^{exp} = \frac{15}{2} \left(\frac{\lambda_L}{2\pi}\right)^4 \int_{-\infty}^\infty I_a(\nu)\, d\nu$[49]. In particular for CH_4 with the potential of Palmer and Anchell we get $M_{0,a}^{DID} = 200\,\text{Å}^9$ which is compatible with the experimental value $M_{0,a}^{exp} = 205\pm14\,\text{Å}^9$ provided in Ref.[50]. For methane, most of available potentials are in agreement with the same criterion (see *e.g.* Refs.[51,52,53]) and generate very similar theoretical contributions. On the contrary, the tetrafluoromethane potential models may yield intensities significantly different. For CF_4, the 12-6 Lennard-Jones potentials given in Refs.[54,55,56] and the potential calculated by Palmer and Anchell[47] provide theoretical momenta compatible with the experimental momentum $M_{0,a}^{exp} = 182 \pm 24\,\text{Å}^9$ measured in Ref.[26] (for the CF_4 potential of Palmer and Anchell, $M_{0,a}^{DID} = 169\,\text{Å}^9$).

In a second step when the potential is chosen, the analysis of CILS spectra and the study of dipole-multipole polarizabilities may go deeply. It is noticeable that Fig.1a (CH_4) and Fig.1b (CF_4) present rather different situations. In the case of the relatively light molecule CH_4 (Fig.1a), the DID contributions cannot be considered as predominant beyond $220\,\text{cm}^{-1}$ whereas the total frequency range scanned goes up to $900\,\text{cm}^{-1}$. This gives a good opportunity to study the scattering intensities generated by dipole-multipole mechanisms and to measure at least the dipole-quadrupole polarizability A. Although a better fit between theory and experiment could be obtained in the $200 - 500\,\text{cm}^{-1}$ frequency range by increasing the A value of about 20%, it can be observed in Fig.1a that the values used for A and E (*ab initio* computed data provided in Ref.[44] and Table 5) generate theoretical DQ and DO intensities compatible with experimental anisotropic spectrum. On the other hand, for gas CF_4 the DID mechanism plays the leading role in the whole frequency range of the CILS anisotropic spectrum presented in Fig.1b.[57] Consequently, the study of the CF_4 dipole-multipole polarizabilities from anisotropic scattering is not accurate. It may only provide maximum values for A and E. Fortunately, the isotropic Rayleigh spectrum of CF_4 presented in Fig.2b offers a safer way to investigate the A and E values. In this spectrum, the DID mechanism is of second order and may be neglected beyond $20\,\text{cm}^{-1}$. Moreover, there is no NL contribution to the spectrum. Therefore, the isotropic spectrum is due to the dipole-multipole mechanisms only. It can be observed in Fig.2b that the main mechanism in the $40 - 120\,\text{cm}^{-1}$ mid frequency range of the CF_4 isotropic spectrum is the dipole-quadrupole one and that the *ab initio* value of A given in Ref.[44]

and Table 5 and used for theoretical calculations is in good agreement with CILS measurements. An even better agreement may be observed for CH_4 in Fig.2a. In the latter case, both *ab initio* values of A and E of Ref.[44], both reported in Table 5, appear to be compatible with experimental measurements.

3.1.2. *CILS within a Raman vibrational band*

For optically isotropic molecules like tetrahedral molecules the ν_1 mode of vibration is totally symmetric and therefore in absence of molecular interactions the depolarization ratio of the ν_1 Raman band is strictly zero. However, for gaseous CF_4 it has been observed that anisotropic intensities induced by molecular interactions are present within the vibrational Raman band ν_1.[58] A detailed analysis of these intensities has shown that multipolar polarizabilities contribute significantly to the wings of this ν_1 band.[23] These polarizabilities concern not only the dipole-quadrupole and the dipole-octopole polarizabilities A and E respectively but also A' and E', their derivatives with respect to the normal coordinate.

In Fig.3 we present the Stokes wing of the binary anisotropic ν_1 Raman spectrum for CF_4 at room temperature[23]. Experimental data are compared with theoretical spectra taking into account DID interaction, dipole-quadrupole and dipole-octopole mechanisms as well as several double transition (QQ, QO, OO) and nonlinear origin (NL) contributions. Theoretical intensities reported in Fig.3 have been computed using i) experimental values of the dipolar polarizability α[43] and of its derivative[59,60]; ii) *ab initio* values of A, E, A' and E'[44,46] reported in Table 5 ; iii) the intermolecular potential of Palmer and Anchell[47] previously chosen for the Rayleigh spectra of Figs. 1b and 2b. Other available potentials of CF_4 compatible with Rayleigh measurements[54,55,56] generate similar theoretical intensities. At low frequencies ($\nu < 25\,cm^{-1}$) contributions from multipolar polarizabilities are negligible and experimental data are well reproduced by the DID intensities. On the other hand, at higher frequencies the theoretical anisotropic spectrum lies lower than the experimental one, probably because of spectral contributions which are not taken into account by the model. However in the $30 - 80\,cm^{-1}$ frequency range, it is noticeable that the *ab initio* values[44,46] reported in Table 5 generate theoretical spectral curves which go through experimental error bars. The quantum chemistry computations of A' and E'[44,46] are then compatible with CILS experiments.

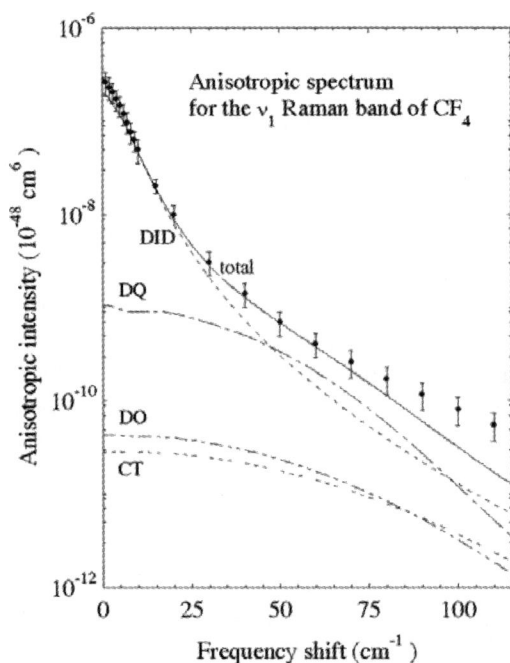

Fig. 3.   Two-body anisotropic scattering spectrum of the ν_1 Raman band of gaseous CF_4 at 294.5 K. Experimental data from Ref. [23] are represented by full circles (•). Partial and total theoretical spectra are computed using the CF_4 potential of Palmer and Anchell [47]: dipole-induced dipole due to free dimers (DID: - - - -); dipole-quadrupole (DQ: - ———); dipole-octopole (DO: – – ———); Cross terms QQ+QO+OO (CT: - - -   - - -); total (solid line). Polarizabilities are: $\alpha = 2.93$ Å^3, $A = 0.972$ Å^4, $E = -1.154$ Å^5, $A' = 1.154$ Å^4 and $E' = 5.53$ Å^5.

The Stokes wing of the binary isotropic spectrum of the ν_1 Raman band of CF_4 has been also measured[39]. However in this spectrum some additional scattering contributions occurs which are not taken into account in the aforementioned theory and contrary to the study of the CILS in the Rayleigh wings no accurate values of multipolar polarizabilities have been deduced from it.

3.1.3. *Set inversion analysis of CILS spectra*

Considering the anisotropic and isotropic CILS spectra in the vicinity of the Rayleigh or the ν_1 Raman line, a "best fitting" procedure is commonly used in order to evaluate A, E and their derivatives at laser frequency for the available intermolecular potentials of any studied molecule.[9,23,26,31,32,34,33] However considering the size of the experimental error bars and the mutual competition between DQ and DO mechanisms, various parameter vectors (A, E, A', E') can generate similar fits. Moreover, local minimization procedures may converge to any local minimum. It is thus preferable to determine the four-dimensional set of the solutions $\{A, E, A', E'\}$ for which each theoretical spectrum lay inside the experimental error bars of the corresponding experimental intensities. This can be carried out through the set inversion method (SIM) used for the first time in this field and described, in particular, in Ref.[61]. It partitions the studied n-dimensional space of parameters into non-overlapping boxes. Then, it tests these boxes in order to know whether they are located inside or outside the solution set. The boxes which are neither inside nor outside are bisected until an accurate bracketing of the solution set is obtained. For CF_4, both isotropic Rayleigh and anisotropic Raman spectra may be studied by SIM in the frequency ranges where DQ and DO mechanisms play the leading role ($0 - 120\,\mathrm{cm}^{-1}$ and $30 - 80\,\mathrm{cm}^{-1}$ for the aforementioned spectra, respectively). Using for example the Palmer and Anchell potential of CF_4, the corresponding set of solutions $\{A, E, A', E'\}$ is a connected box. The projections of this set on the planes (A, E), (A', E'), (A, A') and (E, E') passing by the *ab initio* solution (A_M, E_M, A'_M, E'_M) of Maroulis[44,46] are given in Fig.4. It can be noticed that the latter theoretical solution (represented by a white cross on each figure) belongs to these four planes projections. Overall, the comparison between the set projections provides average values of the multipolar polarizabilities A and E and their derivatives as well as magnitudes of the uncertainties associated to these evaluations. For CF_4 and its Palmer and Anchell potential, the solutions obtained: $A = 0.97 \pm 0.14\,\text{Å}^4$, $A' = 4.38 \pm 0.67\,\text{Å}^3$, $E = 1.38 \pm 0.72\,\text{Å}^5$ and $E' = 7.6 \pm 7.6\,\text{Å}^4$ are fully compatible with the *ab initio* solution (A_M, E_M, A'_M, E'_M) of Maroulis reported in Table 5. In the case of CH_4, the solution set $\{A, E\}$ associated to the isotropic Rayleigh spectra is a connected surface. For the Palmer and Anchell potential of CH_4 used in Fig.2a, the solutions obtained ($A = 0.66 \pm 0.08\,\text{Å}^4$, $E = 0.63 \pm 0.63\,\text{Å}^5$) are both compatible with calculations of Maroulis given in Ref.[46] ($A_M = 0.71\,\text{Å}^4$, $E_M = 0.78\,\text{Å}^5$).[11]

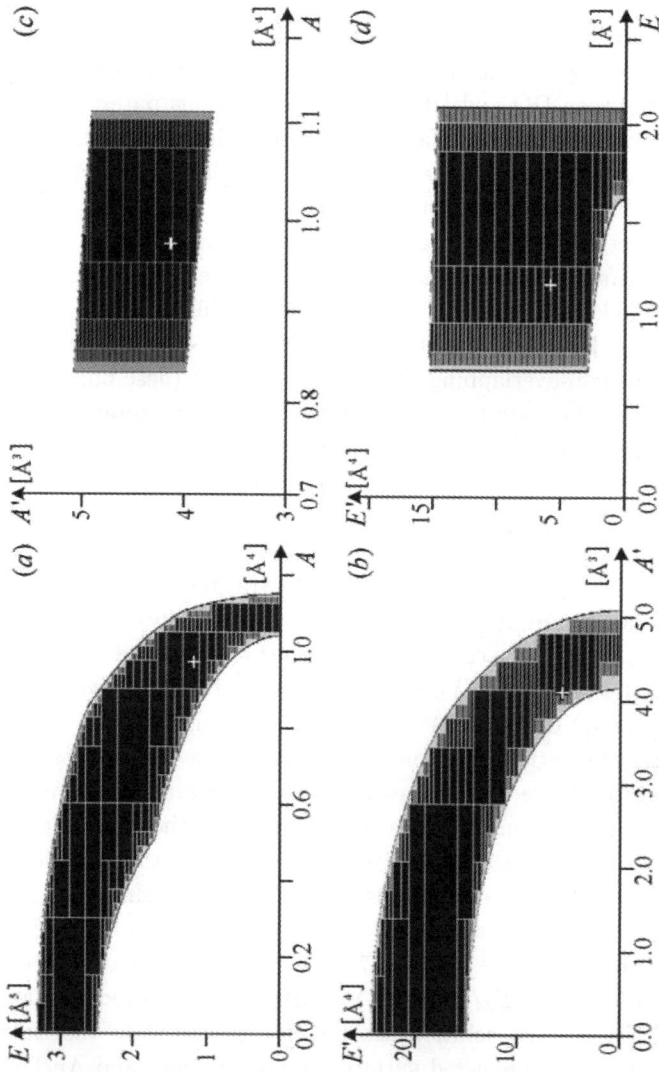

Fig. 4.   Solution areas (black and grey zones) for the multipolar polarizabilities of CF_4 and their derivatives in four projection planes passing by the solution (A_M, E_M, A'_M, E'_M) calculated *ab initio* by Maroulis, reported in Table 4 and represented by the white crosses. (a): (A, E); (b): (A', E'); (c): (A, A'); (d): (E, E'). Computations have been made by using the potential of Palmer and Anchell.

3.2. Octahedral molecules

Octahedral molecules are also optically isotropic and therefore there are suitable molecules for the study of interaction-induced effects in their scattering spectra[10,29,30,33,38,40]. Besides for symmetry reasons, contrary to the case of tetrahedral molecules previously discussed, the dipole-quadrupole polarizability tensor is strictly zero and the lowest order multipolar mechanisms (DID excepted) are due to the dipole-octopole polarizability tensor (Rayleigh wings) and its derivative (Raman vibrational band).

3.2.1. CILS in the Rayleigh wings

Like tetrahedral molecules, light scattering from non-interacting monomers only yields isotropic intensities in the Rayleigh region. On the contrary, if molecular interactions between monomers are taken into account both isotropic and anisotropic light scattering are generated in the Rayleigh wings. For gaseous sulfur hexafluoride, several papers have been published up to 1989 concerning the anisotropic Rayleigh spectrum[6,29,33,62,63] for frequency shifts lower than $70 \, \text{cm}^{-1}$. Nevertheless from these works, due to the narrowness of the frequency range scanned and to the strong influence of translational contributions (DID mainly) to CILS intensities, it is not possible to accurately deduce the rototranslational contribution of the dipole-octopole polarizability tensor. Recently, both anisotropic[38] and isotropic[10,38] Rayleigh spectra of gas SF_6 at room temperature have been reported up to $210 \, \text{cm}^{-1}$. The extension of the frequency range as well as the first recording of the SF_6 isotropic spectrum allows a more complete study of the DO contribution and therefore the measurement of the dipole-octopole polarizability E. The anisotropic $I_{ani}(\nu)$ and isotropic $I_{iso}(\nu)$ binary CILS spectra of SF_6 reported in these works[10,38] are displayed in figures 5a and 5b, respectively. In Fig.5a up to $70 \, \text{cm}^{-1}$, a good agreement can be noticed between experimental data of Ref.[33] and new data reported in Ref.[38]. Below $30 \, \text{cm}^{-1}$, due to too large uncertainties, no isotropic intensities are available[10]. Moreover, the presence of three weak Raman bands at $120 \, \text{cm}^{-1}$, $265 \, \text{cm}^{-1}$ and $335 \, \text{cm}^{-1}$ as well as the feebleness of the CILS intensities make error bars higher at high frequencies[38]. In Figs. 5a-5b, the theoretical contributions have been computed by using the recent SF_6 intermolecular potential of Zarkova[64].

For the value $\alpha = 4.549 \, \text{Å}^3$ of the dipolar polarizability at $\lambda_L = 514.5 \, \text{nm}$[43], this potential provides a first-order DID integrated intensity $M_{0,a}^{DID} = 1219 \, \text{Å}^9$ close to the zero-order spectral moment of the anisotropic

T. Bancewicz, Y. Le Duff, J.-L. Godet

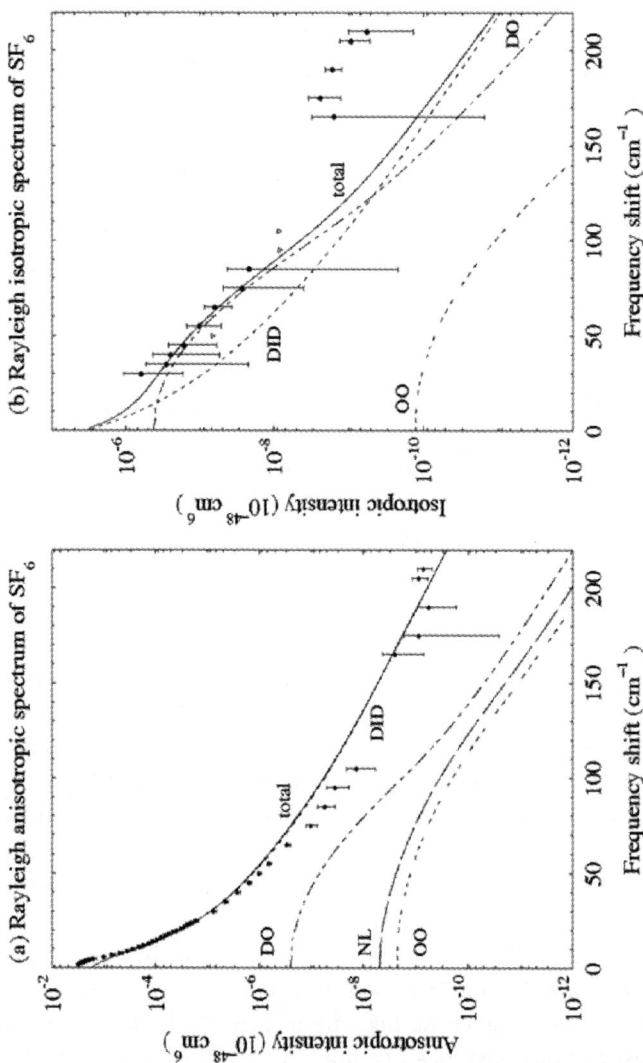

Fig. 5. Two-body scattering spectra in absolute units (cm^6) for SF$_6$ at 294.5 K: a) anisotropic; b) isotropic. Experimental data are those of Ref. [38] for a) and b) (circles (o)) and Ref. [33] for a) (crosses (+)). Partial and total theoretical spectra are computed for the SF$_6$ potential of Zarkova [64], $\alpha = 4.549$ Å^3 and $E = 3.1$ Å^5: dipole-induced dipole due to free dimers (DID: - - -); dipole-octopole (DO: - - - · - - · - - -); Cross term (OO: · · · - · · · - · · -); nonlinear contribution (NL: ———); total (solid line). For the anisotropic spectrum (a), total theoretical and DID intensities are almost the same.

spectrum [49] $M_{0,a}^{exp} = 1239 \pm 134 \text{Å}^9$. Therefore, it may be a good potential to compute the dipole-octopole scattering contribution. Indeed in Fig.5a, like previously for CF_4, the DID mechanism predominates in the whole frequency range of the SF_6 anisotropic spectrum, making this latter spectrum inaccurate to measure the dipole-octopole polarizability E. Nevertheless, E may be extrapolated from the isotropic Rayleigh spectrum provided in Fig.5b. Because the second-order DID mechanism is weak beyond $30\,\text{cm}^{-1}$, the isotropic spectrum recorded from 30 to $85\,\text{cm}^{-1}$ is mainly due to the dipole-octopole mechanism. The fitted value $E = 3.1\,\text{Å}^5$ obtained with the potential of Zarkova is relatively close to the *ab initio* computed value[48] $E = 4.2\,\text{Å}^5$. In Table 6, other values deduced from the isotropic spectrum in this frequency range, taking into account error bars and corresponding to other potentials of SF_6 are listed as well as fitted values previously extrapolated from the anisotropic spectrum. It is noticeable in the same Table 6 that the values of E are correlated with those of $M_{0,a}^{DID}$ for every potential. This is the reason why it is necessary to choose a potential for which $M_{0,a}^{DID}$ is compatible with the measured value $M_{0,a}^{exp}$. Finally, measurements of E made by using the SF_6 isotropic Rayleigh spectrum are by far closer to the *ab initio* computed value [48] $E = 4.2\,\text{Å}^5$ than the one coming from the Rayleigh anisotropic spectrum. This tends to show that the SF_6 isotropic Rayleigh spectrum as well as the theoretical model used (despite of its limitations at high frequencies) are good tools for the measurement of the dipole-octopole polarizability of the molecule SF_6.

Table 6.   Theoretical DID zero-order moment of the anisotropic CILS Rayleigh spectrum $M_{0,a}^{DID}$ and experimental dipole-octopole polarizability E for SF_6 molecule. These results are obtained using several intermolecular potentials and, for E, from either anisotropic Rayleigh spectrum (Ref. [33]) or isotropic Rayleigh spectrum (Ref. [38]).

Potential	$M_{0,a}^{DID}$ (in Å^9)	Anisotropic[33] E (in Å^5)	Isotropic[38] E (in Å^5)
Lennard-Jones[65]	1398	6.1 ± 1.8	2.7 ± 0.6
Lennard-Jones[66]	1016	8.5 ± 0.2	4.0 ± 0.8
MMSV[67]	1492		2.6 ± 0.8
HFD[41]	1013		3.8 ± 0.8
Zarkova[64]	1219		3.1 ± 0.7

3.2.2. *CILS within a Raman vibrational band*

Sulfur hexafluoride molecule belongs to the symmetry group O_h and therefore the monomer contribution to the light scattered by a sample of gas SF_6 in the frequency region of the ν_1 mode of vibration is completely polarized similarly to the CF_4 case. Thus the anisotropic scattering intensities in the vicinity of the ν_1 Raman frequency are completely induced by molecular interactions. In this case, the dipole-octopole polarizability E as well as its derivative E' contribute significantly to scattering intensities as previously shown[40]. In Fig.6 anisotropic CILS spectrum of SF_6 in the vicinity of the ν_1 Raman line is presented[40]. Scattering intensities are given on an absolute scale for the Stokes side of the ν_1 band up to 100 cm-1 where ν is the frequency shift measured from the center of the line.

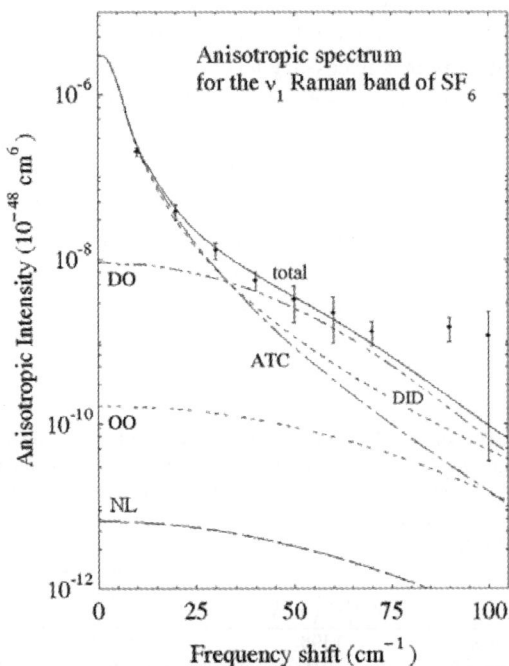

Fig. 6. Two-body anisotropic scattering spectrum of the ν_1 Raman band of gaseous SF_6 at 294.5 K. Experimental data from Ref. [40] are represented by full circles (●). Partial and total theoretical spectra are computed using the CF_4 potential of Zarkova [64], $\alpha = 4.549$ Å^3, $E = 3.1$ Å^5 and $E' = 46$ Å^4: dipole-induced dipole due to free dimers (DID: - - - - -); dipole-octopole (DO: - - ⎯⎯); Cross term (OO: - - -   - - -); nonlinear contribution (NL: ⎯⎯ ⎯⎯); total (solid line).

They are compared with theoretical double differential cross section taking into account translational and rototranslational contribution computed with the SF_6 intermolecular potential of Zarkova[64] used in Figs. 5a-5b. Here, the anisotropic translational contribution (ATC) to the Raman spectrum in the vicinity of the ν_1 frequency is deduced from the study of scattering intensities in the vicinity of the Rayleigh band[40]. Thus an ATC spectrum for the Rayleigh region $I_{ATC}^{Rayleigh}$ is calculated by subtracting theoretical rototranslational intensities to total experimental intensities measured in the Rayleigh wings. Then the intensities of the ATC for the ν_1 Raman band are deduced from the equation

$$I_{ATC}^{Raman}(\nu) = \frac{2}{\alpha^2}\left(\frac{\partial\alpha}{\partial Q_1}\right)^2\left(\frac{\nu_L - \nu_1 - \nu}{\nu_L - \nu}\right)^4 I_{ATC}^{Rayleigh}(\nu) \qquad (45)$$

where ν is the frequency shift measured from the center of either the Rayleigh line for the Rayleigh spectrum or the Raman line for the Raman spectrum. ν_L is the frequency of the laser. For the rototranslational intensities, the dipole-induced octopole (DO), the octopole-induced octopole (OO) and the non-linear contribution are considered. They have been calculated using the value of the dipole-octopole polarizability obtained from the study of CILS in the Rayleigh wings ($E = 3.1\,\text{Å}^5$) and the value of E' ($E' = 46\text{Å}^4$) which yields the best fit of the theoretical intensities with experimental data in the $30 - 70\,\text{cm}^{-1}$ frequency range for which multipole contributions play the leading role. The values of the other molecular parameters are obtained from the calculations reported in Ref.[48]. If the value of E varies in agreement with the error bars given in Ref.[38] and Table 6 ($E = 3.1 \pm 0.7\,\text{Å}^5$), accordingly the value of E' varies from a minimum value $E'_{min} = 44\,\text{Å}^4$ to a maximum value $E'_{max} = 48\,\text{Å}^4$ (these values being obtained with the potential of Zarkova[64]). The intermolecular potential may also influence the value of E' obtained from the analysis of the scattering data. Finally the best/recommended value of E' proposed from the measurement of collision induced anisotropic intensities is [40] $E' = 47 \pm 10\,\text{Å}^4$. It is of the same order of magnitude as the value computed in Ref.[48], $E' = 22.6\,\text{Å}^4$ although it is higher. However, as for the other molecular constants discussed in this paper, it is worth noting that the experimental value is measured at an optical frequency when the theoretical one is a static value.

4. Conclusion

In this short review, we have pointed out that long-range interactions as well as multipolar polarizabilities may contribute significantly to scattering spectra of molecular gases at high frequency shifts. For few optically isotropic molecules (CH_4, CF_4 and SF_6) their high order polarizabilities have been expressed with spherical tensors and corresponding theoretical scattering intensities have been computed. From comparison with Rayleigh CIS data, experimental values of dipole-quadrupole and/or dipole-octopole polarizabilities (A, E) have been obtained and found in good agreement with *ab initio* calculations of these quantities [9,11,38]. Conclusion is similar for the derivative of these polarizabilities deduced from the analysis of Raman vibrational bands (CF_4 and SF_6[23,40]). Finally, we would like to mention that to our knowledge, for the isotropic molecules of this study, the measurement of CIS spectra is the most accurate method to evaluate high order multipolar polarizabilities, especially when set inversion analysis is used. So, this method should be very useful to check the validity of the theoretical models used in quantum chemistry to calculate molecular polarizabilities.

Table 7. Conversion factors from atomic units (au) to SI units and from modified esu (using Å) to SI units. For energy: $E_h = 4.359748 \, 10^{-18}$ J and for length: $a_0 = 0.529177 \, 10^{-10}$ m.

SI units	from au to SI	from esu (using Å) to SI
α and A' (C^2 m^2 J^{-1})	$1.648778 \times 10^{-41} \, e^{-2} a_0^{-2} E_h$	1.11265×10^{-40} Å^{-3}
A and E' (C^2 m^3 J^{-1})	$8.724956 \times 10^{-52} \, e^{-2} a_0^{-3} E_h$	1.11265×10^{-50} Å^{-4}
E (C^2 m^4 J^{-1})	$4.617048 \times 10^{-62} \, e^{-2} a_0^{-4} E_h$	1.11265×10^{-60} Å^{-5}
β (C^3 m^3 J^{-2})	$3.206361 \times 10^{-53} \, e^{-3} a_0^{-3} E_h^2$	3.711401×10^{-49} g$^{0.5}$Å$^{-3.5}$s^{-1}
B (C^3 m^4 J^{-2})	$1.696733 \times 10^{-63} \, e^{-3} a_0^{-4} E_h^2$	3.711401×10^{-59} g$^{0.5}$Å$^{-4.5}$s^{-1}
Ω (C m^3)	$2.374182 \times 10^{-50} \, e^{-1} a_0^{-3}$	3.335641×10^{-52} g$^{-0.5}$Å$^{-4.5}$s
Φ (C m^4)	$1.256363 \times 10^{-60} \, e^{-1} a_0^{-4}$	3.335641×10^{-62} g$^{-0.5}$Å$^{-5.5}$s

References

1. B. J. Berne and R. Pecora, *Dynamic Light Scattering* (John Wiley and sons, New York, 1976).
2. S. Kielich, *Molekularna Optyka Nieliniowa (Molecular Nonlinear Optics)* (PWN, Warszawa-Poznań, 1977), in Polish.
3. L. Frommhold, Adv. Chem. Phys. **46**, 1 (1981).
4. *Collision-and Interaction-Induced Spectroscopy*, Vol. 452 of *NATO ASI Series C: Mathematical and Physical Sciences* (Kluwer Academic Publishers, Dordrecht, 1995), edited by G.C. Tabisz and M.N. Neuman.
5. A. D. Buckingham and G. C. Tabisz, Optics Letters **1**, 220 (1977).
6. A. D. Buckingham and G. C. Tabisz, Mol.Phys. **36**, 583 (1978).
7. F. Barocchi, M. Zoppi, D. P. Shelton, and G. C. Tabisz, Can. J. Phys. **55**, 1962 (1977).
8. H. Posch, Mol. Phys. **37**, 1059 (1979).
9. A. Elliasmine, J.-L. Godet, Y. Le Duff, and T. Bancewicz, Phys. Rev. A **55**, 4230 (1997).
10. J.-L. Godet, F. Rachet, Y. Le Duff, K. Nowicka and T. Bancewicz, J. Chem. Phys. **116**, 5337 (2002).
11. T. Bancewicz, K. Nowicka, J.-L. Godet, and Y. Le Duff, Phys. Rev. A **69**, 062704 (2004).
12. S. Kielich, Proc.Indian Acad.Sci.(Chem.Sci.) **94**, 403 (1985).
13. A. J. Stone, *The Theory of Intermolecular Forces* (Clarendon Press, Oxford, 1996).
14. C. G. Gray and K. E. Gubbins, *Theory of molecular fluids. Vol.1: Fundamentals* (Clarendon Press, Oxford, 1984).
15. S. Kielich, Progress in Optics **XX**, 155 (1983).
16. S. Kielich, J. Physique (Lett.) **43**, L-389 (1982).
17. A. D. Buckingham, Adv.Chem.Phys **12**, 107 (1967).
18. C. G. Gray and B. W. N. Lo, Chem.Phys. **14**, 73 (1976).
19. D. A. Varshalovich, A. N. Moskalev and V. K. Khersonskii, *Quantum theory of angular momentum* (World Scientific, Singapore, 1988).
20. T. Bancewicz, W. Głaz and S. Kielich, Chem. Phys. **128**, 321 (1988).
21. *Phenomena Induced by Intermolecular Interactions*, *NATO ASI Series* (Plenum, New York, 1985), edited by G. Birnbaum.
22. T. Bancewicz, Chem. Phys. Lett. **244**, 305 1995,
23. T. Bancewicz, A. Elliasmine, J.-L. Godet and Y. Le Duff, J. Chem. Phys. **108**, 8084 (1998).
24. R. G. Gordon, Adv. Mag. Resonance **3**, 1 (1968).
25. H. Posch, Mol. Phys. **46**, 1213 (1982).
26. A. Elliasmine, J.-L. Godet, Y. Le Duff and T. Bancewicz, Mol. Phys. **90**, 147 (1997).
27. T. Bancewicz, Y. Le Duff and J. L. Godet, in *Modern Nonlinear Optics, Part 1, Second Edition*, Vol. 119 of *Advances in Chemical Physics*, M.Evans ed. (J. Wiley, New York, 2001), pp. 267-307.
28. S. Wolfram, THE *MATHEMATICA* BOOK, third edition (Cambridge Uni-

versity Press, Cambridge, UK, 1996).

29. D. P. Shelton and G. C. Tabisz, Mol. Phys. **40**, 299 (1980).
30. S. M. El-Sheikh, N. Meinander and G. C. Tabisz, Chem. Phys. Lett. **118**, 151 (1985).
31. S. M. El-Sheikh, N. Meinander and G. C. Tabisz, Mol. Phys. **54**, 479 (1985).
32. F. Barocchi, A. Guasti, M. Zoppi, S. M. El-sheikh, G. C. Tabisz and N. Meinander, Phys. Rev. A **39**, 4537 (1989).
33. S. M. El-Sheikh and G. C. Tabisz, Mol. Phys. **68**, 1225 (1989).
34. U. Bafile, L. Ulivi, M. Zoppi, F. Barocchi and L. Frommhold, Phys. Rev. A **50**, 1172 (1994).
35. N. Meinander, G. C. Tabisz, F. Barocchi and M. Zopii, Mol. Phys. **89**, 521 (1996).
36. U. Hohm, Chem. Phys. Lett. **311**, 117 (1999).
37. U. Hohm, Vacuum **58**, 117 (2000).
38. K. Nowicka, T. Bancewicz, J.-L. Godet, Y. Le Duff and F. Rachet, Mol. Phys. **101**, 389 (2003).
39. J.-L. Godet, A. Elliasmine, Y. Le Duff and T. Bancewicz, J. Chem. Phys. **110**, 11303 (1999).
40. Y. Le Duff, J.-L. Godet, T. Bancewicz and K. Nowicka, J. Chem. Phys. **118**, 11009 (2003).
41. N. Meinander, J. Chem. Phys. **99**, 8654 (1993).
42. A. Gharbi and Y. Le Duff, Physica **90A**, 619 (1978).
43. H. E. Watson and K. L. Ramaswamy, Proc. Roy. Soc. London **A 156**, 144 (1936).
44. G. Maroulis, Chem. Phys. Lett. **259**, 654 (1996).
45. G. Maroulis, Chem. Phys. Lett. **226**, 420 (1994).
46. G. Maroulis, J. Chem. Phys. **105**, 8467 (1996).
47. B. J. Palmer and J. L. Anchell, J. Phys. Chem. **99**, 12239 (1995).
48. G. Maroulis, Chem. Phys. Lett. **312**, 255 (1999).
49. F. Barocchi and M. Zoppi, in *Phenomena Induced by Intermolecular Interactions* (Plenum Press, New York, 1985), edited by G.Birnbaum.
50. N. Meinander, G. C. Tabisz and M. Zoppi, J. Chem. Phys. **84**, 3005 (1986).
51. J. O. Hirschfelder, C. F. Curtis and R. B. Bird, *Molecular Theory of Gases and Liquids* (Wiley, New York, 1954).
52. R. Righini, K. Maki and M. L. Klein, Chem. Phys. Lett. **80**, 301 (1981).
53. N. Meinander and G. C. Tabisz, J. Chem. Phys. **79**, 416 (1983).
54. K. E. MacCormack and W. G. Schneider, J. Chem. Phys. **19**, 849 (1951).
55. A. E. Sherwood and J. M. Prausnitz, J. Chem. Phys. **41**, 429 (1964).
56. G. C. Maitland, M. Rigby, E. B. Smith and W. A. Wakeham, *Intermolecular Forces. Their origin and determination* (Clarendon Press, Oxford, 1981).
57. note that the computation of the theoretical DID contribution presented here is different from this of Ref.[26]. First, the potential used is different. Second the computation made in Ref.[26] was in arbitrary units and for a too narrow frequency range. For the higher frequencies, the extrapolation of the theoretical curve made in Ref.[26] is not satisfactory.
58. Y. Le Duff and A. Gharbi, Phys. Rev. A **17**, 1729 (1978).

59. W. F. Murphy, W. Holzer and H. J. Bernstein, Appl. Spectrosc. **23**, 211 (1969).
60. W. Holzer, J. Mol. Spectrosc. **25**, 123 (1968).
61. L. Jaulin, J.-L. Godet, E. Walter, A. Elliasmine and Y. Le Duff, J. Phys. A **30**, 7733 (1997).
62. P. Lallemand, J. de Phys. (Paris) **32**, 119 (1971).
63. W. Holzer and Y. Le Duff, Phys. Rev. Lett. **32**, 205 (1974).
64. L. Zarkova, Mol. Phys. **88**, 489 (1996).
65. J. C. McCoubrey and N. M. Singh, Trans. Faraday Soc. **55**, 1826 (1959).
66. J. G. Powles, J. C. Dore, M. B. Deraman and E. K. Osae, Mol. Phys. **50**, 1089 (1983).
67. R. A. Aziz, M. J. Slaman, W. L. Taylor and J. J. Hurly, J Chem Phys **94**, 1034 (1991).

CHAPTER 13

POLARIZABILITY FUNCTIONS OF DIATOMIC MOLECULES AND THEIR DIMERS

Michail A. Buldakov[1], Victor N. Cherepanov[2]

[1]*Laboratory of Ecological Instrument Making, Institute of Monitiring of Climatic and Ecological Systems SB RAS, 634055, Tomsk, Russia*
E-mail: bvk@phys.tsu.ru
[2]*Department of Physics, Tomsk State University, 634050, Tomsk, Russia*
E-mail: vnch@phys.tsu.ru

The methods of analytical description of the polarizability functions for homonuclear diatomic molecules and their van der Waals dimers are discussed. The polarizability functions of diatomic molecule obtained have physically correct behavior as at small so at large internuclear separations and agree with the polarizability functions in the vicinity of the molecule nuclei equilibrium position. The polarizability functions of N_2 and O_2 molecules for wide ranges of internuclear separation and frequency of an external electromagnetic field are given. There are considered the polarizabilities of $(N_2)_2$ and $(O_2)_2$ complexes depending on separations and mutual orientation of their components.

1. Introduction

The polarizability function conception of a diatomic molecule for any electronic state (electronic polarizability of the molecule) arises from the adiabatic approximation being used. In this approximation the molecule electronic polarizability turns into a function of the molecule internuclear distance r and the frequency ω of an external electromagnetic field and appears to be more full and important characteristic of the molecule than the polarizability at the molecule equilibrium distance r_e usually given in scientific articles. The molecule electronic polarizability is the second rank tensor that has only two independent components for diatomic molecules, and accordingly there are two polarizability functions $\alpha_{zz}(\omega,r)$ and $\alpha_{xx}(\omega,r) = \alpha_{yy}(\omega,r)$ where axis z is directed along the molecule axis.

Presently, quantum-mechanical and semiempirical methods are used to calculate the polarizability functions of diatomic molecules. The quantum-mechanical calculations of the polarizability functions are usually carried out for relatively small rangs of internuclear distances near r_e (for instance[1-25]), and only for the molecule H_2 such calculations have been carried out for wide r and ω ranges[26-28] using special *ab initio* methods being applicable to two-electron molecules only. Moreover, there are many works on *ab initio* calculations of the polarizability functions for two interacting atoms of noble gases (see the works[29-31] and quoted there references). However, these interacting atoms do not form the molecules with the chemical bonds and in this work are not considered.

Development of semiempirical approaches is caused by wish to obtain the knowledge about polarizability functions of molecules for wide r and ω ranges using simple analytical expressions. Semiempirical calculations are usually based on the empirical or theoretical information about polarizability functions for separate ranges of internuclear distances. So, for the range of small r the polarizability functions have been found in the work[32] using the united atom model. In a vicinity of equilibrium internuclear position the polarizability functions of diatomic molecules are generally given by Taylor series where expansion parameters (derivatives of the polarizability tensor) may be experimentally found from Raman scattering. To construct the polarizability function at large r the classical electrostatic model by Silberstein (DID model) is used where interacting atoms are considered to be a point dipole induced by external electromagnetic field and localized on molecule's nuclei[33,34]. Such approach makes it possible to calculate the molecule polarizability function on a base of known polarizability values for the atoms forming the molecule. Later DID model have been improved by taking into account polarizability changing of these when they come closer[35-37] and multipole and exchange interactions of the atoms[38-40]. Presently, there are only a few works[38-43] where the polarizability functions of diatomic molecules are given for the full range of internuclear distances. First, the polarizability functions for all range of internuclear distances were suggested in the work[41] for N_2 molecule. Unfortunately, these functions don't have physically correct behavior neither at small nor at large internuclear distances. The polarizability functions of the work[42] are based on empirically found behavior regularities of these near their maxima and known asymptotic dependencies at large $r^{33,34,37}$. However, the polarizability functions at $r \rightarrow 0$ in the frame of this method do not agree with ones theoretically predicted in the work[32], and versatility of the correlations found needs to be further studied. The polarizability functions in the work[43] were obtained using their asymptotes at small[32] and large $r^{33,34}$ and the polarizability derivatives empirically found for N_2 and O_2 molecules at their equilibrium positions. Later, in the works[38-40] these functions were improved at the middle range of r by taking into account dispersive, multipole and exchange interactions of the atoms. The behavior of the polarizability functions for the molecule N_2 depending on ω was studied in the work[44].

Recently, there appeared an interest also to the electronic polarizability tensor of interacting molecules in connection with the intensive theoretical and experimental study of the properties of van der Waals molecular complexes. The van der Waals complex is the complex of some molecules bounded by van der Waals interactions. Such complexes have a small bonding energy and the large distances between particles in a complex. For this reason, the components of a complex as a first approximation preserve their individuality. Nevertheless, the properties of a complex are not the simple sum of properties of its components. Other specific feature of van der Waals complexes is their nonrigid structure. So, the quantum-chemical calculations of a surface of the potential energy for complexes (see the reviews[45,46]) show, that the complexes can exist in various configurations, and even the small vibration-rotation excitation can change their configurations. As a result, electronic polarizability of van der Waals complexes depends on the distances between the components of the complex and their mutual orientation. Existing *ab initio* methods allow calculating the polarizability of molecular complexes with a good accuracy, however, these methods can not describe it in an analytical form. So, it is of interest to develop other methods which allow calculating the polarizability tensor in analytical form for the systems of interacting molecules including van der Waals complexes.

The methods of analytical description of the polarizability functions developed by authors of the chapter for diatomic homonuclear molecules and their van der Waals dimers are given below.

2. Polarizability functions of diatomic homonuclear molecules

Finding an analytical form of polarizability function for diatomic molecule is a difficult problem that requires the molecule wavefunction to be computed for arbitrary r values. At present it is not possible to specify the wavefunction at arbitrary r, however, there are three specific ranges of internuclear separations where an analytical form of the wavefunction can be defined: the range of small r values and the range of large r values that, in turn, can be divided into two in accordance with the governing interaction – multipole (at $r \to \infty$) and exchange (at smaller r). In this case, the polarizability functions for a diatomic molecule may be obtained in the form of a piecewise-continuous function that exhibits physically correct behavior at small and large internuclear separations[38-40].

2.1. *The molecule polarizability at small internuclear separations*

The diatomic molecule polarizability at small internuclear distances may be described on the basis of united atom idea[32]. For small internuclear distances the wave function $\Psi_l(r_{el}, r)$ of the electron state l is calculated by the methods of perturbation theory where the wave functions of the united atom $\varphi_m(r_{el})$ are used as unperturbed wave functions. In the first order of the perturbation theory

$$\Psi_l(r_{el}, r) = \varphi_l(r_{el}) + \sum_{m \neq l} \frac{V_{ml}(r)}{E_m^0 - E_l^0} \varphi_m(r_{el}),$$
(1)

where r_{el} is the coordinate set of all electrons of the system and E_k^0 is the kth energy state of the united atom. The index l of the united atom wave function $\varphi_l(r_{el})$ shows the state which the molecule electronic state l goes into when the molecule nuclei unite. The matrix element of the "united atom partition operator" is computed in the work[47] and is a power series of the form

$$V_{ml}(r) = V_{ml}^{(2)}r^2 + V_{ml}^{(3)}r^3 + V_{ml}^{(4)}r^4 + \dots, \qquad (2)$$

where the coefficients $V_{ml}^{(j)}$ are fully specified by the electron density distribution of the united atom at m and l states. The energy $E_l(r)$ of the diatomic molecule for small r may be also represented as the series[47-49]

$$E_l(r) = E_l^0 + \frac{Z_1 \cdot Z_2}{r} + E_l^{(2)}r^2 + E_l^{(3)}r^3 + E_l^{(4)}r^4 + \dots . \qquad (3)$$

Here Z_1 and Z_2 are the charges of the nuclei constituting the united atom and $E_l^{(j)}$ are the constants for the given molecule state.

To calculate the diatomic molecule static polarizability tensor at the electronic state n for small internuclear distances we employ the following well known form for tensor components of the electronic polarizability

$$\alpha_{ii}^{(n)}(0,r) = 2\sum_{m\neq n} \frac{\left|\langle n(r)|d_i|m(r)\rangle\right|^2}{E_m(r) - E_n(r)}, \qquad (4)$$

where the matrix elements of the dipole moment

$$\langle n(r)|d_i|m(r)\rangle \equiv \langle \Psi_n(r_{el},r)|d_i(r_{el})|\Psi_m(r_{el},r)\rangle, \qquad (5)$$

taking into account Eqs. (1) and (2), may be represented as the power series in terms r except the linear term. As a result, Eq. (4) may be written as

$$\alpha_{ii}^{(n)}(0,r) = \alpha_{ii}^{(n)} + A_{ii}^{(n)}r^2 + B_{ii}^{(n)}r^3 + C_{ii}^{(n)}r^4 + \dots, \qquad (6)$$

where $\alpha_{ii}^{(n)}$ are the polarizability tensor components of the united atom at the state which the molecule goes into when its nuclei come closer together, and $A_{ii}^{(n)}$, $B_{ii}^{(n)}$, and $C_{ii}^{(n)}$ are the constants for the given molecule electronic state.

2.2. The molecule polarizability at large internuclear separations

The molecule polarizability at large internuclear distances may be described on the basis of an idea about interacting atoms forming the molecule. There are two types of atom interactions: multipole (long-range) and exchange ones for which an analytical description of the molecule polarizability functions is possible. Each of these is predominant at its own range of internuclear distances: the multipole one – at $r \to \infty$ and exchange one – at smaller r.

Multipole interaction of atoms. To take into account the multipole interactions between two identical atoms we apply a model of two interacting isotropic dielectric spheres[50,51]. In this model the polarizability of two dielectric spheres of radius r_0 located on axis z at distance r each of other and placed into a permanent electric field may be represented in the form[51]:

$$\alpha_{zz}(0,r) = 2\alpha_0 \frac{\varepsilon+2}{\varepsilon-1}(2\sinh\eta_0)^3 \sum_{p=0}^{\infty}\left[\frac{2}{\varepsilon+1}\right]^p \sum_{n=1}^{\infty} n\chi_n^{(p)}(r), \qquad (7)$$

$$\alpha_{xx}(0,r) = 2\alpha_0 \frac{\varepsilon+2}{\varepsilon-1}(2\sinh\eta_0)^3 \sum_{p=0}^{\infty}\left[\frac{2}{\varepsilon+1}\right]^p \sum_{n=1}^{\infty} \frac{n(n+1)}{2}\omega_n^{(p)}(r), \quad (8)$$

where ε is the sphere permittivity, α_0 is the static polarizability of the sphere, $\cosh\eta_0 = r/(2r_0)$, and the coefficients $\chi_n^{(p)}(r)$ and $\omega_n^{(p)}(r)$ are given by cumbersome recurrent relations (27) - (35) in the work[51]. In frames of this model the Eqs. (7) and (8) are accurate in the range of $r\in(2r_0,\infty)$, however, these allow the polarizabilities to be calculated only in a numerical form because of infinite and weakly convergent series.

In the work[38], to represent analytically the polarizability functions $\alpha_{zz}(0,r)$ and $\alpha_{xx}(0,r)$, the power series expansions of Eqs. (7) and (8) with respect to the small parameter $\delta = r_0/r$ have been made and after summation over p and n the following simple asymptotic expressions have been obtained:

$$\alpha_{zz}(0,r) = 2\alpha_0 + 4\alpha_0^2 \frac{1}{r^3} + 8\alpha_0^3 \frac{1}{r^6} + 18\alpha_0^3\left(1+\frac{1}{2\varepsilon+3}\right)\frac{r_0^2}{r^8} +$$

$$+16\alpha_0^4\frac{1}{r^9}+32\alpha_0^3\left(1+\frac{2}{3\varepsilon+4}\right)\frac{r_0^4}{r^{10}}+\cdots,\qquad(9)$$

$$\alpha_{xx}(0,r)=2\alpha_0-2\alpha_0^2\frac{1}{r^3}+2\alpha_0^3\frac{1}{r^6}+6\alpha_0^3\left(1+\frac{1}{2\varepsilon+3}\right)\frac{r_0^2}{r^8}-$$

$$-2\alpha_0^4\frac{1}{r^9}+12\alpha_0^3\left(1+\frac{2}{3\varepsilon+4}\right)\frac{r_0^4}{r^{10}}+\cdots.\qquad(10)$$

Doing so, we used the following relation:

$$\alpha_0=\frac{\varepsilon-1}{\varepsilon+2}r_0^3.\qquad(11)$$

The expressions (9) and (10) include, except well known terms of DID model, additional terms resulting from multipole interaction. Note that at $\varepsilon\to\infty$ the expressions obtained correspond to the "metallic-sphere" model[52].

The model above considered for interacting dielectric spheres does not take into account the anisotropy of atom polarizabilities being at the states with the orbital quantum number $L\geq1$ and also the change of atom polarizabilities due to their drawing together. The polarizability anisotropy may be partially taken into account if in the terms of Eqs. (9) and (10), corresponding to DID model, α_0 appears to be replaced by the components of the polarizability tensor[53] α_{ii}^0 of corresponding atoms, and in the terms which include r_0 the value α_0 appears to be equal to the average atomic polarizability (hereinafter r_0 is the van der Waals radius of the interacting atoms). The change of polarizability for atoms approaching to each other may be taken into account in frames of the method suggested in the works[35,36]. Taking into consideration above notes, Eqs. (9) and (10) may be written as[38]

$$\alpha_{zz}(0,r)=2\alpha_{zz}^0+4\left(\alpha_{zz}^0\right)^2\frac{1}{r^3}+\left[8\left(\alpha_{zz}^0\right)^3+\frac{7\tilde{\gamma}C_6}{9\alpha_0}\right]\frac{1}{r^6}+$$

$$+18\alpha_0^3\left(1+\frac{1}{2\varepsilon+3}\right)\frac{r_0^2}{r^8}+16\left(\alpha_{zz}^0\right)^4\frac{1}{r^9}+32\alpha_0^3\left(1+\frac{2}{3\varepsilon+4}\right)\frac{r_0^4}{r^{10}},\qquad(12)$$

$$\alpha_{xx}(0,r) = 2\alpha_{xx}^0 - 2(\alpha_{xx}^0)^2 \frac{1}{r^3} + \left[2(\alpha_{xx}^0)^3 + \frac{4\tilde{\gamma}C_6}{9\alpha_0} \right] \frac{1}{r^6} +$$

$$+ 6\alpha_0^3 \left(1 + \frac{1}{2\varepsilon + 3} \right) \frac{r_0^2}{r^8} - 2(\alpha_{xx}^0)^4 \frac{1}{r^9} + 12\alpha_0^3 \left(1 + \frac{2}{3\varepsilon + 4} \right) \frac{r_0^4}{r^{10}}, \quad (13)$$

Here the terms containing van der Waals factor C_6 and the average second hyperpolarizability $\tilde{\gamma}$ of an atom show the polarizability change due to dispersion interactions of atoms approaching to each other.

Exchange interaction of atoms. To take into account the exchange of two atoms' electrons one may use asymptotic methods[54] that are applicable to the range of internuclear distances characterized by weak overlapping of the valence electron shells of interacting atoms. In this range, the exchange interaction of atoms can be approximately considered as an exchange interaction of two valence electrons (by one from each atom).

Let us consider the interaction of two valence s-electrons. In this case, the asymptotic radial wavefunction of a valence electron of a neutral atom is determined by the following equation[54]:

$$\varphi(r_{el}) = A r_{el}^{1/\beta - 1} \exp(-r_{el}\beta), \quad (14)$$

where $\beta^2/2$ is the atom ionization potential, and the value of the asymptotic coefficient A depends on the electron distribution in the internal zone of the atom. Then the two-electron (one electron from atom "a" and one from atom "b") molecular wavefunction of the state n can be written in the form:

$$\Psi_n(r_{el1}, r_{el2}, r) = c_n^{(1)} \psi_n^{(1)}(r_{el1}, r_{el2}, R) + c_n^{(2)} \psi_n^{(2)}(r_{el1}, r_{el2}, r), \quad (15)$$

where

$$\psi_n^{(1)}(r_{el1}, r_{el2}, r) = \left[\varphi^{(a)}(r_{el1}, r)\varphi^{(b)}(r_{el2}, r)\chi_I(r_{el1}, r_{el2}, r) \right]_n,$$
$$\psi_n^{(2)}(r_{el1}, r_{el2}, r) = \left[\varphi^{(a)}(r_{el2}, r)\varphi^{(b)}(r_{el1}, r)\chi_{II}(r_{el1}, r_{el2}, r) \right]_n. \quad (16)$$

Here $\varphi^{(a)}(r_{el1}, r)$, $\varphi^{(b)}(r_{el1}, r)$ and $\varphi^{(a)}(r_{el2}, r)$, $\varphi^{(b)}(r_{el2}, r)$ are asymptotic wavefunction of the first and the second electrons located near corresponding atom cores. Equations (15) and (16) are written in the molecular coordinate system in which the interacting atoms are located on the axis z, and the center of the interatomic separation is taken as the origin of coordinates. In this coordinate system r_{el1} and r_{el2} are coordinates of the first and the second electrons. The explicit form of the functions $\chi_I(r_{el1}, r_{el2}, r)$ and $\chi_{II}(r_{el1}, r_{el2}, r)$ accounting for the interaction of electrons with each other and with extraneous nuclei is given in Ref. 54.

The exchange interaction, when the molecule polarizability being calculated, shows itself according to Eq. (4) mainly through its contribution into the matrix element of the electron transition dipole moment. The contribution of the exchange interaction into the matrix elements of ith component of the dipole moment calculated by Eqs. (15) and (16) for a homonuclear molecule can be represented in the form:

$$\langle n(r)|d_i|m(r)\rangle_{ch} \sim \langle \psi_n^{(1)}(r_{el1}, r_{el2}, r)|d_i|\psi_m^{(2)}(r_{el1}, r_{el2}, r)\rangle =$$
$$= \tilde{A}_{nm}^i(\beta_n, \beta_m, r)r^{\delta_i}\exp[-(3\beta_n + \beta_m)r/2], \tag{17}$$

where $\tilde{A}_{nm}^i(\beta_n, \beta_m, r)$ is the function weakly dependent on r in the region of slight overlapping of electron shells of atoms; $\beta_n^2/2$ and $\beta_m^2/2$ are ionization potentials of the atoms in the ground and excited states respectively, and

$$\delta_z = \frac{11}{4\beta_n} + \frac{1}{\beta_m} - \frac{1}{2(\beta_n + \beta_m)} + 1,$$
$$\delta_{x,y} = \frac{11}{4\beta_n} + \frac{1}{\beta_m} - \frac{1}{2(\beta_n + \beta_m)} + \frac{1}{2}. \tag{18}$$

Then, after substitution of Eq. (17) into Eq. (4) and substitution of some effective parameter $\bar{\beta}$ for β_m, the equation for the contribution of the exchange interaction into the molecular polarizability takes the form:

$$\left[\alpha_{ii}^{(n)}(0,r)\right]_{ch} = B_i(\beta_n, \bar{\beta}, r)r^{2\delta_i}\exp[-(3\beta_n + \bar{\beta})r]. \tag{19}$$

Here $B_i(\beta_n, \bar{\beta}, r)$ also weakly depends on r and is further assumed as a parameter B_i.

The result obtained can also be applied to interacting electrons with nonzero orbital moment l. This is connected with the fact that the exchange interaction occurs in the range of electron coordinates near the axis z where the angular wavefunctions of the electrons vary slightly, therefore, they can be substituted by their values on the axis z. As a consequence, the problem of exchange interaction of the valence l-electrons reduces to the problem considered above with the only difference that the coefficient A in Eq. (14) should be multiplied by $\sqrt{2l+1}$ (see the work[55]).

Note that in the considered range of internuclear distances the multipole interactions give a contribution into the molecular polarizability function too.

2.3. *Method for construction of polarizability functions*

Semiempirical method for construction of the polarizability functions of diatomic molecules in the ground electronic state (hereinafter, the index $n = 0$ for the molecular parameters is omitted) comprises three conditions to be satisfied[38-40,43].

1. The polarizability functions $\alpha_{zz}(0,r)$ and $\alpha_{xx}(0,r) = \alpha_{yy}(0,r)$ of the molecule in the range of small r are described by the polynomial like Eq. (6):

$$\alpha_{ii}(0,r) = \alpha_{ii}^{(0)} + a_i^{(2)}r^2 + a_i^{(3)}r^3 + a_i^{(4)}r^4 + \cdots, \qquad (20)$$

where $\alpha_{ii}^{(0)}$ are the components of the static polarizability for the united atom, and the constants $a_i^{(k)}$ $(k \geq 2)$ are found from the known values of the polarizability functions and their derivatives at the point r_e. These coefficients for each component of the polarizability tensor are the solution of the system of linear equations. The system of linear equations is determined by an equality condition at r_e for derivatives of the polynomial (20) and those of the polarizability function in the vicinity of equilibrium internuclear position

$$\alpha_{ii}(0,r) = \left(\alpha_{ii}\right)_e + \left(\alpha_{ii}'\right)_e \xi + \frac{1}{2}\left(\alpha_{ii}''\right)_e \xi^2 + \cdots. \qquad (21)$$

In Eq. (21) the components $(\alpha_{ii})_e$ of the static polarizability tensor and their first derivatives $(\alpha'_{ii})_e$, and second derivatives $(\alpha''_{ii})_e$ with respect to $\xi = (r - r_e)/r_e$ are determined at r_e. Note, that the number of terms in Eq. (20) is determined by the number of known polarizability derivatives of the molecule in Eq. (21). As a result, the polynomials (20) together with the coefficients $a_i^{(k)}$ obtained describe the polarizability functions of a molecule in the range of small r including the vicinity of the equilibrium internuclear position.

2. The polarizability function of a molecule in the range of large r can be presented as sum of contributions from multipole (Eqs. (12) and (13)) and exchange (Eq. 19) interactions:

$$\alpha_{zz}(0,r) = 2\alpha_{zz}^0 + 4\left(\alpha_{zz}^0\right)^2 \frac{1}{r^3} + \left[8\left(\alpha_{zz}^0\right)^3 + \frac{7\tilde{\gamma}C_6}{9\alpha_0}\right]\frac{1}{r^6} + 18\alpha_0^3\left(1 + \frac{1}{2\varepsilon + 3}\right)\frac{r_0^2}{r^8} +$$

$$+ 16\left(\alpha_{zz}^0\right)^4 \frac{1}{r^9} + 32\alpha_0^3\left(1 + \frac{2}{3\varepsilon + 4}\right)\frac{r_0^4}{r^{10}} + B_z r^{2\delta_z} \exp\left[-(3\beta_0 + \overline{\beta})r\right] \quad (22)$$

and

$$\alpha_{xx}(0,r) = 2\alpha_{xx}^0 - 2\left(\alpha_{xx}^0\right)^2 \frac{1}{r^3} + \left[2\left(\alpha_{xx}^0\right)^3 + \frac{4\tilde{\gamma}C_6}{9\alpha_0}\right]\frac{1}{r^6} + 6\alpha_0^3\left(1 + \frac{1}{2\varepsilon + 3}\right)\frac{r_0^2}{r^8} -$$

$$- 2\left(\alpha_{xx}^0\right)^4 \frac{1}{r^9} + 12\alpha_0^3\left(1 + \frac{2}{3\varepsilon + 4}\right)\frac{r_0^4}{r^{10}} + B_x R^{2\delta_x} \exp\left[-(3\beta_0 + \overline{\beta})r\right].(23)$$

The unknown parameters B_i are determined by fitting $\alpha_{ii}(0,r)$ to the *ab initio* data for H_2 molecule[27] in the range of slight overlapping of electron shells of H atoms. Then, taking into account the scaling factor reflecting the size of electron shells of the atoms, these parameters are applied to other homonuclear molecules.

3. The polarizability functions of the molecule in the intermediate range of internuclear separations are determined by joining the polarizability functions of a molecule at small and large r. These joining functions are described by the fifth-order polynomials in terms of r

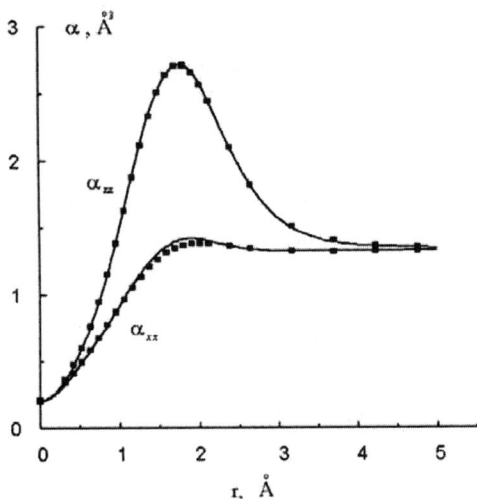

Fig. 1. Static polarizability functions of H_2 molecule: this method (solid curves), *ab initio* calculation[27] (squares).

$$\alpha_{ii}(0,r) = \sum_{j=0}^{5} b_i^{(j)} r^j ,\qquad (24)$$

where the coefficients $b_i^{(j)}$ are determined from the join conditions with an accuracy up to the second derivatives inclusive. The joints r_1 and r_2 are chosen as: $r_1 \approx r_e$, $r_2 \approx 2r_0$. Such a choice of the joints admits some arbitrariness, which does not change significantly the form of the polarizability functions.

2.4. *Static polarizability functions of H_2, N_2 and O_2 molecules*

In this section the method discussed above is applied to H_2, N_2 and O_2 molecules[39-40]. The reliable *ab initio* values of the static polarizability functions of H_2 molecule[27] were used to test the proposed model and to find the unknown parameters B_i in Eqs. (22) and (23). The values of atomic and molecular constants ε, B_i, $\bar{\beta}$, r_1, and r_2 used in the calculation are given in Table 1. The coefficients B_i ($\equiv B_i^{(H)}$) of the exchange interaction of H atoms were determined by the least-square

Table 1. Parameters for calculation of the polarizability of H_2, N_2, and O_2 molecules.

Molecule / Term	H_2 / $^1\Sigma_g^+$	N_2 / $^1\Sigma_g^+$	O_2 / $^3\Sigma_g^-$	
$(\alpha_{xx})_e$, Å^3	0.679[a]	1.53[b]	1.25[b]	
$(\alpha'_{xx})_e$, Å^3	0.669[a]	1.12[b]	0.70[b]	
$(\alpha''_{xx})_e$, Å^3	0.121[a]	0.33[b]	0.80[c]	
$(\alpha_{zz})_e$, Å^3	0.947[a]	2.24[b]	2.33[b]	
$(\alpha'_{zz})_e$, Å^3	1.372[a]	3.35[b]	3.89[b]	
$(\alpha''_{zz})_e$, Å^3	1.221[a]	2.93[b]	8.60[c]	
United atom / Term	He / ^1S	Si / $^1D\,(M_L=0)$	S / $^3P\,(M_L=0)$	
$\alpha_{xx}^{(0)}$, Å^3	0.205[d]	5.62[e]	2.68[e]	
$\alpha_{zz}^{(0)}$, Å^3	0.205[d]	7.50[e]	3.35[e]	
Isolated atom / Term	H / ^1S	N / ^4S	O / $^3P\,(M_L=0)$	$^3P\,(M_L=\pm1)$
α_{xx}^0, Å^3	0.6668[d]	1.101[f]	0.755[f]	0.825[f]
α_{zz}^0, Å^3	0.6668[d]	1.101[f]	0.895[f]	0.755[f]
α_0, Å^3	0.6668	1.101	0.802	
r_0, Å	1.1[g]	1.5[g]	1.4[g]	
ε	4.01[h]	2.45[h]	2.24[h]	
γC_6, Å^{12}	4.18[i]	10.51[i]	5.03[i]	
β_0, a.u.	1.000[j]	1.033[j]	1.000[j]	
$\bar{\beta}$, a.u.	0.41	0.31	0.31	
A, a.u.	2.00[j]	1.49[j]	1.32[j]	
B_x, Å$^{(3-2\delta)}$	32.60	50.16	24.25	
B_z, Å$^{(3-2\delta)}$	108.81	167.44	80.94	
r_1, Å	0.9	1.1	1.2	
r_2, Å	2.2	3.0	2.8	

[a] Calculated using data from Ref. 27; [b] taken from Ref. 56; [c] taken from Ref. 39;.
[d] taken from Ref. 57; [e] taken from Ref. 58; [f] taken from Ref. 43; [g] taken from Ref. 59;
[h] calculated by Eq. (11); [i] the values of γ and C_6 for H were taken from Ref. 36, γ for N and O were taken from Ref. 60, C_6 for N and O were taken from Ref. 61;
[j] taken from Ref. 55.

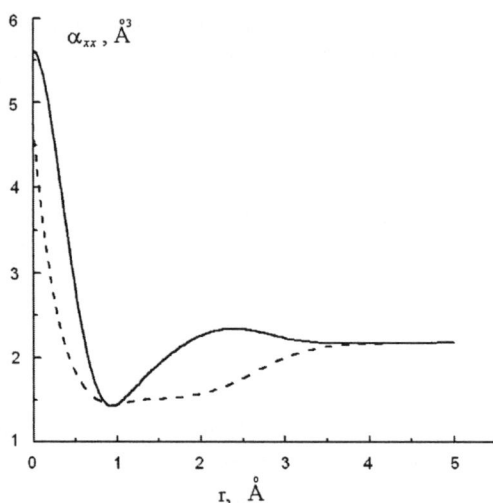

Fig. 2. Static polarizability functions $\alpha_{xx}(0,r)$ of N$_2$ molecule: this method (solid curve), calculation[42] (dashed curve).

method from the data of *ab initio* calculations[27] of the polarizability functions $\alpha_{ii}(0,r)$ in the range of $2.4 \leq r < 4.0$ Å. The coefficients B_i for the atoms N and O were calculated by the equation $B_i^{(N,O)} = 9B_i^{(H)} A_{(N,O)}^6 / A_{(H)}^6$, where $A_{(N,O)}$ is the parameter of the asymptotic radial wavefunction of the valence p-electron of, respectively, N and O atoms, and $A_{(H)}$ is the parameter of the asymptotic radial wavefunction of the valence s-electron of H atom. The parameter $\overline{\beta}$ was estimated on a base of probabilities of radiative transitions in H, N and O atoms.

Figure 1 depicts the polarizability functions of the molecule H$_2$ calculated by both this and *ab initio*[27] methods. It is clearly seen that the polarizability functions calculated by these methods are closely agreed in the entire range of internuclear separations. The use of the single fitting parameter B_i demonstrates the physical adequacy of the method as a whole.

The proposed method of calculation of the polarizability functions was applied to N$_2$ and O$_2$ molecules, and for the oxygen molecule the

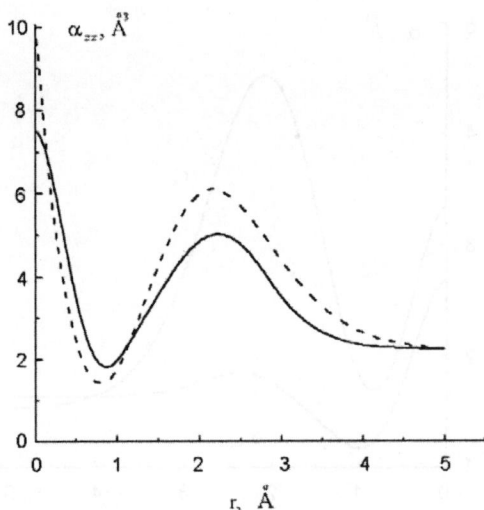

Fig. 3. Static polarizability functions $\alpha_{zz}(0,r)$ of N_2 molecule: this method (solid curve), calculation[42] (dashed curve).

both ways of its decomposition were considered (Figs. 2–5). It is of interest to compare the polarizability functions obtained for N_2 molecule with those calculated by an alternative method[42] in which some found correlations were used. Apart from differences in the polarizability functions at small r due to different values of $\alpha_{xx}^{(0)}$ and $\alpha_{zz}^{(0)}$ chosen for Si atom, there should be noted the low values of $\alpha_{xx}(0,r)$ for $r \approx 1 \div 3$ Å (Fig. 2). This is caused, in our opinion, by the fact that correlations found for the molecules, mainly comprising the first-group atoms of the Periodic Table, were applied to the molecule N_2 without enough foundation. The polarizability functions[41] are not given in Figs. 2–3 because of their only historic interest now.

The calculated polarizability functions for N_2 and O_2 molecules should be considered as approximations because the B_i and $\overline{\beta}$ for N_2 and O_2 are only estimated parameters. Nevertheless, we suppose that these polarizability functions for N_2 and O_2 are more accurate than those from the works[38,41-43].

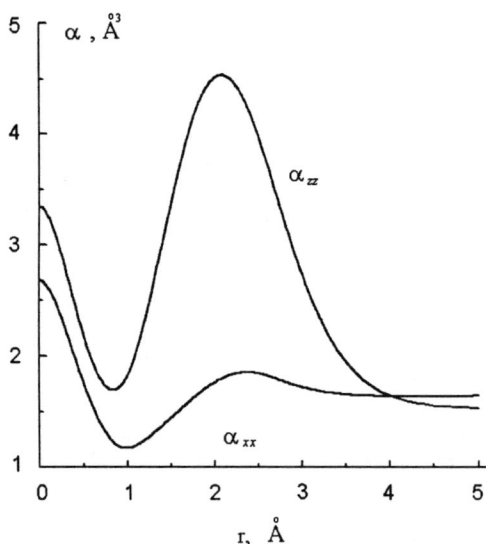

Fig. 4. Static polarizability functions of O_2 molecule (O atoms at the states with $|M_L| = 1$).

2.5. *Dynamic polarizability functions of H₂ and N₂ molecules*

The tensor of dynamic electronic polarizability for a diatomic homonuclear molecule at the electronic state n may be represented as

$$\alpha_{ii}^{(n)}(\omega, r) = 2 \sum_{m \neq n} \frac{[E_m(r) - E_n(r)] |\langle n(R)| d_i | m(r) \rangle|^2}{[E_m(r) - E_n(r)]^2 - (\hbar\omega)^2}. \tag{25}$$

It is obvious that the frequency dependence of the molecule polarizability is fully determined by a structure of electronic energy levels of the molecule and by correspondent probabilities of electric dipole transitions between them. Thus, the frequency dependence of the polarizability tensor is individual as for each molecule so for its different components.

In the first approximation for a molecule in the ground electron state the Eq. (25) can be represented as a product of the static polarizability $\alpha_{ii}(0, r)$ of a molecule and the frequency factor $f_i(\omega, r)$

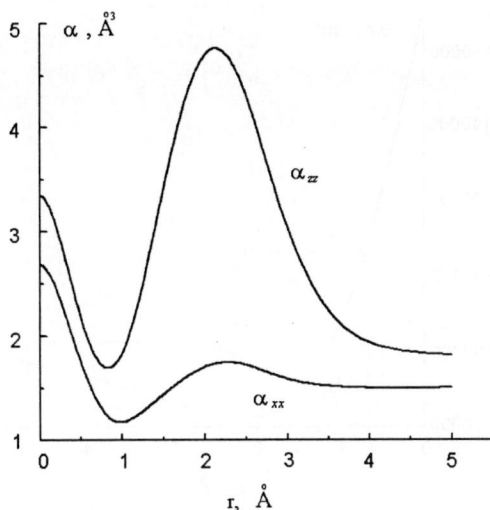

Fig. 5. Static polarizability functions of O_2 molecule (O atoms at the states with $|M_L| = 0$).

$$\alpha_{ii}(\omega,r) \approx \alpha_{ii}(0,r) f_i(\omega,r), \qquad (26)$$

where

$$f_i(\omega,r) = \frac{[E_{1i}(r) - E_0(r)]^2}{[E_{1i}(r) - E_0(r)]^2 - (\hbar\omega)^2}. \qquad (27)$$

Here $E_0(r)$ is an energy of the ground electronic state of a molecule, $E_{1i}(r)$ is an energy of the molecule lower exited electronic state into which electric dipole transition is allowed from the ground state. Note, that in Eq. (27) the energies $E_{1i}(r)$ are different for polarizability tensor components $\alpha_{ii}(\omega,r)$ and they are determined by selection rules according to Λ, where Λ is an eigenvalue of the orbital moment projection of electrons onto the molecule axis. Thus, for the tensor component $\alpha_{zz}(\omega,r)$ the lower exited electronic state (with the energy $E_{1z}(r)$) is determined by the selection rule $\Delta\Lambda = 0$, and for the

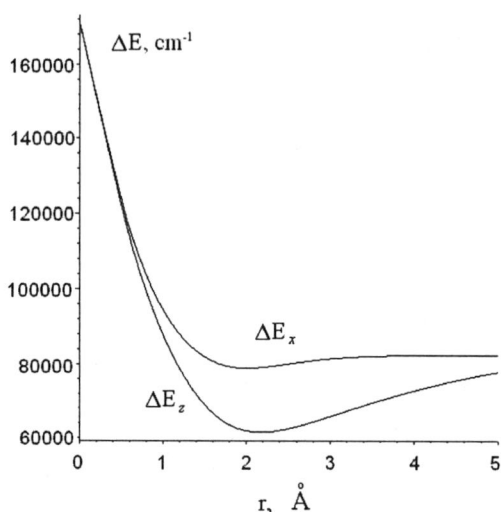

Fig. 6. The functions $\Delta E_z(r)$ and $\Delta E_x(r)$ for H_2 molecule.

components $\alpha_{xx}(\omega,r) \equiv \alpha_{yy}(\omega,r)$ the lower exited electronic state (with the energy $E_{1x}(r)$) is determined by the selection rule $|\Delta\Lambda| = 1$.

For some molecules the functions $E_{1i}(r) - E_0(r)$ in Eq. (27) can be represented as a finite and continuous functions in full range of internuclear separations $r \in [0,\infty)$, and in consequence the frequency factor $f_i(\omega,r)$ is also a continuous function of r. The frequency factor $f_i(\omega,r) = 1$ at $\omega = 0$ and it increases with increasing of the frequency ω of an external electromagnetic field. Note, that Eq. (26) gives an upper boundary for components of the dynamic polarizability tensor $\alpha_{ii}(\omega,r)$ of the molecule in the frequency range $\omega < [E_{1i}(r) - E_0(r)]/\hbar$.

H_2 molecule. The method for calculation $\alpha_{ii}(\omega,r)$ was tested on the hydrogen molecule. For this purpose the dynamic polarizability functions calculated by Eqs. (26) and (27) were compared with the precise *ab initio* values[28] of $\alpha_{ii}(\omega,r)$. To calculate $\alpha_{ii}(\omega,r)$ the energy differences $E_{1i}(r) - E_0(r)$ were formed where the potential energy values $E_0(r)$ for the ground electronic state $X^1\Sigma_g^+$ were taken from Ref. 62 and the

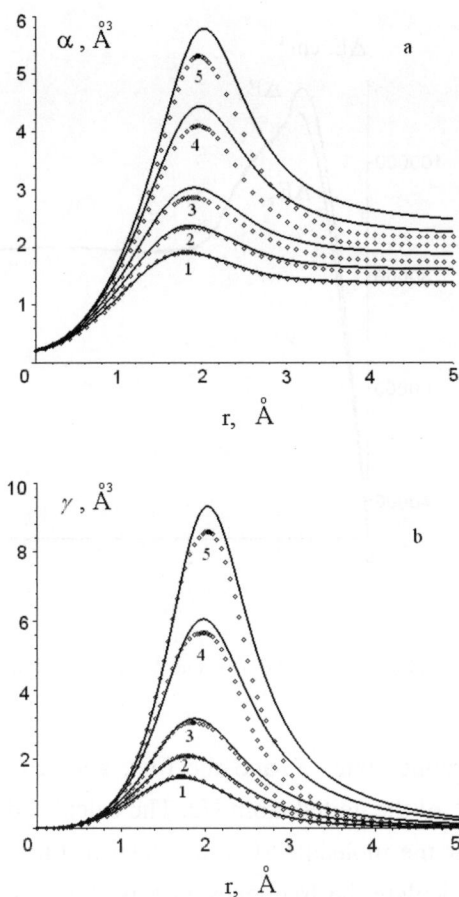

Fig. 7. Average dynamic polarizability (a) and anisotropy of dynamic polarizability (b) of H_2 molecule: this method (solid lines), *ab initio* calculation[28] (diamonds).
1 $-\omega = 15803$ cm^{-1}; 2 $- \omega = 33692$ cm^{-1}; 3 $- \omega = 43424$ cm^{-1}; 4 $- \omega = 51658$ cm^{-1}, 5 $- \omega = 54867$ cm^{-1}.

energy values $E_{1z}(r)$ and $E_{1x}(r)$ for the excited electronic states $B^1\Sigma_u^+$ and $C^1\Pi_u$ – from Ref. 63. In the region of small internuclear separations ($r \leq 0.5$ Å) lacking values of energy differences $\Delta E_z(r) = E_{1z}(r) - E_0(r)$ and $\Delta E_x(r) = E_{1x}(r) - E_0(r)$ were defined by interpolation to the value $\Delta E_z(0) = \Delta E_x(0) = E(^1P^o) - E(^1S)$, where $E(^1S)$ and $E(^1P^o)$ are the

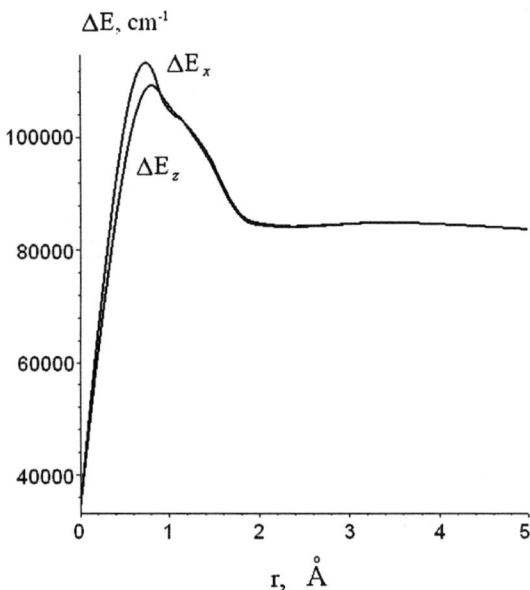

Fig. 8. The functions $\Delta E_z(r)$ and $\Delta E_x(r)$ for N_2 molecule.

terms of the ground state ^1S and the first, allowed by selection rules, exited state $^1P^o$ of the united atom He. The calculated functions $\Delta E_z(r)$ and $\Delta E_x(r)$ for the molecule H_2 are shown in Fig. 6. These functions were used to calculate the frequency factors $f_z(\omega,r)$ and $f_x(\omega,r)$ that subsequently, in combination with the static polarizability[27], made it possible to calculate tensor components of the dynamic polarizability and tensor invariants: the average polarizability

$$\alpha(\omega,r) = \left[\alpha_{zz}(\omega,r) + 2\alpha_{xx}(\omega,r)\right]/3 \qquad (28)$$

and the anisotropy of polarizability

$$\gamma(\omega,r) = \alpha_{zz}(\omega,r) - \alpha_{xx}(\omega,r). \qquad (29)$$

The invariants of dynamic polarizability tensor obtained for the molecule H_2 are shown in Fig. 7 where they are compared with those of *ab initio* calculations[28]. It is clearly seen that the method, despite its simplicity, adequately represents the dependence of $\alpha(\omega,r)$ and $\gamma(\omega,r)$

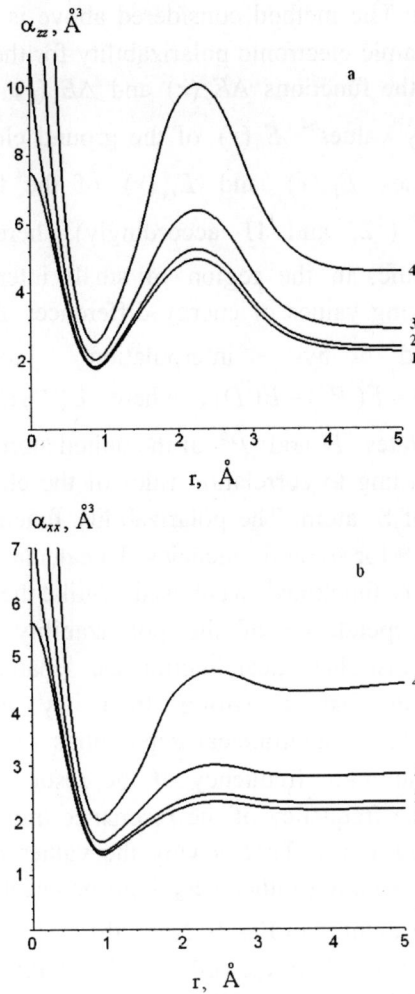

Fig. 9. The components of the dynamic polarizability tensor $\alpha_{zz}(\omega, r)$ (*a*) and $\alpha_{xx}(\omega, r)$ (*b*) of N_2 molecule: $1 - \omega = 0$ cm^{-1}; $2 - \omega = 20000$ cm^{-1}; $3 - \omega = 40000$ cm^{-1}; $4 - \omega = 60000$ cm^{-1}.

on the frequency of an external electromagnetic field for H_2 molecule. Evidently, this conclusion is also valid for individual components of the dynamic polarizability tensor of the hydrogen molecule. Therefore, it is possible to expect that the method will give satisfactory results for other diatomic molecules.

N_2 *molecule*. The method considered above is applied to calculate functions of dynamic electronic polarizability for the nitrogen molecule. Figure 8 shows the functions $\Delta E_z(r)$ and $\Delta E_x(r)$ calculated using the numerical energy values[64] $E_0(r)$ of the ground electronic state $X^1\Sigma_g^+$ and energy values $E_{1z}(r)$ and $E_{1x}(r)$ of the first exited diabatic electronic states[65] ($^1\Sigma_u^+$ and $^1\Pi_u$ accordingly). Therewith, like in case of hydrogen molecule, in the region of small internuclear separations ($r \leq 0.85$ Å) lacking values of energy differences $\Delta E_z(r)$ and $\Delta E_x(r)$ were defined by interpolation to the value $\Delta E_z(0) = \Delta E_x(0) = E(^1P^\circ) - E(^1D)$, where $E(^1D)$ and $E(^1P^\circ)$ are the terms of exited states ^1D and $^1P^\circ$ of the united atom Si. These terms are determined according to correlation rules of the electronic states of N_2 with the states of Si atom. The polarizability functions of N_2 molecule are given in Fig. 9 for some frequencies. To calculate these functions the static polarizability functions[39] were used. Unlike the molecule H_2, where the frequency dependence of the polarizability tensor components appeared to be most sharp near internuclear separation of $r \approx 2$ Å, the nitrogen molecule has a strong frequency dependence of the polarizability at small internuclear separations ($r \approx 0$). In accordance with this, the minimum frequency of the resonance transition of N_2 coincides with the frequency of the resonance transition of the united atom Si (≈ 34700 cm^{-1}). That is why the values of the polarizability tensor components taking either very high or negative values for $r <$ 0.3 Å are not represented in Fig. 9. Note, that the dynamic polarizability functions shown in Fig. 9 agree well with the *ab initio* calculations of the dynamic polarizability of N and Si atoms[66,67], and N_2 molecule[25] at $r = r_e$ (see Ref. 44).

2.6. *Application of the polarizability functions to some problems of molecular physics and spectroscopy*

The polarizability functions can be used to calculate some molecular properties which are difficult to be experimentally determined. Below, there are given a few examples of such calculations for N_2 and O_2 molecules.

2.6.1. *Static polarizability of N_2 and O_2 molecules in excited vibrational-rotational levels*

The polarizability of a molecule in excited vibrational-rotational levels is determined as the vibrational-rotational matrix elements of the components of the polarizability tensor that depend on the vibrational and rotational quantum numbers. These matrix elements can be written to the order of $(B_e/\omega_e)^2$ for the terms dependent only on the vibrational quantum number v and to the order of $(B_e/\omega_e)^3$ for the terms dependent on the rotational quantum number J in the form[68]:

$$
(\alpha_{ii})_{vJ} = \langle vJ|\alpha_{ii}(0,r)|vJ\rangle = (\alpha_{ii})_e + \left[-3a_1(\alpha'_{ii})_e + (\alpha''_{ii})_e\right]\left(v + \frac{1}{2}\right)\left(\frac{B_e}{\omega_e}\right) + .
$$

$$
+ \left\{\left[-\frac{3}{16}a_1^3(30v^2 + 30v + 11) + \frac{1}{8}a_1a_2(78v^2 + 78v + 31) - \right.\right.
$$

$$
\left. -\frac{15}{8}a_3(2v^2 + 2v + 1)\right](\alpha'_{ii})_e + \left[\frac{1}{16}a_1^2(30v^2 + 30v + 11) - \right.
$$

$$
\left.\left. -\frac{3}{8}a_2(2v^2 + 2v + 1)\right](\alpha''_{ii})_e - \frac{1}{48}a_1(30v^2 + 30v + 11)(\alpha'''_{ii})_e\right\}\left(\frac{2B_e}{\omega_e}\right)^2 +
$$

$$
+ J(J+1)(\alpha'_{ii})_e\left(\frac{2B_e}{\omega_e}\right)^2 + \left\{(\alpha'_{ii})_e\left[\frac{27}{8}a_1(1+a_1) + 3(1-a_2)\right] - \right.
$$

$$
\left. -\frac{3}{8}(3a_1+1)(\alpha''_{ii})_e + \frac{1}{4}(\alpha'''_{ii})_e\right\}J(J+1)(2v+1)\left(\frac{2B_e}{\omega_e}\right)^3, \qquad (30)
$$

where a_i are the Dancham anharmonicity constants, B_e is the rotational constant of the molecule, and ω_e is the harmonic frequency of vibrations of its nuclei.

478 M.A. Buldakov, V.N. Cherepanov

Table 2. Polarizabilities (Å³) of vibrationally excited N_2 and O_2 molecules.

v	N_2		O_2	
	$(\alpha_{zz})_{v0}$	$(\alpha_{xx})_{v0}$	$(\alpha_{zz})_{v0}$	$(\alpha_{xx})_{v0}$
0	2.26	1.54	2.35	1.25
1	2.28	1.55	2.39	1.26
2	2.31	1.55	2.43	1.27
3	2.34	1.56	2.48	1.27
4	2.36	1.57	2.52	1.28
5	2.39	1.58	2.57	1.29
6	2.42	1.59	2.61	1.30
7	2.45	1.60	2.66	1.30
8	2.48	1.61	2.71	1.31
9	2.51	1.62	2.76	1.32

The first term of Eq. (30) describes the electronic polarizability of a molecule at its equilibrium separation. The second and third terms take into account the polarizability dependence of the molecule nuclei vibrations. The two last terms in Eq. (30) determine a rotational portion of the molecule polarizability. Calculations show that the contribution of these terms into the polarizability of the molecule is very small. So, for example, for N_2 and O_2 molecules for $J = 20$ this contribution into the polarizability is less than 0.2 %. The polarizability of molecules at their vibrational excitation varies considerably larger. In Table 2 the polarizabilities of vibrationally excited N_2 and O_2 molecules are given. The increase of both polarizability components with v increasing is very noticeable and their v-dependence is more noticeable for $(\alpha_{zz})_{v0}$. The dependence of the polarizabilities on v for $v \leq 9$ is almost linear that indicates the small contribution into the polarizability from the terms of the order $(B_e/\omega_e)^2$.

The dynamic polarizability of the molecules in excited vibrational-rotational states can be calculated by the same Eq. (30) where the polarizability derivatives have to be considered as the functions of the frequency ω of an external electromagnetic field.

Table 3. Parameters for calculating the mean polarizability of molecules.

Parameter	N_2	O_2
$\bar{\alpha}(0,0)$,$\cdot\text{Å}^3$	1.7406[a]	1.5658[a]
$\omega_0 \cdot 10^{-16}$, c^{-1}	2.6049[a]	2.1801[a]
$b \cdot 10^6$, K^{-1}	0.6611[b]	1.146[b]
$c \cdot 10^9$, K^{-2}	0.683[b]	2.01[b]

[a] Taken from Ref. 69; [b] see also Ref. 70.

2.6.2. *Temperature dependence of the mean polarizability of N_2 and O_2 molecules*

In a general case, the temperature dependence of the mean polarizability of a molecule can be represented in the form:

$$\bar{\alpha}(\omega,T) = \sum_{nvJ} f_{nvJ}(T)\langle v,J|\alpha^{(n)}(\omega,r)|v,J\rangle, \tag{31}$$

where $f_{nvJ}(T)$ is the distribution function of molecules over the electronic (n), vibrational (v) and rotational (J) states at the temperature T and $\alpha^{(n)}(\omega,r)$ is the isotropic average polarizability of a molecule. For the gaseous medium at the thermodynamic equilibrium the distribution function $f_{nvJ}(T)$ can be represented as

$$f_{nvJ} = \frac{g_{nJ}(2J+1)\exp[-\Delta E_{nvJ}/kT]}{\sum_{nvJ} g_{nJ}(2J+1)\exp[-\Delta E_{nvJ}/kT]}. \tag{32}$$

Here, g_{nJ} is the statistical weight of J rotational level in nth electronic state and ΔE_{nvJ} is the rovibronic energy of a nonrigid diatomic molecule. The matrix elements of the average polarizability in Eq. (31) can be calculated by Eqs. (28) and (30).

For practical use, the temperature dependence of $\bar{\alpha}(\omega,T)$ in Eq. (31) can be conveniently expressed in the form[69]:

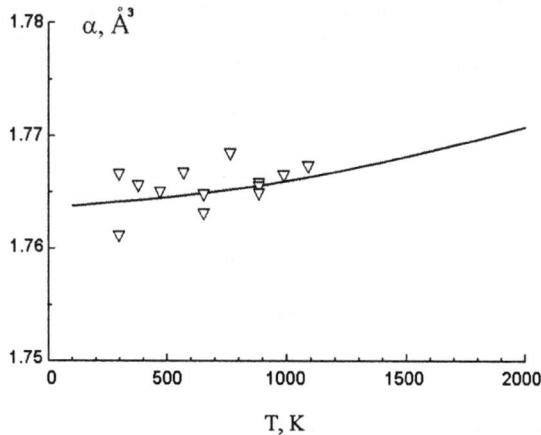

Fig. 10. Temperature dependence of mean polarizability of N_2 molecule at $\lambda = 633$ nm: calculation by Eqs. (30) – (32), triangles – experiment[71].

$$\bar{\alpha}(\omega,T) = \frac{\bar{\alpha}(0,0)}{1-\omega^2/\omega_0^2}(1+bT+cT^2), \qquad (33)$$

where $\bar{\alpha}(0,0)$, ω_0, b and c are some adjustable parameters. Note, that Eq. (33) can't be applied for very low temperatures.

Nitrogen. N_2 molecule has high lying excited electronic states, whose population is negligible at temperature up to 2000 K. Thus, it is possible to neglect the contributions from excited electronic states in Eq. (31) calculating the temperature dependence of the mean polarizability $\bar{\alpha}(\omega,T)$ for N_2 molecule. In Fig. 10 the results of $\bar{\alpha}(\omega,T)$ calculation at $\lambda = 633$ nm are given from which the b and c parameters in Eq. (33) were found. The values of these parameters are listed in Table 3.

Oxygen. Unlike the nitrogen molecule, O_2 molecule has excited electronic states ($^1\Delta_g$ and $^1\Sigma_g^+$) that are not much high over the ground electronic state. In this case, the triplet structure of the ground electronic state and a possibility of populating the excited electronic states should be taken into account in calculating $\bar{\alpha}(\omega,T)$. The calculations of $\bar{\alpha}(\omega,T)$ were carried out in assumption that the polarizability of O_2 molecule to be of the same value for the ground electronic state and

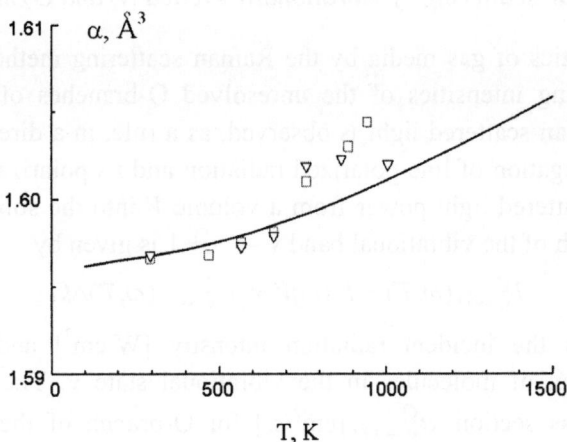

Fig. 11. Temperature dependence of mean polarizability of O_2 molecule at $\lambda = 633$ nm: calculation by Eqs. (30) – (32), the triangles[71] and squares[72] are the experimental results.

excited electronic states $^1\Delta_g$ and $^1\Sigma_g^+$. In Fig. 11 the polarizability $\overline{\alpha}(\omega,T)$ of O_2 molecule at $\lambda = 633$ nm is shown for the temperature range of 100 – 1500 K. The fact that the experimental values $\overline{\alpha}(\omega,T)$ exceed the calculated results at high temperatures can neither be explained by the fine structure of the ground electronic state, nor by the presence of excited $^1\Delta_g$ and $^1\Sigma_g^+$ states, but it is likely to be associated with experimental errors[70].

The results obtained may be useful for calculating the temperature dependence of the refractive index of low-density gases

$$n - 1 = 2\pi N \overline{\alpha}(\omega,T), \qquad (34)$$

where N is the concentration of molecules of the gas. Expression (33) with the obtained values b and c is applicable to the visible spectrum range and yields satisfactory results for a long wavelength range. For example, values of the refractive index calculated at the frequency of 47.7 GHz for nitrogen and oxygen $[(n-1)\times 10^6 = 293.94$ and 264.49, respectively] well agree with the corresponding values 293.81 and 266.95 measured in Ref. 73.

2.6.3. *Raman scattering by vibrationally excited N_2 and O_2 molecules*

Diagnostics of gas media by the Raman scattering method are often realized using intensities of the unresolved Q-branches of vibrational bands. Raman scattered light is observed, as a rule, in a direction at 90° to the propagation of line-polarized radiation and its polarization. In this case the scattered light power from a volume V into the solid angle $\Delta\Omega$ for Q-branch of the vibrational band v → v + 1 is given by

$$I^Q_{v\to v+1}(\omega,T) = I_0(\omega)VN_v\sigma^Q_{v\to v+1}(\omega,T)\Delta\Omega, \qquad (35)$$

where I_0 is the incident radiation intensity [W/cm²] and N_v is the concentration of molecules in the vibrational state v. The differential Raman cross section $\sigma^Q_{v\to v+1}$ [cm²/sr] for Q-branch of the vibrational band can be written as

$$\sigma^Q_{v\to v+1}(\omega,T) = 16\pi^4\left(\omega - \Delta\omega_{v\to v+1}\right)^4 \times$$

$$\times\left[\langle v|\alpha(\omega,r)|v+1\rangle^2 + \frac{7}{45}f_v(T)\langle v|\gamma(\omega,r)|v+1\rangle^2\right], \qquad (36)$$

where ω is the incident light frequency [cm⁻¹], $\Delta\omega_{v\to v+1}$ is Q-branch frequency of the vibrational band v → v + 1 [cm⁻¹], T is the gas temperature. The matrix element of the average polarizability in Eq. (36) can be written within the accuracy up to the terms of $\left(B_e/\omega_e\right)^{3/2}$ inclusively in the form[68]:

$$\langle v|\alpha(\omega,r)|v+1\rangle = \left(\frac{v+1}{2}\right)^{1/2}\left(\frac{2B_e}{\omega_e}\right)^{1/2}\left\{\alpha'_e(\omega) + (v+1)\left(\frac{2B_e}{\omega_e}\right)\times\right.$$

$$\left.\times\left[\left(\frac{11}{16}a_1^2 - \frac{3}{4}a_2\right)\alpha'_e(\omega) - \frac{5}{4}a_1\alpha''_e(\omega) + \frac{1}{4}\alpha'''_e(\omega)\right]\right\}. \qquad (37)$$

The matrix element of the anisotropy of polarizability has the same form. The factor $f_v(T)$ in Eq. (36) for gaseous nitrogen and oxygen being in the thermodynamic equilibrium can be written in an analytical form:

$$f_v(T) = \frac{\sum_J g_J(2J+1)\frac{J(J+1)}{(2J-1)(2J+3)}\exp\left[-\Delta E_{vJ}/kT\right]}{\sum_J g_J(2J+1)\exp\left[-\Delta E_{vJ}/kT\right]}, \qquad (38)$$

where g_J is the degeneracy due to nuclear spin of J state, ΔE_{vJ} is the vibration-rotation energy of a molecule.

The polarization characteristics of the scattering light are determined by the depolarization ratio that for Q-branch of the vibrational band $v \rightarrow v+1$ and for the scattering configuration described above can be written as

$$\rho^Q_{v\rightarrow v+1}(\omega,T) = \frac{3f_v(T)\langle v|\gamma(\omega,r)|v+1\rangle^2}{45\langle v|\alpha(\omega,r)|v+1\rangle^2 + 4f_v(T)\langle v|\gamma(\omega,r)|v+1\rangle^2}. \qquad (39)$$

To calculate the cross sections $\sigma^Q_{v\rightarrow v+1}(\omega,T)$ and the depolarization ratios $\rho^Q_{v\rightarrow v+1}(\omega,T)$ for N_2 and O_2 molecules it's more conveniently to write these in the forms

$$\sigma^Q_{v\rightarrow v+1}(\omega,T) = (v+1)\sigma^{Q,\exp}_{0\rightarrow 1}(\omega,T)\frac{F^\sigma_{v\rightarrow v+1}(T)}{F^\sigma_{0\rightarrow 1}(T)} \qquad (40)$$

and

$$\rho^Q_{v\rightarrow v+1}(\omega,T) = \rho^{Q,\exp}_{0\rightarrow 1}(\omega,T)\frac{F^\rho_{v\rightarrow v+1}(T)}{F^\rho_{0\rightarrow 1}(T)}, \qquad (41)$$

where $\sigma^{Q,\exp}_{0\rightarrow 1}(\omega,T)$ and $\rho^{Q,\exp}_{0\rightarrow 1}(\omega,T)$ are experimental values of the cross sections and the depolarization ratios. The factors $F^\sigma_{v\rightarrow v+1}(T)$ and $F^\rho_{v\rightarrow v+1}(T)$ take into account the intramolecular interactions, weakly dependent on ω, and are written as follows[68]

$$F^\sigma_{vv+1}(T) = 1.018 + 3.99\cdot 10^{-6}T + \left[1.81\cdot 10^{-2} + 8.2\cdot 10^{-8}T\right]v,$$

$$F^\rho_{vv+1}(T) = 1.006 + 1.0\cdot 10^{-6}T + \left[6.0\cdot 10^{-3} + 2.04\cdot 10^{-7}T - 9.8\cdot 10^{-12}T^2\right]v$$

for N_2 molecule and

$$F^\sigma_{vv+1}(T) = 1.028 + 1.163\cdot 10^{-5}T + \left[2.8\cdot 10^{-2} + 3.0\cdot 10^{-7}T\right]v,$$

Table 4. The differential Raman cross sections $\sigma^Q_{v \to v+1}$ ($10^{-30} \cdot cm^2/sr$) and the depolarization ratios $\rho^Q_{v \to v+1} \cdot 100$ for Q-branch of the vibrational bands $v \to v+1$ for N_2 and O_2 molecules at $\lambda = 488$ nm and $T = 300$ K.

v	N₂		O₂	
	$\sigma^Q_{v \to v+1}$	$\rho^Q_{v \to v+1}$	$\sigma^Q_{v \to v+1}$	$\rho^Q_{v \to v+1}$
0	0.55	2.11	0.68	4.27
1	1.12	2.12	1.39	4.33
2	1.71	2.14	2.14	4.38
3	2.31	2.15	2.92	4.44
4	2.94	2.16	3.74	4.49
5	3.59	2.17	4.60	4.55
6	4.26	2.19	5.50	4.61
7	4.94	2.20	6.43	4.66
8	5.65	2.21	7.40	4.72
9	6.37	2.22	8.40	4.77
10	7.12	2.24	9.45	4.83

$$F^\rho_{vv+1}(T) = 1.013 + 1.6 \cdot 10^{-6} T + \left[1.3 \cdot 10^{-2} + 1.19 \cdot 10^{-6} T - 7.0 \cdot 10^{-10} T^2\right] v$$

for O_2 molecule.

The numerical values of the cross sections $\sigma^Q_{v \to v+1}$ and the depolarization ratios $\rho^Q_{v \to v+1}$ calculated for the temperature $T = 300$ K and for $v \le 10$ are given in Table 4. At the calculation the values of $\sigma^{Q,exp}_{0 \to 1}(\omega, T)$ and $\rho^{Q,exp}_{0 \to 1}(\omega, T)$ were taken from Refs. 74 and 75 respectively. As seen from the Table 4 the cross sections $\sigma^Q_{v \to v+1}$ increase nonlinearly as v increases. The depolarization ratios $\rho^Q_{v \to v+1}$ are not changed practically as functions of v. The temperature dependence as of $\sigma^Q_{v \to v+1}$ so of $\rho^Q_{v \to v+1}$ is very weak.

The values of the cross sections and the depolarization ratios given in Table 4 can be used for diagnostics of gas media including vibrationally-excited N_2 and O_2 molecules.

3. Polarizability functions of dimers composed of diatomic molecules

Van der Waals complex may be considered as a molecule with a large amplitude of a movement of its parts and, thus, with an enormous variety of configurations of the complex. As a result, in a gas medium the complexes of different configurations may exist simultaneously. So, to study some polarizability properties averaged on ensemble of the complexes, we need know their polarizability surface as a function of intra- and intermolecular separations and relative orientation of the complex components.

3.1. *Modified dipole-induced-dipole polarizability functions of van der Waals complexes*

The dipole-induced-dipole (DID) theory of the polarizability was formulated by Silberstein for interacting atomic systems[33,34]. This theory may be directly applied to molecular complexes too if only separations between the molecules in a complex are large and each molecule is considered as the atom with the polarizability being equal to the polarizability of the molecule. However, for weakly bounded molecular complexes the separations between the molecules are not much to exceed the sizes of the molecules. For this reason, it is necessary to modify the DID model to take into account their sizes.

Within the modified DID model we consider each molecule of a complex as a set of effective point atoms without any interaction between them[76]. The position of these atoms coincides with those of the nuclei of the molecule. The tensor of the total polarizability of the effective atoms is the same as that of the molecule. For diatomic homonuclear molecule the polarizabilities of the effective atoms are equal to each other. In general, decomposition of the molecular polarizability into effective atomic polarizabilities can be performed, for example, proportionally to the number of the valence electrons or to the polarizability of the atoms forming the molecule. So, the introduced modifications allow using the DID model for calculation of the polarizability of the van der Waals molecular complexes taking into account the sizes of molecules.

Following to the modified DID method the induced dipole moment μ^m of the atom m for the complex may be written as

$$\mu_\alpha^m = \alpha_{\alpha\beta}^m \left(E_\beta^m + \sum_{n \neq m}^N T_{\beta\gamma}^{mn} \mu_\gamma^n \right), \tag{42}$$

where $\alpha_{\alpha\beta}^m$ is the polarizability tensor of the effective atom m, E_β^m is the external electric field applied to atom m, $T_{\beta\gamma}^{mn} \mu_\gamma^n$ is the electric field induced by the atom n in the location of the atom m, and N is the total number of atoms in the complex. The indexes α, β, and γ take the values x, y, and z of the Cartesian coordinate system (here the summation over the Greek repeated indexes is assumed). The tensor $T_{\beta\gamma}^{mn}$ has the form:

$$T_{\beta\gamma}^{mn} = \frac{1}{\left(r^{mn} \right)^5} \left(3 r_\beta^{mn} r_\gamma^{mn} - \left(r^{mn} \right)^2 \delta_{\beta\gamma} \right), \tag{43}$$

where r^{mn} is the distance between the atoms m and n and r_β^{mn} are the components of the vector $\mathbf{r}^{mn}$. Within the modified DID model the components of the tensor $T_{\beta\gamma}^{mn}$ are equal to zero if the atoms m and n belong to the same molecule. Assuming that all the atoms are in a uniform external field ($E_\beta^m = E_\beta$), we have the following expression for the total induced dipole moment of the complex considered:

$$\mu_\alpha = \sum_{m=1}^N \mu_\alpha^m = \alpha_{\alpha\beta} E_\beta, \tag{44}$$

where $\alpha_{\alpha\beta}$ is the polarizability tensor of the complex. Within the modified DID model the exact expressions for $\alpha_{\alpha\beta}$ may be written as in[53,77], however, these are useful for numerical calculations. The exact analytical expressions for $\alpha_{\alpha\beta}$ may be obtained only for preset configurations of the simplest complexes[78]. To describe the polarizability surface of a complex in an analytical form, the iterative procedure has to

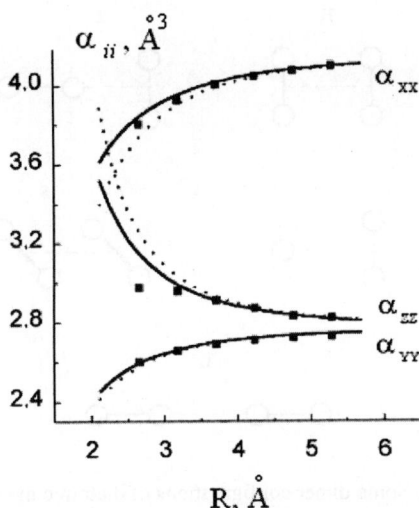

Fig. 12. Tensor components of the static polarizability for the molecular pair N_2-N_2 in the H configuration: modified DID method (solid lines), usual DID method (dotted lines), *ab initio* calculation[79] (squares).

be applied to solve the set of the equations (Eq. (42)). In this case, the polarizability tensor $\alpha_{\alpha\beta}$ of a complex may be written as[76,77]

$$\alpha_{\alpha\beta} = \sum_{m=1}^{N} \alpha_{\alpha\beta}^m + \sum_{m,n=1}^{N} \alpha_{\alpha\delta}^m T_{\delta\gamma}^{mn} \alpha_{\gamma\beta}^n + \sum_{m,n,k=1}^{N} \alpha_{\alpha\delta}^m T_{\delta\gamma}^{mn} \alpha_{\gamma\varepsilon}^n T_{\varepsilon\rho}^{nk} \alpha_{\rho\beta}^k + \cdots . \quad (45)$$

The advantage of the modified DID model is illustrated in Fig. 12. In this figure the polarizability of a pair of interacting N_2 molecules as a result of calculating by *ab initio* method[79], the modified DID model, and the usual DID model in which each interacting molecule is represented as an atom whose polarizability tensor coincides with that of the molecule is given. It is seen that for large intermolecular separations R (R is the distance between the mass centers of interacting molecules) both DID models give the results that are close to *ab initio* calculations. However, for the intermolecular separations $R \approx 3 \div 5$ Å, which correspond to van der Waals X_2-Y_2 complexes, the calculations by modified DID model agree better with the *ab initio* data.

H   T

X   S

L

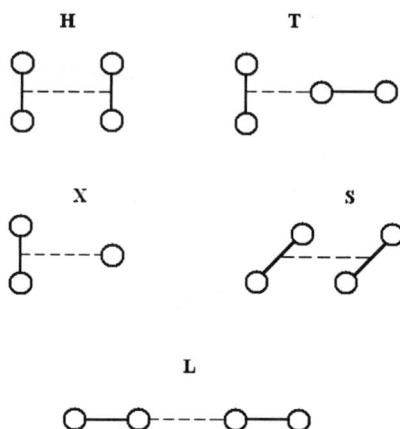

Fig. 13. Some dimer configurations of diatomic molecules.

3.2. Polarizability surface of dimer

Generally, the polarizability tensor of two interacting molecules arbitrarily oriented with respect to each other is a function of Euler angles θ_1, φ_1, χ_1, θ_2, φ_2, and χ_2 of both molecules, the intermolecular separation R, the internuclear separations r_1 and r_2 in the molecules, and the polarizability tensor $\alpha_{ii}(\omega,r)$ of each interacting molecule. Here Euler angles determine the orientation of each molecule with respect to the Cartesian coordinate system XYZ related to both interacting molecules (Z axis passes through the mass center of each molecule). In the case of diatomic molecules, each component of the polarizability tensor of the two interacting molecules represents a surface in the space of variables θ_1, φ_1, θ_2, φ_2, R, r_1, r_2 ($\chi_1 = \chi_2 = 0$), and ω. An analytic equation of this surface in general case is cumbersome one. The components of the polarizability tensor calculated by Eqs. (42) – (45) can be presented in the form:

$$\alpha_{ij} = c_{ij}^{(0)} + \frac{c_{ij}^{(3)}}{R^3} + \frac{c_{ij}^{(5)}}{R^5} + \frac{c_{ij}^{(6)}}{R^6} + \frac{c_{ij}^{(7)}}{R^7} + \frac{c_{ij}^{(8)}}{R^8} + \frac{c_{ij}^{(9)}}{R^9} + \cdots, \qquad (46)$$

Table 5. Polarizability tensor of N_2-N_2 and O_2-O_2 dimers.

Configuration	N_2-N_2			O_2-O_2		
	D_e, cm^{-1}	R_e, Å	$\alpha_{ij}(R_e)$, Å^3	D_e, cm^{-1}	R_e, Å	$\alpha_{ij}(R_e)$, Å^3
T $\theta_1 = 90°$ $\varphi_1 = 0°$ $\theta_2 = 0°$ $\varphi_2 = 0°$	75.97[a]	4.15[a]	$\alpha_{XX} = 3.683$ $\alpha_{YY} = 2.999$ $\alpha_{ZZ} = 3.992$	61.86[b]	4.02[b]	$\alpha_{XX} = 3.492$ $\alpha_{YY} = 3.448$ $\alpha_{ZZ} = 3.782$
X $\theta_1 = 90°$ $\varphi_1 = 0°$ $\theta_2 = 90°$ $\varphi_2 = 90°$	52.01[a]	3.63[a]	$\alpha_{XX} = 3.656$ $\alpha_{YY} = 3.656$ $\alpha_{ZZ} = 3.251$	123.0[b]	3.28[b]	$\alpha_{XX} = 3.450$ $\alpha_{YY} = 3.450$ $\alpha_{ZZ} = 3.655$
H $\theta_1 = 90°$ $\varphi_1 = 0°$ $\theta_2 = 90°$ $\varphi_2 = 0°$	44.30[a]	3.70[a]	$\alpha_{XX} = 4.328$ $\alpha_{YY} = 2.983$ $\alpha_{ZZ} = 3.243$	128.6[b]	3.33[b]	$\alpha_{XX} = 4.444$ $\alpha_{YY} = 2.425$ $\alpha_{ZZ} = 2.652$
L $\theta_1 = 0°$ $\varphi_1 = 0°$ $\theta_2 = 0°$ $\varphi_2 = 0°$	6.83[a]	5.22[a]	$\alpha_{XX} = 3.030$ $\alpha_{YY} = 3.030$ $\alpha_{ZZ} = 4.657$	45.18[b]	4.55[b]	$\alpha_{XX} = 2.459$ $\alpha_{YY} = 2.459$ $\alpha_{ZZ} = 4.968$
S $\theta_1 = 45°$ $\varphi_1 = 0°$ $\theta_2 = 45°$ $\varphi_2 = 0°$	80.79[a]	4.09[a]	$\alpha_{XX} = 3.676$ $\alpha_{YY} = 2.996$ $\alpha_{ZZ} = 3.995$ $\alpha_{XZ} = 0.708$	37.53[b]	4.39[b]	$\alpha_{XX} = 3.510$ $\alpha_{YY} = 2.461$ $\alpha_{ZZ} = 3.726$ $\alpha_{XZ} = 1.087$

[a] Taken from Ref. 80; [b] taken from Ref. 81.

where the coefficients $c_{ij}^{(k)}$ depend on the orientation (θ_1, φ_1, θ_2, φ_2) of two interacting molecules with respect to the *XYZ* coordinate system, on the internuclear separations r_1 and r_2, and the polarizability tensor $\alpha_{ii}(\omega,r)$ of each molecule.

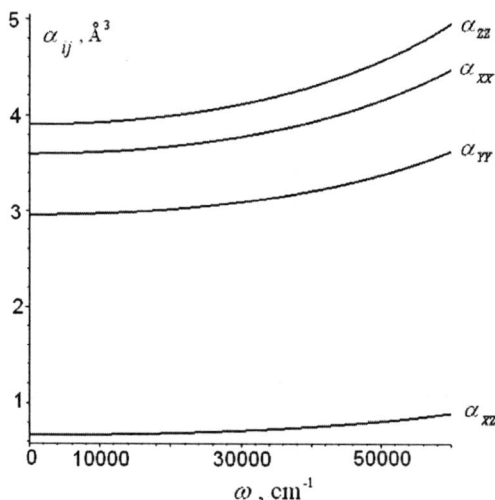

Fig. 14. Dynamic polarizability of N_2-N_2 dimer.

It should be noted that the usual DID model makes it possible to write the components of the polarizability tensor as

$$\alpha_{ij} = c_{ij}^{(0)} + \frac{c_{ij}^{(3)}}{R^3} + \frac{c_{ij}^{(6)}}{R^6} + \frac{\tilde{c}_{ij}^{(9)}}{R^9} + \cdots, \qquad (47)$$

where $c_{ij}^{(0)}$, $c_{ij}^{(3)}$, and $c_{ij}^{(6)}$ coincide with those in Eq. (46) and $\tilde{c}_{ij}^{(9)} \neq c_{ij}^{(9)}$.
Thus, the sizes of interacting molecules being taken into account result in additional terms in Eq. (46) to appear.

Let us consider some particular configurations of N_2-N_2 and O_2-O_2 dimers (Fig. 13). The components of the polarizability tensor of these complexes at $r_1 = r_2 = r_e$, calculated by Eq. (46), are listed in Table 5. This table also contains the dissociation energies D_e and equilibrium intermolecular separations R_e of the dimers, obtained in Refs. 80 and 81 by *ab initio* calculations.

The frequency dependence of the polarizability tensor of the dimer is determined by the polarizability tensors $\alpha_{ii}(\omega, r)$ of the interacting molecules and is calculated by Eq. (46). As an example, the dynamic

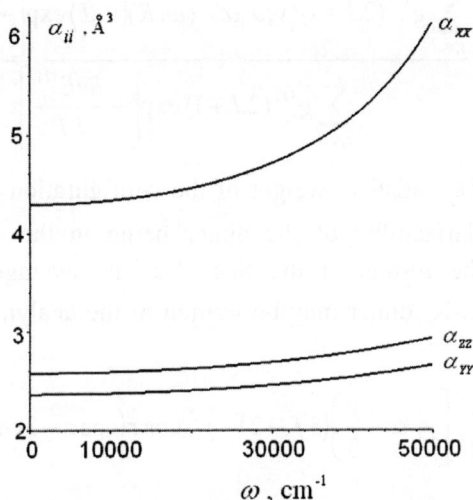

Fig. 15. Dynamic polarizability of O_2-O_2 dimer.

polarizabilities of N_2-N_2 and O_2-O_2 dimers for their most stable configurations are given in Figs. 14 and 15. At the calculation, the *ab initio* dynamic polarizabilities for O_2 and N_2 molecules at $r = r_e$ were taken from Refs. 24 and 25, respectively.

3.3. *Temperature dependence of the polarizability of dimers*

In the real atmosphere the molecular van der Waals complexes of a certain chemical composition are presented by an ensemble of isomers being in the state of dynamic equilibrium with each other. As a result, the complex polarizability, due to statistical averaging over rotational-vibrational states of different complex configurations, is some effective parameter depending on temperature.

Since a significant part of dimers is already in excited rotational-vibrational states of different configurations at low temperatures, the mean values of dimer polarizability invariants are of a practical importance. So, for example, the mean value of the isotropic average polarizability of a dimer may be written as

$$\overline{\alpha}(\omega,T) = \frac{\sum_{i,v,J} g^{(i)}(2J+1)\langle v,J | \alpha^{(i)}(\omega,R) | v,J \rangle \exp\left[-\dfrac{hcE_{vJ}^{(i)}}{kT}\right]}{\sum_{i,v,J} g^{(i)}(2J+1)\exp\left[-\dfrac{hcE_{vJ}^{(i)}}{kT}\right]}, \quad (48)$$

where $g^{(i)}$ is the statistical weight of the configuration i; $\alpha^{(i)}(\omega,R)$ is the average polarizability of the dimer being in the configuration i. Following to the results of the Sec. 3.2, the average polarizability $\alpha^{(i)}(\omega,R)$ of X_2-X_2 dimer may be written in the analytical form (up to R^{-6}):

$$\alpha^{(i)} = 2\alpha + \frac{2}{3R^3}\left[\gamma\left(\alpha - \frac{\gamma}{3}\right)(3f-2) - \gamma^2\cos\tau(\cos\tau - 3\cos\theta_1\cos\theta_2)\right] +$$

$$+ \frac{r^2}{4R^5}\left\{\gamma\left(\alpha - \frac{\gamma}{3}\right)[4\cos\tau(\cos\tau - 10\cos\theta_1\cos\theta_2) + 35f^2 - 40f + 8] +\right.$$

$$+ \gamma^2\cos\tau[(6-15f)\cos\tau - (30-35f)\cos\theta_1\cos\theta_2]\right\} + \frac{1}{R^6}\left[4\alpha^3 - \frac{4}{27}\gamma^3 +\right.$$

$$+ \alpha\gamma\left(\alpha - \frac{\gamma}{3}\right)(3f-2) + \frac{2}{3}\gamma^2\left(2\alpha + \frac{\gamma}{3}\right)(\cos\tau - 3\cos\theta_1\cos\theta_2)^2\right], \quad (49)$$

where

$$\cos\tau = \cos\theta_1\cos\theta_2 + \sin\theta_1\sin\theta_2\cos(\varphi_1 - \varphi_2),$$
$$f = \cos^2\theta_1 + \cos^2\theta_2. \quad (50)$$

Here, $\alpha \equiv \alpha(\omega)$ and $\gamma \equiv \gamma(\omega)$ are the invariants of dynamic polarizability tensor of X_2 molecule at $r = r_e$ (see Eqs. (28) and (29)), and the angles θ_1, φ_1, θ_2, and φ_2 determine the configuration i of the dimer.

The rotational-vibrational energy levels $E_{vJ}^{(i)}$ in Eq. (48) may be written as

$$E_{vJ}^{(i)} = \delta^{(i)} + \sum_{nm} Y_{nm}^{(i)}\left(v + \frac{1}{2}\right)^n J^m(J+1)^m, \quad (51)$$

where $\delta^{(i)} = D_e^{(0)} - D_e^{(i)}$ is the difference in the dissociation energies between the most stable and ith dimer configurations. To calculate the wavefunctions $|v,J\rangle$ and the energies $E_{v,J}^{(i)}$ in Eqs. (48) and (51) the X_2-X_2 dimer is considered as a diatomic homonuclear molecule. To simplify the calculation we simulate the potential energy function of the dimer by the 6:12 Lennard-Jones potential depending only on the two parameters: equilibrium separation $R_e^{(i)}$ and the dissociation energy $D_e^{(i)}$. Then, within this model the molecular constants for each configuration of the dimer may be determined: frequency of harmonic vibrations $\omega_e^{(i)}$, rotational constant $B_e^{(i)}$, and Dunham anharmonic constants a_j, which, in turn, are used to calculate spectroscopic constants $Y_{nm}^{(i)}$, rotational-vibrational energy levels $E_{v,J}^{(i)}$, and matrix elements of polarizability. The matrix elements of polarizability are determined by Eq. (30). The needed polarizability derivatives are determined by expansion of $\alpha^{(i)}(\omega,R)$ into Taylor's series in terms of $x^{(i)} = \left(R - R_e^{(i)}\right)\!\big/R_e^{(i)}$ at $x^{(i)} = 0$.

Calculation of mean value of the isotropic average polarizability of a dimer requires the summation limits in Eq. (48) to be determined. To find the maximum values of the quantum numbers v_{max} and J_{max} (for each $v \le v_{max}$) for different dimer configurations let us use by the technique described in Ref. 82.

The calculated temperature dependence of $\bar{\alpha}(0,T)$ of N_2-N_2 and O_2-O_2 dimers is depicted in Figs. (16) and (17). To calculate $\bar{\alpha}(0,T)$ for each dimer, 169 configurations obtained by changing Euler angles θ_1, φ_1, θ_2, and φ_2 with the step of 45° were taken into consideration. Since only 19 of them are physically different, just those were used in calculations, while the contribution from the others was taken into account by the statistical weights $g^{(i)}$. The values of the parameters $R_e^{(i)}$ and $D_e^{(i)}$ of some configurations are given in Table 5. The lacking values of $R_e^{(i)}$ and $D_e^{(i)}$ for other dimer configurations were estimated by linear interpolation.

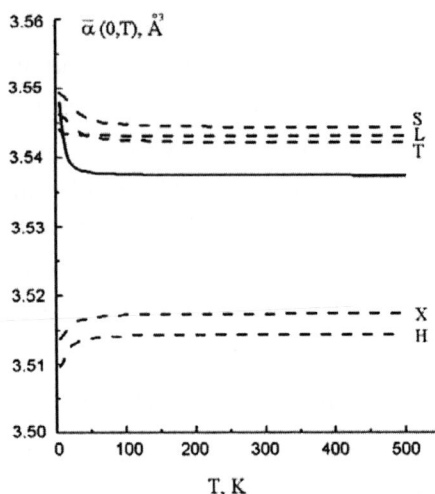

Fig. 16. Mean polarizabilities $\bar{\alpha}(0,T)$ (solid line) and $\bar{\alpha}^{(i)}(0,T)$ (dashed lines) of N_2-N_2 dimer.

For analysis of the results obtained, the mean polarizabilities $\bar{\alpha}^{(i)}(0,T)$ for some dimer configurations are given in Figs. (16) and (17) too. The mean polarizabilities $\bar{\alpha}^{(i)}(0,T)$ are calculated by Eq. (48) without summation over i. Analysis of the temperature dependences leads to the following conclusions[83]:

– the functions $\bar{\alpha}(0,T)$ and $\bar{\alpha}^{(i)}(0,T)$ change significantly at rather low temperatures ($T < 100$ K);

– at $T \to 0$ K $\bar{\alpha}(0,T)$ tends to $\bar{\alpha}^{(i)}(0,T)$, where the index i corresponds to the most stable dimer configuration;

– at $T > 100$ K $\bar{\alpha}(0,T)$ weakly depends on temperature and tends to the mean polarizability of dimer formed by molecules freely rotating around each other. The mean polarizability of such dimer is practically not depend on the temperature and may be calculated by the following equation:

$$\tilde{\alpha}(\omega) = 2\alpha + \frac{4\alpha}{\tilde{R}^6}\left(\alpha^2 + \frac{2}{9}\gamma^2\right) + \frac{5r_e^2}{\tilde{R}^8}\left(\alpha^2 + \frac{2}{9}\gamma^2\right)\left(\alpha + \frac{1}{15}\gamma\right) + \cdots, (52)$$

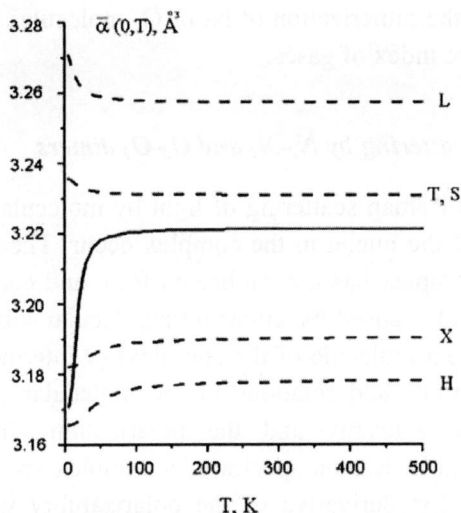

Fig. 17. Mean polarizabilities $\bar{\alpha}(0,T)$ (solid line) and $\bar{\alpha}^{(i)}(0,T)$ (dashed lines) of O_2-O_2 dimer.

where $\tilde{R}$ is an effective intermolecular separation of dimer.

The behavior of the temperature dependences of $\bar{\alpha}(0,T)$ for N_2-N_2 and O_2-O_2 dimers may be explained by the following reasons. At very low temperatures, most of N_2-N_2 and O_2-O_2 dimers are in the lowest rotational-vibrational states of the most stable dimer configurations: S for N_2-N_2 and H for O_2-O_2. As the temperatures increases, other dimer configurations having significantly different polarizabilities appear. This leads to a considerable change in the temperature dependence of $\bar{\alpha}(0,T)$. The character of this change is different for N_2-N_2 and O_2-O_2 dimers because of the different distributions of dimer configurations over the parameter $\delta^{(i)}$. As the temperature reaches $T \approx 100$ K, almost all rotational-vibrational states of all configurations are populated and the further elevation of temperature does not lead to redistribution of dimers over configurations. At such temperatures, the molecules in most dimers are free oriented.

Note that the mean polarizability $\tilde{\alpha}(\omega)$ of a dimer is more than the sum of the average polarizabilities of two noninteracting molecules (see

Eq. (52)). So, the dimerization of N_2 or O_2 molecules leads to increasing of the refractive index of gases.

3.4. *Raman scattering by N_2-N_2 and O_2-O_2 dimers*

Vibrational Raman scattering of light by molecular complex appears if vibrations of the nuclei in the complex occur. The vibrational Raman spectrum of complex has a complicated form and consists of individual vibrational bands caused by either intramolecular vibrations (vibrations of nuclei inside a molecule of the complex) or intermolecular vibrations (relative vibrations and rotations of the molecular components of the complex). The intensity and the polarization characteristics of a vibrational band in Raman spectra of a complex are determined by the tensor of the first derivative of the polarizability with respect to the corresponding coordinate.

The Raman spectrum of N_2-N_2 or O_2-O_2 dimer consists of two groups of vibrational bands: the first group includes two bands caused by the intramolecular vibrations of the nuclei in the molecules of the dimer, while the second one includes four (or five, in the case of linear dimer) bands associated with relative vibrations and rotations of the molecules in the dimer. In the Raman spectrum the vibrational bands of the first group coincide practically with the vibrational bands of free molecules. The Raman vibrational bands of the second group are shifted far from the Raman vibrational bands of the first group and are not considered in this section.

Consider a dimer free oriented in space. Assume that the exciting laser radiation is linearly polarized and the scattered light is observed at the angle of 90° with respect to the electric vector of the exciting radiation. Because the vibrational bands of the first group are almost unresolved, the differential Raman cross section $\sigma_{0 \to 1}$ of the dimer is the sum of the differential Raman cross sections $\sigma_{0 \to 1}^{(k)}$ (here, the index k designates the number of a molecule in the dimer) for these bands:

$$\sigma_{0 \to 1} = \sigma_{0 \to 1}^{(1)} + \sigma_{0 \to 1}^{(2)} . \tag{53}$$

In the harmonic approximation the differential Raman cross section $\sigma_{0\to1}^{(k)}$ may be written as[68]

$$\sigma_{0\to1}^{(k)} = 16\pi^4(\omega - \Delta\omega_{0\to1})^4\left(\frac{B_e}{\omega_e}\right)\left[(\alpha_k')^2 + \frac{7}{45}(\gamma_k')^2\right]. \qquad (54)$$

The degree of depolarization of the Raman scattering light for the dimer may be presented in the form:

$$\rho_{0\to1} = \frac{\sigma_{0\to1}^{\perp(1)} + \sigma_{0\to1}^{\perp(2)}}{\sigma_{0\to1}^{\parallel(1)} + \sigma_{0\to1}^{\parallel(2)}}, \qquad (55)$$

where

$$\sigma_{0\to1}^{\perp(k)} = 16\pi^4(\omega - \Delta\omega_{0\to1})^4\left(\frac{B_e}{\omega_e}\right)\frac{1}{15}(\gamma_k')^2, \qquad (56)$$

and

$$\sigma_{0\to1}^{\parallel(k)} = 16\pi^4(\omega - \Delta\omega_{0\to1})^4\left(\frac{B_e}{\omega_e}\right)\left[(\alpha_k')^2 + \frac{4}{45}(\gamma_k')^2\right], \qquad (57)$$

where $\sigma_{0\to1}$ and $\rho_{0\to1}$ in Eqs. (53) and (55) depend on the dimer configuration (θ_1, φ_1, θ_2, φ_2), the intermolecular separation R, and the frequency ω.

The invariants of the tensor of the first derivative of the dimer polarizability with respect to the vibrational coordinate of kth molecule are expressed through the tensor components $(\alpha_{ij}')_k$ as follows:

$$\alpha_k' = \frac{1}{3}\left[(\alpha_{xx}')_k + (\alpha_{yy}')_k + (\alpha_{zz}')_k\right] \qquad (58)$$

and

$$(\gamma_k')^2 = \frac{1}{2}\left\{\left[(\alpha_{zz}')_k - (\alpha_{xx}')_k\right]^2 + \left[(\alpha_{zz}')_k - (\alpha_{yy}')_k\right]^2 + \left[(\alpha_{yy}')_k - (\alpha_{xx}')_k\right]^2 + \right.$$
$$\left. + 6\left[(\alpha_{xz}')_k^2 + (\alpha_{xy}')_k^2 + (\alpha_{yz}')_k^2\right]\right\}. \qquad (59)$$

Table 6. Invariants $\alpha'(R_e)$ and $\alpha'(R_e)$ for N_2-N_2 and O_2-O_2 dimers at λ =488 nm.

Configu-ration	N_2-N_2		O_2-O_2	
	$\alpha'(R_e)$, Å^3	$(\gamma'(R_e))^2$, Å^6	$\alpha'(R_e)$, Å^3	$(\gamma'(R_e))^2$, Å^6
T^a	1.849	4.286	1.739	9.240
T^b	1.935	7.172	1.858	13.301
X	1.819	4.270	1.700	9.099
H	1.808	3.999	1.666	8.228
L	1.935	6.524	1.939	14.647
S	1.883	5.325	1.778	10.529

[a] Differentiation on ξ_1 (left molecule in Fig.13);
[b] Differentiation on ξ_2 (right molecule in Fig.13).

The components of the tensor of the first derivative of the dimer polarizability $(\alpha'_{ij})_k$ may be obtained from Eq. (46) by differentiation with respect to $\xi_k = (r_k - r_{ek})/r_{ek}$:

$$(\alpha'_{ij}(\omega, \theta_1, \varphi_1, \theta_2, \varphi_2, r_{e1}, r_{e2}, R))_k = (\partial \alpha_{ij} / \partial \xi_k)_{r_k = r_{ek}}, \qquad (60)$$

where r_{e1} and r_{e2} are the equilibrium internuclear separations of the first (left in Fig. 13) and second molecules.

The invariants $\alpha'(R_e)$ and $(\gamma'(R_e))^2$ of N_2-N_2 and O_2-O_2 dimers calculated by Eqs. (58) – (60) are given in Table 6. For H, X, S, and L configurations the invariants $\alpha'(R_e)$ and $(\gamma'(R_e))^2$ obtained by differentiation with respect to ξ_1 and ξ_2 have the same values. For T configuration the values of these derivatives with respect to ξ_1 and ξ_2 differ from each other because of nonequivalent positions of the first and second molecule in dimer and are given in Table 6 both. The values of $\alpha'(R_e)$ and $(\gamma'(R_e))^2$ given in Table 6 differ noticeably from those of ($\alpha'(R_e)$ = 1.859 Å^3 and $(\gamma'(R_e))^2$ = 4.973 Å^6 for N_2; $\alpha'(R_e)$ = 1.759 Å^3 and $(\gamma'(R_e))^2$ = 10.176 Å^6 for O_2) of free N_2 and O_2 molecules. It is also seen, the variations of $(\gamma'(R_e))^2$ for different dimer configurations are more significant than ones of $\alpha'(R_e)$.

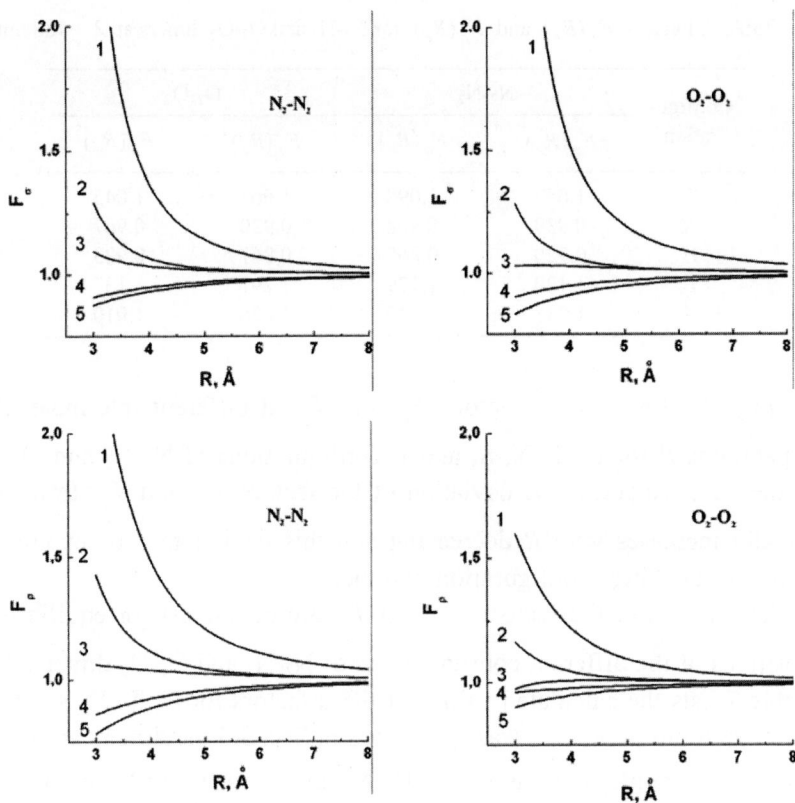

Fig. 18. Factors F_σ and F_ρ for N_2-N_2 and O_2-O_2 dimers at λ = 488 nm.

It is convenient to analyze the Raman scattering parameters of dimers $\sigma_{0\to1}$ and $\rho_{0\to1}$ by comparing them with the similar Raman parameters of free molecules, $\sigma_{0\to1}^0$ and $\rho_{0\to1}^0$. For this purpose let us introduce the factor of Raman cross section

$$F_\sigma = \frac{\sigma_{0\to1}}{2\sigma_{0\to1}^0} \qquad (61)$$

and the factor of depolarization degree of the Raman scattering light

$$F_\rho = \frac{\rho_{0\to1}}{\rho_{0\to1}^0}. \qquad (62)$$

Table 7. Factors $F_\sigma(R_e)$ and $F_\rho(R_e)$ for N_2-N_2 and O_2-O_2 dimers at $\lambda =488$ nm[84].

| Configu- | N_2-N_2 | | O_2-O_2 | |
ration	$F_\sigma(R_e)$	$F_\rho(R_e)$	$F_\sigma(R_e)$	$F_\rho(R_e)$
T	1.057	1.098	1.067	1.045
X	0.939	0.908	0.920	0.967
H	0.920	0.865	0.867	0.923
L	1.120	1.179	1.291	1.137
S	1.033	1.039	1.026	1.010

Figure 18 shows the factors F_σ and F_ρ at different intermolecular separations R for H, T, X, S, and L configurations of N_2-N_2 and O_2-O_2 dimers. As expected, the deviation of the factors F_σ and F_ρ from one usually increases with R decreasing and this deviation is of maximum value for the linear configuration of dimer.

The values of the factors F_σ and F_ρ are of interest for equilibrium positions of the different configurations of N_2-N_2 and O_2-O_2 dimers. The Table 7 lists the calculated values of these factors for H, T, X, S, and L configurations at R_e. As was mentioned in Sec. 3.3 the dimers are in the most stable configurations (S for N_2-N_2 and H for O_2-O_2) only at very low temperatures. Note that the factors $F_\sigma(R_e)$ and $F_\rho(R_e)$ for these low temperatures are more than 1 for N_2-N_2 dimer and less than 1 for O_2-O_2 dimer. As the gas temperature increases, the factors need to be averaged over the dimer configurations. The temperature dependent behavior of these factors is similar to that shown in Sec. 3.3 for the mean polarizability of dimers. At very low temperatures the factors $F_\sigma(R_e)$ and $F_\rho(R_e)$ coincide with the same ones for the most stable dimer configurations, and as gas temperature increases, they tend to the factors for the dimers, in which molecules are practically free oriented.

4. Conclusion

In this chapter the methods of describing the polarizability functions of diatomic molecules and their van der Waals dimers are presented. As the physical properties of these objects considerably differ, the descriptions of their polarizability functions are essentially various.

To describe the polarizability function of diatomic homonuclear molecules the semiempirical method is used here. This semiempirical method allows obtaining in analytical form the polarizability functions in the whole range of internuclear separations. The polarizability functions obtained within the method have correct asymptotic forms as for large so for small internuclear separations. The method takes into account the dispersion, multipole and exchange interactions between the atoms of a molecule that makes it possible to generate the correct idea of physical mechanisms that form the polarizability functions of diatomic molecules with different internuclear separations. Note that this method may be applied to heteronuclear diatomic molecules after simple modification too.

The large intermolecular separations in van der Waals dimers allow using the classical electrostatic theory to describe the polarizability functions of the dimers. The polarizability functions calculated within DID model modified by us take into account sizes of interacting molecules. As a result, nonadditive parts of the polarizability functions from interaction of the molecules in a dimer may be changed considerably. These changes for nondiagonal elements of polarizability tensor exceed ones of its diagonal elements.

The analytical methods to calculate the polarizability functions of molecules and the molecular van der Waals comlexes are not to be considered as the alternative methods to *ab initio* calculations. We assume that the analytical methods along with *ab initio* calculations give more deep insight into the polarizability functions of molecular systems.

References

1. M. A. Morrison and P. J. Hay, *J. Phys.* **B10**, 647 (1977).
2. M. A. Morrison and P. J. Hay, *J. Chem. Phys.* **70**, 4034 (1979).
3. E. N. Svendsen and J. Oddershede, *J. Chem. Phys.* **71**, 3000 (1979).
4. E. N. Svendsen and J. Oddershede, *Chem. Phys.* **64**, 359 (1982).
5. R. D. Amos, *J. Phys.* **B12**, 315 (1979).
6. R. D. Amos, *Mol. Phys.* **39**, 1 (1980).
7. T. Inoue, S. Matsushima and S. Iwata, *Mol. Phys.* **56**, 1097 (1985).

8. S. R. Langhoff, C. W. Bauschlicher Ir. and D. P. Chong, *J. Chem. Phys.* **78**, 5287 (1983).
9. H. Hettema, P. E. S. Wormer, P. Jørgensen, H. J. Aa. Jensen and T. Heldaker, *J. Chem. Phys.* **100**, 1297 (1994).
10. P. A. Hyams, J. Cerratt, D. L. Cooper and M. Raimondi, *J. Chem. Phys.* **100**, 4417 (1994).
11. F. A. Gianturco and U. T. Lamanna, *J. Phys.* **B12**, 2789 (1979).
12. B. O. Roos and A. J. Sadley, *J. Chem. Phys.* **76**, 5444 (1982).
13. A. Hinchliffe, *Chem. Phys. Lett.* **70**, 610 (1980).
14. D. A. Dixon, R. A. Eades and D. G. Truchlar, *J. Phys.* **B12**, 2741 (1979).
15. W. Muller and W. Meyer, *J. Chem. Phys.* **85**, 953 (1986).
16. R. D. Amos, *Chem. Phys. Lett.* **70**, 613 (1980).
17. G. Maroulis, *Mol. Phys.* **77**, 1085 (1992).
18. G. Maroulis, *J. Phys. Chem.* **100**, 13466 (1996).
19. G. Maroulis and C. Makris, *Mol. Phys.* **91**, 333 (1997).
20. G. Maroulis, *J. Chem. Phys.* **108**, 5432 (1998).
21. G. Maroulis, C. Makris, D. Xenides and P. Karamanis, *Mol. Phys.* **98**, 481 (2000).
22. G. Maroulis, *J. Chem. Phys.* **118**, 2673 (2003).
23. C. Jamorski, M.E. Casida and D. R. Salahub, *J. Chem. Phys.* **104**, 5134 (1996).
24. D. Spelsberg and W. Meyer, *J. Chem. Phys.* **109**, 9802 (1998).
25. D. Spelsberg and W. Meyer, *J. Chem. Phys.* **111**, 9618 (1999).
26. W. Kolos and L. Wolniewicz, *J. Chem. Phys.* **46**, 1426 (1967).
27. J. Ruchlewski, *Mol. Phys.* **41**, 833 (1980).
28. J. Rychlevski, *J. Chem. Phys.* **78**, 7252 (1983).
29. B. Fernandez, C. Hatting, H. Koch and A. Rizzo, *J. Chem. Phys.* **110**, 2872 (1999).
30. G. Maroulis, *J. Phys. Chem.* **104**, 4772 (2000).
31. G. Maroulis and A. Haskopoulos, *Chem Phys. Lett.* **358**, 64 (2002).
32. M. A. Buldakov, N. F. Vasiliev and I. I. Matrosov, *Opt. Spektrosk. (Russia)* **75**, 597 (1993).
33. L. Silberstein, *Phil. Mag.* **33**, 92 (1917).
34. L. Silberstein, *Phil. Mag.* **33**, 521 (1917).
35. A. D. Buckingham, *Trans. Faraday Soc.* **52**, 1035 (1956).
36. A. D. Buckingham and K. L. Clarke, *Chem. Phys. Lett.* **57**, 321 (1978).
37. K. L. C. Hunt, B. A. Zilles and J. E. Bohr, *J. Chem. Phys.* **75**, 3079 (1981).
38. M. A. Buldakov, *Atmos. Oceanic Opt.* **15**, 752 (2002).
39. M. A. Buldakov and V. N. Cherepanov, *Atmos. Oceanic Opt.* **16**, 928 (2003).
40. M. A. Buldakov and V. N. Cherepanov, *J. Comp. Meth. Sci. Eng.* **5**, 237 (2004).
41. A. Temkin, *Phys. Rev.* **A17**, 1232 (1978).
42. A. J. C. Varandas and S. P. J Rodrigues, *Chem. Phys. Lett.* **245**, 66 (1995).
43. M. A. Buldakov and I. I. Matrosov, *Optics and Spectroscopy* **78**, 20 (1995).
44. M. A. Buldakov and V. N. Cherepanov, *Atmos. Oceanic Opt.* (in press).

45. A. van der Avoird, P. E. S. Wormer and R. Moszynski, *Chem. Rev.* **94**, 1931 (1994).
46. P. E. S. Wormer and A. van der Avoird, *Chem. Rev.* **100**, 4109 (2000).
47. W. A. Bingel, *J. Chem. Phys.* **30**, 1250 (1959).
48. W. A. Bingel, *J. Chem. Phys.* **49**, 1931 (1968).
49. W. Byers-Brown and J. D. Power, *Proc. Roy. Soc. Lond.* **A317**, 545 (1970).
50. A. Goyette and A. Navon, *Phys. Rev.* **B13**, 4320 (1976).
51. J. L. Godet and B. Dumon, *Phys. Rev.* **A46**, 5680 (1992).
52. H. B. Levine and D. A. McQuarrie, *J. Chem. Phys.* **49**, 4181(1968).
53. J. Applequist, J. R. Carl and Fung Kwok-Keung, *J. Am. Chem. Soc.* **94**, 2952 (1972).
54. B. M. Smirnov, *Asymptotic methods in the theory of atomic collisions* (Atomizdat, Moscow, 1973).
55. E. E. Nikitin and B.M. Smirnov, *Atomic and molecular processes* (Nauka, Moscow, 1988).
56. M. A. Buldakov, I. I. Ippolitov, B. V. Korolev, I. I. Matrosov, A. E. Cheglokov, V. N. Cherepanov, Yu. S. Makushkin and O. N. Ulenikov, *Spectrochimica Acta.* **A52**, 995 (1996).
57. A. A. Radtsig and B. M. Smirnov, *Reference book on atomic and molecular physics* (Atomizdat, Moscow, 1980).
58. K. Andersson and A. J. Sadley, *Phys. Rev.* **A46**, 2356 (1992).
59. L. Pauling and P.Ppauling, *Chemistry* (Freeman and Company, San Francisco, 1975).
60. A. K. Das and A. J. Thakkar, *J. Phys.* **B31**, 2215 (1998).
61. G. D. Zeiss and W. J. Meath, *Mol. Phys.* **33**, 1155 (1977).
62. L. Wolniewicz, *J. Chem. Phys.* **103**, 1792 (1995).
63. L. Wolniewicz and K. Dressler, *J. Chem. Phys.* **88**, 3681 (1988).
64. J. S. Wright and R. J. Buenker, *J. Chem. Phys.* **83**, 4059 (1985).
65. D. Speisberg and W. Meyer, *J. Chem. Phys.* **115**, 6438 (2001).
66. P. K. Mukherjee and K. Ohno, *Phys. Rev.* **A40**, 1753 (1989).
67. H. Hettema and P. E. S. Wormer, *J. Chem. Phys.* **93**, 3389 (1990).
68. M. A. Buldakov, V. N. Cherepanov, B. V. Korolev and I. I. Matrosov, *J. Mol. Spectrosc.* **217**, 1 (2003).
69. K. Kerl, U. Hohm and H. Varchmin, *Ber. Bunsenges. Phys. Chem.* **96**, 728 (1992).
70. M. A. Buldakov, I. I. Matrosov and V. N. Cherepanov, *Optics and Spectroscopy* **89**, 37 (2000).
71. U. Hohm and K. Kerl, *Mol. Phys.* **69**, 803 (1990).
72. U. Hohm and K. Kerl, *Mol. Phys.* **69**, 819 (1990).
73. A. C. Newell and R. C. Baird, *J. Appl. Phys.* **36**, 3751 (1965).
74. H. W. Schrötter and H. W. Klöckner, in *Raman spectroscopy of Gases and Liquids*, Ed. A. Weber (Springer-Verlag, Berlin-Heidelberg-New York, 1979), p. 123.

75. M. A. Buldakov, I. I. Matrosov and T. N. Popova, *Opt. Spektrosk. (Russia)* **47**, 87 (1979).
76. M. A. Buldakov, B. V. Korolev, I. I. Matrosov and V. N. Cherepanov, *Optics and Spectroscopy* **94**, 185 (2003).
77. C. Domene, P. W. Fowler, P. Jemmer and P. A. Madden, *Mol. Phys.* **98**, 1391 (2000).
78. M. A. Buldakov, V. N. Cherepanov and N. S. Nagornova, *Atmos. Oceanic Opt.(Russia)* **18**, 17 (2005).
79. D. G. Bounds, A. Hinchliffe and C. J. Spicer, *Mol. Phys.* **42**, 73 (1981).
80. Akira Wada, Hideto Kanamori, and Suehiro Iwata, *J. Chem. Phys.* **109**, 9434 (1998).
81. B. Bussery and P. E. S. Wormer, *J. Chem. Phys.* **99**, 1230 (1993).
82. A. A. Vigasin, *J. Mol. Spectrosc.* **205**, 9 (2001).
83. M. A. Buldakov, *Atmos. Oceanic Opt.* **15**, 757 (2002).
84. M. A. Buldakov and V. N. Cherepanov, *Atmos. Oceanic Opt.* **17**, 256 (2004).

CHAPTER 14

ATOMIC POLARIZABILITIES AND HYPERPOLARIZABILITIES: A CRITICAL COMPILATION

Ajit J. Thakkar and Concetta Lupinetti

Department of Chemistry, University of New Brunswick, Fredericton, NB E3B 6E2, Canada

Currently available theoretical and experimental estimates of the static dipole polarizabilities (α), second dipole hyperpolarizabilities (γ), quadrupole polarizabilities (α_q), and dipole-dipole-quadrupole hyperpolarizabilities (B) of the ground states of the atoms are critically reviewed. Recommended values are presented in tabular form.

1. Introduction

Theoretical treatments of optical properties of materials, interatomic potentials, electron-atom scattering, and collision-induced spectral shifts all use polarizabilities[1–7] and hyperpolarizabilities.[3,4,8]

It is useful to recall some basic definitions and establish notation. The change in energy of a neutral S-state atom upon introduction of a static, axially-symmetric field F_z, with gradient F_{zz}, is given[3] by

$$\Delta E = -\frac{1}{2}\bar{\alpha}F_z^2 - \frac{1}{8}\bar{\alpha}_q F_{zz}^2 - \frac{1}{4}\bar{B}F_z^2 F_{zz} - \frac{1}{24}\bar{\gamma}F_z^4 \ldots \qquad (1)$$

in which $\bar{\alpha}$ is the dipole polarizability, $\bar{\alpha}_q$ is the quadrupole polarizability, $\bar{B}$ is the dipole-dipole-quadrupole hyperpolarizability, and $\bar{\gamma}$ is the second dipole hyperpolarizability. Note that our $\bar{\alpha}_q$ is identical to Dalgarno's definition[1] but $\bar{\alpha}_q = 2C$ where C is the quadrupole polarizability defined by Buckingham.[3]

An electric field lifts the degeneracy of multiplet states with total and azimuthal angular momentum quantum numbers L and M respectively. States with the same L but different $|M|$ are no longer degenerate and have different polarizability tensors. Moreover, the α tensor is not spherically symmetric when the atom is in a specified L, M state with $L > 0$. There are two independent polarizability components $\alpha_{xx}(L,M) = \alpha_{yy}(L,M)$ and $\alpha_{zz}(L,M)$ in a coordinate system in which the field direction coincides with the atom's axis of high symmetry. Mean or scalar polarizabilities are defined by rotational averages over states with different M:

$$\bar{\alpha} = (2M+1)^{-1} \sum_{M=-L}^{L} \alpha_{aa}(L,M) \tag{2}$$

where a is one of x, y or z. The polarizability anisotropy can be defined by

$$\delta\alpha(L,M) = \alpha_{zz}(L,M) - \alpha_{xx}(L,M) \tag{3}$$

Some authors refer to $\frac{1}{3}\delta\alpha$ as the tensor part of the polarizability.

For P states $(L = 1)$, the polarizability for $M = 0$ determines the polarizability for the $M = \pm 1$ states as follows

$$\alpha_{xx}(1,\pm 1) = \alpha_{yy}(1,\pm 1) = \alpha_{zz}(1,0) \tag{4}$$

$$\alpha_{zz}(1,\pm 1) = \alpha_{xx}(1,0) = \alpha_{yy}(1,0) \tag{5}$$

Hence only two numbers are required to specify the polarizability tensor for an atomic P state. It is not uncommon to use $\alpha_0 \equiv \alpha_{zz}(1,0)$ and $\alpha_{\pm 1} \equiv \alpha_{zz}(1,\pm 1)$. The mean polarizability is then

$$\bar{\alpha} = \frac{1}{3}\left(\alpha_0 + 2\alpha_{\pm 1}\right) \tag{6}$$

as follows from Eq. (2) for any choice of a because of Eqs. (4)–(5). A single number specifies the anisotropy as follows

$$\Delta\alpha \equiv \alpha_{\pm 1} - \alpha_0 = \delta\alpha(1,\pm 1) = -\delta\alpha(1,0) \tag{7}$$

We use $\bar{\alpha}$ and $\Delta\alpha$ to specify the polarizability tensor for a P-state atom. The reader should be aware that some authors define $\Delta\alpha$ with the opposite sign. Exactly analogous averages and anisotropies are defined for the second dipole hyperpolarizability γ. Even though the bars are superfluous, we use $\bar{\alpha}$ and $\bar{\gamma}$ for S-state atoms for the sake of consistency. Note that there is also a simplifying result[2] for D states:

$$\alpha_{zz}(2,\pm 2) = 4\alpha_{zz}(2,\pm 1) - 3\alpha_{zz}(2,0) \tag{8}$$

In spite of their importance, values of polarizabilities and hyperpolarizabilities are not well-established for all the atoms. Dipole polarizabilities have been measured for less than 30 ground-state atoms.[5,7] Experimental dipole hyperpolarizabilities are even more scarce; they have been measured only for five ground-state atoms.[8] There is no experimental information on the anisotropy of an atomic polarizability or hyperpolarizability.

There is relatively more theoretical information available about static polarizabilities and hyperpolarizabilities of atoms. This article critically reviews some of the best theoretical calculations. A compilation of recommended values is presented in tabular form with the hope that it will serve as a handy reference. In most cases a single recommended value is given together with subjective error bars determined on the basis of our experience as well as critical analysis of convergence patterns with respect to methodological variables such as basis set and level of electron correlation included. Moreover, a discussion of the choices made in compiling each table is given with references to competitive calculations that we did not use in the tables. The user who is skeptical of any particular choice that we have made is urged to consult these references. Even though the choice of recommended values and estimated errors is a subjective procedure, we think that it will be an aid to "consumers" of this data. The alternative of simply presenting two or more different values for a property would delegate responsibility for making a choice to the user who may or may not be an expert on the computation of polarizabilities. We refer the reader to standard texts[9, 10] for details of computational and theoretical methods. Hartree atomic units are used throughout. The 2002 CODATA recommended values of the fundamental constants[11] lead to the conversion factors listed in Table 1.

Table 1.   Conversion factors

Quantity	Atomic unit	SI equivalent
E	E_h	$4.359\,744\,17 \times 10^{-18}$ J
F	$e^{-1} a_0^{-1} E_h$	$5.142\,206\,43 \times 10^{11}$ V m^{-1}
α	$e^2 a_0^2 E_h^{-1}$	$1.648\,777\,27 \times 10^{-41}$ C^2 m^2 J^{-1}
γ	$e^4 a_0^4 E_h^{-3}$	$6.235\,380\,74 \times 10^{-65}$ C^4 m^4 J^{-3}
α_q	$e^2 a_0^4 E_h^{-1}$	$4.617\,046\,59 \times 10^{-62}$ C^2 m^4 J^{-1}
B	$e^3 a_0^4 E_h^{-2}$	$1.696\,733\,43 \times 10^{-63}$ C^3 m^4 J^{-2}

We do not refer to many older calculations that have been superseded, and others that were carried out to test methods rather than to provide new polarizability values. Although we have taken pains to search the literature thoroughly for this article and have used the results of searches made by the first author and other collaborators in the past, it nevertheless remains possible that we have missed some important contributions. We apologize for such oversights and urge the authors of omitted references to provide us with the pertinent information so that we may not make the same mistakes in future compilations.

2. Dipole polarizabilities

2.1. *Experiment*

A collection of measured values of the mean dipole polarizability $\bar{\alpha}$ is presented in Table 2. Many other measurements have been reported in the literature for the noble-gas, alkali and alkaline-earth atoms. We have tried to select the most recent and/or most reliable values. The reader should be aware that the literature contains many citations of ostensibly experimental values that, upon closer examination, turn out to be theoretical values taken from one of several well-cited compilations.[5, 12–16] The most striking aspect of Table 2 is that experimental polarizabilities are available only for about a quarter of the periodic table.

Table 2.   Experimental polarizabilities ($\bar{\alpha}$)

Z	Atom	$\bar{\alpha}$
2	^{4}He	1.383223±0.000067[17, 18]
3	Li	164.0±3.4[19]
7	N	7.63±0.40[20]
8	O	5.2±0.4[20]
10	Ne	2.6669±0.0008[21]
11	Na	162.7±0.8[22]
13	Al	45.9±2.0[23, 24]
18	Ar	11.078±0.010,[25] 11.070±0.007[26]
19	K	292.9±6.1[19]
20	Ca	169±17[27]
30	Zn	38.8±0.8[28]
36	Kr	16.782±0.005[21]
37	Rb	319.2±6.1[19]
38	Sr	186±15[29]
48	Cd	49.65±1.62[30]
49	In	68.70±8.10[31]
54	Xe	27.078±0.005[21]
55	Cs	401.0±0.6[32]
56	Ba	268±22[29]
73	Ta	128±20[33]
77	Ir	54±10[33, 34]
80	Hg	33.91±0.34[35]
81	Tl	51±5[a]
92	U	137.0±9.4[36]

[a]Unpublished results of J. A. D. Stockdale, I. Efremov, K. Rubin and B. Bederson as cited by Miller and Bederson[6, 16]

A slightly different value of $\bar{\alpha}$ for He was listed in the original work of Gugan and Michel[17,18] but, as Weinhold[37] pointed out, it was based on an infinite-mass bohr radius rather than a bohr radius appropriate for ^{4}He. The value given in Table 2 is extracted from the experimental molar polarizability[17,18] using the ^{4}He bohr radius.[37] The experimental polarizabilities in Table 2 were obtained via dielectric constant measurements for He,[17,18] Ne,[21] Ar,[25] Kr,[21] and Xe,[21] atomic beam interferometry for Na,[22] a time-of-flight method for Cs,[32] the E–H gradient balance technique for Li,[19] K,[19] Rb,[19] and In,[31] beam deflection for Al,[23,24] Ca,[27] Sr,[29] Ba,[29] In,[31] and Tl,[6,16] refractivity measurements for N,[20] O,[20] Ar,[26] Zn,[28] Cd,[30] and Hg,[35] field ion microscopy for Ta,[34] and Ir,[34] and a light force method for U.[36]

2.2. *Theoretical calculations for light atoms*

We begin with the atoms of the first three periods because accurate polarizability computations have been reported for all of them. Thus, Table 3 lists our choices of the most reliable calculations of the dipole polarizabilities of the atoms from hydrogen ($Z = 1$) to argon ($Z = 18$). Since relativistic corrections are rather small for these light atoms, they have not been reported in all cases. Thus the table includes the non-relativistic (NR) value, the small relativistic correction δ_r when it is known, and the recommended estimate. In the case of He, δ_r is actually the sum of the relativistic, mass-polarization and quantum electrodynamic corrections. Next, we provide a brief discussion for the polarizability of each atom from H to Ar.

As is evident from Table 3, polarizabilities for the few-electron atoms H, He, Li and, to a lesser extent, Be are known quite accurately. Within a year of the birth of quantum mechanics, the exact non-relativistic value of $\bar{\alpha}$ for the hydrogen atom was worked out independently but almost simultaneously by Wentzell,[53] Waller,[38] and Epstein[54] who all used different methods. The relativistic correction was first calculated by Bartlett and Power[39] more than 35 years ago; it has been recalculated with increased accuracy several times[40] and is now known to 15 significant figures or more.

Quantum mechanical calculation of the polarizability of helium began with a simple estimate reported in 1927 by Pauling.[55] Since then, variational perturbation theory with trial functions containing explicit dependence on the interelectronic distance r_{12} has permitted nearly exact calculations of the non-relativistic $\bar{\alpha}$ for the helium atom. The early work of Atanasoff[56] and Hassé[57] culminated in a value accurate to three significant figures.[58] This was subsequently improved to four[59] and six sig-

Table 3. Theoretical polarizabilities ($\bar{\alpha}$)

Z	Atom	$\bar{\alpha}$(NR)	δ_r	$\bar{\alpha}$
1	H	4.5^{38}	$-0.0002485^{39,\,40}$	4.4997515
2	He	$1.383192^{41,\,42}$	-0.000001^{42}	1.383191
3	Li	$164.111^{41,\,43}$	-0.07 ± 0.01^{44}	164.04±0.01
4	Be	37.755^{43}	-0.016 ± 0.003^{45}	37.739±0.03
5	B	20.43^{46}	$\approx 0.00^{47}$	20.43±0.11
6	C	11.67^{46}		11.67±0.07
7	N	7.26^{46}		7.26±0.05
8	O	5.24^{46}		5.24±0.04
9	F	3.70^{46}		3.70±0.03
10	Ne	2.6562^{48}	$+0.0044^{48}$	2.661±0.005
11	Na	163.90^{49}	-1.02^{49}	162.88 ± 0.60^{49}
12	Mg	71.574^{45}	-0.35^{50}	71.22±0.36
13	Al	57.74^{51}	$+0.05^{47}$	57.79±0.30
14	Si	37.17^{51}		37.17±0.21
15	P	24.93^{51}		24.93±0.15
16	S	19.37^{51}		19.37±0.12
17	Cl	14.57^{51}		14.57±0.10
18	Ar	11.065^{51}	$+0.02^{52}$	$11.085±0.06^{51}$

nificant figures.[60] The latest calculations are accurate to more than ten significant figures.[41,42,61,62] Relativistic and mass-polarization corrections were reported first by Weinhold[37] and have been improved recently.[42,61] Curiously enough, the relativistic, mass-polarization and quantum electrodynamic corrections cancel each other out so that the total correction, reported as δ_r in Table 3, is very small.[42]

As in the case of helium, variational perturbation theory with trial functions containing explicit dependence on the interelectronic distances has been the preferred route used to calculate the most accurate values of the non-relativistic dipole polarizability $\bar{\alpha}$ for the lithium atom. The first such calculations were reported by Sims et al.,[63] and improved computations have been reported by several groups.[41,43,64,65] The latest values are accurate to six significant figures. Scalar relativistic corrections and spin-orbit coupling corrections[44] total -0.07 ± 0.01.

Analogous calculations are difficult but possible for the Be atom as well. The first efforts in this direction were made by Sims and Rumble,[66] and much improved calculations were reported by Komasa.[43] We have chosen Komasa's value as the most reliable one. Tunega and coworkers[67] reported a value of 37.73 using both the conventional CCSDT-1a coupled-cluster (CC) method[10] and its CC-R12 variant. In their calculations, an uncontracted (18s13p10d9f8g) basis set of Gaussian-type functions (GTF) was used.[67] Bendazzoli and Monari[68] have reported a full configuration interaction (CI) polarizability of 37.81 using an uncontracted (9s9p5d3f2g) GTF basis set but we think their value is a bit too high. A non-relativistic coupled-cluster calculation[45] of the CCSD(T) variety with a large uncontracted basis set of (14s9p7d6f4g4h3i) GTF led to $\bar{\alpha} = 37.751$ in very close agreement with Komasa's value. Finally, CCSD(T) calculations[45] in an uncontracted (14s9p7d6f4g) GTF basis set led to non-relativistic and 2nd-order Douglas-Kroll[69-71] (DK) scalar relativistic polarizabilities of 37.725 and 37.709 respectively from which $\delta_r = -0.016$ was deduced. Tunega *et al.*[67] suggest a conservative estimate of ± 0.05 for the uncertainty in the polarizability. We are more optimistic and suggest a smaller uncertainty of ± 0.03.

The first systematic sets of polarizability calculations for the second-period atoms were those of Stevens and Billingsley,[72] Werner and Meyer,[73] and Hibbert *et al.*[74] We think the most accurate polarizabilities reported so far for the B, C, N, O and F atoms are the CCSD(T) computations of Das and Thakkar[46] who used optimized basis sets of (14s9p7d6f) GTF contracted to [10s7p7d6f] functions. A similar CCSD(T) value of $\bar{\alpha} = 20.47$ for B was published by Pouchan *et al.*[75] The relativistic correction for $\bar{\alpha}$ was found to be completely negligible, zero to two decimal places,[47] for B and to be[48] +0.0044 for Ne. We estimate that $0 < \delta_r < 0.01$ for the C, N, O and F atoms. The uncertainties shown in Table 3 for B to F are based on an estimate of $\pm 0.5\%$ for the non-relativistic value and ± 0.01 for the relativistic correction. There have been so many computations for Ne that reviewing them all is not worthwhile. Many references to earlier work on Ne can be found elsewhere.[48,76] Table 3 lists the non-relativistic CCSD(T)-R12/B value calculated in an uncontracted (25s19p15d12f10g8h6i4k) GTF basis set,[48] and a δ_r obtained[48] by direct perturbation theory (DPT) at the CCSD(T) level using a q-aug-cc-pCV6Z basis set.

Systematic sets of computed polarizabilities of the third-period atoms were first reported by Reinsch and Meyer,[77] and Hibbert *et al.*[78] Many computed polarizability values have been published for sodium including results from model potentials[79-81] that are particularly suitable for alkali atoms and CCSD(T) results.[44,49,82-84] We have selected the most recent non-relativistic CCSD(T) result of Thakkar and Lupinetti[49] who used a carefully optimized basis set of (20s14p10d8f2g1h1i) GTF contracted to [13s12p10d8f2g1h1i] functions. The choice of δ_r and the error estimate was discussed in detail by them.[49]

We have selected for Table 3 the CCSD(T) polarizabilities of Mg, Al, Si, P, S, Cl and Ar computed by Lupinetti and Thakkar[45,51] who used optimized basis sets of (21s16p7d6f2g) GTF contracted to [13s10p7d6f1g] functions. There are several previous computations for the polarizability of magnesium.[45,50,85–87] CCSD(T) polarizability values have also been published by others for Al,[88] Si,[89] P,[90] and S.[91] For argon, we mention several CCSD(T) studies.[52,92–95] We selected the δ_r obtained with a DK2-CCSD(T) computation by Miadoková *et al.*[50] for Mg, and by Černušák *et al.*[47] for Al, and with a DK3-CCSD(T) calculation by Nakajima and Hirao[52] for Ar. We expect $0.02 < \delta_r < 0.05$ for the Si, P, S, and Cl atoms. Our uncertainty estimates for $\bar{\alpha}$ of the atoms from Mg to Ar have been calculated using $\pm 0.5\%$ for the non-relativistic polarizability and 0.00 ± 0.02 for δ_r of Si to Cl.

Table 2 contains experimental counterparts for eight of the 18 theoretical polarizabilities listed in Table 3. There is satisfactory agreement for seven of the atoms (He, Li, N, O, Ne, Na, Ar) but strong disagreement for Al. As pointed out in previous work,[51,88] the experimental value of $\bar{\alpha}$ for Al is almost certainly in error.

2.3. Reliable theoretical calculations for heavy atoms

Now we move on to the polarizabilities of those heavier atoms with $Z > 18$ for which computations taking adequate account of basis set, electron correlation, and relativistic effects have been published. Table 4 lists our recommended polarizabilities for the heavy elements from the well-studied groups 1, 2, 11, 12, 13, and 18. Relativistic effects are much more important for these atoms than the light ones treated in Section 2.2. All-electron CCSD(T) calculations have been reported for all the elements in Table 4 with inclusion of scalar relativistic effects using the Douglas-Kroll Hamiltonian. Hence Table 4 does not include non-relativistic values.

The polarizabilities of the alkali atoms have been studied intensively. The recommended values selected for Table 4 are the results of all-electron DK2-CCSD(T) calculations.[96] Other reported results of competitive quality include relativistic all-order singles-doubles computations,[100] earlier DK-CCSD(T) values,[44] and pseudopotential (PP) + core-polarization potential (CPP) calculations.[80,81] If non-relativistic polarizabilities are required for methodological purposes, we recommend obtaining them by combining the values in Table 4 with the δ_r corrections from older work.[44]

For the well-studied alkaline-earth atoms of Group 2, Table 4 lists DK2-

Table 4. Theoretical polarizabilities ($\bar{\alpha}$)

Atom	$\bar{\alpha}$	Atom	$\bar{\alpha}$	Atom	$\bar{\alpha}$
Group 1[96]		Group 2[97]		Group 11[98]	
K	291.1±1.5	Ca	157.9±0.8	Cu	46.50±0.35
Rb	316.2±3.2	Sr	199.0±2.0	Ag	52.46±0.52
Cs	396.0±5.9	Ba	273.5±4.1	Au	36.06±0.54
Fr	315.2±6.3	Ra	246.2±4.9		
Group 12[99]		Group 13[47]		Group 18[52]	
Zn	38.35±0.29	Ga	52.91±0.40	Kr	16.80±0.13
Cd	47.55±0.48	In	68.67±0.69	Xe	27.06±0.27
Hg	34.73±0.52	Tl	71.72±1.08	Rn	33.18±0.50

CCSD(T) calculations corrected for spin-orbit effects and basis set incompleteness.[97] Similar values were obtained by a relativistic configuration interaction (CI) + many-body perturbation theory (MBPT) technique.[101] Early PP+CPP calculations should be mentioned as well.[80]

We list in Table 4 the all-electron, spin-adapted DK2-CCSD(T) polarizabilities of Neogrády *et al.*[98] for the coinage-metal atoms of Group 11. Computations using relativistic small-core pseudopotentials have been published as well.[102,103] For the 'battery elements' of group 12, we selected the all-electron DK2-CCSD(T) polarizabilities of Kellö and Sadlej[99] for inclusion in Table 4. Other results to note include CCSD(T) computations using relativistic small-core pseudopotentials.[103,104]

The group 13 heavy atom polarizabilities selected for Table 4 are spin-adapted DK2-CCSD(T) computations in uncontracted GTF basis sets reported by Černušák *et al.*[47] The polarizabilities of the heavy noble-gas atoms have been studied extensively since the first non-relativistic calculations with proper inclusion of electron correlation.[105] The best non-relativistic polarizability for Kr is probably the CCSD(T) computation of Soldán *et al.*[95] We have selected the relativistic DK3-CCSD(T) computations of Nakajima and Hirao[52] for Table 4.

The error estimates in Table 4 are simply ±0.5% for K and Ca, ±0.75% for Cu, Zn, Ga, and Kr, ±1.0% for the fifth-period atoms (Rb, Sr, Ag, Cd, In, and Xe), ±1.5% for the sixth-period atoms (Cs, Ba, Au, Hg, Tl and Rn), and ±2.0% for Fr and Ra.

Table 2 lists experimental values for 13 of these 20 heavy atoms. In 12 of these cases, there is agreement between the theoretical and experimental polarizabilities after the error bars on both values are taken into account. However, there is severe disagreement between theory and experiment for Tl. Moreover, the overlap between the error bars is slim for Cd and Hg.

2.4. Theoretical calculations for the remaining atoms

Finally, we turn to the remaining heavy atoms. In all these cases, the best available calculations lack a completely satisfactory treatment of at least one of the following three aspects: basis set completeness, electron correlation, and relativistic effects.

Electron-correlated $\bar{\alpha}$ values for the fourth-period atoms Ge, As, Se and Br were first reported by Miller and Bederson[5] in an influential 1977 review article in which they stated that these were the preliminary results of PNO-CEPA calculations by E. A. Reinsch and W. Meyer. In 1991, the provenance of these numbers was questioned by Sadlej.[106] However, corrected versions of the numbers were reported again in 1995 by Stiehler and Hinze[107] who cited a 1994 private communication with Reinsch and Meyer.

The non-relativistic PNO-CEPA polarizability for Ge was reported[107] as 40.95. We are unaware of any other electron-correlated or relativistic calculation for Ge. The relativistic corrections δ_r of -0.35 for Ga,[47] and $+0.24$ for Se[108] lead us to estimate that $\delta_r = -0.15 \pm 0.10$ for Ge. Use of this correction leads to the recommended value of 40.80 listed in Table 5.

Table 5. Theoretical polarizabilities ($\bar{\alpha}$)

Z	Atom	$\bar{\alpha}$
32	Ge	40.80 ± 0.82^a
33	As	29.8 ± 0.6^{109}
34	Se	26.24 ± 0.52^{108}
35	Br	21.13 ± 0.42^a
50	Sn	57.30 ± 5.73^{110}
51	Sb	42.2 ± 1.3^{109}
52	Te	38.10 ± 3.81^{110}
53	I	32.98 ± 1.65^a
70	Yb	145.3 ± 4.4^{111}
83	Bi	48.6 ± 1.5^{109}
82	Pb	60.05 ± 6.01^{112}
84	Po	43.62 ± 4.36^{112}
85	At	40.73 ± 2.04^a

^{a}See text for details.

For Se and Br, Medveď *et al.*[108] have reported CASPT2 polarizabilities perturbatively corrected for relativistic effects using the mass-velocity (MV) and Darwin (D) terms. Their non-relativistic results differ from those of Reinsch and Meyer by 2.0% and 1.8% for Se and Br respectively. To decide whether to use the PNO-CEPA or CASPT2 results, we examine results for the isovalent third-period atoms S and Cl. The PNO-CEPA[77] and CASPT2[108] $\bar{\alpha}$ are almost identical and agree to better than 1.1% with CCSD(T) results[51] for S and Cl. However, the CASPT2 $\Delta\alpha$ for S is significantly closer to the CCSD(T) result than its PNO-CEPA counterpart is. For this reason and because the details of the PNO-CEPA calculation are not accessible, we chose to include the relativistically-corrected MVD-CASPT2 polarizabilities[108] in Table 5. Unfortunately, there are inconsistencies for Br. Medved *et al.*'s CASSCF entries[108] indicate that the MVD correction for α_0 is -0.16 but their CASPT2 entries appear to be erroneously based on a correction of $+0.16$. We have corrected this and made corresponding changes to $\bar{\alpha}$ and $\Delta\alpha$. There is also a spin-orbit CI calculation[113] that leads to an M_J averaged value $\alpha_0 = 18.86$ for Br in good agreement with the MVD-CASPT2 value of $\alpha_0 = 18.74$ obtained after correction as described above. Unfortunately, the spin-orbit CI study[113] did not consider any other components and the corresponding $\bar{\alpha}$ cannot be obtained.

In an analogous way, we chose Roos *et al.*'s relativistic DK-CASPT2 value[109] rather than the PNO-CEPA value. Table 5 also lists the DK-CASPT2 polarizabilities reported by Roos *et al.*[109] for Sb and Bi. Table 5 lists the CCSD(T) polarizability for Yb calculated by Wang and Dolg[111] using a relativistic 42-valence-electron pseudopotential for Yb. The uncertainties of the tabulated polarizabilities are estimated as $\pm 2\%$ for the fourth-period atoms Ge, As, Se and Br, and $\pm 3\%$ for Sb, Yb and Bi.

Table 5 lists MVD-HF polarizabilities[110,112] for Sn, Pb, Te and Po. Their uncertainties are estimated at $\pm 10\%$. MVD-HF polarizabilities of 31.87 for I[110] and 37.49 for At[112] are also available. Instead of using the latter, we combined the spin-orbit CI α_0 values[113] of 30.35 for I and 37.28 for At with MVD-HF values of $\Delta\alpha = 3.95$ for I and 5.17 for At to obtain the recommended mean polarizabilities listed in Table 5. Uncertainties of $\pm 5\%$ are estimated for the I and At polarizabilities.

There are no really satisfactory $\bar{\alpha}$ for the first transition series from Sc to Ni. Chandler and Glass[114–116] reported variation-perturbation calculations using small CI wave functions. More recently, numerical HF[107] and modified coupled-pair functional[117] (MCPF) computations[118] have been published. The MCPF calculations used large GTF basis sets and should have led to fairly accurate values. However, comparison of the MCPF $\bar{\alpha}$ for Cu (53.44) and Zn (42.85) with their accurate counterparts in Table 4 suggests that the MCPF values are systematically too large by 10–15%, perhaps because too strong a field was used for the MCPF finite-field computation.[118] Table 6 lists the MCPF $\bar{\alpha}$ with $\pm 15\%$ uncertainties.

Ajit J. Thakkar and Concetta Lupinetti

Table 6.   MCPF^a $\bar{\alpha}$ polarizabilities

Atom	$\bar{\alpha}$	Atom	$\bar{\alpha}$
Sc	142.3±21.3	Mn	75.5±11.3
Ti	114.3±17.2	Fe	63.9±9.6
V	97.3±14.6	Co	57.7±8.7
Cr	94.7±14.2	Ni	51.1±7.7

^{a}From Pou-Amérigo *et al.*[118]

Good $\bar{\alpha}$ values are even more scarce for the rest of the atoms in the periodic table. The best polarizabilities that we can find are based on unpublished calculations made for all the atoms in the periodic table by G. Doolen using a relativistic linear response method.[119] His values were adjusted by Miller[16] to fit accurate experimental values known for some atoms. Miller's recommended values are listed in Table 7 for the reader's convenience, and Miller has assigned uncertainties of ±25% to them. Three values in Tables 5–7 can be compared with their experimental counterparts in Table 2. There is good agreement for Ir, moderate agreement for U, and acceptable overlap between the error bars of the scaled and experimental $\bar{\alpha}$ for Ta.

Table 7.   Scaled $\bar{\alpha}$ polarizabilities^a.

	$\bar{\alpha}$		$\bar{\alpha}$		$\bar{\alpha}$		$\bar{\alpha}$
Y	153	La	210	Lu	148	Ac	217
Zr	121	Ce	200	Hf	109	Th	217
Nb	106	Pr	190	Ta	88	Pa	171
Mo	86	Nd	212	W	75	U	153
Tc	77	Pm	203	Re	65	Np	167
Ru	65	Sm	194	Os	57	Pu	165
Rh	58	Eu	187	Ir	51	Am	157
Pd	32	Gd	159	Pt	44	Cm	155
		Tb	172			Bk	153
		Dy	165			Cf	138
		Ho	159			Es	133
		Er	153			Fm	161
		Tm	147			Md	123
						No	118

^{a}From Miller.[16] Uncertainties are ±25%. A more accurate value of $\bar{\alpha}$ for Yb is given in Table 5.

2.5. *Dipole polarizability anisotropies*

Calculated values of the dipole polarizability anisotropy $\Delta\alpha$ are listed in Table 8 for the atoms in groups 13, 14, 16, and 17 that all have a P ground state. Each $\Delta\alpha$ is obtained from the same sources that the corresponding $\bar{\alpha}$ was obtained for Tables 3–5. The uncertainties are taken as twice the uncertainties for the matching $\bar{\alpha}$ because errors in α_0 and α_1 can be magnified when the latter are subtracted to obtain $\Delta\alpha$. The uncertainties obtained in this way seem reasonable except for some of the heaviest atoms Sn, Te, Po, I, and At where they seem to be too pessimistic.

Table 8.   Theoretical polarizability anisotropies $\Delta\alpha$

Atom	$\Delta\alpha$	Atom	$\Delta\alpha$
Group 13		Group 14	
B	-5.83 ± 0.22[46]	C	$+2.10\pm0.14$[46]
Al	-25.60 ± 0.60[51]	Si	$+8.41\pm0.42$[51]
Ga	-34.66 ± 0.80[47]	Ge	$+13.95\pm1.64$[a]
In	-41.29 ± 1.38[47]	Sn	$+14.21\pm11.46$[110]
Tl	-63.71 ± 2.16[47]	Pb	$+20.46\pm12.02$[112]
Group 16		Group 17	
O	-0.92 ± 0.08[46]	F	$+0.42\pm0.06$[46]
S	-3.63 ± 0.24[51]	Cl	$+1.71\pm0.20$[51]
Se	-7.45 ± 1.04[108]	Br	$+3.59\pm0.84$[a]
Te	-7.26 ± 7.62[110]	I	$+3.95\pm3.30$[110]
Po	-10.98 ± 8.72[112]	At	$+5.17\pm4.08$[112]

[a]See text for details.

3. Second dipole hyperpolarizabilities

3.1. *Experimental hyperpolarizabilities*

Apart from early measurements for Rb vapor near resonance and with an uncertainty of a factor of two or more,[120] experimental hyperpolarizability determinations have been restricted to the noble-gas atoms.[8] Moreover, theoretical values for He have been used to calibrate the most accurate gas-phase hyperpolarizability measurements. Table 9 lists the known experimental static hyperpolarizabilities. Two sets of values are listed for the noble-gas atoms. The more recent data is from the extrapolation to zero frequency of Shelton's electric-field-induced second-harmonic generation (ESHG) measurements.[121, 122] The older $\bar{\gamma}$ values obtained by Buckingham and Dunmur[123] from Kerr effect measurements are also included in Table 9. It is not clear what uncertainties to assign to Shelton's $\bar{\gamma}$ values. We have followed his estimate[122] for Ne. Shelton[121] suggests that the uncertainties for Ar, Kr and Xe could be either $\pm0.5\%$ or $\pm1\%$; we have used uncertainties of $\pm0.5\%$ for Ar, and $\pm1\%$ for Kr and Xe. Rough estimates of $\bar{\gamma} = 1.1 \times 10^5$ and $\bar{\gamma} = 2.2 \times 10^4$ have been obtained for Cd[30] and Hg[35] respectively from refractivity data using some fairly approximate relationships.

Table 9.   Experimental $\bar{\gamma}$ hyperpolarizabilities

	ESHG[121, 122]	Kerr[123]
Ne	108±2	101±8
Ar	1167±6	1171±79
Kr	2600±26	2779±199
Xe	6888±69	7742±596

3.2. Theoretical hyperpolarizabilities for light atoms

Our recommended theoretical $\bar{\gamma}$ and $\Delta\gamma$ hyperpolarizabilities for the light atoms from H to Ar are listed in Table 10. The value for hydrogen is the exact non-relativistic result.[124] An accurate value for He was first obtained by non-relativistic, variation-perturbation computations using wave functions with explicit dependence on the interelectronic separation.[125] Table 10 lists an improved value[61] that should be accurate to all eight significant figures cited.

The calculation of an accurate $\bar{\gamma}$ for the lithium atom is pathologically difficult because of extremely strong electron correlation effects and very stringent basis set requirements arising from the highly diffuse electron density. The first extensive correlated calculations[128] were not even able to decide on the sign of $\bar{\gamma}$. Variation-perturbation computations[64, 65] with Hylleraas-type wave functions, and coupled-cluster calculations[126] validated by extensive testing of convergence with respect to the one-particle basis set have led to the value listed in Table 10. An uncertainty of $\pm10\%$ had been recommended previously[126] but we now think this to be too pessimistic and use an uncertainty of $\pm3\%$.

Coupled-cluster computations of $\bar{\gamma}$ have been reported for all the atoms from Be through Ar. There have been several such calculations[45, 67, 127, 129] of $\bar{\gamma}$ for Be. Tunega *et al.*[67] reported values obtained by the conventional CCSDT-1a coupled-cluster method and its CC-R12 variant. Unlike the case

Table 10. Theoretical $\bar{\gamma}$ and $\Delta\gamma$ hyperpolarizabilities

Atom	$\bar{\gamma}$	$\Delta\gamma$
H[124]	1333.125	
He[61]	43.104227	
Li[64, 65, 126]	$(2.90\pm0.09)\times10^3$	
Be[45, 127]	$(3.09\pm0.03)\times10^4$	
B[46]	$(1.62\pm0.03)\times10^4$	$(-2.86\pm0.06)\times10^4$
C[46]	$(3.48\pm0.07)\times10^3$	$(+3.17\pm0.14)\times10^3$
N[46]	$(9.04\pm0.18)\times10^2$	
O[46]	$(6.98\pm0.14)\times10^2$	$(-9.79\pm0.28)\times10^2$
F[46]	$(2.72\pm0.05)\times10^2$	$(+2.00\pm0.10)\times10^2$
Ne[48]	107.1 ± 0.5	
Na[82]	$(9.56\pm0.48)\times10^5$	
Mg[45, 87]	$(1.05\pm0.02)\times10^5$	
Al[51]	$(2.02\pm0.04)\times10^5$	$(-3.88\pm0.08)\times10^5$
Si[51]	$(4.31\pm0.09)\times10^4$	$(+4.16\pm0.18)\times10^4$
P[51]	$(1.14\pm0.02)\times10^4$	
S[51]	$(6.51\pm0.13)\times10^3$	$(-7.00\pm0.26)\times10^3$
Cl[51]	$(2.73\pm0.05)\times10^3$	$(+1.65\pm0.10)\times10^3$
Ar[51]	1181 ± 24	

of $\bar{\alpha}$, the CC-R12 calculations of $\bar{\gamma}$ were plagued by numerical instabilities and could not be preferred over the CCSD(T) and CCSDT-1a results of 3.07×10^4. Papadopoulos and Sadlej[127] reported a CCSD(T) value of 3.088×10^4, and we[45] obtained a CCSD(T) value of 3.086×10^4. The closeness of these results which have all been obtained with rather elaborate basis sets leads us to assign an uncertainty of $\pm1\%$.

There have not been many calculations of $\bar{\gamma}$ for the open-shell atoms from B to F. We have chosen a set of CCSD(T) computations[46] and assigned uncertainties of $\pm2\%$. A similar CCSD(T) value of $\bar{\gamma} = 1.66 \times 10^4$ for B was published by Pouchan *et al.*,[75] but there are no other comparable computations for C to F. The anisotropies $\Delta\gamma$ have been taken from the same source[46] and assigned uncertainties twice those in $\bar{\gamma}$.

By contrast, there have been very many investigations of $\bar{\gamma}$ for neon stimulated by a discrepancy[130] between theory and experiment that has since been resolved. Reviewing them all does not seem worthwhile. Many references to earlier work on Ne can be found elsewhere.[48, 76] The value in Table 10 was obtained by combining the non-relativistic CCSD(T)-R12/B value of $\bar{\gamma} = 106.53$ calculated in an uncontracted (25s19p15d12f10g8h6i4k) GTF basis set,[48] and a $\delta_r = 0.59$ obtained[48] by direct perturbation theory at the CCSD(T) level using a q-aug-cc-pCV6Z basis set. We have assigned the resulting $\bar{\gamma}$ an uncertainty of $\pm 0.5\%$.

We now turn to atoms of the third period from Na to Ar. Of these atoms, Na is the most difficult to treat because of its extremely diffuse electron density. Maroulis[82] reported a CCSD(T) $\bar{\gamma} = 9.56 \times 10^5$ that is reproduced in Table 10. On the basis of unpublished calculations,[45] we think an uncertainty of $\pm 5\%$ is reasonable.

There is nearly perfect agreement between CCSD(T)[45] and approximate CCSD[87] results for Mg, and between two sets of CCSD(T) computations[51,89] for Si. The calculations of Lupinetti and Thakkar[51] are the only correlated ones available for the $\bar{\gamma}$ and $\Delta\gamma$ of Al, P, S and Cl. The hyperpolarizability of the closed shell atom Ar has been studied much more often. Many references and a comparison of several values can be found elsewhere.[51] Here we note that three recent non-relativistic CCSD(T) calculations with all electrons correlated have led to $\bar{\gamma} = 1159$,[94] 1187[95] and 1172[51] respectively. Combining the last result with a relativistic correction[52] of 9 ± 1 obtained at the DK3-CCSD(T) level leads to the recommended value listed in Table 10. The recommended values for Mg to Ar have all been assigned uncertainties of $\pm 2\%$ in $\bar{\gamma}$ and twice that in $\Delta\gamma$.

Computed relativistic corrections have been discussed above for Ne and Ar. For all the light atoms from Li to Ar except Ne, the relativistic effects are expected to be significantly smaller than the uncertainties in Table 10. The recommended theoretical $\bar{\gamma}$ for Ne and Ar are in fine agreement with the experimental results listed in Table 9.

3.3. *Theoretical hyperpolarizabilities for heavy atoms*

Next we turn to the CCSD(T) values of $\bar{\gamma}$ available[52,92,95] for the heavier noble-gas atoms Kr, Xe and Rn. The best non-relativistic for Kr is probably 2690 obtained by an all-electron CCSD(T) computation[95] using a t-aug-cc-pV5Z basis set. The relativistic correction obtained at the DK3-CCSD(T) level[52] is 28. Combining these two numbers yields the value listed in Table 11 with an uncertainty of $\pm 4\%$. We chose the relativistic, all-electron DK3-CCSD(T) values[52] for Xe and Rn and assigned them uncertainties of $\pm 5\%$ and $\pm 6\%$ respectively. The recommended $\bar{\gamma}$ for Kr and Xe agree with the experimental values in Table 9 within their error bars.

Table 11 lists a fourth-order Møller-Plesset (MP4) perturbation theory calculation[85] for Ca with an estimated uncertainty of $\pm 10\%$. Table 11 includes a scaled CCSD(T) value[28] for Zn with an uncertainty of $\pm 20\%$ because of basis set deficiencies demonstrated by the authors.[28]

Table 11. Theoretical $\bar{\gamma}$ hyperpolarizabilities

Atom	$\bar{\gamma}$	Atom	$\bar{\gamma}$	Atom	$\bar{\gamma}$
K[81]	$(3.6\pm1.1) \times 10^6$	Ca[85]	$(3.83\pm0.38) \times 10^5$	Kr[a]	2718 ± 109
Rb[81]	$(6.2\pm1.9) \times 10^6$	Zn[28]	$(3.1\pm0.6) \times 10^4$	Xe[52]	7037 ± 352
Cs[81]	$(1.1\pm0.3) \times 10^7$			Rn[52]	10030 ± 602

[a]See text for details.

As in the case of $\bar{\alpha}$, the values of $\bar{\gamma}$ for the heavier alkali atoms have been taken from PP+CPP calculations.[81] The method led to $\bar{\gamma} = 6.5 \times 10^4$ for Li and $\bar{\gamma} = 1.03 \times 10^6$ for Na. Comparison with the much more accurate results in Table 10 shows that the PP+CPP value is 22 times too large for Li, and 8% too large for Na. We have estimated uncertainties of $\pm30\%$ for the PP+CPP $\bar{\gamma}$ of K, Rb and Cs.

Numerical Hartree-Fock $\bar{\gamma}$ calculations[107] are listed for the remaining fourth-period atoms in Table 12. Electron correlation reduces the Hartree-Fock value of $\bar{\gamma}$ by a factor of two for Ca[85] and Zn.[28] Similarly large correlation effects are likely for the atoms from Sc to Cu, and hence the Hartree-Fock values of $\bar{\gamma}$ for Sc to Cu are not expected to be very accurate. Electron correlation raises the Hartree-Fock $\bar{\gamma}$ for Kr by about 17%. In Table 12, we have assigned uncertainties of $\pm50\%$ for Sc to Cu and $\pm30\%$ for Ga to Br.

Table 12. Hartree-Fock $\bar{\gamma}$ hyperpolarizabilities[a]

Atom	$\bar{\gamma}$	Atom	$\bar{\gamma}$
Sc	$(5.3\pm2.6) \times 10^5$	Ni	$(8.8\pm4.4) \times 10^4$
Ti	$(3.8\pm1.9) \times 10^5$	Cu	$(1.9\pm1.0) \times 10^5$
V	$(2.9\pm1.5) \times 10^5$	Ga	$(3.0\pm0.9) \times 10^5$
Cr	$(1.9\pm1.0) \times 10^5$	Ge	$(5.9\pm1.8) \times 10^4$
Mn	$(1.8\pm0.9) \times 10^5$	As	$(1.7\pm0.5) \times 10^4$
Fe	$(1.4\pm0.7) \times 10^5$	Se	$(1.0\pm0.3) \times 10^4$
Co	$(1.1\pm0.6) \times 10^5$	Br	$(4.7\pm1.4) \times 10^3$

[a]From Stiehler and Hinze.[107]

4. Dipole-dipole-quadrupole hyperpolarizabilities

There are no reliable experimental values of B hyperpolarizabilities that we are aware of. The number of theoretical calculations is quite limited as well. Table 13 lists our choices of the best available values. The value for hydrogen is the exact non-relativistic result.[131,132] The values for He[125,133] and Li[64,65] were obtained by non-relativistic, variation-perturbation computations using wave functions with explicit dependence on the interelectronic separations and should be accurate to all the digits quoted.

Table 13. Theoretical B hyperpolarizabilities

Atom	B
H[131, 132]	-106.5
He[125, 133]	-7.3267
Li[64, 65]	-5.43×10^4
Be[129]	$-(2.10 \pm 0.06) \times 10^3$
Ne[134]	-18.12 ± 0.54
Na[82]	$-(6.61 \pm 0.20) \times 10^4$
Mg[85]	$-(7.75 \pm 0.78) \times 10^3$
Ar[135]	-159.0 ± 8.0
K[81]	$-(2.19 \pm 0.22) \times 10^5$
Ca[85]	$-(3.29 \pm 0.33) \times 10^4$
Zn[28]	$-(2.37 \pm 0.48) \times 10^3$
Kr[105]	-343 ± 34
Rb[81]	$-(2.74 \pm 0.27) \times 10^5$
Xe[105]	-812 ± 81
Cs[81]	$-(4.50 \pm 0.45) \times 10^5$

Table 13 includes the results of coupled-cluster CCD+ST(CCD) calculations for Be,[129] CCSD(T) values for Ne[134] and Na,[82] and CCSD results for Ar.[135] Uncertainties of $\pm 3\%$ are assigned to the Be, Ne and Na hyperpolarizabilities whereas an uncertainty of $\pm 5\%$ is assigned to the value for Ar because contributions from triple substitutions were not included in the latter.[135] Goebel *et al.*[28] reported a non-relativistic CCSD(T) value of -2960 for Zn and then scaled it for basis set incompleteness to get the value of -2370 listed in Table 13. They assigned it a $\pm 10\%$ uncertainty but we have used $\pm 20\%$ because their scaling procedure is far from unique.

MP4 computations[85] were selected for Mg and Ca. Similar SDQMP4 computations[105] with neglect of triple substitutions are listed in Table 13 for Kr and Xe. Uncertainties of $\pm 10\%$ are assigned to all these MP4 results.

Calculations[81] using a pseudopotential (PP) and core-polarization potential (CPP) are included in Table 13 for K, Rb and Cs. Their uncertainties of $\pm 10\%$ were chosen on the basis of comparisons between PP+CPP values of B for Li and Na and their more accurate coupled-cluster counterparts listed in Table 13, and also taking into account the discussion of the PP+CPP calculations[81] for $\bar{\gamma}$.

5. Quadrupole polarizabilities

There are no reliable experimental values of quadrupole polarizabilities $\bar{\alpha}_q$ that we are aware of. The number of theoretical calculations is quite limited as well. Table 14 lists our choices of the best available values. The value for hydrogen is the exact non-relativistic result.[131, 132] Very accurate values for He, Li and Be have been obtained[41, 43, 60, 64, 65, 133] by non-relativistic, variation-perturbation computations using wave functions with explicit dependence on the interelectronic separations. The latest values[41, 43] quoted in Table 14 should be accurate to all the digits quoted.

Table 14. Theoretical quadrupole polarizabilities $\bar{\alpha}_q$.

Atom	$\bar{\alpha}_q$
H[131, 132]	15
He[41]	2.445083
Li[41]	1423.266
Be[43]	300.96
B[75]	133.3±2.7
C[136]	64.20±6.42
N[136]	30.72±3.07
O[136]	21.20±2.12
F[136]	12.69±1.27
Ne[130, 134, 137, 138]	7.52±0.15
Na[82]	1915±38
Mg[139, 140]	813.9±16.3
Al[141]	695±174
Si[141]	337±84
P[141]	176±44
S[141]	115±29
Cl[141]	74±19
Ar[138]	53.37±1.07

Quadrupole polarizabilities for the open-shell second-period atoms from B to F have not been studied much. Coupled Hartree-Fock calculations[142] were followed by PNO-CEPA computations.[136] A CCSD(T) computation was reported[75] for B. The latter has been selected for inclusion in Table 14 and assigned an uncertainty of ±2%. Comparison shows that the PNO-CEPA value is 9% higher probably because only valence-shell correlation was included in the PNO-CEPA calculations. We have selected PNO-CEPA values[136] for the C, N, O and F atoms and assigned them uncertainties of ±10%. There have been several coupled-cluster computations[130, 134, 137, 138] for neon that all led to the same value of $\bar{\alpha}_q$. It is listed in Table 14 with an uncertainty of $\pm 2\%$.

A CCSD(T) $\bar{\alpha}_q$ for Na[82] is also listed in Table 14 with an uncertainty of $\pm 2\%$. There are several values of $\bar{\alpha}_q$ for Mg. References to older work can be found in a MP4 study[85] whereas more recent work includes an approximate CCSD computation[87] and a variation-perturbation CI calculation[139,140] using a model potential. The latter is included in Table 14 with a $\pm 2\%$ uncertainty. Quadrupole polarizabilities for the open-shell third-period atoms from Al to Cl are available[141] only at the coupled Hartree-Fock level. The latter are listed in Table 14 with uncertainties of $\pm 25\%$. An all-electron CCSD(T) computation[138] for Ar is listed in Table 14 with an uncertainty of $\pm 2\%$. Other CCSD(T) calculations[137] of $\bar{\alpha}_q$ for Ar using both non-relativistic and relativistic pseudopotentials should be mentioned. The latter work reveals a relativistic correction of only -0.07 for the $\bar{\alpha}_q$ of Ar.

Most of the computations of $\bar{\alpha}_q$ for the heavier atoms of Groups 1 and 2 make use of a configuration interaction (CI) treatment with pseudopotentials and a core-polarization potential. Alkali atoms have been examined in this manner by several groups.[81,139,140,143–145] Three of these studies[139,140,143,144] obtain values within 0.1% of $\bar{\alpha}_q = 1424$, 1879 and 5005 for Li, Na and K respectively. Comparison with the accurate values in Table 14 leads us to assign uncertainties of 2% to the $\bar{\alpha}_q$ listed for K and the heavier alkali atoms in Table 15. Similar calculations[139,140] for the Group 2 atoms Ca and Sr are included in Table 15 with uncertainties of $\pm 2\%$. An older MP4 computation[85] of $\bar{\alpha}_q$ for Ca lies within this uncertainty.

Table 15.   Theoretical quadrupole polarizabilities $\bar{\alpha}_q$.

	$\bar{\alpha}_q$			$\bar{\alpha}_q$
Group 1			Group 2	
K[139,140]	5005 ± 100		Ca[139,140]	3063 ± 61
Rb[139,140]	6480 ± 130		Sr[139,140]	4577 ± 92
Cs[144]	$(1.046\pm0.021)\times10^4$			
Other			Group 18	
Zn[28]	324.8 ± 16.2		Kr[137]	97.39 ± 1.95
Pd[146]	268.8 ± 94.1		Xe[137]	209.9 ± 4.2
Hg[146]	337.0 ± 118.0			

The recommended $\bar{\alpha}_q$ for Kr and Xe are CCSD(T) values[137] computed using relativistic pseudopotentials. Comparison with their results obtained using non-relativistic pseudopotentials shows that relativistic effects lower

$\bar{\alpha}_q$ by 1.65 and 9.93 atomic units in Kr and Xe respectively. The uncertainties have been set at $\pm 2\%$. Table 15 lists a non-relativistic CCSD(T) value for Zn that was scaled to correct for basis set deficiencies.[28] Finally, relativistic RPA values have been listed for Pd and Hg, and assigned uncertainties of $\pm 35\%$.

6. Concluding remarks

Clearly, much work remains to be done before the static polarizabilities and hyperpolarizabilities can be considered well-known quantities for all the atoms in the periodic table. The dipole polarizability $\bar{\alpha}$ is known to reasonable accuracy only for the 38 atoms treated in Secs. 2.2–2.3. The heavy atoms of the p-block are the systems that need the most urgent attention, followed by the d- and f-block metal atoms. The situation is even worse for the other properties: $\bar{\gamma}$, $\bar{B}$ and $\bar{\alpha}_q$. We hope that our tables will serve as a handy reference for researchers who need numerical values of these quantities. The gaps in the tables can serve as a challenge for polarizability researchers.

Acknowledgments

We thank Peter Schwerdtfeger for providing us a preprint of his chapter in this book. The Natural Sciences and Engineering Research Council of Canada supported this work.

References

1. A. Dalgarno. *Adv. Phys.* **11**, 281–315 (1962).
2. B. Bederson and E. J. Robinson. *Adv. Chem. Phys.* **10**, 1–27 (1966).
3. A. D. Buckingham. *Adv. Chem. Phys.* **12**, 107–142 (1967).
4. M. P. Bogaard and B. J. Orr. In *MTP International Review of Science, Series Two, Physical Chemistry,* (ed.) A. D. Buckingham (Butterworths, London, 1975), vol. 2, pp. 149–194.
5. T. M. Miller and B. Bederson. *Adv. At. Mol. Phys.* **13**, 1–55 (1977).
6. T. M. Miller and B. Bederson. *Adv. At. Mol. Phys.* **25**, 37–60 (1988).
7. K. D. Bonin and V. V. Kresin. *Electric-dipole polarizabilities of atoms, molecules and clusters* (World Scientific, Singapore, 1997).
8. D. P. Shelton and J. E. Rice. *Chem. Rev.* **94**, 3–29 (1994).
9. C. E. Dykstra. *Ab Initio Calculation of the Structures and Properties of Molecules* (Elsevier, Amsterdam, 1988).
10. T. Helgaker, P. Jørgensen, and J. Olsen. *Molecular Electronic-Structure Theory* (Wiley, New York, 2000).

11. P. J. Mohr and B. N. Taylor. *Rev. Mod. Phys.* **77**, 1–107 (2005).
12. R. R. Teachout and R. T. Pack. *At. Data Nucl. Data Tables* **3**, 195–214 (1971).
13. J. Thorhallsson, C. Fisk, and S. Fraga. *Theor. Chim. Acta* **10**, 388–392 (1968).
14. J. Thorhallsson, C. Fisk, and S. Fraga. *J. Chem. Phys.* **49**, 1987–1988 (1968).
15. S. Fraga, J. Karwowski, and K. M. S. Saxena. *Handbook of Atomic Data* (Elsevier, New York, 1976).
16. T. M. Miller. In *Handbook of Chemistry and Physics*, (ed.) D. R. Lide (CRC Press, Boca Raton, Florida, 2004), chap. 10, pp. 168–171. 85th ed.
17. D. Gugan and G. W. Michel. *Mol. Phys.* **39**, 783–785 (1980).
18. D. Gugan and G. W. Michel. *Metrologia* **16**, 149–167 (1980).
19. R. W. Molof, H. L. Schwartz, T. M. Miller, and B. Bederson. *Phys. Rev. A* **10**, 1131–1140 (1974).
20. R. A. Alpher and D. R. White. *Phys. Fluids* **2**, 153–161 (1959).
21. J. Huot and T. K. Bose. *J. Chem. Phys.* **95**, 2683–2687 (1991).
22. C. R. Ekstrom, J. Schmiedmayer, M. S. Chapman, T. D. Hammond, and D. E. Pritchard. *Phys. Rev. A* **51**, 3883–3888 (1995).
23. W. A. de Heer, P. Milani, and A. Châtelain. *Phys. Rev. Lett.* **63**, 2834–2836 (1989).
24. P. Milani, I. Moullet, and W. A. de Heer. *Phys. Rev. A* **42**, 5150–5154 (1990).
25. D. R. Johnston, G. J. Oudemans, and R. H. Cole. *J. Chem. Phys.* **33**, 1310–1317 (1960).
26. U. Hohm and K. Kerl. *Mol. Phys.* **69**, 819–831 (1990).
27. T. M. Miller and B. Bederson. *Phys. Rev. A* **14**, 1572–1573 (1976).
28. D. Goebel, U. Hohm, and G. Maroulis. *Phys. Rev. A* **54**, 1973–1978 (1996).
29. H. L. Schwartz, T. M. Miller, and B. Bederson. *Phys. Rev. A* **10**, 1924–1926 (1974).
30. D. Goebel and U. Hohm. *Phys. Rev. A* **52**, 3691–3694 (1995).
31. T. P. Guella, T. M. Miller, B. Bederson, J. A. D. Stockdale, and B. Jaduszliwer. *Phys. Rev. A* **29**, 2977–2980 (1984).
32. J. M. Amini and H. Gould. *Phys. Rev. Lett.* **91**, 153001 (2003).
33. M. W. Cole and J. Bardon. *Phys. Rev. B* **33**, 2812–2813 (1986).
34. J. Bardon and M. Audiffren. *J. de Physique* **45**, 245–249 (1984).
35. D. Goebel and U. Hohm. *J. Phys. Chem.* **100**, 7710–7712 (1996).
36. M. A. Kadar-Kallen and K. D. Bonin. *Phys. Rev. Lett.* **72**, 828–831 (1994).
37. F. Weinhold. *J. Phys. Chem.* **86**, 1111–1116 (1982).
38. I. Waller. *Z. Physik* **38**, 635–646 (1926).
39. M. L. Bartlett and E. A. Power. *J. Phys. A* **2**, 419–426 (1969).
40. S. P. Goldman. *Phys. Rev. A* **39**, 976–980 (1989).
41. Z.-C. Yan, J. F. Babb, A. Dalgarno, and G. W. F. Drake. *Phys. Rev. A* **54**, 2824–2833 (1996).
42. K. Pachucki and J. Sapirstein. *Phys. Rev. A* **63**, 012504 (2001).
43. J. Komasa. *Phys. Rev. A* **65**, 012506 (2002).

44. I. S. Lim, M. Pernpointner, M. Seth, J. K. Laerdahl, P. Schwerdtfeger, P. Neogrády, and M. Urban. *Phys. Rev. A* **60**, 2822–2828 (1999).
45. C. Lupinetti and A. J. Thakkar. *To be published* (2005).
46. A. K. Das and A. J. Thakkar. *J. Phys. B* **31**, 2215–2223 (1998).
47. I. Černušák, V. Kellö, and A. J. Sadlej. *Collect. Czech. Chem. Commun.* **68**, 211–239 (2003).
48. W. Klopper, S. Coriani, T. Helgaker, and P. Jørgensen. *J. Phys. B* **37**, 3753–3763 (2004).
49. A. J. Thakkar and C. Lupinetti. *Chem. Phys. Lett.* **402**, 270–273 (2005).
50. I. Miadoková, V. Kellö, and A. J. Sadlej. *Theor. Chem. Acc.* **96**, 166–175 (1997).
51. C. Lupinetti and A. J. Thakkar. *J. Chem. Phys.* **122**, 044301 (2005).
52. T. Nakajima and K. Hirao. *Chem. Lett.* **30**, 766–767 (2001).
53. G. Wentzel. *Z. Physik* **38**, 518–529 (1926).
54. P. S. Epstein. *Phys. Rev.* **28**, 695–710 (1926).
55. L. Pauling. *Proc. R. Soc. Lon. A* **114**, 181–211 (1927).
56. J. V. Atanasoff. *Phys. Rev.* **36**, 1232–1242 (1930).
57. H. R. Hassé. *Proc. Camb. Phil. Soc.* **26**, 542–555 (1930).
58. T. D. H. Baber and H. R. Hassé. *Proc. Camb. Phil. Soc.* **33**, 253–259 (1937).
59. C. Schwartz. *Phys. Rev.* **123**, 1700–1705 (1961).
60. A. D. Buckingham and P. G. Hibbard. *Symp. Faraday Soc.* **2**, 41–47 (1968).
61. W. Cencek, K. Szalewicz, and B. Jeziorski. *Phys. Rev. Lett.* **86**, 5675–5678 (2001).
62. M. Masili and A. F. Starace. *Phys. Rev. A* **68**, 012508 (2003).
63. J. S. Sims, S. A. Hagstrom, and J. R. Rumble, Jr. *Phys. Rev. A* **14**, 576–578 (1976).
64. J. Pipin and D. M. Bishop. *Phys. Rev. A* **45**, 2736–2743 (1992).
65. J. Pipin and D. M. Bishop. *Phys. Rev. A* **53**, 4614 (1996).
66. J. S. Sims and J. R. Rumble, Jr. *Phys. Rev. A* **8**, 2231–2235 (1973).
67. D. Tunega, J. Noga, and W. Klopper. *Chem. Phys. Lett.* **269**, 435–440 (1997).
68. G. L. Bendazzoli and A. Monari. *Chem. Phys.* **306**, 153–161 (2004).
69. M. Douglas and N. M. Kroll. *Ann. Phys. (N.Y.)* **82**, 89–155 (1974).
70. B. A. Hess. *Phys. Rev. A* **33**, 3742–3748 (1986).
71. G. Jansen and B. A. Hess. *Phys. Rev. A* **39**, 6016–6017 (1989).
72. W. J. Stevens and F. P. Billingsley II. *Phys. Rev. A* **8**, 2236–2245 (1973).
73. H.-J. Werner and W. Meyer. *Phys. Rev. A* **13**, 13–16 (1976).
74. A. Hibbert, M. Le Dourneuf, and V. K. Lan. *J. Phys. B* **10**, 1015–1025 (1977).
75. C. Pouchan, M. Rérat, and G. Maroulis. *J. Phys. B* **30**, 167–176 (1997).
76. F. Pawłowski, P. Jørgensen, and C. Hättig. *Chem. Phys. Lett.* **391**, 27–32 (2004).
77. E.-A. Reinsch and W. Meyer. *Phys. Rev. A* **14**, 915–918 (1976).
78. A. Hibbert. *J. Phys. B* **13**, 3725–3731 (1980).
79. F. Maeder and W. Kutzelnigg. *Chem. Phys.* **42**, 95–112 (1979).
80. W. Müller, J. Flesch, and W. Meyer. *J. Chem. Phys.* **80**, 3297–3310 (1984).

81. P. Fuentealba and O. Reyes. *J. Phys. B* **26**, 2245–2250 (1993).
82. G. Maroulis. *Chem. Phys. Lett.* **334**, 207–213 (2001).
83. P. Soldán, M. T. Cvitaš, and J. M. Hutson. *Phys. Rev. A* **67**, 054702 (2003).
84. G. Maroulis. *J. Chem. Phys.* **121**, 10519–10524 (2004).
85. E. F. Archibong and A. J. Thakkar. *Phys. Rev. A* **44**, 5478–5484 (1991).
86. M. A. Castro and S. Canuto. *Phys. Lett. A.* **176**, 105–108 (1993).
87. J. M. Stout and C. E. Dykstra. *J. Mol. Struct. (Theochem)* **332**, 189–196 (1995).
88. P. Fuentealba. *Chem. Phys. Lett.* **397**, 459–461 (2004).
89. G. Maroulis and C. Pouchan. *J. Phys. B* **36**, 2011–2017 (2003).
90. U. Hohm, A. Loose, G. Maroulis, and D. Xenides. *Phys. Rev. A* **61**, 053202 (2000).
91. S. Millefiori and A. Alparone. *J. Phys. Chem. A* **105**, 9489–9497 (2001).
92. J. E. Rice, P. R. Taylor, T. J. Lee, and J. Almlöf. *J. Chem. Phys.* **94**, 4972–4979 (1991).
93. D. E. Woon and T. H. Dunning, Jr. *J. Chem. Phys.* **100**, 2975–2988 (1994).
94. C. Hättig and P. Jørgensen. *J. Chem. Phys.* **109**, 2762–2778 (1998).
95. P. Soldán, E. P. F. Lee, and T. G. Wright. *Phys. Chem. Chem. Phys.* **3**, 4661–4666 (2001).
96. I. S. Lim, P. Schwerdtfeger, B. Metz, and H. Stoll. *J. Chem. Phys.* **122**, 104103 (2005).
97. I. S. Lim and P. Schwerdtfeger. *Phys. Rev. A* **70**, 062501 (2004).
98. P. Neogrády, V. Kellö, M. Urban, and A. J. Sadlej. *Int. J. Quantum Chem.* **63**, 557–565 (1997).
99. V. Kellö and A. J. Sadlej. *Theor. Chim. Acta* **94**, 93–104 (1996).
100. A. Derevianko, W. R. Johnson, M. S. Safronova, and J. F. Babb. *Phys. Rev. Lett.* **82**, 3589–3592 (1999).
101. S. G. Porsev and A. Derevianko. *Phys. Rev. A* **65**, 020701 (2002).
102. P. Schwerdtfeger and G. A. Bowmaker. *J. Chem. Phys.* **100**, 4487–4497 (1994).
103. R. Wesendrup and P. Schwerdtfeger. *Angew. Chem.-Int. Edit.* **39**, 907–910 (2000).
104. M. Seth, P. Schwerdtfeger, and M. Dolg. *J. Chem. Phys.* **106**, 3623–3632 (1997).
105. G. Maroulis and A. J. Thakkar. *J. Chem. Phys.* **89**, 7320–7323 (1988).
106. A. J. Sadlej. *Theor. Chim. Acta* **81**, 45–63 (1991).
107. J. Stiehler and J. Hinze. *J. Phys. B* **28**, 4055–4071 (1995).
108. M. Medveď, P. W. Fowler, and J. M. Hutson. *Mol. Phys.* **98**, 453–463 (2000).
109. B. O. Roos, R. Lindh, P.-Å. Malmqvist, V. Veryazov, and P.-O. Widmark. *J. Phys. Chem. A* **108**, 2851–2858 (2004).
110. A. J. Sadlej. *Theor. Chim. Acta* **81**, 339–354 (1992).
111. Y. Wang and M. Dolg. *Theor. Chem. Acc.* **100**, 124–133 (1998).
112. V. Kellö and A. J. Sadlej. *Theor. Chim. Acta* **83**, 351–366 (1992).
113. T. Fleig and A. J. Sadlej. *Phys. Rev. A* **65**, 032506 (2002).
114. G. S. Chandler and R. Glass. *J. Phys. B* **20**, 1–10 (1987).
115. R. Glass. *J. Phys. B* **20**, 4317–4323 (1987).

116. R. Glass. *J. Phys. B* **20**, 1379–1386 (1987).
117. D. P. Chong and S. R. Langhoff. *J. Chem. Phys.* **84**, 5606–5610 (1986).
118. R. Pou-Amérigo, M. Merchán, I. Nebot-Gil, P.-O. Widmark, and B. O. Roos. *Theor. Chim. Acta* **92**, 149–181 (1995).
119. A. Zangwill and P. Soven. *Phys. Rev. A* **21**, 1561–1572 (1980).
120. J. F. Young, G. C. Bjorklund, A. H. Kung, R. B. Miles, and S. E. Harris. *Phys. Rev. Lett.* **27**, 1551–1553 (1971).
121. D. P. Shelton. *Phys. Rev. A* **42**, 2578–2592 (1990).
122. D. P. Shelton and E. A. Donley. *Chem. Phys. Lett.* **195**, 591–595 (1992).
123. A. D. Buckingham and D. A. Dunmur. *Trans. Faraday Soc.* **64**, 1776–1783 (1968).
124. G. L. Sewell. *Proc. Cambridge Phil. Soc.* **45**, 678–679 (1949).
125. D. M. Bishop and J. Pipin. *J. Chem. Phys.* **91**, 3549–3551 (1989).
126. N. El-Bakali Kassimi and A. J. Thakkar. *Phys. Rev. A* **50**, 2948–2952 (1994).
127. M. G. Papadopoulos and A. J. Sadlej. *Chem. Phys. Lett.* **288**, 377–382 (1998).
128. G. Maroulis and A. J. Thakkar. *J. Phys. B* **22**, 2439–2446 (1989).
129. A. J. Thakkar. *Phys. Rev. A* **40**, 1130–1132 (1989).
130. G. Maroulis and A. J. Thakkar. *Chem. Phys. Lett.* **156**, 87–90 (1989).
131. C. A. Coulson. *Proc. Roy. Soc. Edin.* **A61**, 20–25 (1941).
132. A. D. Buckingham, C. A. Coulson, and J. T. Lewis. *Proc. Phys. Soc. (Lond.)* **A69**, 639–641 (1956).
133. D. M. Bishop and J. Pipin. *Chem. Phys. Lett.* **236**, 15–18 (1995).
134. P. R. Taylor, T. J. Lee, J. E. Rice, and J. Almlöf. *Chem. Phys. Lett.* **163**, 359–365 (1989).
135. S. Coriani, C. Hättig, and A. Rizzo. *J. Chem. Phys.* **111**, 7828–7836 (1999).
136. E.-A. Reinsch and W. Meyer. *Phys. Rev. A* **18**, 1793–1796 (1978).
137. A. Nicklass, M. Dolg, H. Stoll, and H. Preuss. *J. Chem. Phys.* **102**, 8942–8952 (1995).
138. G. Maroulis and A. Haskopoulos. *Chem. Phys. Lett.* **358**, 64–70 (2002).
139. J. Mitroy and M. W. J. Bromley. *Phys. Rev. A* **68**, 052714 (2003).
140. J. Mitroy and M. W. J. Bromley. *Phys. Rev. A* **71**, 019902(E) (2003).
141. P. K. Mukherjee, H. P. Roy, and A. Gupta. *Phys. Rev. A* **17**, 30–33 (1978).
142. A. Gupta, H. P. Roy, and P. K. Mukherjee. *Int. J. Quantum Chem.* **9**, 1–8 (1975).
143. D. Spelsberg, T. Lorenz, and W. Meyer. *J. Chem. Phys.* **99**, 7845–7858 (1993).
144. M. Marinescu, H. R. Sadeghpour, and A. Dalgarno. *Phys. Rev. A* **49**, 982–988 (1994).
145. M. Mérawa, M. Rérat, and B. Bussery-Honvault. *J. Mol. Struct. (Theochem)* **633**, 137–144 (2003).
146. D. Kolb, W. R. Johnson, and P. Shorer. *Phys. Rev. A* **26**, 19–31 (1982).

CHAPTER 15

POLARIZABILITIES OF FEW-BODY ATOMIC AND MOLECULAR SYSTEMS

Zong-Chao Yan, Jun-Yi Zhang, and Yue Li

Center for Laser, Atomic and Molecular Sciences and Department of Physics, University of New Brunswick, P. O. Box 4400, Fredericton, New Brunswick, Canada E3B 5A3

The computational methods for calculating polarizabilities of few-body atomic and molecular systems are reviewed, with particular emphasis on three-body systems using fully-correlated Hylleraas coordinates. The structure of this review is as follows. Sec. I is a short introduction to the importance of polarizabilities. A general theoretical framework is formulated in Sec. II, which includes Hamiltonian, Transition Operator, Generalized Sum Rules, and Basis Set. Sec. III discusses the evaluation of matrix elements for a three-body system in Hylleraas coordinates. Sec. IV concerns with general procedures of calculating polarizabilities. A brief survey of polarizabilities of few-body systems is presented in Sec. V, including He, Li, Be, H_2^+, H_2, and PsH. The final section Sec. VI contains a concluding remark.

1. Introduction

In general, the static 2^ℓ-pole polarizability for an atomic or molecular system is defined in terms of a sum over a complete set of intermediate states, including the continuum [1]:

$$\alpha_\ell = \sum_{n \neq 0} \frac{f_{n0}^{(\ell)}}{(E_n - E_0)^2} , \qquad (1)$$

where $f_{n0}^{(\ell)}$ is the 2^ℓ-pole oscillator strength

$$
\begin{aligned}
f_{n0}^{(\ell)} &= f_{n0}^{(\ell)}(m, M_0, M_n) \\
&= \frac{8\pi}{2\ell+1}(E_n - E_0)|\langle\Psi_0| \sum_i q_i R_i^\ell Y_{\ell m}(\hat{\mathbf{R}}_i)|\Psi_n\rangle|^2 .
\end{aligned}
\qquad (2)
$$

In the above, q_i is the charge of particle i with its position vector $\mathbf{R}_i$ relative to the origin of a laboratory frame, Ψ_0 is the wave function for the state of interest with the associated energy eigenvalue E_0 and the magnetic quantum number M_0, and Ψ_n is the wave function for one of the intermediate states with the associated energy eigenvalue E_n and the magnetic quantum number M_n. In practice, instead of $f_{n0}^{(\ell)}$, an averaged oscillator strength $\bar{f}_{n0}^{(\ell)}$, which is independent of the magnetic quantum numbers m, M_0, and M_n, is used. The $\bar{f}_{n0}^{(\ell)}$ is obtained by averaging over the initial state orientation degeneracy and summing over the final state degeneracy. It is convenient to introduce reduced matrix elements through the Wigner-Eckart theorem [2]

$$\langle \gamma' L' M' | \sum_i q_i R_i^\ell Y_{lm}(\hat{\mathbf{R}}_i) | \gamma L M \rangle = (-1)^{L'-M'} \begin{pmatrix} L' & \ell & L \\ -M' & m & M \end{pmatrix}$$

$$\times \langle \gamma' L' || \sum_i q_i R_i^\ell Y_\ell(\hat{\mathbf{R}}_i) || \gamma L \rangle. \quad (3)$$

With the aid of a sum rule for the $3j$ symbols, we have

$$\bar{f}_{n0}^{(\ell)} = \frac{8\pi}{(2\ell+1)^2(2L_0+1)}(E_n - E_0)|\langle \Psi_0 || \sum_i q_i R_i^\ell Y_\ell(\hat{\mathbf{R}}_i) || \Psi_n \rangle|^2, \quad (4)$$

where L_0 is the total orbital angular momentum for the initial state (the state of interest).

The polarizability of an atom or molecule can be considered as a measure of response of charge cloud to an external electric field. This is best illustrated by the Stark effect. Consider a ground-state atomic hydrogen in a weak external electric field $\mathcal{E}$. According to quantum mechanics, the energy shift due to $\mathcal{E}$ is [3]

$$\Delta E = -\frac{1}{2}\alpha_1 \mathcal{E}^2, \quad (5)$$

which is proportional to the dipole polarizability α_1.

The polarizabilities also appear in long-range interactions between two charged ions, such as H_2^+ and H^-. For example, the long-range interaction energy for two ions a and b in their ground states, accurate to R^{-10}, can be written in the form [4]

$$U(R) = \frac{Q_a Q_b}{R} - \frac{1}{2}Q_a^2 \sum_{\ell=1}^4 \frac{\alpha_\ell^{(b)}}{R^{2\ell+2}} - \frac{1}{2}Q_b^2 \sum_{\ell=1}^4 \frac{\alpha_\ell^{(a)}}{R^{2\ell+2}} - \sum_{n=3}^5 \frac{C_{2n}}{R^{2n}}, \quad (6)$$

where Q_a and Q_b are the total charges of a and b. In the above, C_6, C_8, and C_{10} are the dispersion coefficients which can be expressed in terms of the dynamic polarizabilities [1]

$$\alpha_\ell(\omega) = \sum_{n \neq 0} \frac{\bar{f}_{n0}^{(\ell)}}{(E_n - E_0)^2 - \omega^2}. \tag{7}$$

The dynamic polarizabilities are very important in studying Casimir-Polder effects [5] between two atoms.

The importance of both static and dynamic polarizabilities is also manifested in many areas of physics [6], such as laser trapping and cooling of atoms, optical tweezers, and nanostructure fabrication. In these applications, polarizabilities are crucial quantities in using mechanical properties of light, such as the use of light forces to manipulate atoms and molecules, as well as the use of light to trap and cool atoms.

2. Formulation

2.1. *Hamiltonian*

In this review, atomic units are used throughout. Consider a nonrelativistic Coulombic system consisting of $n + 1$ charged particles. The Hamiltonian [4] is

$$H = -\sum_{i=0}^{n} \frac{1}{2m_i} \nabla_{R_i}^2 + \sum_{i > j \geq 0}^{n} \frac{q_i q_j}{|\mathbf{R}_i - \mathbf{R}_j|}, \tag{8}$$

where q_i and m_i are the charge and mass of the ith particle respectively, and $\mathbf{R}_i$ is its position vector relative to a laboratory frame. The total orbital angular momentum operator, which commutes with H, is

$$\mathbf{L} = -i \sum_{i=0}^{n} \mathbf{R}_i \times \nabla_{R_i}. \tag{9}$$

In order to eliminate the center of mass motion, we introduce the following transformation

$$\mathbf{r}_i = \mathbf{R}_i - \mathbf{R}_0, \quad i = 1, 2, 3, \ldots, n \tag{10}$$

$$\mathbf{X} = \frac{1}{M_T} \sum_{j=0}^{n} m_j \mathbf{R}_j, \tag{11}$$

where $M_T = \sum_{i=0}^{n} m_i$ is the total mass of the system and $\mathbf{X}$ is the position vector for the center of mass relative to the laboratory frame. In the above, we take particle 0 as our reference and $\mathbf{r}_i$ thus represent the internal coordinates of the system. This transformation establishes one to one correspondence between $(\mathbf{X}, \mathbf{r}_1, \mathbf{r}_2, \ldots, \mathbf{r}_n)$ and $(\mathbf{R}_0, \mathbf{R}_1, \mathbf{R}_2, \ldots, \mathbf{R}_n)$. The inverse transformation is

$$\mathbf{R}_0 = \mathbf{X} - \sum_{j=1}^{n} \frac{m_j}{M_T} \mathbf{r}_j, \tag{12}$$

$$\mathbf{R}_i = \mathbf{X} + \left(1 - \frac{m_i}{M_T}\right) \mathbf{r}_i - \sum_{\substack{j \neq i \\ j \geq 1}}^{n} \frac{m_j}{M_T} \mathbf{r}_j. \tag{13}$$

Using the new coordinates $(\mathbf{X}, \mathbf{r}_1, \mathbf{r}_2, \ldots, \mathbf{r}_n)$, the Hamiltonian (8) is transformed into the form of

$$H = -\sum_{i=1}^{n} \frac{1}{2\mu_i} \nabla_i^2 - \frac{1}{m_0} \sum_{i \geq j \geq 1}^{n} \nabla_i \cdot \nabla_j + q_0 \sum_{i=1}^{n} \frac{q_i}{r_i} + \sum_{i \geq j \geq 1}^{n} \frac{q_i q_j}{r_{ij}} - \frac{1}{2M_T} \nabla_X^2, \tag{14}$$

where $r_{ij} = |\mathbf{r}_i - \mathbf{r}_j|$ and $\mu_i = m_i m_0 / (m_i + m_0)$ is the reduced mass between particles i and 0. Similarly, the orbital angular momentum is transformed into

$$\mathbf{L} = -i \sum_{i=1}^{n} \mathbf{r}_i \times \nabla_i - i\mathbf{X} \times \nabla_X. \tag{15}$$

Since the transformed Hamiltonian does not contain $\mathbf{X}$, $\mathbf{X}$ is a cyclic coordinate and can thus be ignored. This means that one works in the center of mass frame where the Hamiltonian and the orbital angular momentum have the forms of

$$H = -\sum_{i=1}^{n} \frac{1}{2\mu_i} \nabla_i^2 - \frac{1}{m_0} \sum_{i \geq j \geq 1}^{n} \nabla_i \cdot \nabla_j + q_0 \sum_{i=1}^{n} \frac{q_i}{r_i} + \sum_{i \geq j \geq 1}^{n} \frac{q_i q_j}{r_{ij}}, \tag{16}$$

and

$$\mathbf{L} = -i \sum_{i=1}^{n} \mathbf{r}_i \times \nabla_i, \tag{17}$$

respectively, which only contain the internal coordinates of the system.

2.2. Transition Operator

The 2^ℓ-pole transition operator [4] for an $(n+1)$-body system, which appears in the definition of the oscillator strength (4), is

$$T_\ell^{(n+1)} = \sum_{i=0}^{n} q_i R_i^\ell Y_{\ell 0}(\hat{\mathbf{R}}_i), \tag{18}$$

which is expressed in terms of the laboratory coordinates $\mathbf{R}_i$. Thus it is necessary to transform it into the center of mass frame by applying (12) and (13), which may be written in the form of

$$\mathbf{R}_i = \sum_{j=1}^{n} \epsilon_{ij} \mathbf{r}_j, \tag{19}$$

with $\epsilon_{ij} = \delta_{ij} - m_j/M_T$, $i = 0, 1, 2, \ldots, n$ and $j = 1, 2, \ldots, n$. Using mathematical induction, one can verify that the following formula is held

$$Y_{\ell m}(\hat{\mathbf{r}}) = \sqrt{\frac{3}{4\pi}} \prod_{i=1}^{\ell-1} \left(\sqrt{\frac{2i+3}{i+1}} \right) \underbrace{(\hat{\mathbf{r}} \otimes \hat{\mathbf{r}} \otimes \cdots \hat{\mathbf{r}})_m^{(\ell)}}_{\ell}, \tag{20}$$

where $\otimes$ denotes the coupling between two irreducible tensor operators. Substituting (20) into (18) yields the transformed transition operator in terms of the internal coordinates $\mathbf{r}_i$

$$T_\ell^{(n+1)} = \sqrt{\frac{3}{4\pi}} \prod_{m=1}^{\ell-1} \left(\sqrt{\frac{2m+3}{m+1}} \right)$$

$$\times \sum_{j_1,\ldots,j_\ell} \left(\sum_{i=0}^{n} q_i \epsilon_{ij_1} \epsilon_{ij_2} \cdots \epsilon_{ij_\ell} \right) (\mathbf{r}_{j_1} \otimes \mathbf{r}_{j_2} \otimes \cdots \mathbf{r}_{j_\ell})_0^{(\ell)}, \tag{21}$$

with the understanding that $\prod_{m=1}^{\ell-1} \sqrt{\frac{2m+3}{m+1}} = 1$ when $\ell = 1$.

For a three-body system where there are only two relative coordinates $\mathbf{r}_1$ and $\mathbf{r}_2$, the transition operators $T_\ell^{(3)}$ with ℓ up to 3 have the following explicit forms:

$$T_1^{(3)} = \sum_{j=1}^{2} \left(\sum_{i=0}^{2} q_i \epsilon_{ij} \right) r_j Y_{10}(\hat{\mathbf{r}}_j), \tag{22}$$

$$T_2^{(3)} = \sum_{j=1}^{2} \left(\sum_{i=0}^{2} q_i \epsilon_{ij}^2 \right) r_j^2 Y_{20}(\hat{\mathbf{r}}_j) + \sqrt{\frac{15}{2\pi}} \left(\sum_{i=0}^{2} q_i \epsilon_{i1} \epsilon_{i2} \right) r_1 r_2 (\hat{\mathbf{r}}_1 \otimes \hat{\mathbf{r}}_2)_0^{(2)},$$

$$\tag{23}$$

$$T_3^{(3)} = \sum_{j=1}^{2} \left(\sum_{i=0}^{2} q_i \epsilon_{ij}^3 \right) r_j^3 Y_{30}(\hat{\mathbf{r}}_j) + \frac{3}{2}\sqrt{\frac{35}{2\pi}} \Big[$$

$$\left(\sum_{i=0}^{2} q_i \epsilon_{i1} \epsilon_{i2}^2 \right) r_2^2 r_1 ((\hat{\mathbf{r}}_2 \otimes \hat{\mathbf{r}}_2)^{(2)} \hat{\mathbf{r}}_1)_0^{(3)}$$

$$+ \left(\sum_{i=0}^{2} q_i \epsilon_{i1}^2 \epsilon_{i2} \right) r_1^2 r_2 ((\hat{\mathbf{r}}_1 \otimes \hat{\mathbf{r}}_1)^{(2)} \otimes \hat{\mathbf{r}}_2)_0^{(3)} \Big]. \qquad (24)$$

It is noted that the masses of particles enter into the transition operator $T_\ell^{(n+1)}$ as a polynomial in m_j/M_T. To be specific, the mass-dependent coeffecients are

$$\sum_{i=0}^{2} q_i \epsilon_{ij} = q_j - Q_T \frac{m_j}{M_T}, \qquad (25)$$

$$\sum_{i=0}^{2} q_i \epsilon_{ij}^2 = q_j - 2q_j \frac{m_j}{M_T} + Q_T \left(\frac{m_j}{M_T} \right)^2, \qquad (26)$$

$$\sum_{i=0}^{2} q_i \epsilon_{i1} \epsilon_{i2} = -\frac{q_1 m_2 + q_2 m_1}{M_T} + Q_T \frac{m_1 m_2}{M_T^2}, \qquad (27)$$

$$\sum_{i=0}^{2} q_i \epsilon_{ij}^3 = q_j - 3q_j \frac{m_j}{M_T} + 3q_j \left(\frac{m_j}{M_T} \right)^2 - Q_T \left(\frac{m_j}{M_T} \right)^3, \qquad (28)$$

$$\sum_{i=0}^{2} q_i \epsilon_{i1} \epsilon_{i2}^2 = -q_2 \frac{m_1}{M_T} + \frac{q_1 m_2^2 + 2q_2 m_1 m_2}{M_T^2} - Q_T \frac{m_1 m_2^2}{M_T^3}, \qquad (29)$$

$$\sum_{i=0}^{2} q_i \epsilon_{i1}^2 \epsilon_{i2} = -q_1 \frac{m_2}{M_T} + \frac{q_2 m_1^2 + 2q_1 m_2 m_1}{M_T^2} - Q_T \frac{m_2 m_1^2}{M_T^3}. \qquad (30)$$

In the above, $Q_T = q_0 + q_1 + q_2$ is the total charge of the system. One can see that the coefficients of $T_\ell^{(n+1)}$ are generally polynomials of degree ℓ in m_j/M_T, except for the case when $Q_T = 0$, ie, a neutral system where the degree of polynomial reduces to $\ell - 1$, as shown above. For an atomic system, the mass ratio m_j/M_T can be considered to be small. In the limit of infinite nuclear mass, $T_\ell^{(n+1)}$ becomes the following simple form

$$T_\ell^{(n+1)} = \sum_{j=1}^{n} q_j r_j^\ell Y_{\ell 0}(\hat{\mathbf{r}}_j). \qquad (31)$$

For a molecular system (more than one center), however, m_j/M_T can no longer be considered as a small parameter. Similar expressions can be obtained for more than three-particle systems with increasing complexity of mass dependency.

2.3. *Generalized Sum Rules*

In order to evaluate the polarizabilities α_ℓ, it is necessary to sum over a complete set of intermediate states, including the continuum. The degree of completeness of the intermediate states determines the quality of the calculated α_ℓ. A direct way of forming such a complete set is to diagonalize the Hamiltonian in a basis set with required angular momentum symmetry. For an atomic system of infinite nuclear mass, the exact dipole oscillator-strength spectrum satisfies the Thomas-Reiche-Kuhn (TRK) sum rule

$$\sum_n f_{n0}^{(1)}(0, M_0, M_n) = N_e \,, \tag{32}$$

where N_e is the total number of electrons. Yan and Drake [7] extended the TRK sum rule to the case of finite nuclear mass. The only modification is the replacement of $f_{n0}^{(1)}$ by $f_{n0}^{(1)}/(1 + N_e m_e/m_0)$ in the above equation, where m_e/m_0 is the mass ratio between electron and nucleus. Of course, the Hamiltonian must include the mass polarization terms $-1/m_0 \nabla_i \cdot \nabla_j$ and the reduced masses, as shown in (16).

The original or the extended TRK sum rule discussed above can not be applied directly to a system of arbitrary masses and charges. A further generalization is needed [8]. Consider a system of $n + 1$ charged particles again, with the Hamiltonian H given by (16). Define the following transition operators in length and velocity gauges respectively

$$\mathbf{d}_0 = \sum_{i=1}^{n} \lambda_i \mathbf{r}_i \,, \tag{33}$$

$$\mathbf{b}_0 = \sum_{i=1}^{n} \xi_i \mathbf{p}_i \,, \tag{34}$$

where

$$\lambda_i = q_i - Q_T \frac{m_i}{M_T} \,, \tag{35}$$

$$\xi_i = \frac{q_0}{m_0} - \frac{q_i}{m_i} \,. \tag{36}$$

It is a straightforward matter to show that

$$[H, \mathbf{d}_0] = i\mathbf{b}_0 \,. \tag{37}$$

Similarly,

$$[\mathbf{d}_0, \mathbf{b}_0] = 3i \sum_{j=1}^{n} \lambda_j \xi_j, \tag{38}$$

with the understanding that $[\mathbf{a}, \mathbf{b}] = \mathbf{a} \cdot \mathbf{b} - \mathbf{b} \cdot \mathbf{a}$. Sandwiching (37) by two eigenstates $|\gamma\rangle$ and $|\gamma'\rangle$ of the Hamiltonian H yields

$$-\omega_{\gamma\gamma'} \langle \gamma | \mathbf{d}_0 | \gamma' \rangle = \langle \gamma | \mathbf{b}_0 | \gamma' \rangle, \tag{39}$$

where $\omega_{\gamma\gamma'} = E_\gamma - E_{\gamma'}$ is the transition energy. Furthermore, consider $\langle \gamma | [\mathbf{d}_0, \mathbf{b}_0] | \gamma' \rangle$. Using (38) and inserting a complete set of intermediate states between $\mathbf{d}_0$ and $\mathbf{b}_0$, one obtains the following sum rule

$$\sum_{\gamma'} \frac{2}{3} \omega_{\gamma\gamma'} |\langle \gamma | \mathbf{d}_0 | \gamma' \rangle|^2 = \sum_{i=1}^{n} \lambda_i \xi_i, \tag{40}$$

with the help of (39). Since

$$\sum_{i=1}^{n} \lambda_i \xi_i = \frac{Q_T^2}{M_T} - \sum_{i=0}^{n} \frac{q_i^2}{m_i}, \tag{41}$$

we finally obtain the generalized Thomas-Reiche-Kuhn sum rule

$$\sum_{\gamma'} \frac{2}{3} \omega_{\gamma'\gamma} |\langle \gamma | \mathbf{d}_0 | \gamma' \rangle|^2 = \sum_{i=0}^{n} \frac{q_i^2}{m_i} - \frac{Q_T^2}{M_T}, \tag{42}$$

in the length gauge. The parallel expression in the velocity gauge is

$$\sum_{\gamma'} \frac{2}{3\omega_{\gamma'\gamma}} |\langle \gamma | \mathbf{b}_0 | \gamma' \rangle|^2 = \sum_{i=0}^{n} \frac{q_i^2}{m_i} - \frac{Q_T^2}{M_T}. \tag{43}$$

In fact, it is easy to see that

$$\frac{2}{3} \omega_{\gamma'\gamma} |\langle \gamma | \mathbf{d}_0 | \gamma' \rangle|^2 = \frac{2}{3\omega_{\gamma'\gamma}} |\langle \gamma | \mathbf{b}_0 | \gamma' \rangle|^2, \tag{44}$$

which establishes the equivalence between the length and velocity gauges of the oscillator strength. One can verify that, for an atomic system, the generalized Thomas-Reiche-Kuhn sum rule reduces to the original one (32) for the case of infinite nuclear mass, as well as to its extension by Yan and Drake [7] for the case of finite nuclear mass. As a numerical example, consider the ground state of H_2^+ with the proton mass taken to be $1836.152701 m_e$. Using the ground state wave function and the intermediate energy spectrum of P symmetry of our previous work [9], we find that

$$\sum_{\gamma'} \frac{2}{3} \omega_{\gamma'\gamma} |\langle \gamma | \mathbf{d}_0 | \gamma' \rangle|^2 = 1.000817000. \tag{45}$$

On the other hand,

$$\sum_{i=0}^{2} \frac{q_i^2}{m_i} - \frac{Q_T^2}{M_T} = 1.000816999, \tag{46}$$

in perfect agreement with (45). For HD$^+$ with the deuteron mass of $3670.483014m_\mathrm{e}$, the corresponding values are 1.000635496 and 1000635494 respectively.

For 2^ℓ-pole oscillator strengths, where ℓ can be larger than 1, a set of sum rules have been developed by us for an atomic system of infinite nuclear mass [10]. The use of the following expression [11]

$$\sum_n (E_n - E_k)|F_{nk}|^2 = \frac{1}{2}\langle k|[F,[H,F]]|k\rangle, \tag{47}$$

where $F(\mathbf{r},\mathbf{p})$ is an arbitrary Hermitian operator, yields the general sum rule

$$2\sum_n (E_n - E_0)|\langle \Psi_n| \sum_{i=1}^{N_e} r_i^\ell P_\ell(\cos\theta_i)|\Psi_0\rangle|^2$$
$$= \ell^2 \langle \Psi_0| \sum_{i=1}^{N_e} r_i^{2\ell-2}\csc^2\theta_i[P_\ell^2(\cos\theta_i) + P_{\ell-1}^2(\cos\theta_i)$$
$$-2\cos\theta_i P_\ell(\cos\theta_i)P_{\ell-1}(\cos\theta_i)]|\Psi_0\rangle. \tag{48}$$

This formula can further be simplified by defining

$$F_\ell(x) = \frac{P_\ell^2(x) + P_{\ell-1}^2(x) - 2xP_\ell(x)P_{\ell-1}(x)}{1 - x^2}. \tag{49}$$

$F_\ell(x)$ can be expanded in the form

$$F_\ell(x) = \sum_{s=0}^{\infty} C_s(\ell)P_s(x), \tag{50}$$

with

$$C_s(\ell) = \frac{2s+1}{2} \int_{-1}^{1} \frac{P_\ell^2(x) + P_{\ell-1}^2(x) - 2xP_\ell(x)P_{\ell-1}(x)}{1 - x^2} P_s(x)dx. \tag{51}$$

The integrand is reducible by applying the following recursion relations [12]:

$$-\ell P_\ell(x) + \ell x P_{\ell-1}(x) = (1 - x^2) P_{\ell-1}'(x), \tag{52}$$

$$\ell P_{\ell-1}(x) - \ell x P_\ell(x) = (1 - x^2) P_\ell'(x). \tag{53}$$

The result is

$$C_s(\ell) = \frac{2s+1}{2\ell} \int_{-1}^{1} (P_{\ell-1}(x)P_\ell'(x) - P_\ell(x)P_{\ell-1}'(x))P_s(x)\,dx. \tag{54}$$

According to the properties of Legendre polynomials,

$$C_s(\ell) = 0, \quad s > 2\ell - 2 \ \text{ or } \ s = \text{odd}. \tag{55}$$

Thus, the sum rule in (48) can finally be recast into the form

$$(2\ell+1)(2L_0+1)\sum_{L_n}\left|\begin{pmatrix} L_0 & \ell & L_n \\ -M_0 & 0 & M_0 \end{pmatrix}\right|^2 S_{Ln}^{(\ell)} = \ell^2 \sum_{m=0}^{\ell-1} C_{2m}(\ell)M_{2\ell-2,2m}(M_0), \tag{56}$$

where

$$M_{mn}(M_0) = \langle \Psi_0 | \sum_{i=1}^{N_e} r_i^m P_n(\cos\theta_i) | \Psi_0 \rangle, \tag{57}$$

and

$$S_L^{(\ell)} = \sum_n \bar{f}_{n0}^{(\ell)}(L). \tag{58}$$

In the above equation, the sum n runs over all the intermediate states for a given angular momentum L. We consider the case of $L_0 = 1$, the P symmetry, as an example. The integral $C_{2m}(\ell)$ can easily be evaluated with the help of a symbolic manipulation program such as Maple. Thus, (56) gives rise to a series of sum rules:

$$3S_S^{(1)} + \frac{6}{5}S_D^{(1)} = N_e, \tag{59}$$

$$\frac{9}{10}S_D^{(1)} + \frac{3}{2}S_P^{(1)} = N_e, \tag{60}$$

$$\frac{6}{7}S_F^{(2)} + \frac{3}{2}S_D^{(2)} + \frac{1}{2}S_P^{(2)} = 2M_{20}(1) + 2M_{22}(1), \tag{61}$$

$$2S_P^{(2)} + \frac{9}{7}S_F^{(2)} = 2M_{20}(0) + 2M_{22}(0), \tag{62}$$

$$\frac{9}{5}S_D^{(3)} + \frac{4}{3}S_G^{(3)} = \frac{24}{7}M_{42}(0) + 3M_{40}(0), \tag{63}$$

$$\frac{3}{5}S_D^{(3)} + \frac{3}{2}S_F^{(3)} + \frac{5}{6}S_G^{(3)} = \frac{24}{7}M_{42}(1) + 3M_{40}(1). \tag{64}$$

Similarly, for $L_0 = 2$, we have

$$2S_P^{(1)} + \frac{9}{7}S_F^{(1)} = N_e, \tag{65}$$

$$\frac{3}{2}S_P^{(1)} + \frac{1}{2}S_D^{(1)} + \frac{8}{7}S_F^{(1)} = N_e, \tag{66}$$

$$2S_D^{(1)} + \frac{5}{7}S_F^{(1)} = N_e, \tag{67}$$

$$5S_S^{(2)} + \frac{10}{7}S_D^{(2)} + \frac{10}{7}S_G^{(2)} = 2M_{20}(0) + 2M_{22}(0), \tag{68}$$

$$\frac{5}{2}S_P^{(2)} + \frac{5}{14}S_D^{(2)} + \frac{5}{7}S_F^{(2)} + \frac{25}{21}S_G^{(2)} = 2M_{20}(1) + 2M_{22}(1), \tag{69}$$

$$\frac{10}{7}S_D^{(2)} + \frac{25}{14}S_F^{(2)} + \frac{25}{42}S_G^{(2)} = 2M_{20}(2) + 2M_{22}(2), \tag{70}$$

$$3S_P^{(3)} + \frac{4}{3}S_F^{(3)} + \frac{50}{33}S_H^{(3)} = 3M_{40}(0) + \frac{24}{7}M_{42}(0) + \frac{18}{7}M_{44}(0), \tag{71}$$

$$S_P^{(3)} + 2S_D^{(3)} + \frac{1}{6}S_F^{(3)} + \frac{5}{6}S_G^{(3)} + \frac{40}{33}S_H^{(3)} =$$
$$3M_{40}(1) + \frac{24}{7}M_{42}(1) + \frac{18}{7}M_{44}(1), \tag{72}$$

$$\frac{1}{2}S_D^{(3)} + \frac{5}{3}S_F^{(3)} + \frac{5}{3}S_G^{(3)} + \frac{35}{66}S_H^{(3)} = 3M_{40}(2) + \frac{24}{7}M_{42}(2) + \frac{18}{7}M_{44}(2). \tag{73}$$

The generalization of (56) to an arbitrary system has not yet been made.

2.4. *Basis Set*

In order to solve the eigenvalue problem for the Hamiltonian H variationally, we seek the common eigenstates of the commuting observables $\{H, \mathbf{L}^2, L_z, \Pi\}$, where Π is the parity operator. For a two-electron atomic system, the basis set can be constructed in Hylleraas coordinates

$$\chi_{ijk}(\alpha, \beta) = r_1^i r_2^j r_{12}^k e^{-\alpha r_1 - \beta r_2} \mathcal{Y}_{\ell_1 \ell_2}^{LM}(\hat{\mathbf{r}}_1, \hat{\mathbf{r}}_2), \tag{74}$$

where α and β are two nonlinear parameters, $\mathbf{r}_1$ and $\mathbf{r}_2$ are the position vectors of the two electrons relative to the nucleus, *ie*, we take nucleus as the reference particle 0, and $\mathcal{Y}_{\ell_1 \ell_2}^{LM}(\hat{\mathbf{r}}_1, \hat{\mathbf{r}}_2)$ is the vector-coupled product of spherical harmonics for the two electrons forming a common eigenstate of $\{\mathbf{L}^2, L_z, \Pi\}$

$$\mathcal{Y}_{\ell_1 \ell_2}^{LM}(\hat{\mathbf{r}}_1, \hat{\mathbf{r}}_2) = \sum_{m_1 m_2} \langle \ell_1 m_1; \ell_2 m_2 | \ell_1 \ell_2; LM \rangle Y_{\ell_1 m_1}(\hat{\mathbf{r}}_1) Y_{\ell_2 m_2}(\hat{\mathbf{r}}_2), \tag{75}$$

with the corresponding eigenvalues $L(L+1)\hbar^2$, $M\hbar$, and $(-1)^{\ell_1+\ell_2}$ respectively. The basis is generated by including all the terms such that

$$i + j + k \leq \Omega, \tag{76}$$

where Ω is an integer. However, some truncations may be applied in order to preserve numerical stability which could be important for large size of basis set. The wave function is expanded according to

$$\Psi(\mathbf{r}_1, \mathbf{r}_2) = \sum_{ijk} a_{ijk} \chi_{ijk}(\alpha, \beta) \pm \text{exchange}, \tag{77}$$

and the expansion coefficients a_{ijk} determined variationally. One can also include more than one set of nonlinear parameters in the wave function so that different areas in the configuration space can be accounted for, and the rate of convergence can thus be enhanced.

The Hylleraas basis set (74) can also be used to three-body molecular systems [9,4,13]. To be specific, let us consider H_2^+. Choosing $m_0 = m_p$, $m_1 = 1$, and $m_2 = m_p$, where m_p is the proton mass, the Hamiltonian (16) becomes

$$H = -\frac{1 + m_p}{2m_p}\nabla_1^2 - \frac{1}{m_p}\nabla_2^2 - \frac{1}{m_p}\nabla_1 \cdot \nabla_2 - \frac{1}{r_1} + \frac{1}{r_2} - \frac{1}{r_{12}}. \tag{78}$$

In (74), r_2 now is the distance between two protons. The vibrational modes, as implied by the Born-Oppenheimer approximation, can be represented by $r_2^j e^{-\beta r_2}$ with the lowest power of r_2 being a nonzero integer. For the case of H_2^+, we found [9] that $j \geq 35$, together with the nonlinear parameters $\alpha \approx 1$ and $\beta \approx 15 \sim 20$. Such modified Hylleraas basis set has generated the ground state energy of H_2^+ to be accurate to a few parts in 10^{19}. The extension to higher angular momentum states has also been made [13]. It is interesting to note that there is no apparent symmetry between $\mathbf{r}_1$ and $\mathbf{r}_2$ in the tranformed Hamiltonian (78). The Hamiltonian, however, is invariant under the following transformation:

$$\mathbf{r}_1 \rightarrow \mathbf{r}_1 - \mathbf{r}_2, \tag{79}$$

$$\mathbf{r}_2 \rightarrow -\mathbf{r}_2, \tag{80}$$

which reflects the exchange symmetry between two protons. Thus, one can always choose a basis which is permutational symmetry adapted. On the other hand, one is also free to choose a basis which has no permutational symmetry imposed explicitly, which can significantly simplify computational efforts. A further discussion along this line can be found in [9,13].

For a three-electron atomic system, the variational basis set can be constructed in a similar way:

$$\Phi(\mathbf{r}_1, \mathbf{r}_2, \mathbf{r}_3) = \phi(\mathbf{r}_1, \mathbf{r}_2, \mathbf{r}_3)\,\chi(1, 2, 3)\,, \tag{81}$$

where the orbital part ϕ can be written in terms of Hylleraas coordinates

$$\begin{aligned}\phi(\mathbf{r}_1, \mathbf{r}_2, \mathbf{r}_3) = &\, r_1^{j_1}\, r_2^{j_2}\, r_3^{j_3}\, r_{12}^{j_{12}}\, r_{23}^{j_{23}}\, r_{31}^{j_{31}}\, e^{-\alpha r_1 - \beta r_2 - \gamma r_3} \\ &\times \mathcal{Y}_{(\ell_1 \ell_2)\ell_{12},\ell_3}^{LM}(\hat{\mathbf{r}}_1, \hat{\mathbf{r}}_2, \hat{\mathbf{r}}_3)\,, \end{aligned} \tag{82}$$

with

$$\begin{aligned}\mathcal{Y}_{(\ell_1 \ell_2)\ell_{12},\ell_3}^{LM}(\hat{\mathbf{r}}_1, \hat{\mathbf{r}}_2, \hat{\mathbf{r}}_3) = &\sum_{\text{all } m_i} \langle \ell_1 m_1; \ell_2 m_2 | \ell_1 \ell_2; \ell_{12} m_{12}\rangle \\ &\times \langle \ell_{12} m_{12}; \ell_3 m_3 | \ell_{12}\ell_3; LM\rangle Y_{\ell_1 m_1}(\hat{\mathbf{r}}_1) Y_{\ell_2 m_2}(\hat{\mathbf{r}}_2) Y_{\ell_3 m_3}(\hat{\mathbf{r}}_3)\end{aligned} \tag{83}$$

being a vector-coupled product of spherical harmonics for the three electrons to form a state of total orbital angular momentum L. The spin part χ can be either

$$\chi(1, 2, 3) = \alpha(1)\beta(2)\alpha(3) - \beta(1)\alpha(2)\alpha(3)\,, \tag{84}$$

for the total electron spin $1/2$, or

$$\chi(1, 2, 3) = = \alpha(1)\alpha(2)\alpha(3)\,, \tag{85}$$

for the total electron spin $3/2$. The variational wave function is a linear combination of the functions Φ antisymmetrized by the three-particle antisymmetrizer

$$\mathcal{A} = (1) - (12) - (13) - (23) + (123) + (132)\,. \tag{86}$$

All the terms in (81) are nominally included such that

$$j_1 + j_2 + j_3 + j_{12} + j_{23} + j_{31} \leq \Omega\,. \tag{87}$$

Encouraged by the success of applying Hylleraas coordinates to three-body molecular systems, we [14] have recently calculated the ground-state energies of muonium hydride and its isotopes MuH, MuD, and MuT variationally in Hylleraas coordinates. These are two-center four-body molecular systems with two positively charged heavy particles and two negatively charged electrons. Convergence to a few parts in 10^9 has been achieved, which has significantly improved the previous results Suffczyński, Kotowski, and Wolniewicz [15]. Other four-body molecular systems that we are currently interested in include H_2, HeH^+, $He^{2+}H^+\mu^-\mu^-$, and $H^+H^+\mu^-e^-$, as well as their isotopes.

3. Matrix Elements

3.1. *Reduction of Hamiltonian*

In order to solve the eigenvalue problem of the Hamiltonian, one needs to evaluate the matrix elements of the Hamiltonian with respect to the basis functions in Hylleraas coordinates. For a three-body atomic or molecular system, they are

$$\{r_1^a \, r_2^b \, r_{12}^c \, e^{-\alpha r_1 - \beta r_2} \mathcal{Y}_{\ell_1 \ell_2}^{LM}(\hat{\mathbf{r}}_1, \hat{\mathbf{r}}_2)\} \, . \tag{88}$$

It is advantageous to separate the radial and angular parts of the operators ∇_i according to

$$\nabla_1 = \frac{\mathbf{r}_1}{r_1} \frac{\partial}{\partial r_1} + \frac{\mathbf{r}_{12}}{r_{12}} \frac{\partial}{\partial r_{12}} + \nabla_1^Y \, , \tag{89}$$

$$\nabla_2 = \frac{\mathbf{r}_2}{r_2} \frac{\partial}{\partial r_2} + \frac{\mathbf{r}_{21}}{r_{21}} \frac{\partial}{\partial r_{21}} + \nabla_2^Y \, , \tag{90}$$

where $\mathbf{r}_{12} = \mathbf{r}_1 - \mathbf{r}_2$ and the operators ∇_1^Y and ∇_2^Y act only on the angular part of the basis function $\mathcal{Y}_{\ell_1 \ell_2}^{LM}(\hat{\mathbf{r}}_1, \hat{\mathbf{r}}_2)$. Applying ∇_1 twice yields the expression for the Laplacian operator

$$\nabla_1^2 = \frac{\partial^2}{\partial r_1^2} + \frac{\partial^2}{\partial r_{12}^2} + \frac{2}{r_1} \frac{\partial}{\partial r_1} + \frac{2}{r_{12}} \frac{\partial}{\partial r_{12}} + \frac{r_1^2 - r_2^2 + r_{12}^2}{r_1 r_{12}} \frac{\partial^2}{\partial r_1 \partial r_{12}}$$
$$- \frac{\ell_1(\ell_1 + 1)}{r_1^2} - \frac{2 r_2}{r_1 r_{12}} \frac{\partial}{\partial r_{12}} (\hat{\mathbf{r}}_2 \cdot \hat{\nabla}_1^Y) \, , \tag{91}$$

where $\hat{\mathbf{r}}_i = \mathbf{r}_i / r_i$ and $\hat{\nabla}_i^Y = r_i \nabla_i^Y$. A similarly expression for ∇_2^2 can be obtained from ∇_1^2 by interchanging $1 \leftrightarrow 2$:

$$\nabla_2^2 = \frac{\partial^2}{\partial r_2^2} + \frac{\partial^2}{\partial r_{12}^2} + \frac{2}{r_2} \frac{\partial}{\partial r_2} + \frac{2}{r_{12}} \frac{\partial}{\partial r_{12}} + \frac{r_2^2 - r_1^2 + r_{12}^2}{r_2 r_{12}} \frac{\partial^2}{\partial r_2 \partial r_{12}}$$
$$- \frac{\ell_2(\ell_2 + 1)}{r_2^2} - \frac{2 r_1}{r_2 r_{12}} \frac{\partial}{\partial r_{12}} (\hat{\mathbf{r}}_1 \cdot \hat{\nabla}_2^Y) \, . \tag{92}$$

Finally

$$\nabla_1 \cdot \nabla_2 = -\left(\frac{\partial^2}{\partial r_{12}^2} + \frac{2}{r_{12}} \frac{\partial}{\partial r_{12}} \right) + \frac{r_1^2 + r_2^2 - r_{12}^2}{2 r_1 r_2} \frac{\partial^2}{\partial r_1 \partial r_2}$$
$$+ \frac{r_2^2 - r_1^2 - r_{12}^2}{2 r_1 r_{12}} \frac{\partial^2}{\partial r_1 \partial r_{12}} + \frac{r_1^2 - r_2^2 - r_{12}^2}{2 r_2 r_{12}} \frac{\partial^2}{\partial r_2 \partial r_{12}}$$
$$+ \frac{r_1}{r_2} \left(\frac{1}{r_1} \frac{\partial}{\partial r_1} + \frac{1}{r_{12}} \frac{\partial}{\partial r_{12}} \right) (\hat{\mathbf{r}}_1 \cdot \hat{\nabla}_2^Y) + \frac{1}{r_1 r_2} \hat{\nabla}_1^Y \cdot \hat{\nabla}_2^Y$$
$$+ \frac{r_2}{r_1} \left(\frac{1}{r_2} \frac{\partial}{\partial r_2} + \frac{1}{r_{12}} \frac{\partial}{\partial r_{12}} \right) (\hat{\mathbf{r}}_2 \cdot \hat{\nabla}_1^Y) \, . \tag{93}$$

One can see that there are four types of integrals that appear in the Hamiltonian: (i) overlap integrals, (ii) integrals involving $\hat{\mathbf{r}}_1 \cdot \hat{\nabla}_2^Y$, (iii) integrals involving $\hat{\mathbf{r}}_2 \cdot \hat{\nabla}_1^Y$, and (iv) integrals involving $\hat{\nabla}_1^Y \cdot \hat{\nabla}_2^Y$.

3.2. *Basic Integrals*

The overlap integral is the basic integral [16] that appears in the variational calculation of three-body Coulombic systems in Hylleraas coordinates

$$I(a,b,c;\alpha,\beta) = \int d\mathbf{r}_1 \, d\mathbf{r}_2 \, \mathcal{Y}_{\ell_1' \ell_2'}^{L'M'}(\hat{\mathbf{r}}_1, \hat{\mathbf{r}}_2)^* \mathcal{Y}_{\ell_1 \ell_2}^{LM}(\hat{\mathbf{r}}_1, \hat{\mathbf{r}}_2)$$
$$\times \, r_1^a r_2^b r_{12}^c e^{-\alpha r_1 - \beta r_2} \,. \tag{94}$$

The coordinate r_{12} can be expanded according to Perkins [17]

$$r_{12}^c = \sum_{q=0}^{L_1} P_q(\cos\theta_{12}) \sum_{k=0}^{L_2} C_{cqk} r_<^{q+2k} r_>^{c-q-2k} \,, \tag{95}$$

where $r_< = \min(r_1, r_2)$, and $r_> = \max(r_1, r_2)$. For $c \geq -1$, for even values of c, $L_1 = \frac{1}{2}c$, $L_2 = \frac{1}{2}c - q$; for odd values of c, $L_1 = \infty$, $L_2 = \frac{1}{2}(c+1)$. Also in (95), the coefficients are given by

$$C_{cqk} = \frac{2q+1}{c+2} \left(\frac{c+2}{2k+1} \right) \prod_{t=0}^{S_{qc}} \frac{2k+2t-c}{2k+2q-2t+1} \,, \tag{96}$$

where $S_{qc} = \min[q-1, \frac{1}{2}(c+1)]$. The angular part can be decoupled using the addition theorem for spherical harmonics

$$P_q(\cos\theta_{12}) = \frac{4\pi}{2q+1} \sum_{m=-q}^{q} Y_{qm}(\hat{\mathbf{r}}_1)^* Y_{qm}(\hat{\mathbf{r}}_2) \,. \tag{97}$$

Thus the basic integral (94) can then be recast into the form

$$I(a,b,c;\alpha,\beta) = \sum_{q,k} C_{cqk} G(q) I_{\mathrm{R}}(a,b,c;\alpha,\beta;q,k) \,, \tag{98}$$

where I_{R} is the radial integral

$$I_{\mathrm{R}}(a,b,c;\alpha,\beta;q,k) = \int_0^\infty d r_1 \int_0^\infty d r_2 \, r_1^{a+2} r_2^{b+2} r_<^{q+2k} r_>^{c-q-2k}$$
$$\times \, e^{-\alpha r_1 - \beta r_2} \,, \tag{99}$$

and $G(q)$ is the angular part

$$
\begin{aligned}
G(q) = \frac{4\pi}{2q+1} \sum_m \sum_{m_1 m_2} \sum_{m'_1 m'_2} (-1)^{\ell_1 - \ell_2 + M + \ell'_1 - \ell'_2 + M'} (L, L')^{1/2} \\
\times \begin{pmatrix} \ell_1 & \ell_2 & L \\ m_1 & m_2 & -M \end{pmatrix} \begin{pmatrix} \ell'_1 & \ell'_2 & L' \\ m'_1 & m'_2 & -M' \end{pmatrix} \\
\times \int d\Omega_1 d\Omega_2 Y_{\ell'_1 m'_1}(\hat{\mathbf{r}}_1)^* Y_{\ell_1 m_1}(\hat{\mathbf{r}}_1) Y_{qm}(\hat{\mathbf{r}}_1)^* Y_{\ell'_2 m'_2}(\hat{\mathbf{r}}_2)^* Y_{\ell_2 m_2}(\hat{\mathbf{r}}_2) Y_{qm}(\hat{\mathbf{r}}_2),
\end{aligned}
\tag{100}
$$

with $(L, L') = (2L+1)(2L'+1)$. Further simplification can be performed by first using the formula

$$
Y_{\ell m}(\hat{\mathbf{r}}) Y_{\ell' m'}(\hat{\mathbf{r}}) = \sum_{LM} \sqrt{\frac{(\ell, \ell', L)}{4\pi}} \begin{pmatrix} \ell & \ell' & L \\ 0 & 0 & 0 \end{pmatrix} \begin{pmatrix} \ell & \ell' & L \\ m & m' & M \end{pmatrix} Y_{LM}(\hat{\mathbf{r}})^*,
\tag{101}
$$

and then using the graphical method of angular momentum [18]. The result is

$$
G(q) = (-1)^{L+q} (\ell_1, \ell_2, \ell'_1, \ell'_2)^{1/2} \begin{pmatrix} \ell'_1 & \ell_1 & q \\ 0 & 0 & 0 \end{pmatrix} \begin{pmatrix} \ell'_2 & \ell_2 & q \\ 0 & 0 & 0 \end{pmatrix} \begin{Bmatrix} L & \ell_1 & \ell_2 \\ q & \ell'_2 & \ell'_1 \end{Bmatrix} \delta_{MM'} \delta_{LL'}.
\tag{102}
$$

It should be noted that the range of q in (98) is restricted by the triangular rule of the above $3j$ symbols

$$
q_{\mathrm{m}} \le q \le q_{\mathrm{M}},
\tag{103}
$$

where

$$
q_{\mathrm{m}} = \max(|\ell_1 - \ell'_1|, |\ell_2 - \ell'_2|),
\tag{104}
$$
$$
q_{\mathrm{M}} = \min(\ell_1 + \ell'_1, \ell_2 + \ell'_2).
\tag{105}
$$

Thus, the summation over q in (98) is always finite even when c is odd.

For the radial part I_{R}, breaking the range of integration at $r_1 = r_2$ yields

$$
\begin{aligned}
I_{\mathrm{R}}(a, b, c; \alpha, \beta; q, k) = \int_0^\infty dr_1 \int_0^{r_1} dr_2 r_1^A r_2^B e^{-\alpha r_1 - \beta r_2} \\
+ \int_0^\infty dr_2 \int_0^{r_2} dr_1 r_1^{A'} r_2^{B'} e^{-\alpha r_1 - \beta r_2},
\end{aligned}
\tag{106}
$$

where

$$A = a + 2 + c - q - 2k,$$
$$B = b + 2 + q + 2k,$$
$$A' = a + 2 + q + 2k,$$
$$B' = b + 2 + c - q - 2k.$$

Using formula 6.5.12 of [19] and formula 4, 7.621 of [20],

$$\int_0^{r_2} r_1^n e^{-\alpha r_1} dr_1 = \frac{r_2^{n+1}}{n+1} e^{-\alpha r_2} {}_1F_1(1; n+2; \alpha r_2), \qquad (107)$$

$$\int_0^{\infty} e^{-st} t^{b-1} {}_1F_1(a; c; kt) dt = \Gamma(b) s^{-b} {}_2F_1(a, b; c; ks^{-1}), \qquad (108)$$

I_R finally becomes

$$I_R(a, b, c; \alpha, \beta; q, k) = \frac{s!}{(\alpha + \beta)^{s+1}} \qquad (109)$$

$$\times \left[\frac{1}{a + 3 + q + 2k} {}_2F_1(1, s+1; a + 4 + q + 2k; \frac{\alpha}{\alpha + \beta}) \right.$$

$$\left. + \frac{1}{b + 3 + q + 2k} {}_2F_1(1, s+1; b + 4 + q + 2k; \frac{\beta}{\alpha + \beta}) \right],$$

where $s = a + b + c + 5$. The above expression is also valid for $a \geq -2$, $b \geq -2$, $c \geq -2$, and $a + b + c + 5 \geq 0$. Note that I_R depends on q and k through $q + 2k$.

3.3. *Integrals involving* $\hat{\nabla}_i^Y$

Integrals involving $\hat{\mathbf{r}}_1 \cdot \hat{\nabla}_2^Y$, $\hat{\mathbf{r}}_2 \cdot \hat{\nabla}_1^Y$, or $\hat{\nabla}_1^Y \cdot \hat{\nabla}_2^Y$ can be evaluated in a uniform way. For ∇ acting only on spherical harmonics, one can use the formula [21]

$$\hat{\nabla}_\mu^Y Y_{\ell m}(\hat{\mathbf{r}}) = \sum_{\lambda \tau} b(\ell; \lambda)(\ell, \lambda)^{1/2} \begin{pmatrix} 1 & \ell & \lambda \\ 0 & 0 & 0 \end{pmatrix} \begin{pmatrix} 1 & \ell & \lambda \\ \mu & m & \tau \end{pmatrix} Y_{\lambda \tau}^*(\hat{\mathbf{r}}), \qquad (110)$$

where $\hat{\nabla}_\mu^Y$ is written in its spherical component form with $\mu = -1$, 0, and 1, and the notation $b(\ell; \lambda)$ is defined by

$$b(\ell; \ell - 1) = \ell + 1, \qquad (111)$$
$$b(\ell; \ell + 1) = -\ell. \qquad (112)$$

On the other hand, since

$$\hat{\mathbf{r}}_\mu = \sqrt{\frac{4\pi}{3}} Y_{1\mu}(\hat{\mathbf{r}}), \qquad (113)$$

one obtains by (101) that

$$\hat{r}_\mu Y_{\ell m}(\hat{\mathbf{r}}) = \sum_{\lambda\tau}(\ell,\lambda)^{1/2}\begin{pmatrix}1 & \ell & \lambda \\ 0 & 0 & 0\end{pmatrix}\begin{pmatrix}1 & \ell & \lambda \\ \mu & m & \tau\end{pmatrix}Y^*_{\lambda\tau}(\hat{\mathbf{r}}). \qquad (114)$$

Comparing (114) with (110), one can see that the action of $\hat{\nabla}^Y_\mu$ on spherical harmonics can be realized by first replacing $\hat{\nabla}^Y_\mu$ by $\hat{r}_\mu$, evaluating the corresponding terms, and then inserting $b(\ell;\lambda)$'s appropriately. Therefore only the following integral needs to be considered

$$J(a,b,c;\alpha,\beta) = \int d\mathbf{r}_1\, d\mathbf{r}_2\, \mathcal{Y}^{L'M'}_{\ell'_1\ell'_2}(\hat{\mathbf{r}}_1,\hat{\mathbf{r}}_2)^*(\hat{\mathbf{r}}_1\cdot\hat{\mathbf{r}}_2)\mathcal{Y}^{LM}_{\ell_1\ell_2}(\hat{\mathbf{r}}_1,\hat{\mathbf{r}}_2)$$
$$\times\ r_1^a r_2^b r_{12}^c e^{-\alpha r_1-\beta r_2}. \qquad (115)$$

Since

$$\hat{\mathbf{r}}_1\cdot\hat{\mathbf{r}}_2 = \sum_\mu(-1)^\mu\,\hat{r}_{1\mu}\cdot\hat{r}_{2-\mu} = \frac{4\pi}{3}\sum_\mu(-1)^\mu Y_{1\mu}(\hat{\mathbf{r}}_1)Y_{1-\mu}(\hat{\mathbf{r}}_2), \quad (116)$$

one can combine the above two spherical harmonics with $Y_{\ell_1 m_1}(\hat{\mathbf{r}}_1)$ and $Y_{\ell_2 m_2}(\hat{\mathbf{r}}_2)$ in $\mathcal{Y}^{LM}_{\ell_1\ell_2}(\hat{\mathbf{r}}_1,\hat{\mathbf{r}}_2)$ according to (101)

$$Y_{1\mu}(\hat{\mathbf{r}}_1)Y_{\ell_1 m_1}(\hat{\mathbf{r}}_1) = \sum_{T_1\tau_1}\sqrt{\frac{(1,\ell_1,T_1)}{4\pi}}\begin{pmatrix}1 & \ell_1 & T_1 \\ 0 & 0 & 0\end{pmatrix}\begin{pmatrix}1 & \ell_1 & T_1 \\ \mu & m_1 & \tau_1\end{pmatrix}$$
$$\times(-1)^{\tau_1}Y_{T_1-\tau_1}(\hat{\mathbf{r}}_1), \qquad (117)$$

$$Y_{1-\mu}(\hat{\mathbf{r}}_2)Y_{\ell_2 m_2}(\hat{\mathbf{r}}_2) = \sum_{T_2\tau_2}\sqrt{\frac{(1,\ell_2,T_2)}{4\pi}}\begin{pmatrix}1 & \ell_2 & T_2 \\ 0 & 0 & 0\end{pmatrix}\begin{pmatrix}1 & \ell_2 & T_2 \\ -\mu & m_2 & \tau_2\end{pmatrix}$$
$$\times(-1)^{\tau_2}Y_{T_2-\tau_2}(\hat{\mathbf{r}}_2). \qquad (118)$$

After some simplification, the J integral can be reduced to the basic integral type with the angular coefficient $G(\hat{\mathbf{r}}_1\cdot\hat{\mathbf{r}}_2)$

$$J(a,b,c;\alpha,\beta) = \sum_{q,k}C_{cqk}G(\hat{\mathbf{r}}_1\cdot\hat{\mathbf{r}}_2)I_R(a,b,c;\alpha,\beta;q,k), \qquad (119)$$

where

$$G(\hat{\mathbf{r}}_1\cdot\hat{\mathbf{r}}_2) = (\ell_1,\ell_2,\ell'_1,\ell'_2)^{1/2}\sum_{T_1T_2}(T_1,T_2)\begin{pmatrix}\ell'_1 & T_1 & q \\ 0 & 0 & 0\end{pmatrix}\begin{pmatrix}\ell'_2 & T_2 & q \\ 0 & 0 & 0\end{pmatrix}$$
$$\times\begin{pmatrix}1 & \ell_1 & T_1 \\ 0 & 0 & 0\end{pmatrix}\begin{pmatrix}1 & \ell_2 & T_2 \\ 0 & 0 & 0\end{pmatrix}G_0, \qquad (120)$$

and

$$
G_0 = \sum_{m_1' m_2'} \sum_{m_1 m_2} \sum_{\tau_1 \tau_2} \sum_{\mu m} (-1)^{\ell_1' - \ell_2' + M' + \ell_1 - \ell_2 + M + \mu + \tau_2 + m_2'} (L, L')^{1/2}
$$
$$
\times \begin{pmatrix} \ell_1 & \ell_2 & L \\ m_1 & m_2 & -M \end{pmatrix} \begin{pmatrix} \ell_1' & \ell_2' & L' \\ m_1' & m_2' & -M' \end{pmatrix} \begin{pmatrix} 1 & \ell_1 & T_1 \\ \mu & m_1 & \tau_1 \end{pmatrix}
$$
$$
\times \begin{pmatrix} 1 & \ell_2 & T_2 \\ -\mu & m_2 & \tau_2 \end{pmatrix} \begin{pmatrix} \ell_1' & T_1 & q \\ m_1' & \tau_1 & m \end{pmatrix} \begin{pmatrix} \ell_2' & T_2 & q \\ -m_2' & -\tau_2 & m \end{pmatrix}. \tag{121}
$$

After using the graphical method [18], G_0 becomes

$$
G_0 = \delta_{LL'} \delta_{MM'} (-1)^{\ell_1 + \ell_1'} \begin{Bmatrix} T_2 & \ell_2 & 1 \\ \ell_1 & T_1 & L \end{Bmatrix} \begin{Bmatrix} \ell_2' & T_2 & q \\ T_1 & \ell_1' & L \end{Bmatrix}. \tag{122}
$$

The final expression for the angular coefficient $G(\hat{\mathbf{r}}_1 \cdot \hat{\mathbf{r}}_2)$ is therefore

$$
G(\hat{\mathbf{r}}_1 \cdot \hat{\mathbf{r}}_2) = \delta_{LL'} \delta_{MM'} (\ell_1, \ell_2, \ell_1', \ell_2')^{1/2} (-1)^{\ell_1 + \ell_1'} \sum_{T_1 T_2} (T_1, T_2) \begin{pmatrix} \ell_1' & T_1 & q \\ 0 & 0 & 0 \end{pmatrix}
$$
$$
\times \begin{pmatrix} \ell_2' & T_2 & q \\ 0 & 0 & 0 \end{pmatrix} \begin{pmatrix} 1 & \ell_1 & T_1 \\ 0 & 0 & 0 \end{pmatrix} \begin{pmatrix} 1 & \ell_2 & T_2 \\ 0 & 0 & 0 \end{pmatrix} \begin{Bmatrix} T_2 & \ell_2 & 1 \\ \ell_1 & T_1 & L \end{Bmatrix} \begin{Bmatrix} \ell_2' & T_2 & q \\ T_1 & \ell_1' & L \end{Bmatrix}. \tag{123}
$$

The angular coefficients containing $\hat{\nabla}_i^Y$ are obtained by the following replacements:

$$
\hat{\mathbf{r}}_1 \cdot \hat{\nabla}_2^Y \longrightarrow \hat{\mathbf{r}}_1 \cdot \hat{\mathbf{r}}_2 \, b(\ell_2; T_2), \tag{124}
$$
$$
\hat{\mathbf{r}}_2 \cdot \hat{\nabla}_1^Y \longrightarrow \hat{\mathbf{r}}_1 \cdot \hat{\mathbf{r}}_2 \, b(\ell_1; T_1), \tag{125}
$$
$$
\hat{\nabla}_1^Y \cdot \hat{\nabla}_2^Y \longrightarrow \hat{\mathbf{r}}_1 \cdot \hat{\mathbf{r}}_2 \, b(\ell_1; T_1) \, b(\ell_2; T_2), \tag{126}
$$

where $b(\ell_1; T_1)$ and $b(\ell_2; T_2)$ are understood to be inserted into the summation sign $\sum_{T_1 T_2}$ in (123).

3.4. *Matrix Elements of Transition Operators*

From (22) to (24), one can see that in order to calculate the polarizabilities α_ℓ with ℓ up to 3, one also needs to evaluate the matrix elements of operators $Y_{\ell m}(\hat{\mathbf{r}}_i)$, $(\hat{\mathbf{r}}_1 \otimes \hat{\mathbf{r}}_2)^{(2)}$, $((\hat{\mathbf{r}}_2 \otimes \hat{\mathbf{r}}_2)^{(2)} \otimes \hat{\mathbf{r}}_1)^{(3)}$, and $((\hat{\mathbf{r}}_1 \otimes \hat{\mathbf{r}}_1)^{(2)} \otimes \hat{\mathbf{r}}_2)^{(3)}$. The angular coefficients of these matrix elements may be obtained using the graphical method. The final results are listed below [4].

$$\langle \mathcal{Y}_{\ell'_1 \ell'_2}^{L'M'} R' \| Y_{\ell m}(\hat{\mathbf{r}}_1) \| \mathcal{Y}_{\ell_1 \ell_2}^{LM} R \rangle = \sum_{qk} g_1^{(1)}(q) C_{\bar{c}qk} I_{\mathrm{R}}(\tilde{a}, \tilde{b}, \tilde{c}; \tilde{\alpha}, \tilde{\beta}; q, k),$$

$$g_1^{(1)}(q) = \sqrt{\frac{2\ell+1}{4\pi}} \varpi (-1)^{L+L'+\ell+\ell'_2} \sum_{\Lambda} (2\Lambda+1) \begin{pmatrix} \ell & \ell_1 & \Lambda \\ 0 & 0 & 0 \end{pmatrix} \begin{pmatrix} \ell'_1 & \Lambda & q \\ 0 & 0 & 0 \end{pmatrix}$$

$$\times \begin{pmatrix} \ell'_2 & \ell_2 & q \\ 0 & 0 & 0 \end{pmatrix} \begin{Bmatrix} L & \ell_1 & \ell_2 \\ \Lambda & L' & \ell \end{Bmatrix} \begin{Bmatrix} \ell'_2 & \ell_2 & q \\ \Lambda & \ell'_1 & L' \end{Bmatrix}. \tag{127}$$

$$\langle \mathcal{Y}_{\ell'_1 \ell'_2}^{L'M'} R' \| Y_{\ell m}(\hat{\mathbf{r}}_2) \| \mathcal{Y}_{\ell_1 \ell_2}^{LM} R \rangle = \sum_{qk} g_2^{(1)}(q) C_{\bar{c}qk} I_{\mathrm{R}}(\tilde{a}, \tilde{b}, \tilde{c}; \tilde{\alpha}, \tilde{\beta}; q, k),$$

$$g_2^{(1)}(q) = \sqrt{\frac{2\ell+1}{4\pi}} \varpi (-1)^{\ell'_1} \sum_{\Lambda} (2\Lambda+1) \begin{pmatrix} \ell & \ell_2 & \Lambda \\ 0 & 0 & 0 \end{pmatrix} \begin{pmatrix} \ell'_2 & \Lambda & q \\ 0 & 0 & 0 \end{pmatrix} \begin{pmatrix} \ell'_1 & \ell_1 & q \\ 0 & 0 & 0 \end{pmatrix}$$

$$\times \begin{Bmatrix} L & \ell_2 & \ell_1 \\ \Lambda & L' & \ell \end{Bmatrix} \begin{Bmatrix} \ell'_1 & \ell_1 & q \\ \Lambda & \ell'_2 & L' \end{Bmatrix}. \tag{128}$$

$$\langle \mathcal{Y}_{\ell'_1 \ell'_2}^{L'M'} R' \| (\hat{\mathbf{r}}_1 \otimes \hat{\mathbf{r}}_2)^{(2)} \| \mathcal{Y}_{\ell_1 \ell_2}^{LM} R \rangle = \sum_{qk} g_1^{(2)}(q) C_{\bar{c}qk} I_{\mathrm{R}}(\tilde{a}, \tilde{b}, \tilde{c}; \tilde{\alpha}, \tilde{\beta}; q, k),$$

$$g_1^{(2)}(q) = \sqrt{5}\, \varpi (-1)^{\ell'_1 + \ell_2 + L' + 1} \sum_{\Lambda_1 \Lambda_2} (\Lambda_1, \Lambda_2) \begin{pmatrix} 1 & \ell_1 & \Lambda_1 \\ 0 & 0 & 0 \end{pmatrix} \begin{pmatrix} 1 & \ell_2 & \Lambda_2 \\ 0 & 0 & 0 \end{pmatrix}$$

$$\times \begin{pmatrix} \ell'_1 & \Lambda_1 & q \\ 0 & 0 & 0 \end{pmatrix} \begin{pmatrix} \ell'_2 & \Lambda_2 & q \\ 0 & 0 & 0 \end{pmatrix} \begin{Bmatrix} \ell'_2 & \Lambda_2 & q \\ \Lambda_1 & \ell'_1 & L' \end{Bmatrix} \begin{Bmatrix} L & \ell_1 & \ell_2 \\ 2 & 1 & 1 \\ L' & \Lambda_1 & \Lambda_2 \end{Bmatrix}. \tag{129}$$

$$\langle \mathcal{Y}_{\ell'_1 \ell'_2}^{L'M'} R' \| ((\hat{\mathbf{r}}_2 \otimes \hat{\mathbf{r}}_2)^{(2)} \otimes \hat{\mathbf{r}}_1)^{(3)} \| \mathcal{Y}_{\ell_1 \ell_2}^{LM} R \rangle$$

$$= \sum_{qk} g_1^{(3)}(q) C_{\bar{c}qk} I_{\mathrm{R}}(\tilde{a}, \tilde{b}, \tilde{c}; \tilde{\alpha}, \tilde{\beta}; q, k),$$

$$g_1^{(3)}(q) = \sqrt{\frac{14}{3}}\, \varpi (-1)^{\ell_1 + \ell'_2 + L' + 1} \sum_{\Lambda_1 \Lambda_2} (\Lambda_1, \Lambda_2) \begin{pmatrix} 1 & \ell_1 & \Lambda_1 \\ 0 & 0 & 0 \end{pmatrix} \begin{pmatrix} 2 & \ell_2 & \Lambda_2 \\ 0 & 0 & 0 \end{pmatrix}$$

$$\times \begin{pmatrix} \ell'_1 & \Lambda_1 & q \\ 0 & 0 & 0 \end{pmatrix} \begin{pmatrix} \ell'_2 & \Lambda_2 & q \\ 0 & 0 & 0 \end{pmatrix} \begin{Bmatrix} \ell'_2 & \Lambda_2 & q \\ \Lambda_1 & \ell'_1 & L' \end{Bmatrix} \begin{Bmatrix} L & \ell_1 & \ell_2 \\ 3 & 1 & 2 \\ L' & \Lambda_1 & \Lambda_2 \end{Bmatrix}. \tag{130}$$

$$\langle \mathcal{Y}^{L'M'}_{\ell'_1 \ell'_2} R' || ((\hat{\mathbf{r}}_1 \otimes \hat{\mathbf{r}}_1)^{(2)} \otimes \hat{\mathbf{r}}_2)^{(3)} || \mathcal{Y}^{LM}_{\ell_1 \ell_2} R\rangle$$

$$= \sum_{qk} g_2^{(3)}(q) C_{\bar{c}qk} I_R(\tilde{a}, \tilde{b}, \tilde{c}; \tilde{\alpha}, \tilde{\beta}; q, k),$$

$$g_2^{(3)}(q) = \sqrt{\frac{14}{3}} \, \varpi(-1)^{\ell'_1 + \ell_2 + L' + 1} \sum_{\Lambda_1 \Lambda_2} (\Lambda_1, \Lambda_2) \begin{pmatrix} 2 & \ell_1 & \Lambda_1 \\ 0 & 0 & 0 \end{pmatrix} \begin{pmatrix} 1 & \ell_2 & \Lambda_2 \\ 0 & 0 & 0 \end{pmatrix}$$

$$\times \begin{pmatrix} \ell'_1 & \Lambda_1 & q \\ 0 & 0 & 0 \end{pmatrix} \begin{pmatrix} \ell'_2 & \Lambda_2 & q \\ 0 & 0 & 0 \end{pmatrix} \begin{Bmatrix} \ell'_2 & \Lambda_2 & q \\ \Lambda_1 & \ell'_1 & L' \end{Bmatrix} \begin{Bmatrix} L & \ell_1 & \ell_2 \\ 3 & 2 & 1 \\ L' & \Lambda_1 & \Lambda_2 \end{Bmatrix}. \tag{131}$$

In the above expressions, $\tilde{a} = a' + a$, etc.,

$$R = r_1^a r_2^b r_{12}^c e^{-\alpha r_1 - \beta r_2}, \tag{132}$$

$$R' = r_1^{a'} r_2^{b'} r_{12}^{c'} e^{-\alpha' r_1 - \beta' r_2}, \tag{133}$$

$$\varpi = (\ell'_1, \ell'_2, L', \ell_1, \ell_2, L)^{1/2}. \tag{134}$$

4. Calculations of Polarizabilities

4.1. *Computational Procedures*

As pointed before, the quality of calculated polarizabilities depends on the degree of completeness of intermediate states. It is therefore essential to have an adequate representation of the whole spectrum of the Hamiltonian. For atomic systems with more than one electron, the central problem is the inclusion of electron-electron correlations. The simplest approach for generating intermediate states is to diagonalize the Hamiltonian in a basis set constructed using Hylleraas coordinates. In this approach, the discrete eigenstates with positive eigenvalues represent the continuum. As the size of basis increases progressively, one can monitor the convergence pattern of the calculated polarizability. For excited states, one may also use the projection operator technique [22] to project all the low-lying states out. To be specific, let Ψ_0 be an excited state of interest and all the intermediate states with correct angular momentum symmetry below Ψ_0 be $\{\phi_i\}_{i=1}^g$. The contributions to the polarizability α_ℓ from all the low-lying states can be calculated explicitly and the contributions from the remaining part of the spectrum can be handled by diagonalizing the Hamiltonian in a restricted basis set

$$\left(1 - \sum_{i=1}^g |\phi_i\rangle\langle\phi_i|\right) \times \text{(basis function)}. \tag{135}$$

The resulting eigenvalues are all above the energy of state of interest and the problem is reduced to the problem for the ground-state case. The projection operator method is suited for low-lying states. For example, consider the $1s3s\,^1S$ state of helium. For the calculation of dipole polarizability α_1, the only state of P symmetry, which is below $1s3s\,^1S$, is the state of $|\phi_1\rangle = |1s2p\,^1P\rangle$. The contribution to α_1 from ϕ_1 can be calculated directly and the value is $-37.155\,811\,9(5)$. After diagonalizing the Hamiltonian in a restricted functional space formed by the projection operator $1 - |\phi_1\rangle\langle\phi_1|$, one obtains the total contribution from all the states of P symmetry above $1s3s\,^1S$ to be $16924.31(2)$. Combining these two contributions together yields the total α_1 of $16\,887.15(2)$, which is in agreement with the one calculated directly by diagonalizing the Hamiltonian in the whole functional space. Nevertheless, the projection operator method becomes more difficult to apply for high-lying Rydberg states. This is because that, as $\{\phi_i\}_{i=1}^g$ becomes more complete, the gram matrix becomes more singular, leading eventually to numerical instabilities, unless multiple precision arithmetic is adoped [23] beyond the regular quadruple precision (about 32 decimal digits). Therefore, the strategy we have adopted is to diagonalize the Hamiltonian directly without using the projection operator technique. The reliability of the method can be judged by the convergence pattern as the size of basis set for a given intermediate symmetry increases progressively.

4.2. *Symmetries*

The allowed angular symmetries of intermediate states that contribute to the 2^ℓ-pole polarizability α_ℓ can be determined by the triangle rule of angular momenta among (L_0, ℓ, L_n), as well as by the parity selection rule

$$\text{Parity } \Psi_0 \times \text{Parity } T_\ell^{(n+1)} \times \text{Parity } \Psi_n = 1, \qquad (136)$$

where Ψ_0 is the state of interest, $T_\ell^{(n+1)}$ is the transition operator, and Ψ_n is one of the intermediate states. For a two-electron system, for instance, the parity selection rule is

$$L_0 + \ell + \ell_1 + \ell_2 = \text{even}, \qquad (137)$$

where L_0 is the total angular momentum of the singly excited state Ψ_0 and (ℓ_1, ℓ_2) are the angular momenta of the two electrons forming the total angular momentum L_n of the intermediate states. Therefore, Ψ_n and Ψ_0 should have the same parity for even ℓ and the opposite parity for odd ℓ.

For example, suppose the state of interest is an even parity D state. For the dipole polarizability α_1, the parity of the intermediate states must be odd. On the other hand, from the triangle rule, the intermediate symmetries are P, D, and F. Thus P and F can be generated by the odd-parity configurations $(nsn'p)P$ and $(nsn'f)F + (npn'd)F$ respectively, whereas the D symmetry can be generated by the odd-parity configuration $(npn'd)D$. All other configurations are included implicitly through Hylleraas coordinates r_{12}^k.

4.3. An Example

Consider the dipole polarizability for the $1s3d\,^1D$ state of heliumlike ion Li^+. The wave function of $1s3d\,^1D$ can be constructed according to Drake [24]

$$
\begin{aligned}
\Psi(\mathbf{r}_1, \mathbf{r}_2) = &\sum_{i+j+k\leq\Omega_A} a_{ijk}^{(A)} r_1^i r_2^j r_{12}^k e^{-\alpha_A r_1 - \beta_A r_2} \mathcal{Y}_{02}^{2M}(\hat{\mathbf{r}}_1, \hat{\mathbf{r}}_2) \\
&+ \sum_{i+j+k\leq\Omega_B} a_{ijk}^{(B)} r_1^i r_2^j r_{12}^k e^{-\alpha_B r_1 - \beta_B r_2} \mathcal{Y}_{02}^{2M}(\hat{\mathbf{r}}_1, \hat{\mathbf{r}}_2) \qquad (138)\\
&+ \sum_{i+j+k\leq\Omega_C} a_{ijk}^{(C)} r_1^i r_2^j r_{12}^k e^{-\alpha_C r_1 - \beta_C r_2} \mathcal{Y}_{11}^{2M}(\hat{\mathbf{r}}_1, \hat{\mathbf{r}}_2) \pm \text{exchange}\,,
\end{aligned}
$$

where Ω_A, Ω_B, and Ω_C are three integers controlling the lengths of expansions in each blocks. One can see that the basis functions containing the configuration $(\ell_1, \ell_2) = (0, 2)$ is doubled into two blocks: A and B. One might think that such kind of wave function would contain near linear dependence terms and would lead to numerical problems. In fact, a complete optimization of the nonlinear parameters leads to well-separated values for the parameters. For block A, the optimum values of α_A and β_A are close to their screened hydrogenic values $\alpha_A \approx Z$ and $\beta_A \approx (Z-1)/n$, where n is the principal quantum number of the outer electron. The physical meaning is that this block describes the asymptotic behavior of the wave function. The nonlinear parameters for block B tends to be much larger, thus describing the complex inner correlation effects. The wave functions for the intermediate states of P, D, and F symmetries can also be constructed in a similar way. Table 1 presents a convergence study for the contributions to α_1 from the three intermediate symmetries ^1P, ^1D, and ^1F, where ^1D has a basic doubly-excited configuration of (pd), as the sizes of basis sets increase progressively. One can see that the overall convergence is excellent. A good way of insuring the completeness of basis sets is to check sum rules numerically. A neglection of an important configuration would make sum rules unbalanced. For the sum rule (67), the contribution from the configuration $(pd)D$ is $S_D^{(1)} = 0.3341998$, and the contribution from the configuration $(sf)F + (pd)F$ is $S_F^{(1)} = 1.864240$. Thus $2S_D^{(1)} + \frac{5}{7}S_F^{(1)} = 1.999999$, which is very close to the exact value of 2.

Table 1: Convergence of the contributions to the dipole polarizability of the heliumlike ion Li$^+$ in $1s3d\,^1D$ from the P, $(pd)D$, and F symmetries with respect to the sizes of basis sets N_D, N_P, $N_{(pd)D}$, and N_F, where N_D is for $1s3d\,^1D$, and N_P, $N_{(pd)D}$, and N_F are for the three intermediate symmetries. Units are atomic units.

N_D	N_P	$N_{(pd)D}$	N_F	$\alpha_1(^1P)$	$\alpha_1((pd)^1D)$	$\alpha_1(^1F)$
425	440	455	360	3053.95836	0.018656066	119.9695135
565	572	560	495	3053.95857	0.018656097	119.9695180
733	728	680	660	3053.95881	0.018656101	119.9695190
931	910	969	858	3053.95885	0.018656106	119.9695194

5. Review of Polarizabilities

5.1. *Helium*

The dipole polarizability α_1 of the ground state $(1s)^2\,^1S$ of atomic helium has been calculated by many authors. An extensive tabulation for α_1, together with computational techniques, has been made by Bonin and Kresin [6] for the period of 1968 to 1996. Since then, some important progress has been made towards higher precision, as well as towards relativistic and quantum electrodynamic (QED) effects. Table 2 highlights some of the most accurate calculations for the nonrelativistic value of α_1 in the past 25 years. Thakkar [25] evaluated α_1 using the all exponential basis where the correlation coordinate r_{12} is contained in

$$\exp(-\alpha r_1 - \beta r_2 - \gamma r_{12}), \qquad (139)$$

developed by Thakkar and Smith [26]. The accuracy he achieved is 50 ppm (parts per million). Using the same method of Thakkar, Bishop and Pipin [27] were able to push the accuracy in α_1 to 70 ppb (parts per billion). A similar result was also obtained by Jamieson et al. [28] using Hylleraas bases. Johnson and Cheng [29] calculated α_1 using the method of configuration interaction. Yan et al. [1] significantly improved the best previous results of Bishop and Pipin, and of Jamieson et al., by four more significant figures with more extensive Hylleraas basis sets. The Hylleraas coordinates were also used by Bhatia and Drachman [30] to evaluate α_1 to high precision. A further improvement upon the result of Yan *et al.* was made by Pachucki and Sapirstein [31] using all exponential basis with the nonlinear parameters α, β, and γ chosen in a random fashion. The dipole polarizability of helium was also calculated to high precision by Cencek *et al.* [32] using the basis of Gaussian-type

$$\exp(-\alpha|\mathbf{r}_1 - \mathbf{A}|^2 - \beta|\mathbf{r}_2 - \mathbf{B}|^2 - \gamma|\mathbf{r}_1 - \mathbf{r}_2|^2). \tag{140}$$

The nonrelativistic value of the dipole polarizability of the ground state of helium has now become a benchmark for other theoretical methods. The relativistic correction to the ground-state dipole polarizability of helium was evaluated by Johnson and Cheng [29] using the method of relativistic configuration interaction, as well as by Bhatia and Drachman [33] using Hylleraas coordinates. Pachucki and Sapirstein [31] also evaluated the corrections to the helium dipole polarizability due to the finite nuclear mass, leading relativistic and QED effects. Similar calculations were also performed by Cencek *et al.* [32] and by Lach *et al.* [34]. The important conclusion from these calculations is that the the finite nuclear mass, relativistic, and QED effects almost completely cancel out, leading to the final establishment of the dipole polarizability of the ground state of helium [31]

$$\alpha_1 = 1.383191(2) \, a_0^3, \tag{141}$$

accurate to 1.5 ppm. The experimental measurement of α_1 of helium has been performed by several groups [6], where the most accurate determination of $\alpha_1 = 1.38376(7) \, a_0^3$ was made by Gugan and Michel [35] using the dielectric constant technique. Comparing with the theoretical value (141) one can see that the discrepancy between theory and experiment is at the level of 0.04%.

In addition to α_1, the quadrupole and octupole polarizabilities of the ground state of helium α_2 and α_3 were also calculated nonrelativistically by several authors, including Thakkar [25], Bishop and Pipin [27], Chen and Chung [36], Yan *et al.* [1], and Bhatia and Drachman [30].

Table 2: The dipole polarizability of the ground state of helium for the case of infinite nuclear mass, in atomic units.

Author (year)	α_1	Ref.
Thakkar (1981)	1.383 12	25
Bishop and Pipin (1993)	1.383 192	27
Jamieson *et al.* (1995)	1.383 192	28

Johnson and Cheng (1996)	1.383 199(8)	[29]
Yan et al. (1996)	1.383 192 174 40(5)	[1]
Bhatia and Drachman (1997)	1.383 192 18	[30]
Pachucki and Sapirstein (2000)	1.383 192 174 455(1)	[31]
Cencek et al. (2001)	1.383 192 174 17	[32]

Although most of the theoretical work has been limited to the ground and the metastable states of helium, the extension to Rydberg states of helium has recently been made by Yan. The static dipole, quadrupole, and octupole polarizabilities have been calculated precisely by Yan [22,37] for the Rydberg series $1snL\,^1L$ and $1snL\,^3L$ with L from 0 to 9 and n up to 10 $(n \geq L + 1)$. One of the important applications of polarizabilities for an atom in Rydberg states is to understand the threshold behavior of photodetachment cross sections for negative ions such as He^-, Li^-, Na^-, and K^-. The recent experiment by Kiyan et al. [38] on the doubly excited states of He^- has shown that the polarizabilities of excited states of the parent atom He are crucial in the determination of the number of bound states below a particular excited state of the parent atom. For heliumlike ions, the precise values of polarizabilities are available [30,39,40,10] for $(1s)^2\,^1S$, $1s2s\,^1S$, $1s2s\,^3S$, $1s2p\,^1P$, and $1s2p\,^3P$ for Z up to 10, as well as [41] $1s3d\,^1D$ and $1s3d\,^3D$ for Z up to 15. Finally, the polarizability of the two-electron negative hydrogen ion H^- in $(1s)^2\,^1S$, the only true bound state that this system can have [42], was calculated nonrelativistically to high precision by Golver and Weinhold [43], Pipin and Bishop [44], and Bhatia and Drachman [45] using Hylleraas-type bases.

5.2. Lithium

Lithium, a three-electron atomic system, is the simplest atom with both an open and closed shell. It also serves as a prototype for other alkali-metal atoms. Compared to atomic helium, precision calculation of spectroscopic properties of lithium, including relativistic and QED effects, is still a challenging problem, even though significant theoretical advances have been made recently [46,47]. Nevertheless, the polarizabilities of the ground state $(1s)^2 2s\,^2S$ of lithium have been calculated by many authors [48,49,50,51,52,53,54,55,56,57,58,59,60,1]. The works by Pipin and Bishop [55], by Kassimi and Thakkar [59], and by Yan et al. [1] are among the most accurate ones. Pipin and Bishop [55] used the CI-Hylleraas method, where each term in the basis set contains only one factor of the interelectronic coordinate r_{ij}^k. In the CI-Hylleraas method, the three-electron integrals become tractable [61]. The work by Kassimi and Thakkar [59] was performed using the finite-field coupled-cluster at the single and double substitution level with perturbative correction for triple substitutions. In the calculations of Yan *et al.* [1], fully-correlated Hylleraas basis sets were adopted, yielding the most precise value for the dipole polarizability of the ground state lithium $164.111(2)$ a_0^3. Yan *et al.* [1] also calculated the quadrupole and octupole polarizabilities of the ground state lithium to high precision.

The dipole polarizability of lithium in $(1s)^2 2p\,^2P$ was evaluated by Pipin and Bishop [62] using CI-Hylleraas method. In their calculation, however, the contribution from the doubly excited intermediate states of $(pp')P$ symmetry was ignored. More precise calculation has recently been done by Zhou *et al.* [63] using Hylleraas coordinates, where all the intermediate symmetries of S, $(pp')P$, and D have been included explicitly. Our value of α_1 is $126.948(4)$ a_0^3, in comparison with the value of Pipin and Bishop 126.844 a_0^3. The polarizabilities of low-lying S, P, and D states of lithium were also investigated by Ashby and van Wijngaarden [64] using the method of Coulomb approximation. Finally, the reader is referred to the review article by van Wijngaarden [65] on the developments of measurements of polarizabilities up to 1996. Recently Ashby *et al.* [66] performed an experimental study of Stark shifts of the lithium $(1s)^2 3d\,^2D_{3/2,5/2}$ states and determined the corresponding dipole polarizabilities.

5.3. *Beryllium*

High precision calculation for beryllium is an extremely challenging problem in atomic few-body physics. So far, there has been no fully-correlated Hylleraas-type calculation even for the ground state energy, although a few investigations [67,68,69,70,71,72,73] have been published using restricted basis functions in order to avoid some intractable four-electron integrals [74]. Nevertheless, important progress has been made by Komasa and collaborators [75,76] in high precision variational calculations of beryllium energy levels using the method of exponentially correlated Gaussian (ECG) bases. For the ground state, the lowest-order relativistic and QED corrections have also been evaluated [77]. In ECG, the orbital part of basis function is written in the form

$$\exp[-\mathbf{r}\mathbf{A}\mathbf{r}^T], \qquad (142)$$

where $\mathbf{r}$ is the all electron position vectors and $\mathbf{A}$ is a positive definite matrix determined variationally. Recently Komasa [78,79] has applied the method of

ECG to calculate the dipole and quadrupole polarizabilities of the ground state of beryllium to high precision. The reported values in atomic units for α_1 and α_2 are 37.755 and 300.96 respectively, which represent the most accurate ones in the literature. A detailed comparison among various calculations can be found in [78].

5.4. *The hydrogen molecular ion*

The hydrogen molecular ion H_2^+ is another fundamental three-body system, which can be considered as an extreme case of "heliumlike" atom. Like helium, H_2^+ has been used as a benchmark molecule for other molecular computational methods. Recently, the nonrelativistic energy eigenvalues of H_2^+ have been calculated to high precision nonadiabatically, using various methods, including the traditional molecular physics method [80,81,82], perimetric coordinates [83], and Hylleraas or Hylleraas-type coordinates [84,85,86,9,87]. For example, the nonrelativistic ground state energy has been calculated to about 24 significant figures [87], which is more than sufficient for the calculation of higher-order (up to α^5 a.u.) relativistic and QED corrections, although such corrections have not yet been derived.

Recent theoretical studies of the dipole polarizabilities of the hydrogen molecular ions have been stimulated by the measurements of Jacobson *et al.* [88,89] for H_2^+ and D_2^+. Using the finite element method, Shertzer and Greene [90] evaluated the polarizabilities of the ground state H_2^+ and D_2^+. Bhatia and Drachman [84,91] calculated the polarizabilities of H_2^+, D_2^+, and HD^+ using second-order perturbation theory in Hylleraas coordinates. Taylor *et al.* [92] obtained the polarizabilities of H_2^+ and D_2^+ variationally within the framework of the traditional molecular physics method. Similar approach was also used by Moss and Valenzano [81,93] who calculated the polarizabilities of H_2^+, D_2^+, and HD^+ not only for the ground states but also for many vibrational states. Hilico *et al.* [94] evaluated the polarizabilities of H_2^+ and D_2^+ using perimetric coordinates. Finally, Korobov [95] calculated the lowest-order relativistic corrections to the ground state dipole polarizabilities of H_2^+ and D_2^+. The most accurate nonrelativistic values of the dipole polarizabilities of H_2^+, D_2^+, T_2^+, HD^+, HT^+, and DT^+ are due to Yan *et al.* who used Hylleraas coordinates [9]. For instance, the dipole polarizability of H_2^+ has been established to be $\alpha_1 = 3.16872580267(1)\, a_0^3$. However, there is an existing discrepancy between theory and experiment [88,89] for the dipole polarizability of H_2^+ even though the relativistic correction of the order α^2 has been considered [95]. More studies are needed on both theory and experiment. Very recently Zhang and Yan [4] have extended their calculations to α_ℓ with ℓ up to 4 for the systems H_2^+, D_2^+, T_2^+, HD^+, HT^+, and DT^+ in their ground states.

5.5. *The hydrogen molecule*

With one more electron added to H_2^+, the hydrogen molecule H_2 is much more difficult to calculate. Most of the studies on H_2 have been within the framework of the Born-Oppenheimer approximation until recently [96,97,98,99]. Kinghorn and Adamowicz [96] performed variational calculation on the ground state energy of H_2 nonadiabatically, using exponentially correlated Gaussian (ECG) basis sets. Based on the convergence study they gave, the ground state energy seems to converge to about 7 to 8 significant figures, although the energy they presented has 11 digits. Later, they [97] extended their calculation of H_2 to other isotopomers D_2, T_2, HD, HT, and DT with similar accuracy. Furthermore, Cafiero and Adamowicz [98] performed the first nonadiabatical calculations of polarizabilities of H_2 and its isotopomers. For example, the reported nonadiabatical value of the dipole polarizability of H_2 is 6.74 a_0^3. However, since neither uncertainty nor convergence study is given in their paper, it is difficult to see how many significant figures have been obtained in their values. It would be very interesting to see how much improvement one can made using Hylleraas coordinates, as for the case of H_2^+.

5.6. *Positronium hydride*

Positronium hydride PsH is a four-body Coulombic system containing a positronium and a hydrogen. This exotic system has recently been a focus of some theoretical attention since its discovery experimentally by Schrader [100] *et al.*. This system is believed to have only one bound state which has been calculated to high precision, including the lowest-order relativistic corrections [101]. The polarizabilities of PsH had been controversial [102,103,104] until the work by Yan [105] was published in 2002. Yan calculated the polarizabilities of PsH using Hylleraas coordinates, where the dipole polarizability was determined with an accuracy of one part in 10^5, representing an improvement of three orders of magnitude over the best previous result of Mella *et al.* [104].

6. A Concluding Remark

Hylleraas coordinates have been demonstrated to be powerful for calculating properties, such as energy eigenvalues and polarizabilities, of three- and four-body atomic systems. For few-body molecular systems, however, much better convergence of these properties can be expected if one includes all the inter-particle distances on the shoulder of exponential functions. For H_2^+ for example, instead of using (74), one could use

$$\chi_{ijk}(\alpha, \beta, \gamma) = r_1^i\, r_2^j\, r_{12}^k\, e^{-\alpha r_1 - \beta r_2 - \gamma r_{12}} \mathcal{Y}_{\ell_1 \ell_2}^{LM}(\hat{\mathbf{r}}_1, \hat{\mathbf{r}}_2)\,, \qquad (143)$$

due to the fact that there are two hydrogen formation channels in H_2^+. The explicit use of $e^{-\gamma r_{12}}$ will enhance the rate of convergence dramatically, as demonstrated by Cassar and Drake [87]. For four-body molecular systems, such as H_2, $H^+H^+\mu^-e^-$, $H^+\mu^+e^-e^-$, and HeH^+, one needs to evaluate the following integral:

$$\int d\mathbf{r}_1 d\mathbf{r}_2 d\mathbf{r}_3 r_1^{j_1}\, r_2^{j_2}\, r_3^{j_3}\, r_{12}^{j_{12}}\, r_{23}^{j_{23}}\, r_{31}^{j_{31}}\, e^{-\alpha_1 r_1 - \alpha_2 r_2 - \alpha_3 r_3 - \alpha_{12} r_{12} - \alpha_{23} r_{23} - \alpha_{31} r_{31}}\,.$$

$$(144)$$

However, efficient evaluation of this type of integrals has not yet been found, although there is an analytic expression derived first by Fromm and Hill [106], which is rather complicated involving subtle tracking of the branches of the multi-valued complex functions. Some attempts of simplification of the original Fromm and Hill formula were made [107,108]. For a recent review, see the article by Harris [109]. Once this bottleneck has been removed, a large number of interesting four-body molecular systems can be accessible to high precision study.

Acknowledgments

Research support by the Natural Sciences and Engineering Research Council of Canada, by the ACRL of the University of New Brunswick, by SHARCnet, and by WestGrid is gratefully acknowledged.

References

1. Z.-C. Yan, J. F. Babb, A. Dalgarno, and G. W. F. Drake, Phys. Rev. A **54**, 2824 (1996).
2. A. R. Edmonds, *Angular Momentum in Quantum Mechanics* (Princeton University Press, Princeton, 1985).
3. L. I. Schiff, *Quantum Mechanics* (McGraw-Hill Book Company, New York, 1968), p. 265.

4. J.-Y. Zhang and Z.-C. Yan, J. Phys. B **37**, 723 (2004).
5. Z.-C. Yan, A. Dalgarno, and J. F. Babb, Phys. Rev. A **55**, 2882 (1997).
6. K. D. Bonin and V. V. Kresin, *Electric-Dipole Polarizabilities of Atoms, Molecules and Clusters* (World Scientific, Singapore, 1997).
7. Z.-C. Yan and G. W. F. Drake, Phys. Rev. A **52**, R4316 (1995).
8. B.-L. Zhou, J.-M. Zhu, and Z.-C. Yan (to be submitted).
9. Z.-C. Yan, J.-Y. Zhang, and Y. Li, Phys. Rev. A **67**, 062504 (2003).
10. Z.-C. Yan, J.-M. Zhu, and B.-L. Zhou, Phys. Rev. A **62**, 034501 (2000).
11. J. Y. Zeng, *Quantum Mechanics* (Science Press, Beijing, 1995, in Chinese), Vol. I, p. 254.
12. G. B. Arfken and H. J. Weber, *Mathematical Methods for Physicists, 5th edition* (Harcourt/Academic Press, San Diego, 2001).
13. Z.-C. Yan and J.-Y. Zhang, J. Phys. B **37**, 1055 (2004).
14. B.-L Zhou, J.-M. Zhu, and Z.-C. Yan, J. Phys. B **38**, 305 (2005).
15. M. Suffczyński, T. Kotowski and L. Wolniewicz, Acta Phys. Pol. A **102**, 351 (2002).
16. Z.-C. Yan and G. W. F. Drake, Chem. Phys. Lett. **259**, 96 (1996).
17. J. F. Perkins, J. Chem. Phys. **48**, 1985 (1968).
18. R. N. Zare, *Angular Momentum* (John Wiley & Sons, New York, 1988).
19. M. Abramowitz and I. A. Stegun, *Handbook of Mathematical Functions* (Dover, New York, 1972).
20. I. S. Gradshteyn and I. M. Ryzhik, *Table of Integrals, Series, and Products* (Academic, San Diego, 1980).
21. D. M. Brink and G. R. Satchler, *Angular Momentum*, second edition (Clarendon, Oxford, 1968).
22. Z.-C. Yan, Phys. Rev. A **62**, 052502 (2000).
23. D. H. Bailey, Comput. Sci. Eng. **2**, 24 (2000).
24. G. W. F. Drake, in *Long-Range Casimir Forces: Theory and Recent Experiments on Atomic Systems*, edited by F. S. Levin and D. A. Micha (Plenum, New York, 1993).
25. A. J. Thakkar, J. Chem. Phys. **75**, 4496 (1981).
26. A. J. Thakkar and V. H. Smith, Jr., Phys. Rev. A **15**, 1 (1977).
27. D. M. Bishop and J. Pipin, Int. J. Quantum Chem. **45**, 349 (1993).
28. M. J. Jamieson, G. W. F. Drake, and A. Dalgarno, Phys. Rev. A **51**, 3358 (1995).
29. W. R. Johnson and K. T. Cheng, Phys. Rev. A **53**, 1375 (1996).
30. A. K. Bhatia and R. J. Drachman, Can. J. Phys. **75**, 11 (1997).
31. K. Pachucki and J. Sapirstein, Phys. Rev. A **63**, 012504 (2000).
32. W. Cencek, K. Szalewicz, and B. Jeziorski, Phys. Rev. Lett. **86**, 5675 (2001).
33. A. K. Bhatia and R. J. Drachman, Phys. Rev. A **58**, 4470 (1998).
34. G. Łach , B. Jeziorski, and K. Szalewicz, Phys. Rev. Lett. **92**, 233001 (2004).
35. D. Gugan and G. W. Michel, Mol. Phys. **39**, 783 (1980).
36. M.-K. Chen and K. T. Chung, Phys. Rev. A **53**, 1439 (1996).
37. Z.-C. Yan, J. Phys. B **35**, 2713 (2002).
38. I. Yu. Kiyan, U. Berzinsh, D. Hanstorp, and D. J. Pegg, Phys. Rev. Lett. **81**, 2874 (1998).

39. J.-M. Zhu, B.-L. Zhou, and Z.-C. Yan, Chem. Phys. Lett. **313**, 184 (1999).
40. J.-M. Zhu, B.-L. Zhou, and Z.-C. Yan, Mol. Phys. **98**, 529 (2000).
41. Y. Li, M.Sc. thesis, University of New Brunswick (2004) (unpublished).
42. R. N. Hill, Phys. Rev. Lett. **38**, 643 (1977).
43. R. M. Glover and F. Weinhold, J. Chem. Phys. **65**, 4913 (1976).
44. J. Pipin and D. M. Bishop, J. Phys. B **25**, 17 (1992).
45. A. K. Bhatia and R. Drachman, J. Phys. B **27**, 1299 (1994).
46. Z.-C. Yan and G. W. F. Drake, Phys. Rev. Lett. **91**, 113004 (2003).
47. K. Pachucki, Phys. Rev. A **71**, 012503.
48. F. Maeder and W. Kutzelnigg, Chem. Phys. **42**, 95 (1979).
49. J. Muszyńska, D. Papierowska, J. Pipin, and W. Woźnicki, Int. J. Quantum Chem. **22**, 1153 (1982).
50. J. Pipin and W. Woźnicki, Chem. Phys. Lett. **95**, 392 (1983).
51. C. Pouchan and D. M. Bishop, Phys. Rev. A **29**, 1 (1984).
52. W. Müller, J. Flesch, and W. Meyer, J. Chem. Phys. **80**, 3297 (1984).
53. P. J. Knowles and W. J. Meath, Chem. Phys. Lett. **124**, 164 (1986).
54. G. Maroulis and A. J. Thakkar, J. Phys. B **22**, 2439 (1989).
55. J. Pipin and D. M. Bishop, Phys. Rev. A **45**, 2736 (1992).
56. D. V. Ponomarenko and A. F. Shestakov, Chem. Phys. Lett. **210**, 269 (1993).
57. Z-W Wang and K. T. Chung, J. Phys. B **27**, 855 (1994).
58. M. Mérawa, M. Rérat, and C. Pouchan, Phys. Rev. A **49**, 2493 (1994).
59. N. E. Kassimi and A. J. Thakkar, Phys. Rev. A **50**, 2948 (1994).
60. C. Laughlin, J. Phys. B **28**, L701 (1995).
61. G. W. F. Drake and Z.-C. Yan, Phys. Rev. A **52**, 3681 (1995).
62. J. Pipin and D. M. Bishop, Phys. Rev. A **47**, R4571 (1993).
63. B.-L. Zhou, J.-M. Zhu, and Z.-C. Yan (to be submitted).
64. R. Ashby and W. A. van Wijngaarden, J. Quant. Spectrosc. Radiat. Transfer **76**, 467 (2003).
65. W. A. van Wijngaarden, Adv. At. Mol. Opt. Phys. **36**, 141 (1996).
66. R. Ashby, J. J. Clarke, and W. A. van Wijngaarden, Eur. Phys. J. D **23**, 327 (2003).
67. L. Szasz and J. Byrne, Phys. Rev. **158**, 34 (1967).
68. R. F. Gentner and E. A. Burke, Phys. Rev. **176**, 63 (1968).
69. J. S. Sims and S. A. Hagstrom, Phys. Rev. A **4**, 908 (1971).
70. J. F. Perkins, Phys. Rev. A **8**, 700 (1973).
71. J. S. Sims and S. A. Hagstrom, Int. J. Quantum Chem. **9**, 149 (1975).
72. G. Büsse and H. Kleindienst, Phys. Rev. A **51**, 5019 (1995).
73. G. Büsse, H. Kleindienst, and A. Lüchow, Int. J. Quantum Chem. **66**, 241 (1998).
74. F. W. King, J. Chem. Phys. **99**, 3622 (1993).
75. J. Komasa, W. Cencek, and J. Rychlewski, Phys. Rev. A **52**, 4500 (1995).
76. J. Komasa and J. Rychlewski, Chem. Phys. Lett. **342**, 185 (2001).
77. K. Pachucki and J. Komasa, Phys. Rev. Lett. **92**, 213001 (2004).
78. J. Komasa, Phys. Rev. A **65**, 012506 (2001).
79. J. Komasa, Chem. Phys. Lett. **363**, 307 (2002).
80. R. E. Moss, J. Phys. B **32**, L89 (1999).

81. R. E. Moss, Chem. Phys. Lett. **311**, 231 (1999).
82. J. M. Taylor, Z.-C. Yan, A. Dalgarno, and J. F. Babb, Mol. Phys. **97**, 25 (1999).
83. L. Hilico, N. Billy, B. Grémaud, and D. Delande, Eur. Phys. J. D **12**, 449 (2000).
84. A. K. Bhatia and R. J. Drachman, Phys. Rev. A **59**, 205 (1999).
85. V. I. Korobov, Phys. Rev. A **61**, 064503 (2000).
86. A. M. Frolov, J. Phys. B **35**, L331 (2002).
87. M. M. Cassar and G. W. F. Drake, J. Phys. B **37**, 2485 (2004).
88. P. L. Jacobson, D. S. Fisher, C. W. Fehrenbach, W. G. Sturrus, and S. R. Lundeen, Phys. Rev. A **56**, R4361 (1997); **57**, 4065(E) (1998).
89. P. L. Jacobson, R. A. Komasa, W. G. Sturrus, and S. R. Lundeen, Phys. Rev. A **62**, 012509 (2000).
90. J. Shertzer and C. H. Greene, Phys. Rev. A **58**, 1082 (1998).
91. A. K. Bhatia and R. J. Drachman, Phys. Rev. A **61**, 032503 (2000).
92. J. M. Taylor, A. Dalgarno, and J. F. Babb, Phys. Rev. A **60**, R2630 (1999).
93. R. E. Moss and L. Valenzano, Mol. Phys. **100**, 1527 (2002).
94. L. Hilico, N. Billy, B. Grémaud, and D. Delande, J. Phys. B **34**, 491 (2001).
95. V. I. Korobov, Phys. Rev. A **63**, 044501 (2001).
96. D. B. Kinghorn and L. Adamowicz, Phys. Rev. Lett. **83**, 2541 (1999).
97. D. B. Kinghorn and L. Adamowicz, J. Chem. Phys. **113**, 4203 (2000).
98. M. Cafiero and L. Adamowicz, Phys. Rev. Lett. **89**, 073001 (2002).
99. M. Cafiero, S. Bubin, and L. Adamowicz, Phys. Chem. Chem. Phys. **5**, 1491 (2003).
100. D. M. Schrader, F. M. Jacobsen, N.-P. Frandsen, and Ulrik Mikkelsen, Phys. Rev. Lett. **69**, 57 (1992).
101. Z.-C. Yan and Y. K. Ho, Phys. Rev. A **60**, 5098 (1999).
102. C. Le Sech and B. Silvi, Chem. Phys. **236**, 77 (1998).
103. M. Mella, G. Morosi, D. Bressanini, and S. Elli, J. Chem. Phys. **113**, 6154 (2000).
104. M. Mella, D. Bressanini, and G. Morosi, Phys. Rev. A **63**, 024503 (2001).
105. Z.-C. Yan, J. Phys. B **35**, L345 (2002).
106. D. M. Fromm and R. N. Hill, Phys. Rev. A **36**, 1013 (1987).
107. F. E. Harris, Phys. Rev. A **55**, 1820 (1997).
108. F. E. Harris, A. M. Frolov, and V. H. Smith, Jr., J. Chem. Phys. **119**, 8833 (2003).
109. F. E. Harris, Adv. Quantum Chem. **47**, 129 (2004).

CHAPTER 16

NONLINEAR OPTICAL PROPERTIES OF TRANSITION-METAL CLUSTERS

Kechen Wu

State Key Laboratory of Structural Chemistry, Fujian Institute of Research on the Structure of Matter, Chinese Academy of Sciences, Fuzhou, Fujian 350002, China
E-mail: wkc@fjirsm.ac.cn

In this mini-review we presented the recent theoretical advances in calculations of the electric hyperpolarizability of transition-metal clusters with the emphasis on the investigations in our laboratory including the first hyperpolarizability, IR transparent spectra and UV-vis absorption spectra of a series of inorganic transition-metal cluster compounds. The perspectives of transition-metal clusters applied in IR second harmonic generation (SHG) devices were demonstrated. The adapted far/medium IR optical transparency, distinctive stereo charge-transfer character, diverse structures and flexible coordinated organic/inorganic ligand make transition-metal cluster compounds great potential for nonlinear optical materials. We predicted several potential IR SHG transition-metal materials with expectation to be further verified by experiment. The study may benefit to the efforts that have been made to search novel IR nonlinear optical materials.

1. Introduction

The colorful nonlinear optical (NLO) phenomena are displayed via the special materials called NLO materials. In 1961, when Franken and coworkers firstly detected the second harmonic generation (SHG) of ruby laser light,[1] they discovered the first NLO material as well, a quartz single crystal. In the past four decades, novel NLO materials developed very rapidly and kept in great demand due to the critical role that they are playing in the contemporary photonic devices.[2] At present stage, the key NLO materials applied in the different regions of optical spectrum could be listed a long table. Some excellent inorganic mineral oxides NLO materials including KH_2PO_4 (KDP), $LiNbO_3$, β-BaB_2O_4 (BBO), and LiB_3O_5 (LBO) have already been commercialized. Efforts have also been extensively devoted to the studies of semiconductor, conjugated polymers and organic crystals.[2, 3] In recent years, much attention has also been paid to fullerenes,[4] and organometallic compounds.[5] Inorganic transition-metal coordinated cluster compounds, on the other hand, seem to have been remained unexplored until very recently.[6] In fact, transition-metal clusters offer a great variety of structural configurations and a diversity of electronic properties shaped by the metal-characterized core. They may comprise a promising family of NLO materials, besides its special roles demonstrated in catalytic reactions,[7] biological chemical processes[8] and magnetic materials.[9]

The development of NLO theory accelerates the process of discovery of novel NLO materials. The successful D-π-A theory[10] for organic molecules has been extended to understanding the NLO nature of hybrid and even inorganic NLO materials. Recently, the theory is developing towards the octopolar theory.[11] Organometallic NLO materials are developed rationally by introducing heavy transition-metal elements to organic NLO-active moieties to enhance the electronic accepting/donating ability in order to enhance NLO responses. At molecular level, transition-metal coordinated A/D effectively enlarge the hyperpolarizability, β. For example, ferrocenyl organometallic complexes directed to the second-order NLO properties having been the hot topic since the powders of cis-1-ferrocenyl-2-(N-Nitrophenyl)ethylene were reported to possess remarkable macroscopic second-order nonlinear optical (NLO) susceptibility.[12] Most recently Robinson and coworkers reported a series of ferrocenyl nonlinear optical chromophores.[13] The transition-metal coordinated NLO chromophores with Fc groups as the electron donor moieties have been measured to have large hyperpolarizabilities and electrooptic coefficients as well. The study has demonstrated the congruence of the theory and experiments in the discovery of novel NLO materials. The first static hyperpolarizabilities of some ferrocenyl complexes have been well predicted at MP2 quantum chemical level.[14] We analyzed the contribution of electronic excitation to the first hyperpolarizability through the combination of the two-level model with the calculated linear optical parameters.

Some reports investigated the transition-metal coordinated cluster complexes for NLO properties. For example, Coe and coworkers investigated various Ru[II] complexes with divers substituted ligands.[15] They employed Hyper-Rayleigh Scattering (HRS) measurement and electronic Stark effect spectroscopy combined with theoretical calculations to study detailed structure-property correlations for the molecular hyperpolarizability. Coe pointed out that the key contribution for β came from metal-to-ligand charge transfer (MLCT) for this type of mononuclear TM complexes. As a result, the modification of the ligands could tune the NLO response of the chromophores. Another example is the transition-metal-based Schiff base complexes. Di Bella and coworkers presented many reports on this topic.[16] In the most recent report, they found that Ni[II] complex crystal exhibits SHG efficiency 1.2 times that of urea crystal.[17] This is an exciting result suggesting the brightness along this direction. Di Bella and coworkers also employed theoretical methods to elucidate and predict molecular NLO response in this series of Ru[II]/Zn[II] mononuclear complexes and pointed out the significant octopolar contribution to optical nonlinearity.

Strictly speaking, in the above works D-π-A theory for organic molecule were employed in analyzing NLO nature of the transition-metal coordinated mononuclear clusters. The colorfulness of transition-metal cluster exhibits in multi-nuclear compounds. Mo/W-Cu/Ag-S cluster is a representative system reported to have NLO property.[18] In the most recent report, Xin and coworkers discovered a nest-shaped Mo/W-Ag-S hetero-bimetallic cluster with significant third-order NLO property.[19] Actually, Xin and his coworkers reported many types of Mo/W-Cu/Ag sulfur transition-metal cluster with NLO properties. Their exciting works emphasized on the third-order NLO properties. In our laboratory the experimental and theoretical works are emphasized on the discovery of the IR SHG materials in transition-metal clusters.

It is believed that metal clusters may combine some advantages of inorganic materials (such as colorful structures and high stabilities) and organic materials (such as easy to be structurally tailed for large nonlinear optical polarizabilities). However it still has a long way to go to establish the theory of NLO property of transition-metal cluster. We listed some attractive figures of transition-metal cluster for NLO materials. (1) Diversity of configurations. It is possible to tune the molecular hyperpolarizability by modifying and even designing the ancillary ligands of transition-metal clusters due to the key contribution of the metal-ligand joint-effect to NLO response. (2) Distinct CT characteristics. Many evidences suggested the special stereo CT characteristics in transition-metal clusters. In some circumstance, three-dimension CT may result in very large NLO response. The efforts that people made on searching novel IR NLO materials in transition-metal cluster will definitely have rewards and it will be gigantic.

In the past decade, a great improvement has been made in computational NLO material science within the first principle framework. The in-depth understanding of the relationship between the microscopic structure and NLO functionality, the pre-designed tailing/modifying at atomic/molecular level and the modeling/simulation of specified NLO properties become the key and efficient way in developing novel NLO materials. Maroulis performed numerous high-level first-principle calculations on molecular polarizability and hyperpolarizability.[20] The contributions are great to the accurate computations and hence the thorough understanding of molecular optical nonlinear.

However, the high-level first-principle computational study of the NLO property of the inorganic transition-metal clusters remains less unexplored. The special characteristics of a transition-metal cluster include the large structural size, large number of electrons and non-negligible relativistic effect. The traditional high-level accurate computational methods such as coupled-cluster method (CC) and analytical coupled-perturbed HF method (CPHF) would encounter great difficulties in calculating the electronic polarizability and hyperpolarizability of transition-metal cluster due to the limited computer resources, particularly when the cluster has no symmetry. But in screening potential NLO materials the efficiency and timesaving are critically required. To this end the finite-field (F-F) approach, which may track back to the Buckingham expansion,[21] shows its value. The electronic polarizability and hyperpolarizability could be obtained by calculating the differences of total energies or dipole moments in the weak external electric fields. The computational requirements for F-F calculations is not that critical as CC or CPHF methods because it may perform single-point total energy calculation at SCF level (with external electric field being applied). Consequently, F-F method is a suitable choice for a complex system such as transition-metal cluster, which is usually with low symmetry or even without any symmetry. The example of applying *ab initio*/F-F method to NLO polarizability calculations of transition-metal clusters has been reported.[22]

Density functional theory (DFT) has become very popular in recent years. The developments of DFT approach makes the theoretical evaluation of NLO property for transition-metal clusters more practical.[23, 24] Some recent reports give the efficiency and reliability of DFT calculations in transition-metal clusters for optical absorption spectra and NLO property as well.[25-31] This is justified based on the fact that it is less computationally intensive than an MP2 calculation with similar computational accuracy.

2. Computational Models and Methods

2.1 *ab initio/F-F approach*

The calculation of the static polarizabilities and hyperpolarizabilities of a molecule is straightforward in a finite-field approach.[32-34] Suppose that an uncharged molecule is placed in weak homogeneous static electric fields. The energy can be expressed by the Buckingham type expansion:

$$\Delta E(F_i, F_j, F_k) = E(F_i, F_j, F_k) - E_0 = -\mu_i F_i - \tfrac{1}{2}\alpha_{ij}F_i F_j$$
$$-\tfrac{1}{6}\beta_{ijk}F_i F_j F_k - \tfrac{1}{24}\gamma_{ijkl}F_i F_j F_k F_l \tag{1}$$

where E is the energy of the molecule under the electric field F, E_0 is the unperturbed energy of the free molecule, F_i is the vector component of the electric field along the i direction, and μ_i, α_{ij}, β_{ijk} and γ_{ijkl} are the dipole moment, linear polarizability, first-order hyperpolarizability, and second-order hyperpolarizability, respectively. In Eq. (1), each Greek subscript of i, j, k and l denote the indices of the Cartesian axes x, y or z, and a repeated subscript means a summation over its corresponding index.

The independent tensor components of μ_i, α_{ij}, β_{ijk}, and γ_{ijkl} of a specified molecule depend on its molecular symmetry. As the molecules concerned in this paper are all of C_l symmetry, three, six, ten and fifteen independent components are needed to specify μ_i, α_{ij}, β_{ijk}, and γ_{ijkl}, respectively. Accordingly, a set of 34 appropriate energy differences, $\Delta E(F_i, F_j, F_k)$, are required to resolve these tensor components. The general formulation was published by Dupuis *et al.*[35] and the general formulations for molecules with major molecular point group symmetries were also summarized in our previous paper.[36]

In a uniform approach, the explicit expressions for the diagonal tensor components, μ_i, α_{ii}, β_{iii}, γ_{iiii} and γ_{iijj} are:

$$
\begin{aligned}
\mu_i &= \frac{1}{2(\sigma-\sigma^3)F}\{\sigma^3[\Delta E(F_i)-\Delta E(-F_i)]-[\Delta E(\sigma F_i)-\Delta E(-\sigma F_i)]\} \\
&= \frac{1}{2(\sigma-\sigma^3)F}\{[\sigma^3\Delta E(F_i)+\Delta E(-\sigma F_i)]-[\sigma^3\Delta E(-F_i)+\Delta E(\sigma F_i)]\}
\end{aligned}
\tag{2}
$$

$$
\alpha_{ii} = \frac{1}{(\sigma^2-\sigma^4)F^2}\{\sigma^4[\Delta E(F_i)+\Delta E(-F_i)]-[\Delta E(\sigma F_i)+\Delta E(\sigma F_i)]\}
\tag{3}
$$

$$
\begin{aligned}
\beta_{iii} &= \frac{3}{(\sigma^3-\sigma)F^3}\{\sigma[\Delta E(F_i)-\Delta E(-F_i)]-[\Delta E(\sigma F_i)-\Delta E(-\sigma F_i)]\} \\
&= \frac{3}{(\sigma^3-\sigma)F^3}\{[\sigma\Delta E(F_i)+\Delta E(-\sigma F_i)]-[\sigma\Delta E(-F_i)+\Delta E(\sigma F_i)]\}
\end{aligned}
\tag{4}
$$

$$
\gamma_{iiii} = \frac{12}{(\sigma^4-\sigma^2)F^4}\{\sigma^2[\Delta E(F_i)+\Delta E(-F_i)]-[\Delta E(\sigma F_i)+\Delta E(\sigma F_i)]\}
\tag{5}
$$

$$\gamma_{iijj} = \frac{1}{F^4}\{2[\Delta E(F_i) + \Delta E(-F_i) + \Delta E(F_j) + \Delta E(-F_j)]$$
$$-[\Delta E(F_i,F_j) + \Delta E(-F_i,F_j) + \Delta E(F_i,-F_j) + \Delta E(-F_i,-F_j)]\} \tag{6}$$

where σF ($\sigma \neq 0,1$) is a multiple of F. The suitable values of electric field strengths are crucial in finite-field computations.[37] We set median values so that $F = 0.0050$ a.u. and $\sigma F = 0.0030$ a.u. with $\sigma = 0.6$.

The isotropic scalar values for μ, α, β, and γ averaged from their tensor components are defined as,[38]

$$\alpha_{av} = \frac{1}{3}\sum_i \alpha_{ii}, \quad i = x, y, z \tag{7}$$

$$\beta_{i,av} = \frac{1}{3}\sum_j \beta_{ijj} + \beta_{jij} + \beta_{jji}, \quad i = x, y, z \tag{8}$$

$$\gamma_{av} = \frac{1}{5}\sum_{ij} \gamma_{iijj}, \quad i = x, y, z \tag{9}$$

2.2 DFT/TD-DFT approach

2.2.1 Density functional theory (DFT)

The premise behind DFT is that the energy of a molecular can be determined from the simpler electron density instead of the complicated N-electron wave function.[42] This theory use the variational principle and the self-consistent field procedure (SCF) to solve the electron density $\rho(\mathbf{r})$ and the system energy $E[\rho(\mathbf{r})]$. This theory originated authentically with the theorems that were developed by Hohenberg and Kohn.

Theorem I: For any system of interacting particles in external potential $V_{ext}(\mathbf{r})$, the external potential $V_{ext}(\mathbf{r})$ is determined uniquely, except for a constant, by the ground state particle density $\rho(\mathbf{r})$.[39] That is to say, the energy is the functional of electron density $\rho(\mathbf{r})$,[40]

$$E = E[\rho(r)] \tag{10}$$

Then the system energy can be expressed as

$$E_V[\rho] = T[\rho] + V_{ne}[\rho] + V_{ee}[\rho] = \int \rho(r)v(r) + F_{HK}[\rho] \tag{11}$$

where

$$F_{HK}[\rho] = T[\rho] + V_{ee}[\rho] \tag{12}$$

The term $V_{ee}[\rho]$ may be divided into two term

$$V_{ee}[\rho] = J_{ee}[\rho] + E_{XC}[\rho] \tag{13}$$

where $J_{ee}[\rho]$ is the classical repulsion

$$J_{ee}[\rho] = \frac{1}{2} \iint \frac{1}{2} \rho(r_1)\rho(r_2) dr_1 dr_2 \tag{14}$$

$E_{xc}[\rho]$ is the exchange-correlation energy.

Theorem II: A universal functional for the energy $E[\rho]$ in terms of the density $\rho(\mathbf{r})$ can be defined, valid for any external potential $V_{ext}(\mathbf{r})$. For any particular $V_{ext}(\mathbf{r})$, the exact ground state energy of the system is global minimum value of this functional, and the density $\rho(\mathbf{r})$ that minimizes the functional is the exact ground state density. The second theorem provides the energy variational principle. Assuming a trial density $\widetilde{\rho}(\mathbf{r})$, which is controlled by conditions of $\widetilde{\rho}(\mathbf{r}) \geq 0$ and $\int \widetilde{\rho}(r)dr = N$, this theorem can be written

$$E_0 \leq E_V[\widetilde{\rho}] \tag{15}$$

where the form of the energy functional $E_V[\widetilde{\rho}]$ is the same as formula (11).

$F_{HK}[\rho]$ is defined independently of the external potential $V_{ext}(\mathbf{r})$. As shown in the formulas (13) and (14), it is concluded that $F_{HK}[\rho]$ also is the functional of $\rho(r)$. Thus if we knew the exact $F_{HK}[\rho]$, we would obtain the exact ground state election density, then calculate the system energy.

Hohenberg-Kohn theorems only can solve the problems caused by the ground state electron density and calculate the molecular properties of ground state. However, the theorems don't show how to ascertain the functions of election density $\rho(\mathbf{r})$ and kinetic energy $T[\rho]$, and the exchange-correlation energy $E_{xc}[\rho]$. So Kohn and Sham provided the Kohn-Sham theorem. The three KS equations can be written as follows.

$$\hat{h}_{eff}\varphi_i = \left[-\frac{1}{2}\nabla_2 + \upsilon_{eff} \right]\varphi_i = \varepsilon_i\varphi_i \tag{16}$$

$$\upsilon_{eff}[\rho](\vec{r}) = v(\vec{r}) + \int \frac{\rho(r')}{|\vec{r} - \vec{r}'|}d\vec{r}' + v_{xc}(\vec{r}) \tag{17}$$

$$\rho(\vec{r}) = \sum_i^N \sum_s \left| \varphi_i(\vec{r},s) \right|^2 \tag{18}$$

The forms of KS equations are similar to that of Hartree-Fock. Once we have an explicit form of $v_{xc}(\vec{r})$, we can obtain the system energy of the ground state and other properties by iterative computations. So the key of the problem is how to express the term $v_{xc}(\vec{r})$.

2.2.2 *Time-dependent density functional theory*

Time-dependent density functional theory (TDDFT) extends the basic ideas of the ground state density functional theory to the treatment of excitations and of more general time-dependent phenomena. The basic variable is the time-dependent electron density, $\rho(\mathbf{r}, t)$. We may write time-dependent Kohn-sham equation

$$\left\{-\frac{1}{2}\nabla^2 + \upsilon_s[\rho](\vec{r},t)\right\}\varphi_i^\sigma(\vec{r},t) = i\frac{\partial}{\partial T}\varphi_i^\sigma(\vec{r},t) \qquad \sigma = \alpha, \beta$$

The time-dependent electron density is approximate to

$$\rho(\tilde{r},t) = \sum_{i=1}^{N}\left|\varphi_i(\vec{r},t)\right|^2$$

Where N is the number of electronic occupation, and $v_s(\vec{r},t)$ is called the time-dependent KS potential. It is can be defined from

$$v_s[\rho](\vec{r},t) = v(\vec{r},t) + \int\frac{\vec{\rho}(r',t)}{\left|\vec{r}-\vec{r}'\right|}d\vec{r}' + v_{xc}(\vec{r},t)$$

In the expression $v(\vec{r},t)$ is the external potential, and $v_{xc}(\vec{r},t)$ is the exchange-and-correlation potential.

2.2.3 *The exchange-and-correlation functional*

The Density Functional, also called the exchange-and-correlation (XC) functional, consists of an LDA and a GGA part. LDA stands for the Local Density Approximation, which implies that the XC functional in each point in space depends only on the (spin) density in that same point. GGA stands for Generalized Gradient Approximation and is an addition to the LDA part, by including terms that depend on derivatives of the density.[41]

The present the correction about GGA are as follows.[41]

For the exchange part the options are:

Becke: the gradient correction proposed in 1988 by Becke.

PW86x: the correction advocated in 1986 by Perdew-Wang.

PW91x: the exchange correction proposed in 1991 by Perdew-Wang.

mPWx: the modified PW91 exchange correction proposed in 1998 by Adamo-Barone.

PBEx: the exchange correction proposed in 1996 by Perdew-Burke-Ernzerhof.

RPBEx: the revised PBE exchange correction proposed in 1999 by Hammer-Hansen-Norskov.

RevPBEx: the revised PBE exchange correction proposed in 1998 by Zhang-Wang.

OPTX: the OPTX exchange correction proposed in 2001 by Handy-Cohen.

For the correlation part the options are:

Perdew: the correlation term presented in 1986 by Perdew.

PBEc: the correlation term presented in 1996 by Perdew-Burke-Ernzerhof.

PW91c: the correlation correction of Perdew-Wang in 1991.

LYP: the Lee-Yang-Parr 1988 correlation correction

Some GGA options define the exchange and correlation parts in one stroke. These are:

PW91: this is equivalent to pw91x + pw91c together.

mPW: this is equivalent to mPWx + pw91c together.

PBE: this is equivalent to PBEx + PBEc together

RPBE: this is equivalent to RPBEx + PBEc together

revPBE: this is equivalent to revPBEx + PBEc together

Blyp: this is equivalent to Becke (exchange) + LYP (correlation).

Olyp: this is equivalent to OPTX (exchange) + LYP (correlation).

OPBE: this is equivalent to OPTX (exchange) + PBEc (correlation).

LB94: this refers to the XC functional of Van Leeuwen and Baerends.

There are no separate entries for the Exchange and Correlation parts respectively of LB94.

SAOP: this refers to the statistical averaging of (model) orbit potentials proposed in 1999 by Gritsenko, Shipper and Baerends.

2.3 *Theoretical calculations*

The systems we have studied are mono-nuclear organometallic compounds and bi-nuclear and multi-nuclear transition metal clusters, so it is strictly necessary to include the relativistic effects and electron correlation effects in order to provide accurate calculations of the electronic structures and accurate descriptions of the properties. So the selectivity of methods, basis sets, and the exchange-and-correlation potentials has great influence upon the calculation results.

2.3.1 *Mono-nuclear coordinated compounds*

2.3.1.1 *Co(NH₃)₂(L-ala–gly–gly)*

The second-order polarizability β was calculated by using the finite-field approach and Møller–Plesset perturbation correction to the second-order (MP2) method. The time-dependent Hartree–Fock (TDHF) method is used to calculate the UV-vis spectrum.

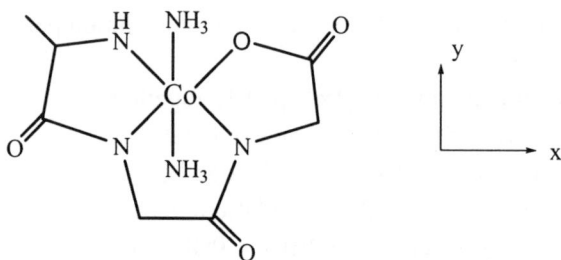

Fig. 1 Molecular structure and orientation of Co(NH₃)₂(L-ala–gly–gly).

As shown in Fig.1, the tripeptide was almost planar and the other two NH3 groups were perpendicular to this plane. We let the *xy* plane coincide with the tripeptide plane and used the crystal structure data such as bond lengths and angles in the calculations of the β values. The choice of the basis is as following: for metal Co, the Los Alamos ECP plus DZ basis set (Lanl2dz) is used. For atom C, N, O and H, the Lanl2dz extended by diffuse and polarization function (p, d function for C, H and O; s, p for H) is used: C (P, 0.0311; D, 0.587), N (P, 0.0533; D, 0.736), O (P, 0.0673; D, 0.961), H (S, 0.0498; P, 0.356) (denoted as Lanl2dz1pd).[42] All *ab initio*/F-F computations were performed by Gaussian 98 program package.[43]

2.3.1.2 *cis-[Rh(CO)₂Cl(4-Nme₂-C₅H₄N)]*

We optimized the crystal structure using DFT/B3LYP method, then the optimized structure acted as the initio configuration for calculation (Fig.2). We carried out time-dependent Density Functional Theory (TDDFT) to study the UV-*vis* spectra, and DFT/B3LYP to compute IR spectra and the second-order polarizability β. It is necessary to use the basis sets with diffuse functions because of the close relationship between the nonlinear optical properties and the excited behavior of outer electrons.[44-46] So the 6-31+G* basis set was chosen for C, H, N, O and Cl. For metal Rh, the Stuttgart/Dresden ECP basis set, ECP28MHF,[47, 48] was used. The calculations were performed by Gaussian 03 Package.[49]

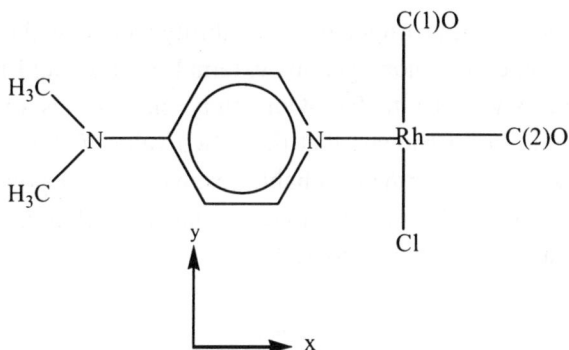

Fig. 2 Molecular structure and orientation of *cis*-[Rh(CO)₂Cl(4-Nme₂-C₅H₄N)]

2.3.2 Di-nuclear transition metal cluster, [Mo₂S₃(CO)₆(C₆H₁₁)₃]·N(C₂H₃)₄

The crystal structure belongs to Pna_{21}, a non-centric symmetric space group. A trigonally tetrahedral structure was formed by metal cluster core [Mo₂S₃], which was linked to organic ligands. The molecular coordination was set up with the z direction along the connection line of Mo₁ and Mo₂. This molecule was used as model cluster **I** (Fig. 3) In order to analyze the role of the ionic group, [N(C₂H₃)₄]⁺, played in the NLO response, we set another theoretical model, model cluster **II**, which was constructed only by anionic metal cluster [Mo₂S₃(C₆H₁₁)₃]⁻, with the coordinate unchanged.

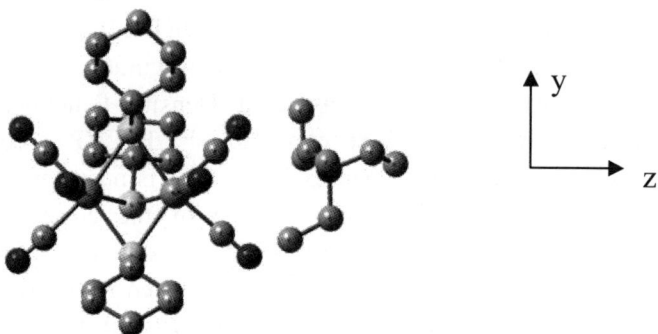

Fig. 3 The molecular structure of model cluster (hydrogen atoms are not shown).

The linear and nonlinear optical polarizability were calculated by using finite-field method. The homogenous external electric field had a value of 0.005 a.u., σ was set to 0.6. Our study on a series of tri-nuclear transition-metal cluster showed that these two value, $F = 0.005$ a.u. and $\sigma F = 0.003$ a.u., could provide reliable results of NLO polarizability.[50] The ECP/Lanl2dz basis set was used to take the relativistic effect of transition metal molybdenum into account.

2.3.3 Tri-nuclear transition metal cluster

2.3.3.1 MS₄(M′PPh₃)₂(M′PPh₃) (M=Mo, W; M′ = Cu, Ag, Au)

Six molecular clusters in Mo(W)-S(Se)-Cu(Ag, Au) series had been modeled and studied. They included, (a) MoS₄Cu(PPh₃)₂Cu(PPh₃)

[5140], (b). $MoS_4Ag(PPh_3)_2Ag(PPh_3)$,[52] (c) $WS_4Cu(PPh_3)_2Cu(PPh_3)$,[53]
(d) $WS_4Ag(PPh_3)_2Ag(PPh_3)$,[53] (e) $MoS_4Au(PPh_3)Au(PPh_3)$[54]
(f) $WS_4Au(PPh_3)Au(PPh_3)$.[55] They were categorized into two groups
because of the similarities of the crystal and molecular structures.
Group 1 includes the isomorphous compounds (a), (b), (c) and (d),
while (e) and (f) belong to group 2. Each of the crystals of group 1 is in
monoclinic symmetry containing 4 molecules in a unit cell, which are
in orientation nearly parallel with the b axis. In group 1, two M′
(M′ = Cu, Ag) atoms are bridged by an essentially tetrahedral MS_4^{2-}
(M = Mo, W) moiety, with one (M′$_1$) distortedly tetrahedral coordinated
and the other (M′$_2$) trigonally planar. Whereas the two crystals of group
2 have a triclinic unit cell, which is composed of two molecules. In
clusters (e) and (f), the gold atoms have triangle geometry and that
atoms P-Au-M(W)-Au-P form an approximately linear chain. Thus the
two Au atoms are nearly identical.

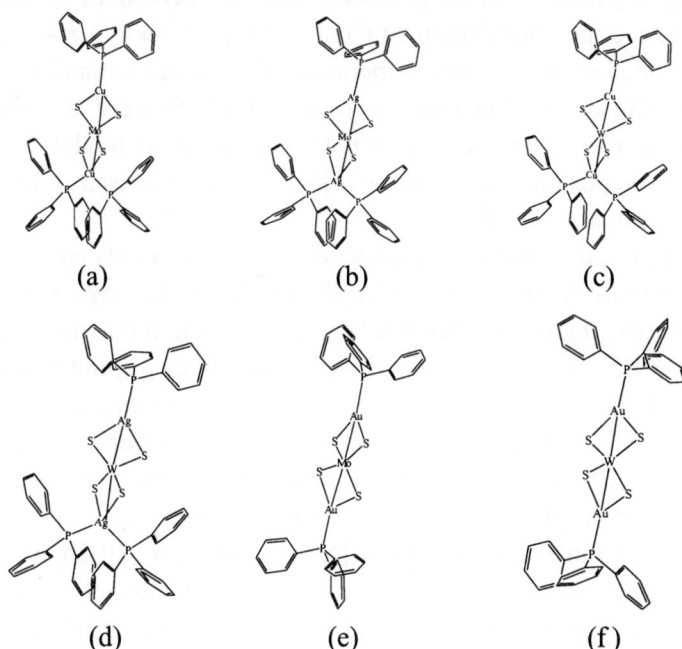

(a)   (b)   (c)

(d)   (e)   (f)

Fig. 4 Structural depictions of clusters (a) $MoS_4Cu(PPh_3)_2Cu(PPh_3)$,
(b) $MoS_4Ag(PPh_3)_2Ag(PPh_3)$ (c) $WS_4Cu(PPh_3)_2Cu(PPh_3)$, (d) $WS_4Ag(PPh_3)_2Ag(PPh_3)$,
(e) $MoS_4Au(PPh_3)Au(PPh_3)$ and (f) $WS_4Au(PPh_3)Au(PPh_3)$.

The optimization was performed at the DFT/B3LYP/LANL2DZ level. The density was converged to 10^{-6} for geometry optimizations and 10^{-8} for energy computations. The calculation of the static polarizabilities and hyperpolarizabilities of a molecule is straightforward in a finite-field approach.

2.3.3.2   *[MAg₂X₄C₅H₅NS](PPh3)₂· CH₂Cl₂ (M = Mo, W; X = S, Se)*

A series of tri-nuclear transition metal clusters $[MAg_2X_4C_5H_5NS](PPh_3)_2 \cdot CH_2Cl_2$ (M = Mo, W; X = S, Se)[56,57] with incomplete cubane-like structures (model **1-3**) had been investigated by DFT method. The crystal structures of three cluster complexes were similar to each other and crystallized in $P2_12_12_1$ non-central symmetry space group. Each neutral metal cluster contains a metal-core $[MAg_2X_4]$ with an incomplete cubane-like configuration and coordinated organic ligands, a pyridine-2-thiol (C_5H_5NS) and two neutral PPh3 ligands. In order to see the contribution of CH_2Cl_2 molecule to hyperpolarizability, we introduced the isolated supermolecules as the computing models. Additional two simulated models (model **4** and **5**) were designed in the following way: removing the CH_2Cl_2 molecule in model **1** result in model **4**; replacing the pyridine-2-thiol ligand in cluster **1** by a CH_2S ligand result in model **5**.

The selected geometric parameters of these two simulated models have been optimized at a DFT/LDA/SVWN level. In hyperpolarizability calculations, the generalized gradient approximation potential (GGA) by LB94 XC potentials was used. In both optimizations and hyperpolarizability calculations, we made use of an unrestricted Triple-ξ STO basis set extended by one polarization function for non-metal elements, P, S, Se, N, O, C and H, and for transition metal elements, an unrestricted triple-ξ STO basis set extended by two additional polarization functions. The cores (N, O, C: 1s; P, S, Cl: 2p; Se: 3d; Mo, W, Ag: 4p) were kept frozen. The calculation of the optical spectrum was carried out only for cluster **1** because of the similarity of three prototype clusters. GGA/LB94 XC potentials and the same basis sets mentioned above were used in the excitation calculations. In all calculations we

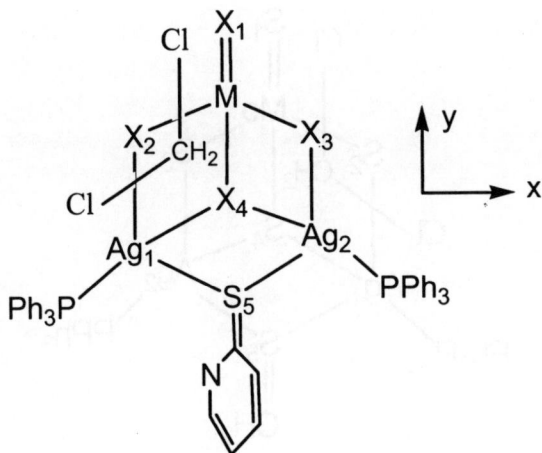

1 M=Mo, X=S   2 M=W, X=S
3 M=W, X=Se

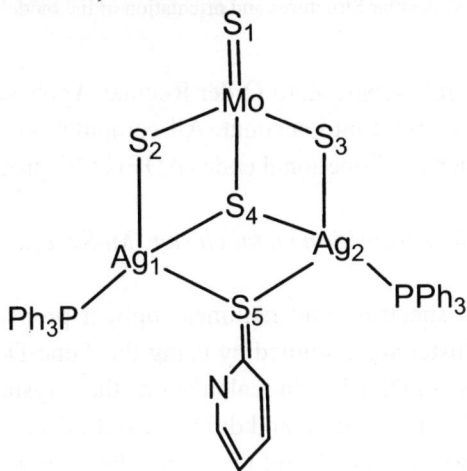

4

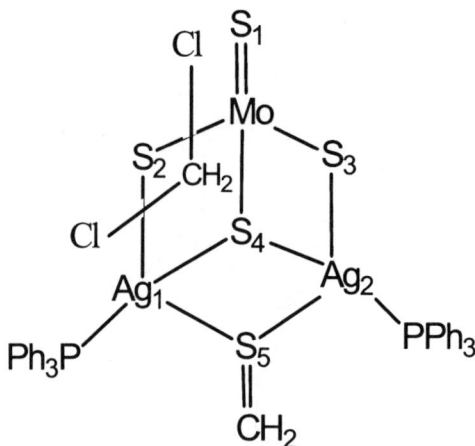

5

Fig. 5 Molecular Structures and orientation of the model clusters.

Adopted a scalar relativistic Zero Order Regular Approximation (ZORA) to take relativistic effect into account. All computations were performed by Amsterdam Density Functional code (ADF) 2003 program package.[58]

2.3.4 *Penta-nuclear transition metal cluster,* $MoS_4Cu_4Br_2(py)_6$[59]

The electronic spectrum and nonlinear optical properties of a penta-nuclear planar cluster were studied by using the Time-Depended Density Function Theory (TDDFT). In calculation the crystal structure was chosen the initial configuration and didn't be optimized. The space group of crystal structure was Fdd2 and the molecular symmetry was *C2*. The *C2* axis was perpendicular to the molecular plane and transited the molybdenum atom. So both crystal and molecule had second-order polarizabilities.

A scalar relativistic ZORA was used to take into account the relativistic effect. For metal elements, the standard basis set, STO-TZP was chosen. When taking into account the excited states and the

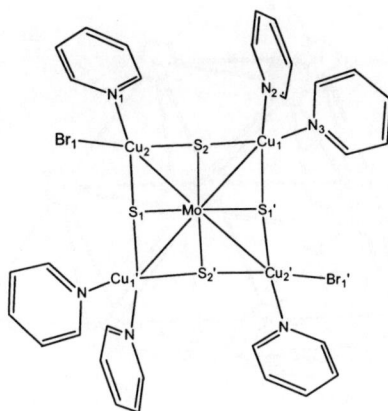

Fig. 6 Molecular Structure of $MoS_4Cu_4Br_2(py)_6$ cluster

nonlinear response, the behavior of the outer electron becomes very dispersionless. So for non-metal elements we adopted the TZP or DZ basis sets, and at the same time we added corresponding assistant basis for TZP and DZ. The cores of all atoms except for Hydrogen were kept frozen, namely, Mo: 3d; Cu: 3p; S: 2p; C: 1s; N: 1s; Br: 3d. Becke-Perdew potential was used to calculate the electron structure and UV-vis spectrum, while LB94 was adopted to compute the nonlinear optical properties. All calculations were performed by ADF 2002 program package.

2.3.5 *Nano-sized Au_{20} cluster*[60]

The dipole polarizability, static first hyperpolarizability and UV-vis spectrum of the recently identified nano-sized tetrahedral cluster of Au_{20} have been investigated by using time-dependent density functional response theory.

A scalar relativistic ZORA was adopted in all the calculations to account for the scalar relativistic effects. The generalized-gradient approximation (GGA) was employed in the geometry optimizations by

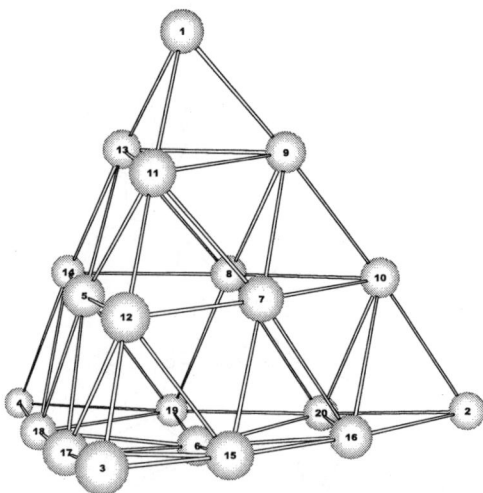

Fig. 7 The molecular structure of the tetrahedral Au_{20} cluster with T_d symmetry

using the Perdew-Wang 1991 (PW91) exchange-correlation (XC) functional. In the frozen core calculations we used the basis set of triple-zeta plus p- and f-type polarization functions (TZ2P) for Au atoms, while in the all-electron (AE) calculations the basis set of quadruple-zeta plus 2p- and 2f-type polarization functions (QZ4P) was used. The LB94 XC potential was chosen to calculate the static polarizability and hyperpolarizability, also to calculate the excited states and UV-vis spectrum.

3. Results and Discussion

3.1 $Co(NH_3)_2(L\text{-}ala\text{-}gly\text{-}gly)$

3.1.1 *Hyperpolarizabilities*

It is well known that in order to obtain reliable β values, the electron correlation effect and the more diffuse basis function are critical. Table 1 showed that the β component values of MP2 calculations were larger than the correspondent HF result, which were the general phenomena in treating the organic molecules. At both HF and MP2 levels, the implemented diffuse functions led most of b components to decrease, especially for the $\beta_{xxx}, \beta_{xxy}, \beta_{xyy}, \beta_{yyy}$ components.

Table 1 The values of β components with various methods and basis sets (10^{-30} esu)

Method/basis set	β_{xxx}	β_{xxy}	β_{xyy}	β_{yyy}	β_{xxz}	β_{xyz}	β_{yyz}	β_{xzz}	β_{yzz}	β_{zzz}
HF/LANL2DZ	-215	66.9	-100	38.5	-0.52	5.27	1.37	-14.8	12.4	-4.25
HF/LANL2DZ+pd[a]	-173	53.2	-88.6	15.7	6.87	5.27	-0.36	-10.7	-2.97	-5.62
MP2/LANL2DZ	-298	86.8	-164	86.9	-5.96	11.7	-4.49	-17.3	13.5	-6.39
MP2/LANL2DZ+pd	-262	75.3	-157	53.2	10.0	10.5	-5.69	-14.4	-0.88	-7.79
New orientation[b]	-368	-0.03	-46.8	-145	-0.03	8.64	3.91	-10.1	-8.34	-8.66

[a]Co: LANL2DZ, C, H, O, N: LANL2DZ+pd ECP Polarization
[b]*The result of MP2/LANL2DZ+pd, but the molecule is rotated around z axis with 38 degree in anticlockwise direction.*

From Table 1, the β components with the z subscript, i.e. β_{xxz}, β_{xyz}, β_{yyz}, β_{xzz}, β_{yzz}, β_{zzz} were about ten times less than the components only including the x and y subscript. This result indicated that the hyperpolarizability is mainly caused by the quadratic chelate tripeptide.

3.2.2 *UV-vis-IR spectra*

The UV and IR spectra were shown in Fig. 8 and Fig. 9 respectively. In the UV region, there were several intense absorptions near 180 nm, 200 nm, 235 nm and 327 nm, which the corresponding experimental values were 190 nm, 208 nm, 270 nm and 356 nm. The energy gap between the highest occupied molecular orbital and lowest unoccupied molecular orbital was 547 nm, which was good agreement with the experimental value 540 nm. However, the IR spectra showed that this molecule has a transparent region from 2000 cm^{-1} (5 μm) to 3200 cm^{-1}. The region from 3200 cm^{-1} to 3820 cm^{-1} (2.62 μm) was the low absorption of C–H and N–H stretching vibration. In the visible and near infrared region from 0.55 μm to 2.62 μm, there were no obvious absorptions in the UV and IR spectra. These results indicated that this molecule may be used as a near infrared NLO material, especially to be used a second harmonic generation material in the region from 0.55 μm to 5 μm.

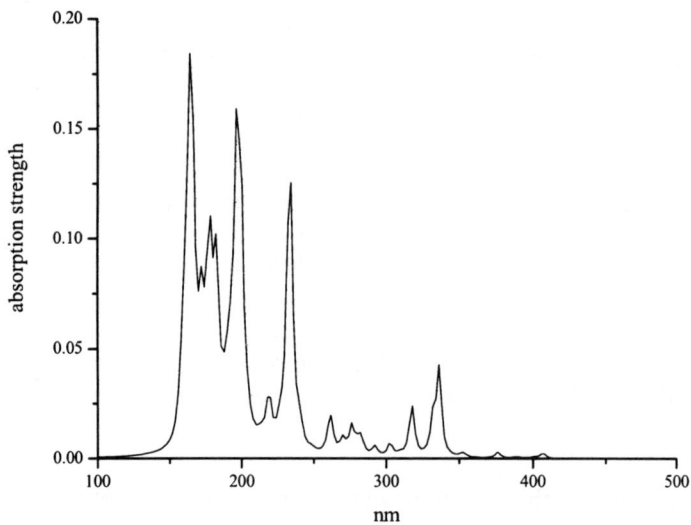

Fig. 8 UV-vis spectrum of Co(NH$_3$)$_2$(L-ala-gly-gly)

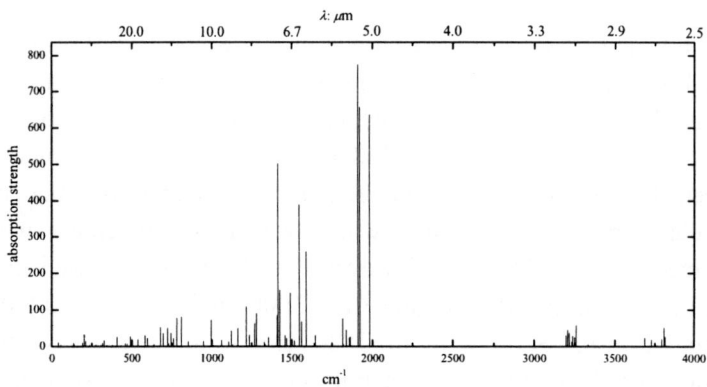

Fig. 9 IR spectrum of Co(NH$_3$)$_2$(L-ala-gly-gly)

These result showed that the organometallic complex, $Co(NH_3)_2$(L-ala-gly-gly) had a relative large first-order hyperpolarizability β value and a transparent IR region from 0.55 μm to 5 μm. It may be use as second harmonic generation materials in the near infrared region.

3.2 *cis-[Rh(CO)₂Cl(4-Nme₂-C₅H₄N)]*

We listed the calculated values of the first hyperpolarizability in Table 2.

Table 2 The β components of *cis*-[Rh(CO)₂Cl(4-NMe₂-C₅H₄N)] (unit: 10^{-30}esu)

	β_{xxx}	β_{xxy}	β_{xyy}	β_{yyy}	β_{xxz}	β_{xyz}	β_{yyz}	β_{xzz}	β_{yzz}	β_{zzz}
B3LYP/ MHF28	9.892	-0.001	-0.454	3.301	-0.034	-0.195	1.170	-1.154	0.027	-0.202

As shown in Fig. 10, there was very intensive adsorption near 284nm, which was good agreement with the experimental value 289nm. The adsorption band was ascribed mainly to the inter-ligand charge transfer (ILCT), primarily by the electron transfer form NMe_2 to pyridine.

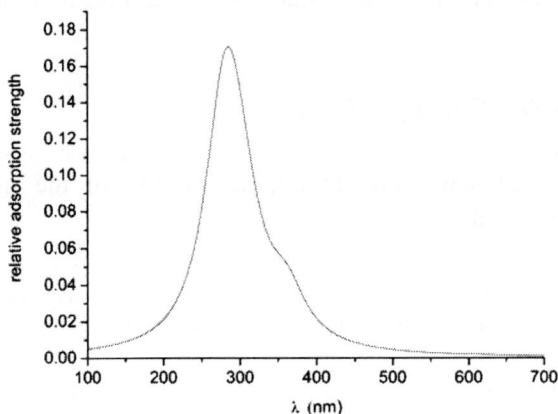

Fig. 10 UV-visible spectra of *cis*-[Rh(CO)₂Cl(4-NMe₂-C₅H₄N)]

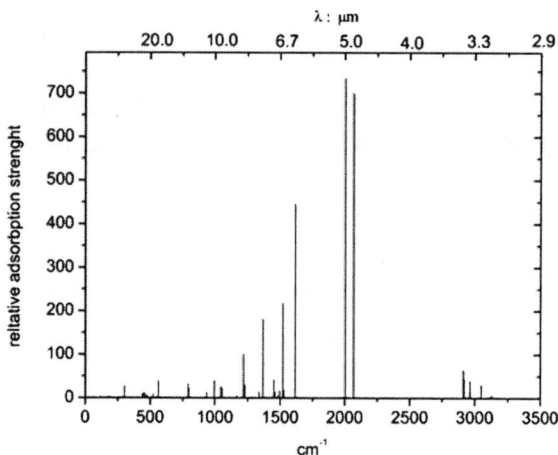

Fig. 11 The IR spectrum of *cis*-[Rh(CO)$_2$Cl(4-NMe$_2$-C$_5$H$_4$N)]

Combining the Fig. 10 with Fig. 11, there were no obvious adsorptions from 0.7 μm to 4.83 μm. On the other hand, the computed results of the second-order polarizabilities were shown in Table 2. It is obvious that β_{xxx} was great than any other components and determined the β_{vec}. So it may be approximately considered that the hyperpolarizability was one dimension. This complex also may act as a potential near infrared material.

3.3 [Mo$_2$S$_3$(CO)$_6$(C$_6$H$_{11}$)$_3$]·N(C$_2$H$_3$)$_4$

The calculated static first hyperpolarizability of the model **I** and model **II** were listed in Table 3.

Table 3 The calculated independent components of the static first hyperpolarizability and the spatial average values β_i of model **I** and **II** (unit: 10^{-30}esu)

	β_{xxx}	β_{xxy}	β_{xxz}	β_{xyy}	β_{xzz}	β_{yyy}	β_{yyz}	β_{yzz}	β_{zzz}	β_x	β_y	β_z
Model I	-0.2	-2.6	0.2	0.3	0.2	4.5	0.2	-1.4	13.5	0.4	0.5	13.8
Model II	-0.2	-2.7	0.6	0.2	0.6	3.6	0.1	1.9	-2.6	0.5	2.8	-3.1

The calculated results provided the valuable information to understand the NLO nature of the compound. The largest spatial average β value was in z direction, which was about three times larger than β_x and β_y. And in z direction we could find the connection of Mo_1-Mo_2, and the tetraethyl amine ionic group (Fig. 3). The charge transfer between the metals would be the origin the hyperpolarizability of the metal cluster. And in addition, the inter-group interaction between the metal cluster $[Mo_2S_3(C_6H_5)_3]^-$ and the positively charged organic group $[N(C_2H_3)_4]^+$ obviously play a role in the enhancement of the β_z value. The β_z value of model **II**, although kept to be the largest one among the three β_i values, reduced to about one fifth of that of model **I**. The electrostatic interaction between $[N(C_2H_3)_4]^+$ and $[Mo_2S_3(C_6H_5)_3]^-$ positively affect the CT inside the metal core.

The diagonal components of the polarizability and second hyperpolarizability of model **I** were listed in Table 4. Being similar to β_i, both the largest component of α and γ occurred in z direction. It was noticeable that γ_{zzzz} had a quite large value of -150×10^{-36} esu, which means that, in addition to the practical potential in SHG application, the third-order NLO property of this compound, such as self-defocusing (γ_{zzzz} is negative) is a very interesting subject for further studies.

The electronic excitation properties provided another way to help us to understand the origin of NLO effect. We calculated the electronic excitation spectrum of model **I** by using the time-dependent HF (TDHF) method at Lanl2dz level. The results in Fig. 12 showed that there were two absorption bands in the region of 200 nm - 500 nm. The low-energy (LE) absorption band centered at 431 nm was contributed by the excitation transition HOMO-2$\rightarrow$LUMO. On the other hand, the high-energy (HE) absorption band centered at 230 nm was dominated by the excitation from the inner occupied orbitals HOMO-7$\rightarrow$LUMO.

Table 4 The calculated diagonal components of polarizability ($\times 10^{-24}$ cm^3) and second hyperpolarizability ($\times 10^{-36}$ esu) of model **I**

α_{xx}	α_{yy}	α_{zz}	γ_{xxxx}	γ_{yyyy}	γ_{zzzz}
62.1	64.2	99.8	12	11	-150

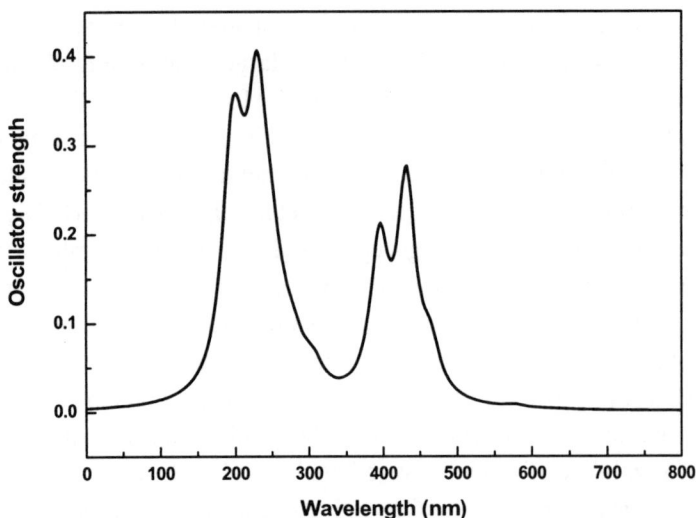

Fig. 12 UV-visible spectra of molder **I**

3.4 $MS_4(M'PPh_3)_2(M'PPh_3)$ $(M = Mo, W; M' = Cu, Ag, Au)$

The selected results of calculated dipole moments and the first-order hyperpolarizabilities were given in Table 5. It could also be seen that to some extent, the magnitude of β_{iii} will roughly follow that of μ_i, the dipole moment in the corresponding direction. In Table 5, model V and VI of group 2, which are nearly centrosymmetric, had negligible dipole moments and the first hyperpolarizabilities, while those of group 1, with negligible μ_y and μ_z but larger μ_x values, had negligible β_{yyy} $(\beta_{y,av})$ and β_z $(\beta_{z,av})$, but much considerable β_{xxx} and hence $\beta_{x,av}$. Though the relative order of μ_i would not coincide with that of β_{iii} or $\beta_{i,av}$ due to nonlinearity, it could be expected that much larger first-order hyperpolarizabilities would be obtained by construction of unsymmetrical clusters with larger dipole moments. The small first-order hyperpolarizabilities of gold compounds V and VI are due to symmetric structures.

Table 5 The calculated results of dipole moments (unit: Debye) and the first-order hyperpolarizabilities (unit: au) of the six clusters

	Group 1				Group 2	
	I	II	III	IV	V	VI
μ_x	−0.1492	−0.1559	−0.1308	−0.1330	0.0026	0.0019
μ_y	0.0208	0.0241	0.0222	−0.0038	−0.0348	0.0603
μ_z	0.0215	0.0312	0.0176	−0.0233	0.0173	0.0395
β_{xxx}	−579.66	−798.76	−523.74	−753.29	1.83	1.53
β_{yyy}	5.94	6.27	−6.50	8.70	−2.67	5.03
β_{zzz}	−10.86	−7.80	−12.76	5.43	4.42	6.59
$\beta_{x,av}$	−467.36	−705.87	−392.74	−634.14	33.56	−55.54
$\beta_{y,av}$	300.70	297.57	313.08	311.19	94.47	71.79
$\beta_{z,av}$	−88.88	−105.47	−87.82	−96.12	−80.57	−86.78

Table 6 The calculated results of the linear polarizabilities (unit: au) and the second-order hyperpolarizabilities (unit: au) of the six model clusters

	Group 1				Group 2	
	I	II	III	IV	V	VI
α_{xx}	342.114	351.160	330.573	341.878	393.979	383.060
α_{yy}	205.633	220.322	203.452	217.669	175.192	173.122
α_{zz}	177.981	187.918	176.214	184.979	175.356	173.182
α_v	241.909	253.134	236.747	248.175	248.176	243.121
γ_{xxxx}	325710	469310	292397	426557	409873	393967
γ_{yyyy}	35370	44693	35247	45053	4480	4707
γ_{zzzz}	7643	7613	7343	6947	3953	3717
γ_{xxyy}	43602	55400	40280	52152	5080	6253
γ_{yyzz}	3309	3731	3027	3397	2360	1898
γ_{zzxx}	8411	15003	9570	16208	5258	6322
γ_{av}	95873	133977	88148	124414	88740	86267

On the other hand, the second-order hyperpolarizability components γ_{iiii}, as well as those of linear polarizability α_{ii} were list in Table 6. Models in group 1 had similar structures and impressively, the similar

NLO properties. The components with x subscripts are much larger than those with only y or z. In each model cluster, $\gamma_{xxxx} > \gamma_{yyyy} > \gamma_{zzzz}$ ladder with one order of magnitude and $\gamma_{xxyy} \gg \gamma_{zzxx} > \gamma_{yyzz}$. Second, the two terminal phosphine ligands coordinated to M'_1 in group 1, though important to γ_{yyyy}, contribute much less to γ_{av} due to the dominance of γ_{xxxx}. Models V and VI in group 2 had two nearly identical moieties $M(\mu\text{-S})_2Au(M = Mo, W)$ perpendicular to each other. The two sides of the central M atoms were chemically indistinguishable, but the sulfur-bridged metal cores again were the main source of NLO response. Therefore, in the group 2 the orders were as follows, $\gamma_{xxxx} \gg \gamma_{yyyy} \sim \gamma_{zzzz}$ and $\gamma_{xxyy} \sim \gamma_{zzxx} > \gamma_{yyzz}$ ($\gg$ means larger with two orders of magnitude, and $\sim$ means comparable).

The analysis of the frontier orbitals shows that HOMOs do not change much. In Figure 13(b), it is the p orbitals of μ-S atoms that contribute mainly to HOMO of the molecules. This is very similar to the cases of group 1. On the other hand, Figure 13 (f)-(h) shows there are significant deviations among the extensive LUMOs in Case 2 and 3 from the free status in a similar way to that in group 1. However, the actual symmetry on two sides of M determines the symmetric shapes of LUMOs in Cases 2 and 3.

Several conclusions could be drawn from the study. The model clusters, $M\text{-}(\mu\text{-S})_2\text{-}M'(M = Mo, W; M' = Cu, Ag, Au)$, were joined by sharing the bridge metal atom M. It was the charge transfers from one of these moieties to the other in these characteristic sulfido-transitional metal cores that were responsible for polarizabilities and hyperpolarizabilities. Secondly, the structural effects on properties are important.

3.5 $[MAg_2X_4C_5H_5NS](PPh3)_2 \cdot CH_2Cl_2$ (M = Mo, W; X = S, Se)

3.5.1 *Hyperpolarizabilities of model 1, 2, 3*

Due to the $C1$ molecular symmetry, the hyperpolarizability (β) tensor of each cluster thus contains 27 non-zero components. For simplicity, we only listed in Table 7 spatial average values along i direction $\beta_{av,i}$ ($I = x, y, z$)

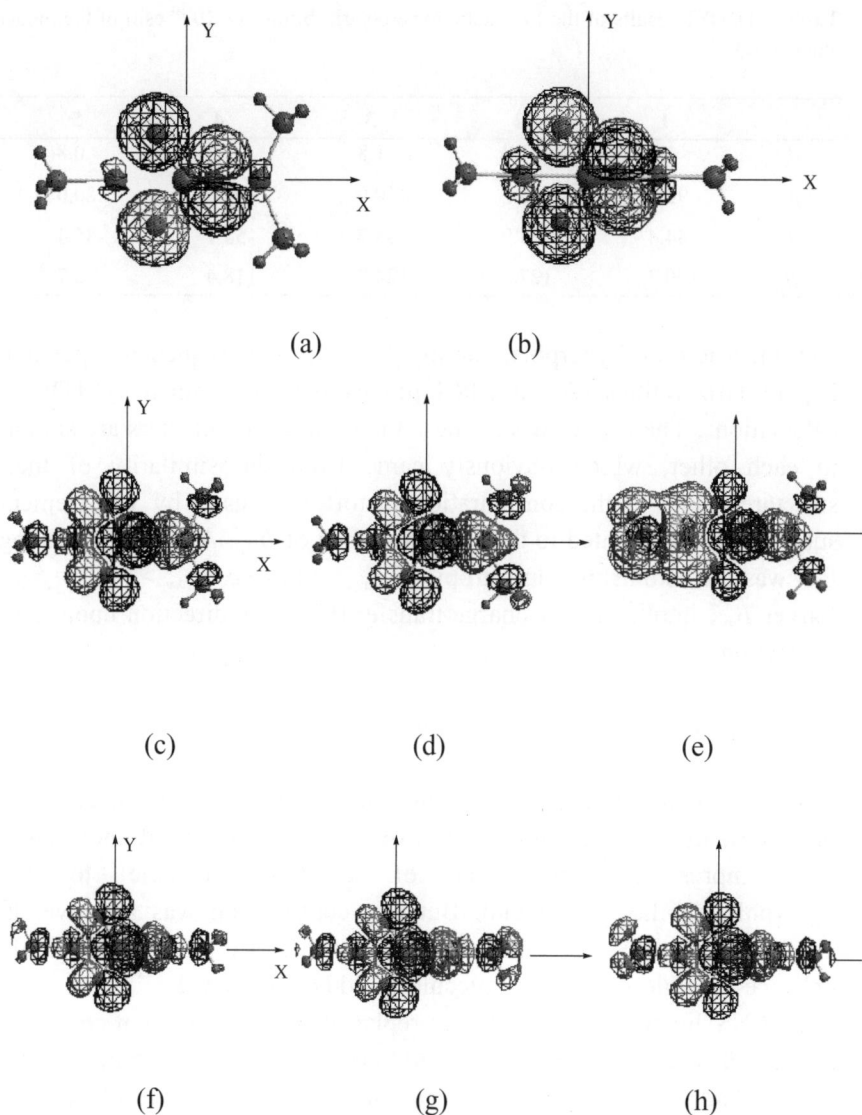

(a)   (b)

(c)   (d)   (e)

(f)   (g)   (h)

Fig. 13 Contour graphs of frontier orbitals (a) HOMO of molecule II, (b) HOMO of molecule V, (c) LUMO of molecule II in free state, (d) LUMO of molecule II with $F_x = 0.005$ au, (e) LUMO of molecule II with $F_x = -0.005$ au. (f) LUMO of molecule V in free state, (g) LUMO of molecule V with $F_x = 0.005$ au and (h) LUMO of molecule V with $F_x = -0.005$ au.

Table 7 TDDFT results of the first static hyperpolarizabilities ($\times 10^{-30}$ esu) of the model clusters **1-5**.

	1	2	3	4	5
$\beta_{av,x}$	1.5	1.3	1.8	7.9	0.8
$\beta_{av,y}$	90.0	115.2	110.9	104.5	20.0
$\beta_{av,z}$	-44.4	-53.9	-55.7	-55.1	-35.4
β_{tot}	100.3	127.2	124.2	118.4	40.7

and intrinsic total hyperpolarizability β_{tot}. The first frequency-dependent hyperpolarizabilities (β_d) at 1.064 μm were also obtained by TDDFT calculations. The $\beta_{av,i}$ values of these three molecular clusters are similar to each other, which obviously came from the similarity of their structures. The slight configuration distortion caused by the element substitution contributed to the small variance of $\beta_{av,i}$. Another interesting fact was the astonishing anisotropy of $\beta_{av,i}$ values *i.e.* $\beta_{av,y} > \beta_{av,z} >> \beta_{av,x}$. Larger $\beta_{av,y}$ implies larger charge-transfer (CT) in y direction upon laser irradiation.

3.5.2 *Hyperpolarizability of model 4, 5*

The molecule CH_2Cl_2 was removed in model **4**. Comparing to prototypal model **1**, all three values of $\beta_{av,\,i}$ of **4** enhanced about 20%. The impressive contribution of CH_2Cl_2 molecule to the hyperpolarizability was found. But this contribution was negative. In another word, the NLO response of prototypal cluster compounds would enhance if dichloromethane molecule could be eliminated.

C_5H_5NS ligand in model **1** was replaced by a simple non-conjugate CH_2S in **5**. The $\beta_{av,\,y}$ of **5** was reduced to more than one fourth of that of **1** due to "bad" electron acceptor located in y direction. The results clearly indicated the important role of the p-π planar conjugate organic ring in enhancing the β values of prototypal clusters.

The enhanced β_{tot} values are originated from various means of charge transfer (CT). In these multi-nuclear transition metal clusters (TMC), the CT from metal-core to coordinate ligands was similar to the common MLCT. However in addition to above CT, there was a special CT in this series of metal clusters that strongly relied on the asymmetry of the structural configuration, *i.e.* the incomplete cubane-like structure. This distinct intra-metal-cluster CT made great contributions to β_{tot} values.

We found in this study the completely different contributions of $(PPh3)_2$ ligands and pyridine-2-thiol ligand to β_{tot} and CH_2Cl_2 molecule reduce the hyperpolarizability.

3.6 $MoS_4Cu_4Br_2(py)_6$

The calculated results of the excited states and UV-*vis* spectrum were listed in Table 8.

Table 8 Calculated excitation energies under the experimental geometry (Mo-S 2.178Å) with different basis sets and the others results

	v_1		v_2		v_3	
Expt.	$1t_1$-2e	2.64 eV	$3t_2$-2e	3.88	t_1-$4t_2$	5.11
TDDFT/IV	$1t_1$-2e	2.59	$4t_2$-2e	3.96	$4t_2$-$5t_2$	5.01
			$1t_1$-$5t_2$			
TDDFT/II	$1t_1$-2e	2.80	$1t_1$-$5t_2$	4.30	$4t_2$—$5t_2$	5.43
			$4t_2$-2e		$3a_1$-$5t_2$	
IV+	$1t_1$-2e	2.48	$4t_2$-2e	3.57	t_1-$4t_2$	4.57
					$4t_2$-$5t_2$	
II+	$1t_1$-2e	2.52	$4t_2$-2e	3.68	t_1-$4t_2$	4.83
					$4t_2$-$5t_2$	
LDT	$1t_1$-2e	2.51	$3t_2$-2e	3.50	$3t_2$-$4t_2$	4.90
			$1t_1$-$4t_2$			
SW-X_α	$3t_2$-2e	2.70	$1t_1$-$4t_2$	3.40	$1e$-$4t_2$	5.51
SCCC	$1t_1$-2e	3.20	$1t_1$-$4t_2$	4.20	$3t_2$-2e	4.90
CIS	$2t_1$-3e	3.43	$2t_1$-$9t_2$	3.93	$8t_2$-3e	4.63
ZINDO	$2t_1$-3e	3.14	$2t_1$-$9t_2$	4.12	$8t_2$-3e	5.03

Two different basis sets resulted in great difference, and the calculation results with TZP for S were good agreement with the experimental value. The values using DZ for S were 0.2-0.3eV greater than those of the experimental. We calculated the former three hundred excited states of this penta-nuclear cluster. Due to the $C2$ symmetry, there were two kinds of transition modes, a and b irreducible representations. It was found that all excited states belonging to a irreducible representations had very small oscillator strength. Whereas the excited states adopting b mode could include all excited states completely. So we fitted the values of the excited states with b irreducible representation. The results of TDDFT were basically agreement with experimental values. But the calculated were greater than the experimental. The possible reasons were that the calculation was not use greater basis sets, the advanced exchange-and-correlation potentials and was shot of the solvent effects. The static polarizabilities, first- and second- hyperpolarizabilities were listed in Table 9.

Table 9 Calculated static polarizabilities, first- and second-order hyperpolarizabilities

Polarizability (au)		Hyperpolarizability (10^{-30}esu)		Second hyperpolarizability (10^{-34}esu)	
α_{xx}	662.39	β_{zxx}	11.78	γ_{xxxx}	13.33
				γ_{yyyy}	17.41
α_{yy}	716.05	β_{zyy}	54.38	γ_{zzzz}	2.65
				γ_{xxyy}	5.67
α_{zz}	493.32	β_{zzz}	29.50	γ_{xxzz}	3.27
				γ_{yyzz}	5.07
α_{av}	623.92	β_{vec}	95.66	γ_{ave}	12.28

The penta-nuclear planar cluster had large first-hyperpolarizability, 95.7×10^{-30} esu. The reasons might be as follows. First of all, the penta-nuclear planar cluster had lower symmetry. Secondly, there were two ligands with different properties, namely, one was the electron donor and another was the electron accepter. More importantly, there were charge transfer between cluster core and ligands, which might produce large nonlinear responses.

3.7 *Nano-sized Au₂₀ cluster*[60]

The tetrahedral structure of the Au_{20} cluster is shown in Fig.7. The twenty gold atoms in this tetrahedral cluster can be divided into three categories: four equivalent vertex atoms Au_A, four equivalent face-centered atoms Au_B, and twelve edged atoms Au_C. Due to the T_d symmetry of the tetrahedral Au_{20} cluster, only the diagonal dipole polarizability tensors α_{ii} ($i = x, y, z$) are nonzero, with the averaged polarizability being equal to α_{zz}. Similarly, there is only one nonzero component of the first hyperpolarizability, β_{xyz}, which corresponds to the d_{14} coefficient of a macroscopic SHG effect. The α_{zz} and β_{xyz} values calculated at the LB94 level with the DZ, TZP, TZ2P, and QZ4P basis sets and with different level of frozen core approximation are listed in Table 10.

Table 10 The calculated LB94 dipole polarizability and static first hyperpolarizability of tetrahedral Au_{20} Cluster[*]

Basis/Core	α_{zz}		β_{xyz}	
	au	10^{-24} esu	au	10^{-30} esu
DZ/Au.4f	657.76	97.47	-1656.1	14.31
TZP/Au.4f	679.64	100.71	-1656.2	14.31
TZ2P/Au.4f	682.02	101.07	-1656.9	14.32
DZ/Au.4d	652.72	96.72	-1648.0	14.24
TZP/Au.4d	678.53	100.55	-1652.3	14.28
TZ2P/Au.4d	681.01	100.92	-1645.9	14.22
DZ/Au	648.00	96.02	-1616.5	13.97
TZP/Au	677.33	100.37	-1653.8	14.29
TZ2P/Au	679.85	100.74	-1647.2	14.23
QZ4P/Au	685.20	101.54	-1520.9	13.14

[*] Due to the tetrahedral symmetry, $\alpha_{xx} = \alpha_{yy} = \alpha_{zz}$.

When frozen core approximation is imposed all the calculated hyperpolarizability values show little dependency on the basis sets or the level of the frozen core approximation. In the AE calculations, the augmentation of the basis sets with more polarization functions indeed affected the calculated polarizability and hyperpolarizability. The considerably large β value of Au_{20} cluster is very interesting, particularly when considering that no π-electron conjugation is present in the gold clusters.

Table 11 listed the major MO transitions and assignments of these seven states and the calculated UV-Vis spectrum of Au_{20} is shown in Fig. 14. The calculated UV-vis spectrum of Au_{20} cluster indicates several lower-energy (LE) absorption peaks around 370 nm (band **A** in Fig. 14) and a few high-energy (HE) peaks B, C and D. The major charge transfers that contribute to the β value of the cluster were those from the edged atoms to the vertex atoms and those between the face-centered Au atoms and vertex Au atoms.

Table 11 Calculated excitation wavelengths (λ_{ng}, nm), oscillator strengths (f), corresponding dominant MO transitions, and CT assignments of the optically-allowed excited states

State	λ_{ng} (nm)	f	MO transition [a]	CT assignment
2 T_2	527	0.044	29 $t_2 \rightarrow$ 30 t_2^*	face-centered atoms $\rightarrow$ vertex atoms
8 T_2	417	0.136	14 $a_1 \rightarrow$ 31 t_2^*	vertex atoms $\rightarrow$ face-centered atoms
9 T_2	403	0.030	18 $t_1 \rightarrow$ 30 t_2^*	edged atoms $\rightarrow$ vertex atoms
			14 $a_1 \rightarrow$ 31 t_2^*	vertex atoms $\rightarrow$ face-centered atoms
10 T_2	388	0.539	17 $t_1 \rightarrow$ 30 t_2^*	edged atoms $\rightarrow$ vertex atoms
11 T_2	374	0.383	17 $t_1 \rightarrow$ 30 t_2^*	edged atoms $\rightarrow$ vertex atoms
			18 $t_1 \rightarrow$ 19 t_1^*	
15 T_2	354	0.230	15 $e \rightarrow$ 30 t_2^*	edged atoms $\rightarrow$ vertex atoms
			18 $t_1 \rightarrow$ 19 t_1^*	face-centered atoms $\rightarrow$ vertex atoms
20 T_2	329	0.066	26 $t_2 \rightarrow$ 30 t_2^*	vertex atoms $\rightarrow$ vertex atoms

[a]HOMO: 16e; HOMO-1: 29t_2; HOMO-2: 14a_1; HOMO-3: 18t_1; HOMO-4: 17t_1; HOMO-6: 15e; HOMO-10: 26t_2 and LUMO: 30t_2^*; LUMO+1: : 31t_2^*; LUMO+2: 19t_1^*.

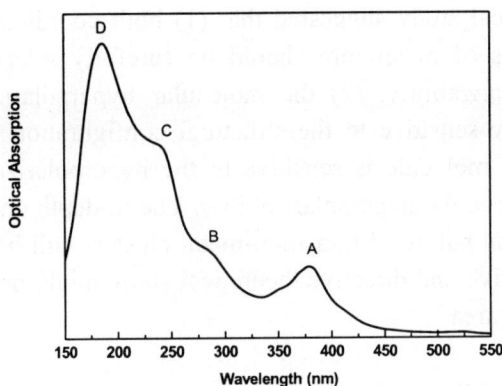

Fig. 14 Calculated UV-vis spectrum of the Au_{20} cluster

4. Conclusion

The theoretical results, although not yet supported by the experimental show great potential of the transition-metal cluster compounds to be IR SHG nonlinear optical materials. They may possess large first hyperpolarizability with non-central space symmetry, favorable cluster orientation, compact packing pattern, applicable IR transparent window and unique charge-transfer capacity.

The combined analytical results indicate that the enhanced first hyperpolarizability are originated from various charge-transfer means. Based on the concepts of traditional D-π-A charge-transfer in organic molecules and MLCT in organometallic systems, it is understandable the most charge-transfer found in transition-metal clusters. For instance, the charge-transfer from metal-core to coordinate ligands is similar to the common MLCT in organometallics described in many literatures.[28] However in addition to above known charge-transfer, we found a special charge-transfer that may exist in some transition-metal clusters. It is that strongly relied on the asymmetry of the structural configuration, *i.e.* the incomplete cubane-like structure. This distinct intra-metal-cluster charge-transfer may make great contributions to the first hyperpolarizability values.

Our theoretical study suggested that (1) both coordinate ligands and coordinate sites of metal-core should be carefully selected to archive large hyperpolarizability; (2) the molecular hyperpolarizabilities of a cluster are very sensitive to the structural configuration of metal-core; (3) the solvent molecule is sensitive to the hyperpolarizability. It may reduce or enhance the hyperpolarizability. The in-depth understanding of nonlinear optical nature of transition-metal clusters still has long way to go. The predictive and directive theoretical study might be helpful to the research in this area.

Acknowledgement

The works were continually supported by National Science Foundation of China (NSFC, 69978021, 20173064 and 90203017).

References

1. P. A. Franken, A. E. Hill, C. W. Peters, G. Weinreich, *Phys. Rev. Lett.*, **7**, 118 (1961)
2. S. P. Karna, *J. Phys. Chem.* **A104**, 4671 (2000).
3. D. J. Williams, ed., *Nonlinear Optical Properties of Organic and Polymeric Materials*,
   ACS Symposium Series, Vol.233, American Chemical Society, Washington (1983).
4. 3a. L. W. Tutt,; A. Kost, *Nature* **356**, 225 (1992).
   3b. D. G. Mclean, R. L.Sutherland, M. C. Brant, D. M. Brandelik, P. A. Fleitz, T. Pottenger, *Opt. Lett.*, **18**, 858 (1993).
5. a. N. Matsuzawa, J. Seto, D. A. Dixon, *J. Phys .Chem.*, **A101**, 9391 (1997).
   b. I. R. Whittall; A. M. Mcdonagh; M. G. Humphrey; M. Samoc, *Adv. Organomet. Chem.* 42, 291(1998).
6. S. Shi, *Nonlinear Optical Properties of Inorganic Clusters,* Chapter 3 of *Optoelectronic Properties of Inorganic Compounds*, Roundhill, D. M.; Fackler, J. P. Jr. eds. Plenum Press, New York, 1998. And the references therein.
7. A. Bose; C. R. Saha, *J. Mol. Catal.* **49**, 271 (1989).
8. G. Mclendon, A. E. Martell, *Coord. Chem. Rev.*, **19**, 1 (1976).
9. D. Gatteschi,, O. Kahn, J. S. Miller, F. Palacio, *Magnetic Molecular Materials*, Reidel, Dordrecht, 1991.
10. D. S. Chemla, J. Zyss, *Nonlinear Optical Properties of Organic Molecules and Crystals* Academic Press: London (1986).
11. J. Zyss, I. Ledoux, *Chem. Rev.* **94**, 77 (1994).
12. a. M. L. H. Green, S. R. Marder, M. E. Thompson, J. A. Bandy, D. Bloor, P. V. Kolingskey, R. J. Jones, *Nature* 330, 360(1987);
    b. E. Peris, *Coord. Chem. Rev.* **248**, 279 (2004) and the references herein.

13. Y. Liao, B. E. Eichinger, K. A. Firestone, M. Haller, J. Luo, W. Kaminsky, J. B. Benedict, P. J. Reid, A. K-Y Jen, L. R. Dalton,B. H. Robinson, *J. Am. Chem. Soc.* **127**, 1758 (2005).

14. C. Mang, K. Wu, M. Zhang, T. Hong, Y. Wei, *J. Mol. Struct.* (THEOCHEM) **674**, 77 (2004).

15. a. B. J. Coe, J. A. Harris, B. S. Brunschwig, *J. Phys. Chem.* **106**, 897 (2002).

    b. B. J. Coe, J. A. Harris, B. S. Brunschwig, J. Garin, J. Orduna, S. J. Coles, M. B. Hursthouse, *J. Am. Chem. Soc.* **126**, 10418 (2004).

    c. B. J. Coe, J. L. Harries, J. A. Harris, B. S. Brunschwig, S. J. Coles, M. E. Light, M. B. Hursthouse, *Dalton Trans.* **18**, 2935(2004). And the references herein.

16. a. S. Di Bella, I. Fragala, I. Ledoux, J. Zyss, *Chem. Eur. J.* **7**, 3738 (2001).

    b. S. Di Bella, *Chem. Soc. Rev.* **30**, 355 (2001).

    c. S. Di Bella, I. Fragala, *New. J. Chem.* **26**, 285 (2002).

    d. S. Di Bella, I. Fragala, *Eur. J. Inorg. Chem.* 2606 (2003) and the references herein.

17. S. Di Bella, I. Fragala, A. Guerri, P. Dapporto, K. Nakatani, *Inorg. Chim. Acta.,* **357**, 1161 (2004).

18. a. S. Shi, W. Ji, S.H. Tang, J.P. Lang, X.Q. Xin, *J. Am. Chem. Soc.* **116**, 3615 (1994).

    b. S. Shi, H.W. Hou, X.Q. Xin, *J. Phys. Chem.* **99**, 4050 (1995).

    c. D.X. Zeng, W. Ji, W.T. Wong, W.Y. Wong, X.Q. Xin, *Inorg. Chim. Acta* **279**, 172 (1998).

    d. P. Ge, S.H. Tang, W. Ji, S. Shi, H.W. Hou, D.L. Long, X.Q. Xin, S.F. Lu, Q.J. Wu, *J. Phys. Chem.* **B101**, 27 (1997).

    e. Q. F. Zhang, Y. N. Xiong, T. S. Lai, W. Ji, X. Q. Xin, *J. Phys. Chem.,* **B104**, 3446(2000).

    f. C. Zhang, G. C. Jin, J. X. Chen, X. Q. Xin, K. P. Qian, Coord. Chem. Rev. **213**, 51 (2001) and the references herein.

19. Y. Wang, Y. L. Song, M. F. Lappert, X. Q. Xin, A. Usman, H. K. Fun, H. G. Zheng, *Inorg. Chim. Acta.* **358**, 2217 (2005).

20. a. G. Maroulis, *J. Chem. Phys.* **94**, 1182 (1991).

    b. G. Maroulis, *J. Chem. Phys.* **96**, 6048 (1992).

    c. G. Maroulis, *J. Chem. Phys.* **101**, 4949 (1994).

    d. G. Maroulis, *J. Phys. Chem.* **100**, 13466 (1996).

    e. G. Maroulis, C. Pouchan, *Phys. Rev. A* **57**, 2440 (1998).

    f. G. Maroulis, *J. Chem. Phys.* **113**, 1813 (2000).

    g. G. Maroulis, *J. Comput. Chem.* **24**, 443 (2003).

    h. G. Maroulis, *J. Chem. Phys.* **121**, 10519 (2004). And the references herein.

21. A. D. Buckingham, *J. Chem. Phys.* **30**, 1580(1959).

22. X. Chen, C. Lin, K. Wu, B. Zhuang, R. Sa, S. Peng and Z. Zhou, *Chin. J. Struct. Chem.* **20**, 369 (2001).

23. N. Matsuzawa, J. Seto and D. A. Dixon, *J. Phys. Chem.,* **A 101**, 9391 (1997).

24. E. R. Davidson, Ed., Special Issue: Computational Transition Metal Chemistry, *Chem. Rev.* 100, 351-818 (2000)..

25. G. Ricciardi, A. Rosa, S. J. A. van Gisbergen and E. J. Baerends, *J. Phys. Chem.* A **104**, 635 (2000).
26. C. Lin, K. Wu, J. G. Snijders, R. Sa and X. Chen, *Acta Chimica Sinica.* **60**, 664 (2000).
27. G. Ricciardi, A. Rosa, E. J. Baerends and S. A. J. van Gisbergen, *J. Am. Chem. Soc.*, **124**, 12319 (2002).
28. V. Cavillot and B. Champagne, *Chem. Phys. Letts.*, **354**, 449 (2002).
29. P. J. Hay, *J. Phys. Chem.* A **106**, 1634 (2002).
30. Y. Yamaguchi, *J. Chem. Phys.* **117**, 9688 (2002).
31. Y. Q. Qiu, Z. M. Su, L. K. Yan, Y. Liao, M. Zhang and R. S. Wang, *Synth.* **137**, 1523 (2003).
32. H.D.Cohen, C.C.J.Roothaan, *J. Chem. Phys.* **43**, S34 (1965).
33. E.Perrin, P.N.Prasad, P.Mougenot, M.Dupuis, *J. Chem. Phys.* **91**, 4728 (1989).
34. H.A.Kurtz, J.P.Stewart, K.M.Dieter, *J. Comput. Chem.* **11**, 82 (1990).
35. See the numerous works by M. Dupuis et al. For example, F. Sim, S. Chin, Dupuis, M.; Rice, J. E.; *J. Phys. Chem.*, **97**, 1158 (1992).
36. X.Chen, C.Lin, K.Wu, *Chin. J. Strcut. Chem.* **20**, 369 (2001).
37. J.Guan, P.Duffy, J.T.Carter, D.P.Chong, K.C.Casida, M.E.Casida, M.Wrinn, *J. Chem. Phys.*, **98**, 4753 (1993).
38. D.C.Young, *Computational Chemistry: A Practical Guide for Appling Techniques to Real-World Problems*, John Wiley & Sons. Inc. (2001)
39. R.M.Martin, *Electronic Structure: Basic Theory and Practical Methods,* Cambridge University Press (2004).
40. R.G. Parr and W.T.Yang, *Density-Functional Theory of Atoms and Molecules*, Oxford University Press (1989).
41. ADF User's Guide, 2004.01
42. C.E. Check, T.O. Faust, J.M. Bailey, B.J. Wright, T.M. Gilbert, L.S. Sunderlin, *J. Phys. Chem. A* **105**, 8111 (2001).
43. Gaussian 98, Revision A.11.3, Frisch, M. J. *et al.* Gaussian, Inc., Pittsburgh, PA, 2002.
44. E. R. Davidson and D. Feller, *Chem. Rev.*, **86**, 681 (1986).
45. G. D. Purvis, R. J. Bartlett, *Phys. Rev. A*, **23**, 1549 (1981).
46. H. Sekino and R. J. Bartlett, *J. Chem. Phys.* **84**, 2726 (1986).
47. Æleen Frisch, M. J. Frisch, and G. W. Trucks, *Gaussian 03 User's Reference,* Gaussian, Inc. (2003).
48. D. Andrae, U. Häußermann, M. Dolg, H. Stoll, and H. Preuß, *Theo. Chim. Acta*, **77**, 123 (1990).
49. Gaussian 03, Revision A.1, M. J. Frisch et al. Gaussian, Inc., Pittsburgh PA, 2003.
50. X. Chen, K. Wu and C. Lin, *Inorg. Chem.* **42**, 532 (2003).
51. a. A.Müller, H.Bögge, H.G.Tölle, R.Jostes, U.Schimanski, M.Dartmann, *Angew. Chem. Int .Ed. Engl.* **19**, 654 (1980);
    b. A.Müller, H.Bögge, U.Schimanski, U., *Inorg. Chim. Acta* **45**, L249 (1980).
52. A.Müller, H.Bögge, U.Schimanski, *Inorg.Chim.Acta* **69**, 5 (1983).
53. A.Müller, H.Bögge, E.Koniger-Ahlborn, *Z. Naturforsch. Teil B* **34**, 1698 (1979).
54. J.M.Charnock, S.Bristow, J.R.Nicholson, C.D.Garner, W.Clegg, *J. Chem. Soc. Dalton Trans.* 303 (1987).

55. R.G.Pritchard, L.S.Moore, R.V.Parish, C.A.McAuliffe, B. Beagley, *Acta Crystallography. C*, **44**, 2022 (1988).
56. Q. Wang, X. Wu, Q. Huang, T. Sheng and P. Lin, *Polyhedron*, **16**, 1543 (1997).
57. Q. Wang, X. Wu, Q. Huang and T. Sheng *J. Cluster. Sci.*, **7**, 371 (1996).
58. a. C. Fonseca Guerra, O. Visser, J. G. Snijders, G. te Velde and E. J. Baerends, In *Methods and Techniques for Computational Chemistry*, ed. E. Clementi and G.Corongiu, STEF, Calgary, 1995, pp. 303-395.
   b. G. te Velde, F. M. Bickelhaupt, E. J. Baerends, S. J. A. van Gisbergen, C. Fonseca Guerra, J. G. Snijders and T. J. Ziegler, *J. Comput. Chem.* **22**, 931 (2001).
59. C. Zhang, Y. L. Song, F. E. Kuhn, Y. X. Wang, H. Fun, X. Q. Xin. *J. Mater. Chem.*, **12**, 239 (2002).
60. K. Wu, J. Li, C. Lin, Chem. Phys. Lett. **388**, 353 (2004).

CHAPTER 17

INTERACTION (HYPER)POLARIZABILITY IN N_2-He, CO_2-He, H_2O-He, $(H_2O)_2$-He AND O_3-He.

George Maroulis[1], Anastasios Haskopoulos

[1]Department of Chemistry, University of Patras, GR-26500 Patras, Greece
E-mail: maroulis@upatras.gr

We have used high-level ab initio quantum chemical methods to investigate the interaction polarizability and hyperpolarizability in the complexes of helium with dinitrogen, carbon dioxide, water, water dimer and ozone. We rely on flexible basis sets that yield accurate electric properties for all subsystems. We find that the interaction (hyper)polarizability is consistently negative for all studied systems. Our best values for the mean interaction dipole polarizability are α_{int} (N_2-He) = -0.0218, α_{int} (CO_2-He) = -0.0464, α_{int} (H_2O-He) = -0.0031, α_{int} (($H_2O)_2$-He) = -0.0182 and α_{int} (O_3-He) = -0.1178 $e^2a_0^2E_h^{-1}$. For the mean of the second dipole hyperpolarizability we find γ_{int} (N_2-He) = -18.17, γ_{int} (CO_2-He) = -10.39, γ_{int} (H_2O-He) = -18.93, γ_{int} (($H_2O)_2$-He) = -62.25 and γ_{int} (O_3-He) = -62.72 $e^4a_0^4E_h^{-3}$.

1. Introduction

Interaction electric polarizabilities and hyperpolarizabilities are fundamental properties of atomic and molecular systems. The total (hyper)polarizability of two interacting systems is, as intuition dictates, different from the sum of the respective properties of the two moieties. The determination of this "excess" polarizability is of certain importance for the analysis and understanding of phenomena induced by intermolecular interactions [1,2] and consequently for the rigorous analysis of observations in collision and interaction induced spectroscopies [3]. It is easily seen that the determination of the interaction or excess polarizability can also be used to understand bonding in complex architectures. An elegant quantification of the effect is achieved by the introduction of the **differential-per-atom (hyper)polarizability**. For a homoatomic cluster consisting of n interacting M atoms this property is defined as [4,5]:

$$\overline{\alpha}_{diff}(M_n)/n = [\overline{a}(M_n) - n\overline{a}(M)]/n$$

or

$$\overline{\alpha}_{diff}(M_n)/n = \overline{a}(M_n)/n - \overline{a}(M) \tag{1}$$

It is obviously a measurable quantity. Experimentally deduced values exist for important classes of molecules e.g. lithium clusters. It has been recently successfully applied to studies of clusters [6-8].

An overview of early efforts to determine the interaction polarizability has been given in two important review papers by Hunt [9]. It is worth noticing that various models have been proposed for the theoretical determination of interaction properties for atomic species [10,11]. These models rely on the knowledge of the electric properties of the moieties. Quantum chemical methods offer a more general and direct approach to the calculation of interaction properties. A variety of methods has been applied in recent years to important studies of interaction (hyper)polarizability [12-27].

In this study we present an investigation of interaction (hyper)polarizability effects in a sequence of systems of increasing complexity. These are the complexes of helium with molecules of primary importance to environmental science: dinitrogen, carbon dioxide, water, water dimer and ozone. The reduction of symmetry for the partner molecule increases considerably the computational cost of the investigation. In cases where the geometry of the complex is known from previous high-level theoretical investigations we adopt the relevant findings. In other cases, we have fully or partially explored the potential energy surface in order to locate the most stable configuration of the complex. Our computational approach to the calculation of the interaction properties relies on the finite-field method [28]. We have used in all cases large, flexible, purpose-oriented basis sets of Gaussian-type functions (GTF). The construction/choice of such basis sets represents always a major part of our effort. They have been especially designed for the systems of interest (He, N_2, CO_2, H_2O, $(H_2O)_2$ and O_3) following closely a computational strategy developed by us and tested successfully on various systems. Particular aspects of this strategy may be found in previous work [29].

2. Theory and Computational Strategy

The theory of electric polarizability [30,31] is a powerful tool for the rigorous description of the perturbation of an atomic/molecular system by an external electric field. Following Buckingham [30] and McLean and Yoshimine [31], we write the energy (E^P) and electric dipole (μ_α^P), quadrupole ($\Theta_{\alpha\beta}^P$) and octopole ($\Omega_{\alpha\beta\gamma}^P$) moment of a system perturbed by a weak, static electric field as an expansion:

$$
\begin{aligned}
E^P \equiv{}& E^P(F_\alpha, F_{\alpha\beta}, F_{\alpha\beta\gamma}, F_{\alpha\beta\gamma\delta}, \ldots) \\
={}& E^0 - \mu_\alpha F_\alpha - (1/3)\Theta_{\alpha\beta}F_{\alpha\beta} - (1/15)\Omega_{\alpha\beta\gamma}F_{\alpha\beta\gamma} - (1/105)\Phi_{\alpha\beta\gamma\delta}F_{\alpha\beta\gamma\delta} \\
& + \ldots \\
& - (1/2)\alpha_{\alpha\beta}F_\alpha F_\beta - (1/3)A_{\alpha,\beta\gamma}F_\alpha F_{\beta\gamma} - (1/6)C_{\alpha\beta,\gamma\delta}F_{\alpha\beta}F_{\gamma\delta} \\
& - (1/15)E_{\alpha,\beta\gamma\delta}F_\alpha F_{\beta\gamma\delta} + \ldots \\
& - (1/6)\beta_{\alpha\beta\gamma}F_\alpha F_\beta F_\gamma - (1/6)B_{\alpha\beta,\gamma\delta}F_\alpha F_\beta F_{\gamma\delta} + \ldots \\
& - (1/24)\gamma_{\alpha\beta\gamma\delta}F_\alpha F_\beta F_\gamma F_\delta + \ldots
\end{aligned}
\tag{2}
$$

$$
\begin{aligned}
\mu_\alpha^P ={}& \mu_\alpha + \alpha_{\alpha\beta}F_\beta + (1/3)A_{\alpha,\beta\gamma}F_{\beta\gamma} + (1/2)\beta_{\alpha\beta\gamma}F_\beta F_\gamma \\
& + (1/3)B_{\alpha\beta,\gamma\delta}F_\beta F_{\gamma\delta} + (1/6)\gamma_{\alpha\beta\gamma\delta}F_\beta F_\gamma F_\delta + \ldots
\end{aligned}
\tag{3}
$$

$$\Theta_{\alpha\beta}{}^P = \Theta_{\alpha\beta} + \mathbf{A}_{\gamma,\alpha\beta}E_\gamma + \mathbf{C}_{\alpha\beta,\gamma\delta}\,F_{\gamma\delta} + (1/2)\mathbf{B}_{\gamma\delta,\alpha\beta}\,F_\gamma F_\delta + \dots \qquad (4)$$

$$\Omega_{\alpha\beta\gamma}{}^P = \Omega_{\alpha\beta\gamma} + \mathbf{E}_{\delta,\alpha\beta\gamma}\,F_\delta + \dots \qquad (5)$$

The coefficients of the above expansions (**in bold**) are the permanent electric moments, polarizabilities and hyperpolarizabilities of the molecule. F_α, $F_{\alpha\beta}$, etc. are the field, field gradient, etc. at the origin. E^0 is the energy of the unperturbed molecule. The number of independent components needed to specify the electric moment and (hyper)polarizability tensors in the above expansions is strictly regulated by symmetry [30]. In addition to the Cartesian components of the above tensors, in this work we are especially interested in the total dipole moment (μ), the mean ($\overline{\alpha}$) and the anisotropy ($\Delta\alpha$) of the dipole polarizability and the mean second dipole polarizability ($\overline{\gamma}$), defined as:

$$\mu = (\mu_x{}^2 + \mu_y{}^2 + \mu_z{}^2)^{1/2}$$

$$\overline{\alpha} = (\alpha_{xx} + \alpha_{yy} + \alpha_{zz})/3$$

$$\Delta\alpha = 2^{-1/2}[(\alpha_{xx} - \alpha_{yy})^2 + (\alpha_{yy} - \alpha_{zz})^2 + (\alpha_{zz} - \alpha_{xx})^2 + 6(\alpha_{xy}{}^2 + \alpha_{yz}{}^2 + \alpha_{zx}{}^2)]^{1/2}$$

$$\overline{\gamma} = (\gamma_{xxxx} + \gamma_{yyyy} + \gamma_{zzzz} + 2\gamma_{xxyy} + 2\gamma_{yyzz} + 2\gamma_{zzxx})/5 \qquad (6)$$

For very weak fields, the expansions of Eqs. (2)-(5) converge rapidly enough. This allows the determination of the molecular properties either from the induced multipole moments [32-36] or the perturbed energies [37-41]. Our finite-field approach extends easily to the calculation of the electric properties of molecular complexes of the type considered here.

The interaction electric properties of the MOL···He (where MOL = N_2, CO_2, H_2O, $(H_2O)_2$ and O_3) system are obtained via the well-tested Boys-Bernardi counterpoise-correction (CP) method [42]. The interaction quantity $P_{int}(MOL···He)$ at a given geometric configuration is computed as

$$P_{int}(MOL\cdots He) = P(MOL\cdots He) - P(MOL\cdots X) - P(X\cdots He) \qquad (7)$$

The symbol $P(MOL\cdots He)$ denotes the property P for $MOL\cdots He$. $P(MOL\cdots X)$ is the value of P for subsystem MOL in the presence of the ghost orbitals of subsystem He. By virtue of Eq. (7), the interaction mean polarizability or hyperpolarizability is a linear combination of the Cartesian components of $\alpha_{\alpha\beta}$ or $\gamma_{\alpha\beta\gamma\delta}$. The interaction anisotropy $\Delta\alpha_{int}$ at a given level of theory is computed by inserting in Eq. (6) the respective interaction quantities for the Cartesian components of $\alpha_{\alpha\beta}$. The same method obviously applies to the total dipole moment. It should be noted here that the elimination of basis set superposition error (BSSE) effects is a major problem in interaction property calculation. This quantity is basis set dependent and certainly extremely basis set sensitive for large classes of systems. Nevertheless, it can be reduced and made arbitrarily small by a suitable choice of purpose-oriented basis set. This has been clearly born out in our study on the interaction (hyper)polarizability of two water molecules [43].

Theoretical arguments about the determination of electron correlation may be found in standard textbooks [44,45]. In this study, electron correlation effects on the molecular properties are accounted for via MP2 and MP4 (second and fourth order Møller-Plesset perturbation theory), CCSD (single and double excitation coupled cluster theory) and CCSD(T) which includes an estimate of connected triple excitations by a perturbational treatment. We lean heavily on the predictive capability of coupled-cluster methods [46]. Last, we define the electron correlation correction (ECC) for each system as

$$ECC = CCSD(T) - SCF \qquad (8)$$

Atomic units are used throughout this paper. Conversion factors to SI units are, Energy, $1\ E_h = 4.3597482 \times 10^{-18}$ J, Length, $1\ a_0 = 0.529177249 \times 10^{-10}$ m, dipole moment (μ), $1\ ea_0 = 8.478358 \times 10^{-30}$ Cm, dipole polarizability ($\alpha_{\alpha\beta}$), $1\ e^2a_0^2E_h^{-1} = 1.648778 \times 10^{-41}$ $C^2m^2J^{-1}$, first hyperpolarizability ($\beta_{\alpha\beta\gamma}$), $1\ e^3a_0^3E_h^{-2} = 3.206361 \times 10^{-53}$ $C^3m^3J^{-2}$ and second hyperpolarizability ($\gamma_{\alpha\beta\gamma\delta}$), $1\ e^4a_0^4E_h^{-3} = 6.235378 \times 10^{-65}$ $C^4m^4J^{-3}$.

Property values are in most cases given as pure numbers, i.e. μ/ea_0 or $\alpha_{\alpha\beta}/e^2a_0^2E_h^{-1}$.

All calculations in this study were performed with Gaussian 94 [47] and Gaussian 98 [48].

3.1. N_2-He

The interaction of N_2 with He has attracted considerable attention in late years [49].

The potential energy surface (PES) of the N_2-He system constructed in this work is shown in Fig. 1. It was obtained at the MP2(Full) level of theory, with the N-N bond length fixed at the experimental value of 2.07432 a_0 [50] using a large [7s5p4d2f] basis set for N_2 [51] and a [6s4p3d] one for He [17]. The PES shows clearly three minima. The global one corresponds to a T-shaped configuration with the He centre located at R = 6.562 a_0 from the N_2 midpoint and two linear-shaped minima with the He centre at R = 7.681 a_0.

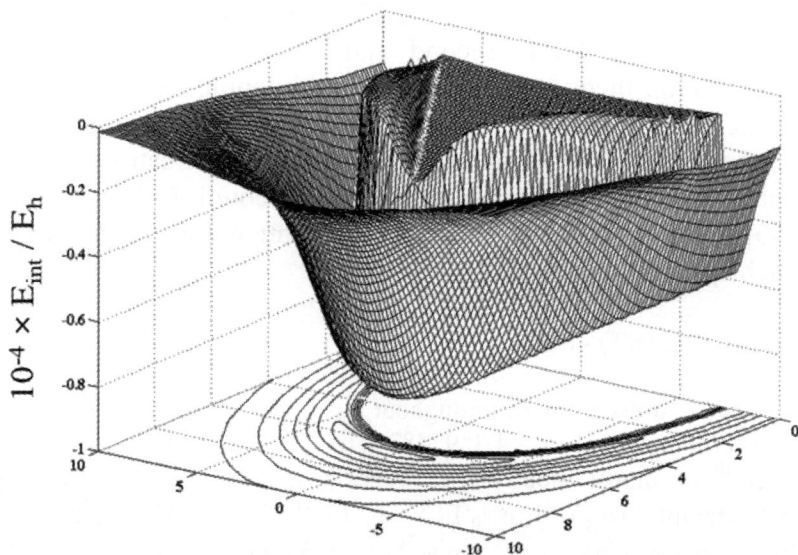

Fig. 1. Potential energy surface of the N_2-He interaction

The two molecular configurations considered in the interaction property calculations are shown in Fig. 2.

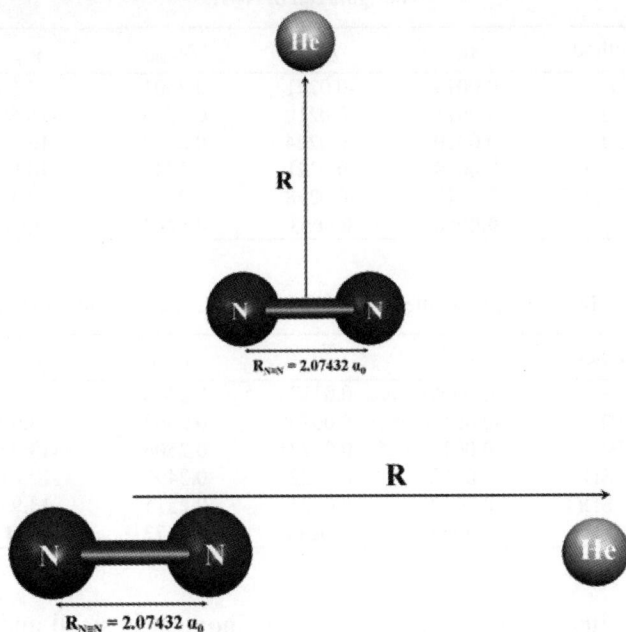

Fig. 2. T- and L-shaped configurations of N_2-He considered in this work.

We have used the [7s5p4d2f/6s4p3d] basis set to calculate the interaction properties of the complex. The results are shown in Tables 1 and 2.

At the T-shaped configuration the complex is characterized by a small dipole moment. Its SCF value is $\mu_x = 0.0019$ and is little affected by electron correlation. The SCF interaction mean dipole polarizability is small and negative while the anisotropy is considerable larger and positive. Electron correlation has an appreciable effect on both properties. The magnitude of $\overline{\alpha}_{int}$ decreases at post-Hartree-Fock levels of theory while that of $\Delta\alpha_{int}$ increases. We find a negative SCF $\overline{\gamma}_{int}$. The CCSD(T) value of this property is $\overline{\gamma}_{int} = -18.17$, which corresponds to an increase of magnitude by 16.8%.

Table 1. Interaction properties for the most stable T-shaped (global minimum) configuration of N_2-He.

Method	μ_x	$\overline{\alpha}_{int}$	$\Delta\alpha_{int}$	$\overline{\gamma}_{int}$
SCF	0.0019	-0.0271	0.2501	-15.56
MP2	0.0021	-0.0210	0.2656	-16.60
MP4	0.0020	-0.0224	0.2757	-16.97
CCSD	0.0019	-0.0221	0.2740	-16.89
CCSD(T)	0.0019	-0.0218	0.2770	-18.17
ECC	**0.0000**	**0.0053**	**0.0269**	**-2.61**

Table 2. Interaction properties for the most stable L-shaped configuration of N_2-He.

Method	μ_z	$\overline{\alpha}_{int}$	$\Delta\alpha_{int}$	$\overline{\gamma}_{int}$
SCF	-0.0009	0.0137	0.2343	-7.39
MP2	-0.0011	0.0086	0.2363	-13.90
MP4	-0.0012	0.0122	0.2504	-13.82
CCSD	-0.0012	0.0122	0.2491	-12.61
CCSD(T)	-0.0012	0.0123	0.2515	-13.93
ECC	**-0.0003**	**-0.0014**	**0.0172**	**-6.54**

At the linear configuration the dipole moment is small and negative. Its size increases slightly with electron correlation. Both components of the interaction polarizability are positive for this configuration. Electron correlation reduces the magnitude of the mean (with the notable exception of the MP2 value) and increases the anisotropy of the dipole polarizability. We find a SCF $\overline{\gamma}_{int}$ = -7.39. The CCSD(T) value is -13.93. In absolute terms this represents a rather small increase in magnitude but in relative terms this represents an impressive 88.5%.

3.2. CO_2-He

The interaction of carbon dioxide with helium is relevant to wide range of experimental investigations, including the Raman spectroscopy of CO_2 perturbed He [52], CO_2 solvated with He atoms [53] and collisional line broadening in CO_2-He [54]. We note also a recent theoretical study of PES CO_2-He [55].

In a previous paper [18] we reported the construction of a MP2(FULL)/[5s3p3d1f/6s4p3d] PES for this system. The C-O bond length was kept fixed at the experimental value of 2.192 a_0 [56]. An aspect of this surface is shown in Fig. 3. The global minimum (T-shaped) and the two local minima (L-shaped) have been reported elsewhere [18].

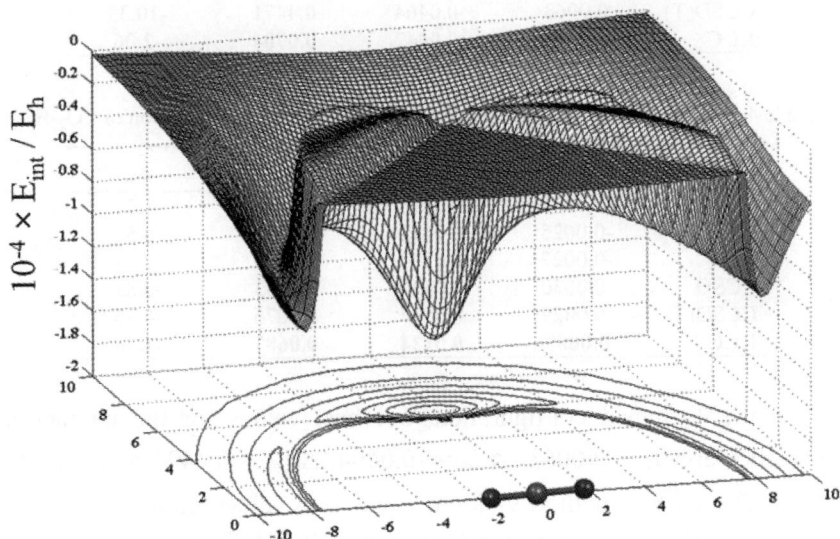

Fig. 3. Potential energy surface of the CO_2-He interaction

The basis set used in our study are [5s3p3d1f] for CO_2 [57] and [6s4p3d] for He [17]. The [5s3p3d1f] basis yields SCF values $\bar{\alpha} = 15.81$ and $\Delta\alpha = 11.78$, quite close to the reference 15.86 and 11.68 obtained with a very large (19s14p8d6f1g) basis set [58].

The calculated interaction properties for the two configurations are given in Tables 3 (T-shaped) and 4 (L-shaped). The SCF and MP2 values for all properties have been presented and discussed in previous work [18,27].

G.Maroulis, A.Haskopoulos

Table 3. Interaction properties for the most stable T-shaped (global minimum) configuration of CO_2-He.

Method	μ_x	$\overline{\alpha}_{int}$	$\Delta\alpha_{int}$	$\overline{\gamma}_{int}$
SCF	0.0070	-0.0417	0.3767	-8.33
MP2	0.0063	-0.0494	0.4375	-11.66
MP4	0.0063	-0.0483	0.4463	-11.19
CCSD	0.0064	-0.0465	0.4426	-9.47
CCSD(T)	0.0063	-0.0464	0.4471	-10.39
ECC	**-0.0007**	**-0.0047**	**0.0704**	**-2.06**

Table 4. Interaction properties for the most stable L-shaped configuration of CO_2-He.

Method	μ_z	$\overline{\alpha}_{int}$	$\Delta\alpha_{int}$	$\overline{\gamma}_{int}$
SCF	-0.0032	0.0393	0.3389	-3.39
MP2	-0.0025	0.0512	0.4014	-6.58
MP4	-0.0027	0.0524	0.4073	-7.29
CCSD	-0.0030	0.0520	0.4033	-5.83
CCSD(T)	-0.0029	0.0527	0.4076	-6.15
ECC	**0.0003**	**0.0134**	**0.0687**	**-2.76**

In the T-shaped configuration, our best values for the interaction properties are μ_x = 0.0063, $\overline{\alpha}_{int}$ = -0.0464, $\Delta\alpha_{int}$ = 0.4471 and $\overline{\gamma}_{int}$ = -10.39. The CCSD method yields μ_x and $\overline{\alpha}_{int}$ values quite close to the above. The difference between CCSD and CCSD(T) is more important for $\Delta\alpha_{int}$ and $\overline{\gamma}_{int}$, indicating a sizeable triples correction for these properties. We also note the reasonable agreement between the MP2 and CCSD(T) values. The calculated ECC shows that electron correlation affects the magnitude of the SCF values by 10.0% (μ_x), 11.3% ($\overline{\alpha}_{int}$), 18.7% ($\Delta\alpha$) and 24.8% ($\overline{\gamma}_{int}$).

We observe a change of sign for the interaction dipole moment and mean polarizability for the L-shaped configuration. Electron correlation has qualitatively the same effect on the magnitude of the properties for both configurations. For the linear case the effect is 9.4% (μ_z), 34.1% ($\overline{\alpha}_{int}$), 20.2% ($\Delta\alpha$) and 81.4% ($\overline{\gamma}_{int}$).

3.3. *H₂O-He*

For the H_2O-He we adopted the molecular (C_s) geometry determined by Calderoni et al. [59] and shown in Fig. 4. The geometry of the H_2O moiety is the experimental one [60], with R_{OH} = 0.9572 Å and $\angle HOH$ = 104.52°. The reduction of symmetry, compared to N_2-He and CO_2-He, increases considerably the computational effort. The basis sets used on this system are [6s4p3d1f/4s3p1d] for H2O [43] and [6s4p3d] for He [17]. The basis set used for the water monomer yields $\overline{\alpha}$ (SCF) = 8.48, which is just 0.6% lower than the 8.53 obtained with a (18s13p8d5f/12s7p3d2f) basis [61].

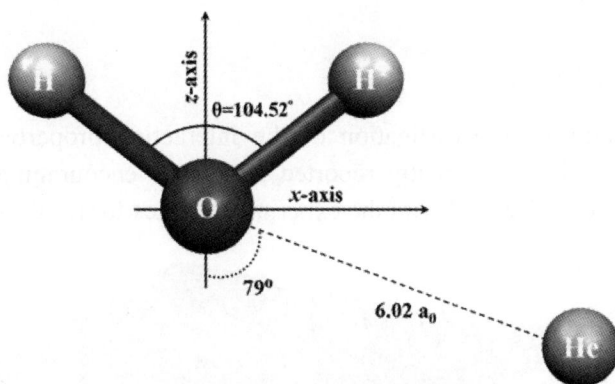

Fig. 4. Equilibrium configuration of the H₂O-He system.

Table 5. Interaction properties for the most stable (C_s) configuration of H₂O-He. Basis set [6s4p3d1f/4s3p1d/6s4p3d].

Method	μ_{int}	$\overline{\alpha}_{int}$	$\Delta\alpha_{int}$	$\overline{\gamma}_{int}$
SCF	0.0072	-0.0067	0.2853	-14.19
MP2	0.0074	-0.0072	0.3332	-22.26
MP4	0.0073	-0.0044	0.3440	-22.07
CCSD	0.0073	-0.0037	0.3389	-18.18
CCSD(T)	0.0073	-0.0031	0.3448	-18.93
ECC	**0.0001**	**0.0036**	**0.0595**	**-4.74**

The calculated interaction properties are given in Table 5. The magnitude of the interaction dipole moment (μ_{int}) is very small. The SCF value is $\mu_{int} = 0.0072$. This value changes very little at the post-Hartree-Fock levels of theory. It is rather remarkable that the mean $\overline{\alpha}_{int}$ is very small while the anisotropy $\Delta\alpha_{int}$ is considerably larger. Electron correlation, with the notable exception of the MP2 method, decreases the magnitude of the mean and increases that of the anisotropy. In relative terms, the ECC is 0.0036 for $\overline{\alpha}_{int}$ which corresponds to a decrease of the magnitude of the SCF value by 53.7%. For $\Delta\alpha_{int}$ ECC = 0.0595 or an increase of the SCF value by 20.9%. The most notable effect for the interaction hyperpolarizability is the disagreement of the MP2, MP4 and CCSD, CCSD(T) values for this property. Electron correlation increases the magnitude of $\overline{\gamma}_{int}$ by 33.4%.

3.4. $(H_2O)_2$-He

A preliminary investigation of the interaction properties for the $(H_2O)_2$-He has been recently reported [62]. It is encouraging that our findings agree quite well with the experimental ones for $(H_2O)_2$-Ar [63].

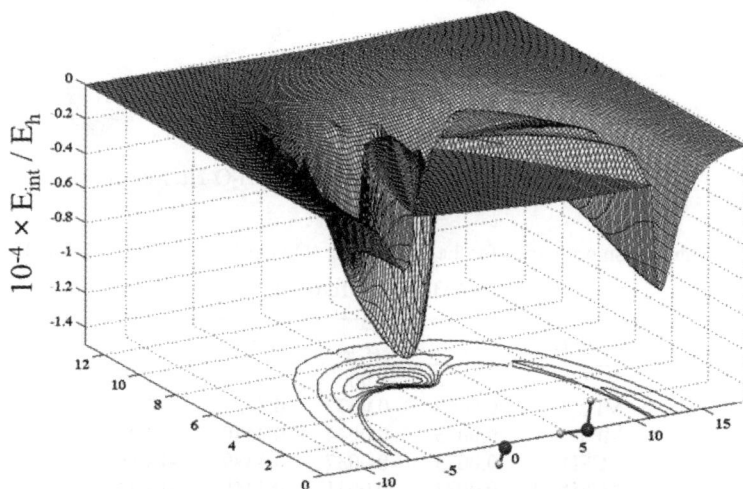

Fig. 5. Potential energy surface of the $(H_2O)_2$-He interaction

Adopting the theoretical geometry of Frisch et al. [64] for the water dimer, we explored this fundamental interaction as much as possible. The results presented here were obtained at the MP2(Full) level of theory with a [6s4p3d1f/4s3p1d] basis set for H_2O [43] and [6s4p3d] for He [17]. A part of the PES and the adopted configuration is shown in Figs. 5 and 6.

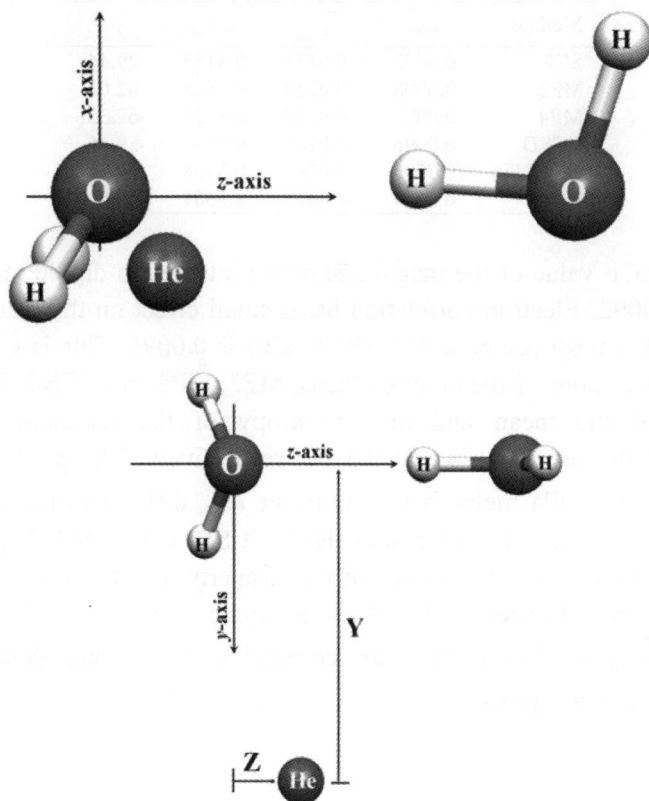

Fig. 6. Orientation of the $(H_2O)_2$-He system.

The surface in Fig. 5 shows clearly that the minimum of the interaction energy is very well defined. The most stable geometry obtained from this investigation is characterized by the position of the helium atom at the side of the water dimer. The Cartesian coordinates of this position are (-0.010, 5.985, 1.299), in a_0. Relying on this

G.Maroulis, A.Haskopoulos

configuration for the complex, we obtained the interaction properties for
$(H_2O)_2$-He shown in Table 5.

Table 5. Interaction properties for the most stable configuration of $(H_2O)_2$-He. Basis set
[6s4p3d1f/4s3p1d/6s4p3d].

Method	μ_{int}	$\overline{\alpha}_{int}$	$\Delta\alpha_{int}$	$\overline{\gamma}_{int}$
SCF	0.0092	-0.0174	0.3157	-29.44
MP2	0.0098	-0.0230	0.3628	-62.08
MP4	0.0097	-0.0218	0.3725	-69.35
CCSD	0.0096	-0.0188	0.3701	-58.66
CCSD(T)	0.0096	-0.0182	0.3758	-65.25
ECC	**0.0004**	**-0.0008**	**0.0601**	**-35.81**

The SCF value of the magnitude of the interaction dipole moment is
μ_{int} = 0.0092. Electron correlation has a small effect on this value. Our,
presumably most accurate, CCSD(T) value is 0.0096. This is very close
to the predictions of the other methods, MP2, MP4 and CCSD. The SCF
values of the mean and the anisotropy of the interaction dipole
polarizability are -0.0174 and 0.3157, respectively. The predictions of
the MP2 and MP4 methods for the mean $\overline{\alpha}_{int}$ differ significantly from
those of the coupled cluster ones. Both CCSD and CCSD(T) predict a
small increase of the magnitude of this property. For the anisotropy, the
predicted post-Hartree-Fock values are quite close. The SCF method
yields $\overline{\gamma}_{int}$ = -29.44. Electron correlation more than doubles the
magnitude of this property.

3.5. O_3-He

We are not aware of previous experimental work on the structure of
the O_3-He system. The geometry of the similar O_3-Ar was investigated
by radio frequency and microwave molecular beam spectroscopy
sometime ago by DeLeon et al. [65]. According to their findings, the
argon atom is located approximately above the center of mass of the O_3
moiety. We relied on their work in order to investigate the structure of
the O_3-He complex. We used a [6s4p3d1f] basis set for O_3 [66] and

[6s4p3d] for He [17]. The O-O bond length and the ∠O-O-O angle of O_3 were kept fixed at the experimental values of 1.2717 Å and 116.783° , respectively. The PES in Fig. 7 (xz as the molecular plane with z as the C_2 axis and He on the yz plane) was obtained at the MP2(Full) level of theory shows clearly that the total minimum is very well defined. In the adopted configuration of Fig. 8, the He atom is located at (y,z) = (5.371,1.052), in a_0.

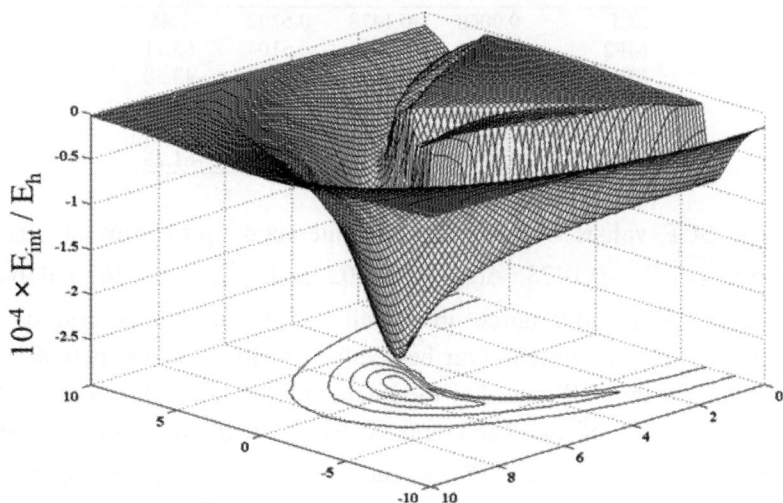

Fig. 7. Potential energy surface of the O_3-He interaction with the He atom on the yz plane. The molecule is on the xz plane (see text).

Fig. 8. Most stable configuration of the O_3-He system. O_3 is on the xz plane.

The interaction properties calculated with the same basis set as above are shown in Table 6. This basis gives $\overline{\alpha}$ (SCF) = 20.88, very close to the 20.93 reported elsewhere [67].

Table 6. Interaction properties for the most stable configuration of O_3-He. Basis set [6s4p3d1f/6s4p3d].

Method	μ_{int}	$\overline{\alpha}_{int}$	$\Delta\alpha_{int}$	$\overline{\gamma}_{int}$
SCF	0.0060	-0.1478	0.6202	-1.46
MP2	0.0074	-0.0895	0.5104	-63.71
MP4	0.0075	-0.1189	0.5678	-43.50
CCSD	0.0070	-0.1327	0.6063	-23.97
CCSD(T)	0.0072	-0.1178	0.5682	-62.72
ECC	**0.0008**	**0.0300**	**-0.0520**	**-61.26**

The SCF values obtained for the interaction properties are μ_{int} = 0.0060, $\overline{\alpha}_{int}$ = -0.1478, $\Delta\alpha_{int}$ = 0.6202 and $\overline{\gamma}_{int}$ = -1.46. All post-Hartree-Fock methods agree quite well and predict an increase of the interaction dipole moment. Our best value for this property is 0.0072 or 13.3% above the SCF one. For the interaction polarizability we note the good agreement of the MP4 and CCSD(T) methods. The CCSD(T) method gives $\overline{\alpha}_{int}$ = -0.1178 and $\Delta\alpha_{int}$ = 0.5682, which corresponds to the reduction of the SCF values by 20.3 and 8.4%, respectively. The property most affected by electron correlation, in relative terms, is the hyperpolarizability. Although the MP2 is close enough to the CCSD(T) for this property, the other methods predict a significantly lower magnitude for $\overline{\gamma}_{int}$.

4. Conclusions

We have collected in this review the findings of our investigation of the interaction electric properties of the complexes MOL-He, where MOL = N_2, CO_2, H_2O, $(H_2O)_2$ and O_3. The interaction dipole moment, dipole polarizability and second dipole hyperpolarizability were obtained from finite-field RHF, Møller-Plesset and coupled cluster calculations. We have relied exclusively on especially designed, purpose-oriented, basis sets of Gaussian-type functions. All basis sets yield dipole polarizabilities very close to the respective Hartree-Fock limit.

Overall, the complexes are characterized by a small interaction dipole moment at the SCF level, a value that is little affected by electron correlation. The mean polarizability $\bar{\alpha}_{int}$ is invariably smaller than the anisotropy. The magnitude of the second hyperpolarizability $\bar{\gamma}_{int}$ is rather small for N_2, CO_2, H_2O but somewhat larger for $(H_2O)_2$ and O_3. The calculated electron correlation effects for the interaction polarizability and hyperpolarizability depend largely on the partner molecule.

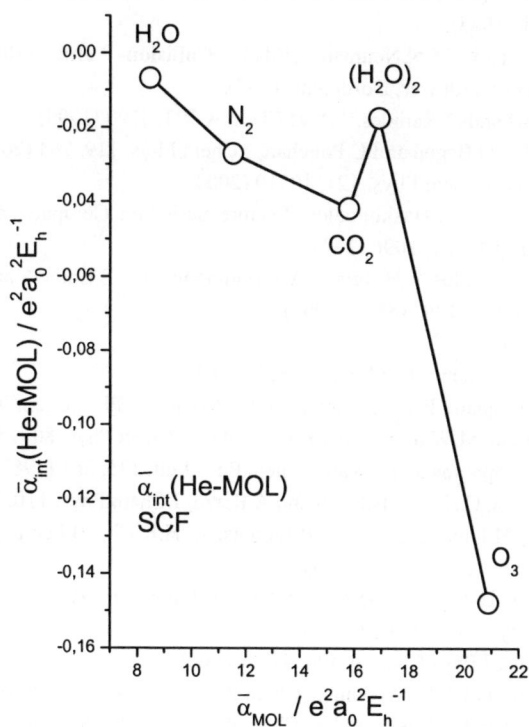

Fig. 9. $\bar{\alpha}_{int}(He - MOL)$ versus $\bar{\alpha}(MOL)$ for MOL = H_2O, N_2, CO_2, $(H_2O)_2$ and O_3.

We have plotted in Fig. 9 the evolution at the SCF level of the mean $\overline{\alpha}_{int}$ (MOL-He) with $\overline{\alpha}$ (MOL). With the notable exception of the water dimer, the interaction property $\overline{\alpha}_{int}$ (MOL-He) becomes increasingly negative.

References

1. G.C.Maitland, M.Rigby, E.B.Smith and W.A.Wakeham, *Intermolecular forces: their origin and determination* (Clarendon, Oxford, 1981)
2. G.Birnbaum (Ed.), *Phenomena induced by intermolecular interactions* (Plenum, New York, 1985).
3. G.Tabisz and M.N.Neuman (Eds.), **Collision- and interaction induced spectroscopy** (Kluwer, Dordrecht, 1995).
4. G.Maroulis and D.Xenides, J.Phys.Chem.A 103, 4590 (1999).
5. G.Maroulis, D.Begué and C.Pouchan, J.Chem.Phys. 119, 794 (2003).
6. G.Maroulis, J.Chem.Phys. 121, 10519 (2004).
7. G.Maroulis and A.Haskopoulos, Lecture Series on Computer and Computational Science (LSCCS) 1, 1096 (2004).
8. M.G.Papadopoulos, H.Reis, A.Avramopoulos, S.Erkoc and L.Amirouche, J.Phys.Chem. B 109, 18822 (2005).
9. See reference 2.
10. K.L.C.Hunt, Chem.Phys.Lett. 70, 336 (1980).
11. A.D.Buckingham, E.P.Concannon and I.D.Hands, J.Phys.Chem 98, 10455 (1994).
12. E.W.Pearson, M.Waldman and R.G.Gordon, J.Chem.Phys. 80, 1543 (1984).
13. M.G.Papadopoulos and J.Waite, Chem.Phys.Lett. 135, 361 (1987).
14. B.Fernandez, C.Hättig, H.Koch and A.Rizzo, J.Chem.Phys. 110, 2872 (1999).
15. C.Hättig, H.Larsen, J.Olsen, P.Jørgensen, H.Koch, B.Fernandez and A.Rizzo, J.Chem.Phys. 111, 10099 (1999).
16. H.Koch, C.Hättig, H.Larsen, J.Olsen, P.Jørgensen, B.Fernandez and A.Rizzo, J.Chem.Phys. 111, 10108 (1999).
17. G.Maroulis, J.Phys.Chem.A 104, 4772 (2000).
18. G.Maroulis and A.Haskopoulos, Chem.Phys.Lett. 349, 335 (2001).
19. G.Maroulis and A.Haskopoulos, Chem.Phys.Lett. 358, 64 (2002).
20. R.J.Li, Z.R.Li, D.Wu, X.Y.Hao, B.Q.Wang and C.C.Sun, J.Phys.Chem. A 107, 6306 (2003).
21. X.Y.Hao, Z.R.Li, D.Wu, Y.Wang, Z.S.Li and C.C.Sun, J.Chem.Phys. 118, 83 (2003).
22. G.Maroulis, A.Haskopoulos and D.Xenides, Chem.Phys.Lett. 396, 59 (2004).

23. W.Chen, Z.R.Li, D.Wu, F.L.Gu, X.Y.Hao, B.Q.Wang, R.J.Li and C.C.Sun, J.Chem.Phys. 121, 10489 (2004).
24. J.Lopez-Cacheiro, B.Fernandez, D.Marchesan, S.Coriani, C.Hättig and A.Rizzo, Mol.Phys. 102, 101 (2004).
25. A.Haskopoulos, D.Xenides and G.Maroulis, Chem.Phys. 309, 271 (2005).
26. P.Karamanis and G.Maroulis, Comp.Lett. 1, 117 (2005).
27. A.Haskopoulos and G.Maroulis, Lecture Series on Computer and Computational Science (LSCCS) 3, 93 (2005).
28. H.D.Cohen and C.C.J.Roothaan, J.Chem.Phys. 43S, 34 (1965).
29. G.Maroulis, J.Chem.Phys. 108, 5432 (1998) and references therein.
30. A.D.Buckingham, Adv.Chem.Phys. 12, 107 (1967).
31. A.D.McLean and M.Yoshimine, J.Chem.Phys. 47, 1927 (1967).
32. G.Maroulis and D.M.Bishop, Chem.Phys.Lett. 114, 182 (1985).
33. D.M.Bishop and G.Maroulis, J.Chem.Phys. 82, 2380 (1985).
34. G.Maroulis an D.M.Bishop, Chem.Phys. 96, 409 (1985).
35. G.Maroulis, Chem.Phys.Lett. 199, 250 (1992).
36. G.Maroulis, Chem.Phys.Lett. 312, 255 (1999).
37. G.Maroulis and A.J.Thakkar, J.Chem.Phys. 88, 7623 (1988).
38. G.Maroulis, J.Chem.Phys. 94, 182 (1991).
39. G.Maroulis, J.Chem.Phys. 111, 583 (1999).
40. D.Xenides and G.Maroulis, 115, 7953 (2001).
41. G.Maroulis, J.Phys.Chem. A 107, 6495 (2003).
42. S.F.Boys and F.Bernardi, Mol.Phys. 19, 55 (1970).
43. G.Maroulis, J.Chem.Phys. 113, 1813 (2000).
44. A.Szabo and N.S.Ostlund, *Modern Quantum Chemistry* (McMillan, New York, 1982).
45. T.Helgaker, P.Jørgensen and J.Olsen, *Molecular Electronic Structure Theory* (Wiley, Chichester, 2000).
46. J.Paldus and X.Li, Adv.Chem.Phys. 110, 1 (1999) and references therein.
47. M.J.Frisch, G.W.Trucks, H.B.Schlegel et al., GAUSSIAN 94, Revision E.1 (Gaussian, Inc, Pittsburgh PA, 1995).
48. M.J.Frisch, G.W.Trucks, H.B.Schlegel et al., GAUSSIAN 98, Revision A.7 (Gaussian, Inc., Pittsburgh PA, 1998).
49. L. Beneventi, P.Casavecchia, G.Volpi, C.C.Wong, F.R.W.McCourt and D.Lemoine, J.Chem.Phys. 95, 5827 (1991).
50. K.P.Huber and G.Herzberg, *Molecular Spectra and Molecular Structure: IV. Constants of Diatomic Molecules* (Van Nostrand, New York, 1979).
51. G.Maroulis, J.Chem.Phys. 118, 2673 (2003).
52. C.Boulet, J.P.Bouanich, J.M.Hartmann, B.Lavorel and A.Deroussiaux, J.Chem.Phys. 111, 9315 (1999).
53. J.Tang and A.R.W.McKellar, J.Chem.Phys. 121, 181 (2004).
54. J.Buldyreva, S.V.Ivanov and L.Nguyen, J.Raman Spectrosc. 36, 148 (2005).

55. F.Negri, F.Ancilotto, G.Mistura and F.Toigo, J.Chem.Phys. 111, 6439 (1999).
56. G.Graner, C.Rossetti and D.Bailly, Mol.Phys. 58, 627 (1986).
57. G.Maroulis, Chem.Phys. 291, 81 (2003).
58. G.Maroulis, Chem.Phys.Lett. 396, 66 (2004).
59. G.Calderoni, F.Cargnoni and M.Raimondi, Chem.Phys.Lett. 370, 233 (2003).
60. W.S.Benedict, N.Gailar and E.K.Plyler, J.Chem.Phys. 24, 1139 (1956).
61. G.Maroulis, Chem.Phys.Lett. 289, 403 (1998).
62. A.Haskopoulos and G.Maroulis, MATCH Comm.Math.Comp.Chem. **53**, 253-260 (2005).
63. E.Arunan, T.Emilsson and H.S.Gutowsky, J.Am.Chem.Soc. 116, 8418 (1994).
64. M.J.Frisch, J.A.Pople and J.E.Del Bene, J.Phys.Chem. 89, 3664 (1985).
65. R.L.DeLeon, K.M.Mack and J.S.Muenter, J.Chem.Phys. 71, 4487 (1979).
66. G.Maroulis, unpublished results.
67. G.Maroulis, J.Chem.Phys. 111, 6846 (1999).

CHAPTER 18

THEORETICAL STUDIES ON POLARIZABILITY OF ALKALI METAL CLUSTERS

K. R. S. Chandrakumar, Tapan K. Ghanty and Swapan K. Ghosh
Theoretical Chemistry Section, RC & CD Division, Chemistry Group,
Bhabha Atomic Research Centre, Mumbai 400 085, India.
Email: skghosh@magnum.barc.ernet.in

In this article, we will present a brief review on the polarizability of alkali metal clusters, particularly for the cases of sodium and lithium clusters, calculated by various theoretical methods. The role of electron correlation in determining the polarizability of these clusters has been discussed elaborately. The significance of the issues related to the use of density functional and other ab initio approaches is highlighted and the need for proper inclusion of electron correlation effects through better exchange-correlation density functionals or other approaches is also emphasized. Some of the possible sources of error present in the experimental measurements as well as in the different theoretical methods have been highlighted. The existence of a good linear correlation between the dipole polarizability and the ionization potential and softness (reactivity parameter) of the neutral clusters has been discussed.

I. Introduction

The response of atomic and molecular systems to an external or internal electric or magnetic field perturbation has been of immense importance in a variety of problems in physics and chemistry.[1-6] Depending on the nature of perturbations, one can have many response properties. Although most of these response properties are important and useful in understanding various phenomena exhibited by the atomic and molecular systems, the focus of our attention here will be only on polarizability and that too for metal clusters for many laudable reasons. The dipole polarizability is one of the fundamental electronic properties and provides a measure of the extent of distortion of the electron density on the application of external electric field.[2,3,7-10] In recent years, efficient techniques have been developed for the determination of such response properties using *ab initio* or density functional methods of electronic structure theory in the area of computational chemistry.[1-5]

The study of polarizability has been considered to be one of the challenging and interesting tasks for many decades and has a long history.[3,10,11] It has also emerged as an active research field in recent years due to the usefulness of the information contained in the polarizability property for a variety of important systems and phenomena.[7,8,10,11,12-25] Typical applications of polarizability in different areas include the interpretation of molecular electronic structure,[12] reactivity,[13-15] structure-activity relationships[16-18] and long-range interactions.[10,19,21] In modeling the solvent effects, the polarizability is used in the reaction field model as an important parameter. This property as well as its derivatives form the basis for an understanding of many linear and nonlinear optical phenomena.[3,7,8,12,22] The intrinsic properties namely, refraction, absorption, and scattering of electromagnetic waves in bulk materials can be associated with nonzero values of polarizability.[10] Intermolecular interactions are also strongly influenced by the magnitude of the polarizability tensor elements.[21,23-26] The role of polarizability for modeling the ionic and nonionic solvation has also been recently emphasized.[26-30] Moreover, the interaction between slow electrons and atoms or molecules resulting in scattering or even binding of these electrons, is strongly dependent on polarizabilities.[31] This property can also be related linearly with the reactivity and ionization potential for various systems.[32-36] An accurate knowledge of the response properties also plays important role for the development of nonlinear optical materials.

Detailed review articles and books on the polarizability of atomic and molecular systems are available[1,2,9,11,12,21,22,37-48] and therefore our attention will be focused here to the case of newly emerging system of alkali metal clusters. This article is, however, not aimed at providing an exhaustive and complete review of this area of research and we highlight here only some of the important aspects of our own work along with some other the recent developments. One of the objectives is to analyze the performance of density functional theory (DFT) with various exchange-correlation density functionals in determining the polarizability of the alkali metal clusters by comparing the predicted results with those from experiments and wave function based methods.

In this article, we provide an overview on the recent results on theoretically calculated electric dipole polarizability of alkali metal clusters. We begin this review with a brief outline of the metal clusters and their significance in Section II. The experimental techniques and theoretical methods for the determination of polarizability of the clusters are discussed in Section III. Next, in Section IV and V, the structure and cluster size effects on the polarizability of these clusters have been discussed. The applicability of the different DFT as well as *ab initio* methods in determining the polarizability has been discussed in Section VI. Further, a detailed discussion for the electron correlation effect on the polarizability of the Na and Li clusters has been illustrated in Section VII. The relationship between the polarizability and various quantities, such as, ionization potential, reactivity parameter (softness), and binding energy of the clusters has been discussed in Section VIII. In Section IX, we provide a brief note on the effect of temperature on the cluster polarizabilities. Finally, we present a few concluding remarks in the last section.

II. Metal Clusters and their Significance

In recent years, the study of small metal clusters has been the subject of intense interest from both experimental and theoretical points of view, mainly because of the interesting dependence of their electronic structure and properties on their size and geometry.[11,45,46-48,49-61] Clusters are the aggregates of atoms (or molecules) consisting of two to few thousand atoms, with the size ranging from 1 to 10 nm and can thus be called super atoms or giant atoms.[46,47,51,57,59] The properties of clusters are intermediate between those of the corresponding atoms and the bulk phase. Thus, clusters form a bridge between the well-established fields of atomic and solid-state physics (gradual transition from molecular to condensed matter physics). Besides metal clusters, other clusters are also possible viz., van der Waals, molecular, covalent, ionic and hydrogen bonded clusters, with the forces involved varying from very weak to very strong ones.[50,51,56,57,61]

The primary goals and interests of many physicists and chemists in the area of metal clusters span over many problems, ranging from the formation of the clusters and its mechanism;[62-65] to the influence of the particle size on various aspects, such as, catalysis (due to high surface/volume ratio), electron transfer, nucleation process, etc.[58-61,66-71] More importantly, the optical, electrical and magnetic properties of these clusters have emerged as important research areas.[56] The electronic properties and structures of small metal clusters primarily depend on the size (number of atoms) of the clusters and their physical and chemical properties can even be tuned in desired manner by simply changing the cluster size and the charge on the system.[57,61,71] Most of these properties often show non-monotonic and oscillating dependence due to the quantum size effects for small clusters, thus behaving quite differently, as compared to the bulk.[46-48,72,73,74] As the size of the clusters becomes large, the properties of the clusters, however, smoothly saturate to yield the bulk properties. Hence, an understanding of the size sensitive properties is very essential, particularly, in view of their potential applications. It, has, however, been realized that it is difficult to understand the behavior of these clusters under different circumstances and hence, many fundamental questions are yet to be addressed.

This rapidly developing area of research on the size dependent physical and chemical properties of clusters has made significant progress in the direction of many experimental and theoretical studies.[56] Although the exploration of interest in this subject appears to be more in recent years, Kubo predicted long back in the sixties that the properties of all systems will change when the size of the systems reduces to the nano regime.[75] The theoretical aspects of the various novel and basics aspects of the clusters and the corresponding cluster size effects have been elaborately reviewed by Jortner[57,61] and other leading research workers in this field.[8,47-50] From the experimental point of view, the remarkable progress in the areas of supersonic jets and cluster beams has made significant advances for the study of isolated clusters.[38-40,46,76-79] Parallel to these advancements, a variety of theoretical and computational methods have also provided a framework for the understanding of the evolution of the electronic structural,[62-65] and thermodynamic properties[80-85] and also quantum size effects[51,52,56,57,75,86-92] in clusters. The physical and chemical aspects of these clusters were explored in a remarkable manner by the experiments conducted by Knight et al.[48,93-96] especially for the direct measurement of the polarizability of sodium and potassium clusters[93] by electron deflection of a molecular beam. It has been observed that the properties are, generally, size-dependent in nature, reflecting definite electronic shell effects, with minima in the polarizability values at n = 2, 8, 18, and 20 for alkali metal clusters.[46,48,93-96,97-105]

In the initial periods, the theoretical understanding of these clusters mostly relied on the Jellium model of clusters wherein the properties of the clusters are determined by their electronic properties with the effect of the nuclear positions smeared out.[47,48,106-115] The simple three-dimensional harmonic oscillator potential or a square well or the "Woods-Saxon" potential have largely been used.[47,48] The major success of the Jellium model is the prediction of the shell structure for the electronic states similar to the shell structure in atoms and the concept of magic numbers.[46-48] The concepts of shell structure and magic numbers had been very much familiar to the nuclear physics community.[116] It should be noted that the observed shell structure is not due to the packing of atoms, or ions, but is produced by the occupancy of the `free' metallic electrons, otherwise responsible for the electric conductivity of metals. Hence, the observed shell structure in alkali and other metal clusters appears to be of electronic origin. Experimental measurements, at low temperatures, on alkali and metal clusters have shown that the behavior of some of their electronic properties is dominated by an odd-even alternation, with the alternation magnitude depending strongly on the element of which the cluster is formed.[46-48]

Stable clusters, which contain so called "magic numbers" of atoms, are formed upon closing a shell, analogous to the formation of rare gases and are more stable with lower polarizabilities and higher ionization potentials than their immediate neighbors.[48] The number of electrons corresponding to closed shells is 2, 8, 18, 20, 34, 40, 58, etc. The shell model has accounted for such diverse characteristics as the polarizabilities and the ionization potentials of alkali and other type of clusters.[48,100,101,117-119] The major features in the mass spectra of metal clusters also identify clusters of special stability associated with the closing of electronic shells at the particular number of electrons corresponds to the magic numbers.

III. Theoretical and Experimental Determination of Polarizability: A Brief Outline

III.1 Theoretical methods and computational details

In an *ab initio* calculation of the electronic structure of metal clusters, the geometry optimization can be carried out using suitable methods. In our recent work, the geometry optimization of sodium clusters (Na_n, with n=2-10) has been performed through DFT with the B3-LYP exchange-correlation functional using the 6-311+G(d) basis set.[120] These globally optimized geometries have then been used in all the *ab initio* and DFT calculations of polarizability using Sadlej basis set. The post HF methods, second and fourth order Möller-Plesset (MP2 and MP4) calculations have been performed using the same geometry. DFT methods using different functionals, namely, PBE-LYP,[121] Becke-LYP, B3-LYP,[122] BHH-LYP,[123] H-LYP, H-OP, B-OP and PBE-OP have been employed to include the exchange and correlation effects, with the latter treated through the Lee-Yang-Parr (LYP)[124] functional and the One-parameter Progressive (OP)[125,126] functional. The restricted HF method has been used for the calculation of polarizability of the even member of the clusters while for the odd member clusters, the restricted open-shell HF method has been used. Both *ab initio* and the DFT calculations have been performed using the GAMESS system of programs.[127] We have used the grid-based DFT in GAMESS that employs a typical grid quadrature to compute the integrals. During the SCF procedure, the grid used consists of 96 radial shells with 12 and 24 angular points.

In general, there are different methods (derivative techniques, perturbation theory and propagator methods) for the calculation of the effect of an external field perturbation on many electron systems. Here, we will discuss the most common approach for calculating the response properties, namely the derivative formulation. Herein, the static response properties of a molecule are defined by expanding the field-dependent energy E(F) as a series in the components of a uniform electric field F, viz.

$$E(F) = E(0) + \sum_{i} \mu_i F_i + \sum_{ij} \alpha_{ij} F_i F_j$$

where E(0) is the energy of the system in the absence of the electric field, μ is the dipole moment and α_{ij} is the dipole polarizability tensor. The components of the polarizability tensor are obtained as the second-order derivatives of the energy with respect to the Cartesian components (i,j = x,y,z) of the electric field. The derivatives are evaluated numerically using the finite field method and the mean polarizability is calculated from the diagonal elements of polarizability tensor as

$$\alpha = (\alpha_{xx} + \alpha_{yy} + \alpha_{zz})/3$$

Besides the finite field method, one can also use the approaches like coupled perturbed Hartree-Fock (CPHF) method for the calculation of polarizabiliy.[2,4] The finite field method is known to be one of the very efficient ways of calculating the polarizability within the framework of Hartree-Fock as well as density functional theory. Although the analytical approach with the DFT based coupled perturbed method is not developed, the hybrid solution has recently been proposed.[128] We have, however, followed here the finite field method using the various DFT based exchange-correlation functionals and other post post-HF methods to evaluate the mean polarizability of the alkali metal clusters.

III.2 Experimental Determination of Polarizabilities for the Alkali Metal Clusters

A brief outline on the experimental setup, focusing on the determination of polarizability is discussed here. The experimental methods for producing free nano clusters have been discussed in details by de Heer *et al*[46,48] and Rayane[79] and coworkers.   The static polarizability measurements are, in general, carried out by deflecting a well-collimated beam through a static inhomogeneous transverse electric field.   The experiment consists of a laser vaporization source coupled to an electric deflector and a mass spectrometer.   The first step is the production of the free clusters and many different types of sources have been developed to produce clusters of various sizes and properties. The cluster sources basically rely either on condensation/aggregation or on break up, and sometimes on both. The most widely used cluster source is based on the supersonic jet cluster source.   The resulting beam is collimated by two round skimmers with diameters of 0.4 mm and 1.5 mm.   A voltage of 30 kV is applied across the gap with no detected leakage current. Clusters are ionized one meter after the deflector in the extraction region of a position sensitive time of flight mass spectrometer. The mass of the cluster and the profile of the beam are obtained from the arrival time at the detector. The beam profile is measured as a function of the electric field in the deflector. The polarizability is proportional to the measured deflection in the z direction, $d = \alpha C V^2 / nmv^2$ where, α is the rotationally averaged static electric dipole polarizability, m is the mass of a alkali metal atom, C is a factor defined by the deflection plate geometry and apparatus dimensions, V is the plate voltage, and v is the cluster velocity. Here, the velocity is determined by means of a coaxial time-of-flight   measurement.   The   precision   of   the   measured polarizabilities is estimated to be at most 10%, in which the main source of error is the velocity measurement. Other descriptions for the experimental setup can be referred elsewhere.[38,39,46,48,76-79,129-131]

IV. Equilibrium Structures of Sodium and Lithium Clusters

One of the important steps for the understanding of the cluster properties is to study their structures. For a given size and composition, the clusters can have different geometries which are energetically close to each other.   In recent years, the structural properties of the alkali metal clusters have been extensively studied by different theoretical methods.[132-143] In Figure 1, the equilibrium geometries of the lithium and sodium clusters obtained through optimization at the level of B3LYP/6-311+G(d) are shown.   It can be seen that the cluster size effects on the geometries of these clusters is important. In general, it is noted that the shape and structure of the Li and Na clusters are very similar to each other. However, it is observed that the average Li-Li bond distance in the lithium clusters is shorter than the Na-Na bond length in sodium clusters. For instance, the Li-Li bond length in Li_2 is 2.709Å whereas the Na-Na bond length is 3.055Å, and the

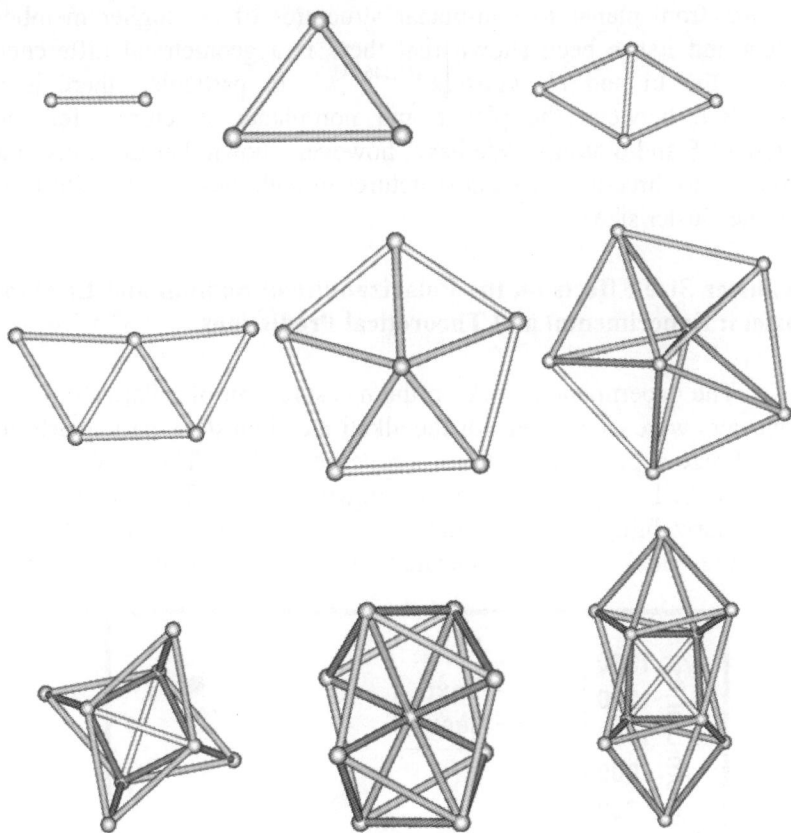

Fig. 1. The Equilibrium geometry of sodium and lithium clusters

corresponding experimental values are 2.673Å and 3.08Å, respectively.[144] A similar trend is observed for other clusters also. In the case of trimer, the equilibrium geometry is isosceles obtuse angled triangle and our results for the bond length for the shorter side is comparable with the experiment as well as other theoretical values while the longer side of the triangle is found to be longer by 0.7Å, compared to the experimental value of 4.157Å.[145]   This kind of trend has been observed in many of the earlier theoretical works and in general, the B3-LYP based bond lengths are little higher than the B88-P86 values and the LDA results are considerably smaller than the results from other functionals.   There have been other studies, focusing exclusively on the transition from planar to non-planar structures of the higher member clusters and it has been shown that there is a geometrical difference between the Li and Na clusters.[142,146,147]   In particular, there is a competition between the planar and non-planar structures, for the clusters of 5 and 6 atoms. We have, however, shown that the transition from two– to three-dimensional structures in both these clusters starts at the same cluster size of 6.[120]

V. Cluster Size Effects on the Polarizability of Sodium and Lithium Clusters: Experimental and Theoretical Predictions

The experimental work on the measurement of polarizability of the clusters were carried out for the alkali metal clusters, particularly to study of size-dependent electronic properties.[46-48,120]   The experimental values for the Li and Na clusters are shown in Figure 2 and the minima in the polarizability values for the cluster size 2 and 8 show that there are definite shell effects, associated with special stability. Several

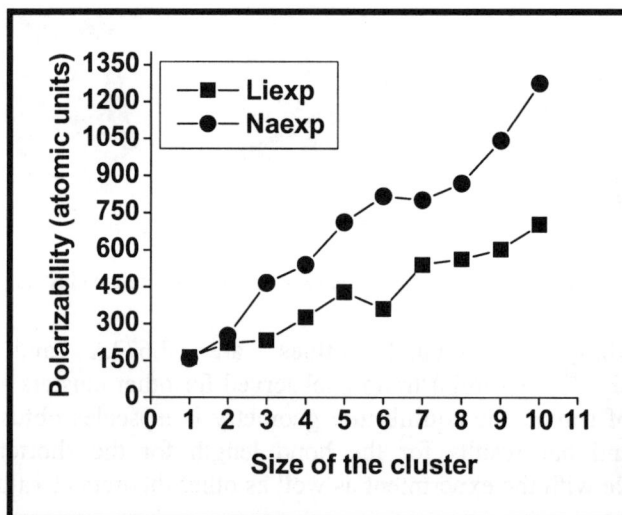

Fig. 2. Comparison of the experimental polarizabilities of sodium and lithium clusters.

theoretical calculations have been carried out for the static dipole polarizability of alkali metal clusters.[91,96,100,101,106-115,132-135,140] Our calculated results on the polarizability of clusters, Na_n and Li_n (for $1 \leq n \leq 10$), by using different theoretical methodologies and large extensive Sadlej basis set are shown in Figs. 2-6. It is clear that there is considerable disagreement among the various experimental results and reproducible experimental results have been reported only for few clusters. The values provided by Knight *et al*[46,48,93] are higher than those of Rayne *et al*.[79,131] for lower member clusters, (n < 3 - 8), while for the higher member clusters (n > 8 - 20), these experimental values are quite different from each other and sometimes even the generally observed trend (odd-even oscillations) differs. The experimental error in the earlier work was in the range of 12% whereas the error present in the recent experimental values is estimated to be in the range of few atomic units per atom and hence, the accuracy of the results will change drastically as the cluster size grows. It can be seen from Fig. 2 that the size evolution of the electronic properties is mostly characterized by the odd-even oscillations and this feature is more prominent for lithium clusters as compared to the sodium clusters. The experimental values of polarizability of sodium and lithium atoms are already available in the literature.[148-150] Although the atomic polarizabilities of Li and Na are close to each other, it is interesting to note that the difference in the polarizability becomes very significant as the cluster size grows. For the atomic cases, the temperature effect is insignificant and hence the comparison of the theoretical and experimental values would pave a way to assess the behavior (or accuracy) of different DFT functionals and the MP2 method. There are three experimental polarizability values available in the literature for the sodium atom, viz., 159.3±4.3, 164.7±11.5 and 162.7±0.8au though different methods.[184-150] For the lithium atom, the best available value is 164.0±3.4 au.[148] Based on the finite-field coupled cluster methods with large, systematically optimized basis sets of Gaussian-type functions, Maroulis[151] has reported the polarizability value of sodium atom, 165.06au which is fairly larger than the available experimental results. From figs. 3 and 4, it is clear that all the pure DFT functionals significantly underestimates the polarizability values of Na and Li atoms (figs. 5 and 6). The hybrid DFT functionals also underestimates the polarizability while DFT methods with full HF exchange yield results, which are slightly higher than the experimental results, but are close to the MP2 values. In case of sodium atom case, the PBE-LYP functional predicts the polarizability with 12.5% error, which is reduced to 6.7% if the OP correlation functional is used instead of LYP functional. A Similar trend is also observed for the lithium atom, and the underestimation in the calculated polarizability is more than 15% except the values predicted by the HOP and MP2 methods, which are very close to the experimental result. This is an important observation, which suggests that the effect of correlation is the most important factor in determining the polarizability of such systems. It can also be mentioned that among the methods considered here, the MP2 can be considered the best available theoretical methods for the polarizability calculations. Analyzing the polarizability trend for the Na and Li clusters, some interesting features can be revealed from Figure 2. In case of Li clusters, the polarizability of the clusters measured by the experimental method gradually increases from Li_1 to Li_5, passes through a minimum at Li_6 and is then followed by a steep rise for the rest of the clusters.[129,131] It is interesting to note that none of the theoretical methods used in the present work, is able to reproduce this trend of polarizability evolution with respect to the size of the clusters. However, the MP2 method shows that there is a deep minimum in polarizability for the cluster Li_7. On the other hand, the experimental polarizability values for the case of sodium clusters provided by Knight et al., gradually increase from the cluster size 1 to 6, then pass through a minimum at 7, and then increase for other higher clusters.[93] Such minimum in the polarizability trend has been reproduced only by the MP2 method and not by any DFT methods, which rather predict that there is a continuous monotonic increase in the polarizability values. Although both the clusters have the same number of electrons in the valence shell and similar cluster geometry, there is a difference in the polarizability evolution of the two clusters, particularly at the cluster size of 6 and 7. The origin of this anomaly has not been understood clearly. This problem has, however, been explained to some extent in terms of the planar and non-planar structures of these two clusters.[140-143] The recent study of Benichou et al. also confirms that there is a competition between the planar and the non-planar structures for the Li_5 and Li_6 clusters.[141]

VI. Polarizabilities of Na and Li Clusters Calculated by Different Methods: Applicability of Post Hartree-Fock and Density Functional Theory Based Approaches

The early calculations, which were based on variants of the jellium model within the framework of density functional theory (DFT), have been reasonably successful in explaining qualitatively the trend in the size dependence of polarizability.[46-48,106-115] However, the calculated polarizabilities were found to be less than the corresponding experimental values, due to the fact that the jellium model does not incorporate the effects of discrete atoms in the clusters. Subsequently, calculations have been carried out by considering the actual geometrical arrangement of the metal atoms in the clusters using DFT with different types of pseudo potentials as well as all-electron correlated methods.[132-143] These studies have mainly demonstrated that the theoretically evaluated polarizability values deviate from the experimental polarizability values significantly (often underestimated by an amount of 20-30%). It has been speculated that the observed discrepancy may be due to the effect of temperature on the structure of the cluster, since the calculations have been performed at zero temperature.[152-155] It has also been noted that the measured polarizability values could be in error due to the experimental setup and the accuracy of the measurement, which is dependent on many parameters. For instance, the recent experiments carried out by Rayne et al[131] and Tikhonov *et al.*[130] reveal that for some of the sodium clusters, the observed polarizability values differ even by ~100 au from the earlier experimental results. Although the general qualitative trend is found to be similar in all the experimental results, the reproducibility of the values is rather poor in experimental as well as theoretical results. Hence, a reliable and systematic all electron calculation of the polarizability by accurate theoretical methods is essential to resolve the ambiguity present in the experimental as well as theoretical results. The large discrepancy present in the previous theoretical calculations is due to the fact that the effect of electron correlation has not been taken into account effectively and most of the reported values were obtained by the DFT based methods.[106-115, 132-143] Keeping all these factors in mind, let us now discuss the recent results on the polarizability of sodium and lithium clusters calculated by various theoretical methods. The calculated polarizability values obtained by the HF and MP2 methods using the Sadlej basis set are listed in Table 1 along with the available experimental results. We have noted that

Table 1: Polarizability values of sodium and lithium clusters obtained by experimental, Hartree-Fock and MP2 (in atomic units)

Cluster size	Sodium		Lithium		Experimental results		
					Sodium		Lithium
	MP2	HF	MP2	HF	Knight et al.[a]	Rayan et al.[b]	
1	166.88	191.01	167.59	167.86	--	--	163.87
2	255.44	270.58	205.79	203.45	251.90	265.24	221.35
3	431.72	464.50	236.20	299.83	464.70	444.83	232.82
4	508.56	530.17	346.24	347.38	538.62	565.58	326.62
5	635.34	637.30	432.36	414.74	710.92	630.03	428.52
6	699.34	728.61	510.45	503.39	816.62	754.42	360.36
7	655.47	827.02	402.22	659.10	800.69	808.34	538.52
8	797.65	851.62	611.84	602.10	868.75	901.14	561.47
9	949.89	1042.31	659.81	841.96	1042.49	1062.98	601.28
10	1067.96	1098.57	747.77	719.51	1274.14	1309.33	701.83

[a] Ref.93   [b] Ref. 131

inclusion of the electron correlation effects by more sophisticated methods, such as, MCQDPT2 and CCSD(T), does not change the polarizability value significantly from the values calculated by the MP2 method.   The calculated polarizabilities for the higher clusters, even by the more sophisticated electron correlation methods, in general, deviate considerably from the experimental results. The HF method overestimates the polarizability due to the complete neglect of correlation effects.   First, we discuss the performance of DFT using various hybrid, non-hybrid and other XC functionals for sodium clusters. It can be seen from Figures 3 and 4 that the performance of a functional depends mostly on whether it is a hybrid or non-hybrid functional.   Interestingly, the polarizability values predicted by the DFT functionals (except H-OP functional), are less than the MP2 values for most of the sodium clusters. In order to compare the performance of different methods of incorporating the correlation effects, we first consider methods where full HF exchange is used, viz. MP2, H-OP and H-LYP. The H-OP values of polarizability are found to be better than the MP2 values for Na_3, Na_8 and Na_9 and far better for Na_7, while the predictions of the two methods for other clusters are quite

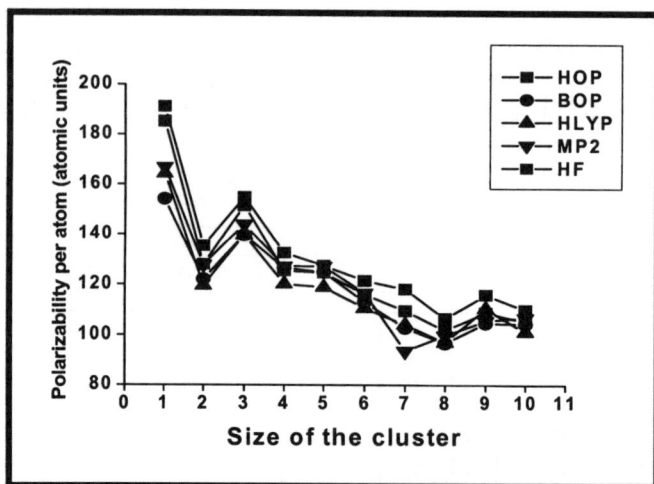

Fig. 3. Comparison of the sodium cluster polarizabilities calculated by MP2, HF and various DFT exchange-correlation functionals.

close to each other. It is also imperative to note that the H-OP value for Na_3 is very close to the experimental values, being only 2% lower than the experimental value of Knight et al.[93] and 2% higher than that of Rayane et al[131]. Although the previous reports[131-134] have suggested that the polarizability value for the trimer can be improved by the inclusion of temperature effect in the calculations, we argue that the discrepancy appeared in the previous calculations is completely due to the partial neglect of the electron correlation effects. It also suggests that the temperature effect might be rather small. Even for the higher clusters, Na_8 and Na_9, the H-OP values deviate by only ~6% when it is compared with the earlier experimental values. However, when the correlation effect is treated by the LYP method rather than the OP functional, even the H-LYP method consistently underestimates the polarizability in all cases (Figure 3). This shows a better performance of the OP correlation functionals than the LYP functional for such calculations. It is also interesting to note that even after introducing the 100% HF exchange effect along with the LYP correlation functional in H-LYP, it does not reach the accuracy of MP2 or any other post HF methods (Fig.4). This discrepancy is clearly due to the effect of dynamical correlation corrections included in MP2 method, but not effectively taken care of by the LYP correlation functional. This particular aspect may probably signify that there is a need to improve the LYP correlation functional.

However, it should be noted that the correlation contribution introduced by the OP functional through the H-OP scheme, is almost close to same through the MP2 method.

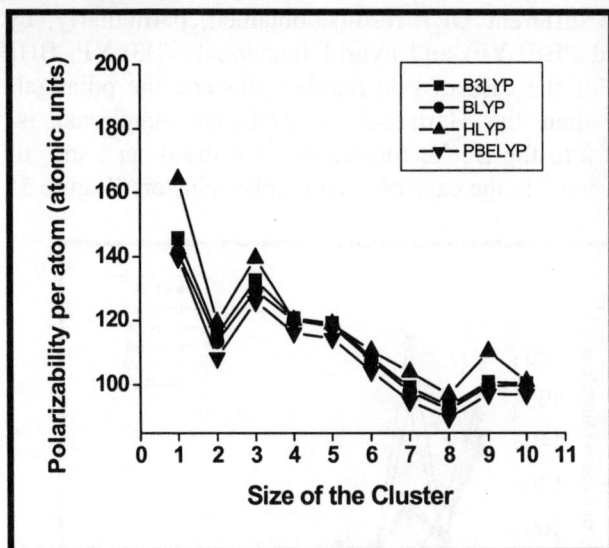

Fig. 4. The performance of pure and hybrid DFT exchange functionals along with LYP correlation functional for the determination of sodium cluster polarizabilities.

Considering the effect of changing the exchange functionals from the pure DFT to hybrid ones, we observe that the incorporation of the HF exchange improves the polarizability results significantly (Figure 4). For instance, the BHH-LYP that incorporates 50% of HF exchange yields polarizabilities slightly better than those predicted by B3-LYP, consisting of 20% HF exchange. However, the BHH-LYP values themselves are lower than those from the full HF exchange functional H-LYP. When the exact exchange effect is parameterized through the Becke and PBE functionals (B-LYP and PBE-LYP), the calculated polarizability values are further deteriorated. The functionals B-OP and PBE-OP, however, show some improvement in the polarizability values. The deficiency shown by different exchange functionals (Becke, PBE and Slater) can be attributed to the incorrect asymptotic behavior of these functionals viz. exponential decay instead of $-1/r$ decay as shown by the exact HF exchange potential.[156-159]
    In case of lithium clusters also, similar interesting remarks can be made by analyzing the polarizability values calculated by different

DFT exchange-correlation functionals. Although we observe many discrepancies between the theoretically calculated and experimental results, some uniformity and consistency in trends can be observed among the different DFT results, obtained, particularly, by the pure (BLYP and PBELYP) and hybrid functionals (B3LYP, BHHLYP and HLYP). For the case of odd number clusters, the polarizability value increases when the Hartree-Fock exchange functional is gradually introduced into the Becke functional. On the other hand, this trend is exactly reversed in the case of even number clusters (Figure 5).

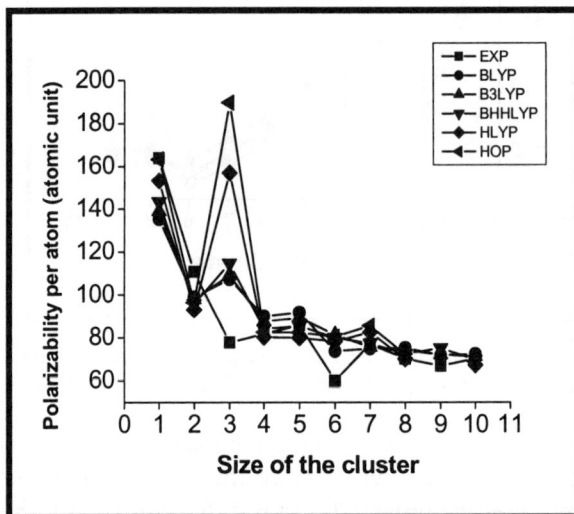

Fig.5.Lithium cluster polarizabilities calculated by various DFT exchange-correlation functionals.

In addition, it can also be noticed that when the correlation effect is treated by the OP method rather than the LYP functional in HLYP, the polarizability values are consistently overestimated for all the clusters. While comparing the HF and MP2 values also, a similar type of method dependent odd-even oscillations in the polarizability values is noticed particularly for the higher member clusters. Although the present analysis shows that the exchange-correlation effects are very important in determining the polarizability values, it is still not possible to understand how the polarizabilities obtained from different methods depend on the type (odd or even) of clusters. Apart from these interesting results obtained for lithium clusters, we have specifically made an attempt to discuss the effect of electron correlation for the determination of polarizability for sodium and lithium clusters in the next section.

VII Comparison of Polarizability of Sodium and Lithium Clusters: Effect of Electron Correlation

Although the Na and Li clusters have similar electronic and geometrical structures, a number of theoretical studies on lithium clusters have shown that the lithium clusters are not very simple and some of them have unusual spin states with the structures involving complicated dynamics, thermal plasmon broadening, etc.[136,139,140,147] It has also been noted that the properties derived from the simple jellium models were less successful for lithium clusters than for the sodium clusters.[140] The electron density distributions in lithium clusters are also known to exhibit unusual topological properties and it has been shown that the electron density accumulation is maximum at the non-nuclear sites in the lithium molecule or clusters.[159-161] Thus, it can be seen that the lithium clusters are more complicated than sodium or any other alkali metal cluster and they merit detailed calculations to understand the factors that govern their size evolution properties.[162] In this section, we will indeed show some of the distinctive features exhibited by the lithium clusters in their polarizability trends.

On comparing the polarizability of sodium and lithium clusters, calculated by different theoretical methods, it is observed that the polarizability trend is different from each other at the intermediate sizes. Let us recall from the earlier section that the experimental polarizabilities of sodium clusters are found to be more than the theoretical results computed at zero temperature for all the clusters with the aggregation number, $n > 4$. In addition, the pure (BLYP and PBELYP) and hybrid (B3LYP, BHHLYP and HLYP) exchange-correlation density functionals, except the HOP functional, are found to considerably underestimate the calculated polarizability values as compared to the MP2 and experimental results. The HF values for the sodium clusters are always higher than the MP2 values. From Figures 3 and 4, it can be seen that the largest deviation in the polarizability values for sodium clusters is observed for the PBELYP functional and the MP2 values are close to the experimental values with about 10% deviation.[120]

On the contrary, the trend observed for the lithium cluster polarizabilities, as shown in Fig. 6, is completely different from the sodium clusters and in particular, the calculated values for most of the higher member clusters are larger than the corresponding experimental results, which is a rather surprising and somewhat unusual behavior in these types of systems.[162] It is also pertinent to note that the values

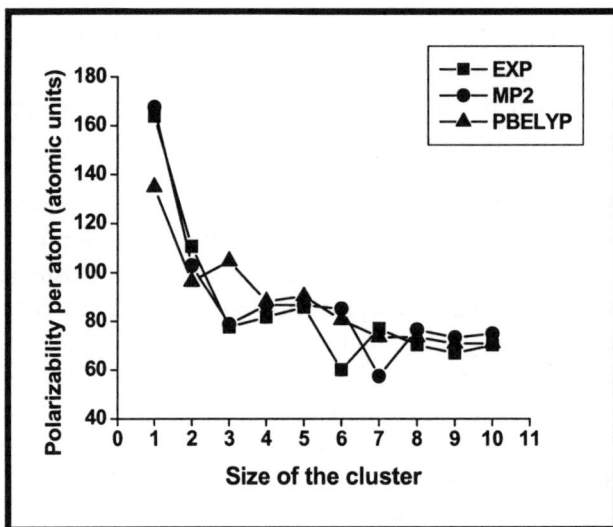

Fig. 6. Comparison of the lithium cluster polarizabilities calculated by MP2, PBELYP with experimental values.

obtained by various methods differ significantly among themselves and hence, it is not possible to draw any conclusion from the above results on the systematic inclusion of the electron correlation effects. For instance, it can be seen that even the PBELYP values are higher than the MP2 values for some of the lower member clusters (n=5 and 7). Moreover, the HF values for lithium clusters are found to be less than the MP2 values except for Li_7 and Li_9 cases. Maroulis and Xenides have also noticed a similar type of observation from highly accurate *ab initio* calculations and this study is limited only to the cluster size of 4.[163]

The above observations raise some important issues, viz., whether the electron correlation has any significant role in the evaluation of polarizability of lithium clusters or not? Even if we include the electron correlation effect very systematically starting from the uncorrelated level of calculations (HF methods) to the approximate pure (Becke-LYP and PBE-LYP) and hybrid (B3-LYP, BHH-LYP and HLYP) DFT exchange-correlations to *ab initio* MP2 methods, we can not draw any consistent or systematic conclusions on the electron correlation effects from the polarizability values of the lithium clusters. Thus, the effect of electron correlation has been found to be somewhat unpredictable in the calculation of polarizability of lithium clusters. Detailed further calculations are required to clarify this particular issue. One can even argue that the geometry of the clusters optimized through the higher order correlated level of theory, will have some impact on the polarizability calculations. The earlier report from our group[162] suggests that the effect of geometry of the lithium clusters obtained through optimization by higher order sophisticated methods does not resolve the issues of the discrepancy between the polarizability values obtained by the experimental and theoretical methods.

VIII. Polarizability and Its Relation with Ionization Energy, Chemical Hardness/Softness and Binding Energy

In this section, we will discuss how this important response property, polarizability, is related to other quantities, such as, ionization potential and reactivity of these clusters in terms of the chemical hardness or softness parameters. As it has been mentioned earlier, the static dipole polarizability is a measure of the distortion of the electronic density and carries the information about the response of the system under the effect of an external static electric field. However, the ionization potential reveals how tightly an electron is bound within the nuclear attractive field of the system. Thus, it has been qualitatively expected that these two parameters should be inversely related to each other.[32-36] Such a relationship has been first derived by Dmitrieva and Plindov[36] and has been shown to be valid for atomic systems using a statistical model. Fricke has shown that polarizability of neutral atoms correlates very well with their first ionization potential (in a logarithmic scale) within the groups of elements with the same angular momentum of the outermost electrons.[35]

Similarly, the concept of hardness or softness parameters, introduced by Pearson to rationalize the reactivity of the molecular systems,[13] has been related with the polarizability of the systems in a qualitative way.

The softness is defined as the inverse of the difference between the ionization potential and electron affinity of the clusters. One of the key aspects in these conceptual developments lies in the fact that these reactivity parameters are not experimentally observable quantities and hence further work has often been necessary to explore the possible relationships of these descriptors with polarizability. In view of this, few studies have been carried out to show the correlation between the softness of the systems with the polarizability[164-167] While such relationships have been studied for atomic and selected molecular systems with reasonable success, they have not been explored for the metal clusters.

The relationship between the polarizability α and the ionization potential of the sodium and lithium clusters calculated by the MP2 method shows that there is a good correlation between these two quantities ($\alpha^{1/3}/n$ and IP^{-1}/n, where n is the number of atoms present in the cluster) with a linear correlation coefficient 0.999.[32]   This is an important observation because the above results can provide a way to calculate the polarizability of the clusters of larger size from the values of their ionization potential. This success in relating the polarizabilities to the ionization potential of these clusters would also be of interest in relating the local ionization energies with local polarizability, as suggested by Politzer et al.[33,34]

Now, let us analyze the relationship between the softness and polarizability in these metal clusters. For this case also, a linear relation has been observed for these two quantities, namely, the average softness (S/n) and the average of the cubic root of polarizability ($\alpha^{1/3}/n$) of the sodium and lithium clusters with linear correlation coefficient values of 0.995 calculated by MP2 method.[32]   Such relationship can be rationalized by the following factors. The electron affinity values of these clusters are much less than the ionization potentials and hence one can approximate the global softness parameter in terms of the inverse of the ionization potential alone. In addition, it is also known that the softness of any system, which represents essentially the charge capacity of a species, is expected to be proportional to its size.[164] Similarly, for a sphere of radius R, the polarizability can be approximated[166] as R^3. Hence, it has also been possible to extend a linear relationship between the cubic root of polarizability and the softness from the relationship between the softness and the size of the system. This may explain the reason why the polarizability correlates well with the softness parameter as well as inverse ionization potential. Based on the Fermi-Amaldi model, Dmitrieva and Plindov have shown that ionization potential is inversely proportional to the radius of the atomic systems and this relation has further connected the cubic root of polarizability to the inverse of the ionization potential.[36] From the above arguments, the physical basis for the relationship between the cube root of polarizability and the softness as well as the inverse of ionization potential can be rationalized. Herein, it has, however, been shown that this relationship can be validated for the metal clusters as well. In addition, a better correlation between these parameters has been observed when these quantities, viz., polarizability, softness, and the inverse of the ionization potential, are divided by the cluster aggregation number (n).

We will now discuss another important relationship between the polarizability with binding energy of the clusters, known as "Minimum Polarizability Principle".[15,169,170] The binding energy of the clusters is defined as the difference between the energy of the clusters and that of isolated atoms. It has been observed that higher value of binding energy of the clusters (more stable cluster) is, in general, accompanied by relatively lower value of the polarizability. This is consistent with the recently proposed "minimum polarizability principle", according to which, the natural direction of evolution of any system is towards a state of minimum polarizability.[169,170] As a consequence of this principle, it has been mentioned that the stability of system is always accompanied by a lowering of the polarizability of the systems as has been demonstrated earlier from a study of polarizability in exchange, dissociation, and isomerization types of reactions.[169] The condition of minimum polarizability has also, in general, been associated with the energetically more stable situations or the maximum hardness.[15]

IX. A Critical Remarks on the Effect of Temperature on Polarizability of Alkali Metal Clusters

In spite of using improved correlated methods and better non-local XC functionals for the calculation of polarizability, even the best set of values is still found to be lower than the actual experimental results especially for the higher member clusters, which indicates that the remaining difference could be due to the neglect of the temperature effect or it can be even due to the experimental error. Attempt has been made to improve the level of agreement between the theoretical and experimental results through inclusion of temperature effect in the calculations.[152-155]   Blundell *et al.* have reported the temperature dependence of the polarizability of sodium clusters in the size range N= 8-139 by averaging over large statistical ensembles of ionic configurations and found that there is an increment in the polarizability (about 15% at 300 K) which is consistent with the level of discrepancy between theory and experiment.[153] Kümmel *et al.* have also studied the polarizability of these clusters by different theoretical approaches and have explained the observed discrepancy in terms of the average expansion coefficients of these clusters at higher temperatures.[152] Their calculations show that the bond length increase is in the order of 0.48Å for Na_8 and 0.87Å for Na_{10} at 500 K, which in turn leads to an increase in the polarizability of 12Å^3 and 23Å^3, respectively. The resulting polarizabilities of 130Å^3 for Na_8 and 172Å^3 for Na_{10} compare favorably with the experimental values 134Å^3 and 190Å^3, respectively. Although these studies have shown that the temperature effect is significant for the polarizability calculations, these calculations involve the extended Thomas-Fermi description. Later, the pseudopotential based *ab initio* calculations have been performed by Chelikowsky and co-workers[154,155] within the framework of gradient corrected density functional theory and shown that by taking into account the effect of finite temperature on the cluster structure, the quantitative accuracy can be achieved for the experimental and theoretical polarizabilites.[154,155] Although the effect of temperature has been ascribed to be one of the reasons for the deviation of the calculated values from the corresponding experimental values, we argue that the accurate estimate of the polarizability values at zero temperature using different theoretical methods is essential to assess the effect of temperature on the polarizability measurements. Since these calculations involving finite temperature methods, used approximate DFT exchange-correlation functionals, such as PBE and extended Thomas-Fermi models, which generally underestimates the calculated polarizabilities significantly, no definite conclusion regarding the temperature effect could be reached in absence of better calculations at finite temperature, particularly for lower member clusters.

Since the source of experimental errors can be due to many factors, the temperature factor alone may not be able to explain the discrepancy between the experimental and theoretical results. In this context, it is interesting to note that the theoretical results obtained at zero temperature for lithium clusters have been found to be higher than the experimental results. [79,129,131] As we have stated earlier, none of these factors (temperature, thermal expansion or average bond length increase in the clusters) can explain the unusual trend shown by the lithium clusters. Several other factors may also contribute to the discrepancy in the polarizability trend. In general, the experimental static dipole polarizability is normally measured by the electric field deflection of a supersonic cluster beam and the deflection is characteristic of the number of sodium atoms in the cluster. It should, however, be noted that the measured deflection in the actual experiment is not only due to a single cluster, but also due to the mixture of other higher and lower member clusters which are normally present in the supersonic beam. Another main source of error might be due to the cluster beam velocity. Since the measured experimental polarizability is directly proportional to the deflection of the beam as well as the cluster beam velocity, any error introduced in these two quantities will change the value of the polarizability considerably. Although the recent work of Tikhonov *et al.*[130] has taken care of the above aspects, the accuracies of the measured values are still disputable due to the cluster temperature as well as the applied inhomogeneous electric field. Yang and Jackson have recently studied the polarizabilities of small and intermediate sized copper clusters[171] and shown that the effects of basis set, electron correlation, and equilibrium geometry are too small to account for the theory-experiment gap and thermal expansion of the clusters also leads to very small changes in polarizability. On the other hand, it has been suggested that the presence of permanent dipoles in the clusters could account for the experimental observations. However, in our opinion, electronic structure of the clusters, reliable ground-state structures and accurately measured experimental temperatures are the important factors in determining accurate polarizabilities of the clusters with a good quantitative agreement between the calculated and experimentally observed polarizabilities.

X. Concluding Remarks

In this review, we have made an attempt to present an overview on the recent studies on polarizability of alkali metal clusters, particularly for sodium and lithium clusters determined by various DFT and post Hartree-Fock methods. In particular, it has been shown that highly accurate computational methods, which incorporate the electron correlation effects, are necessary to obtain the polarizability of these clusters in quantitative agreement with the experimental results. All the pure (BLYP and PBELYP) and hybrid (B3LYP, BHHLYP and HLYP) DFT functionals are found to considerably underestimate the calculated polarizability values as compared to the MP2 results. The significance of these issues has been highlighted and the need of highly accurate DFT exchange-correlation functionals and *ab initio* methods in the study of the electronic properties of these clusters is emphasized. For the case of lithium clusters, the calculated polarizability values by various methods at zero Kelvin temperature are found to be higher than the experimental results measured at higher temperature and this trend is entirely in contradiction with the sodium cluster cases. In view of these, we have also discussed the issue regarding the role of electron correlation for these alkali metal clusters. In addition, some of the issues relating to the cluster ionization potential, reactivity and their relationship with the polarizability of these clusters have been discussed. Unless we have a proper understanding of the physics of the metal clusters, it is rather difficult to get a quantitative agreement with the experimental trend even if we perform any advanced theoretical methods in the calculation

Acknowledgement   It is a pleasure to thank Dr. T. Mukherjee for his kind interest and encouragement.

References
1. P. Jørgensen, J. Simons, *Geometrical Derivatives of Energy Surfaces and Molecular Properties* Eds. D. Reidel, Dordrecht, 1986.
2. C. E. Dykstra, *Ab initio Calculations of the Structures and Properties of Molecules,* Elsevier Science, New York, 1988.
3. T. Helgaker, P. Jorgensen, and J. Olsen, *Molecular Electronic-Structure Theory,* Wiley, Chichester, 2000.
4. R. D. Amos, in Advances in Chemical Physics, *Ab initio Methods in Quantum Chemistry*, Vol. I, edited by K. P. Lawley, Wiley, New York, 1987.
5. McWeeny, R. *Methods of Molecular Quantum Mechanics*, Academic Press: New York, 1992.

6. A. Szabo and N. S. Ostlund, Modern Quantum Chemistry, MacMillan, New York, 1982.
7. The Optical Properties of Materials, Mater. Res. Soc. Symp. Proc., Vol. 579, edited by J. R. Chelikowsky, S. G. Louie, G. Martinez, and E. L. Shirley, Materials Research Society, Warrendale, PA, 2000.
8. U. Kreibig and M. Vollmer, *Optical Properties of Metal Clusters, Springer Series in Materials Sciences*, Springer-Verlag, Berlin, 1995.
9. G. D. Mahan and K. R. Subbaswamy, *Local Density Theory of Polarizability*, Plenum Press, New York, 1990.
10. A. D. Buckingham, Adv. Chem. Phys. **12**, 107 (1967).
11. U. Hohm, Vacuum, **58**,117 (2000).
12. K. D. Bonin and V. V. Kresin, *Electric-Dipole Polarizabilities of Atoms, Molecules and Clusters*, World Scientific, Singapore, 1997.
13. R. G. Pearson, J. Am. Chem. Soc., **85**, 3533 (1963).
14. R. G. Parr and R. G. Pearson, J. Am. Chem. Soc., **105**, 7512 (1983).
15. K. R. S. Chandrakumar, *"Theoretical Studies on Some Aspects of Chemical Reactivity Using Density Based Descriptors"*, Ph. D. Thesis, Pune University, 2003.
16. M, Staikovaa, F. Waniab and D. J. Donaldson, Atmospheric Environment **38**, 213 (2004).
17. R. P. Verma, A. Kurup and C. Hansch Bioorganic & Medicinal Chemistry, **13**, 237S (2005).
18. S. Gupta, Chem. Rev. **87**, 1183 (1987).
19. V. V. Kresin, G. Tikhonov, V. Kasperovich, K. Wong, and P. Brockhaus J. Chem. Phys. **108**, 6660 (1998).
20. V. V. Kresin, V. Kasperovich, G. Tikhonov, and K. Wong Phys. Rev. A **57**, 383 (1998).
21. G. C. Maitland, M. Rigby, E. B. Smith, and W. A. Wakeham, *Intermolecular Forces: Their Origin and Determination*; Oxford University Press: New York, 1981.
22. C. E. Dykstra and P. G. Jasien, Chem. Phys. Lett. **109**, 388 (1984).
23. C. Hansch, W. E. Steinmetz, A. J. Leo, S. B. Mekapati, A. Kurup and D. Hoekman, J. Chem. Inf. Comput. Sci.; **43**, 120 (2003).
24. G. Maroulis J. Chem. Phys. **121**, 10519 (2004).
25. C. G. Gray and K. E. Gubbins. *Theory of Molecular Fluids*, vol. I. Oxford: Clarendon Press, 1984.
26. G. Ahn-Ercan and H. Krienke Current Opinion in Colloid & Interface Science **9**, 92 (2004).
27. S. Yoo, Y. A. Lei, and X. C. Zeng, J. Chem. Phys. **119**, 6083 (2003).
28. A. Perera and F. Sokolic J. Chem. Phys. **121** 11272, (2004).
29. M. Pretti and C. Buzano, J. Chem. Phys. **121**, 11856 (2004).
30. E. M. Knipping, M. J. Lakin, K. L. Foster, P. Jungwirth, D. J. Tobias, R. B. Gerber, D. Dabdub, and B. J. Finlayson-Pitts, Science **228**, 301 (2000).
31. N. F. Lane Rev Mod Phys **29**, 52 (1980).
32. K. R. S. Chandrakumar, T. K. Ghanty and S. K. Ghosh, J. Phys. Chem A **108**, 6661 (2004).
33. P. Politzer, P. Jin, and J. S. Murray, J. Chem. Phys. **117**, 8197 (2002).

34. P. Politzer, J. S. Murray, M. E. Grice, T. Brinck and S. Ranganathan, J. Chem. Phys. **95**, 6699 (1991).
35. B. Fricke, J. Chem. Phys. **84**, 862 (1986).
36. I. K. Dmitrieva and G. I. Plindov, Phys. Scr., **27**, 402(1983).
37. A. Dalgarno, Adv. Phys. 11, 281(1962).
38. T. M. Miller and B. Bederson, Adv. At. Mole. Phys. **13**, 1 (1977).
39. T. M. Miller and B. Bederson, Adv. At. Mole. Phys. **25**, 37 (1988).
40. H. D. Cohen and C. C. J. Roothaan J. Chem. Phys. **43**, S34 (1965).
41. D. M. Bishop Rev. Mod. Phys. **62**, 343-374 (1990).
42. D. M. Bishop, L. M. Cheung, A. D. Buckingham, Mol. Phys. **41**,1225 (1980).
43. C. E. Dykstra, J. Mol. Str. THEOCHEM, 573, 63 (2001).
44. M. J. Stott and E. Zaremba Phys. Rev. A **21**, 12 (1980).
45. K. D. Bonin and M. A. Kadar-Kallen, Int. J. Mod. Phys.B **8**, 3313 (1994).
46. W.A. de Heer, Rev. Mod. Phys. **65,** 611 (1993).
47. M. Brack, Rev. Mod. Phys. **65**, 677 (1993).
48. W.A. deHeer, W. D. Knight, M. Y. Chou and M. L. Cohen, *Solid State Physics*, **40**, 93 (1987).
49. *Physics and Chemistry of Finite Systems: From Clusters to Crystals*: Ed. By P. Jena *et al.*, Kluwar Academic, Netherlands, **1992**, Vol. I and II.
50. V. Bonacic-Koutecky, P. Fantucci, and J. Koutecky, Chem. Rev. **91**, 1035 (1991).
51. M. Moskovits, Annu. Rev. Phys. Chem. *42*, 465 (1991).
52. A. Nitzan, Annu. Rev. Phys. Chem., **52**, 681(2001).
53. A. P. Alivistaos, J. Phys. Chem. **100**, 13226 (1996).
54. A. T. Bell, Science, 299, 1688 (2003).
55. M. Valden, X. Lai, D. W. Goodman, Science, **281**, 1647(1998).
56. F. Baletto and R. Ferrando, Rev. Mod. Phys.**77,** 371 (2005).
57. J. Jortner, Z. Phys. D: At., Mol. Clusters 24, 247 (1992).
58. J. A. Alonso, Chem. Rev. **100,** 637 (2000).
59. M. A. El-Sayed, Acc. Chem. Res. 37, 326 (2004).
60. M. A. El-Sayed, Acc. Chem. Res. 34, 257 (2001).
61. J. Jortner, Clusters as a key to the understanding of properties as a function of size and dimensionality, In *Physics and Chemistry of Finite Systems: From Clusters to Crystals*: Ed. By P. Jena *et al.* Kluwar Academic, Netherlands, 1992, Vol. 1, p1-17.
62. T. P. Martin Physics Reports, **273**, 199 (1996).
63. T. P. Martin Surface Science, **106**,79 (1981).
64. D. Babikov, E. Gislason, M. Sizun, F. Aguillon, and V. Sidis J. Chem. Phys. **112**, 9417 (2000).
65. S. J. Oh, S. H. Huh, H. K. Kim, J. W. Park, and G. H. Lee J. Chem. Phys. 111, 7402 (1999).
66. R. Narayanan, M. A. El-Sayed, J. Phys. Chem. B. 2005, **In Press,** DOI: 10.1021/jp051066p.
67. B. F. G. Johnson Coord. Chem. Rev., **190-192**, 1269 (1999).
68. M. C. Daniel and D. Astruc Chem. Rev. **104**, 293 (2004).
69. Y. K. Kim, Int. J.Mass Spec., **238**, 17 (2004).
70. C. N. R. Rao, V. Vijayakrishnan, A. K. Santra, M. W. J. Prins, Angew. Chem. Int. Ed. Engl. **31**, 1062 (1992).
71. C. N. R. Rao, G. U. Kulkarni, P. John Thomas and P.P. Edwards Chem. Eur. J. **8**, 29 (2002)

72. V. V. Kresin, Phys.Rep. **220**, 1 (1992).
73. A. Sanchez, S. Abbet, U. Heiz, W. –D. Schneider, H. Hakkinen, R. N. Barnett and U. Landman, J. Phys. Chem. A. **103** 9573(1999).
74. H. Nishioka, K. Hansen, and B. R. Mottelson Phys. Rev. B **42**, 9377 (1990).
75. R. Kubo, J. Phys. Jpn. **17**, 975 (1962).
76. B. D. Leskiw and A. W. Castleman Jr. Comptes Rendus Physique **3**, 251 (2002).
77. H. Schaber and T. P. Martin Surf. Sci. **156**, 64 (1985).
78. T. P. Martin Surf. Sci., **156**, 584 (1985).
79. M. Broyer, R. Antoine, E. Benichou, I. Compagnon, P. Dugourd and D. Rayane *Comptes Rendus Physique*, **3**, 301(2002) .
80. D. J. Wales, Energy Landscapes and with Applications to Clusters, Biomolecules and Glasses, Cambridge University, Cambridge, England, 2003.
81. T. L. Hill, Thermodynamics of Small Systems, Parts I and II, Benjamin, Amsterdam, 1964.
82. M. Bixon and J. Jortner J. Chem. Phys. **91**,1631 (1989).
83. R. S. Berry and B. M. Smirnov J. Exp. Theo. Phys. **98**, 366 (2004).
84. O. Mulken and H. Stamerjohanns and P. Borrmann Phys. Rev. E **64**, 047105 (2001).
85. K. Koga, T. Ikeshoji, and K. Sugawara Phys. Rev. Lett. **92**, 115507 (2004).
86. W. Ekardt, Surf. Sci. **152**, 180 (1985).
87. M. M. Kappes, M. Schär, P. Radi, and E. Schumacher, J. Chem. Phys. **84**, 1863 (1986).
88. Feng Liu, S. N. Khanna, and P. Jena, J. Appl. Phys. **67**, 4484 (1990).
89. D. A. Evans, M. Alonso, R. Cimino, and K. Horn, Phys. Rev. Lett. **70**, 3483 (1993).
90. C. W. J. Beenakker Phys. Rev. B **50**, 15170 (1994).
91. R. Schäfer, S. Schlecht, J. Woenckhaus, and J. A. Becker Phys. Rev. Lett. **76**, 471 (1996).
92. G. K. Gueorguiev, J. M. Pacheco, and D. Tománek, Phys. Rev. Lett. **92**, 215501 (2004).
93. W. D. Clemenger, K.; W.A de Heer and W.A Saunders, Phys. Rev. B **31**, 2539 (1985).
94. W. D. Knight, K. Clemenger, W. A. de Heer, W. A. Saunders, M. Y. Chou, and M. L. Cohen, Phys. Rev. Lett. **52**, 2141 (1984).
95. W. A. Saunders, K. Clemenger, W. A. de Heer, and W. D. Knight Phys. Rev.B **32**, 1366-1368 (1985).
96. W. D. Knight, W. A. De Heer, K. Clemenger and W. A. Saunders, Solid State Comm. **53**, 445 (1985).
97. K. I. Peterson, P. D. Dao, R. W. Farley, and A. W. Castleman, Jr. J. Chem. Phys. **80**, 1780 (1984).
98. M. Y. Chou, A. Cleland and M. L. Cohen, Solid State Comm **52**, 645 (1984).
99. A. I. Yanson, I. K. Yanson and J. M. van Ruitenbeek, Phys. Rev. Lett. 87, 216805 (2001).
100. M. L. Cohen, In Micro Clusters, Ed. by S. Sugano, Y. Nishina and S. Ohnishi, Springer-Verlag, Berlin, p.2-9, 1987.
101. W. A. de Heer and W. D. Knight, In Elemental and Molecular Clusters, Ed. by G. Benedek, T. P. Martin and G. Pacchioni, Springer-Verlag, Berlin,1988.
102. H. Gohlich, T. Lange, T. Bergmann, and T. P. Martin, Phys. Rev. Lett. **65**, 748 (1990).

103. H. Nishioka, K. Hansen, and B. R. Mottelson , Phys. Rev. B **42**, 9377 (1990).
104. C. Yannouleas and U. Landman, Phys. Rev. Lett. **78**, 1424 (1997).
105. V. Kresin Phys. Rev. B **38**, 3741 (1988).
106. B. B. Dasgupta and R. Fuchs, Phys. Rev. B **24**, 554 (1981).
107. D. R. Snider and R. S. Sorbello, Phys. Rev. B **28**, 5702 (1983).
108. W. Ekardt, Phys. Rev. Lett. **52**, 1925 (1984).
109. D. E. Beck, Phys. Rev. B **30** 6935 (1984).
110. M. J. Puska, R. M. Nieminen and M. Manninen, Phys. Rev. B **31**, 3487 (1985).
111. M.Manninen, Phys. Rev. B **34**, 6886 (1986).
112. M. Y. Chou, and M. L. Cohen, Phys. Lett. A **113**, 420 (1986).
113. M. Manninen, Phys. Rev. B **34**, 6886 (1986).
114. I. Moullet, J. L. Martins, F. Reuse and J. Buttet, Phys. Rev. B **42**, 11598 (1990).
115. M. K. Harbola, Solid State Commun. **98**, 629 (1996).
116. M. G. Mayer and J. H. D. Jensen, Elementary Theory of Nuclear Shell Structure, Wiley, New York, 1955.
117. U. Rothlisberger and W. Andreoni, J. Chem. Phys. **94**, 8129 (1991).
118. J. Blanc, V. Bonacic-Koutecky, M. Broyer, J. Chevaleyre, Ph. Dugourd, J. Koutecky, C. Scheuch, J.P. Wolf and L. Woste, J. Chem. Phys. **96** 1793 (1992).
119. T. K. Ghanty, K. R. S. Chandrakumar and S. K. Ghosh, J. Chem. Phys. **120**, 11363 (2004).
120. K. R. S. Chandrakumar, T. K. Ghanty and S. K. Ghosh, J. Chem. Phys. **120**, 6487 (2004).
121. J. P. Perdew, K. Burke and M. Ernzerhof, Phys. Rev. Lett. **77**, 3865 (1996).
122. A. D. Becke, J. Chem. Phys. **98**, 5648 (1993).
123. A. D. Becke, J. Chem. Phys. **98**, 1372 (1993).
124. C. Lee, W. Yang and R. G. Parr, Phys. Rev. B **37**, 785 (1988).
125. T. Tsuneda and K. Hirao, Chem. Phys. Lett. **268**, 510 (1997).
126. T. Tsuneda, T. Suzumura and K. Hirao, J. Chem. Phys., **110**, 10664 (1999).
127. M. W. Schmidt, K. K. Baldridge, J. A. Boatz, S. T. Elbert, M. S. Gordon, J. H. Jensen, S. Koseki, N. Matsunga, K. A. Nguyen, S. J. Su, T. L. Windus, M. Dupuis, J. A. Montgomery, (Department of Chemistry, North Dakota State University and Ames Laboratory, Iowa State University). GAMESS, General Atomic and Molecular Electronic Structure System. *J. Comput. Chem.* **14**, 1347 (1993) (GAMESS version 2002).
128. K. B. Sophy and S. Pal. J. Chem. Phys. **118**, 10861 (2003).
129. E. Benichou, R. Antoine, D. Rayane, B. Vezin, F. W. Dalby, Ph. Dugourd, M. Broyer, C. Ristori, F. Chandezon, B. A. Huber, J. C. Rocco, S. A. Blundell, and C. Guet, Phys. Rev.A **59**, R1 (1999).
130. G. Tikhonov, V. Kasperovich, K. Wong, and V. V. Kresin Phys. Rev. A **64**, 063202 (2001).
131. D. Rayane, A. R. Allouche, E. Benichou, R. Antoine, M. Aubert-Frecon, Ph. Dugourd, M. Broyer, C. Ristori, F. Chandezon, B. A. Huber and C. Guet, Eur. Phys. J. D **9**, 243 (1999).
132. J. Guan, M. E. Casida, A. M. Köster, and D. R. Salahub, Phys. Rev. B **52**, 2184 (1995).
133. P. Calaminici, K. Jug, and A. M. Köster, J. Chem. Phys. **111**, 4613 (1999).
134. S. Kümmel, T. Berkus, P.-G. Reinhard, and M. Brack, Eur. Phys. J. D **11**, 239 (2000).

135. I. A. Solov'yov, A. V. Solov'yov and W. Greiner, Phys. Rev. A **65**, 53203 (2002).
136. S. E. Wheeler, K. W. Sattelmeyer, P. v. R. Schleyer, H. F. Schaefer III J. Chem. Phys. **120**, 4683 (2004).
137. R. Rousseau and D. Marx, Chem. Eur. J. **6**, 2982 (2000).
138. R. Rousseau and D. Marx, Phys. Rev. Lett., **80**, 2574 (1998).
139. R. O. Jones, A. Lichtenstein and I. J. Hutter, J. Chem. Phys. **106**, 4566(1997).
140. G. Gardet, F. H. Rogenmond and H. Chermette, J. Chem. Phys. **105**, 9933 (1996).
141. E. Benichou, A. R. Allouche, M. Aubert-Frecon, R. Antoine, M. Broyer, Ph. Dugourd and D. Rayane. Chem. Phys. Lett, **290**, 171 (1998).
142. P. Fuentealba, O. Reyes, J. Phys. Chem. A **103**, 1376 (1999).
143. V. Bonacic-Koutecky, J.Gaus, M. F. Guest, L. Cespiva and J. Koutecky, J. *Chem. Phys. Lett.* **206**, 528 (1993).
144. K. P. Huber and G. Herzberg, *Molecular Spectra and Molecular Structure IV: Constants of Diatomic Molecules,* Van Nostrand Reinhold, New York, **1979**.
145. H. A. Eckel, J. M. Gress, J. Biele and W. Demtroder, J. Chem. Phys. **98**, 135 (1992).
146. E. Benichou, A. R. Allouche, M. Aubert-Frecon, R.Antoine, M. Broyer, Ph. Dugourd and D. Rayane, *Chem. Phys. Lett.,* **290**, 171 (1998).
147. B. K. Rao, P. Jena, and A. K. Ray, *Phys. Rev. Lett.* **1996**, *76*, 2878.
148. R. W. Molof, H. L. Schwartz, T. M. Miller and B. Bederson, *Phys. Rev. A 10*, 1131, (1974).
149. W. D. Hall and J. C. Zorn, Phys. Rev. A. **10**, 1141 (1974).
150. C. R. Ekstrom, J. Schmiedmayer, M. S. Chapman, T. D. Hammond and D. E. Pritchard, Phys. Rev. A **51**, 3883, (1995).
151. G. Maroulis, Chem. Phys. Lett. **334**, 207 (2001).
152. S. Kümmel, J. Akola, and M. Manninen, Phys. Rev. Lett. **84**, 3827 (2000).
153. S. A. Blundell, C. Guet, and R. R. Zope, Phys. Rev. Lett. **84**, 4826 (2000).
154. L. Kronik, I. Vasiliev, and J. R. Chelikowsky, Phys. Rev. B **62**, 9992 (2000).
155. L. Kronik, I. Vasiliev, M. Jain, and J.R. Chelikowsky J. Chem. Phys. **115**, 4322 (2001).
156. D. J. Tozer and N. C. Handy, J. Chem. Phys. **109**, 10180 (1998).
157. R. van Leeuwen and E. J. Baerends, Phys. Rev. A **49**, 2421 (1994).
158. F. Della Sala and A. Görling, Phys. Rev. Lett. **89**, 33003 (2002).
159. V. Luana, P. M. Sanchez, A. Costales, M. A. Blanco and A. M. Pendas, J. Chem. Phys. **119**, 634 (2003).
160. R. F. W. Bader, T. T. Nguyen-Dang and Y. Tal, Rep. Prog. Phys. **44**, 893 (1981).
161. T. Baruah, D. G. Kanhere and R. R. Zope, Phys Rev A **63**, 063202 (2001).
162. K. R. S. Chandrakumar, T. K. Ghanty and S. K. Ghosh, Int. J. Quan. Chem. **In Press** (2005)
163. Maroulis, G.; Xenides, D. J Phys Chem A **103**, 4590 (1999).
164. P. Politzer, J. Chem. Phys. **86**, 1072 (1987).
165. A.Vela and J. L. Gazquez, J. Am. Chem. Soc. **112**, 1490 (1990).
166. T. K. Ghanty and S. K. Ghosh, J. Phys. Chem. **100**, 17429 (1996).
167. T. K. Ghanty and S. K. Ghosh, J. Phys. Chem. **97**, 4951(1993)
168. L. Komorowski, Chem. Phys. Lett. **134**, 536 (1987).
169. T. K. Ghanty and S. K. Ghosh J. Phys. Chem. **100**, 12295 (1996).
170. P. K. Chattaraj and S. Sengupta, J. Phys. Chem. **100**, 16126 (1996).
171. Y. Mingli and K. A. Jackson J. Chem. Phys. **122**, 184317 (2005).

CHAPTER 19

CHARGE DISTRIBUTION AND POLARISABILITIES OF WATER CLUSTERS

Patrick Senet[1], Mingli Yang[2] and Christian Van Alsenoy[3]

[1] *Théorie de la matière condensée, CNRS-UMR 5027, Laboratoire de Physique, Univ. de Bourgogne, 9 Avenue Alain Savary - BP 47870, F-21078 Dijon Cedex, France*
E-mail: psenetu-bourgogne.fr

[2] *Department of Physics, Central Michigan University, Mt. Pleasant, MI 48859, USA*
E-mail: yang2mcmich.edu

[3] *Structural Chemistry Group, Department of Chemistry, University of Antwerp, Universiteitsplein 1, B-2610 Antwerp, Belgium*
E-mail:kris.vanalsenoyua.ac.be

By combining Density Functional Theory calculations, the Hirshfeld partitioning of the electronic density, and a physical point dipole model, one tries to answer the following question: how varies the polarisability of a water molecule in a cluster ? The structures and electrostatic properties of small- and mediate-sized water clusters (up to 20 molecules) are computed *ab initio* using a supermolecule approach and B3LYP and B3PW91 exchange-correlation functionals. The dielectric properties and geometries of the clusters are also evaluated using the second order Moller-Plesset perturbation theory (for the clusters up to 15 molecules) for comparison with the DFT results. Based on these data and a simple model, one establishes the influence of the intermolecular interactions on the effective dipole and polarisability of *a molecule in a cluster*. In particular, one demonstrates the correlation between charge transfer occurring in hydrogen bonding and molecular polarisability.

1. Introduction

Water is an ubiquitous molecule in physical chemistry and biochemistry and liquid water is one of the most studied solvent. The polar character of the molecule has two well known major consequences on the interactions between solutes and between solutes and water. First, the oscillations of the molecular dipole due to thermal agitation screen the electrostatic interactions in the liquid and provide the major contribution to the high water dielectric constant, which is around 80 at room temperature. Second, the dipolar electric field of the molecule is responsible, at short distances, of polarisation and strong electrostatic interactions with molecular species (including water itself). According to the Pauling point of view, is the electrostatic interaction which drives the hydrogen bonding between water molecules. Dipole, polarisability and hydrogen bonds are intimately related.

The hydrogen bond network formed in liquid water is dynamic and can be thought in terms of metastable water clusters.[1,2,3] On the other hand, stable water clusters are observed at low temperature in gas clathrates[4,5,6] or around amino acids of hydrophobic proteins,[7] for instance. Isolated water clusters exist naturally in the atmosphere[8] and can be produced experimentally.[9] These clusters have been the subject of an intense experimental and theoretical research since many years because they offer the possibility to understand in deeper details the interactions between water molecules and between water and a solute.[10] Varying the size of clusters allows to "build up the liquid one step at a time".[11] Indeed, small clusters form well-defined hydrogen-bond networks[12,13,14,15,16] which "mimic" the short-live structure in liquid with molecules having different coordination numbers. As suggested by Weinhold, water clusters might be the appropriate building block needed to describe the thermodynamical properties of the condensed phases.[17] On the other hand, existence of molecules in different environments within a cluster may elucidate some aspects of bulk water which are otherwise hidden by symmetry in the homogeneous phases. Moreover, the clusters provide various simple water/vacuum interfaces which can help to understand more complicated heterogeneous interfaces like water/biomolecules, for instance.

In order to get insight on the interactions and reactivity of water molecules in various clusters, one focuses here on the dipolar properties: the dipole moment and the polarisability of a "*molecule in a cluster*". These quantities can be simply related to the energy of a cluster of n molecules to which we apply a non uniform electric field $\mathbf{E}$. The total energy of a cluster W_n can be decomposed formaly in two parts

$$E(n) = E_0(n) - \sum_{I=1}^{n}\sum_{\alpha} E_{I\alpha}\delta P_{I\alpha} + \frac{1}{2}\sum_{I=1}^{n}\sum_{\substack{J=1 \\ J \neq I}}^{n}\sum_{\alpha}\sum_{\beta} \Phi_{I\alpha J\beta}\delta P_{I\alpha}\delta P_{J\beta} + ...,$$
(1)

where the first part, E_0, is the energy of a reference state (to be defined) and where the second part represents the response of its reference state, defined by the vectors $\delta\mathbf{P}_I$, to the applied field. More precisely, $\delta P_{I\alpha}$ represents the α Cartesian component of the variation of the dipole moment of the *Ith* water molecule within the cluster when the state of the cluster is changed from the reference to the perturbed state. The quantities $E_{I\alpha}$ and $\Phi_{I\alpha J\beta}$ are of course

$$E_{I\alpha} \equiv -\left[\frac{\partial E(n)}{\partial P_{I\alpha}}\right]_0,$$
(2)

$$\Phi_{I\alpha J\beta} \equiv \left[\frac{\partial E(n)}{\partial P_{I\alpha}\,\partial P_{J\beta}}\right]_0.$$
(3)

The reference state is defined as a fictitious cluster, having same geometry, but for which the molecules are *not* polarisable. The dipole moments of the water molecules of the reference state have their ground-state value, noted $\mathbf{P}^0$. Then $\delta P_{I\alpha}$ is interpreted as the modification of the molecular dipole of an individual molecule induced by the variation of the local electric field. $\delta P_{I\alpha}$ depends on the molecule position and of its number of hydrogen bonds (see below).

If we apply the variational principle of Density Functional Theory (DFT) to the model Eq. (1) truncated to the second order, one finds an equation for the induced dipoles

$$\left[\frac{\partial E(n)}{\partial P_{I\alpha}}\right]_0 = 0,$$
(4)

$$E_{I\alpha} = \sum_{J=1}^{n}\sum_{\beta} \Phi_{I\alpha J\beta}\delta P_{J\beta},$$
(5)

the formal solution of which is

$$\delta P_{I\alpha} = P_{I\alpha} - P_{I\alpha}^0 = \sum_{J=1}^{n} \sum_{\beta} \Phi_{I\alpha J\beta}^{-1} E_{J\beta} \ . \tag{6}$$

From Eq. (6), one interprets Φ_{IJ}^{-1} as a nonlocal polarisability $(3nx3n)$tensor representing the dipole moment induced on molecule I by a unit electric field applied to the molecule J.[18] The polarisability Ω of a molecule in a cluster can be defined by using the Lorentz relation[18]

$$\delta P_{I\alpha} = \sum_{\beta} \Omega_{I\alpha\beta} e_{I\beta}, \tag{7}$$

where e_I is a local field. According to Eq. (5), one finds that the molecular polarisability is the inverse of the (3x3) tensor $\Phi_{I\alpha I\beta}$

$$\Omega_{I\alpha\beta} = \Phi_{I\alpha I\beta}^{-1}, \tag{8}$$

and that the local field is

$$e_{I\beta} = E_{I\beta} - \sum_{\substack{J=1 \\ J\neq I}}^{n} \sum_{\beta} \Phi_{I\alpha J\beta} \delta P_{J\beta}. \tag{9}$$

$\Omega_{I\alpha\beta}$ is the (effective) polarisability of the Ith water molecule in the cluster and $\mathbf{P}_I$ in Eq. (6) is its (effective) dipole moment. The effective dipole moment of a water molecule in a given phase depends on the partitioning of the polarisation of the phase and is not therefore uniquely defined. However the value calculated for a molecule in the liquid phase[19] and in ice[20] is always found greater than the value in gas phase. Several authors have predicted also an increase of the effective dipole moment of water molecules in clusters.[21,22,23,24,25] The *average* increase of the effective dipole moment with water condensation may be reproduced using an appropriate induction model based on the polarisabilities of the water molecule.[20] The evaluation of the polarisabilities of water clusters and of molecules within the clusters are thus essential to understand intermolecular interactions. Water cluster polarisabilities calculations using first principles began rather recently. Kim et al.[26] and Maroulis[27] have reported *ab initio* evaluations of the dimer polarisability. Otto et al.[28] have computed the polarisabilty of polymeric water with coupled Hartree-Fock method. Polarisability of some larger clusters W_n $(n = 4, 5$ and $13)$ have been calculated by Jensen et al.[29] using a point dipole model and time-dependent density functional theory. These authors concluded that the *mean polarisability* of a water molecule, defined as the cluster polarisability divided by the aggregation number n, depends on the cluster size for small clusters and is lower than in gas phase. In their study of small clusters ($n < 5$), Tu and Laaksonen[30] found also that the polarisability of a water molecule decreases with n using the second-order Møller-Plesset perturbation theory (MP2). The size-dependence of polarisability of water clusters was first studied by Rodriguez et al.[31] for $W_2 - W_8$ with DFT and Generalised Gradient Approximations of exchange-correlation functionals. More recently, Ghanty and Ghosh[32] reported in a systematic study, a surprisingly linear size-increase of the cluster W_n polarisability for $n = 1$ to 20 using Hartree-Fock (HF). Calculations performed at DFT-B3LYP and MP2 levels for the clusters W_1 to W_{10}, show that this independence of the mean water molecule polarisability with n does not depend on the approximation used for exchange and electron correlation at least for $n \leq 10$.[32]

The present paper is based on a systematic study of electronic polarisabilities of all water clusters from 1 to 20 molecules using DFT-B3LYP and DFT-B3PW91 exchange-correlation functionals.[33] Our aim is to address the evaluation of the polarisabilities of the individual molecules within the clusters. In particular, we study the possible dependence of the polarisability of a water molecule with its environment in a cluster of a given size. Using a partitioning of the electronic density, we will describe how the charge distribution varies within a cluster and influences the molecule polarisation. The importance of intermolecular charge transfer, which is debated in the recent literature,[34,35,36] is emphasized. A preliminary account of the relations between hydrogen bonding and polarisabilities is also reported for water cluster isomers in the case of W_6 (a more detailed study will be reported elsewhere).

The chapter is organised as follows. The methodology is explained in next Section. The dependence of dipole moment with cluster size is presented in the third Section. In the fourth Section, the variation of the polarisability of a molecule with the size of the cluster and its location is established and discussed. The fifth Section is devoted to the correlation between hydrogen bonding and molecular polarisability for isomeric structures of W_6. A résumé and discussion end the chapter.

2. Methodology

The initial Hartree-Fock optimised geometries for W_1 to W_{20}[37] are taken from the Cambridge Cluster Database (CCD)[38] and optimised using the B3LYP functional and the 6-31G(d,p) basis set. The stability of the equilibrium structures is tested by performing a normal mode analysis. All frequencies are real confirming that these geometries are local minima of the corresponding potential energy surfaces. It is worth noting that these B3LYP-optimised structures are not very different from the initial Hartree-Fock structures. In addition, one finds no major differences between the present DFT geometries and those computed at the MP2/aug-cc-pvDZ level for the smallest clusters.

The electronic polarisabilities are calculated using a finite field method[39] at the HF and DFT levels using the B3LYP-optimised structures. The applied field strength is set to 0.001au and the SCF convergence criteria are tighten to 10^{-9}. Only the trace of the polarisability tensor $\alpha = \frac{1}{3} [\alpha_{xx} + \alpha_{yy} + \alpha_{zz}]$ is presented and discussed. For the calculation of the dipole moments and polarisabilities, one chooses a basis set 6-31++G(d,p), often used in the study of the (hyper)polarizabilities of conjugated molecules.[40,41] Larger basis sets become computationally demanding for water clusters as large as W_{20}. In order to achieve a full picture of the variations of water polarizabilities with cluster size, the basis set 6-31++G(d,p) is selected as a compromise between reliability of the results and the computational efficiency. This choice is based on a careful numerical analysis of the basis set effects for cluster size up to $n = 6$ which demonstrates that the values obtained with the 6-31++G(d,p) calculations reproduce about 70-80% of the values obtained with large basis sets such as aug-cc-pVDZ and aug-cc-pVTZ as shown by the Fig. 1.[42] For the smallest clusters, one observes that large basis sets have a tendency to reduce the variations of the mean polarisability with the cluster size. More details can be found in Ref. [42].

The dependence of the present results on the approximations used to treat exchange and electronic correlation is tested by comparing the values of the polarisabilities obtained using the functionals B3LYP and B3PW91 (all n) and the MP2 approximation ($n \leqslant 15$).

We define the *mean polarisability of a water molecule* in a cluster as stated above by the cluster polarisability divided by the number of molecules: it is a well defined quantity. The polarisability of *"a water molecule in a cluster"* (Eqs. (7) and (8)) has no unique definition, on the contrary. We

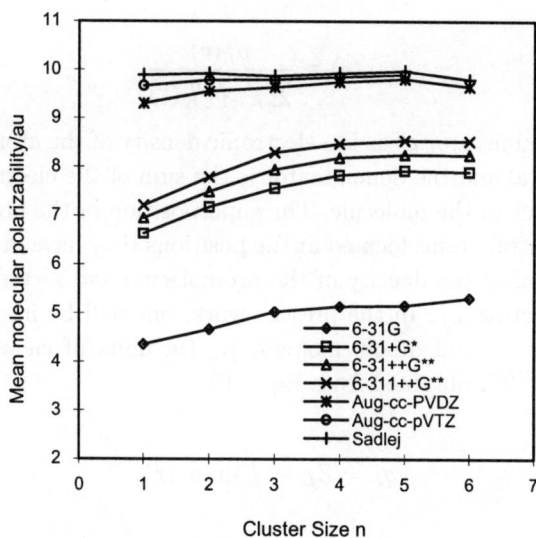

Fig. 1. Basis set dependence of mean molecular polarisability calculated at MP2 level. All values are in a.u.

named it molecular polarisability. Several approaches in quantum chemistry are available to partition the cluster polarisability in molecular ones.[43, 44] Stockholder scheme[45,46] is used in the present study to partition the cluster electron density into local contributions and then in molecular polarisabilities as follows.

In the stockholder or Hirshfeld scheme, the electronic density of an atom in a molecule ρ_L may be defined as the expectation value of an atomic charge operator $\hat{\omega}_L$ by the following equation

$$\rho_L(\mathbf{r}) = \left\langle \hat{\omega}_L \right\rangle = \omega_L(\mathbf{r})\rho(\mathbf{r}), \tag{10}$$

$$\hat{\omega}_L \equiv \sum_{i=1}^{N} \omega_L(\mathbf{r})\delta(\mathbf{r} - \mathbf{r}_i), \tag{11}$$

where $\rho(\mathbf{r})$ is the molecular electronic density and $N = \int d\mathbf{r}\,\rho(\mathbf{r})$ its integral whereas $\mathbf{r}_i$ is the *ith* electron coordinate. The scalar function $\omega_L(\mathbf{r})$ is defined by

$$\omega_L(\mathbf{r}) \equiv \frac{\rho_L^0(\mathbf{r})}{\sum_{K=1}^M \rho_K^0(\mathbf{r})}, \tag{12}$$

in which the numerator ρ_L^0 is the electronic density of the atom L when it is free and neutral and the denominator is the sum of the electronic densities of all the atoms of the molecule. The superposition of the atomic densities of the free neutral atoms located at the positions they have in the molecular geometry is called the density of the promolecule and serves to define the weighting function ω_L. In the present work, one will be interested by the atomic charge q_L and dipole moment $\mathbf{p}_L$ (in units of elementary charge $e = 1.602 \ 10^{-19}C$) obtained from Eq. (10)

$$q_L = Z_L - \int d\mathbf{r} \ \rho_L(\mathbf{r}), \tag{13}$$

$$\mathbf{p}_L = - \int d\mathbf{r} \ \mathbf{r} \ \rho_L(\mathbf{r}) + Z_L \mathbf{R}_L, \tag{14}$$

where Z_L and $\mathbf{R}_L$ are the atomic number and position, respectively. From these atomic quantities, one computes the effective dipole moment of the water *molecule I* in the cluster

$$\mathbf{P}_I = \sum_{K(I)} \mathbf{p}_K, \tag{15}$$

where the sum is restricted to the three atoms K of the molecule I. Assuming that each molecule has an effective polarisability given by the Lorentz approximation, Eq. (7), the variation of the molecular dipole moment due to the application of a uniform external field $\mathbf{E}$ is simply

$$\delta\mathbf{P}_I = \mathbf{P}_I(\mathbf{E}) - \mathbf{P}_I(\mathbf{E}=0) = \Omega_I \ \mathbf{e}_I, \tag{16}$$

where the electrostatic local field $\mathbf{e}_I$[47] is

$$\mathbf{e}_I = \mathbf{E} + \sum_{J \neq I} \frac{3(\delta\mathbf{P}_J.\mathbf{R}_{JI})\mathbf{R}_{JI} - R_{JI}^2\delta\mathbf{P}_J}{R_{JI}^5}, \tag{17}$$

in which J runs over all the molecules but I in the cluster and $\mathbf{R}_{JI} = \mathbf{R}_J - \mathbf{R}_I$.

The induced dipole moment evaluated from Eq. (16) is coordinate-dependent because water molecules are in general charged in the cluster due to small intermolecular charge transfers.[34,35,36] To remove the coordinate-dependent terms, $\delta \mathbf{P}_I$ in Eqs. (16) and (17) must be replaced by a "corrected" induced dipole defined by

$$\delta \mathbf{P}_I^c = \delta \mathbf{P}_I - \frac{Q_I}{3} \sum_{K(I)} \mathbf{R}_K = \Omega_I^c \, \mathbf{e}_I^c, \qquad (18)$$

$$Q_I \equiv \sum_{K(I)} q_K, \qquad (19)$$

where the net charge of molecule is Q_I. From Eq. (18), one evaluates a average dipolar polarisability of the molecule I (trace of Ω_I^c) in the cluster that we called the *"corrected molecular polarisability"*. Eq. (18) is independent of the choice of coordinates because it defines the induced dipole moment relative to the geometrical center of the molecule. Only Ω_I^c is expected to be a transferable property of a water molecule in a condensed phase. On the contrary, Ω_I deduced from Eq. (16) is coordinate-dependent because it contains a charge dependent contribution which is a cluster property and is thus not expected to be a transferable molecular property. We call here Ω_I as *"uncorrected molecular polarisability"*.

The calculations of polarisabilities and dipole moments described in this section are performed with Gaussian98 program[48] and the stockholder multipolar expansion of the electronic density is computed with the program STOCK[46] which is part of the BRABO package as in Ref. [49].

3. Structures and dipoles moments of the clusters

Fig. (2) presents the optimised structures of water clusters studied here. They are similar to the structures first reported by[37] and deposited in CCD. The onset of the formation of three dimensional clusters occurs for the cage isomer of the water hexamer, which is the most stable isomer.[50] For smaller clusters ($n = 3, 4$ and 5) two-dimensional rings structures are preferred in agreement with experiments.[51,52,53,54] Addition of water molecules to the hexamer tend to favor structures which can be viewed as the stacking of 4 and 5-member water rings. For instance W_{10} corresponds to two 5-members water rings parallel to each other, W_{16} and W_{20} are piles of respectively four and five 4-members water rings. By consequence, for these cluster sizes, and for a zero temperature, the molecules do not adopt an "ice-like" structure

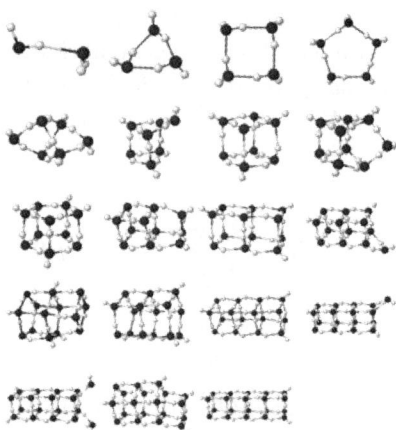

Fig. 2.   Structures of the water clusters.

which can be viewed as a stacking of *hexagonal* water sheets. The number of hydrogen bonds per water molecule in the clusters varies: the molecules at the corners of the cuboïd structures are the less coordinated having 1, 2 or 3 bonds whereas the molecule in the middle of the stacked structures are tetra-coordinated. It is interesting to note that the number of "dangling hydrogen bonds" is equal to 3 or 4 in nearly all the structures: there is a tendency to minimize the number of hydrogen atoms unbounded.

An analysis of the optimised structures up to $n = 6$, shows that the average length of the covalent O-Hb bond, in which Hb is an atom involved in the hydrogen bond, is larger than the O-H bondlength of the isolated molecule (0.960 Å). It increases from the dimer to the pentamer (W_3 =0.984 Å, W_4 =0.993 Å, W_5 =0.995 Å, W_6 =0.984 Å). On the other hand, the O...Hb distance decreases with the cluster size (W_2 =1.93 Å, W_3 =1.838 Å, W_4 =1.714 Å, W_5 =1.684 Å, W_6 =1.798 Å). The later finding is in agreement with experimental data.[55] For larger clusters, we do not find any systematic variation of the average O-Hb bond lenght with the cluster size, its value is comprised between 0.985 Å (W_{18}) and 0.992 Å (W_8, W_9 and W_{10}). The average hydrogen bond length is around 1.8 Å for all water clusters being slightly smaller for W_8, W_9 and W_{10}.

Table (1) shows the calculated dipole moment of water clusters for $n = 1$ to 20. All the methods (HF, DFT-B3LYP, DFT-B3PW91 and MP2) give similar results. Calculations with larger basis sets for small clusters (n <

Table 1.   Calculated dipole moments of water clusters. Results are performed with 6-31++G(d,p) basis set. All values reported are in Debye.

n	HF	DFT-B3LYP	DFT-B3PW91	MP2
1	2.308	2.219	2.224	2.255
2	2.913	2.880	2.883	2.862
3	1.517	1.454	1.457	1.477
4	0	0	0	0
5	1.357	1.289	1.294	1.309
6	2.427	2.359	2.367	2.364
7	1.240	1.246	1.246	1.215
8	0.008	0.008	0.008	0.008
9	1.863	1.784	1.792	1.797
10	2.867	2.781	2.786	2.783
11	1.939	1.858	1.853	1.876
12	0.008	0.008	0.008	0.008
13	2.290	2.204	2.211	2.214
14	5.178	5.170	5.181	5.112
15	4.802	4.672	4.675	4.667
16	0.008	0.008	0.008	
17	3.050	2.900	2.899	
18	1.792	1.703	1.700	
19	5.142	4.995	4.998	
20	0.005	0.005	0.005	

7)[22] gave the same trend but slightly smaller values. The dipole moment of the cluster is the sum of the effective dipole *vectors* Eq. (15) of the individual molecules and depends mainly on the orientation of the molecules and of their arrangement in the cluster. Thus, the dipole moment of the dimer W_2 is larger than the monomer W_1. On the other hand, W_4, W_8, W_{12}, W_{16} and W_{20} have no dipole moment because of the symmetry of their geometry while W_6, W_{10}, W_{14} and W_{19} have relatively large dipole moments. Nigra and Kais[56] suggested to separate the water clusters into four groups according to their structures. The fluctuations of dipole moments in Table (1) reflect approximately the $4n$ rule suggested by these authors.

4. Size and geometry dependence of the molecular polarisabilities

The computed mean polarisabilities of water clusters are listed in Table (2). To the best of our knowledge, there is no experimental values to compare excepted for the monomer (9.92 au,[58]). MP2 with extended basis set (aug-cc-pVDZ) reproduces 94% of the polarisability of a single molecule

Table 2. ^a Calculated mean polarisabilities of water clusters. Results are performed with 6-31++G(d,p) basis set. All values reported are in a.u. Other works are given for comparison: [a] MP2/aug-cc-pVDZ, ref. [57], [b] Experimental value, ref. [58], [c] Static TDDFT/TZ2P+, ref. [29], [d] B3LYP/Sadlej, ref. [32], [e] MP2/Sadlej, ref. [32].

n	HF	DFT-B3LYP	DFT-B3PW91	MP2	Other works		
1	6.340	7.210	7.030	6.940	9.290[a]	9.920[b]	9.150[c]
2	6.650	7.800	7.600	7.405	9.820[d]	9.610[e]	
3	7.023	8.403	8.217	7.957	9.787[d]	9.540[e]	
4	7.232	8.642	8.447	8.215	9.832[d]	9.592[e]	9.167[c]
5	7.318	8.740	8.540	8.312	9.868[d]	9.624[e]	9.376[c]
6	7.268	8.725	8.550	8.290	9.668[d]	9.415[e]	
7	7.317	8.778	8.601	8.350	9.708[d]	9.460[e]	
8	7.342	8.804	8.632	8.389	9.685[d]	9.437[e]	
9	7.399	8.870	8.690	8.455	9.725[d]	9.475[e]	
10	7.420	8.894	8.715	8.482	9.717[d]	9.465[e]	
11	7.332	8.817	8.654	8.391			
12	7.345	8.822	8.663	8.406			
13	7.338	8.827	8.669	8.402			9.140[c]
14	7.354	8.851	8.679	8.423			
15	7.434	8.918	8.753	8.512			
16	7.362	8.856	8.703				
17	7.353	8.863	8.710				
18	7.340	8.857	8.704				
19	7.424	8.919	8.761				
20	7.372	8.881	8.731				

in steam.[42] Our best estimation in the present work using a moderate 6-31++G(d,p) basis set is only 73 % of this experimental value and is obtained with DFT-B3LYP. This inaccuracy compared to experiment does not modify the main conclusions described below as they concern not the absolute values of the cluster polarisabilities but their variations at a given approximation level, with the size and the geometry.

One observes in fact no qualitative differences between the various methods and approximations used here. All give a variation of the mean polarisability with the size of the cluster which is similar. As reported elsewhere, an accurate evaluation of the polarisabilities for the smallest water clusters ($n < 7$) give also a trend similiar to the results presented in Table (2).[42] The only differences are in the absolute values of the mean polarisabilities which are systematically larger in DFT approach. For the same basis set, one finds that the cluster polarisability follows the order HF < MP2 < DFT-B3PW91 < DFT-B3LYP.

One remarkable property deduced from the values listed in Table (2) is that the mean polarisability is rather constant.[33] It means that the polarisability of water clusters scales linearly with their size. This has been found also by Ghanty and Ghosh[32] in a recent study. For the smallest size ($n < 6$), the mean polarisability varies with the size but this variation depends also on the basis set as shown at Fig. (1).

Although the mean polarisability is nearly constant for cluster size $n > 5$, the molecular polarisabilities are different for the different water molecules in a given cluster. In Fig. (3), one finds indeed fluctuations of the molecular polarisability (Eq. (18)) in W_{15} and in W_{20}. Similar variations are observed in all the stable clusters studied here and their isomers (see next Sec.). The variations can be as large as 1.0 a u. The "uncorrected molecular polarisabilities" obtained directly from Eq. (16) are larger than the correspondingly corrected ones but their variations within a cluster are qualitatively similar.[33]

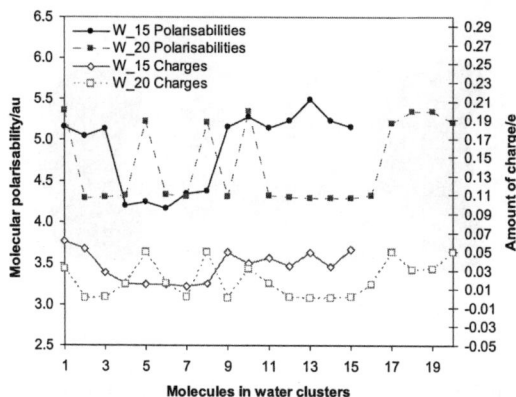

Fig. 3. Molecular polarisabilities in water clusters W_{15} and W_{20}. Calculations are performed at DFT-B3LYP level with basis set 6-31++G(d,p). All values are in a.u.

One may also note that the molecular polarisabilities shown in Fig. (3) are either around 5.3 au or 4.3 au. Eight molecules in W_{20} and ten molecules in W_{15} have larger polarisabilities than the others. This reflects in fact two types of water molecules depending on their location in the cluster. Indeed, W_{20} has a fused cube structure with five layers in which eight molecules occupy the first and fifth layers while the other twelve molecules share the other three layers in the body (Fig. (2)). Similarly, there are three layers in W_{15} and each layer contains five molecules, and thus ten molecules are in the first and third layers and the other five molecules in the body (Fig.

(2)). Fig. (4) shows that the molecules at both ends in W_{20} possess larger polarisabilities than those in the body of the cluster. There are 36 hydrogen bonds in the W_{20}. Molecules at the ends are less bonded than those in the body. In contrast to ice where each molecule bonds to other four molecules and acts as both electron donor and acceptor, molecules in water clusters are unsaturated in their bonding, especially for the molecules at corners or at surface which might serve only as either electron donor or acceptor. Therefore the molecules at corners are charged highest. This can be seen in Fig. (5), where the molecules at end possess greater amount of net charge (absolute value) than the molecules in body which indicates a close relation between the molecular polarisability and net charge (compare Figs. (4) and (5)).

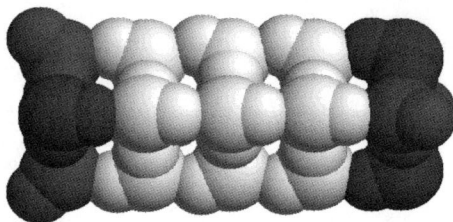

Fig. 4.   Molecular polarisability in W_{20}. Polarisability increases from white (low values) to black (high values). Calculations are performed using B3LYP/6-31++G(d,p).

A correlation between molecular charge distribution and molecular polarisability is found for other water clusters. Fig. (6) summarizes the variations of the molecular polarisabilities and of molecular charges for all the molecules in the stable geometries studied here (Fig. (2)). It is obvious that molecular polarisabilities fluctuate in accordance with molecular net charge for all molecules in a cluster independent of its size. For fused cube geometries (see Fig. (2)), there is clearly two groups of molecules as in W_{15} and W_{20}.

Though the molecular polarisability (Eq. (18)) is related closely to the location of the molecule in the cluster, more exactly, to its net charge, the average of the molecular polarisabilities in a cluster W_n varies simply

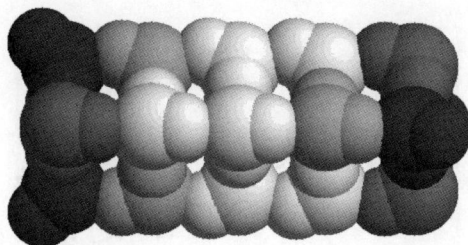

Fig. 5.   Molecular charge distribution in W_{20}. The amount of net charge (in absolute value) increases from white (low values) to black (high values). Calculations are performed using B3LYP/6-31++G(d,p).

with n. Table (3) lists the average corrected molecular polarisabilities. The smallest values are obtained using HF. The DFT calculations overestimate both the cluster and "corrected" molecular polarisabilities compared to the MP2 results, but the deviations are quite small (less than 6% and 4% with DFT-B3LYP, and 4% and 2% with DFT-B3PW91). All these data show that the corrected molecular polarisability decreases with an increase of n and reaches saturation at the largest cluster size.

We may extrapolate from the DFT-B3LYP results in Table (3) that the corrected polarisability per molecule in large water clusters is approximately 35% smaller than the corrected polarisability of an isolated molecule. The reduction of polarisability of water due to hydrogen bonding has been also predicted by het in Panhuis et al.[59] for the water dimer by using the Stone's distributed polarisabilities analysis.[43] For the scheme and the accuracy of the method used here, one may predict from the Table (3) that the average molecular polarisability should approach about 4.3 au if the framework like W_{20} grows unidimensionally since the number of molecules in the body (saturated with H-bonds) increases on the contrary of the molecules at the end (unsaturated with H-bonds and having a larger polarisability of 5.3 a.u., as discussed above).   However, it should be noted that the molecules are all at the surface of the clusters. According to our analysis, the molecular polarisability of water could be therefore different in very large systems in which most molecules are in the bulk like in ice. Further studies on the dielectric properties of water in larger clusters and crystals are now being undertaken.

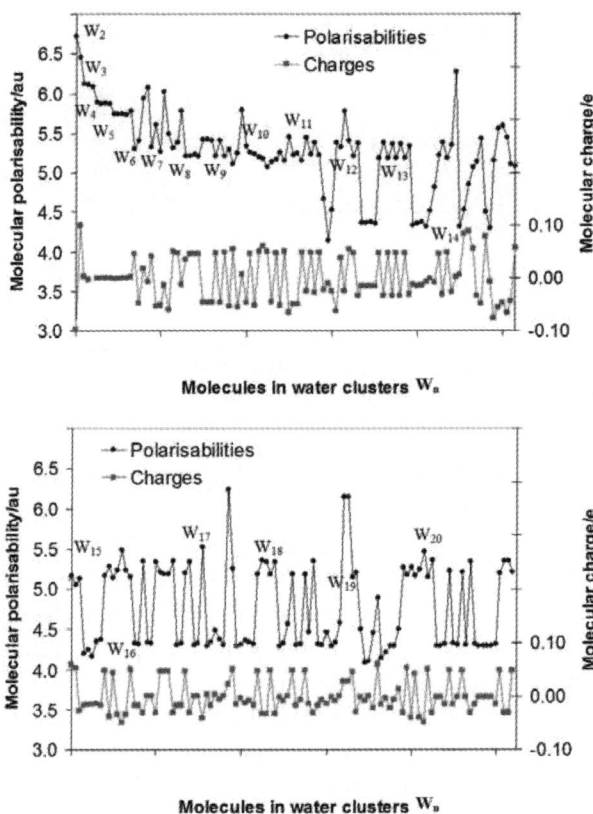

Fig. 6.   Correlation between molecular polarisability and molecular net charge in water clusters. Calculations are performed at B3LYP level with basis set 6-31++G(d,p).

5. Polarisability and hydrogen bonding

According to the so-called Bernal and Fowler rules,[1] each molecule in ice donates its two protons and receives two protons of the surrounding molecules, acting as both proton donor and acceptor at the same time. This is not the case for the molecules in water clusters as shown in the previous section. The relation shown above between the water molecular charge and the polarisability is a consequence of these different local electronic prop-

Table 3. Average of the corrected polarisabilities in water clusters.Calculations are performed at DFT-B3LYP level with basis set 6-31++G(d,p). All values are in a.u.

n	HF	DFT-B3LYP	DFT-B3PW91	MP2
1	6.34	7.21	7.030	6.94
2	5.87	6.65	6.49	6.42
3	5.46	6.15	6.01	5.98
4	5.28	5.95	5.81	5.79
5	5.21	5.86	5.73	5.72
6	4.99	5.57	5.46	5.46
7	4.90	5.46	5.35	5.36
8	4.74	5.26	5.16	5.17
9	4.75	5.27	5.16	5.11
10	4.69	5.19	5.09	5.05
11	4.64	5.11	5.02	4.99
12	4.52	4.96	4.87	4.83
13	4.54	4.96	4.88	4.82
14	4.58	5.01	4.93	4.85
15	4.49	4.90	4.82	4.77
16	4.41	4.80	4.72	
17	4.42	4.80	4.73	
18	4.42	4.79	4.72	
19	4.41	4.78	4.71	
20	4.34	4.70	4.62	

erties, i.e. of these different water-water hydrogen bonds. One can expect therefore an influence of hydrogen bonding on the molecular polarisability.

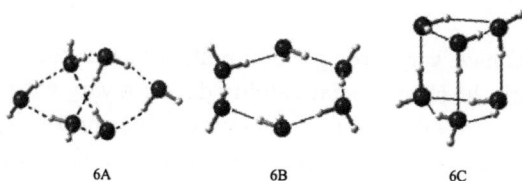

Fig. 7. Structure of the cage (A), cyclic (B) and prism (C) isomers of W_6.

In order to get further insight in the relations between hydrogen bond network and molecular polarisability, we carried on a study of the isomers of water clusters. Only the preliminary results for the hexamer will be discussed here. One emphasize that the isomers studied here are among the most stable conformations predicted by theoretical calculations[37] : the cage (6A), the cyclic (6B), and the prism (6C) hexamers (Fig. (7)). For these structures, the DFT-B3LYP calculated cluster polarizabilities of W_6

Table 4. Corrected polarisabilities for water molecules in W_6 isomers. Calculations are performed at DFT-B3LYP level with basis set 6-31++G(d,p). All values are in a.u. [a] Number of molecules, [b] number of molecules accepting protons (A), [c] number of molecules donating protons (D), [d] molecule type, [e] average corrected polarizability of molecules in the cluster and of the corresponding type, [f] mean polarizability (cluster polarisability/n).

N^a	n_A^b	n_D^c	Type^d	Molecular polarisability (charge)			Aver.^{e}	Mean^f
6A							5.61	8.72
2	1	1	AD	5.94 (0.017)	6.08 (-0.007)		6.01	
2	1	2	ADD	5.41 (-0.047)	5.60 (-0.052)		5.31	
2	2	1	AAD	5.30 (0.046)	5.32 (0.043)		5.31	
6B							5.65	8.77
6	1	1	AD	5.66 (0.00)	5.64 (0.00)	5.65 (0.00)		
				5.68 (0.00)	5.66 (0.00)	5.63 (0.00)		
6C							5.58	8.69
3	1	2	ADD	5.90 (-0.064)	5.61 (-0.023)	5.81 (-0.015)	5.77	
3	2	1	AAD	5.47 (0.023)	5.41 (0.045)	5.28 (0.034)	5.39	

are presented in Table (4^f). The accuracy of the results can be measured by comparing these data with those obtained with the aug-cc-pVDZ basis sets and the MP2 method. The reported cluster polarizability of 6A accounts for 88% and 91% of the DFT-B3LYP and MP2/aug-cc-pVDZ results,[42] respectively. The corresponding ratios are 86% and 88% for 6B and 88% and 91% for 6C.[42] Calculations at all levels give the same order of magnitude, 6B > 6A > 6C.

6A, 6B and 6C have very close polarizabilities with a standard deviation of 0.43%. Although the cluster polarisability does not vary with the isomeric structure, the polarisability of a molecule within the cluster depends on its location and more precisely on its hydrogen bonds with the other molecules. The computed polarizabilities of the individual molecules in the W_6 isomers are summarized in Table (4). There are three kinds of molecules in 6A. The two molecules donating one proton and receiving one proton have greatest polarizabilities, followed by the two donating two protons and receiving one proton. The left two molecules donate one proton and receive two protons and have the smallest polarizabilities. All the six molecules in 6B are almost equivalent. Each of them donates one proton and receives one proton, and therefore have nearly equivalent polarizabilities. The molecules in 6C form two groups. Three of them with two protons donated and one proton received have greater polarizabilities than the other three molecules with one proton donated and two protons received. On concludes that the polarizability of a water molecule is directly related to its bonding with the surrounding molecules. Molecules with more HBs tend to have smaller polarizabilities. In addition, one observes that the molecules donating a proton tend to be charged positively and the molecules receiving a proton tend to be charged negatively. The net charge seems to depend on the difference between the number of bonds "donor" and "acceptor". However the correlation observed between the charge and polarisability shown in Fig. (6) is not sufficient to explain the relation between hydrogen bonding and molecular polarisability. For instance in 6A, the greatest polarisabilities are obtained for molecules receiving and donating a proton (they should be neutral water molecules). Considering the first neighboring shell of a given water molecule is therefore not enough to explain all the effects on its molecular charge and polarisability. This reflects the well known effect of donor-acceptor cooperativity of hydrogen bonding.[60] Its relation to charge distribution might be elucidated by the careful study of hydrogen bond networks of larger isomers. A such study is being undertaken.

6. Résumé and discussion

Using DFT, one has demonstrated that the mean polarisability of a water molecule in a cluster is nearly constant (Table (2)), in agreement with recent Hartree-Fock calculations.[32] In addition, the study of isomers of W_6 shows that the mean polarisabilty of a cluster of a given size does not vary significantly with the hydrogen bond network. On the contrary, the polarisability of *"a water molecule in a cluster"* (Eq. (18)) varies according to its net charge due to charge redistributions in hydrogen bonds and depends thus on the hydrogen bond network of the cluster (Figs. (3-6)). In résumé, although the global properties of water clusters scale simply with the cluster size, their local properties, charges and polarisabilities, vary in a complicated manner.

The cluster polarisability suffices only to describe the interaction of the cluster as a whole and at large distances. On the opposite, the chemical reactivity of the water cluster and its strongest interactions with solutes occur at short distances for which the inhomogenities of static and induced multipoles (beyond the dipole) are important. However, the fact that all water molecules are not equivalent regarding their (electrostatic) interaction with other species, as expected from the present results (Fig. (6)), is not considered in the modelisation of intermolecular interactions.

Building an accurate water force field is still an open important problem in physical chemistry (See for instance, Refs. [61,62,63,64] and references therein) Indeed, in spite of the progress made in first principles methodologies and computer power, one still needs simple potential models to describe complicated systems like solvated proteins. The present findings might serve to improve the existing water force fields. One may note that one major shortcoming of most if not all force fields for water is the absence of intermolecular charge transfers. The polarisation is thus represented by an effective intramolecular polarisability (empirical or fitted on *ab initio* data), identical for all molecules. This is possible even if there is actual charge transfer between the water molecules because the macroscopic bulk polarization has not a unique representation. In other words, the value of the average intramolecular polarisability used in force fields, larger than the present "corrected" one, must account in some average way for the intermolecular part of the polarization found. However this "mean field" representation is not expected to be transferable because the intermolecular charge transfer dependent on the hydrogen bond network as shown above and an intramolecular effective polarisability does not. Very recent published works have shown that polarisable force fields are indeed not easily transferable[65,66] and some attempts to reintroduce charge transfers have been done.[34,66,67] We believe that a better transferability of the electronic polarisability of a water molecule in a condensed phase could be achieved by assigning an intramolecular "corrected" polarisability to a molecule in a force field and by treating separately the effects of the local bonding (intermolecular polarisability). Such model, based on the partitioning of the electronic density, is presently under development.

Acknowledgments

M. Yang thanks the "Conseil Régional de Bourgogne" for his postdoctoral fellowship during his stay in Dijon where the present work was done. The authors acknowledge the use of the computer resources of the "Centre de Ressources Informatiques" of the "Université de Bourgogne". Part of this work was also supported by the University of Antwerp under grant GOA-BOF-UA Nr.23.

References

1. J. D. Bernal and H. R. Fowler, *J. Chem. Phys.* **1**, 515 (1933).
2. R. C. Dougherty and N. L. Howard, *J. Chem. Phys.* **109**, 7379 (1998).
3. M. F. Chaplin, *Biophys. Chem.* **83**, 211 (2000).

4. E. D. Sloan, *Nature (London)* **426**, 353 (2003).
5. J. H. Van der Waals and J. C. Plateeuw, *Adv. Chem. Phys.* **2**, 1(1959).
6. R. K. Mc Mullan and G. A. Jeffrey, *J. Chem. Phys.* **42**, 2725 (1965).
7. M. M. Teeter, *Proc. Natl. Acad. Sci. USA* **81**, 6014 (1984).
8. K. Pfeilsticker, A. Lotter, C. Peters, H. Bösch, *Science* **300**, 2078 (2003).
9. U. Buck, F. Huisken, J. Schleusener, J. Schaefer, *J. Chem. Phys.* **72**, 1512 (1980).
10. W. H. Robertson, E.G. Diken, E. A. Price, J.-W. Shin, M. A. Johnson, *Science* **299**, 1367 (2003).
11. R. J. Saykally, Proceedings of the Nobel Symposium 117: The Physics and Chemistry of Clusters, Campbell, E. E. B., Laarson, M. Eds, World Scientific: Singapore, 2001, p206.
12. J. K. Gregory, D. C. Clary, *J. Phys. Chem.* **100**, 18014 (1996).
13. J. Kim, K. S. Kim, *J. Chem. Phys.* **109**, 5886 (1998).
14. J. M. Ugalde, I. Alkorta, J. Elguero, *Angew. Chem. Int. Ed.* **39**, 717 (2000).
15. K. Müller-Dethlefs, P. Hobza, *Chem. Rev.* **100**, 143 (2000).
16. F. N. Keutsch, J. D. Cruzan, R. J. Saykally, *Chem. Rev.* **103**, 2533 (2003).
17. F. Weinhold, *J. Chem. Phys.* **109**, 373 (1998).
18. P. Senet, L. Henrard, Ph. Lambin, and A. A. Lucas in "Electronic Properties of Novel Materials", Eds. H. Kuzmani, J. Fink, M. Mehring, and S. Roth, World Scientific Publishing, Singapore, 393 (1994).
19. P. L. Silvestrelli, M. Parrinelo, *Phys. Rev. Lett.* **82**, 3308 (1999).
20. E. R. Batista, S. S. Xantheas, H. Jonssón, *J. Chem. Phys.* **109**, 4546 (1998).
21. C. Gatti, B. Silvi, F. Colonna, *Chem. Phys. Lett.* **247**, 135 (1995).
22. J. K. Gregory, D. C. Clary, K. Liu, M. G. Brown, R. J. Saykally, *Science* **275**, 814 (1997).
23. Y. Tu, A. Laaksonen, *J. Chem. Phys.* **111**, 7519 (1999).
24. T. D. Poulsen, P. R. Ogilby, K. V. Mikkelsen, *J. Chem. Phys.* **116**, 3730 (2002).
25. K. Coutinho, R. C. Guedes, B. J. C. Cabral, S. Canuto *Chem. Phys. Lett.* **369**, 345 (2003).
26. K. S. Kim, B. J. Minh, U. S. Choi, K. Lee, *J. Chem. Phys.* **91**, 6649 (1992).
27. G. Maroulis, *J. Chem. Phys.* **113**, 1813 (2000).
28. P. Otto, F. L. Gu, J. Ladik, *J. Chem. Phys.* **110**, 2717 (1999).
29. L. Jensen, M. Swart, P.T. van Duijnen, J. G. Snijders, *J. Chem. Phys.* **117**, 3316 (2002).
30. Y. Tu, A. Laaksonen, *Chem. Phys. Lett.* **329**, 283 (2000).
31. J. Rodriguez, D. Laria, E. J. Marceca, D. A. Estrin, *J. Chem. Phys.* **110**, 9039 (1999).
32. T. K. Ghanty, S. K. Ghosh, *J. Chem. Phys.* **118**, 8547 (2003).
33. M. Yang, P. Senet and C. Van Alsenoy, *Int. J. of Quant. Chem.* **101**, 535 (2005).
34. R. Chelli, M. Pagliai, P. Procacci, G. Cardini, and V. Schettino, *J. Chem. Phys.* **122**, 74504 (2005).
35. A. van der Vaart, and J. K. M. Merz, *J. Chem. Phys.* **116**, 7380 (2001).
36. O. Gálvez, P. C. Gómez, and L. F. Pacios, *J. Chem. Phys.* **115**, 11166 (2001);

ibid *J. Chem. Phys.* **118**, 4878 (2003).

37. S. Maheshwary, N. Patel, N. Sathyamurthy, A. D. Kulkarni, S. R. Gadre, *J. Phys. Chem. A* **105**, 10525 (2001).
38. D. J. Wales, J. P. K. Doye, A. Dullweber, M.P. Hodges, F. Y. Naumkin, F. Calvo, J. Hernández-Rojas, T. F. Middleton, The Cambridge Cluster Database.
39. H. A. Kurtz, J. J. P. Stewart , K. M. Dieter, *J. Comput. Chem.* **11**, 82 (1990).
40. D. Jacquemin, B. Champagne, C. Hättig, *Chem. Phys. Lett.* **319**, 327 (2000).
41. M. Yang, B. Champagne, *J. Phys. Chem. A* **107**, 3942 (2003).
42. M. Yang, P. Senet and C. Van Alsenoy, to be published.
43. A. J. Stone, *Chem. Phys. Lett.* **83**, 233 (1981).
44. R. F. W. Bader, *Chem. Rev.* **91**, 893 (1991).
45. F. L. Hirshfield, *Theoret. Chim. Acta (Berl)* **44**, 129 (1977).
46. B. Rousseau, A. Peeters, C. Van Alsenoy, *Chem. Phys. Lett.* **324**, 189 (2000).
47. R. Kubo, T. Nagamya, Solid State Physics, Mc Graw-Hill: New York, (1969).
48. M. J. Frisch, G. W. Trucks, H. B. Schlegel, G. E. Scuseria, M. A. Robb, J. R. Cheeseman, V. G. Zakrzewski, J. A. Montgomery Jr., R. E. Stratmann, J. C. Burant, S. Dapprich, J. M. Millam, A. D. Daniels, K. N. Kudin, M. C. Strain, O. Farkas, J. Tomasi, V. Barone, M. Cossi, R. Cammi, B. Mennucci, C. Pomelli, C. Adamo, S. Clifford, J. Ochterski, G. A. Petersson, P. Y. Ayala, Q. Cui, K. Morokuma, D. K. Malick, A. D. Rabuck, K. Raghavachari, J. B. Foresman, J. Cioslowski, J. V. Ortiz, A. G. Baboul, B. B. Stefanov, G. Liu, A. Liashenko, P. Piskorz, I. Komaromi, R. Gomperts, R. L. Martin, D. J. Fox, T. Keith, M. A. Al-Laham, C. Y. Peng, A. Nanayakkara, M. Challacombe, P. M. W. Gill, , B. Johnson, W. Chen, M. W. Wong, J. L. Andres, C. Gonzalez, M. Head-Gordon, E. S. Replogle, J. A. Pople, 1998, Gaussian 98, Revision A.11, Gaussian, Inc., Pittsburgh PA.
49. C. Van Alsenoy, A. Peeters, *J. Mol. Struct. (THEOCHEM)* **286**, 19 (1993).
50. K. Liu et al., *Nature (London)* **381**, 501 (1996).
51. T. R. Dyke and J. S. Muenter, *J. Chem. Phys.* **57**, 5011 (1972); ibid, *J. Chem. Phys.* **59**, 3125 (1973).
52. B. D. Key and A. W. Castleman, *J. Phys. Chem.* **89**, 4867 (1985).
53. M. F. Vernon et al., *J. Chem. Phys.* **77**, 47 (1982).
54. F. N. Keutsch, J. D. Cruzan, and J. R. Saykally, *Chem. Rev.* **103**, 2533 (2003).
55. K. Liu, M. G. Brown, J. D. Cruzan, and R. J. Saykally, *Science* **271**, 62 (1996).
56. P. Nigra, S. Kais, *Chem. Phys. Lett.* **305**, 433 (1999).
57. S. S. Xantheas and T. H. Dunning Jr, *J. Chem. Phys.* **99**, 8774 (1993).
58. W. F. Murphy, *J. Chem. Phys.* **67**, 5877 (1977).
59. M. in het Panhuis, P. L. A. Popelier, R. W. Munn, and J. G. Ángyán, *J. Chem. Phys.* **114**, 7951 (2001).
60. A. E. Reed, L. A. Curtiss, and F. Weinhold, *Chem. Rev.* **88**, 899 (1988).
61. B. Guillot, *J. Mol. Liq.* **101**, 219 (2002).
62. C. J. Burham and S. S. Xantheas, *J. Chem. Phys.* **116**, 1479 (2002).

63. S. S. Xantheas, C. J. Burham, and R. J. Harrison, *J. Chem. Phys.*, **116**, 1493 (2002).
64. C. J. Burham and S. S. Xantheas, *J. Chem. Phys.* **116**, 1500 (2002).
65. T. J. Giese and D. M. York, *J. Chem. Phys.* **120**, 9903 (2004).
66. R. Chelli, V. Schettino, and P. Procacci, *J. Chem. Phys.* **122**, 234107 (2005).
67. Z.-Z. Yang, Y. Wu and D.-X. Zhao, *J. Chem. Phys.* **120**, 2541 (2004).